Springer Water

More information about this series at http://www.springer.com/series/13419

Lothar Mueller · Askhad K. Sheudshen
Frank Eulenstein
Editors

Novel Methods for Monitoring and Managing Land and Water Resources in Siberia

Springer

Editors
Lothar Mueller
Leibniz Centre for Agricultural Landscape Research (ZALF) e.V.
Müncheberg
Germany

Frank Eulenstein
Leibniz Centre for Agricultural Landscape Research (ZALF) e.V.
Müncheberg
Germany

Askhad K. Sheudshen
Kuban State Agrarian University
Krasnodar
Russia

ISSN 2364-6934 ISSN 2364-8198 (electronic)
Springer Water
ISBN 978-3-319-24407-5 ISBN 978-3-319-24409-9 (eBook)
DOI 10.1007/978-3-319-24409-9

Library of Congress Control Number: 2015950904

Springer Cham Heidelberg New York Dordrecht London

Printed on acid-free paper

Springer International Publishing AG Switzerland is part of Springer Science+Business Media (www.springer.com)

Preface

Inducement for the Book

A breakthrough in soil evaluation was achieved in the year 2006. The working group, Soil Classification of the International Union of Soil Sciences, published a new version of the World Reference Base of Soil Resources (WRB) for improving international communication about soils. It enabled an apposite name to be allocated to all soils of the globe. This name characterises typical features of soils and the processes behind them.

At the same time discussion about the need to improve public awareness about the functions of soils for humankind had reached a climax. It seemed to be difficult or even impossible to characterise soil functions using globally valid evaluation schemes. This was a momentous occasion for the editors and some of the authors of this book to remember a vision of the Russian geographer V.V. Dokuchaev (1846–1903), the father of modern soil science. He pointed out that the findings of soil science needed to improve the food security of the population. According to his vision, the characterisation and classification of soils should focus on characterising and maintaining or even improving their fertility.

Interesting work was carried out by scientists from the United States, Australia and New Zealand to evaluate the performance of soil for cropping and grazing. Russia, too, has long-standing traditions in such work. The approaches had a clear focus on soil structure but did not closely correlate with crop yields outside of their home areas, because other soil and climate factors dominated, such as drought.

Based on all these considerations the framework of the Muencheberg Soil Quality Rating (M-SQR) was born: a system which allocates fertility numbers to soils. Now we had a combination of two tools available to characterise soils by meaningful names and by their fertility potential. We were keen to test the feasibility of the M-SQR along with the WRB in very different climate and soil regions. The Russian partners and the German Federal Office for Agriculture and Food (BLE) were very open for this idea. In 2007 the project 05/07 "Indicators of fertility and function of agricultural soils" got started. It enabled joint fieldwork about soil classification and evaluation on numerous agricultural sites in both countries. The main partners were the Institute of Soil Science and Agrochemistry (ISSA) from Novosibirsk, Russia, and the Leibniz Centre of Agricultural Landscape Research (ZALF) from Müncheberg, Germany. Two years later the Pryanishnikov All Russian Research Institute of Agrochemistry (VNIIA) Moscow joined this project. Many sites studied and sampled were in Siberia.

Another small-budget project "Effect of climate change in boreal and sub-arctic ecosystems on water quality and soil functions, code 01DJ12058" was supported by the German Federal Ministry of Education and Research (BMBF) and enabled this cooperation between ZALF Müncheberg and ISSA Novosibirsk to be deepened jointly with IWEP Barnaul during 2012–2013. Overall, more than 15 joint publications appeared as an outcome of both projects.

Contacts deepened, and the network of researchers became broader. Gaps in the knowledge became clear and interest in closer cooperation accrued. However, first it was necessary to combine the numerous results, to interlink them with other running projects and extremely innovative activities, and to make them available for application in research and practice. That is the intention behind editing this book.

The focus on Siberia was obvious because of the exciting, remote landscapes and their great potential for the future of the population and of our whole planet. Land degradation and desertification are threats there which will have implications not only for food security but also for water resources and their quality, for biodiversity, the livelihood of the population and other crucial targets of landscape development and evolution. We put the emphasis on the unity and interactions between land and water, their resources, functions and quality. Reliable data are required about the status of land and water resources based on advanced, internationally proven and acknowledged methods.

One of the most important recognitions of our project activities was that many scientific and practical solutions for monitoring of land resources are available in Europe but not yet in Siberia.

All this is to be revealed in detail and some more eyes must be opened to see the potential improvements and the need for them. Learning from neighbours and cooperating with them helps to avoid same mistakes that they have already made, and it helps to save time and money. Based on the knowledge and technologies presented here, it is the responsibility of the current generation of scientists, decision-makers and other stakeholders that practical measures of monitoring and sustainable resources have to be taken.

Purpose of the Book

This book summarises the outcomes of the above-mentioned projects and of a number of other recent studies related to the topic of land and water monitoring and management. It is intended to be a source of information for all those dealing with its subject: methods for the characterisation and wise utilisation of land and water resources in Siberia. Besides information, it aims to deliver motivation for thinking about applications and new site-adapted solutions. It will also provide understanding and confidence that those better solutions are feasible based on the power of scientific-technical innovations and people's creativity and efforts in handling them.

The book will not overfeed readers with facts and data. The main intended innovation of the book is its focus on transferrable novel methods. Scientific tools will be proposed for measuring, evaluating, modelling and controlling processes in the landscapes of Siberia, especially in rural landscapes. The application of these new scientific tools requires not only open minds but also high levels of motivation and education. In some cases investments are needed. Thus, outreach and the adaption of new methods can only be realistically carried out in the framework of pilot studies based on further strengthened international scientific cooperation. The book is to serve as an advanced platform for new and more sustainable research cooperation between inventors and protagonists of new methods coming from different leading research institutions of Russia, Germany and other regions of the globe.

Content and Structure

The book offers a broad array of methods to measure, assess, forecast and control land and water resources: laboratory and field measurement methods of water and soil quality, methods of resource evaluation, functional mapping and remote sensing methods for monitoring and modelling large areas. It contains methods for ecosystem modelling, and the field monitoring of soils, and methods and technologies for optimising land use systems.

The book has 32 individual chapters in four parts and seven thematic clusters. In order to focus on the scientific value of individual chapters and the expertise of their authors, the editors have decided to keep the structure on a flat level of hierarchy and to allocate the chapters to four parts of the book only. These are

Part I, "Environmental and Societal Framework for Monitoring and Managing Land and Water Resources". It analyses the status of land and water resources in Asian Russia, evaluates the agri-environmental research and points out gaps in the knowledge.

Part II "Methods and Case Studies for Understanding and Monitoring the Landscapes of Siberia", presents further advanced research studies from Siberia about water and land monitoring and their methodologies.

Part III "Novel Approaches and Technologies of Application Potentials for Siberia" offers methods developed outside Siberia, but with great potential to be applied there in the near future.

Another Part IV "Synopsis and Overall Conclusions" consists of the chapter "Potential of Applying Novel Monitoring and Management Methods to Siberian

Landscapes" It reviews all thematic clusters and their individual chapters, summarises the overall book and draws conclusions for the application of novel methods.

Readers and Authors

Our addressees are scientists, planners, teachers, students, decision-makers and all readers who feel responsible for initiating the sustainable use of resources by scientific-technical innovations. Readers will gain some information and inspiration for their own work from this book. Based on this, they are encouraged to find their individual optimum when drawing conclusions and acting imaginatively.

Readers are also encouraged to contact the authors for more information. The chapter authors are pioneers behind novel methods, as well as being innovative and experienced scientists. Most of them come from Russia and Germany, others from different regions of the globe. Possible divergences between the findings, conclusions and statements of individual authors are natural. Data given in the various chapters of this book may include slight uncertainties, biases and inconsistencies. The editors have made no attempt to harmonise them because this is natural and reflects the different sources and local and temporal scales of the data. The chapter authors' conclusions do not necessarily need to coincide with the particular opinion of the editors. Chapters reflect the views of their author, and editors cannot be held responsible for any interpretation which may be made based on the information contained therein.

It is important to mention that in some chapters, trade names are used to provide specific information about proven technologies applied in the study. Mentioning a trade name does not constitute a guarantee of the product by the authors or editors. It does also not mean a preference for, or recommendation of this product.

Acknowledgements

Many people and institutions provided the basis for this publication. We would like to thank the German Federal Office for Agriculture and Food (BLE) for travel funding as part of the project 05/07 of the German–Russian list of agricultural research cooperation. The International Bureau of the German Federal Ministry of Education and Research also provided travel funding for research work in project 01DJ12058.

Ms. Anne Koth (Dresden) proofread the chapters with prudence and expertise. The Springer publishing house ensured that the editorial and printing process was smoothly managed and completed. The editors would like to thank all funding bodies and other supporters for their help and engagement.

It was our pleasure to serve as editors of this book by coordinating and reviewing the written concepts and findings of motivated, enthusiastic scientists. We hope that our book can contribute to initiating the sustainable use of the land and water resources in Siberia.

Letschin
Krasnodar
Müncheberg
April 2015

Lothar Mueller
Askhad K. Sheudshen
Frank Eulenstein

Contents

Editors and Contributors

About the Editors

Lothar Mueller is a guest scientist at the Leibniz Centre for Agricultural Landscape Research (ZALF) in Müncheberg, Germany. He studied Agriculture at the Humboldt University in Berlin (Agricultural Engineer, 1970) and graduated from the Academy of Agricultural Science of the GDR (Dr. agr.) in 1974. He has worked in Müncheberg at the Research Centre of Soil Fertility (1974–1991), and the ZALF Institutes of Soil Landscape Research and Landscape Hydrology. His main fields of work and expertise are in the areas of soil hydrology, soil quality, agricultural soil and water management, drainage engineering and subsoiling technology.

Lothar Mueller has published more than 350 papers, amongst them more than 100 peer-reviewed publications and book chapters. Most of this work was done in cooperation with his colleague Uwe Schindler.

Since 2013 he has been an Honorary Professor at the Kuban State Agrarian University in Krasnodar (Russian Federation) and an Honorary Doctor of the Russian Research Institute of Rice in Krasnodar. Since 2015 he is an Honorary Doctor of the Uspanov Institute for Soil Science and Agrochemistry in Almaty, Kazakhstan.

Over the past 20 years Lothar Mueller has built up international cooperation with leading experts around the world on the topic of agricultural soil and water monitoring and management. Many of them are contributors to this book.

Askhad K. Sheudshen (Шеуджен Асхад Хазретович) is a researcher and professor at the Kuban State Agrarian University in Krasnodar, Russia. He is a corresponding member of the Russian Academy of Science. Askhad K. Sheudshen is also the deputy head of the Scientific Research Institute of Rice in Krasnodar and fulfils a number of public tasks in scientific research, education and advisory.

He studied agronomy at the Agronomic Faculty of the Kuban Agricultural Institute, worked for three years in the Tulsk Oblast and became an aspirant of the All Russian Institute for Rice in Krasnodar. In 1985 he defended his Ph.D. dissertation and in 1992 his doctoral dissertation (doctor of biological sciences), both about the physiology and nutrition of rice. His main specialisations are in Biogeochemistry, Agrochemistry and Ecology.

Askhad K. Sheudshen's current research interests are the efficiency of crops' productivity, their qualities and their safety and soil fertility depending on the conditions of fertilisation.

He has published more than 200 research papers, including 40 textbooks and monographs about plant nutrition. Askhad K. Sheudshen has educated numerous young scientists in Russia. He has been honoured as a merited scientist of the Russian Federation, the Kuban Region and the Republic of Adygaya.

Frank Eulenstein is a scientist at the Leibniz Centre for Agricultural Landscape Research (ZALF) in Müncheberg, Germany. He studied Agriculture at the Georg August University of Göttingen and graduated from the Institute for Soil Science in 1992 (Dr. sc. agr.), his thesis being about groundwater quality related to fertilisation in agriculture. From 1992 to 1993 he was head of the department soil hydrology and soil physics at the institute.

Since 1993 he has worked as scientist at the ZALF Institute of Land Use Systems. His fields of interest are water, matter and energy fluxes in agricultural systems and the practical questions of cropping and fertilisation.

He has worked on some national and international research projects, and headed a number of them. Frank Eulenstein has published more than 250 scientific papers. He is active in the international scientific community and a member of scientific organisations and working groups dealing with soil science, land management and water quality issues.

Since 2013 he has been an Honorary Professor of the Kuban State Agrarian University in Krasnodar (Russian Federation) and an Honorary Doctor of the Russian

Research Institute of Rice in Krasnodar. Since 2015 he is an Honorary Doctor of the Uspanov Institute for Soil Science and Agrochemistry in Almaty, Kazakhstan.

Frank Eulenstein has also practical experience as a business manager of two companies in the segments of the development, production and sale of organic fertilisers and farm supplies, and the trade and storage of agricultural products.

Contributors

Kanat Akshalov Scientific Production Center of Grain Farming named after A.I. Barayev, Shortandy, Kazakhstan

Jürgen Augustin Leibniz Centre for Agricultural Landscape Research (ZALF), Institute for Landscape Biogeochemistry, Müncheberg, Germany

Vladimir L. Badenko State Polytechnical University St. Petersburg, St. Petersburg, Russia

Dmitry Balykin Institute for Water and Environmental Problems, Siberian Branch of the Russian Academy of Sciences (IWEP SB RAS), Barnaul, Russia

Pavel Barsukov Siberian Branch, Institute of Soil Science and Agrochemistry (ISSA), Russian Academy of Sciences, Novosibirsk, Russia

Axel Behrendt Leibniz Centre for Agricultural Landscape Research (ZALF), Müncheberg, Germany

Natalia L. Belkova Siberian Branch, Limnological Institute (LIN), Russian Academy of Sciences, Irkutsk, Russia; Irkutsk State University (ISU), Irkutsk, Russia

Vladimir I. Belyaev Faculty Agricultural Engineering, Barnaul Altai State Agrarian University, Barnaul, Altai Krai, Russia

Michael Berg Leibniz-Centre for Agricultural Landscape Research (ZALF), Müncheberg, Germany

Norbert Bischoff Institute for Soil Sciences, Leibniz University Hannover (LUH), Hannover, Germany

Bogdan H. Chojnicki Department of Meteorology, Poznan University of Life Sciences, Poznań, Poland

Alexander S. Chumbaev Siberian Branch, Institute of Soil Science and Agrochemistry (ISSA), Russian Academy of Sciences, Novosibirsk, Russia

Darya A. Chupina V.S. Sobolev Institute of Geology and Mineralogy, Siberian Branch of Russian Academy of Sciences, Novosibirsk, Russia

Valentina V. Chuprova Krasnoyarsk State Agricultural University, Krasnoyarsk, Russia

Philippe Ciais CEA Centre de Saclay, Laboratoire des Sciences Du Climat et de l'Environnement (LSCE), Gif-sur-Yvette, France

Nils Cremer Erftverband, Bergheim, Germany; DWA, German Association for Water, Wastewater and Waste, Hennef, Germany

Alina Danielewska Department of Meteorology, Poznan University of Life Sciences, Poznań, Poland

Ralf Dannowski Leibniz Centre for Agricultural Landscape Research (ZALF), Müncheberg, Germany

Valentina M. Domysheva Siberian Branch, Limnological Institute (LIN), Russian Academy of Sciences, Irkutsk, Russia

Hartwig Drechsler Dr. Drechsler Ingenieurdienst, Göttingen, Germany; DWA, German Association for Water, Wastewater and Waste, Hennef, Germany

Wolfgang Durner Braunschweig Technical University, Institute of Geoecology, Brunswick, Germany

Serafima Dvurechenskaya Novosibirsk Department, Institute for Water and Environmental Problems (IWEP), SB RAS, Novosibirsk, Russia

Jonas Eberle Department for Earth Observation, Institute for Geography, Friedrich-Schiller-University, Jena, Germany

Frank Eulenstein Leibniz Centre for Agricultural Landscape Research (ZALF), Müncheberg, Germany; Kuban State Agrarian University, Krasnodar, Russia

Johann Fank Joanneum Research, Graz, Austria

Manfred Frühauf Martin-Luther-University Halle-Wittenberg, Institute of Geosciences and Geography, Halle, Germany

Roger Funk Leibniz Centre for Agricultural Landscape Research (ZALF), Müncheberg, Germany

Nadezhda V. Glushkova V.S. Sobolev Institute of Geology and Mineralogy, Siberian Branch of Russian Academy of Sciences, Novosibirsk, Russia; Novosibirsk State University, Novosibirsk, Russia

Liudmila P. Golobokova Siberian Branch, Limnological Institute (LIN), Russian Academy of Sciences, Irkutsk, Russia

Lars-Christian Grunwald Amazone Company, Hasbergen, Germany

Malte Hamann Amazone Company, Hude, Germany

Katharina Helming Leibniz Centre for Agricultural Landscape Research (ZALF), Müncheberg, Germany

Volker Hennings Hannover, Germany

Martin Herold Centre for Geo-Information, Wageningen University, Wageningen, The Netherlands

Soeren Hese Department for Earth Observation, Institute of Geography, Friedrich-Schiller-University, Jena, Germany

Anna Homolka Department for Earth Observation, Institute for Geography, Friedrich-Schiller-University, Jena, Germany

Heinrich Höper State Authority for Mining, Energy and Geology (LBEG), Hannover, Germany

Christian Hüttich Department for Earth Observation, Institute for Geography, Friedrich-Schiller-University, Jena, Germany

Patrick Illiger Martin-Luther-University Halle-Wittenberg, Institute of Geosciences and Geography, Halle, Germany

Radosław Juszczak Department of Meteorology, Poznan University of Life Sciences, Poznań, Poland

Zheksenbai Kaskarbayev Scientific Production Center of Grain Farming named after A.I. Barayev, Shortandy, Kazakhstan

Tamara Khodzher Siberian Branch, Limnological Institute (LIN), Russian Academy of Sciences, Irkutsk, Russia

Natalia P. Kosykh Siberian Branch, Institute of Soil Science and Agrochemistry (ISSA), Russian Academy of Sciences, Novosibirsk, Russia

Natalia Kowalska Department of Meteorology, Poznan University of Life Sciences, Poznań, Poland

Nikita A. Kozhanov Farm OOO KX “Partner”, Michailovskiy Rayon, Altai Krai, Russia

Barbara Köstner Department of Meteorology, Technical University Dresden, Tharandt, Germany

Marcos Lana Leibniz Centre for Agricultural Landscape Research (ZALF), Müncheberg, Germany

Jacek Leśny Department of Meteorology, Poznan University of Life Sciences, Poznań, Poland

Sergey Lukin Institute for Organic Fertilization and Peat (VNIIOU), Vyatkino, Vladimir Oblast, Russia

Viktoriya A. Lyamina V.S. Sobolev Institute of Geology and Mineralogy, Siberian Branch of Russian Academy of Sciences, Novosibirsk, Russia; Novosibirsk State University, Novosibirsk, Russia

Galina I. Lysanova V.B. Sochava Institute of Geography, Siberian Branch, Russian Academy of Sciences, Akademgorodok, Irkutsk, Russia

Walter H. Mayer PROGIS Software GmbH, Villach, Austria

Tobias Meinel Amazone Company in Kazakhstan, Saryarka District, Astana, Kazakhstan

Ralph Meissner Department Soil Physics, Helmholtz Centre for Environmental Research—UFZ, Lysimeter Station, Falkenberg, Germany

Irina V. Mikheeva Siberian Branch, Institute of Soil Science and Agrochemistry (ISSA), Russian Academy of Sciences, Novosibirsk, Russia

Nina P. Mironycheva-Tokareva Siberian Branch, Institute of Soil Science and Agrochemistry (ISSA), Russian Academy of Sciences, Novosibirsk, Russia

Wilfried Mirschel Leibniz-Centre for Agricultural Landscape Research (ZALF), Müncheberg, Germany

Lothar Mueller Leibniz Centre for Agricultural Landscape Research (ZALF), Müncheberg, Germany; Kuban State Agrarian University, Krasnodar, Russia

Jürgen Müller Johann Heinrich v. Thünen-Institute, Institute of Forest Ecosystems, Eberswalde, Germany

Claas Nendel Leibniz Centre for Agricultural Landscape Research, Institute of Landscape Systems Analysis, Müncheberg, Germany

Janusz Olejnik Department of Meteorology, Poznan University of Life Sciences, Poznań, Poland; Department of Matter and Energy Fluxes, Global Change Research Centre, Czech Academy of Sciences, Brno, Czech Republic

Ludmila Onishenko Kuban State Agrarian University, Krasnodar, Russia

Konstantin Pachikin Kazakh Research Institute of Soil Science and Agrochemistry named after U.U. Uspanov, Almaty, Kazakhstan

Klara G. Pankratova Pryanishnikov All-Russian Research Institute of Agrochemistry, Moscow, Russia

Elena I. Parfenova V.N. Sukachev Institute of Forest, Russian Academy of Sciences, Krasnoyarsk, Russia

Valentina V. Parfenova Siberian Branch, Limnological Institute (LIN), Russian Academy of Sciences, Irkutsk, Russia

Olga N. Pavlova Siberian Branch, Limnological Institute (LIN), Russian Academy of Sciences, Irkutsk, Russia

Denis V. Pchelnikov V.S. Sobolev Institute of Geology and Mineralogy, Siberian Branch of Russian Academy of Sciences, Novosibirsk, Russia

Anna Peregon CEA Centre de Saclay, Laboratoire des Sciences Du Climat et de l'Environnement (LSCE), Gif-sur-Yvette, France

Olga S. Pestunova Siberian Branch, Limnological Institute (LIN), Russian Academy of Sciences, Irkutsk, Russia

Stefan Poecking Department for Earth Observation, Institute of Geography, Friedrich-Schiller-University, Jena, Germany

Oleg S. Pokrovsky Geoscience and Environment Toulouse, UMR 5563 CNRS, Toulouse, France; Institute of Ecological Problems of the North, Ural Branch of Russian Academy of Sciences, Arkhangelsk, Russia; BIO-GEO-CLIM Laboratory, Tomsk State University, Tomsk, Russia

Aleksandr Puzanov Institute for Water and Environmental Problems, Siberian Branch of the Russian Academy of Sciences (IWEP SB RAS), Barnaul, Russia

Vladimir A. Romanenkov Pryanishnikov All-Russian Research Institute of Agrochemistry, Moscow, Russia

Nikolay V. Rudev Farm OOO KX "Partner", Poluyamki, Michailovskiy Rayon, Altai Krai, Russia

Olga V. Rukhovich Pryanishnikov All-Russian Research Institute of Agrochemistry, Moscow, Russia

Karolina Sakowska Department of Meteorology, Poznan University of Life Sciences, Poznań, Poland; Department of Sustainable Agro-Ecosystems and Bioresources, Research and Innovation Centre—Fondazione Edmund Mach, San Michele all'Adige, TN, Italy

Abdulla Saparov Kazakh Research Institute of Soil Science and Agrochemistry named after U.U. Uspanov, Almaty, Kazakhstan

Peter Schad Lehrstuhl für Bodenkunde, Technische Universität München, Freising-Weihenstephan, Germany

Roland Schindler NEW Niederrhein Wasser GmbH, Viersen, Germany; DWA, German Association for Water, Wastewater and Waste, Hennef, Germany

Uwe Schindler Leibniz Centre for Agricultural Landscape Research (ZALF), Müncheberg, Germany

Gerd Schmidt Martin-Luther-University Halle-Wittenberg, Institute of Geosciences and Geography, Halle, Germany

Christiane Schmullius Department for Earth Observation, Institute of Geography, Friedrich-Schiller-University, Jena, Germany

Larisa S. Sharaya Institute of Ecology of the Volga Basin, RAS, Togliatti, Samara Region, Russia

Peter A. Shary Institute of Physicochemical and Biological Problems in Soil Science, Pushchino, Moscow Region, Russia

Vladimir I. Shchelokov Pryanishnikov All-Russian Research Institute of Agrochemistry, Moscow, Russia

Askhad K. Sheudshen Kuban State Agrarian University, Krasnodar, Russia; Russian Scientific Research Institute of Rice, Krasnodar, Russia

Paweł Siedlecki Department of Meteorology, Poznan University of Life Sciences, Poznań, Poland

Cristian Siewert Faculty of Agriculture/Landscape Management, University of Applied Sciences, Dresden, Germany

Elena N. Smolentseva Siberian Branch, Institute of Soil Science and Agrochemistry (ISSA), Russian Academy of Sciences, Novosibirsk, Russia

Amber J. Soja NASA Langley Research Center, Hampton, VA, USA

Larisa M. Sorokovikova Siberian Branch, Limnological Institute (LIN), Russian Academy of Sciences, Irkutsk, Russia

Eckart Stephan Martin-Luther-University Halle-Wittenberg, Institute of Geosciences and Geography, Halle, Germany

Galina A. Stupakova Pryanishnikov All-Russian Research Institute of Agrochemistry, Moscow, Russia

Mekhlis Suleimenov Scientific Production Center of Grain Farming named after A.I. Barayev, Shortandy, Kazakhstan

Mariya Yu. Suslova Siberian Branch, Limnological Institute (LIN), Russian Academy of Sciences, Irkutsk, Russia

Victor G. Sychev Pryanishnikov All-Russian Research Institute of Agrochemistry, Moscow, Russia

Aleksandr Syso Siberian Branch, Institute of Soil Science and Agrochemistry (ISSA), Russian Academy of Sciences, Novosibirsk, Russia

Anatoly A. Tanasienko Siberian Branch, Institute of Soil Science and Agrochemistry (ISSA), Russian Academy of Sciences, Novosibirsk, Russia

Marion Tauschke Leibniz Centre for Agricultural Landscape Research (ZALF), Müncheberg, Germany

Nadezhda M. Tchebakova V.N. Sukachev Institute of Forest, Russian Academy of Sciences, Siberian Branch (SIF SB RAS), Krasnoyarsk, Russia

Vitaly V. Terleev State Polytechnical University St. Petersburg, St. Petersburg, Russia

Alexandre G. Topazh Agrophysical Research Institute, St. Petersburg, Russia

Aibek Tulegenov Scientific Production Center of Grain Farming named after A.I. Barayev, Shortandy, Kazakhstan

Georg von Unold UMS GmbH, Munich, Germany

Marcel Urban Department for Earth Observation, Institute of Geography, Friedrich-Schiller-University, Jena, Germany

Marek Urbaniak Department of Meteorology, Poznan University of Life Sciences, Poznań, Poland

Michael Voltersen Department for Earth Observation, Institute of Geography, Friedrich-Schiller-University, Jena, Germany

Karl-Otto Wenkel Leibniz-Centre for Agricultural Landscape Research (ZALF), Müncheberg, Germany

Ralf Wieland Leibniz-Centre for Agricultural Landscape Research (ZALF), Müncheberg, Germany

Yoshiki Yamagata Center for Global Environmental Research (CGER), National Institute for Environmental Studies (NIES), Tsukuba, Japan

Evgeny N. Yefremov Pryanishnikov All-Russian Research Institute of Agrochemistry, Moscow, Russia

Nadezhda Yermolaeva Novosibirsk Department, Institute for Water and Environmental Problems (IWEP), SB RAS, Novosibirsk, Russia

Klaudia Ziemblińska Department of Meteorology, Poznan University of Life Sciences, Poznań, Poland

Ivan D. Zolnikov V.S. Sobolev Institute of Geology and Mineralogy, Siberian Branch of Russian Academy of Sciences, Novosibirsk, Russia; Novosibirsk State University, Novosibirsk, Russia

Part I
Environmental and Societal Framework for Monitoring and Managing Land and Water Resources

Chapter 1
Land and Water Resources of Siberia, Their Functioning and Ecological State

Lothar Mueller, Askhad K. Sheudshen, Aleksandr Syso, Pavel Barsukov, Elena N. Smolentseva, Tamara Khodzher, Victor G. Sychev, Vladimir A. Romanenkov, Olga V. Rukhovich, Aleksandr Puzanov, Ralf Dannowski, Uwe Schindler and Frank Eulenstein

Abstract Siberia is the backbone of the economy of modern Russia due to huge reserves of gas, oil, land and water. Not only resource extracting and processing industries, but also forestry and agriculture capitalize these resources with implications for local and global processes of nature and society. We analysed the state of land and water resources with regard to the impacts of human activity and climate change. The environmental status of forests, agricultural lands and inland water bodies was evaluated based on our own research and the recent literature. The focus was on agro-ecosystems. Our synthetic review revealed that peatlands and

L. Mueller (✉) · R. Dannowski · U. Schindler · F. Eulenstein
Leibniz Centre for Agricultural Landscape Research (ZALF) e.V., Eberswalder Strasse 84, 15374 Müncheberg, Germany
e-mail: mueller@zalf.de

A.K. Sheudshen
Kuban State Agrarian University, 13 Kalinin Str, Krasnodar 350044, Russia
e-mail: kubagrohim@mail.ru

A. Syso · P. Barsukov · E.N. Smolentseva
Siberian Branch, Institute of Soil Science and Agrochemistry (ISSA), Russian Academy of Sciences, Ac. Lavrentieva Av 8, Novosibirsk 630090, Russia
e-mail: syso@issa.nsc.ru

T. Khodzher
Siberian Branch, Limnological Institute (LIN), Russian Academy of Sciences, 3 Ulan-Batorskaya St, Irkutsk 664033, Russia
e-mail: khodzher@lin.irk.ru

V.G. Sychev · V.A. Romanenkov · O.V. Rukhovich
Pryanishnikov All-Russian Research Institute of Agrochemistry (VNIIA), Pryanishnikova St. 31a, Moscow 127550, Russia
e-mail: geoset@yandex.ru

A. Puzanov
Institute for Water and Environmental Problems, 1 Molodyoznaya St, Barnaul 656038, Altai Krai, Russia
e-mail: puzanov@iwep.ru

L. Mueller et al. (eds.), *Novel Methods for Monitoring and Managing Land and Water Resources in Siberia*, Springer Water,
DOI 10.1007/978-3-319-24409-9_1

Tundra ecosystems are endangered by resource-extracting industries and industrial air pollution. Mining and industrial activity damage soil and vegetation and accelerate thermokarst processes. Forest ecosystems suffer increasingly from fires, insect outbreaks and improper management. Past and recent mining and industrial activity has polluted soils and water seriously in many regions. Permafrost melting could expose cases of old and inherited pollution. The impact of agriculture on water quality is still low but will increase. Agriculture is in a recession and operates inefficiently, destroying the soil. There is largely a lack of any agri-environmental monitoring in many regions. The rural infrastructure is on the verge of collapse, in the High North and the Far East in particular. State natural reserves (zapovedniks) are endangered by illegal activities and lack integration into scientific monitoring. Overall, monitoring programmes on the status of land and water resources lack consistency and modern technology. Climate change will put a great deal of additional pressure on Siberian landscapes, but hard data are required, and monitoring systems need to be modernized. Siberian landscapes have great potential for the mitigation of climate change through carbon sequestration and for improving people's livelihoods. Environmentally friendly business activities such as organic food production, environmental tourism and recreational fishing are still underdeveloped. We conclude that the status of food production and the disintegration of rural areas are risks for Russian food security and national security. Modern technologies for monitoring and research ecosystems are needed to generate sustainable developments in managing the land and water resources of Siberia.

Keywords Siberia · Land · Water · Quality · Environment · Ecosystems · Monitoring

1 Attributes and Resources of Siberia

Siberia is the region beyond the Urals, the Asian part of Russia, from a European perspective. It is a region of extremes and superlatives. The extremely harsh climate and massive stocks of natural resources—land, water, oil, gas, and minerals—determine its status of recognition and further development. Siberia is the coldest populated area on the planet, characterised by permafrost soil, snow and ice. The Siberian Taiga is one of the largest global ecosystems, comprising about one third of the total forest land on our planet. Siberia includes vast and impassable bogs and swamps, amongst them the Vasyugan bog, the largest global peatland. Forest Steppe and steppe regions in the south are part of the longest agricultural belt in the world. Yenisei, Lena, Ob-Irtysh and Amur rank amongst the top twelve global rivers, and Lake Baikal is the deepest freshwater lake, containing the largest available freshwater volume of the planet. Siberia's huge dimensions are impressive, always creating the illusion of the exhaustlessness of its resources.

Siberia is the treasure chamber and a basis of the economy of modern Russia. More than 70 % of the oil resources and more than 80 % of the gas resources of the country are stored in the north-west of West Siberia (Adam and Mamin 2001). The exploitation of gas, oil and minerals has been the driver for the increasing welfare of the urban population over the past years.

Siberia has much potential but faces some risks. Human exploration and climate change will induce a dynamic of ecosystem alterations with impacts for global cycles (Shvidenko et al. 2013). Some aspects and trends of climate change have been already recognized. These are global warming and the release of greenhouse gases into the atmosphere. Further resource exploration and utilisation and accelerated global warming may pose threats to nature including the human population. When thinking in terms of "environmental currencies" such as carbon sequestration potentials, Russia will be a leading global player due to Siberia's resources (Lioubimtseva 2010).

Food security is another issue of importance for Russia, but agricultural land and water management are in a recession (Gordeev and Romanenko 2008). The sustainable production of traditional goods such as local food or timber, which are based on healthy soil and water bodies, requires more scientific data and public awareness. We are at the beginning of understanding the complexity of these processes and re-evaluating land and water resources. Decisions about possible controlling and mitigating mechanisms need a permanent scientific background. What are the consequences of global warming and resource utilisation for soil and water quality, for the agriculture, forestry, human health and biodiversity of Siberia? There is a high risk that our understanding of these processes cannot keep pace with the speed of alterations, making them more and more uncontrollable. Trends in changes to the terrestrial and aquatic ecosystem need to be monitored permanently.

Are our available methods and tools for research and monitoring landscape processes doing well and do they meet future requirements? One aim of this book is to help maintain human society's capability to control processes and minimise the risks for future generations. This will be done by explaining some scientific, technological methods for understanding, monitoring, forecasting and controlling landscape processes in Siberia. The focus is on the main resources of land and water, on terrestrial and aquatic ecosystems. This introductory chapter analyses some basic settings and drivers of ecosystem change.

2 Geographical Regions and Landscapes in a Nutshell

From its physical geography, Siberia can be subdivided into several geomorphological areas: the West Siberian Plain, the Central Siberian Plateau, the Mountains of South Siberia and the Uplands of Northeast Siberia (Fig. 1). In-between the upland areas are further large lowlands such as the North Siberian Lowland, the Central Yakutia Lowland and the East Siberian Lowland.

Fig. 1 Sketch of geomorphological landscape units. *Map* Ralf Dannowski

The West Siberian Plain begins east of the Urals and mainly covers the Ob River basin, along with some western parts of the Yenisei basin. In a world without glaciers much of this land and of the North Siberian Lowland would fall under the elevated sea level. The Yamal Peninsula and the land south of the Ob mouth would disappear. On a sketch drawn by M.G. Groswald and M.N. Sverkova one can recognise it as the largest closed area on the Northern Hemisphere which the ocean would claim back (Kotlyakov 1986). The West Siberian Lowland comprises more than a thousand lakes and ponds with an area of more than 100 thousand square kilometres (Antipov 2006).

The Central Siberian Plateau stretches roughly from the Yenisei River to the Lena River. It is bounded on the south by the Baikal mountain system and on the north by the North Siberian Lowland. The South Siberian Mountains system comprises a number of mountains such as the Altai, Sayan and Baikal Mountains. It is the spring and headwater area of the Ob, Yenisei and Lena rivers. The Northeast Siberian Uplands consist of some average and high mountain chains east of the Lena River.

From a historical European perspective, Siberia is considered the Non-European part of Russia. As Europe and Asia have no clear natural border, Siberia is also not an exactly defined geographical region. Based on administrative units of the Russian Federation the designation of Siberia is associated with the Federal District of Siberia (administrative centre: Novosibirsk). Other Federal Districts (Fig. 2) east of Europe are Ural (adm. centre: Yekaterinburg) and the Far East (adm. centre: Khabarovsk). Adding the area and population of these three Federal Districts (Siberia in a wider sense) reveals an area of 13.1 million square kilometres and a population of 37.6 million (Wikipedia 2013a). This is roughly comparable with Canada, the second largest country in the world, covering an area of about

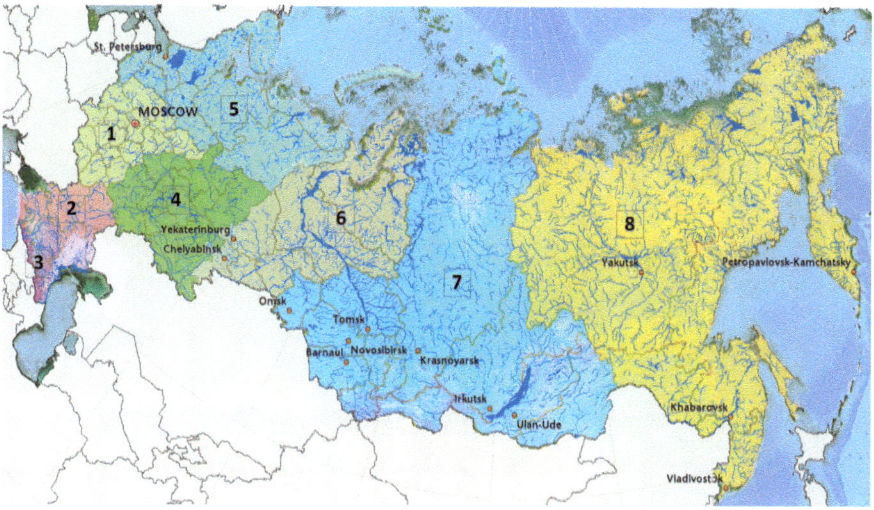

Fig. 2 Federal Districts (FD) of the Russian Federation. *1* Central, *2* Southern, *3* North Caucasian, *4* Volga, *5* Northwestern, *6* Ural, *7* Siberian, *8* Far Eastern. The map demonstrates the huge dimensions of the FD in Asian Russia (*7*, *8* and partly *6*). Note that FD's are administrative units and not the Federal subjects. Detailed information is given in Wikipedia (2013a). *Map* Ralf Dannowski

10 million square kilometres and inhabited by about 35 million people. Other sources exclude areas of the Russian Far East and the Ural districts from the definition of Siberia, then referring to an area of about 10 million square kilometres (Antipov 2006; Groisman and Gutman 2013).

Siberia is extremely sparsely populated but cannot be considered as a rural region. The region has a relatively short history. The Russian colonization started in about 1750. Siberia's exploration is well set out in the Museum of History in Irkutsk (Fig. 3). The emergence and growth of cities has been associated with the Trans-Siberian Railroad and industrial complexes. The population is concentrated in some large cities of the south. About 70 % of the entire population lives there whilst large regions are virtually uninhabited (Groisman and Gutman 2013). Cultural landscapes with stable rural centres, traditions in soil and water management and owners aware of how to sustainably handle land and water resources are still underdeveloped or not even found in Siberia. This has implications for the perception and evaluation of landscape values by the population.

Novosibirsk, the third largest city of Russia, is the political, social, scientific and cultural centre of Siberia. It lies about 1400 km south-west from the geographical centre of Russia (Lake Vivi, Evenki district of the Krasnoyarsk Krai). The distance from Novosibirsk to locations in the Far East such as Petropavlovsk Kamchatsky, or to the Bering Strait, is greater than the distance to the North Sea at Hamburg.

Fig. 3 The historical museum in Irkutsk presents the exploration and development of Siberia. *Photo* Ralf Dannowski

3 Climate

3.1 Temperature

Siberia is cold and dry, with particularly severe and long winters. Towards the south, mountain ranges protect it from the direct influx of water vapour and heat from the tropics and also largely from the moderate Atlantic climate system. The dominating climate-forming factor of Central Siberia is the Arctic Ocean. Cold and dry air masses form over the Arctic and dictate the climate of most parts of Siberia over the winter and the northern parts all year round (Richter 1963; Groisman and Gutman 2013). Most cities of Siberia are located between 52° and 56° Northern Latitude. This is comparable with some West and North European cities. Ulan-Ude has about the same latitude as London, Irkutsk that of Amsterdam or Berlin, Novosibirsk and Omsk that of Copenhagen or Moscow. Yekaterinburg, Tomsk and Krasnoyarsk lie slightly further south than Oslo, Stockholm and St. Petersburg. The position of Yakutsk is comparable with that of Trondheim. Large differences in winter temperatures between Europe and Siberia can largely be explained by the warm water heating system of the Gulf Stream which Europe benefits from. January temperatures in North-Western Europe are not far from the freezing point. They are clearly less than −15 °C in most parts of Siberia (Table 1). Summer temperatures in South Siberia are comparable with or even slightly higher than those in the above-mentioned European cities. Most climate stations have negative mean annual values indicating permafrost soil conditions. The cities of the Far East lie further south. The latitude of Khabarovsk is comparable with that of Paris or Vienna, and Vladivostok with Marseille or Sofia. Much colder winters are found in the Far East than in Europe at the same latitudes.

Table 1 Main climate data of some stations in Siberia and the Far East

No.	Location	Latitude °N	Longitude °E	Altitude masl	Koeppen zone	Temp ° C	Temp January	Temp July	P mm	ETp mm	Aridity
1	Cape Chelyuskin	77.70	104.30	13	ET	−15.3	−28.9	1.3	247	233	Humid
2	Ostrov Dikson	73.50	80.40	47	ET	−12.1	−26.9	4.4	365	223	Humid
3	Khatanga	71.98	102.46	33	Dfc	−13.3	−33.8	12.5	270	234	Humid
4	Chocurdakh	70.62	147.91	61	ET	−14.3	−34.6	9.5	232	240	Humid
5	Olenyok	68.51	112.48	127	Dfc	−12.0	−37.5	14.8	291	307	Subhumid
6	Verkhoyansk	67.55	133.38	137	Dfd	−15.3	−47.0	15.1	177	319	Dry subhumid
7	Igarka	67.46	86.56	30	Dfc	−8.8	−28.9	14.8	675	237	Humid
8	Beryozovo	65.05	63.33	27	Dfc	−3.6	−22.5	16.7	530	336	Humid
9	Tarko-Sale	64.91	77.81	27	Dfc	−6.4	−25.8	16.2	488	290	Humid
10	Markovo	64.68	170.41	3	Dfc	−9.7	−24.0	13.8	407	278	Humid
11	Tura	64.16	100.06	186	Dfc	−9.3	−36.3	16.5	379	322	Humid
12	Oymyakon	63.26	143.15	726	Dfd	−16.4	−47.0	13.8	220	309	Subhumid
13	Omsukchan	62.51	155.78	803	Dfc	−9.7	−29.5	12.8	414	335	Humid
14	Suntar	62.15	117.65	124	Dfc	−7.7	−33.5	17.5	276	398	Subhumid
15	Yakutsk	62.08	129.75	103	Dfd	−10.0	−41.0	18.7	239	541	Dry subhumid
16	Baykit	61.66	96.36	179	Dfc	−6.5	−30.8	17.0	517	322	Humid
17	Surgut	61.25	73.50	44	Dfc	−2.3	−21.8	17.7	675	438	Humid
18	Kolpasevo	58.30	82.90	76	Dfc	−1.3	−20.0	18.3	495	434	Humid
19	Kirensk	57.76	108.11	258	Dfc	−4.4	−27.7	17.7	387	391	Subhumid
20	Tyumen	57.15	65.50	104	Dfb	1.0	−17.3	18.3	524	543	Subhumid
21	Chara	56.91	118.27	711	Dwc	−7.8	−33.3	15.8	357	372	Subhumid
22	Yekaterinburg	56.83	60.63	310	Dfb	1.4	−15.4	17.5	565	467	Humid

(continued)

Table 1 (continued)

No.	Location	Latitude °N	Longitude °E	Altitude masl	Koeppen zone	Temp ° C	Temp January	Temp July	P mm	ETp mm	Aridity
23	Ayan	56.45	138.15	9	Dwc	−2.87	−18.9	11.3	892	345	Humid
24	Krasnoyarsk	56.03	92.75	275	Dfc	0.7	−17.0	19.0	485	496	Subhumid
25	Kurgan	55.46	65.40	74	Dfb	1.9	−17.3	19.7	388	560	Subhumid
26	Novosibirsk	55.03	82.90	162	Dfb	0.0	−19.2	18.8	514	462	Humid
27	Omsk	54.93	73.40	94	Dfb	1.3	−17.3	19.7	390	508	Subhumid
28	Bomnak	54.71	128.85	357	Dwc	−4.8	−31.1	17.7	584	424	Humid
29	Slavgorod	53.96	78.65	115	Dfb	−0.2	−20.0	19.0	313	531	Dry subhumid
30	Romanovka	53.21	112.76	923	Dwc	−5.53	−29.3	16.7	365	432	Subhumid
31	Petropavlovsk-Kamchatsky	52.96	158.75	7	Dfc	1.5	−8.2	11.1	995	400	Humid
32	Biysk Zonalny	52.68	84.95	224	Dfb	0.5	−18.5	18.7	625	482	Humid
33	Irkutsk	52.26	104.35	485	Dwc	0.0	−18.8	17.5	466	475	Subhumid
34	Ulan-Ude	51.80	107.43	51	BSk	−1.0	−24.0	18.8	266	559	Semiarid
35	Kyzyl	51.71	94.50	628	Dfb	−4.3	−34.0	19.7	253	555	Semiarid
36	Nerchinskiy Savod	51.31	119.61	619	Dwc	−2.8	−26.9	18.0	435	457	Subhumid
37	Kosh Agash	50.01	88.68	1758	BSk	−6.5	−31.5	13.6	126	511	Semiarid
38	Khabarovsk	48.51	135.16	72	Dwb	1.8	−20.8	21.2	710	587	Humid
39	Ekaterino-Nikolskoye	47.73	130.96	74	Dwb	1.9	−20.3	21.1	619	589	Humid
40	Vladivostok	43.11	131.90	138	Dfb	4.2	−13.2	17.2	786	590	Humid
	Minimum					−16.4	−47.0	1.3	126	223	
	Maximum					4.2	−8.2	21.2	995	590	

Source Database LocClim 1.10 (FAO 2006). Stations were sorted by latitude. Masl = meters above sea level, P = precipitation, ETp = potential evapotranspiration, Koeppen zones: ET = Tundra climate, Dfc = boreal, humid, cool continental/subarctic, Dfd = boreal, humid, cold continental/subarctic, Dfb = boreal, temperate humid continental with warm summers, Dwb = boreal, dry winters, temperate continental/humid continental, Dwc = boreal, dry winters, cool continental/subarctic, BSk = semi-arid cold Steppe

Fig. 4 Snowbound landscape of the Siberian Forest Steppe. Snow cover is shallow and unevenly distributed. *Photo* Alexander S. Chumbaev, ISSA Novosibirsk

3.2 Precipitation

In West Siberia the annual rainfall is moderate to low, ranging from about 300 to 600 mm depending on latitudes and altitudes. Central and East Siberia receive low to moderate precipitation heights of 300 to 400 mm. Continental Yakutia is not only extremely cold but also an extremely dry region with precipitation of less than 300 mm. As part of the precipitation falls as snow, it cannot be used by the vegetation completely (Fig. 4). Snowmelt above a partly frozen soil causes surface and subsurface lateral discharge processes. In the Far East region belonging to the monsoon climate zone, precipitation is high.

3.3 Range of Climate Data

The wide range of basic climate data at stations east of the Urals is visible from Table 1. The mean annual temperatures of the stations range from −16.4 °C (Oymyakon) to 4.2 °C (Vladivostok). January temperatures range from −47 °C (Verkhoyansk and Oymyakon) to −8.2 °C (Petropavlovsk-Kamchatsky). July temperatures range from 1.3 °C (Cape Chelyuskin) to 21.2 °C (Khabarovsk). Annual precipitation varies between 126 mm (Kosh Agash) and 995 mm (Petropavlovsk-Kamchatsky). Potential evaporation ranges between 223 mm (Ostrov Dikson) and 590 mm (Vladivostok). All positive extremes were measured in humid and maritime regions of the Far East. The continental climate of most parts of Siberia leads to practically two main seasons only: winter and summer. Spring and autumn are only short transition phases between them.

3.4 Short Description of the Climate Station Locations

The characterisation of climate stations refers to Table 1 and is based on different sources. The position of the stations is given in Fig. 5.

1. **Cape Chelyuskin** (Мыс Челюскин) on the Taimyr Peninsula is both the northernmost point of the Siberian mainland and the northernmost continental place on earth. It belongs to the Krasnoyarsk Krai.
2. **Ostrov Dikson** (Ди́ксон) is a very small island and settlement near the mouth of the Yenisei River in the Taimyrsky Dolgano-Nenetsky District of the Krasnoyarsk Krai.
3. **Khatanga** (Хатанга) is a rural settlement on the Taimyr Peninsula, in the Krasnoyarsk Krai. It is located on the river of the same name.
4. **Chokurdakh** (Чокурда́х) is a settlement and the administrative centre of the Allaikhovsky District of the Sakha Republic. It is located on the Indigirka River.
5. **Olenyok** (Оленёк) is a settlement and the administrative centre of the Olenyoksky District in the Sakha Republic. It is located on the river of the same name.
6. **Verkhoyansk** (Верхоянск) is a small town in the Sakha Republic, located on the Yana River.
7. **Igarka** (Игарка) is a town on the Yenisei River in the Krasnoyarsk Krai.

Fig. 5 Permafrost zones of the Russian Federation and some representative climate stations in Asian Russia (numbers acc. to Table 1.1). Permafrost zones after Roshydromet (2008b). *1* Zone where permafrost may occur on less than 50 % of the land, *2* zone where permafrost may occur on 50–90 % of lands, *3* complete permafrost zone (more than 90 % of lands) *4* zone of seasonal freezing (no permafrost). *Map* Ralf Dannowski

8. **Beryozovo** (Берёзово) is an urban settlement and the administrative center of Beryozovsky District of Khanty-Mansi Autonomous Okrug, located on the Ob River.
9. **Tarko-Sale** (Та́рко-Сале́) is a town in the Yamalo-Nenets Autonomous Okrug. It is located on the Pyakupur River near its confluence with the Ayvasedapur.
10. **Markovo** (Марково) is a settlement in the Chukotka Autonomous Okrug, Anadyrsky District, Far East, close to the Anadyr River.
11. **Tura** (Тура) is a settlement in the Krasnoyarsk Krai and the centre of the Ewenkiyskiy Rayon. It is located on the Nizhnyaya Tunguska River.
12. **Oymyakon** (Оймякон) is a settlement in the Republic of Sakha (Yakutia), considered as the coldest place of all inhabited areas on earth.
13. **Omsukchan** (Омсукчан) is an urban settlement and the administrative centre of the Omsukchansky District of the Magadan Oblast, Far East. It is located on the river of the same name, a tributary of the Kolyma River.
14. **Suntar** (Сунта́р) is a rural settlement on the Vilyuy River. It is located in the Republic of Sakha and the administrative centre of the Suntarsky District.
15. **Yakutsk** (Якутск) is the capital city of the Sakha Republic (3.08 million square kilometres) located on the Lena River.
16. **Baykit** (Байкит) is a rural settlement in the Evenkiysky District of the Krasnoyarsk Krai. It is located on the Podkamennaya Tunguska River.
17. **Surgut** (Сургут) is a city in the Khanty-Mansi Autonomous Okrug, Russia, located on the Ob River near its confluence with the Irtysh River.
18. **Kolpashevo** (Колпа́шево) is a town and the administrative centre of the Kolpashevsky District in the Tomsk Oblast, located on the Ob River.
19. **Kirensk** (Киренск) is a town and the administrative centre of the Kirensky District in the Irkutsk Oblast. It is located on the confluence of the Kirenga River into the Lena River.
20. **Tyumen** (Тюмень) is a city and the administrative centre of the Tyumen Oblast, located on the Tura River.
21. **Chara** (Чара) is a settlement in the north Trans-Baikal Region, known for the Chara Sands, a big dune area within the larch Taiga. The settlement is located on the river of the same name.
22. **Yekaterinburg** (Екатеринбург) is the second-largest city east of the Urals and the administrative center of the Sverdlovsk Oblast.
23. **Ayan** (Аян) is a rural settlement and the administrative centre of the Ayano-Maysky District of the Khabarovsk Krai, Far East. It is located on the shore of the Sea of Okhotsk.
24. **Krasnoyarsk** (Красноярск) is a city and the administrative centre of the Krasnoyarsk Krai (2.34 million square kilometres), located on the Yenisei River.
25. **Kurgan** (Курган) is a city and the administrative center of the Kurgan Oblast in South-West Siberia, located on the Tobol River.
26. **Novosibirsk** (Новосибирск) is the largest city of Siberia, located on the Ob River. It is the administrative centre of the Siberian Federal District and the Novosibirsk Oblast.

27. **Omsk** (Омск) is a city and the administrative center of the Omsk Oblast. It is located at the confluence of the Irtysh and Om Rivers.
28. **Bomnak** (Бомнак) is a rural settlement in the Zeysky District of the Amur Oblast, located at the Zeyskiy reservoir.
29. **Slavgorod** (Сла́вгород) is a town in the Altai Krai in the Kulunda Steppe. The region is largely an inland watershed, has extremely low or no discharge.
30. **Romanovka** (Романовка) is a rural settlement in the Vitimsky Selsoviet of the Bauntovsky Evenkiysky District, Republic of Buryatia. It is located on the Vitim River, a tributary of the Lena River.
31. **Petropavlovsk-Kamchatsky** (Петропа́вловск-Камча́тский) is a city and the administrative centre of the Kamchatka Krai, Far East.
32. **Biysk** (Бийск) is a city in the Altai Krai located on the Biya River close to its confluence with the Katun River (from then on called Ob River).
33. **Irkutsk** (Иркутск) is a city and the administrative centre of the Irkutsk Oblast. It is located on the Angara River, close to its outflow from Lake Baikal.
34. **Ulan-Ude** (Улан-Удэ) is the capital city of the Republic of Buryatia, located about 100 km southeast of Lake Baikal, close to the Selenga River.
35. **Kyzyl** (Кызыл) is the capital of the Tuva Republic on the Yenisei River.
36. **Nerchinsk** (Не́рчинск) is a town and the administrative centre of the Nerchinsky District of the Zabaykalsky Krai (Trans-Baikal Region). The climate station lies close to the Chinese border, which is the Argun River, the southern headwater stream of the Amur.
37. **Kosh-Agach** (Кош-Ага́ч) is a rural settlement and administrative centre of the Ten Rayon in the Altai Republic.
38. **Khabarovsk** (Хабаровск) is a city of the Far East at the confluence of the Amur and Ussuri Rivers close to the Chinese border. It is the administrative centre of the Khabarovsk Krai.
39. **Ekaterino-Nikolskoye** (Екатерино-Никольское) is a settlement in the Jewish Autonomous Oblast, Far East. It is located on the banks of the Amur River, directly at the Chinese border.
40. **Vladivostok** (Владивосток) is a city and the administrative centre of the Primorsky Krai, Far East. It is located on the southern tip of the Muravyov-Amursky Peninsula close to the borders with China and North Korea.

4 The Siberian Permafrost Zone

4.1 Extent and Features

About 80 % of the Siberian area is composed of permafrost in the subsurface (Suzuki 2011). Permafrost soil formed in the latest glacial period (Sartanian = Weichselian = Wisconsinan) and remained largely because the temperature was 8–10 K (°C) higher, though still negative, in most parts of Siberia during the

Holocene (Anisimov et al. 2002). Permafrost depths range between a few metres at its current southern boundary in West Siberia (≈60–62°N) and up to about one kilometre in the coldest parts of Yakutia (Baulinym and Belopukhovoy 1963; Suzuki 2011). South of 60°N it may exist as a relict at greater depth. At very cold climates in the north (< −8 °C, French 1999) active permafrost formation occurs. Within the permafrost zone, non-frozen parts of the ground (Taliki) exist beneath lakes and rivers. Mounds of earth-covered ice (Pingos, Bulgunniakhs) are typical landscape elements in the permafrost region. Detailed maps of permafrost have been developed for West Siberia (Shpolyanskaya 1971; Pavlov 2008) and general maps for whole Russia (Roshydromet 2008b; Fig. 5).

Permafrost landscapes are characterised by thermokarst processes and resulting landforms. Thermokarst expresses the ablation of excess ice in permafrost and the subsequent consolidation of the soil or bedrock (Murton 2009). A review by the same author shows that modes of thermokarst activity comprise (1) deepening of the freezing-thawing layer (active layer), (2) ice-wedge melting, (3) thaw slumping, (4) groundwater flow, (5) shoreline thermokarst and (6) basin thermokarst (Murton 2009). Alases are also typical landscape elements of the permafrost zone formed by thermokarst in an aridic climate of the type which prevails in Yakutia. They are shallow depressions which are formed by soil subsidence because of repeated melting and refreezing. They have been developed from former temporary shallow meltwater lakes in depressions. Trees can no longer grow under these bad drainage conditions and are replaced by grasses and herbs, which have a steppe-like feature and can be used as pasture and hayland.

Soil formation takes place in the active layer. It comprises sorting of soil material, shrinkage and swelling, frost wedge formation and polygonal patterns (French 1976).

4.2 *Permafrost Thawing*

The permafrost area and thickness was reduced during the Holocene. Warming of the Arctic region continues (Climate Committee 2002; ACIA 2004; Fedotov et al. 2012). Frozen and non-frozen grounds differ crucially in important properties and processes such as mechanical stability and biological activity. As thawing starts from the soil surface, the increasing thickness of the active layer is of most interest and concern. The depth of the active layer is commonly 1–2 m depending on latitude and soil. It is shallow on organic soils in the Tundra zone and deep on sandy soils in the Taiga zone (Baulinym and Belopuchovoy 1963). Active-layer deepening has many consequences for landscape functions and properties. Frozen ground is stable and hard. At the beginning of thawing the soil is over-saturated and shows quasi-plastic or liquid behaviour; it is prone to translocation and bodies or structures sinking in. Distortions and destruction of buildings, roads, railways, pipelines and other structures are a consequence with a huge economic impact. Solifluction, erosion, changes in hydrology (Woo 2012) and vegetation, and other processes lead to a transformation of existing landforms (Anisimov and Reneva 2006).

A large amount of stored carbon is being lost by the soil and released to the atmosphere, mostly in the form of methane (Tarnocai et al. 2009; Zimov et al. 2006). Pleistocene Yedoma permafrost contains nearly a third of all organic matter (OM) stored in circum-Arctic permafrost. Yedoma OM is highly biologically available (biolabile) upon thawing (Vonk et al. 2013). Its decomposition after thawing will boost nitrogen and phosphorus cycles with consequences for stocks, fluxes and the whole ecosystem (Mack et al. 2010).

Some feedback processes triggered by climate warming, such as the expansion of deciduous shrubs, may reduce fast summer permafrost thawing (Blok et al. 2010). Monitoring of the permafrost zone (cryolithic zone) in Siberia includes some fundamental processes such as groundwater hydrology, the thermal regime of the active layer, and numerous cryogenic geological processes such as ground deformation and translocation, solifluction, thermoabrasion of shorelines, thermokarst, thermoerosion and ice dynamics (Pavlov 2008).

Human impacts have particularly big consequences in permafrost landscapes. Measures of industrial and urban activity lead to damage or the destruction of the vegetation. This leads to fast melting, active layer deepening, erosion and solifluction (Chuprova 2006). Successional re-vegetation is too slow in this region, thus immediate re-greening by landscaping (biological recultivation) will be required.

5 Bio-geographical Zones

Concept of bio-geographical zones: landscape bio-geographical zones (natural zones, geographical zones, vegetation zones, eco-zones) were defined and parameterised based on the ideas and work of Dokuchaev at the end of the nineteenth century (in Dokuchaev 1949), and Berg (1947). The main zones relevant to Siberia are Tundra, Taiga, Forest Steppe and Steppe (Gerasimov et al. 1971; Lydolph 1990). Further transition zones such as the Forest Tundra and the aspen-birch forest, or sub-divisions of the Taiga into northern, middle and southern regions are based on temperatures in the vegetation period but also associated with a winter climate (Shvareva 1963; Table 2). The values in this table are averages over long observation periods in the first half of the twentieth century, not classification thresholds.

Current geographical zoning can be done based on the scheme of Milkov (1977), Table 3. This scheme holds for all zones of the former USSR and is based on thermal conditions for the Tundra and Taiga zones and on thermal and moisture conditions of the temperate zones. It sub-divides the Tundra zone into Arctic desert, Arctic Tundra and common Tundra. This scheme allocates many parts of South Siberia to the temperate forest zone. There is no more southern Taiga. A number of locations in the south of West Siberia characterised by southern Taiga vegetation

Table 2 Temperatures of the warmest month and length of snow cover in different natural zones of West Siberia (Shvareva 1963 p. 96, excerpt)

Natural zone	Temp. °C warmest month	Permanent snow cover	
		Begin	End
Tundra	10–13	10 October	1–10 June
Northern Taiga	15–16	10–20 October	10–20 May
Middle Taiga	17	20 October–1 November	20 April–1 May
Southern Taiga	17–18	1 November	20 April
Aspen-birch forest (Sub-Taiga)	17–18	1–10 November	20 April
Forest Steppe	19–20	10 November	10–20 April
Steppe	20	10 November	10 April

Table 3 Geographical zones of Russia relevant to Siberia (Milkov 1977, shortened[a])

Zone	Radiation balance kcal/cm^2	Group of zones				
		Desert	Semi-desert	Steppe	Forest Steppe	Forest
Polar	0 to 30–35	Thermal conditions: average air temperature of the warmest month				
		<5°	5–6°	6–11°	11–14°	14–17°
		Arctic desert	Arctic Tundra	Tundra	Forest Tundra	Taiga (northern and middle)
		Moisture conditions: Moisture coefficient and radiation index of dryness (in parentheses)				
		0.12–0.00 (>3)	0.29–0.13 (2–3)	0.59–0.30 (1.5–2)	0.99–0.60 (1.5–1)	>1 (<1)
Temperate	30–35 to 55–60	Desert	Semi-desert	Steppe	Forest Steppe	Broadleaf mixed forest

[a]Subtropical and tropical zones omitted; Moisture coefficient after Vysotsky-Ivanov: $K = R/E_p$, where R = sum of precipitation and E_p is potential evapotranspiration, K is considered as Aridity Index (AI) in the FAO climate database LocClim 1.10 (FAO 2006); note that values of that K can differ from AI because of different methodology of E_p computation. Radiation index of dryness after Budyko $R_i = R/L_r$, where R_s is the radiation balance and L_r is the heat sum required for the evaporation of precipitation. R_i < 0.45 excessive, R_i = 0.45–1.00 sufficient R_i = 1.00–3.00 insufficient. Both indexes are given in the LocClim database (FAO 2006)

with admixtures of aspen and birch but having a July temperature of 17–19 °C fall into the broadleaf forest or Forest Steppe class of the temperate zone.

There are further systems for classifying ecological zones. The Ecological Atlas of Russia (Isachenko 1990) published on the website of the Lomonosov Moscow State University, Faculty of Geography, contains further detailed information about geographical landscape zoning. Geographical zones are associated with typical vegetation and soil forming and landscape structure and processes (Richter 1963).

More and more data are interpreted in the context of landscape zones: for example, patterns of lake water elements in the landscape are like the fingerprints of geographical zones and regions (Kremleva et al. 2012).

Tundra The Tundra zone stretches south of the Arctic Ocean and the Arctic desert. It is a roughly 200–800-km-wide zone north of the Arctic Circle or north of 70° northern latitude, if located in the Central Siberian Upland. This zone has been formed by glacial processes since the last ice age. The Tundra can be subdivided into the Arctic Tundra (coast regions of the Arctic Ocean, southern part of Novaya Zemlya), the typical or moss-lichen Tundra, and the subzone of the southern Tundra (shrub Tundra). July temperatures may give an orientation. They are less than 5–6 °C in the Arctic Tundra, about 6–9 °C in the typical Tundra and 9–11 °C in the southern Tundra (Milkov 1977). Typical zonal soils are Tundra-Gley Soils (Gleyic Cryosols). Due to the Arctic climate, biogeochemical processes occur slowly and with low intensity. Soils are characterised by a permafrost regime, poor drainage and wet humus accumulation. Plant growth is limited by a short vegetation period, low biochemical weathering intensity of rock, wind intensity and other factors. Mosses, lichens, grasses and dwarf shrubs prevail (Fig. 6). Photosynthesis and plant growth benefit from the permanent solar radiation in the short summer whilst plant decomposition is limited by low temperatures. This leads to a very slow but

Fig. 6 Mountain Tundra (**a**) and Forest Tundra (**b**). Reindeer (**c**) are the most important domestic animals and the living basis of the indigenous peoples in the High North. *Photos* Pavel Barsukov

permanent accumulation of organic material. Half of this material is carbon, and the Tundra is a carbon sink.

The occurrence of previously accumulated loess-like soil material in the Tundra zone is a typical feature of the Siberian subarctic region. Soils of Yedoma complexes (Schirrmeister et al. 2008) are rich in nutrients and provide higher ecosystem productivity. This ecosystem is the Tundra Steppe (Schirrmeister et al. 2008) or Mammoth Steppe (Zimov et al. 1995). Not only global climate change at the end of the Pleistocene, but also human activity in remaining Mammoth Steppe areas could have led to the extinction of this characteristic mammal (Zimov et al. 1995). The main current human activities in the Siberian Tundra are reindeer farming and oil and gas extraction.

Forest Tundra The Forest Tundra is a roughly 20–200-km-wide transition zone between Tundra and Taiga, characterised by very sparse stands of coniferous vegetation. Forest Tundra and the northern treeline fluctuate with climate change. Six thousand years ago, the northern treeline on the Taimyr Peninsula was located about 150 km further north than at present due to the warmer climate (Juday et al. 2009).

Taiga This is characterised by coniferous forests. The Taiga is the main vegetation zone of Russia, stretching south of the Tundra over the whole of Siberia and westwards to Scandinavia as a closed belt. It is an approximately 1000–2000-km-wide zone south of a line connecting Ust Kamchatsk, Verchoyansk, Olenek, Igarka, Salechard, Mesen and Murmansk and north of a line connecting Sakhalin, Irkutsk, Gorno-Altaisk, Kemerovo, Tomsk, Tyumen, Yekaterinburg, Izhevsk, Nizhniy Novgorod, Tula and St. Petersburg. Taiga patches also exist along mountain ranges. Typical zonal soils of the Taiga are Podsols (northern and middle Taiga) and Derno-Podsols (southern Taiga). The main coniferous tree species are larch, spruce, fir, cedar, and pines (Fig. 7). Larch, which is a deciduous conifer, prevails in the coldest parts of the Taiga zone. In the southern Taiga, admixtures of small-leaved trees such as birch, aspen and poplar are typical. The Taiga zone also includes large or small patchy wetlands free of trees where willow bushes or reed

Fig. 7 Pine-Taiga in the Altai Mountains (**a**). Near villages land was cleared for grazing milking cows (**b**)

vegetation and mosses occur depending on the drainage conditions and availability of plant nutrients. Taiga vegetation is well adapted to permafrost conditions and protects the soil from heating up and thawing. However, even in the coldest parts of Siberia it suffers, sometimes irreversibly, from thermokarst processes which are induced by fire, lodging and insect outbreaks. In semiarid regions (Yakutia) it features numerous alases, and in the humid north of West Siberia numerous lakes and ponds.

Taiga vegetation is also a carbon sink, accumulating organic material in vegetation and soil. However, the system is vulnerable to fire and non-sustainable management. The timber industry utilizes the vegetation of this zone. Also, most mining and industrial activity in Siberia is located there and has led to large urban complexes. Most Siberian cities are located in the southern Taiga zone or in the transition zone of the mixed forest and Forest Steppe (sub-Taiga). The state and tendencies of forest ecosystem dynamics are negatively impacted by increasing anthropogenic pressure and insufficient governance of natural resources (Groisman and Gutman 2013).

Forest Steppe This is a roughly 200-km-wide transition zone between forest and Steppe. As in a parkland, patches of coniferous trees, mainly pines, admixed by birches and aspen, are interrupted by Steppe areas in Southern Siberia (Fig. 8a).

Located on fertile loess sediments, it forms the northern part of the Chernozem belt. Typical zonal soils are leached Chernozems (Phaeozems). This zone has been largely transformed by humans over the past centuries because of their relatively ideal conditions for settlement and cropping agriculture (Fig. 8b). Ploughing has induced accelerated soil erosion by water and wind has diminished the soil quality in some parts of this zone. Large fields of 100 ha (100 ha = 1 km^2) and more surrounded by windbreaks have been created in the last century.

Steppe This covers the southernmost part of Siberia and is a relative small part of the Eurasian Steppe landscape, extending roughly from North China across Mongolia, Kazakhstan, southern Russia, and Ukraine to Hungary. A lack of precipitation formed these largely treeless grasslands.

Deep rooting grass, herb and shrub vegetation have developed on silty and fine sandy aeolian loess or loess-like sediments, forming soils rich in humus. Loess is extensive in southern Siberia between the Irtysh basin in the west and the Angara basin in the east (Chlachula 2003). Large areas of salinized lands and saltwater lakes are found in the Steppe landscape of southwest Siberia between the Ob and the Irtysh rivers (Bulatov et al. 2006).

Typical Steppe soils in Siberia are Chernozems, Southern Chernozems and Dark Kastanozems. The Steppe zone provides favourable conditions for pastoral agriculture. Other subsistence agriculture by cropping cereals, potatoes, vegetables and other food have also been practised since the time of human settlement. Tillage-based cropping systems face problems of soil fertility being degraded by wind and water erosion, and humus loss in the Steppe zone. Crop yields are limited by drought. The Steppe biome is threatened by agricultural conversion into cropland, overgrazing, fossil hydrocarbon extraction, over-exploitation and poaching (Smelansky and Tishkov 2012).

Fig. 8 Spring aspect (end of May) of the vegetation in the Baraba Forest Steppe, located south of Novosibirsk. **a** *Anemone* species in *Stipa* grasslands. **b** Soil tillage for wheat sowing. Though the land is not ploughed but tilled by cultivators and harrows in this case, soil degradation by wind erosion cannot be prevented. The field size is 100 ha, which is too large for windbreaks to function. Many windbreaks have disintegrated

6 Bio-climatic Conditions for Agriculture and Forestry

The vegetation period of plants is largely limited by frost and days of temperatures less than 5 °C, whilst conditions for good growth of grasses and trees require daily temperatures higher than 10 °C. Water availability at this time is another important bioclimatic factor. The situation for a subset of climate stations is given in Table 4. Stations are also sorted according to their latitude from North to South. Except Cape Chelyuskin in the Arctic desert region, all other locations in this table are sites of agriculture. The data show that the months of April and October are not growth months all over Siberia because of a high probability of frost. The same holds for the first half of May and the second half of September. The long frost period in combination with harsh temperatures and sometimes shallow protecting snow cover does not allow for the growth of winter crops such as winter wheat or barley in most parts of Siberia. This is in contrast to Europe where, despite some late frosts, grass

Table 4 Agro-climate data of some locations in Siberia

	C.Chelyuskin 77.7° N	Verchoyansk 67.55° N	Tarko-Sale 64.91° N	Surgut 61.25° N	Omsk 54.93° N	Slavgorod 53.96° N	Ulan-Ude 51.8° N	Ekaterino-Nik.47.73° N
Ground frost probability %								
April	100	100	100	90	58	67	73	53
May	97	69	72	40	10	15	32	10
June	27	6	10	2	0	0	1	0
July	1	1	0	0	0	0	0	0
Aug	9	3	3	0	0	0	0	0
September	46	55	22	15	8	12	21	2
October	100	100	85	66	52	54	74	45
Days of t > 0 °C	63	133	137	168	192	189	188	200
Growing days (t > 5 °C)	0	110	107	135	168	162	156	178
Good growing days (t > 10°C)	0	79	75	98	132	129	120	144
Growing degree days								
0° base	57	1510	1496	1691	2634	2391	2240	2626
5° base	0	698	816	1134	1378	1584	1404	1721
10° base	0	292	311	500	832	746	636	959
Prec 4 warmest months [mm]	111	103	237	300	203	155	183	441
WB def. 4 warmest months [mm]	88	167	-1	0	221	247	219	−79

(continued)

Table 4 (continued)

	C.Chelyuskin 77.7° N	Verchoyansk 67.55° N	Tarko-Sale 64.91° N	Surgut 61.25° N	Omsk 54.93° N	Slavgorod 53.96° N	Ulan-Ude 51.8° N	Ekaterino-Nik.47.73° N
NPP_{Temp} [g/m^2]	***124***	***125***	***334***	***511***	721	625	575	***758***
NPP_{Rain} [g/m^2]	454	332	832	1084	***684***	***563***	***469***	1011
Geographical zone	Arctic desert	Northern Taiga	Northern Taiga	Southern Taiga[a]	Forest Steppe	Steppe	Steppe	Broadleaf mixed forest

Growing degree days = sum of daily temperatures if daily mean exceeds the threshold of 0 or 5 °C, WB def. = Water balance deficit (*P-ET*p). A high deficit implies drought risk and requires a high soil water storage capacity, negative values indicate an excess of water. NPP = Net Primary Production, NPP_{Temp} and NPP_{Rain} data were also given by LocClim 1.10, calculated by the Miami model developed by Lieth (In: FAO 2006). Limited NPP, either by temperature or by precipitation, is marked in bold and italics. [a]Transition between southern Taiga and mixed forest zone growth starts in most regions in April and is still full in action in October, and the greening of leaf trees starts in April/May.

From the total number of months and days above 0 °C the latitudinal zonation is clearly recognizable. About 130–200 days > 0 °C and 110–180 growing days (> 5 °C) provide some opportunities for growing spring crops. However, this can be managed with much less risk if growth periods are not interrupted by frosts. In this respect, agricultural activity is less limited in the south part of the Far East (Ekaterino-Nikolskoe) and of West Siberia (Omsk) but much more limited at Verchoyansk and Tarko-Sale. On these locations, grasses and other robust fodder crops can be grown, but all crops requiring a flowering and maturity phase are grown at risk. Nevertheless, as long as some 100 years ago Laschtschenkow (2012) reported on barley cropping in Verchoyansk and wheat cropping up to 63.5° North in Yakutia. Stations located further south provide long periods free of frost and acceptable conditions for growing a broader variety of annual crops. In most cases, not only in Southern Siberia but in Yakutia too, drought is a severe limiting cropping factor. The water balance deficit in summer is so high that only considerable amounts of water stored in soils (>150 mm) or fed by shallow groundwater or irrigation can provide acceptable crop yields. Practically, only loess or lowland soils of a shallow water table are suited for dryland cropping under these conditions. The situation in the humid Far East (location Ekaterino-Nikolskoe) is completely different and characterised by a significantly higher cropping potential.

Rough estimates of the Net Primary Production (NPP), provided by the database Loc-Clim 1.10 (FAO 2006) confirm the tendency for NPP to increase from North to South. Temperature limited NPP estimates range from 1.2 to 7.7 tonnes per hectare and year dry matter for the sites in Table 4. The values are significantly lower than those estimated for locations in Europe (Müncheberg/Berlin 12.8 t/ha; Shebantsevo/Moscow 11.0 t/ha). Also, the number of growing days (t > 5 °C) is higher in the forestry and agricultural regions of Europe (Müncheberg/Berlin 226; Shebanzevo/Moscow 186) but in most parts of Siberia less than 170. This shows clearly lower yield potentials of forest trees and crops in Siberia mainly due to low temperatures.

A comparison of Table 4 with growing degree requirements of some annual crops (Table 5) for the stations of Verchoyansk and Tarko-Sale confirms the possibility of cropping spring barley, cereals or potatoes in the Taiga zone of the Extreme North. Data on the summer period length (Table 6) also indicate that there is a sufficient period for growing vegetables and crops in all natural zones except Tundra. However, there is a considerable frost risk for agricultural and horticultural plants in the northern Taiga. Boreal forest is frost-resistant and can use the considerable number of summer days (Table 6) for biomass production.

Growing days (t > 5 °C) or days of t > 10 °C seem to be a suitable criterion of natural vegetation zoning. The northern boundary of tree growth (polar treeline) is characterised by more than 64 days > 5 °C and the northern boundary of broadleaf mixed forest more than 160 days > 5 °C (Weischet 1960).

Table 5 Growing degree-day requirements for some annual crops (data by Juday et al. 2009)

	Growing degree-days	
	5 °C base	0 °C base
Peas (green)	700–800	1000
Spring barley	700–900	1200–1500
Peas (seed)	800–1150	1500–1700
Oats		1300–1700
Canola	950–1050	1350–1550
Potatoes	1000–1100	
Spring wheat	1000–1200	400–1650

Table 6 Duration of the summer period and temperature sums in different natural zones of West Siberia (Shvareva 1963, excerpt, modified)

Zone or sub-zone	Begin and end of the summer period		Number days > 10 °C
Tundra	10 July	15 August	40–50
Northern Taiga	20 June	25 August	60–80
Middle Taiga	5 June	5 September	100
Southern Taiga	30 May	10 September	100–150
Sub- Taiga (Aspen-birch forest)			115–120
Forest Steppe	15 May	15 September	120–140
Steppe			125–140
Dry Steppe	5 May	25 September	

7 Climate Change

From predicted trends, Russia is warming faster than the global average, and Siberia and the Arctic region of Siberia are warming exceedingly fast (Climate Committee 2002; Table 7; Roshydromet 2008a). The temperature trend coefficient over the past 30–40 years has been a roughly 0.3–0.5 °C increase per decade. The depth of the active layer could deepen by 15–50 %, in Yakutia and Chukotka in particular (Roshydromet 2008b). Under the ground, at a depth of 10–20 m, the temperature could increase by 1 K by the end of this century (Pavlov 2008).

Climate change could mean an increase in rainstorm probability of about 7 % by 2030 and 16 % by 2060 in West Siberia, and 10 % by 2030 and 19 % by 2060 in East Siberia, respectively (Roshydromet 2008b). River discharges (Table 8) and the probability of flood disasters are predicted to increase, especially in the Lena River catchment (Roshydromet 2008b; Shiklomanov and Georgievsky 2009).

Increasing freshwater discharge of Siberian rivers is of concern because of consequences for the circulation in the Arctic Ocean and global climate. Rawlins et al. (2010) observed a significant intensification of the Arctic freshwater cycle: increasing precipitation, evapotranspiration and river discharge. Discharge shows a

Table 7 Predicted temperature increase by 2050 at the soil surface in permafrost regions (Climate Committee 2002)

Region	Temperature increase °C		
	Tundra	Forest Tundra	Taiga
European North	1.6		
West Siberia	2.4	2.6	2.4
Yakutia	2.8	2.6	2.5
North East		1.5	1.5

Table 8 Predicted increase of river discharges in the 21st century (ROSHYDROMET 2008b)

River catchment, region	Increase of river discharges (%)		
	2011–2030	2041–2060	2080–2099
Ob	5 ± 5	7 ± 7	13 ± 10
Yenisei	5 ± 4	10 ± 6	21 ± 8
Lena	8 ± 5	17 ± 7	34 ± 15
Amur	3 ± 8	11 ± 10	23 ± 17
West Siberia	5 ± 3	11 ± 4	22 ± 5
East Siberia	6 ± 2	14 ± 5	32 ± 10
Russia	5 ± 2	11 ± 3	23 ± 4

significant positive trend of 0.31 mm/year^2 over Eurasia, which may have bad consequences for the global water circulation including the Gulf Stream as the heat pump of Europe. This increasing freshwater input to the Arctic is dominated by Siberian rivers (Rawlins et al. 2010). Increasing discharge is associated with increased sediment loads (Syvitski 2002).

All these trends will have implications for nature and humans of Siberia and around the globe.

The latest climate report (Roshydromet 2014) addresses some possible negative biophysical consequences of climate change. Those examples are the decrease of agroclimatic potentials in Steppe regions of Siberia due to drought, the spread of crop pests like the Colorado beetle (*Leptinotarsa decemlineata*) and the spread of hosts for pathogenic organisms like tics (*Ixodes spec.*) (Roshydromet 2014). However, scenarios are uncertain and need to be improved by better data and models.

8 Land Resources and Their Quality

8.1 Quantity of Forest Land

The Russian land for forestry covers about 809 million ha according to FAO criteria (FRA 2005). It is by far the largest forest land in the world. According to original Russian statistics the forest land including non-stocked lands and other land users was higher, about 891 million ha in 2008 (Filipchuk and Moiseev 2010). The Russian Forest Fund even counted 1.18 billion ha of land (Nilsson and Svidenko 1998).

Table 9 Forest land (2008) in the Siberian and Far East Federal Districts (data by Alexeenko 2012)

	Siberian FD	Far East FD[a]	Total for Siberia and Far East	Share of Russian total volume, %
Forest land area, M ha	347.4	496.2	843.6	75.4
Stocked area, M ha	257.9	290.2	548.1	73.4
Growing stock, billion m^3	31.3	20.3	51.6	67.5
Mature and overmature stands, M ha	122.8	122.3	245.1	74.5
Total average annual increment, M m^3	338.1	234.4	572.5	60.4
Annual allowable cut, M m^3	196.5	90.9	287.4	52.2
Area of protected forest, M ha	91.0	78.9	169.9	68.1

M ha = million hectares, M m^3 = million cubic metres. [a]The Far East Federal District includes the Sakha Republic

About three quarters of the Russian forest land belongs to the Taiga region of the Siberian and Far East Federal districts (Table 9).

Only part of the forest land is stocked by forest (Fig. 9). Bogs cover one third of the land in the Siberian and Far East Federal districts, which are considered as forest lands but not stocked by forests (Alexeenko 2012). The proportion of stands with aged trees is relatively high. This is favourable for the natural reproduction and the overall ecosystem.

The primary designated function of about the half of forest land is production (416 million ha in 2010); the rest is designated for reserves, the protection of soil and water conservation, of biodiversity and other purposes (Filipchuk and Moiseev 2010). The key principles of legislation and management of forests as a resource are

Fig. 9 Dense stands of *Picea* in the Southern Taiga (**a**) and mixed forest in the SubTaiga (**b**)

ruled by the Forest Code of the Russian Federation (Forest Code 2006). More than 90 % of forest land is in public (state) ownership and under the responsibility of the Ministry of Natural Resources of the Russian Federation (Filipchuk and Moiseev 2010). Forest management rights lie to about 80 % in the hands of the public administration. 20 % of land is leased by private corporations and institutions (Filipchuk and Moiseev 2010). The allowable cut amounts to about half of the annually growing forest mass.

About 250 million ha of forest and Tundra land is designated area for reindeer farming in the Siberian and Far East FD. The Republic of Sakha (80 M ha forest land), the Chukot Autonomous Region (44 M ha) and the Taimyr Autonomous Region (40 M ha) are regions containing the largest lands for this purpose (Alexeenko 2012).

8.2 *State of Forest Land*

Most forest land of Siberia is in a good ecological state and stable or very resilient. Some soils of the Russian Federation, mainly in subarctic and Arctic regions or in mountain regions, underlie significant risks of disturbances or degradation. Stolbovoi and Fischer (1997) estimated that about 10 million ha of Russian forest land suffers from disturbances to the soil organic horizon due to cuttings, and another 15.5 million ha disturbances to the soil organic horizon due to conflagration. In Taiga and Tundra regions about 31.2 million ha are damaged by thermokarst due to industrial activities, and 60.2 million ha show signs of surface corrosion from overgrazing (Stolbovoi and Fischer 1997). Forest logging in mountain regions has triggered an irreversible loss of soils by erosion. Vast forest areas in Siberia and the Far East suffer from air pollution. They are at risk of damage from sulphur and nitrogen depositions. Sulphur deposition may stress 210 million ha and nitrogen deposition 87 million ha of forest land in Asian Russia (Nilsson and Shvidenko 1998). Those depositions aggravate natural processes of soil acidification, podsolization and leaching of nutrition elements and carbon, leading to irreversibly diminished soil and habitat quality.

The south Siberian mountains and some other mountainous regions are rich in uranium ore and thus have high natural radioactivity (Utkin 1997 quoted in Adam et al. 2013; Puzanov et al. 2013). The processing industries and radioactive waste disposal and the related dumping of materials have caused serious contamination in Siberia. Much is accumulated in the active layer and will disperse later (Fukuda 1997). The Altai territory and Republic of Buryatia are particularly damaged by fallout from nuclear tests (Baranovskaya et al. 2012). Elevated levels of radioactive Caesium (Cs-137) and/or Polonium were measured in soils adjacent to centres of the nuclear industry (Seversk, Zheleznogorsk) and at sites of accident fallout at underground nuclear explosions in Yakutia (Baranovskaya et al. 2012). The large region around Lake Baikal which is mainly under forest is a zone of elevated soil radioactivity levels due to natural and human-induced sources (Chernyago et al. 2012).

8.3 Quantity of Agricultural Land

Russia has agricultural lands covering about 180–402 million ha. Statistical data for Russia in 2007 were 220.6 million ha of agricultural land, of which 121.6 million ha were cropping land. This is about 13 and 7 %, respectively, of the total land (1709.8 million ha) of the Russian Federation (Gordeev and Romanenko 2008).

The Russian Land Fund considers 401.6 million ha of land as designated for primary agricultural purposes, and 69 % of this land is in Federal and Municipal ownership (FAO 2013a). From FAO data sources (FAO 2013b), agriculture in the Russian Federation is based on 215.25 million ha of cultivated land (arable land and permanent crops).

Schepaschenko et al. (2012) refer to 220.5 million ha of agricultural land of Russia in 2009. Cropland covered 121.6 million ha. Of the total cropland, only 77.8 million ha were under cropping, the rest was set aside or abandoned (Schepaschenko et al. 2012). This 77.8 million ha is a low level and corresponds to about the level at the end of the 1920s and late 1940s (Lyuri et al. 2010). In Russia, about 35.7 million ha of land are potentially available for wheat production (Petrick et al. 2013). Much of this land is located in South Siberia.

Most agricultural land and cropping land in Russia is located in Europe. From data gathered by Adam and Mamin (2001) less than a quarter of Russia's arable land is located in the Asian part of the country (Table 10), most of that being in the southern part of West Siberia. About 10 million ha of good agricultural lands are located in the Forest Steppe zone of West Siberia (Egorova Egorova 2014), and also in the Steppe zone. Of the administrative units, the Altai Krai comprises the largest land areas for agriculture in Siberia. Though bioclimatic conditions in the Far East are better than in the Steppe regions of Siberia, less land is under agriculture there.

The agricultural land area has experienced great fluctuations over the past century. According to Lyuri et al. (2010) between 1990 and 2007 about 13.5 million ha of agricultural lands in Asian Russia (48 million ha in Russia) were taken out of production. This was more than that gained during the huge land reclamation campaign in the 1950s. The area of potential (not yet used) cropland of Russia amounts to 39–40 million ha (Lambin et al. 2013) and is about in the magnitude of the abandoned land. Land abandonment was correlated with a loss of rural population, which was about 30 % in most regions of Siberia.

Table 10 Agricultural land in Siberia in million hectares (Adam and Mamin 2001, modified)

Region	Agricultural land	Arable land	Pastures and hayland	Others
Russian Federation	178.6	117.8	58.3	2.5
West Siberia	29.0	17.8	10.9	0.3
East Siberia	18.2	7.3	10.0	0.9
Far East	4.7	2.4	2.1	0.2

8.4 *State of Agricultural Land*

8.4.1 Industrial Pollution of Soil

The industrial pollution of soil is a serious problem worldwide and also in Siberia. The air quality in Siberian cities and their industrial areas is very bad (Adam and Mamin 2001; Kashapov et al. 2008). The urban air quality in several Siberian cities (e.g. Norilsk, Barnaul, and Novokuznetsk) is considered among the worst in Russia (Baklanov et al. 2013). Siberian ecosystems have begun to show stress from the accumulation of pollution depositions that come from cities and industrial plants (Belozertseva 2013; Gutman et al. 2013). Pollutants are being dispersed over soils of the region. The mining, transport and storage of industrial minerals or products have also led to point source pollution. Many areas around major metallurgical, chemical and energy enterprises have been found to be polluted by toxic substances such as heavy metals, oil and oil products, sulphur oxides and chemical wastes. Soil quality is also adversely affected by soil contamination with radionuclides and other pollutants. The pollution of soils poses a risk to food production and biodiversity. In the Irkutsk region a historical atmospheric source of polychlorinated biphenyl (PCB) has led to the widespread contamination of soil with PCBs which were ingested by milking cows and contaminated the milk in this region (Mamontova et al. 2007).

About 3.6 million ha of agricultural land in Russia are contaminated with radionuclides and heavy metals (FAO 2013a). Under permafrost conditions industrial pollution is a particular threat because the damage to vegetation initiates thermokarst processes (Chuprova 2006; Baranov et al. 2010).

8.4.2 Soil Quality for Agriculture

For environmental monitoring, both soil and water and quality can be measured using sets of chemical, biological and physical data. In the case of soils, there is a lack of conventions and international standards on the parameters required for this kind of monitoring. Some approaches do not meet one of the basic requirements described by Dokuchaev (1951): that soil quality assessment on agricultural land should reflect crop yield potentials. It is useful to measure and evaluate soil quality in terms of its functions for society. For example, the specific role of soil and land in producing plant biomass for humans (productivity function, Mueller et al. 2011) remains crucial. Consequently, higher soil quality means the land has a higher crop yield potential. Land rating approaches of the former Soviet Union (Gavrilyuk 1974; Vostokova and Yakushevskaya 1979) meet these requirements in general. However, they are based on yield data which are 60 and more years old and do not consider soil functional properties and climate conditions. More recent approaches (Krupkin and Toptygin 1999), based on important functional properties of soils such as humus content or nutrient stocks, have more regional meaning and cannot be transferred to other parts of Russia.

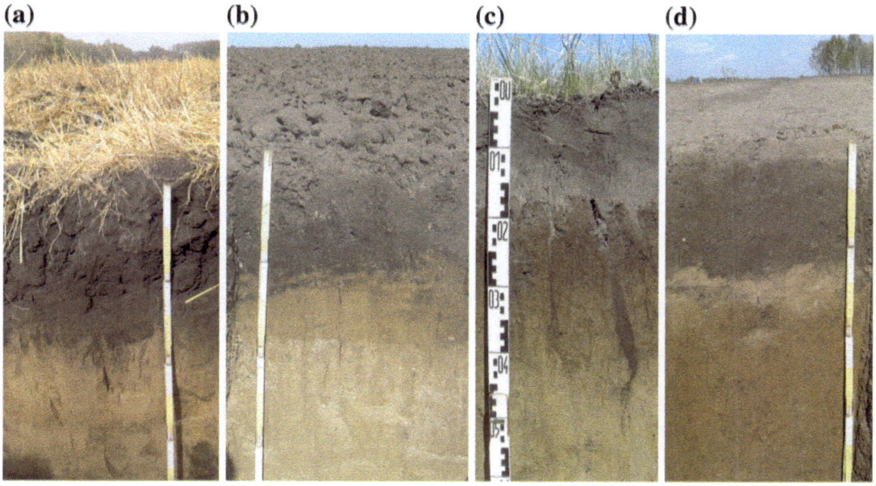

Fig. 10 Examples of agricultural soils in Siberia and their quality for cereal cropping: **a** Ordinary Chernozem of the research station Krasnoobsk, Forest Steppe, rating points of M-SQR (Mueller et al. 2011): 98 (Upgraded Basic Rating)/42 (Final Rating, including climate factors). **b** Southern Chernozem of the Grushevka field near Bagan, Kulunda Steppe, degraded by wind erosion, rating points 78/20. **c** Solonetz near Bagan, Kulunda Steppe, rating points: 55/14. **d** Luvisol of the Kochky site, Baraba Forest Steppe, rating points: 74/24. Classes of rating points are 0–20 very poor, 20–40 poor, 40–60 moderate, 60–80 good, 80–100 very good

The overall situation of soil quality in Asian Russia and future trends in the context of the Eurasian and global situation does not seem to be clear yet but could be found out by using the Muencheberg Soil Quality Rating (M-SQR, Mueller et al. 2011) to create a strategy of assessing food security for Eurasia or the world. Work by Smolentseva et al. (2014) showed that loess-borne soils of South Siberia have excellent basic soil properties in terms of texture, structure, rooting potential and water and nutrient storage capacity. However, climatic conditions such as too-late warming and drought are serious yield-declining factors. This leads to low overall rating values on a global scale (Smolentseva et al. 2014). Some examples are given in Fig. 10.

8.4.3 Land Degradation by Agriculture

Degradation of agricultural land by wasteful and unsustainable management is one of the most important socioeconomic problems worldwide. This is also true for the lands of Russia including Siberia. It poses a threat to the country's ecological, economic and national security. Data are inconsistent and differ depending on their assessment methodology but demonstrate the huge dimensions of the problem.

Table 11 Degraded agricultural soils in Siberia (data by Gordeev and Romanenko 2008)

	Agricultural land (thousand ha)	Degraded land in %			
		Wetness	Water erosion	Wind erosion	Salinization and alkalinisation
West Siberia	34,434	20.3	6.6	12.9	35.1
East Siberia	23,196	7.8	9.8	14.3	3.8
Far East	7932	36.5	7.0	0.8	4.3

The worst damage to Russian soils is caused by water and wind erosion (50 million ha), waterlogging (40 million ha), droughts (up to 170 million ha), land salinization and alkalinisation (40 million ha) (FAO 2014a).

Cropping agriculture, if associated with soil tillage, is risky for soil fertility and associated with a permanent loss of soil humus and fertility. Schepaschenko et al. (2012) calculated that the arable land of Russia was a carbon source (carbon loss of 0.8 t/ha per year), whilst pasture and hayfields were a sink of 0.29 t/ha per year. Zhulanova (2013) found negative carbon balances of the Dry Steppe (0.76 t/ha per year) and Steppe (0.19 t/ha per year), whilst carbon of the Forest Steppe was about balanced under agriculture in South Siberia. Table 11 shows the percentage of degraded agricultural land in Siberia.

Wetness of agricultural land is a permanent yield-declining issue in West Siberia and the Far East. Salinization and wind erosion are also typical of Steppe regions. Degradation means an irreversible loss of soil productivity potential. Drought, the most severe productivity-limiting factor leading to desertification in current Dry Steppe and Steppe regions (Schreiner and Meyer 2014), is not listed here. Data about other soil degrading factors such as permanent humus loss and soil compaction are also not included in Table 11.

When tilled, soil loses its protective vegetation and becomes prone to wind erosion. In the Steppe regions of Asia alone, wind erosion may reach global dimensions in future due to the mismanagement of soils by ploughing (Suleimenov et al. 2014). Halting anthropogenically induced land degradation by introducing more sustainable land management is a challenge for Asian Russia. This also has implications for the sector of agri-environmental research.

8.4.4 State of Grasslands

Grasslands and rangelands are an underestimated resource for biochemical cycles and for human welfare. There is a lack of reliable data on the state of pasture or rangeland degradation in Siberia. The situation is similarly unclear to that in the landscapes of Kazakhstan and other countries of Central Asia (Mueller et al. 2014). Long-term succession studies in grasslands done by the Sochava Institute of Geography (Nechaeva et al. (2010) along with chemical soil analyses are important for understanding local grassland ecosystems. However, those studies lack linkages with modern diagnostic and monitoring methods. A loss of plant and wild animal diversity, an increase in unpalatable or toxic plants, a loss of soil fertility and productivity and a decline in livestock production are examples of possible indicators. Rangeland recovery may comprise palatable biomass, biodiversity and rare species.

The biodiversity of grasslands is influenced or threatened by several disturbances such as habitat loss, fragmentation of natural communities, over-exploitation such as overgrazing, penetration of non-native species, environmental pollution, climate change, and other elements. Local overgrazing or periodic ploughing of grasslands are crucially negative impact factors. Both overgrazing and underutilisation decrease the natural potential of Steppe soils (Kandalova and Lysanova 2010). Desertification tendencies of Siberian ecosystems, grasslands in particular, have been already detected (Meyer et al. 2008; Plyusnin and Danko 2011).

9 Peatlands and Their Significance for a Functioning Landscape

Almost 370 million ha of peatlands are located in the Russian Federation, the majority of them in the Taiga zone of the Asian part (Table 12; Fig. 11). They are of crucial importance for biodiversity, carbon storage, hydrology and other environmental functions (Liss et al. 2001; Kremenetski et al. 2003).

The West Siberian lowlands are the world's largest high-latitude wetland region, including vast and deep peatlands. Sheng et al. 2004 estimated the total area of peatlands in West Siberia at 59.2 million ha and the total carbon pool at 70.21 Pg (Petagrams = billion tonnes). The Vasyugan Mire is the largest of them, covering an area of 5.3 million ha (Inisheva et al. 2011).

Humid Siberian Taiga and Tundra are regions of permanent peat accumulation and thus a permanent natural carbon sink. From recent models based on numerous measurements, sphagnum mosses grow about 12 mm/year in the northern Taiga, 16 mm/year in the middle Taiga and 12 mm/year in the southern Taiga. The annual carbon accumulation ranges from 90 to 160 g/m^2 (Dyukarev et al. 2011). Inisheva and Berezina (2013) report on a peat layer increasing by about 0.4–0.7 mm/year

Table 12 Peatland areas in the Russian Federation (Climate Committee 2002)

Territory	Area million ha
Russia as a whole	369.1
Of which: In the European part	58.8
In the Asian part	310.3
In different zones	
Tundra and Forest Tundra	106.2
Taiga and other zones	262.9
In the permafrost zone	270.6
In the West Siberian lowland	99.1

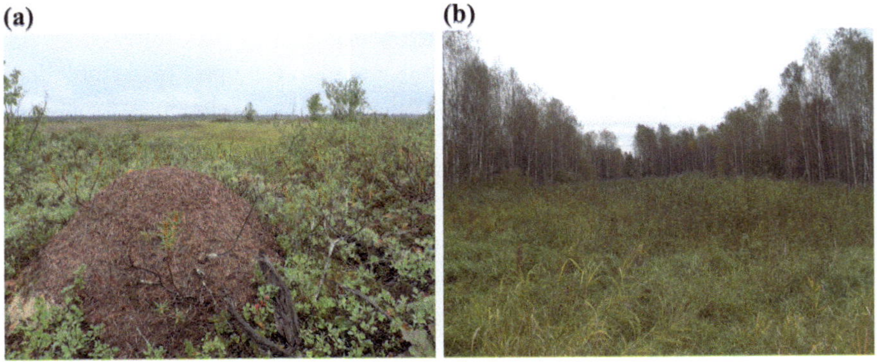

Fig. 11 Peatlands in the Southern Taiga. **a** Natural bog with *Formica spec.* anthill. These ants are very important for preventing insect outbreaks through biological means. **b** Cultivated and abandoned peatland

and quote an enlargement of the peat area of 92 km^2/year based on data collected by Neistadt (1971). In the warmer Sub-Taiga of South Siberia, Larin and Guselnikov (2011) also found an increase of wetland areas by 4.6 km^2 per year.

NPP is a crucial parameter for characterising the biological productivity of sites and a starting point for carbon and ecosystem balances. It has been measured by traditional methods (harvesting and analysing above-ground and below-ground plant biomass). Table 13 shows data on peatland sites in West Siberia.

The NPP in eutrophic swamps of the Forest Steppe was about 600–700 g/m^2 per year of dry mass (Kosykh 2009). It was higher (1285 g/m^2) in the peatlands of the Altai mountains, where the roots of wetland grasses contributed 80 % of the NPP (Kosykh et al. 2010). About half of the dry biomass is carbon. Inisheva et al. (2011) measured the NPP of peat sites at 171–296 g C/m^2 per year. Lapshina and Bleuten (2011) report on an NPP of 150–500 g C/m^2 per year in the peatlands of West Siberia. Carbon storage (NPP minus heterotrophic respiration) was 30–70 g C/m^2 per year, and methane emissions were about 0–5 g C/m^2 per year.

Table 13 Net primary production (NPP, g/m^2 per year dry matter) of two peat sites in West Siberia (Vasilyev 2007, shortened)

NPP fractions	NPP g/m^2 per yr dry matter			
	Northern Taiga (Noyabrsk site, 63.2°N, 75.5°E)		Southern Taiga (Plotnikovo site, 56.7°N, 83.0°E)	
	Ridges	Depressions	Ridges	Depressions
Vascular plants	208	360	592	584
Mosses	200	241	265	295
Dwarf shrubs	88	10	124	86
Herbs	10	41	5	45
Total	506	652	986	1010

100 g/m^2 = 1 tonne/ha

The peatlands of Siberia face threats different from those that are known from peatland regions in Western Europe. In Western Europe, peatlands are a carbon source due to drainage and agriculture in a temperate, warm and subhumid climate (Drösler et al. 2008; Mueller et al. 2008). Land drainage, which is responsible for the carbon loss of west European peatlands, has had no significant influence on the carbon balance in West Siberia (Inisheva et al. 2011). Loss of carbon due to heterotrophic respiration higher than NPP seems not to be a threat to Siberian peatlands. However, more reliable data are required for assessing climate change. In general, measuring NPP and respiration by different methods, as was done in most cited studies on Siberian peatlands, may lead to biased carbon balance results. Today, more sophisticated and accurate methods of measuring carbon balances in situ are available (Chojnicki et al. 2010; Juszczak et al. 2013).

Threats to Siberian peatlands are more associated with industrial pollution. The process of conversion of natural land into industrial land goes on. There are 13.7 million ha of licensed sites for oil exploration and processing in the Khanty-Mansi Autonomous Okrug of West Siberia, the majority of them located in peatlands (Lopatin 2011). Peat contamination by heavy metals around oil and gas extraction and processing industrial complexes, degradation of the natural vegetation, thermokarst and erosion processes are common in those areas (Opekunova et al. 2011). Secondary salinization of lands is another detrimental by-effect of oil and gas exploration leading to destruction of the peat vegetation (Ermolov et al. 2011). Industrial burning processes distribute pollutants such as heavy metals and products of incomplete combustion over peat landscapes. Wetland monitoring and the development of proper wetland management practices in Siberia are needed (Robarts et al. 2013). The pollution status of peatlands should be part of those programmes.

10 Status of Zapovedniks and Other Protected Areas

Based on the research and great foresight of Russian and Soviet scientists of the nineteenth and twentieth centuries, Russia has always developed ambitious nature protection projects. Scientific nature reserves (zapovedniks) are strictly protected areas without any economic use of natural resources such as soil, water, minerals, flora and fauna. It is the highest category of protection restricted to the public. They were created for nature and ecosystem research almost about 100 years ago. The degree of human disturbance in these areas is extremely low. In Europe and most parts of the world, such regions practically no longer exist. Russia has about 100 zapovedniks covering an area of 31,000 km^2 (Juday et al. 2009). Many of them are located in Siberia and the Far East. Unique data about rare and threatened species have been gathered in zapovedniks, for example Gilg et al. (2000), Ebmer (2006), Goodrich et al. (2008).

Zapovedniks are a national treasure of Russia and of great importance for studying largely natural landscape processes. However, they face problems of destruction and risks for survival. Their protection status and management are ruled by law (Government RF 1994). However, goals and reality have differed. The monitoring and protection of zapovedniks cannot be guaranteed because of financial hardship and a local population which does not back the concept (Oostergreen and Shvarts 2000; Lieske 2011).

Zapovedniks have great potential for environmental education and scientific tourism. The peripheral zones could be more open for scientific and eco-tourism and the environmental education of the young generation (Luzhkova 2013). This would make the region more economically and ecologically stable and the population more environmentally aware (Oostergreen and Shvarts 2000). On the other hand this would mean a certain weakening of the zapovednik concept and require investment to prevent the outer zapovednik zone from fast destruction. One of the basic purposes of zapovedniks—research for understanding natural landscape processes as a baseline of landscape research—is still underdeveloped. Research activities in and about zapovedniks need to be increased.

Besides zapovedniks, a system of protected areas and sites of particular ecological, historical and aesthetic value has been created. More than 40 national parks have been funded since 1991 (Wikipedia 2014a). They are open for regulated tourism. One of the largest of these national parks is the Tunkinsky National Park in the Buryat Republic (1183 km^2). Most national parks are located in Forest Fund areas and are managed by the state forestry authorities. These areas are managed for a range of scientific and biodiversity values (Juday et al. 2009). Wildlife refuges (zakazniki) are another category of protected areas; amongst them are many Ramsar sites. Russia has designated 35 wetlands for the Ramsar List; the total area of these sites is over 10.3 million ha. 20 of them are in Siberia and the Far East (Norbalwet 2013).

Ethno-ecological museums (ecomuseums) are a new type of multidisciplinary open-air museum explaining the natural and cultural environment of the Siberian peoples (Kimeev 2008).

Some Siberian zapovedniks and national parks are natural World Heritage Sites (WHS), which are associated with enhanced international (UNESCO) and national obligations for their protection. These are Lake Baikal WHS (includes Barguzin Zapovednik); Sikhote-Alin WHS (includes Sikhote-Alin Zapovednik); Golden Mountains of Altai WHS (includes Altai and Katun zapovedniks); Volcanoes of Kamchatka WHS (includes Kronotski Zapovednik and three national parks); Uvs Nuur basin WHS (includes Uvs Nuur Nature Reserve) and Wrangel Island (includes Wrangel Island Zapovednik) (Wikipedia 2014b). In individual cases the monitoring of recreational impacts has started (Zavadskaya 2011).

All protected areas, and zapovedniks in particular, need to be better involved in research and monitoring programmes.

11 Water Resources and Quality

11.1 Water Resources

11.1.1 Fresh Water Resources

Most of the freshwater resources of the Russian Federation are contained in the permafrost ground of Siberia and the Far East region (FAO 2014b). This freshwater-like water from Siberian glaciers is a valuable resource which largely does not underlie any management. The Russian Federation also has large renewable water resources and a corresponding dependency ratio of 4.3 (%). This ratio means that only a small amount of water flows in from other countries, for example the Irtysh from Kazakhstan. Russia is thus largely independent from freshwater inflow from other countries (FAO 2012). Most Russian water resources occur in Siberian river catchments which drain into the Arctic Ocean. Table 14 shows the catchment and discharge of the largest rivers and watersheds, respectively, east of the Urals. The large catchments of the three biggest rivers of Siberia, Yenisei, Lena and Ob, provide more than a third of the renewable overall Russian freshwater resources. Due to the low precipitation in these areas, the annual

Table 14 Renewable water resources of major river basins east of the Urals

Name of river/catchment	Major region within the RF	Area of basin 1000 km^2		Internal RSWR (km^3/year)	Inflow (km^3/year)	Total RSWR (km^3/year)	Outflow to
		Total	In the RF				
Arctic Ocean							
Ob	Ural, W. Siberia	2990	2330	364.0	38.0[a]	402.0	Kara Sea
Yenisei	Central Siberia	2580	2180	605.0	25.0[b]	630.0	Kara Sea
Pyasina	East Siberia	182	182	82.0	–	82.0	Kara Sea
Lena	E. Siberia, F. East	2470	2470	532.0	–	532.0	Laptev Sea
Khatanga	East Siberia	422	422	88.0	–	88.0	Laptev Sea
Olenek	Far East	219	219	34.0	–	34.0	Laptev Sea
Indigirka	Far East	360	360	55.0	–	55.0	East Siberian Sea
Kolyma	Far East	647	647	126.0	–	126.0	East Siberian Sea
Pacific Ocean							
Amur	E. Siberia, F. East	1855	780	225.0	100.0[c]	325.0	Sea of Okhotsk
Kamchatka	Far East	56	56	33.0	–	33.0	Pacific
Anadyr	Far East	191	191	53.0	–	53.0	Bering Sea
Total RF		20,057	17,075	4 036.7	185.54	4222.24[d]	Total RF

Data from AQUASTAT data basis, country profile on Russian Federation (FAO 2012), *RSWR* Renewable Surface Water Resources, *RF* Russian Federation, [a]from Kazakhstan, [b]from Mongolia, [c]from Mongolia and China, [d]other estimates 4508 km^3/year (2001 data, FAO 2012)

discharge of these three rivers (≈224 mm) is slightly lower than the average over the whole Russian territory (≈236 mm from 460 mm of precipitation) (FAO 2012).

11.1.2 Peculiarities of the Landscape Water Balance

The water cycle of cold regions has some specific features. A high proportion of surface runoff is typical, and even subsurface drainage has specific pathways because of impeding frost layers in the ground. Surface runoff amounts to 70 % of the total runoff and almost one third of the total precipitation in Siberia (Table 15). This discharge component is responsible for fast river discharge, floods, soil erosion and the eutrophication of open waters. It is highest in the Tundra and lowest in the forest. Subsurface runoff (deep drainage) is responsible for groundwater formation, the base flow of rivers, nutrient leaching of soils and nitrate and other contamination of aquifers. It can be relatively high in the forest and extremely low or non-existent in Steppe and Forest Steppe regions.

This kind of discharge formation and drainage is very different from the temperate zones of Europe, where deep subsurface drainage dominates and is an important process for landscape water balance, groundwater formation and water quality (Dyck and Peschke 1983; Müller et al. 1996; Schindler et al. 1998).

On the other hand, high surface runoff does not necessarily generate fast river discharge in Siberia. Thermokarst depressions may have a powerful storage function, thus very effectively reducing river discharge. Surface runoff is predominantly redistributed over the land surface. Low precipitation and high surface runoff in Siberia considerably limit the water available for evapotranspiration and vegetation biomass formation. The figures for precipitation, total runoff and surface runoff are based on measurements (precipitation samplers, river discharge measurements and surface runoff weirs, respectively). Subsurface drainage, transpiration and gross soil moisture are balancing terms in this table. The calculated gross soil moisture of Table 15 is a theoretical or orientation value for the amount of water that would be potentially available for the growth of vegetation under the assumption that subsurface drainage water could be used by plant roots. Practically, only some of the

Table 15 Landscape water balance in West Siberia in mm/year (Kemmerikh et al. 1963)

Zone or region	Precipitation	Runoff			Transpiration[b]	Gross soil moisture[c]
		Total	Surface runoff	Subsurface drainage[a]		
Tundra	341	260 (76 %)	231(89 %)	29	81	110
Forest	480	223 (46 %)	142 (64 %)	81	257	338
Forest Steppe and Steppe	329	43 (13 %)	37 (86 %)	6	286	292
Mountain areas	664	356 (54 %)	268 (75 %)	88	308	396
West Siberia total	453	201 (44 %)	140 (70 %)	61	252	313

[a]Total runoff minus surface runoff, [b]Precipitation minus runoff, [c]Subsurface drainage plus transpiration

Table 16 Dates of ice drift and ice cover for some rivers in West Siberia (Kemmerikh et al. 1963, p. 124, excerpt)

River	Station	Natural zone	Begin of ice drift	End of ice drift	Days with ice cover
Shchuchya	Shchuchye	Tundra	8 October	8 June	229
Pyaku-Pur	Tarko-Sale	Forest Tundra	15 October	31 May	218
Ob	Salechard	Forest Tundra	22 October	30 May	207
Vasyugan	Vasyugan	Taiga	27 October	6 May	180
Ob	Surgut	Southern Taiga	30 October	11 May	185
Ob	Barnaul	Forest Steppe	4 November	26 April	160
Aley	Khabazino	Steppe	4 November	20 April	156
Kulunda	Shimolino	Steppe	5 November	18 April	168

subsurface drainage water is available for plants. Meanwhile, modern lysimeters and other technologies are available for exact measurements of landscape water and solute balances (Meissner et al. 2014; Schindler 2014).

Freezing conditions determine river discharge processes (Table 16). Most rivers are frozen from November until mid-April. Frozen rivers are still important temporary transportation routes in Siberia.

11.1.3 Monitoring Lakes and Reservoirs

Natural lakes and dammed rivers form a system of large water reservoirs for multi-purpose utilisation. The general rules of their water use and management are set out in the water code of the Russian Federation (Water Code 2006). Monitoring all water bodies including lakes and reservoirs is an important basis for management decisions. The State Hydrological Institute in St. Petersburg is responsible for the monitoring programme all over Russia (State Hydrological Institute 2013). Several scientific institutes of the Siberian Branch of the Russian Academy of Sciences (SB-RAS), such as the Institute for Water Resources and Ecological Problems (IWEP) in Barnaul and the Limnological Institute (LIN) in Irkutsk, provide the scientific basis and conduct monitoring programmes of waters and the surrounding landscapes in Siberia (Figs. 12 and 13). To develop and adjust the methodological basis of monitoring programmes, their scientists cooperate with experts at regional universities, other All-Russian and regional institutes and leading scientists from abroad. Lake Baikal (Fig. 13) attracts the international scientific community particularly. Since the 1990s, 249 international expeditions have been carried out on Lake Baikal with the participation of 456 Russian and 1353 foreign scientists from 36 countries (LIN Irkutsk 2012).

Lake Baikal This is the oldest, deepest (1637 m) and purest natural lake on earth. Having a water volume of 23.0 thousand km^3 it contains the largest global available open freshwater resources (20 % of the world, and more than 85 % of Russia). The lake is very pure; visibility depths of up to 40 m are evident. It has an excellent

Fig. 12 Landscapes of the Ob river system. They are subject to monitoring by the Institute for Water and Environmental Problems in Barnaul. **a** Katun river in the Altai mountains, a headwater stream of the Ob river. **b** Middle Ob near Barnaul. The river has cut into a 70-m-deep loess sediment layer

Fig. 13 Lake Baikal near Irkutsk. The lake is the largest global open freshwater reservoir. The water has excellent drinking water quality. *Photo* Ralf Dannowski

drinking water quality (Grachev 2002). Its ecosystem is extremely stable. However, significant changes have been monitored over the past 60 years: an increase in the water temperature in the shore regions (1.21 K since 1946), enhanced amounts of phyto- and zooplankton and long-term changes in the basal food web have been recorded (Shimaraev and Domysheva 2002; Hampton et al. 2008; Moore et al. 2009). The lake starts to freeze later (Shimaraev et al. 2002; Todd and Mackay 2003), and the ice-free season has lengthened by 16 days over the past 137 years (Magnuson et al. 2000). The release of methane from Lake Baikal and other Siberian lakes is another consequence of warming and has a positive effect on climate warming (Walter et al. 2006).

Representative monitoring of Lake Baikal's water quality is particularly challenging because of its huge water volume. For example, a pollutant input of 23,000 tonnes would be required to detect a concentration shift by 1 microgram per

Table 17 Largest water reservoirs in the Asian part of Russia (*Source* UNEP/IETC 2000)

Name	*V* (km³)	*R* (years)	CP (μg/l)	*B* (g/m³)		Chl (μg/l)	SD (m)
				Mean	Max		
Bratsk	169.3	1.82	15	2.21	12.8	26	5.6
Krasnoyarsk	73.3	0.81	30	0.8	46.7	93	3.3
Ust-Ilimsk	58.9	0.58	36	2.2	245	490	3.4
Vilyuy	35.9	1.92	32	–	–	–	–
Kolyma	14.4	1.01	16	–	–	–	–
Novosibirsk	8.8	0.16	44	–	60	36	2.3

V is the volume of water; *R* is the residence time; CP is the concentration of total phosphorus; *B* is the biomass of phytoplankton in the vegetation period, $B = 0.5$–0.9 g/m³ (weak), $B = 1.0$–9.9 g/m³ (moderate), $B = 10$–99 g/m³ (intensive), $B > 100$ g/m³ (hyperintensive); Chl is a maximum concentration of chlorophyll-a; SD is the visible depth, a measure of water transparency

litre (Grachev 2002). Monitoring has thus to include methodical analytical work and balances of the whole lake catchment (Shimaraev 2009, IW-LEARN 2010).

Water reservoirs Russia has more than 2200 man-made water bodies. They store 793 km³ of water and are of great importance for municipal water supply, hydropower generation, local industries and other purposes. 417 reservoirs are located east of the Urals, with a storage capacity of 489 km³ (UNEP/IETC 2000). In terms of their volume, the Bratsk reservoir (Angara River) and the Krasnoyarsk reservoir (Yenisei River) are some of the largest in the world. Table 17 gives some parameters of the largest Siberian reservoirs along with some water quality data from the 1990s.

The siltation, pollution and eutrophication of reservoirs are of great concern and need to be monitored and controlled permanently (UNEP/IETC 2000).

11.2 Water Quality

11.2.1 General Situation

Water bodies in the Asian part of Russia are cleaner than in European Russia, where many rivers and reservoirs are heavily polluted. However, rivers and other open waters of industrial and/or more densely populated regions indicate moderate to severe water pollution. These are mainly the transboundary rivers of the Ob-Irtysh and the Amur basins. The 8th international Conference "Rivers of Siberia and the Far East" in Irkutsk, June 2013, summarised the critical water quality status of many Siberian rivers. The resolution of this conference addressed numerous critical issues of water quality and management in Asian Russia. Amongst them are the damage to the unique ecosystem of the Gulf of the Ob by the gas industry, the pollution of the trans-boundary Selenga River, which is the main tributary to Lake Baikal, and the violation of forest logging rules in the catchment of Lake Baikal associated with pollution loads (Proceedings Irkutsk 2013). Radioactive pollutants have accumulated in the sediments of Siberian rivers and the Arctic Ocean (Fukuda 1997).

High and excessive levels of persistent organic pollutants such as PCB and DDT were detected in the Ob and Yenisei river deltas and in sediments of the Indigirka and Pechora rivers (AMAP 1997). Considerable amounts of heavy metals (especially copper and nickel) and radioactive materials are found in the sediments of the Yenisei River (AMAP 1997). Mining and industrial activities during the past are the main sources of water pollution. However, some of the pollution centres, such as Norilsk Nikel, which is among the global top ten worst polluters (Bergen et al. 2013; Blacksmith Institute 2014), are still active. The quality of the Yenisei River in the upper and middle reaches is acceptable and stable (UNECE 2007).

The contribution of agriculture to water pollution is still low. However, it is already significant in the Ob drainage basin, where cropland covers about one third of the surface (Yang et al. 2004).

11.2.2 Ob-Irtysh Basin

Some of the Siberian rivers, such as the Ob-Irtysh system, show moderate to severe pollution (Koronkevitch 2002). The Irtysh is already heavily polluted when it enters Russia, but the high loads of pollutants are a result of mining, industrial and military activity during Soviet times (UNECE 2007), when Kazakhstan was part of the Soviet Union. Open-cut mining and developed oil-refining, power plants and other industries lead to the contamination of the environment including the water by heavy metals and other contaminants (Gordeev et al. 2004). Deposition and re-mobilisation of river sediments make these problems long-lasting. The current classification of the Irtysh in Kazakhstan is class 3 (moderately polluted) (UNECE 2007). The overall water quality of the Tobol, another transboundary river in Kazakhstan, upstream of the border with the Russian Federation, was 6 (heavily polluted) in 2006. The Ishim, a tributary of the Irtysh, was in quality class 2 (clean) at the same time before entering Russia (UNECE 2007).

11.2.3 Amur Basin

At the border between China and the Russian Federation, the Argun River is classified as "polluted" or "very polluted" (UNECE 2007). The Sungari (Songhua) river ranks among the five most polluted Chinese watercourses and is the most significant pollution source in the middle part of the Amur basin. Elevated mercury loads were measured in the bottom sediments of the Amur river due to poor industrial waste management practices before 1990 (Kot et al. 2009). The water quality of the Ussuri varies between classes 3 (moderately polluted) and 4 (polluted) (UNECE 2007).

Table 18 Some environmentally relevant elements in the water of Lake Baikal (Grachev 2002, modified)

Element	Concentration in Lake Baikal	Concentration in Selenga River	Inventory in Lake Baikal, thou tonnes	Annual input with Selenga River thou tonnes
Chloride	0.4 mg/l	2.1 mg/l	9200	63.0
Nitrate N	0.1 mg/l	0.43 mg/l	2300	12.3
Iron	40 μg/l	0.89 mg/l	920	27.03
Phosphate P	15 μg/l	40 μg/l	345	1.2
Copper	0.3 μg/l	0.98 μg/l	7	0.5

11.2.4 Lake Baikal

The pollution levels of Lake Baikal, the cleanest lake in the world, are very low (Table 18) and will probably stay low over the next 100 years. However, Table 18 shows that element concentrations of input waters are significantly higher. Concentrations of Cl could be double soon. Moreover, input balances are still incomplete, not even considering the re-mobilisation of particulate elements of the bottom sediment and local aspects. Melting permafrost in the lake catchment will probably exacerbate the effects of industrial pollution and cultural eutrophication in the catchment. Though the lake has an extremely high buffer capacity, this could have consequences for ecosystem functioning, in the shore region in particular (Moore et al. 2009). There will be a need for careful ecosystem monitoring and modelling. The pollution levels of this river are of particular importance (Bazhenova and Kobylkin 2013). The soil degradation processes in the Selenga River basin may contribute to the pollution level, as they have been significant since the first half of the 19th century (Bazhenova and Kobylkin 2013).

11.2.5 Other Lakes and Reservoirs

Other Siberian lakes and reservoirs are more susceptible to disturbances. In some dry areas, such as Yakutia, drinking water quality and water scarcity are related problems because of permafrost conditions and/or salinity. Permafrost melting will affect the hydrological regime of all water bodies greatly, both in quantity and quality. As some current pollutants may have penetrated into the permafrost layer, further permafrost melting will disclose the whole extent of water pollution caused by past industrial activity throughout Siberia.

Most underground waters, some of which are a source of drinking water, still remain clean (Koronkevitch 2002). From their hydrochemical data, lakes in the Tundra, Forest Tundra and Northern Taiga zones are pure on average (Kremleva et al. 2012). In the denser populated southern Taiga and Forest Steppe zones, increased mineralisation and loads of elements are evident (Table 19).

Table 19 Lake water micro-elements in natural zones of West Siberia (Kremleva et al. 2012, modified)

Concentration range (μg/l)	Tundra and Forest Tundra	Northern Taiga	Middle Taiga	Southern Taiga	Forest Steppe
>100		Fe	Fe	Sr	Sr > Ba > B
10–100	Fe > Al > B	Al > B	Al > B	Fe > Ba > Mn > B>Al	Mn > Al > Fe > Zn > Li
1–10	Mn > Sr > Zn > Ba > Cu	Zn > Mn > Ba > Cu > Sr	Mn > Zn > Ba > Sr > Cu > Li	Zn > Li > Cu > Rb > As	Cu > Rb > As > V>Mo
0.1–1	Mo > Ni > Li > Bi > As > Rb > Ti > Pb > Cr > Sb > V	Bi > Mo > Ni > Li > Rb > Pd > As > Ti > Cr > Sb	Mo > Rb > As > Pb > Ni > Ti > Bi > Cr > V>Sb	Mo > Bi > Pb > Ti > U>Cr > Sb > Ni	U > Pb > Ti > Bi > Sb > Cr > Ni

Table 19 is based on 11–48 samples per zone/sub-zone in the Yamalo-Nenets Autonomous Okrug and Tyumen Oblast. It shows that element distribution in the north largely reflects the natural conditions of the surrounding soils and rock whilst, towards the south, anthropogenic effects become more significant. Fe, Al and Mn are typical indicators of soil processes (acidification, gleyzation). Sr, Ba and B are also very reactive elements and common in the earth. Increased overall element concentrations and their dominance indicate anthropogenic impacts and air pollution by industrial and military-industrial activity, in particular.

12 Land- and Water-Resource-Based Industries and Natural Services

12.1 *Preface*

Forestry, agriculture and inland fishery utilize the land and water resources of Siberia for human livelihood. They provide food for the population and raw materials for industries and handicrafts. Compared with other regions of the world, this occurs with modest intensity in terms of inputs to the ecosystems such as fertilisers, herbicides, pesticides or veterinary drugs.

Resource management is based on different land rights, laws and programmes.

12.2 *Forestry*

12.2.1 Importance of the Forestry Sector

The forest sector is of great social, economic and environmental importance (IIASA 2011; FAO 2014a). The composition of growing stocks of Russia consists mainly of larch (*Larix* sp., 25.3 billion cubic metres), pine (*Pinus* sp., 16.3), birch (*Betula* sp., 11.0), spruce (*Picea* sp. 10.9), Siberian stone pine (*Pinus sibirica,* 8.4) and aspen (*Populus tremula*, 3.2) (FRA 2005). Timber stocks of the Siberian and Far East Federal Districts are 51.6 billion m^3, which is about two thirds of the Russian timber stock (Alexeenko 2012).

Industries based on the biological resources of Siberia's forests have always played an important role for the economy of Russia. About 200 years ago this was the hunting of fur-bearing animals. Later, the timber industry became prosperous. Annually harvested timber amounts to about 60–70 % of the annual increase in timber biomass in Russian forests (Aurenhammer 2003). This proportion is even much lower in Siberia and the Far East. The annual allowable cut there is about 50 % of the estimated annual increase of growing stocks, and current utilisation of this allowable cut is less than 40 % (Alexeenko 2012).

Those overall data sound sustainable, but regional disproportions are great and may conflict with other landscape functions. The forest sector makes up about 1–3 % of the industrial production in the administrative units of the Siberian and the Far East Federal Districts (Alexeenko 2012). In Siberia it is highest in the region around Lake Baikal (Irkutsk Oblast 21 %, Republic of Buryatia 5 %). In the Far East the share of the forest sector is highest in the Jewish Autonomous Oblast (20 %) and in the Khabarovsk Oblast (13 %) (Alexeenko 2012). Also, the proportion of the annually allowable cut used is particularly high in these regions. However, forests in these regions are hotspots of a unique biodiversity. Lake Baikal is a World Heritage Site. As forest ecosystems are a guarantee for top-quality water resources but logging and deforestation inevitably lead to eutrophication and pollution of waters (Neary et al. 2009). This will be a threat to Lake Baikal in the long run. The forests of the Jewish Autonomous Oblast are inhabited by the endangered Himalayan bear (*Ursus thibetanus*), and the forests of the Khabarovsk Krai by the endangered Siberian tiger (*Pantheratigris altaica*) (Newell 2004).

12.2.2 Forest Products

Timber: In 2012, 179 million cubic metres of industrial roundwood were produced, the majority in Siberia and the Far East. About three quarters of the roundwood (about 15 % of the production) is exported to China (UNECE 2010). Sawnwood, plywood and pulp and paper also play an important economic role (FAO 2014a). Russian production of soft sawnwood was 19.6 million m^3, along with 2.6 million m^3 of plywood, 2.0 million m^3 of fibreboard, 7.2 million tonnes of wood pulp and 7.7 million tonnes of paper and paperboard in 2008 (UNECE 2010). Total forest product exports were worth 6.4 billion USD in 2004 (Killmann and Whiteman 2006).

The area of annual timber harvesting in Siberia and the Far East was 763 thousand ha and 65 million m^3 per year over the first decade of this century (Alexeenko 2012). Timber production data for Siberia and the Far East Federal Districts in 2008 were 38.3 million m^3 of roundwood, 8.3 million m^3 of sawnwood and 0.2 million m^3 of plywood (Alexeenko 2012).

Table 20 Biological stock of edible products in Russian forests (Alexeenko 2012, modified)

Type of product	Biological stock, thousand tonnes		
	Total for Russia	In Federal Districts (FD)	
		Siberian FD	Far East FD
Wild berries, total	8840.5	4257.2	1186.3
Cranberries (*Vaccinium oxycoccus*)	1600.0	390.1	281.8
Cowberries (*Vaccinium vitis-idaea*)	3010.2	1328.6	507.6
Whortleberries (*Vaccinium uliginosum*)	2618.7	1723.5	57.2
Blueberries (*Vaccinium myrtillus*)	1013.8	626.8	258.2
Raspberries (*Rubus arcticus, Rubus idaeus*)	144.3	30.4	70.2
Cloudberries (*Rubus chamaemorus*)	453.5	157.8	11.3
Nuts from *Pinus sibirica, Pinus koraiensis* and *Pinus pumila*	3592.7	1098.5	2288.6
Mushrooms	4325.4	1089.6	2151.7
Birch sap	875,504.6	420,041.5	52,794.0

Roundwood and sawnwood are timber products at a low processing stage. The production of more processed and higher value-added timber goods is relatively poorly developed. In Siberia and the Far East the disproportion of harvesting volumes and timber processing capacities is particular high (CIBC 2007, Simeone 2014). Liberalisation measures for the forest industry based on the Forest Code of 2006 promise investments in the forest sector, mainly for timber-processing industries (Alexeenko 2012). This will strengthen the forest economy but put more pressure on ecosystems and require better monitoring and control.

Other products and services of forest ecosystems The gathering of wild fruits, mushrooms, nuts, medical plants, birch bark, honey and other natural products from forests has always been important for the livelihood of the local population (Filipchuk and Moiseev 2010). Biological stocks of these resources are huge (Table 20). Other benefits of forest ecosystems are spiritual and cultural values (Krankina et al. 1997).

Natural medical plants are a largely unexplored resource of forests in Siberia and the Far East. About 970 plant species are known to be suitable for medical purposes. About 20 % of them are officially considered to be medicinal (Alexeenko 2012). The Forest Tundra and other forest land of the north are important for deer farming and hunting.

12.2.3 Vulnerability of Forestry

Forests are a major stabilisation component of the natural landscapes of Siberia (IIASA 2011). They affect regional and global cycles positively by preventing land erosion and watercourse eutrophication, and are a great carbon store and buffer. Carbon stock in living forest biomass (including shrubs and understory) in Siberia and the Far East is 24.1 billion tonnes (75 % of the Russian stock) (Nilsson and Shvidenko 1998). Forest understory vegetation such as mosses, lichens and shrubs protects the soil from extreme heating and cooling (Blok et al. 2010).

Forest ecosystems are vulnerable to climate and human impacts (Krankina et al. 1997). Russian forests face disturbances by fire (1.3 million ha in the year 2000), insects (4.9 million ha), and diseases (1 million ha) (FRA 2005, Shvidenko and Schepaschenko 2013). About three quarters of these disasters occur in Siberia and the Far East. Such calamities are indicators of disturbed ecosystems and indicate risks for the survival of useful or protected species.

On average 1.5 million ha of forest in the Siberian and Far East Federal Districts burn down every year, causing an annual loss of 29 million m^3 of timber (Alexeenko 2012). Forest fires are frequently triggered by wrong human behaviour and have gone on for a long time. Eichler et al. 2011 found that most forest-fire activity in the Altai region occurred from 1600 to 1680 following an extremely dry period in the sixteenth century. No significant increase in biomass burning has occurred in the Altai region during the last 300 years.

The latter finding seems to contradict data collected by Groisman et al. 2007, who observed significant increases in temperatures and fire risk indices that characterise the weather conditions conducive to forest fires in the twentieth century. In the same region Bezuglova et al. (2012) found temperature increases ranging from 0.19 to 0.53 K/10 years in the period 1940–2008, and the most significant increase was registered during the cold seasons.

The consequences of forest fires are manifold. Economic loss of wood biomass, threats to life and property, carbon (and particularly black carbon) release to the atmosphere. Minerals released from burned areas underlie risks of drainage. In sloped areas the surface runoff and erosion are significantly intensified by forest fires (Evdokimenko 2013). Illegal logging results in an increase in fire hazard and higher carbon emissions than legal logging (Kukavskaya et al. 2014).

12.3 Agriculture

12.3.1 Agriculture and Food Security in Russia

Agriculture is an important component of the Russian state economy and a business for part of the population. The basic food supply is of strategic importance. The potentials and status of agriculture are determined by several natural and socio-economic constraints such as climate, soil quality, macro- and microeconomics, infrastructure and human entrepreneurship.

FAO data from 2007 show that the Russian agricultural sector provides 12 % of the gross national product and over 15 % of national income, and accumulates 16 % of capital assets (FAO 2010). The regions of highest agricultural intensity and productivity are located in the south of European Russia. Average Russia wheat yields were 2.3 t/ha in 2011 and 1.8 t/ha in 2012 (FAO 2014a). The highest cereal yields, of 4–5 t/ha, have been achieved over the past years in the Krasnodar Krai, located in the southeast of European Russia (FAO 2009), and the lowest yields in Siberia.

In spite of an extensive agricultural sector, food security in Russia still cannot be considered as given. The overall production of cereals in 2007 (78 million tonnes) was clearly lower than the production during Soviet times (102 million tonnes) (FAO 2014a). The droughts of 2010 and 2012 showed the vulnerability of Russian agriculture and led to an export ban and state regulations on the grain trade (Kiselev et al. 2012). The combination of instable yields with fluctuations of lands under cropping results in an insecure situation for harvested grain. Due to climate change the risk of food production shortfalls will probably triple by the year 2070 (Alcamo et al. 2007).

On the demand side of food security, an increase in meat consumption in Russia can be expected and, as a result, an increase in demand for cereals. Almost one third of the Russian domestic supply of pork and poultry meat was imported in 2009 (FAO 2014a), but the country is trying to increase the domestic supply. There is a need for strengthening regional reproductive processes in agriculture and for developing customer loyalty to local markets (Belyayev et al. 2014b). Overall food security and safety in the northern regions of Siberia and the Far East are at particular risk (Dudarev et al. 2013a).

12.3.2 Agriculture in Siberia

Meat and dairy husbandry based on grasslands, hay lands and forage crops, and cereal cropping are well developed in Siberia. The main crops are cereals (wheat), buckwheat and potatoes. In the lowlands of the Far East, even soybean (15 % of cultivated land) and rice cropping (5 %) is possible (Blagoveshchensky et al. 2002). The largest agricultural regions for cereal production are the Altai Krai (3.6 million ha), the Omsk Oblast (1.9 million ha) and the Novosibirsk Oblast (1.6 million ha) (IASA 2012). The intensity of the agricultural production is clearly dependent on the distance to urban centres and to the Trans-Siberian railway. The Siberian Federal District is an important grain-producing region for the Russian and world markets but has comparatively low agricultural productivity (Table 21). The effect of mineral fertiliser is also relatively low, mainly because of a lack of moisture and insufficient application technologies. Rotations that include black fallow after several years, or even every other year (Dry Steppe) are common. The grain production in the Siberian FD was 14.6 million tonnes on 9.8 million ha in 2011 (1.49 t/ha). This productivity level is still low and insecure. Belyayev et al. (2014a) emphasize the role of innovative technologies for enhancing the agricultural productivity and maintaining the soil fertility. Sufficient fertilisation and plant management could help to raise yield levels of wheat to more than two tonnes per hectare on the best soils (Gamzikov and Nozov 2010; Gamzikov et al. 2014).

Table 21 Long-term average effect of mineral fertiliser use on the grain yield of spring wheat on Siberian soils (Gamzikov and Nozov 2010, modified)

Soils[a]	Yield without fertilisers, t/ha	Yield increase with fertilisers[b], t/ha			
		N	P	NP	NPK
Soddy-podzolic soil (Retisols)	1.06	0.46	0.32	0.57	0.79
Grey forest soil (Phaeozems, Luvisols)	1.57	0.41	0.30	0.60	0.67
Chernozem (Chernozems, Phaeozems)	1.68	0.33	0.22	0.49	0.52
Chestnut soil (Kastanozems, Calcisols)	1.14	0.16	0.18	0.31	0.31

[a]Reference soil group of WRB (2014) in parentheses, [b]60–140 kg/ha of nitrogen (N), phosphorus (P_2O_5), and/or potassium (K_2O) fertiliser

12.3.3 Historical Development of Agriculture in West Siberia

Unlike the evolution of European agriculture, harsh climate and low demands for agricultural products have long restricted the development of a self-supplying and prospering market production of agricultural goods in most regions of the sparsely populated Siberia. The cheapest way for the rural population has always been the exploitation of natural resources and the utilisation of services which ecosystems provide. After the installation of the Trans-Siberian railroad, cropping agriculture developed notably in the Forest Steppe and Steppe regions, where the climate is appropriate and soils are fertile.

The enhanced exploitation of Siberia in the last century and the development of mining, industrial and military complexes and large cities induced a higher local demand for agricultural products. More lands were ploughed and intensive cropping systems were developed in the Forest Steppe and Steppe zones. Even some fruit like different kinds of apple, pear, black current, sea buckthorn, plum, sour cherry, strawberry and others could successfully bred and cultivated in Siberia since the 1930 years (Kalinina 1985).

The Altai Krai developed as a leading agricultural region of modern Russia. This region is part of the globally largest agricultural belt. The future of the grain production in this belt is likely to have a significant impact on the global and regional food security over the next decades (Lioubimtseva et al. 2013).

The Kulunda Steppe The development of agriculture can be demonstrated from the example of the Kulunda Steppe. The Kulunda is part of the Eurasian Steppe belt, located in the south-east of the West Siberian lowland between the Ob and Irtysh rivers and a main agricultural region of Siberia. The largest areas belong to the Altai Krai, with other parts in the Novosibirsk Oblast and Kazakhstan. Table 22 shows the human impact over the past 4.5 thousand years as re-constructed from pollen and mollusc analyses of lake sediments, and supplemented by recent historical data. Sedentary settlement and agricultural activity in this region started late, about 150 years ago (Rudaya et al. 2012). The Kulunda was subject to a huge land reclamation campaign starting in 1954/1955 (Meinel 2002). Land was ploughed

Table 22 Human activity and environmental conditions in the Kulunda Steppe during the Holocene

Chrono-logy	Archaeological period or historical event	Human activities	Environment
≈2440–1795 BC	Bronze Age, Elunino culture	Animal breeding	Dry climate, semi-desert Steppe, Birch kolki
≈1795–710 BC	Bronze Age, Onset of Iron Age, Andronovo culture	Cattle breeding Transition to a nomadic lifestyle Increase in sheep, goats and horses	More humid climate, grass and forb Steppes, conifers started to spread
≈710 BC–AD 580	Iron Age Early Scythian	Animal breeding with sheep, goats and horses which can forage under the snow	Relatively humid climate Spread of conifers
After AD 500	First and Second Turkic Khaganates	Nomadic lifestyle	Maximum of forest distribution and reduction in Steppe plant communities
AD 1200–1700	Mongolian Golden Horde	Increase of population	Relatively humid climate
After 1750 AD	Start of Russian colonisation		
After ≈ 1860 AD	Settlements of Russian peasants	Sedentism of population Development of industry and agriculture	Sharp decrease in woodland, aridisation of climate
After ≈ 1955 AD	Soviet land reclamation campaign	Ploughing Steppe land, wheat monoculture, planting poplar windbreaks, installation of rural infrastructure	Wind erosion, soil fertility degradation, loss of biodiversity
After ≈ 1990 AD	Disintegration of the Soviet Union, break-down of Kholkhozes and Sovkhozes	Decline of infrastructure, shrinking of rural population	Industrial and agricultural fallowing, re-vegetation of barren land
After ≈ 2000 AD	Partial recovery of the agricultural sector	Accelerated shrinking of rural population, consolidation of infrastructure, modernization of agriculture	Dry period, shrinking of lakes, loss of soil fertility, land desertification, litter pollution of landscapes

This table is based on data by Rudaya et al. (2012), Meinel (2002), Meyer et al. (2008) and our own studies

and wheat was grown. The human impact to this region accelerated with much benefit for the rural population at the cost of the environment and future generations. Later, after the decline of the Soviet Union, most of this land was abandoned. Since the beginning of this century the agriculture has partly recovered.

Nowadays, a great deal of wheat is grown in monoculture on ploughed Steppe soils and exported for feeding pigs. Despite some modernisation, current agricultural practice is not sustainable and will lead to serious damage to soils, air and water if no consistent measures are taken. Conservation cropping and no-till technologies (Derpsch et al. 2010) could help to make agriculture more sustainable.

The Eurasian Steppe region is an important target for investment by international agroholdings (Petrick et al. 2013). This will be associated with improvements for the rural population but risks for biodiversity and landscape functioning. Agri-environmental monitoring is needed.

12.3.4 Food Security and Agriculture in the High North

Food security and safety: Analyses by Dudarev et al. (2013a, b) found a particularly critical situation of food security and safety in regions of the High North. Food costs in remote regions were high and food was unsafe. Food costs amounted to about one third of the household income. High levels of biological and chemical contaminants were found in many cases, mainly in Chukotka, the Evenkiysky District of the Krasnoyarsk Krai and in the Magadan Oblast (Dudarev et al. 2013a). Drinking water as a basic human food was also not safe but seriously contaminated by biological and chemical agents (bacteria, spores, cysts, virus, pesticides, metals) both in centralised and non-centralised water sources (Dudarev et al. 2013b). Incidences of high rates of infectious and parasitic food- and waterborne diseases were the consequence (Dudarev et al. (2013c).

Agriculture Reindeer farming is the most important branch of the rural economy of the Extreme North. It has always been a domain of the indigenous peoples of Siberia such as Evenks, Nenets or Chukchis. This branch of the economy is closely associated with their lifestyle and survival. Their population and reproduction correlates with the reindeer population (Alexeenko 2012). During Soviet times, reindeer breeding was an important branch of agriculture in the Arctic and subarctic region. It was stabilised by state support programmes and economic regulations. Currently reindeer breeding is permanently declining, though some provincial government programmes are trying to stabilise the situation. Marketing aspects and the decay of sedentary infrastructure in the north are a burden and existential threat both for the indigenous population and for reindeer breeding as their economic basis. The infrastructure needs to be maintained for marketing of local products and for children's education, for example.

In principle, nomadic or semi-nomadic pastoral farming in Taiga and Tundra zones is environmentally friendly. Problems and conflicts with resource-exploring industries which block migration routes, or claim and pollute large areas of land, have arisen in the recent past (Degteva and Nellemann 2013).

The High North has also potentials for regular farming in order to supply the local population with basic food (Tichanovsky 2004). All ecological zones may provide forage for grazing domestic animals. Though the vegetation period is short and yields are low, forage production and the breeding of cows, or growing potatoes and some vegetables, are possible over the whole Taiga zone. Of the cereals, spring barley, being a "long day plant", grows well in the North. Additionally, suitable bio-climatic conditions can be created under adverse climatic conditions with higher inputs of energy and costs. Greenhouse production for vegetables is possible in all regions but not yet well developed.

Current crop production in the North is often limited more by non-climatic factors such as lack of infrastructure, a small population base, remoteness from markets and land ownership issues (Juday et al. 2009). There was some advanced knowledge about agricultural systems in the High North (Polyakov and Ivanov 1986; Tichanovsky 2004) but both practical agriculture in this region and research into it have declined over recent years.

12.3.5 The Recession of Agricultural Land and Water Management

Land, crop and water management are a precondition for productive and sustainable agriculture. This includes important activities such as soil tillage, fertilisation, drainage, irrigation and others. Those management activities have been seriously disregarded since the nineties. The recession persists and overall management intensity is minor.

Russian agriculture operates at a low average level of input intensity. The nutrient application in Russia in 2011 was 10 kg N, 3 kg P_2O_5 and 2 kg K_2O per hectare of cultivated land (FAO 2013b). Comparable data for Germany are 136, 20 and 32 kg/ha. From data collected by Bezuglov and Gogmachadze (2008) the mineral fertilisation in the cereal cropping regions of Siberia was clearly below the already low Russian average. The application of NPK fertiliser in Siberia amounts to less than 10 % of the application rates in the 1990s (Gamzikov and Nozov 2010). The annual balance deficit of main nutrients was 27 kg/ha N, 8.6 kg/ha P_2O_5 und 22.4 kg/ha K_2O in 2006–2009 (Gamzikov and Nozov 2010). This was due to nutrient mining from soils, diminishing their fertility.

Other soil fertility maintaining or improving measures were disregarded too. Organic fertilisation, the liming of acid soils and the amelioration of Solonetz soils were at levels of 12, 7 and 4 %, respectively, if compared with levels in the 1980s (Gordeev and Romanenko 2008).

About 29–74 million ha of land in Russia would require irrigation. Water availability has become a central issue for social and ecological sustainability (Groisman and Gutman 2013). In fact, the proportion of irrigated land has declined to a level of 2 % and the proportion of drained land to 4 % of the cultivated land (FAO 2014a). The agricultural land of Siberia is located in the forest and Steppe zones, where about 2–5 and 5–8 % of land, respectively, would require irrigation (FAO 1997). There is a great demand for the reconstruction of irrigation and drainage systems, which have undergone severe decay over the past 25 years (Table 23). Due to these deficits, cropland became more prone to frequent droughts and wetness.

Table 23 Drained and irrigated agricultural lands in the Siberian and Far East Federal Districts (data from Gordeev and Romanenko 2008)

		Total area (thousand ha)	Area of insufficient function	
			Thousand ha	%
Drainage	Siberian FD	230	77	33
	Far East FD	645	174	27
Irrigation	Siberian FD	515	48	9
	Far East FD	122	28	23

About 20 % of formerly drained areas in the Siberian FD and more than half of areas in the Far East FD are completely out of function, which has led to these lands being given up for agriculture (Gordeev and Romanenko 2008).

There is virtually no plant management on most agricultural land in Siberia. The land is tilled, wheat is seeded and harvested only. Conservation tillage, which is the only method to avoid land degradation by wind erosion, has almost no practical significance. Land quality is permanently declining and wheat yields are lower than 1 t/ha in most regions. Land degradation is delayed by the fact that much land was set aside and may recover during this time of fallow. 18 % of agricultural land in the Siberian FD and almost half of land in the Far East FD (including Yakutia) has been abandoned (Gordeev and Romanenko 2008). Setting land aside from agricultural production has different effects depending on natural zones. In Steppe regions, set-aside land can be returned to cropping without problems. However, in the Forest Steppe or Taiga, the growth of bushes or trees has prevented this for a few years and cements the status of land abandonment.

12.3.6 Implications of Climate Change for Agriculture

Russia's agriculture may have already experienced the initial consequences of climate change. In 2010 and 2012, drought caused a significant drop in grain production in the country, as well as a consequent increase in grain prices (Safonov and Safonova 2013). Agriculture in Siberia is highly dependent on the weather and climate. High springtime temperatures and a lack of precipitation, which is very typical of continental climates, affect cereal yields negatively (Kharlamova and Silantyeva 2011).

There is a high risk that mitigation and adaption efforts such as investments in soil fertility and agricultural technologies will continue to fail (Kiselev et al. 2012). Adaption strategies depend on the economy and infrastructure in specific regions. There are large inequalities within Russia. Regions of Russia without any large industrial or mining complexes are relatively poor and underdeveloped. This holds just for the agricultural and forestry regions of Siberia and the Far East such as the Republic of Buryatia, the Amurskaya Oblast, the Trans-Baikal region, the Altai Republic and Altai Krai, the Jewish Autonomous Oblast and the Tuva Republic

Table 24 Ranks of human potential in Asian Russia (from data by Aysan and Bobolev 2011)

Region	Federal District	HDI 2009	HDI rank
Tyumenskaya Oblast	Ural	0.882	3
Sakhalinskaya Oblast	Far East	0.855	5
Tomskaya Oblast	Siberia	0.850	7
Average HDI value of the Russian Federation		0.840	
Sakha Republic (Yakutia)	Far East	0.836	8
Krasnoyarsk Krai	Siberia	0.834	9
Omskaya Oblast	Siberia	0.834	10
Sverdlovskaya Oblast	Ural	0.828	15
Novosibirskaya Olast	Siberia	0.828	17
Magadanskaya Oblast	Far East	0.817	30
Kemerovskaya Oblast	Siberia	0.812	34
Irkutskaya Oblast	Siberia	0.811	37
Average HDI rank of the Russian Federation			40
Chukot Autonomous Okrug	Far East	0.809	43
Chakassia Republic	Siberia	0.809	44
Khabarovskiy Krai	Far East	0.804	51
Primorskiy Krai	Far East	0.804	52
Kamchatskiy Krai	Far East	0.798	58
Altaiskiy Krai	Siberia	0.796	59
Kurgan Oblast	Ural	0.796	60
Republic of Buryatia	Siberia	0.791	66
Amurskaya Oblast	Far East	0.789	68
Trans-Baikal Krai	Siberia	0.782	73
Altai Repubic	Siberia	0.763	77
Jewish Autonomous Okrug	Far East	0.762	79
Tuva Republic	Siberia	0.732	80

HDI Human Development Index, which comprises indicators of the gross domestic product, purchase power parity, life expectancy and education

(Table 24; Fig. 14). These regions rank clearly below average in the Human Development Index.

Most of these regions have an increasing risk of soil degradation, desertification and off-site damage. Forest belts around agricultural land are important in Steppe regions to prevent wind erosion and dust storms. Over the past 20 years, the area of forest belts has decreased by approximately half in the Altai region, and these belts are low in quality and function poorly (Safonov and Safonova 2013). Kharlamova and Silantyeva (2011) found that dust storms were at a minimum (20 days a year) between 1980–1990 and have increased ever since. The reconstruction and creation of new forest belts will be required as part of mitigation and adaption strategies.

Possible positive consequences of climate change for crop production in Siberia are, for example, longer vegetation periods, a potential increase in areas appropriate for farming; advantages for permanent crops due to milder winters, the improvement

Fig. 14 Life in villages is difficult as most people have low incomes and rural infrastructure is obsolete

of farming operations due to warmer periods. Maksimova et al. (2014) found that over the 1964–2009 period the growing season duration increased by 6 days as compared to the average for the agro-climatic areas of the Altai Region. Kirsta et al. (2013) modelled a trend of cereal crop yields in the range of −0.012 to +0.032 % per year for different counties of the Altai Krai until 2020.

The negative consequences are, however. drought stress to crops because of more, longer hot and dry periods (Kiselev et al. 2012). Balancing all positive and negative effects of climate change on agriculture will result in negative consequences for Russia (Götz 2009, Dronin and Kirilenko 2011, FAO 2013a). Also, relative advantages compared with other food-producing countries are improbable (Aysan and Bobolev 2011).

12.4 Inland Fishery

Extensive inland water resources of 71.7 million ha (Filipchuk and Moiseev 2010) are a basis for the fishing industry, subsistence and recreational fishing. The inland fish catch of Russia was 246 thousand tonnes in 2009 (Welcomme 2011). A lot of wild fish comes from Siberia and the Far East (Table 25). More than a quarter of the total Russian inland wild fish catch comes from the Ob-Irtysh catchment. Other

Table 25 Important fish species of inland fishery in Siberia and the Far East

Common name	Scientific name	Catch, tonnes	Remarks
Chum salmon	*Oncorhynchus keta*	41,913 (17.0)	Seafish with spawning habitats in rivers, main occurrence in the Far East and East Siberia
Pink salmon	*Oncorhynchus gorbuscha*	30 632 (12.4)	
Sockeye salmon	*Oncorhynchus nerka*	15,403 (6.3)	
Freshwater bream	*Abramis brama*	27 886 (11.3)	Freshwater fish, belonging to Cypriniformes order (carp-like fish), main catch in the Ob-Irtysh catchment
Roaches	*Rutilus* spec. (mainly)	12,527 (5.1)	
Orfe	*Leuciscus* spec.	4837 (2.0)	
Tench	*Tinca tinca*	2622 (1.1)	
Other cyprinids	*Cypriniformes* order	26,192 (10.6)	
Whitefishes	*Coregonus* Genus (mainly)	3750 3381 (5.4)	Including Omul (*Coregonus migratorius*) of Lake Baikal
Northern pike	*Esox Lucius*	15,988 (6.5)	
European perch	*Perca fluviatilis*	8565 (3.5)	
Pike-perch	*Sander lucioperca*	2515 (1.0)	

Table is based on data by Welcomme (2011), catch refers to 2009, numbers in parentheses are percentages of the Russian inland catch

important sources of inland fish in Siberia are Lake Baikal and the Yenisei catchment (Wikipedia 2013b). Freshwater aquaculture is less developed east of the Urals due to climatic constraints.

Table 25 shows that salmon from the Far East and cyprinids from West Siberia dominate the catch. The salmon is not a freshwater fish but has its spawning habitats in rivers of good water quality. Salmon is in high demand by consumers because of its high-quality flesh and caviar.

Russian fisheries are currently unable to meet the domestic demand for fish (FAO 2007). The poverty of the local population, lack of public awareness and public involvement in the management of local resources contribute to fish poaching and overfishing (Newell 2004; FAO 2007). In the Far East regions, ocean fishing and fishery industry are major employers of the population. The fleet is obsolete. Fish is exported to neighbouring Asian countries whilst the demands of the Russian market cannot be satisfied (FAO 2007).

12.5 Landscape Potentials for Emerging Alternative Activities

The ecosystems of Asian Russia are of great functional, cultural, aesthetic and recreational importance (Fig. 15). Landscapes offer largely underdeveloped potential for environmentally friendly land and water management concepts, for

Fig. 15 Siberian landscapes have excellent potential for leisure activities and education. **a** At the World Heritage Site of Lake Baikal. **b** Watching wild animals such as squirrels, **c** and **d** Gathering wild fruits (wortleberries, cowberries) in the Forest Tundra. **e** learning about the culture of indigenious people: Nenets women in festive dress (*Photo* Pavel Barsukov). **f** Lichens are a main food for reindeer

recreation, leisure activities and tourism. These potentials should be developed and preserved. All activities in these fields could provide a major contribution to maintaining wildlife and biodiversity. The rural population must be a main stakeholder in all processes of landscape development.

Organic agriculture More and more people mistrust conventionally produced food, which could be polluted and risk their good health. Organic (biological) agriculture is an alternative option. Sales of organic products could reach up to $225 million in 2015, indicating an increasing demand from more affluent Russian consumers (Kolchevnikova 2013).

Soils of the Forest Steppe have particularly good potentials for organic agriculture, as their nutrient levels and storage and buffer capacity are high. Soil drainage and leaching rates are low and would allow biologically fixed nitrogen to be maintained for the nutrition of other crops in the rotation. These conditions need to be underpinned by experiments and analyses (Kislov and Dolmatov 2012). Organic agriculture requires proximity to consumers. Both these conditions are given in the vicinity of large Siberian cities.

Hunting Hunting is a traditional human activity for the indigenous peoples of the North, in particular. Whilst the economic importance of hunting is declining, its recreational importance is increasing. About 600 million ha of land (mainly forest land) were given to organisations or individual hunters in 2003. Since 2006, hunting areas are leased based on the new Forest Code. There are 31 species of fur-producing animals, 81 species of ungulates and around 70 species of birds for hunting in Asian Russia (Alexeenko 2012). If properly organised and strictly controlled, hunting can contribute to the sustainable management of wild animal stocks.

Recreational fishing The Far East and Siberia have much potential for developing recreational fishing. Recreational fishery both by local populations and by tourists is beneficial for the local economy. It generates income, for example from sale of fishing licences and equipment, and is beneficial for boat renters, lodge owners, travel agencies, and restaurants (Welcomme 2011). It is also beneficial for fish stocks and the status of aquatic ecosystems as it stimulates the sustainable management of natural resources by all people involved in this activity or business.

Recreation of the urban population Most people in Asian Russia work and live in cities. They consider and require rural and remote areas to be a source of recreation and inspiration. This is an opportunity for the inhabitants of rural areas and communities to improve their income and to keep pace with the increasing living standards in cities. There will be an increasing demand for farm holidays or vacancies in foresters' lodges. Short-term and individual tourism will also emerge, but requires the development of infrastructure and rules for keeping landscapes green and clean.

Tourism, ecotourism, scientific tourism General tourism is already developed in most exciting regions. Some popular tourist areas are Lake Baikal, the mountain areas of Altai and Sayan, Kamchatka, the Amur River valley and some locations in Yakutia (Alexeenko 2012; Evstrop'eva 2013; Luzhkova 2013). Natural landscapes have great potential for environmental tourism (eco-tourism), adventure tourism or sports tourism. In other regions agritourism could help to improve the farm economy (Kundius et al. 2011). Developing tourism must be supported by promotional programmes, as other stakeholders such as local people and authorities have less interest in the touristic development of their region (Shekhovzhova 2013). Scientific tourism is another important branch (Ilyina and Mieczkowski 1992). It can be an important step towards genuine scientific cooperation. The outer zone of reserves and national parks or planned protected areas are preferred objects for scientific studies. It is important that those studies are headed by acknowledged scientists.

13 Overall Assessment of the State of the Environment

Large ecosystems have a large buffer capacity against disturbances. Due to the huge dimensions and low population density of Siberia, the overall status of the environment is still acceptable. In many areas it is even excellent. Moreover, the region has exciting, largely undisturbed landscapes and many natural potentials and benefits for the local and global population. However, during the analysis some deficits and distressing trends regarding the status of the environment in Siberia and the Far East became clear, such as

- wasteful mining, forest, fishing and other industries
- air, water and soil pollution
- man-induced accelerated permafrost melting
- ineffective and soil-destroying agriculture
- recession of rural areas
- uncontrolled fires in peatlands, forests and on agricultural land
- lack of environmental awareness and illegal behaviour in protected areas
- lack of reliable data and monitoring systems
- lack of innovation in research and technology

(Newell 2004; Juday et al. 2009). A great deal of landscape pollution is a burden from the USSR period, when industrial environmental standards were low. However, the introduction of the market economy over the past 25 years has not diminished but exacerbated negative developments (Shaw and Oldfield 2007). There is largely a lack of consistent data and contemporary monitoring and technologies. This would be motivation enough to tackle some of the above-mentioned problems by introducting more innovations and technologies, some of which are available for free (Mueller et al. 2014). Climate change will put a lot of additional pressure on Siberian landscapes (Groisman and Gutman 2013) and urge responsible people to take action towards initiating sustainable developments. Modern methods of monitoring, assessing and controlling the status of land and water resources, as presented in the following chapters, could help to start this process.

References

ACIA (2004) Impacts of a Warming Arctic—Arctic climate impact assessment. Cambridge University Press, New York, 1042 p. http://www.acia.uaf.edu/pages/scientific.html. Accessed 30 Mar 2015

Adam AM, Mamin RG (2001) Natural resources and ecological safety of West Siberia (Природные ресурсы и экологическая безопасность Западной Сибири), 2nd edn. Publishing house "NIA Priroda" Moscow 2001, 164 p. http://window.edu.ru/resource/477/70477/files/Mamin_kniga.pdf. Accessed 30 Mar 2015

Adam AM, Gromov YuA, Elagin VB, Koniashkin VA (2013). Radon hazard estimation in Tomsk Oblast (Оценка радоноопасности в Томской Области) In: Radioactivity and radioactive elements in environment. Proceedings of IV international conference, Tomsk. June 4–8, 2013, pp. 38–39, (in Russian, English abstracts). http://portal.tpu.ru/files/conferences/radioactivity/book-light.pdf. Accessed 30 Mar 2015

Alcamo J, Dronin N, Endejana M, Genady Golubev G, Kirilenko A (2007) A new assessment of climate change impacts on food production shortfalls and water availability in Russia. Glob Environ Change 17(2007):429–444

Alexeenko A (2012) Russian Federation Forestry outlook study. Asia-Pacific Forestry Sector outlook study II. Working paper series. Working paper no. APFSOS II/ WP/ 2012/ 35. Food and Agriculture Organization of the United Nations. Regional Office for Asia and the Pacific. Bangkok, 2012. http://www.fao.org/docrep/016/an002e/an002e00.pdf. Accessed 30 Mar 2015

AMAP (Arctic Monitoring and Assessment Programme) (1997). Arctic Pollution Issues: A State of the Arctic Environment Report ISBN 82-7655-060-6, 1997 published by AMAP, Oslo, Norway. www.amap.no/documents/download/79. Accessed 30 Mar 2015

Anisimov O, Reneva S (2006) Permafrost and changing climate: the Russian perspective. Ambio 35(4):169–175. Royal Swedish Academy of Sciences 2006. http://permafrost.su/sites/default/files/Ambio.pdf. Accessed 30 Mar 2015

Anisimov OA, Velichko AA, Demchenko PF, Eliseev AV, Mokhov II, Nechaev VP (2002) Effect of climate change on permafrost in the past, present, and future. Izvestiya, Atmospheric and oceanic physics, vol 38, Suppl. 1, pp. 25–39. http://ifaran.ru/old/mokhov/AnisimovEtAl02_FAOs.pdf. Accessed 30 Mar 2015

Antipov AN (2006) Landscape-hydrological analysis of a territory, pp 52–64. In: Antipov AN (ed) Geography of Siberia. Research India Publications New Delhi, 235 p

Aurenhammer P (2003) Russlands Forstwirtschaft. Wald und Holz 4/03, pp 44–48, http://www.waldwissen.net/lernen/weltforstwirtschaft/wsl_russlands_forstwirtschaft/wsl_russlands_forstwirtschaft_originalartikel.pdf. Accessed 30 Mar 2015

Aysan AA, Bobolev SN (2011) UNDP report. Human Development in the Russian Federation Report 2011: Modernisation and Development of Human Potential (Доклад о развитии человеческого потенциала в Российской Федерации 2011. Модернизация и развитие человеческого потенциала). Moscow, 147 p http://www.undp.ru/documents/nhdr2011rus.pdf . Accessed 30 Mar 2015

Baklanov AA, Penenko VV, Mahura AG, Vinogradova AA, Elansky NF, Tsvetova EA, Rigina OYu, Maksimenkov LO, Nuterman RB, Pogarskii FA, Zakey A (2013) Aspects of atmospheric pollution in Siberia. In: Groisman PJ, Gutman G (eds) Regional environmental changes in Siberia and their global consequences (Chap. 8), Springer Environmental Science and Engineering 2013, ISBN: 978-94-007-4568-1 (Print) 978-94-007-4569-8 (Online) http://link.springer.com/chapter/10.1007/978-94-007-4569-8_8. Accessed 30 Mar 2015

Baranov AV, Grigoriev VYa, Yakushev NL, Unanyan KL (2010) Degradation and conservation of soil in the areas of developing of fields of hydrocarbons of the far north (Деградация и Охрана Почв в Районах Освоения Месторождений Углеводородов Крайнего Севера). Institute for Problems of Oil and Gas of the Russian Academy of Sciences 2010. Georesources, Geo-Energetics, Geopolitics (Георесурсы, геоэнергетика, геополитика). http://oilgasjournal.ru/vol_2/articles/7.html. Accessed 30 Mar 2015

Baranovskaya NV, Rikhvanov LP, Matveenko IN, Strakhovenko VD, Malikova IN, Shcherbakov BL, Sukhorukov FV, Aturova VP (2012) Radioactive elements in soils of Siberia (Russia) EGU General Assembly 2012, held 22–27 April, 2012 in Vienna, Austria, p. 14198. http://adsabs.harvard.edu/abs/2012EGUGA..1414198B. Accessed 30 Mar 2015

Baulinym VV, Belopuchovoy EB (1963) Permafrost (Многолетняя мерзлота) p. 44. In: Richter GD (ed) Natural conditions and existing resources. West Siberia, in Russian (Природные Условия и Естественные Ресурсы. Западная Сибирь). USSR Academy of Sciences. Moscow, 488 p. http://www.uriit.ru/dzz/rffi/book/richter.pdf. Accessed 30 Mar 2015

Bazhenova OI, Kobylkin DV (2013) The dynamics of soil degradation processes within the Selenga basin at the agricultural period. Geogr Nat Res 34(3):221–227

Belozertseva IA (2013) The degradation and pollution of soils in the area of gas condensate field development. J Earth Sci Eng 2:131–137. file:///C:/Users/Ilse/Downloads/245492013062006509362.pdf. Accessed on May 22, 2014

Belyayev VI, Dobrynin VI, Sopko II, Yefremov VV (2014a) The change of the Russian economy development paradigm (Смена Парадигмы Развития Российской Экономики). J Altai State Agric Univ Barnaul 11(121):144–148 (in Russian). http://www.asau.ru/files/vestnik/2014/11/144-148.pdf. Accessed 30 Mar 2015

Belyayev VI, Belyayev VV, Bezgodova OI, Chernysheva YeV, Kosheleva NA, Novikova MG, Shumilova AYu (2014b) Territorial marketing in local markets of territorial industry's enterprises and strategic management of regional reproduction, case study of the Zao "Molochnaya Skazka" (Маркетинг Территорий на Локальных Рынках Редприятий Местной Промышленности и Стратегическое Управление Региональным Воспроизводством, на Материалах Зао « Молочная Сказка »). J Altai State Agric Univ Barnaul 2(112):160–171 (in Russian). http://www.asau.ru/files/vestnik/2014/2/160-171.pdf. Accessed 30 Mar 2015

Berg LS (1947) Geograficheskie zony Sovetskogo Soyuza (Geographical zones of the Soviet Union), 3rd edn. OGIZ, Moscow

Bergen KM, Hitztaler SK, Kharuk VyI, Krankina ON, Loboda TV, Zhao T, Shugart HH, Sun G (2013) Human dimensions of environmental change in Siberia, In: Groisman PJ, Gutman G (eds) Regional environmental changes in Siberia and their global consequences (Chap. 7). Springer Environmental Science and Engineering. ISBN: 978-94-007-4568-1 (Print) 978-94-007-4569-8 (Online). http://link.springer.com/chapter/10.1007%2F978-94-007-4569-8_7. Accessed 30 Mar 2015

Bezuglov VG, Gogmachadze GD (2008) Application of fertilisers in agriculture (Применение удобрений в сельском хозяйстве). AgroEcoInfo of the Russian Federation http://agroecoinfo.narod.ru/journal/STATYI/2008/2/st_18.doc. Accessed 30 March 2015

Bezuglova NN, Zinchenko GS, Malygina NS, Papina TS, Barlyaeva TV (2012) Response of high-mountain Altai thermal regime to climate global warming of recent decades. Theor Appl Climatol 110:595–605. doi:10.1007/s00704-012-0710-2

Blacksmith Institute (2014) Top ten toxic threats in 2013. http://www.worstpolluted.org/projects_reports/display/116. Accessed 30 Mar 2015

Blagoveshchensky G, Popovtsev V, Shevtsova L, Romanenkov V, Komarov L (2002) Country pasture/forage resource profiles. Russian Federation. http://www.fao.org/ag/AGP/AGPC/doc/Counprof/russia.htm. Accessed 30 Mar 2015

Blok D, Heijmans MMPD, Schaepman-Strub G, Kononov AV, Maximov TC, Berendse F (2010) Shrub expansion may reduce summer permafrost thaw in Siberian Tundra. Global Change Biol 16(4):1296–1305. doi:10.1111/j.1365-2486.2009.02110.x

Bulatov VI, Rotanova IN, Chernykh DV (2006) Landscape ecology and cartographical analysis of natural salt complexes in the south west Siberia Basins of Lake Chany and Lake Kulundinskoye. Sabkha Ecosyst Tasks Veg Sci 42:233–247. http://link.springer.com/chapter/10.1007/978-1-4020-5072-5_19. Accessed 30 Mar 2015

Chernyago BP, Nepomnyashchikh AI, Medvedev VI (2012) Current radiation environment in the central ecological zone of the Baikal Natural Territory. Russ Geol Geophys 53(9):926–935

Chlachula J (2003) The Siberian loess record and its significance for reconstruction of the Pleistocene climate change in north-central Asia. In: Derbyshire E (ed) Dust indicators and records of terrestrial and marine palaeoenvironments (DIRTMAP). Quat Sci Rev 22(18–19): 1879–1906

Chojnicki BH, Michalak M, Acosta M, Juszczak R, Augustin J, Droesler M, Olejnik J (2010) Measurements of carbon dioxide fluxes by chamber method at Rzecin wetland ecosystem in Poland. Polish J Environ Stud 19:283–291

Chuprova IL (2006) Optimization of technogene landscapes in the High North (Оптимизация техногенных ландшафтов крайнего севера, Норильский промышленный район, п-ов Таймыр). Diss. Petrozavodsk 2006. http://earthpapers.net/optimizatsiya-tehnogennyh-landshaftov-kraynego-severa. Accessed 30 Mar 2015

CIBC (2007) CIBC world markets. 22 Feb 2007. http://conservation-economics.com/pdf_pubs/presentation/CIBC_RussiaLogExportTax_022207.pdf. Accessed 30 March 2015

Climate Committee (2002) Commission of the Russian Federation for problems of climate change (Межведомственная комиссия Российской Федерации по проблемам изменения климата. Третьенациональное Сообщениероссийской Федерации представленное в соответствии со статьями 4 и 12 рамочной Конвенции. Организации Объединенных Наций об изменении климата). Moscow, 158 p. http://unfccc.int/resource/docs/natc/rusncr3.pdf. Accessed 30 Mar 2015

Degteva A, Nellemann C (2013) Nenets migration in the landscape: impacts of industrial development in Yamal peninsula. Russia Pastoralism: Res Policy Practice 3:15. doi:10.1186/2041-7136-3-15. http://www.pastoralismjournal.com/content/3/1/15. Accessed 30 Mar 2015

Derpsch R, Friedrich T, Kassam A, Li H (2010) Current status of adoption of no-till farming in the world and some of its main benefits. Int J Agric Biol Eng 3(1):1. Open Access at http://www.ijabe.org. http://www.fao.org/ag/ca/CA-publications/China_IJABE.pdf. Accessed 30 Mar 2015

Dokuchaev VV (1949) K izucheniyu o zonakh prirody (Towards the study of the zones of nature). Izbrannye sochineniya (Selected essays) vol III, pp 317–322. Gosudarstvennoe izdatel'stvo Sel'skokhozyaistvennoi Literatury, Moscow (Orig. pub. 1898–1899)

Dokuchaev VV (1951) Izbranniye sochineniya (Selected essays). vol VI. Gosudarstvennoe izdatel'stvo Sel'skokhozyaistvennoi Literatury Moscow, 515 p. (Orig. pub. 1898–1899)

Dronin N (2011) Kirilenko A (2011) Climate change, food stress, and security in Russia. Reg Environ Change 11(Suppl 1):167–178. doi:10.1007/s10113-010-0165-x

Drösler M, Freibauer A, Christensen T., Friborg T (2008) Observation and status of peatland greenhouse gas emission in Europe. In: Dolman H, Valentini R, Freibauer A (eds) The continental-scale greenhouse gas balance of Europe. Ecological studies, 203, 237–255

Dudarev AA, Dushkina EV, Sladkova YuN, Alloyarov PR, Chupakhin VS, Dorofeyev VM, Kolesnikova TA, Fridman KB, Evengard B, Nilsson LM (2013b) Food and water security issues in Russia II: water security in general population of Russian Arctic, Siberia and Far East, 2000–2011. Int J Circumpolar Health 72. 10.3402/ijch.v72i0.22646. Published online 9 Dec 2013. doi:10.3402/ijch.v72i0.22646. http://www.ncbi.nlm.nih.gov/pubmed/24350065. Accessed 30 Mar 2015

Dudarev AA, Dorofeyev VM, Dushkina EV, Alloyarov PR, Chupakhin VS, Sladkova YuN, Kolesnikova TA, Fridman KB, Nilsson LM, Evengard B (2013c) Food and water security issues in Russia III: food and waterborne diseases in the Russian Arctic, Siberia and the Far East, 2000–2011. Int J Circumpolar Health 72. doi:10.3402/ijch.v72i0.21856. Published online 9 Dec 2013. http://www.ncbi.nlm.nih.gov/pmc/articles/PMC3860330/. Accessed 30 Mar 2015

Dyck S, Peschke G (1983) Grundlagen der Hydrologie. VEB Verlag für Bauwesen Berlin, 386 p

Dyukarev EA, Lapshina ED, Pologova NN (2011) Sphagnum mosses growth dynamics and its response to variations in climatic characteristics (in Russian), p 16–17. In: Vompersky SE (ed) West Siberian peatlands and carbon cycle: past and present. In: Proceedings of the third international field symposium (Khanty-Mansyisk June 27–July 5, 2011), Novosibirsk 2011, 220 p. ISBN 978-5-9611-066-2

Ebmer AW (2006) Daten zur Aculeaten-Fauna der Ussuri-Region unter Berücksichtigung der angrenzenden Gebiete – 2 Arten der Gattungen Halictus, Lasioglossum, Dufourea, Macropis aus dem Lazovski Zapovednik - Naturreservat Laso (Insecta: Hymenoptera: Apoidea: Halictidae, Melittidae) Linzer biol. Beitr. 38/1 541-593 21.7.2006
Egorova NI (2014). Ecological peculiarities of the agro-phytocenosis of brewing barley in the southern forest steppe of West Siberia. (Экологические особенности состояния агрофитоценоза пивоваренного ячменя в южной лесостепи Западной Сибири) Dissertation Omsk, 188 p
Eichler A, Tinner W, Brütsch S, Olivier S, Papina T, Schwikowski M (2011) An ice-core based history of Siberian forest fires since AD 1250. Quat Sci Rev 30(2011):1027–1034
Ermolov YV, Makhatkov ID, Milyaeva EV (2011) Application of conductance measuring method for studying of self-cleaning of a salted peatbog. p 168–170. In: Vompersky SE (ed) West Siberian peatlands and carbon cycle: past and present. Proceedings of the third international field symposium (Khanty-Mansyisk June 27–July 5, 2011), Novosibirsk 2011, 220 p. ISBN 978-5-9611-066-2
Evdokimenko MD (2013) Pyrogenic disturbances of the hydrothermic regime of cryogenic soils under light coniferous forests in southeastern Siberia. Eurasian Soil Sci 46(2):117–126. http://link.springer.com/article/10.1134%2FS106422931302004X. Accessed 30 March 2015
Evstrop'eva OV (2013) Ethno-recreational potential of the Baikal region. Geogr Nat Res 34 (1):61–68. http://www.turismulresponsabil.ro/wp-content/uploads/2012/11/IJRT-2.3_full1.pdf. Accessed 22 May 2014
FAO (1997) Russian Federation National water report Nr 15, 1997 http://www.fao.org/nr/water/aquastat/countries_regions/RUS/print1.stm. Accessed 30 Mar 2015
FAO (2006) New_LocClim 1.10. Local climate estimator. ftp://ext-ftp.fao.org/SD/Reserved/Agromet/New_LocClim/. Accessed 30 Mar 2015
FAO (2007) Fishery and aquaculture. Country profiles. National fishery sector overview. The Russian Federation. ftp://ftp.fao.org/fi/document/fcp/en/FI_CP_RU.pdf. Accessed 30 Mar 2015
FAO (2009) Analysis of the agribusiness sector in Southern Russia FAO Investment Centre/EBRD Cooperation Programme Report Series, No. 13. http://www.fao.org/fileadmin/user_upload/tci/docs/RS13-Russian%20Federation_Analysis%20of%20the%20Agribusiness%20Sector%20in%20Southern%20Russia.pdf. Accessed 30 Mar 2015
FAO (2010) Country report on the state of plant genetic resources for food and agriculture. Russian Federation. http://www.fao.org/docrep/013/i1500e/Russian%20Federation.pdf. Accessed 30 Mar 2015
FAO (2012) AQUASTAT database, food and agriculture organization of the United Nations (FAO). http://www.fao.org/nr/water/aquastat/data/wrs/readPdf.html?f=WRS_RUS_en.pdf. Accessed 30 Mar 2015
FAO (2013a) The second national report on plant genetic resources for food and agriculture in the Russian Federation. Country report on the state of plant genetic resources for food and agriculture: Russian Federation. http://www.fao.org/docrep/013/i1500e/Russian%20Federation.pdf. Accessed 30 Mar 2015
FAO (2013b) FAOSTAT database, country profiles. http://faostat.fao.org/CountryProfiles/Country_Profile/Direct.aspx?lang=en&area=185. Accessed 30 Mar 2015
FAO (2014a) FAOSTAT database. http://faostat3.fao.org/faostat-gateway/go/to/home/E. Accessed 30 Mar 2015
FAO (2014b) AQUASTAT database. http://www.fao.org/nr/water/aquastat/main/index.stm. Accessed 30 Mar 2015
Fedotov AP, Phedorin MA, Enushchenko IV, Vershinin KE, Melgunov MS, Khodzher TV (2012) A reconstruction of the thawing of the permafrost during the last 170 years on the Taimyr Peninsula (East Siberia, Russia). Global Planet Change 98–99(2012):139–152
Filipchuk A, Moiseev B (2010) Global forest resources assessment 2010. Country Report Russian Federation. http://www.fao.org/docrep/013/al608E/al608E.pdf. Accessed 30 Mar 2015

Forest Code (2006) Forest Code of the Russian Federation Adopted by the State Duma on November 8, 2006 http://webarchive.iiasa.ac.at/Research/FOR/forest_cdrom/downloads/Forest_code/ForestCode-3rdReading-061108-eng.txt. Accessed 30 Mar 2015

FRA (2005) Global forest resources assessment 2005. FAO Forestry Paper 147. http://www.fao.org/dOCrEP/008/a0400e/a0400e00.htm. Accessed 30 Mar 2015

French HM (1976) The Periglacial Environment. Longman, London 309 p

French HM (1999) Past and present permafrost as an indicator of climate change. Polar Res 18:269–274

Fukuda M (1997) Radioactive and other hazardous contamination in Arctic Siberia. http://src-h.slav.hokudai.ac.jp/sympo/97summer/fukuda.html. Accessed 30 March 2015

Gamzikov GP, Nozov V (2010) Role of crop nutrition in narrowing the yield gap for spring wheat in Siberia. Better Crops Plant Food 94(4):9–11 [A Publication of the International Plant Nutrition Institute (IPNI)]

Gamzikov GP, Gamzikova OI, Barsukov PA, Lapuchin TP, Dmitriev NN, Sychev VG, Loshakov VG, Romanenkov VA (2014) Effectivity of long-term fertiliser applications on croplands in Siberia (Эффективность длительного применения удобрений в полевых агроценозах Сибири). Bull Geograph Netw Fertilisation Trials 16:48 p. (Бюллетень Географической сети опытов с удобрениями. Выпуск 16. М.: ВНИИА, 2014.—48 с). ISBN: 978-5-9238-0170-5

Gavrilyuk FY(1974) Bonitirovka pochv. Moskva.: Vysshaya shkola, 1974, 270 p

Gerasimov IP, Armand DI, Yefron KM (eds) (1971) Natural resources of the Soviet Union: their use and renewal. English edition: Douglas Jackson WH, Freeman WH (1971) San Francisco, 349 p

Gilg O, Sané R, Solovieva DV, Pozdnyakov VI, Sabard B, Tsanos D, Zöckler C, Lappo EG, Syroechkovski Jr EE, Eichhorn G (2000) Birds and mammals of the Lena delta nature reserve, Siberia. Arctic vol 53(2), pp 118–133

Goodrich JM, Kerley LL, Smirnov EN, Miquelle DG, McDonald L, Quigley HB, Hornocker MG, McDonald T (2008) Survival rates and causes of mortality of Amur tigers on and near the Sikhote-Alin Biosphere Zapovednik. J Zool 276(4):323–329. doi:10.1111/j.1469-7998.2008.00458

Gordeev AV, Romanenko GA (2008) (eds) Problems of agricultural land degradation and rehabilitation of its productivity in Russia (Проблемы деградации и восстановления продуктивности земель сельскохозяйственного назначения в России). Ministry of Agriculture of the Russian Federation and Russian Academy of Agricultural Sciences 2008. Moscow Rosinformagrotex, 67 p. http://agro.geonet.ru/publications/degradation.pdf. Accessed 30 Mar 2015

Gordeev VV, Rachold V, Vlasova IE (2004) Geochemical behaviour of major and trace elements in suspended particulate material of the Irtysh river, the main tributary of the Ob river. Siberia Appl Geochem 19(4):593–610. http://www.sciencedirect.com/science/article/pii/S0883292703001707. Accessed 30 Mar 2015

Götz R (2009) Klimawandel in Russland. Diskussionspapier SG8, 2009/06 SWP Berlin. http://www.swp-berlin.org/fileadmin/contents/products/arbeitspapiere/2009_DisP_gtz_ks.pdf. Accessed 30 March 2015

Government RF (1994) Правительство Российской Федерации. Распоряжение от 23 апреля 1994, г. N 572-р г. Москва. http://pravo.gov.ru/proxy/ips/?docbody=&prevDoc=102047764&backlink=1&&nd=102029398. Accessed 10 Apr 2014

Grachev MA (2002) On the present state of the ecological system of Lake Baikal. http://www.lin.irk.ru/grachev/eng/index.htm. Accessed 30 Mar 2015

Groisman PY, Gutman G (ed) (2013) Regional environmental changes in Siberia and their global consequences. Springer, Berlin. http://link.springer.com/book/10.1007%2F978-94-007-4569-8. Accessed 30 Mar 2015

Groisman PY, Sherstyukov BG, Razuvaev VN, Knight RW, Enloe JG, Stroumentova NS, Whitfield PH, Førland E, Hannsen-Bauer I, Tuomenvirta H, Aleksandersson H, Mescherskaya AV, Karl TR (2007) Potential forest fire danger over Northern Eurasia: changes during the 20th century. Global Planet Change 56(2007):371–386

Gutman G, Groisman PYa, Gordov EP, Shiklomanov AI, Shiklomanov NI, Shvidenko AZ, Bergen KM, Baklanov AA (2013) Summary and outstanding scientific challenges for research of environmental changes in Siberia. In: Groisman PJ, Gutman G (eds) Regional environmental changes in Siberia and their global consequences (Chap. 9). Springer Environmental Science and Engineering 2013. ISBN: 978-94-007-4568-1 (Print) 978-94-007-4569-8 (Online). http://link.springer.com/chapter/10.1007/978-94-007-4569-8_9. Accessed 30 Mar 2015

Hampton SE, Izmest'eva IR, Moore MV, Katz SE, Dennis B, Silow EA (2008) Sixty years of environmental change in the world's largest freshwater lake—Lake Baikal. Siberia Global Change Biol 14:1947–1958. doi:10.1111/j.1365-2486.2008.01616.x

IASA (2012) Interregional Association of the Economic Cooperation of the Constituent Entities of the Russian Federation "Siberian Accord" (IASA). http://www.sibacc.ru/upload/iblock/5d4/03.pdf. Accessed 30 Mar 2015

IIASA (2011) Russian forests & forestry, data and analysis. http://webarchive.iiasa.ac.at/Research/FOR/forest_cdrom/home_en.html. Accessed 30 Mar 2015

Ilyina L, Mieczkowski Z (1992) Developing scientific tourism in Russia. Tourism Manage 13 (3):327–331. doi:10.1016/0261-5177(92)90106-H. Accessed 30 Mar 2015

Inisheva LI, Berezina NA (2013) Origin and development of the peat formation process in the West Siberian lowland (Возникновение И Развитие Процесса Заболачивания На Западно-Сибирской Равнине). Scientific notes of the Tomsk State University 2013. No. 366, pp 172–179. http://ltorf.tspu.ru/files/Inisheva_Berezina.pdf. Accessed 30 Mar 2015

Inisheva LI, Zemtsov AA, Novikov SM (2011) Vasyugan Mire. Natural Conditions, Structure and Functioning. Tomsk State Pedagogical University Press, 162 p. http://ltorf.tspu.ru/files/Vasyugan_Mire-Tekst.pdf, accessed on March 30, 2015

LIN Irkutsk (2012) http://www.lin.irk.ru/eng/about.htm. Accessed 30 Mar 2015

Isachenko AG (1990) Ecological atlas of Russia. Map of landscapes (Экологический атлас России: карта ландшафтов). http://www.landscape.edu.ru/images/maps/fgr/ecorusatlas_landscapes.jpg. Accessed 20 Jan 2015

IW-LEARN (2010) Integrated natural resource management in the baikal basin transboundary ecosystem. http://iwlearn.net/iw-projects/4029. Accessed 30 Mar 2015

Juday GP, Barber V, Duffy P, Linderholm H, Rupp S, Sparrow S, Vaganov E, Yarie J, Berg E, D'Arrigo R, Eggertsson O, Furyaev VV, Hogg EH, Huttunen S, Jacoby G, Kaplunov VYa, Kellomaki S, Kirdyanov AV, Lewis CE, Linder S, Naurzbaev MM, Pleshikov FI, Runesson UT, Savva YuV, Sidorova OV, Stakanov VD, Tchebakova NM, Valendik EN, Vedrova EF, Wilmking M (2009) Forests, land management, and agriculture in the Arctic. In: Draggan S et al. (ed) Arctic climate impact assessment (Chap. 14). http://www.eoearth.org/view/article/152826/. Accessed 30 Mar 2015

Juszczak R, Humphreys E, Acosta M, Michalak-Galczewska M, Kayzer D, Olejnik J (2013) Ecosystem respiration in a heterogeneous temperate peatland and its sensitivity to peat temperature and water table depth. Plant Soil 366(1–2):505–520

Kalinina IK (1985) Die Gärten Sibiriens. pp 105–119. In: Wissenschaft und Menschheit 1985. Wissenschaftliches Jahrbuch. Snanye Moscow and Urania Leipzig-Jena-Berlin. 1st ed. 400 pp. ISSN 0201-680X

Kandalova GT, Lysanova GI (2010) Rehabilitation of steppe pastures of Khakassia. Geograph Nat Res 31(4):356–361. http://dx.doi.org/10.1016/j.gnr.2010.11.009. Accessed 30 Mar 2015

Kashapov NG, Lukicheva TA, Kuchma VF (2008). Hygienic evaluation of the influence of environmental factors on adolescents' health in a gas-and-oil producing region. Gig Sanit 4:15–18. Article in Russian, English abstract at http://www.ncbi.nlm.nih.gov/pubmed/19102014. Accessed 30 Mar 2015

Kemmerikh AO, Kupriyanova EH, Albyl CP, Malik LK (1963) Water. In: Richter GD (ed) Natural conditions and existing resources. West Siberia (Природные Условия и Естественные Ресурсы. Западная Сибирь). USSR Academy of Sciences. Moscow 1963, 488 p. http://www.uriit.ru/dzz/rffi/book/richter.pdf. Accessed 30 Mar 2015

Kharlamova NF, Silantyeva MM (2011) Dependence of the cereal cultures productivity in the Kulunda areas on climatic factors (Зависимость урожайности зерновых культур в районах Кулунды от климатических факторов). Earth Sci. http://izvestia.asu.ru/2011/3-2/geos/TheNewsOfASU-2011-3-2-geos-03.pdf. Accessed 30 Mar 2015

Killmann W, Whiteman A (2006) The Russian forestry sector in the global forest products market: trends, outlook and opportunities for development. Presentation to the VII international forestry forum, 10–13 Oct 2006, St. Petersburg, Russian Federation. ftp://ftp.fao.org/docrep/fao/009/ah231e/ah231e00.pdf. Accessed 30 Mar 2015

Kimeev VM (2008) Ecomuseums in Siberia as centers for ethnic and cultural heritage preservation in the natural environment. Archaeology, Ethnology and Anthropology of Eurasia, 35:119–128

Kirsta YB, Kurenina NY, Lovzkaya OV (2013) Prognosis of the climate and agroclimatic potential of the Altai krai until 2020 (Прогноз климата и агроклиматического потенциала Алтайского края до 2020 года). J Altai State Agric Univ Barnaul 1(99):27–32 (in Russian). http://www.asau.ru/files/vestnik/2013/1/Agroecology_Kirsta.pdf. Accessed 30 Mar 2015

Kiselev S, Romashkin R, Bobylev S, Solovieva S (2012) The adaptation challenge key: issues for crop production and agricultural livelihoods under climate change in the Russian Federation, pp 7–21 In: OXFAM research reports October 2012. http://growcampaign.clicr.ru/attach_files/file_public_320.pdf. Accessed 30 Mar 2015

Kislov AV, Dolmatov AP (2012) Maintenance and rehabilitation of soil fertility in organic farming in the frame of ecologization of crop rotations in the South Ural region (Сохранение и Восстановление Почвенного Плодородия в Биологическом Земледелии при Экологизации Севооборотов на Южном Урале), pp 302–310. In: Научное издание. Теоретические и Технологические Основы Воспроизводства Плодородия Почв и Урожайность Сельскохозяйственных Культур. Materials of the International Scientific-practical conference. Moscow, Publishing House of the Timiryasev Academy (РГАУ-МСХА) 2012, 633 p. http://www.pole-st.ru/Sbornik_100.pdf. Accessed 30 Mar 2015

Kolchevnikova O (2013) Russian organic market continues to grow. GAIN report number RSATO1302. Global Agriculture Information Network of the United States Department of Agriculture (USDA), Washington, 25 Feb 2013. http://gain.fas.usda.gov/Recent%20GAIN%20Publications/Russian%20Organic%20Market%20Continues%20to%20Grow_Moscow%20ATO_Russian%20Federation_2-15-2013.pdf. Accessed 30 Mar 2015

Dudarev AA, Alloyarov PR, Chupakhin VS, Dushkina EV, Sladkova YN, Dorofeyev VM, Kolesnikova, TA, Fridman KB, Nilsson LM, Evengård B (2013a) Food and water security issues in Russia I: food security in the general population of the Russian Arctic, Siberia and the Far East, 2000–2011. Int J Circumpolar Health 72(1). http://www.ncbi.nlm.nih.gov/pubmed/24471055. Accessed 30 Mar 2015

Koronkevitch N (2002) Water resources of Russia, In: Stolbovoi V et al (ed) Land resources of Russia. Version 1.1 Copyright © 2002, IIASA & RAS. http://webarchive.iiasa.ac.at/Research/FOR/russia_cd/hydro_des.htm#top. Accessed 30 Mar 2015

Kosykh NP (2009) Biological productivity of bog in forest-Steppe (in Russian). http://vestnik.tspu.edu.ru/files/PDF/articles/kosih_n._p._87_90_3_81_2009.pdf. Accessed 30 Mar 2015

Kosykh NP, Mironycheva-Tokareva NP, Kirpotina LV (2010) Bogs in the Altai Mountains (Осоковые Болота Горного Алтая). http://e-lib.gasu.ru/konf/biodiversity/2010/52.pdf. Accessed 30 Mar 2015

Kot FS, Bakanov KG, Goryachev NA (2009) Mercury in bottom sediments of the Amur River, its flood-plain lakes and estuary, Eastern Siberia Environ Monit Assess (2010) 168:133–140. doi:10.1007/s10661-009-1097-0 © Springer Science+Business Media B.V. 2009

Kotlyakov VM (1986) Die natürlichen Eismassen des Planeten Erde. pp 91–121. In: Wissenschaft und Menschheit 1986. Publishers Snanye, Moscow, and Urania Verlag Leipzig Jena Berlin 1986 ISSN 0201-680X, 400 p

Krankina O, Dixon R, Kirilenko A, Kobak K (1997) Global climate change adaptation: examples from Russian boreal forests. Clim Change 36(1–2):197–215

Kremenetski KV, Velichko AA, Borisova OK, MacDonald GM, Smith LC, Frey KE, Orlova LA (2003) Peatlands of the West Siberian lowlands: current knowledge on zonation, carbon content and Late Quaternary history. Quatern Sci Rev 22(2003):703–723

Kremleva TA, Moiseenko TI, Khoroshavin VYu, Shavnin AA (2012) Geochemical specifics of natural waters in West Siberia: composition of microelements (Геохимические особенности природных вод Западной Сибири: микроэлементный состав). Vestnik Tjumenskogo Gosudarstvennogo Universiteta, No. 12, pp 80–89. http://www.utmn.ru/docs/8245.pdf. Accessed 30 Mar 2015

Krupkin PI, Toptygin VV (1999) Refinement of approaches of soil bonitation, example of the Krasnoyarsk Krai (Совершенствование способов бонитировки почв на примере Красноярского края). Soil Science (Почвоведение). Moscow 1990. vol 12, pp 1481–1491

Kukavskaya EA, Buryak LV, Ivanova GA, Conard SG, Kalenskaya OP, Zhila SV, McRae DJ (2014) Influence of logging on the effects of wildfire in Siberia. Environ Res Lett 8(4):11 p. doi:10.1088/1748-9326/8/4/045034. http://iopscience.iop.org/1748-9326/8/4/045034/pdf/1748-9326_8_4_045034.pdf. Accessed 30 Mar 2015

Kundius VA, Chermyanina VV, Kudinova MG, Balashova SP, Santalova VN (2011) Agritourism in the Altai as an alternative kind of activity and diversification of the farm economy (Сельский Туризм на Алтае как Альтернативный Вид Деятельности в Стратегии Диверсификации Сельской Экономики). J Altai State Agric Univ Barnaul 9(83), 201, 1, 113–119 (in Russian). http://www1.asau.ru/doc/nauka/vestnik/2011/9/Recreation_Kundius.pdf. Accessed 30 Mar 2015

Lambin EF, Gibbs HK, Ferreira L, Grau R, Mayaux P, Meyfroidt P, Morton DC, Rudel TK, Gasparri I, Munger J (2013) Estimating the world's potentially available cropland using a bottom-up approach. Glob Environ Change 23(2013):892–901

Lapshina ED, Bleuten W (2011) Biospheric importance of peat bogs of West Siberia: Topicality and lines of investigation, pp 36–37. In: Vompersky SE (ed) West Siberian peatlands and carbon cycle: past and present. Proceedings of the third international field symposium (Khanty-Mansyisk June 27–July 5, 2011), Novosibirsk 2011, 220 p. ISBN 978-5-9611-066-2

Larin SI, Guselnikov VL (2011) Age and development tendencies of wetlands in the south of West Siberia, pp 40–41, In: Vompersky SE (ed) West Siberian peatlands and carbon cycle: past and present. Proceedings of the third international field symposium (Khanty-Mansyisk June 27–July 5, 2011), Novosibirsk 2011, 220 pp. ISBN 978-5-9611-066-2

Laschtschenkow P (2012) Das Getreide des Gebietes von Jakutsk. Zeitschrift für Hygiene und Infektionskrankheiten 1912, 71(1):157–160

Lieske B (2011) Selbstverständnis des russischen Naturschutzes und die Entwicklung der Sapowedniks – eine Interviewstudie. Waldökologie, Landschaftsforschung und Naturschutz Heft 11 (2011) S. 63–79. http://www.afsv.de/download/literatur/waldoekologie-online/waldoekologie-online_heft-11-2.pdf. Accessed 30 Mar 2015

Lioubimtseva E (2010) Russia's role in the post-2012 climate change policy: key contradictions and uncertainties. Forum on public policy. http://forumonpublicpolicy.com/spring2010.vol2010/spring2010archive/lioubimtseva.pdf. Accessed 30 Mar 2015

Lioubimtseva E, De Beurs KM, Henebry GM (2013) Grain production trends in Russia, Ukraine and Kazakhstan in the context of the global climate variability and change. In: Younos Tamim; Grady, Caitlin A. (Eds.) Climate change and water resources, the handbook of environmental chemistry (Chap. 6), vol. 25, XVIII. Springer, Berlin, 221 p

Liss OL, Abramova LI, Avetov NA, Berezina NA, Inisheva LI, Kurnishkova TV, Sluka A, Tolpysheva TYu, Shvedzhikova HK (2001) Peatland systems of West Siberia and their nature stabilizing importance (Болотные Системы Западной Сибири и их Природоохранное Значение). Moscow 2001, 584 p. http://vasyugan.sokik.ru/files/vasyugan/publikacii/Bolota_Zapadnoi_Sibiri.pdf. Accessed 30 Mar 2015

Lopatin KI (2011) Optimization of the arrangement of oil-industrial plants in forest-peat landscapes, pp 186–189. In: Vompersky SE (ed) West Siberian peatlands and carbon cycle: past and present. Proceedings of the third international field symposium (Khanty-Mansyisk June 27–July 5, 2011), Novosibirsk 2011, 220 pp. ISBN 978-5-9611-066-2

Luzhkova N (2013) Hiking tourism development in protected areas and nature hazard prevention in the Lake Baikal Natural Territory. Prev Educ Nat Disasters 165 UDC: 338.8:502.7. doi:10.2298/IJGI1303165L

Lydolph PE (1990) Geography of the U.S.S.R. Fifth edition. Misty Valley Publishing 1990. ISBN 0-9624933-0-9

Lyuri DI, Goryachkin SV, Karavaeva NA, Denisenko EA, Nefedova TG (2010) Dynamics of agricultural lands of Russia in XX century and postagrogenic restoration of vegetation and soils (Динамика сельскохозяйственных земель России в XX веке и постагрогенное восстановление растительности и почв) GEOS Moscow 2010 426 p. ISBN: 9785891185005

Mack MC, Finlay JC, Demarco J, Chapin F, Schuur EA, Neff JC, Zimov SA (2010) Nitrogen and phosphorus in Yedoma soils of Northeast Siberia: stocks, fluxes and the ecosystem consequences of nutrient release from permafrost thaw American Geophysical Union, Fall Meeting 2010. Abstract #GC52A-05 http://adsabs.harvard.edu/abs/2010AGUFMGC52A.05M. Accessed 30 Mar 2015

Magnuson JJ, Robertson DM, Benson BJ, Wynne RH, Livingstone DM, Arai T, Assel RA, Barry RG, Card V, Kuusisto E, Granin NG, Prowse TD, Stewart KM, Vuglinski VS (2000) Historical trends in lake and river cover in the Northern Hemisphere. Science 289:1743–1746

Maksimova NB, Arnaut DV, Morkovkin GG (2014) The evaluation of growing season duration change in the agro-climatic areas of the Altai region (Оценка Изменения Продолжительности Вегетационного Периода по Агроклиматическим Районам Алтайского Края). J Altai State Agric Univ Barnaul 10(120):49–53 (in Russian). http://www.asau.ru/files/vestnik/2014/10/049-053.pdf. Accessed 30 March 2015

Mamontova EA, Tarasova EN, Mamontova AA, Kuzmina MI, McLachlan MS (2007) The influence of soil contamination on the concentrations of PCBs in milk in Siberia. Chemosphere 67(9):S71–S78. doi:10.1111/j.1600-0889.2004.00102. http://www.sciencedirect.com/science/article/pii/S0045653506015979. Accessed 30 Mar 2015

Meinel T (2002) Die geoökologischen Folgewirkungen der Steppenumbrüche in den 50-er Jahren in Westsibirien. Diss. Halle 2002. http://sundoc.bibliothek.uni-halle.de/diss-online/02/03H082/t1.pdf. Accessed 30 Mar 2015

Meissner R, Rupp H, Seyfarth M (2014) Advanced technologies in Lysimetry. In: Mueller L, Saparov A, Lischeid G (eds) Novel measurement and assessment tools for monitoring and management of land and water resources in agricultural landscapes of Central Asia. Springer International Publishing, pp 159–173. http://link.springer.com/chapter/10.1007/978-3-319-01017-5_8. Accessed 30 Mar 2015

Meyer BC, Schreiner V, Smolentseva EN, Smolentsev BA (2008) Indicators of desertification in the Kulunda steppe in the south of West Siberia. Arch Agron Soil Sci 54(6):585–603

Milkov FN (1977) Natural zones of the USSR (Природные зоны СССР) Mysl, Moscow 1977, 296 p. Quoted in http://www.geo-site.ru/index.php/2011-01-11-14-47-10/133-2011-09-26-18-56-26/682-geographechesky-pojas.html. Accessed 30 Mar 2015

Moore MV, Hampton SE, Izmest'eva LR, Silow EA, Peshkova EV, Pavlov BK (2009) Climate change and the world's "Sacred Sea"—Lake Baikal, Siberia. Biological Sciences Faculty Scholarship. Paper 2. http://repository.wellesley.edu/biologicalsciencesfaculty/2. Accessed 8 April 2014

Mueller L, Dannowski R, Schindler U, Eulenstein F, Meissner R (1996) Gebietsabflüsse aus Agrarlandschaften Nordost- und Mitteldeutschlands. Arch Agron Soil Sci 40(5):345–362. doi:10.1080/03650349609365963

Mueller L, Drösler M, Höper H, Zinecker F, Behrendt A, Sommer M (2008) Properties of peatsoils and their spatial variability at some representative sites in Germany. In: After wise use—the future of peatlands: Proceedings of the 13th international peat congress; Tullamore, Ireland 8–13 June 2008, International Peat Society, Jyväskylä, vol 2, pp 214–217

Mueller L, Schindler U, Mirschel W, Shepherd TG, Ball BC, Helming K, Rogasik J, Eulenstein F, Wiggering H (2011) Assessing the productivity function of soils. Sustainable agriculture, vol 2. Springer Science+Business Media B.V, Dordrecht, pp 743–760

Mueller L, Suleimenov M, Karimov A, Qadir M, Saparov A, Balgabayev N, Helming K, Lischeid G. (2014) Land and water resources of Central Asia, their utilisation and ecological status. In: Mueller L, Saparov A, Lischeid G (eds) Novel measurement and assessment tools for monitoring and management of land and water resources in agricultural landscapes of Central Asia. Springer International Publishing, Cham, pp 3–59. http://link.springer.com/chapter/10.1007%2F978-3-319-01017-5_1. Accessed 30 Mar 2015

Murton JB (2009) Global warming and thermokarst. In: Margesin R (ed) Permafrost soils. Soil biology, vol 16, pp 185–203. http://link.springer.com/chapter/10.1007%2F978-3-540-69371-0_13. Accessed 30 Mar 2015

Neary DG, Ice GG, Jackson CR (2009) Linkages between forest soils and water quality and quantity. For Ecol Manage 258:2269–2281

Nechaeva EG, Belozerntseva IA, Naprasnikova EV, Vorobyeva IB, Davydova ND, Dubynina SS, Vlasova NV (2010) Monitoring and forecasting of the substance-dynamical state of geosystems in the Siberian regions (Мониторинг и прогнозирование вещественно-динамического состояния геосистем сибирских регионов). Novosibirsk, Nauka, 360 p. ISBN 978-5-02-023315-7. http://irigs.irk.ru/docs/Nechaeva/Nechaeva_et_al_2010.pdf. Accessed 30 Mar 2015

Neistadt MI (1971) About the lower boundary of the Holocene (О нижней границе голоцена). Palynology of Holocene (Палинология голоцена). Moscow, АН СССР, pp 7–17

Newell J (2004) The Russian Far East: A Reference Guide for Conservation and Development. McKinleyville, CA: Daniel & Daniel. 466 p http://urbansustainability.snre.umich.edu/wp-content/uploads/2011/04/RFE.04.pdf. Accessed 30 Jan 2015

Nilsson S, Shvidenko A (1998) Is sustainable development of the Russian forest sector possible? International Union of Forestry Research Organizations. IUFRO Occasional Paper No. 11 ISSN 1024-414X, 77 p. file:///C:/Users/Ilse/Downloads/op11.pdf. Accessed 30 Mar 2015

NORBALWET (2013) Nordic-baltic wetlands initiative for the Ramsar convention on wetlands. http://www.norbalwet.org/. Accessed 30 Mar 2015

Oostergreen D, Shvarts E (2000) Russian Zapovedniki in 1998: recent progress and new challenges for Russia's strict nature preserves. USDA forest service proceedings RMRS-P-14, pp 209–213. http://www.fs.fed.us/rm/pubs/rmrs_p014/rmrs_p014_209_213.pdf. Accessed 30 Mar 2015

Opekunova M, Kokushkin S, Dotsenko T (2011) Assessment of the oil and gas complex impact on peat mires in the north of West Siberia, pp 194–196. In: Vompersky SE (ed) West Siberian peatlands and carbon cycle: past and present. Proceedings of the third international field symposium (Khanty-Mansyisk June 27–July 5, 2011), Novosibirsk 2011, 220 pp. ISBN 978-5-9611-066-2

Pavlov AV (2008) Monitoring of the cryolithic zone (in Russian, Engl subtitles) Novosibirsk. Acad. Publisher Geo, 229 p. ISBN 978-5-9747-0128-3

Petrick M, Wandel J, Karsten K (2013). Rediscovering the Virgin Lands: agricultural investment and rural livelihoods in a Eurasian Frontier Area, World Development, vol 43, pp 164–179

Plyusnin VM, Danko LV (2011) Global change and regional landscape response and desertification in Siberia. J Resour Ecol 2(3):277–279. doi:10.3969/j.issn.1674-764x.2011.03.012. www.jorae.cn, NE Asia. file:///C:/Users/Ilse/Downloads/Global%20Change%20and%20Regional%20Landscape%20Response%20and%20Desertification%20in%20Siberia.pdf. Accessed 22 May 2014

Polyakov VA, Ivanov VG (1986) Dairy cattle pasture use in the extreme northern USSR. Grazing Research at Northern Latitudes. NATO ASI Series, vol 108, pp 153–155

Proceedings Irkutsk (2013) Proceedings of the 8th international conference, rivers of Siberia and the Far East Irkutsk, 6–7 June 2013, 273 p. ISBN 978-5-91344-536-0. http://www.wwf.ru/resources/publ/book/845. Accessed 30 Mar 2015

Puzanov AV, Balykin SN, Saltykov AV, Alekseev IA, Balykin DN (2013) Natural radionuclides (238U, 232Th, 40 K) in soil of the Vostochny Cosmodrome and adjacent territory (the Amur region) (Естественные Радионуклиды (238U, 232Th, 40 K) в почвах космодрома « Восточный » и прилегающих территорий (Амурская Область). In: Radioactivity and radioactive elements in environment. Proceedings of IV international conference, Tomsk. 4–8 June 2013, pp 426–427 (in Russian, English abstracts). http://portal.tpu.ru/files/conferences/radioactivity/book-light.pdf. Accessed 30 Mar 2015

Rawlins MA, Steele M, Holland MM, Adam JC, Cherry JE, Francis JA, Groisman PY, Hinzman LD, Huntington TG, Kane DL, Kimball JS, Kwok R, Lammers RB, Lee CM, Lettenmaier DP, Mcdonald KC, Podest E, Pundsack JW, Rudels B, Serreze MC, Shiklomanov A, Skagseth Ø, Troy TS, Vö CS, Smarty R, Wensnahan M, Wood EF, Woodgate R, Yang D, Zhang K, Zhang T (2010) Analysis of the arctic system for freshwater cycle intensification: observations and expectations. J Clim 23:5715–5737. doi:10.1175/2010JCLI3421.1. http://journals.ametsoc.org/doi/pdf/10.1175/2010JCLI3421.1. Accessed 30 Mar 2015

Richter GD (1963) (ed) Natural conditions and existing resources. West Siberia (Природные Условия и Естественные Ресурсы. Западная Сибирь). USSR Academy of Sciences. Moscow 1963, 488 p. http://www.uriit.ru/dzz/rffi/book/richter.pdf. Accessed 30 Mar 2015

Robarts RD, Zhulidov AV, Pavlov DF (2013) The state of knowledge about wetlands and their future under aspects of global climate change: the situation in Russia. Effects Clim Change Wetlands Aquat Sci 75:27–38. doi:10.1007/s00027-011-0230-7

Roshydromet (2008a) Federal service for hydrometeorology and environmental monitoring (Roshydromet). Assessment report on climate change and its consequences in Russian Federation. General Summary Moscow 2008 24 p. http://climate2008.igce.ru/v2008/pdf/resume_ob_eng.pdf. Accessed 30 Mar 2015

Roshydromet (2008b) Federal service for hydrometeorology and environmental monitoring (Roshydromet). Assessment report on climate change and its consequences in Russian Federation (in Russian) Moscow 2008 87 p. http://climate2008.igce.ru/v2008/pdf/resume_teh.pdf. Accessed 30 Mar 2015

Roshydromet (2014) Second assessment report of Roshydromet about climate change and its consequences on the territory of the Russian Federation. Overall summary (Второй оценочный доклад росгидромета об изменениях климата и их последствиях на территории Российской Федерации. общее резюме) Moscow Planeta 57 pp. ISBN 978-5-904206-13-0. http://voeikovmgo.ru/download/2014/od/od2.pdf. Accessed 30 Mar 2015

Rudaya N, Nazarova L, Nourgaliev D, Palagushkina O, Papin D, Frolova L (2012) Mid-late Holocene environmental history of Kulunda, southern West Siberia: vegetation, climate and humans. Quat Sci Rev 48:32–42. doi:10.1016/j.quascirev.2012.06

Safonov G, Safonova Yu (2013) Economic Analysis of the impact of climate change on agriculture in Russia. National and regional aspects. OXFAM research reports April 2013. http://www.oxfam.org/sites/www.oxfam.org/files/rr-economic-impacts-climate-change-agriculture-russia-010413-en.pdf. Accessed 30 Mar 2015

Schepaschenko D, Shvidenko A, Schepaschenko M (2012) Carbon balance of Russian agricultural land. Presentation on the EGU meeting 2012. http://neespi.org/web-content/meetings/EGU_2012/Schepaschenko.pdf. Accessed 30 Mar 2015

Schindler U (2014) A field method for quantifying deep seepage and solute leaching. In: Mueller L, Saparov A, Lischeid G (eds) Novel measurement and assessment tools for monitoring and management of land and water resources in agricultural landscapes of Central Asia. Springer International Publishing, pp 185–198. http://link.springer.com/chapter/10.1007/978-3-319-01017-5_10. Accessed 30 Mar 2015

Schindler U, Steidl J, Eulenstein F, Mueller L (1998) Simulation of water balance at a catchment of the young moraine landscape—estimation and testing of soil data. Arch Agron Soil Sci 43:373–384

Schirrmeister L, Meyer H, Wetterich S, Siegert C, Kunitsky VV, Grosse G, Kuznetsova TV, Derevyagin AYu (2008) The Yedoma Suite of the North East Siberian shelf region:

characteristics and concept of formation. Ninth international conference on permafrost, pp 1596–1600. http://epic.awi.de/18721/1/Sch2008r.pdf. Accessed 30 Mar 2015

Schreiner V, Meyer BC (2014) Indicators of Land Degradation in Steppe Regions: Soil and Morphodynamics in the Northern Kulunda. In: Mueller L, Saparov A, Lischeid G (eds) Novel measurement and assessment tools for monitoring and management of land and water resources in agricultural landscapes of Central Asia. Springer International Publishing, pp 535–548. http://link.springer.com/chapter/10.1007%2F978-3-319-01017-5_33. Accessed 30 Mar 2015

Tarnocai C, Canadell JG, Schuur, EAG, Kuhry P, Mazhitova G, Zimov S (2009) Soil organic carbon pools in the northern circumpolar permafrost region. Global Biogeochem Cycles 23: GB2023. doi:10.1029/2008GB003327

Shaw DJB, Oldfield J (2007) Landscape science: a Russian geographical tradition. Ann Assoc Am Geogr 97(1):111–126. ISSN 0004-5608

Shekhovzhova TN (2013) Life quality of the population and search for a regional strategy of economic development (Качество жизни населения и поиск региональной стратегии экономического развития (на примере Ольхонского района Иркутской области). Naukovedeniye (Науковедение) 6(19). http://naukovedenie.ru/PDF/60EVN613.pdf. Accessed 22 May 2014

Shiklomanov IA, Georgievsky VYu (2009) Influence of climate changes on the hydrological regime and water resources of the Russian Rivers (Влияние изменений климата на гидрологический режим и водные ресурсы рек России) pp 143, In: Vasiliev OF, Arnell NW (ed) Hydrological impact of climate change. Proceedings of British-Russian conference, 13–15 June 2007, Novosibirsk, Russia. http://www.iwep.ru/ru/bibl/books/matkonf/Hydrological_Impact_2009.pdf. Accessed 22 May 2014

Shimaraev MN (2009) Tendencies of hydrological processes occurring in Lake Baikal under climate changes: review (Тенденции гидрологических процессов на Байкале в условиях изменения климата: обзор) pp 152, In: Vasiliev OF, Arnell NW (eds) Hydrological impact of climate change proceedings of British-Russian conference, 13–15 June 2007, Novosibirsk, Russia. http://www.iwep.ru/ru/bibl/books/matkonf/Hydrological_Impact_2009.pdf. Accessed 22 May 2014

Shimaraev MN, Domysheva VM (2002) Dynamics of dissolved silica in Lake Baikal. Geologiya i Geofizika (Russian Geology and Geophysics) 40:1502–1505

Shimaraev MN, Kuimova LN, Sinyukovich VN, Tsekhanovsky VV (2002) Manifestation of global climatic changes during the twentieth century. Dokl Earth Sci 383:288–291

Shpolyanskaya NYa (1971) Basic principles of the occurrence of permafrost and stages of its development in West Siberia (in Russian) pp 102–123. In: Popov AI (ed) Natural conditions of landscapes in West Siberia. Part 1. Moscow State University, Moscow, 238 p

Shvareva OO (1963) Climate (Климат) pp 70–99 In: Richter GD (ed) Natural conditions and existing resources. West Siberia, in Russian (Природные Условия и Естественные Ресурсы. Западная Сибирь). USSR Academy of Sciences, Moscow, 488 p

Shvidenko AZ, Schepaschenko DG (2013) Climate change and wildfires in Russia. Contemp Probl Ecol 6(7):683–692

Shvidenko AZ, Gustafson E, McGuire AD, Kharuk VI, Schepaschenko DG, Shugart HH, Tchebakova NM, Vygodskaya NN, Onuchin AA, Hayes DJ, McCallum I, Maksyutov S, Mukhortova LV, Soja AJ, Belelli-Marchesini L, Kurbatova JA, Oltchev AV, Parfenova EI, Shuman JK (2013). In: Groisman, PJ and Gutman G (eds) Regional environmental changes in Siberia and their global consequences (Chap. 6). Springer Environmental Science and Engineering, pp 171–249. ISBN: 978-94-007-4569-8

Simeone J (2014). Russia's Forest Sector and International Trade in forest products: export taxes on roundwood, priority investment projects, and WTO accession. Vestnik, J Russian Asian Stud. http://www.sras.org/russia_forest_sector_wto. Accessed 30 Mar 2015

Smelansky IE, Tishkov AA (2012) The Steppe biome in Russia: ecosystem services, conservation status, and actual challenges Eurasian Steppes. Ecol Probl Livelihoods Changing World Plant Veg 6(2012):45–101

Smolentseva EN, Smolentsev B, Pachikin K, Mueller L (2014) Assessing the soil quality and crop yield potentials of some soils of Eurasia. In: Mueller L, Saparov A, Lischeid G (eds) Novel measurement and assessment tools for monitoring and management of land and water resources in agricultural landscapes of Central Asia. Springer International Publishing, pp 505–517. http://link.springer.com/chapter/10.1007%2F978-3-319-01017-5_31. Accessed 30 Mar 2015

State Hydrological Institute (2013) http://www.hydrology.ru/main/. Accessed 30 Mar 2015

Stolbovoi V, Fischer G (1997) A new digital georeferenced database of soil degradation in Russia. IIASA Laxenburg, INTERIM REPORT IR-97-084/November. http://citeseerx.ist.psu.edu/viewdoc/download?doi=10.1.1.56.689&rep=rep1&type=pdf. Accessed 30 Mar 2015

Suleimenov M, Kaskarbayev Z, Akshalov K, Yushchenko N (2014) Conservation agriculture for long-term soil productivity. In: Mueller L, Saparov A, Lischeid G (eds) Novel measurement and assessment tools for monitoring and management of land and water resources in agricultural landscapes of Central Asia. Springer International Publishing, pp 441–454. http://link.springer.com/chapter/10.1007/978-3-319-01017-5_26. Accessed 30 Mar 2015

Suzuki K (2011) SIBERIA In: Singh VP, Singh P, Haritashya UK (2011) Encyclopedia of snow, ice and glaciers, encyclopedia of earth sciences series, pp 1028–1031, © Springer Science +Business Media B.V. 10.1007/978-90-481-2642-2_486

Syvitski JPM (2002) Sediment discharge variability in Arctic rivers: implications for a warmer future. Polar Res 21(2):323–330. doi:10.1111/j.1751-8369.2002.tb00087.x

Tichanovsky AN (2004) Optimization of fertilization on soils of the severe north of West Siberia (Оптимизация применения удобрений на почвах Крайнего Севера Западной Сибири). Doctoral dissertation, Tyumen, 358 p

Todd MC, Mackay AW (2003) Large-scale climatic controls on lake Baikal ice cover. J Clim 16:3186–3199

UNECE (2007) United Nations, Economic Commission for Europe, convention on the protection and use of transboundary watercourses and international lakes. First assessment of transboundary rivers, lakes and groundwaters. United Nations, New York and Geneva. http://www.unece.org/fileadmin/DAM/env/water/blanks/assessment/assessment_full.pdf. Accessed 30 Mar 2015

UNECE (2010) Production, consumption and trade in forest products in the Russian Federation in 2008–2011, Moscow 2010. http://www.unece.org/fileadmin/DAM/timber/country-info/Russia2010.pdf. Accessed 30 Mar 2015

UNEP/IETC (2000) Planning and management of lakes and reservoirs. An integrated approach to eutrophication. Technical Publications series [11] http://www.unep.or.jp/ietc/publications/techpublications/techpub-11/. Accessed 30 Mar 2015

Utkin VI (1997) Gas respiration of lands (Газовое дыхание земли). Sorosovskiy Educ J (Соросовский Образовательный Журнал). 1:57–64 (Quoted in Adam et al 2013)

Vasilyev SV (2007) Forests and wetlands of West Siberia (Лесные и болотные ландшафты Западной Сибири), Tomsk: NTL, 276 p. ISBN 978-5-89503-334-0

Vonk JE, Mann PJ, Dowdy KL, Davydova A, Davydov SP, Zimov N, Spencer RGM, Bulygina EB, Eglinton TI, Holmes RM (2013) Dissolved organic carbon loss from Yedoma permafrost amplified by ice wedge thaw. Environ Res Lett 8(3). doi:10.1088/1748-9326/8/3/035023

Vostokova LB, Yakushevskaya IV (1979) Soil bonitation (Bonitirovka povc) Publising House of the Moscow State University 1979, 1st edn, 102 p

Walter KM, Zimov SA, Chanton JP, Verbyla D, Chapin FS (2006) Methane bubbling from Siberian thaw lakes as a positive feedback to climate warming. Nature 443:71–75

Water Code (2006) June 3rd, 2006 No 74-FZ Russian Federation Federal Law, water code of the Russian Federation. Adopted by the State Duma on April 12th, 2006. Approved by the Federation Council on May 26th, 2006. http://dinrac.nowpap.org/documents/law/Russia/Water_Code_Russia.pdf. Accessed 30 Mar 2015

Weischet W (1960) Zum Problem der Stabilität der Klimabedingungen in Westsibirien während der Glaziale und Interglaziale. http://quaternary-science.publiss.net/system/articles/pdfas/165/original_vol11_no1_a09.pdf?1284107429. Accessed 30 Mar 2015

Welcomme R (2011) Review of the state of the world fishery resources: inland fisheries. FAO Fish Aquac Circular 942(2):97 pp. Rome, FAO. http://www.fao.org/docrep/015/i2484e/i2484e.pdf. Accessed 30 Mar 2015

Wikipedia (2013a) Federal districts of Russia. http://en.wikipedia.org/wiki/Federal_districts_of_Russia. Accessed 30 Mar 2015

Wikipedia (2013b) Fishing industry in Russia. http://en.wikipedia.org/wiki/Fishing_industry_in_Russia#cite_note-federal_2006-5. Accessed 30 Mar 2015

Wikipedia (2014a) National parks of Russia. http://en.wikipedia.org/wiki/List_of_national_parks_of_Russia. Accessed 30 Mar 2015

Wikipedia (2014b) Zapovednik. http://en.wikipedia.org/wiki/Zapovednik. Accessed 30 Mar 2015

Woo MK (2012) Permafrost hydrology, vol XII. Springer, Berlin, 563 p

WRB (2014) World reference base for soil resources. World soil resources report no. 106, 181 p. ISBN: 978-92-5-108369-7. http://www.fao.org/3/a-i3794e.pdf. Accessed 14 Nov 2014

Yang D, Baisheng Ye, Shiklomanov A (2004) Discharge characteristics and changes over the Ob River watershed in Siberia. J Hydrometeor 5:595–610. doi:10.1175/1525-7541(2004)005<0595:DCACOT>2.0.CO;2. Accessed 30 Mar 2015

Yo Sheng, Smith LC, MacDonald GM, Kremenetski KV, Frey KE, Velichko AA, Lee M, Beilman DW, Dubinin P (2004) A high-resolution GIS-based inventory of the West Siberian peat carbon pool. Global Biogeochem Cycles 18(GB3004):2004. doi:10.1029/2003GB002190

Zavadskaya AV (2011) Recreation monitoring of resource conditions in the Kronotsky State natural biosphere preserve (Kamchatka): an initial assessment. Geogr Environ Sustain (4):47–58. http://int.rgo.ru/wp-content/uploads/2011/12/GES2-11.pdf. Accessed 30 Mar 2015

Zhulanova VN (2013) Agro-ecological asssessment of soils in Tuva (Агроэкологическая Оценка Почв Тувы). Doctoral dissertation, Moscow. http://www.timacad.ru/catalog/disser/referat/JulanovaVN.doc. Accessed 30 Mar 2015

Zimov SA, Chuprynin VI, Oreshko AP, Chapin FS III, Reynolds JF, Chapin MC (1995) Steppe-Tundra transition: a herbivore-driven biome shift at the end of the Pleistocene. Am Nat 146(5):765–794

Zimov SA, Schuur EAG, Chapin FST (2006) Permafrost and the global carbon budget. Science 312:1612–1613

Chapter 2
Status Report About Understanding, Monitoring and Controlling Landscape Processes in Siberia

Lothar Mueller, Askhad K. Sheudshen, Victor G. Sychev, Aleksandr Syso, Pavel Barsukov, Elena N. Smolentseva, Cristian Siewert, Ralph Meissner, Ralf Dannowski, Vladimir A. Romanenkov, Olga V. Rukhovich, Katharina Helming, Uwe Schindler and Frank Eulenstein

Abstract Siberia has experienced significant transformations over the past 70 years and particularly since the introduction of the market economy 25 years ago. This has caused implications for landscape processes and for the status of terrestrial and aquatic ecosystems. We review the role of science and technology in monitoring, understanding and developing Siberian landscapes. Data sources were international literature and own expeditions and studies. Russia has great traditions

L. Mueller (✉) · R. Dannowski · K. Helming · U. Schindler · F. Eulenstein
Leibniz Centre for Agricultural Landscape Research (ZALF) e.V.,
Eberswalder Strasse 84, 15374 Müncheberg, Germany
e-mail: mueller@zalf.de

A.K. Sheudshen
Kuban State Agrarian University, 13 Kalinin Str, 350044 Krasnodar, Russia
e-mail: kubagrohim@mail.ru

V.G. Sychev · V.A. Romanenkov · O.V. Rukhovich
Pryanishnikov All-Russian Research Institute of Agrochemistry (VNIIA), Pryanishnikova St. 31a, 127550 Moscow, Russia
e-mail: geoset@yandex.ru

A. Syso · P. Barsukov · E.N. Smolentseva
Siberian Branch, Institute of Soil Science and Agrochemistry (ISSA), Russian Academy of Sciences, Ac. Lavrentieva av. 8, Novosibirsk 630090, Russia
e-mail: syso@issa.nsc.ru

C. Siewert
Faculty of Agriculture/Landscape Management, University of Applied Sciences, Pillnitzer Platz 2, 01326 Dresden, Germany
e-mail: cs@csiewert.de

R. Meissner
Department Soil Physics, Helmholtz Centre for Environmental Research—UFZ, Lysimeter Station, Dorfstrasse 55, 39615 Falkenberg, Germany
e-mail: ralph.meissner@ufz.de

L. Mueller et al. (eds.), *Novel Methods for Monitoring and Managing Land and Water Resources in Siberia*, Springer Water,
DOI 10.1007/978-3-319-24409-9_2

in landscape research disciplines such as geography, soil science, hydrology and agronomy. Substantial progress has been achieved in all these fields over the past 25 years. We found particular progress in landscape research based on international projects in the fields of Arctic research, climate change and carbon cycle. Other fields such as agricultural research remained traditional and underdeveloped. In the 1990s there was a great shift of knowledge and technology in the better-interlinked English-speaking European scientific community. In Russia, at the same time, the introduction of the market economy accelerated environmental problems, caused a greater discrepancy between the livelihoods of urban and rural populations, created new knowledge gaps and enlarged the gap between theory and practice in landscape research. The decay of infrastructure in rural landscapes produced an inhospitable environment for science and technology. In view of this, landscape research in Siberia and in the Far East remained very traditional. Other deficits were based on a lack of communication with the international community due to language barriers. Cooperation between leading Russian and European scientists is still poorly developed and funded. The Russian academic scientific system was highly organized until 2013. However, efficiency was low and scientific outputs did not meet the requirements of decision-makers. The ongoing reform of the academic system entails the risk that precisely the opposite to the desired effects of higher efficiency could come true, such as accelerated brain drain and loss of objectivity. We conclude that Trans-Eurasian research cooperation is becoming very important in the current critical transition phase. Modern analytical methods, sophisticated technologies, models and evaluation schemes for landscape research and environmentally friendly soil management technologies are available in the English-speaking community. Substantial progress in monitoring, understanding and controlling landscape processes in the framework of international research projects could be achieved by applying new research methods in Siberia. We present some of them in the following chapters of this book.

Keywords Landscape research · Soil · Water · Russia · Siberia · Academic system · Cooperation

1 Recent Progress in Landscape Research

1.1 Overview

There are several reasons to identify the processes that affect land and water resources. Human activity, land management and changing climate may degrade or upgrade their functions for nature and society. Underlying processes need to be quantified in order to forecast significant effects and to take measures of prevention, control, mitigation or adaption. Scientists have achieved substantial progress in researching into these functions of Siberian landscapes over the past 25 years. Inputs have come from different scientific disciplines such as geography, soil science, hydrology, climatology, agriculture and others. They have come from scientists working in Siberia, from scientists at leading research institutes in Moscow, European Russia, and from cooperating scientists abroad.

There has been a major focus on Arctic and boreal ecosystems in view of climate change, in monitoring vegetation, water and soil quality. Agricultural scientists have developed technologies for optimizing agricultural systems in different climatic zones and regions. Some interesting and important areas of research and examples are mentioned below.

1.2 Arctic Research, Permafrost Processes

Early analyses of Arctic processes and orientation data on Arctic warming were reported by Pavlov (1994), Pavlov and Moskalenko (2002) and Anisimov and Reneva (2006) from polar desert and tundra regions. Romanovsky et al. (2007) also provided a complex analysis of past and more recent changes in air and permafrost temperatures. Ananicheva et al. (2011) measured recent and forecasted changes to Siberian glaciers as indicators of the ongoing climate change. Climate models of thermal and hydrological soil regimes of the Arctic region were developed to make forecasts more reliable (Arzhanov et al. 2008), and the impact of climate warming on vegetation cover and permafrost in the northern Taiga was quantified using deep ground measurements (Moskalenko 2013). These measured a deepening of the active layer by 10 cm between 1970 and 2010 and an increase in the temperature at a depth of 10 cm by about 1 K (°C) in 40 years.

Polar research has been an object of active international cooperation for many years. The Circumpolar Active Layer Monitoring (CALM) programme was implemented (Shiklomanov et al. 2008), and new permafrost observatories were developed and installed (Boike et al. 2012). The evolution of thermokarst in ice-rich permafrost was explained by Morgenstern et al. (2013), and Yedoma complexes were defined and characterized (Schirrmeister et al. 2013). The Yedoma coastline with the Arctic Ocean in the vicinity of the Lena River Delta retreated at a mean rate of 0.59 m/year between 1951 and 2006 (Lantuit et al. 2011).

Various original approaches were developed; for example, it was possible to reconstruct the 500,000-year history of Siberian permafrost by analysing speleothem growth in caves (Vaks et al. 2013). Areawide digital maps of permafrost temperature and active layer thickness data were developed for modelling climate change over key regions such as Yakutia (Beer et al. 2013). Zakharova et al. (2014) identified the variability in hydrological conditions in Siberian wetlands in terms of water level fluctuations and water storage capacities by satellite radar altimetry. Seasonal amplitudes were 0.7–1.5 m for lakes and 0.2–0.5 m for bogs.

1.3 Carbon Inventory and Cycle, GHG Emissions

In order to identify carbon pools in the soils of the Arctic region, the Northern Circumpolar Soil Carbon Database was developed in international cooperation (Tarnocai et al. 2009). From these data the northern permafrost region contains approximately 1672 Pg (billion tonnes) of organic carbon, accounting for approximately 50 % of the estimated global below-ground organic carbon pool.

Permafrost degradation leads to carbon decomposition and greenhouse gas (GHG) emissions. They were measured in a thermokarst depression (alas) using closed-chamber methods (Takakai et al. 2008). The results showed that the vegetation zone around the pond was an important source of methane (CH_4) and nitrous oxide (N_2O) but a sink of carbon dioxide (CO_2) during the summer time. Khvorostyanov et al. (2008) constructed a new model to study the sensitivity of permafrost carbon stocks to future climate warming. The one-dimensional model solved an equation for diffusion of heat penetrating from the overlying atmosphere and took into account additional in situ heat production by active soil microorganisms. The stability, storage, decomposition and mobility of different soil carbon fractions of permafrost soils (Gleyic Cryosols) is of great importance during permafrost thawing (Rusalimova and Barsukov 2006; Guggenberger et al. 2008). The mobility of black carbon (BC) in its dissolved and colloidal phase is an important export pathway from catchments. Guggenberger et al. (2008) concluded that this transport mechanism may explain the high BC concentrations found in sediments of the Arctic Ocean. This fact was reinforced and made more precise by flux analyses of carbon in rivers (Prokushkin et al. 2011). Zech et al. (2011) analysed the deuterium/hydrogen isotopic ratios (δD) of alkanes in a permafrost loess–paleosol sequence in north-east Siberia and found that maintaining permafrost conditions is most important for the formation and preservation of soil organic matter. Quegan et al. (2011) estimated the carbon balance of central Siberia using different methods, amongst them a landscape–ecosystem approach, atmospheric inversion and dynamic global vegetation models.

Semiletov et al. (2012) determined CO_2 and CH_4 fluxes from the East Siberian Arctic Shelf (ESAS) to the atmosphere. Carbon from degrading terrestrial and subsea permafrost and from coastal erosion contributes to the carbon pool of the ESAS. This affects hydrological and biogeochemical parameters of the Arctic region. Mi et al. (2014) improved the wetland CH_4 emission model Peatland-VU by including an improved hydrological module, incorporating a gross primary productivity (GPP) module, and employing a more realistic soil-freezing scheme.

Shirokova et al. (2013) analysed the biogeochemistry of organic carbon, CO_2, CH_4 and trace elements in shallow and small water bodies in the discontinuous permafrost zones. Upon future permafrost thawing, dissolved organic carbon (DOC) and colloidal metal stocks in the surface will increase in aquatic systems, but CO_2 and CH_4 fluxes from the water surface to the atmosphere will also rise, leading to much higher overall fluxes than previously assumed.

Fig. 1 **a** Fires in peatland and Taiga ecosystems initiate vegetational shifts and altered fluxes of greenhouse gasses. **b** Part of the measuring station in the south of the Vasyugan bog, West Siberia. Tomoko Nakano and colleagues measured the soil/atmosphere methane exchange there after a fire (Nakano et al. 2006)

Glagolev et al. (2011) measured CH_4 emission from a mire in West Siberia using a static chamber method. Similar methods had been developed and tested by Nakano et al. (2006), Fig. 1. Projecting these data to West Siberia by standard model estimates reveals a CH_4 flux from West Siberian mires of 2.93 ± 0.97 million tonnes CH_4–C/year, accounting for about 2.4 % of the total methane emission from all mires or 0.7 % of the global CH_4 emission from all sources. Vasileva and Moiseenko (2013) focused on methane emissions from wildfires using MODIS satellite data. They found most emissions in the southern part of the boreal forest zone (48°–55°N) contributing 5–20 % of the ecosystem methane emissions of the regions, which come mainly from wetlands. Schepaschenko et al. (2013) have developed an automated information system including the carbon pool of the 1-m-deep soil layer, which makes it possible to estimate spatial distributions of soil organic carbon pools with a resolution of about 1 km^2 for Russia as a whole.

Back in the 1990s, scientists already recognized that organic carbon emissions from cropping soils contribute considerably to GHG emissions. In a continuous fallow in grey forest soils of the Baikal forest steppe, the total carbon release into the atmosphere varied from 558 to 1880 kg/ha per year. This was associated with permanent and irreversible humus losses from soils (Pomazkina et al. 1996). Titlyanova et al. (2001) estimated that within 200 years of soil cultivation in West Siberia the stocks of soil organic matter (0–20 cm) decreased by 30 %. This is very different from the largely undisturbed vegetational succession. Vorobyeva (2012) monitored a dark-grey wood soil and weakly leached Chernozem on slopes with forest steppe of South Siberia and found the contents of total carbon, humic acids and insoluble residues had increased up to threefold, indicating a shift from the forest steppe towards the grassland steppe.

1.4 Soil Hydrological Processes and Runoff Generation

Scientists found and predicted a significant alteration of hydrological processes in all landscape zones. Frozen soil acts more or less as a hydrological barrier depending on thermal and other conditions. This has different effects on runoff pathways and consequences for soil processes. Those interactions were studied by a research team of the Institute of Soil Science and Agrochemistry (ISSA) Novosibirsk (Fig. 2). Snow ablation and spring runoff generation depend on the autumn soil moisture and the maximum snow water equivalent (Suzuki et al. 2006). Blome (2014) analysed the influence of permafrost processes on large-scale energy and water cycles over Siberia by applying the regional climate model REMO (Jacob and Podzun 1997) and found that the freezing-induced reduction of water infiltration is the most sensitive process to alter near-surface climate and fluxes of water, matter and energy. Being largely impermeable for snowmelt water, frozen soil favours the development of shallow runoff and erosion processes. On the other hand, when snowmelt water is prevented from infiltrating beyond the soil profile, this stop Chernozems from forming a percolative water regime, which would lead to the fast leaching of minerals (Ollesch et al. 2008; Tanasienko and Chumbaev 2010). During snowmelt infiltration and runoff generation, in combination with enhanced biological and biochemical processes in the active layer, the overall discharge regime of subarctic rivers is altered in its quantity and quality. Increased river loads of elements have already been monitored (Pokrovsky et al. 2012), and more extreme hydrological events are probable (Korytny et al. 2007).

Fig. 2 Russian–German field expedition team for measuring landscape hydrological processes in West Siberia in the frame of the "Soil quality indicators" project. Participants came from the Institute of Soil Science and Agrochemistry (ISSA) Novosibirsk and from the Leibniz Centre of Agricultural Landscape Research Müncheberg (ZALF) in Germany

Research has been carried out into the processes and regulators of lake and river chemistry, as well as ecology, to create better forecasts. Bazhenova and Kobylkin (2013) analysed the dynamics of soil erosion as a source of water pollution in the Selenga basin. Parham et al. (2013) studied the combined effect of permafrost and fires on the chemistry of watercourses, focussing on specific concentrations of DOC and cations. Those empirical data are important for modelling biogeochemical cycles in this region.

The chemistry of surface water is largely associated with salinization processes, mainly in the steppes, but also in semiarid Taiga regions. Zolnikov et al. (2011) monitored the humidification dynamics in the south of West Siberia by means of ground-based field observations and space-acquired images of moderate spatial resolution (Landsat, SPOT). The areas of lakes and solonchaks are aridization indicators. Romanov et al. (2014) used the SMOS satellite with a MIRAS radiometer for monitoring waterlogging, soil salinization and salt lake dynamics. In ground studies, Lebedeva and Lopukhina (2006) analysed the chemical and mineralogical composition of salts, classified salt lakes and surrounding solonchak soils in the Kulunda steppe. Seasonal changes of salt and soil moisture distribution in the active layer of undisturbed larch forest and a thermokarst depression (alas) were studied by Lopez et al. (2007) to explain how forest growth is hindered by surface salinization. The technogenic salinization of soils is also widespread in forest–tundra and northern Taiga ecosystems of areas that are polluted by oil and gas mining (Solntseva and Sadov 2000).

1.5 Analysing the Ecosystem's Response to Climate Change

Research has provided some new findings on the Arctic's response to climate change. Zimov et al. (1995) analysed the impact of grazing herbivores on the kind of vegetation. Results are based on observations that large mammals trampling and grazing tundra regions cause a vegetational shift in dominance from mosses to grasses. Either grass-dominated steppe or moss-dominated Tundra can exist under Pleistocene, current and future climates. Water and nutrient cycles are important under those conditions. Nitrogen (N) in soils can be lost by denitrification and leaching from the system in thawing permafrost, or it can be acquired by plants and boost the N-cycle (Mack et al. 2010). Those processes and ecosystem alterations will probably cause a decline in the populations of Arctic animal species and the expansion of the ranges of some southern species into the Arctic (Callaghan et al. 2011). Overall plant species richness will probably increase at high latitudes (Venevskaia et al. 2013). Many alien species could invade new territories, but this issue has not received much attention (Olonova and Zhang 2013). Models and assessment frameworks have been developed for individual aspects of climate–vegetation interactions. Khomutova et al. (2007) computed the climate-driven limitation of biomass at high latitudes using the Lund-Potsdam-Jena Dynamic Global Vegetation Model (LPJ-DGVM), which is a process-based biogeochemistry/biogeography model. Vasil'evskaya et al. (2012) developed a procedure for calculating and assessing the degradation status of Tundra soil and vegetation based on the properties of natural and disturbed soil cover. Potential alterations to forest, forest steppe and steppe ecosystems have also been the subject of intensive empirical disciplinary studies conducted by international teams over recent years. Interesting examples of these research topics in Siberia are the ongoing German–Russian interdisciplinary projects "Sustainable land management and adaptation strategies to climate change for the Western Siberian corn-belt" (abbreviation "SASCHA") and "How to prevent the next "Global Dust Bowl"?—Ecological and economic strategies for sustainable land management in the Russian steppes" (KULUNDA project 2014; Illiger et al. 2014).

1.6 Development of Landscape Information Systems and Databases

To understand the complexity of landscape processes, research and monitoring programmes of higher complexity and interdisciplinarity are required and in progress (Moskalenko 2013). The creation of Russian soil geography databases (Shoba et al. 2008; Rozhkov et al. 2010) and their correlation with international standards were important steps towards monitoring, modelling and understanding landscape processes. Stolbovoi and Fischer (1997) developed a digital geo-referenced database of soil degradation in Russia, and Stolbovoi (2000) correlated Russian soil types with the legend of the FAO Soil Map of the World and World Reference Base for Soil Resources (WRB). Schepaschenko et al. (2011) developed a new hybrid land cover dataset for Russia which is very important for modelling landscape processes. Datasets of local or regional importance and validity, such as soil survey databases (Mikheeva 2013), vegetation databases (Chepinoga 2012) and others, have been created based on and for the purpose of further local and regional studies. Koneva and Batuev (2014) developed a biogeographical map series for Asian Russia. Most of those databases also have the potential to become integrated into databases and information systems of higher complexity. An information system of this kind has been developed by Gordov et al. (2013). It shall serve as a basis for environmental research in Siberia and is a crucial component of the Northern Eurasia Earth Science Partnership Initiative (NEESPI).

1.7 Development of Research Equipment and Methodology

The focus of research methodologies has changed crucially over the past decades. Whilst most textbook knowledge about landscape structures and processes is based on very detailed ground studies at individual locations, a lot of current knowledge has been generated by studies over large areas utilizing and elaborating on remote sensing methods. Examples are the application of remote sensing methods for vegetation and land cover mapping (Golubeva et al. 2010; Urban et al. 2010), thermokarst monitoring (Fedorov and Konstantinov 2003), glacier observation (Ananicheva et al. 2011), monitoring soil and surface hydrological processes such as salinization (Zolnikov et al. 2011), or environmental pollution and their consequences (Bergen et al. 2013; Gutman et al. 2013). BC emissions from agricultural burning in Russia have been quantified using methodologies based on remote sensing data (Romanenkov et al. 2014).

Particular progress in remote sensing methods was possible because of the change in the political situation in Europe at the beginning of the 1990s, when satellite data and other geo-referenced airborne data became available for civil purposes. At the same time new generations of information technologies have emerged and boomed. Those geo-referenced spatiotemporal data enable more sophisticated simulations of the land and water surface and atmospheric processes (Arzhanov et al. 2008). Data handling by interdisciplinary teams is becoming more reliable and convenient thanks to soft- and middleware tools developed by Eberle et al. (2013).

If the focus is not on the spatial scale of observations alone but on the timescale in high resolution, advanced technologies are also required. Examples of those ground-based process studies and monitoring are GHG flux measurement systems, advanced lysimeters and automatic stations for measuring hydrological and aerial matter fluxes, and sophisticated ecosystem models. In this field of research and monitoring the progress needed for the landscapes of Siberia can be only provided by better international cooperation.

1.8 Advances in Research on Forest and Agro-Ecosystems

Forest ecosystems: The research on forest ecosystems has focused on climate change issues. Outstanding work has been carried out quantifying carbon budgets over the whole area of Russia. The current soil carbon could be estimated from soil maps (Stolbovoi 2006) but possible alterations to this pool remain the largest uncertainty in carbon budgeting (Shvidenko et al. 2011). Aspects of forest acclimation (Lapenis et al. 2005; Tchebakova et al. 2006) and of adaption to stressors such as fire (Ivanova et al. 2011; Kharuk et al. 2011; Krasnoshchekov et al. 2013) or pollution (Sorokin 2009) have become better understood over recent years.

Agro-ecosystems, agronomy: Based on an extended system of education and research in soil science, agronomy, agrochemistry and related disciplines, agro-ecosystems have been well studied over recent years. The greatest focus was on the optimization of local and regional crop production systems. Much progress has been achieved, and numerous dissertations and other publications reflect this best. Enhancing and stabilizing cereal yields by breeding, soil tillage and fertilization and plant management have been of main importance (Gamzikov and Nozov 2010; Trubacheva et al. 2011). Multifactorial trials containing factors of fertilization and other agrochemistry, soil tillage and rotations have been very typical for dissertations. Crop yields and yield quality have been the most frequently tested parameters of success, tested using classical statistical methods. Some examples of the latest work are studies about the effects of organo-mineral fertilization (Sorokin 2011), about the crop yield of a cereal–fallow rotation at different primary tillage (Rzaeva and Fedotkin 2013), about the influence of cultivars and seeding time on yield and grain quality of barley (Puzyreva 2013), about improving the agrotechnology for barley (Shtro 2013) and about optimizing spring wheat yielding capacity in the forest steppe (Skomoroshchenko et al. 2013). Experiments have revealed the site-specific optimum selection of cultivars, fertilization, tillage and plant management. They showed that systems which are more biological have an advantage over those on cropping sites in Europe: local cultivars, extended rotations and low to moderate fertilization levels. A few Siberian experimental sites are part of the Russian agro-ecological monitoring network (Fig. 3). Remote sensing methods for controlling the state of agricultural crops are emerging (Pugacheva et al. 2010).

Modelling agro-ecosystems and agroclimatic potentials: Some progress has been achieved in this emerging topic. Concepts and initial results of modelling agroclimatic potentials were developed by Kirsta (2006) and Tchebakova et al. (2011). Dronin and Kirilenko (2011) developed climate change and food stress scenarios by including socio-economic criteria in models.

Fig. 3 Monitoring stations coordinated by the Geo-Network Department of the Pryanishnikov All-Russian Institute of Agrochemistry (VNIIA) in Moscow. Only a few stations are located in Asian Russia

Soil science related to agronomy: Besides the classical work of soil mapping and studies of pedogenetic processes under different land use, the valuation of soil fertility and productivity, and testing interactions of cropping systems with soil fertility have been an important topic of experimental research work in soil science. Examples of such work include studies about the fertility, rational use and preservation of Chernozems (Abramov and Salova 1998; Eremina 2009; Khmelev and Tanasienko 2009), about acid soils (Tandelov 2012), about the migration of chemical elements between adjacent ecosystems along a geochemical catena (Titlyanova 2008) and about assessing the fertility of typical soils used for agriculture (Smolentseva et al. 2014; Fig. 4). Afonin et al. (2008) developed an interactive agricultural ecological atlas of Russia, which contains important pedologic criteria such as soil types, humus contents and rooting depths. Another important topic of soil scientific research is studies about the status and rehabilitation of technogenic soils (Ermolov et al. 2011). This is related to agronomic

Fig. 4 Assessment of soil conditions in a Solonetz–Solonchak soil association of the Kulunda Steppe (**a**) and a Phaeozem–Chernozem association in the Pri–Ob Forest steppe (**b**). The underlying cooperation between the Institute of Soil Science and Agrochemistry (ISSA) in Novosibirsk and the Leibniz Centre for Agricultural Landscape Research in Müncheberg (Germany) was aimed at harmonizing methods for the taxonomic and functional evaluation of soils. **c** Young scientists measured soil and water quality parameters as part of this project

questions in so far as the cropping of perennials is often the best rehabilitation solution (Chuprova 2006).

A great deal of research not cited here has been done in numerous local institutes of the Russian Academies of Sciences (RAS) and Agricultural Sciences (RAAS).

2 Factors Promoting or Constraining Scientific Progress

2.1 *Traditions in Landscape and Agri-environmental Research*

Russian science, including landscape research, is based on great and largely stand-alone traditions (Antipov et al. 2006; Shaw and Oldfield 2007). Only two outstanding persons from the last century shall be mentioned here: the geographer V.V. Dokuchaev, who recognized he natural laws of soil formation and was the originator of the modern soil science, and the geographer L.S. Berg, who developed the Landscape–Geographical Zones of the USSR. Until now, national traditions and scientific schools play a particular role in the Russian research field of landscape science including contributions from physical geography, soil science, agricultural science, biological science and others. The emphasis on Russian research traditions is based on and associated with some political and social developments in the country, in particular with those of the recent past, the former USSR. The huge dimensions of the country, backward information flow and technologies, and the continued use of the Russian language may have had some negative side effects. It is still a deficit of many research papers and dissertations in agriculture, soil and environmental sciences that some progress achieved by the international research community was not known or not seen as worth consideration. On the other hand, much important work by Siberian scientists is not known abroad because of language barriers. For example, the Sochava Institute of Geography in Irkutsk is conducting a long-term (>50 years) geo-biochemical monitoring programme over numerous sites in Siberia and neighbouring countries (Antipov et al. 2006; Nechaeva et al. 2010).

2.2 *Institutional Research*

The Russian Federation operates a huge scientific network for understanding processes and monitoring land and water resources. It is the largest national network worldwide dealing with landscape research. The Russian Academy of Sciences (RAS) and the Russian Academy of Agricultural Sciences (RAAS) comprise several hundred research institutes under general Russian or local control. Additionally, based on the federal structure of Russia, a number of universities and academies deal

Fig. 5 International communication and cooperation is key to scientific progress in landscape research. The photograph shows participants of an international symposium about water quality monitoring, organized by the Limnological Institute Irkutsk in November 2013. *Photo* Marco Natkhin

with research and education in science and agricultural science. The Siberian branches of the RAS and RAAS are responsible for landscape and agri-environmental research over Siberia. Environmental problems in Russia have grown over the past 25 years (Shaw and Oldfield 2007). Landscape scientists therefore have a lot to do to monitor and combat these problems in their landscapes.

Some examples of particularly active scientific groups in the field of landscape and agri-environmental research of Siberia are the Limnological Institute in Irkutsk, Fig. 5, the Sukachev Institute of Forest in Krasnoyarsk, the Institute for Water and Environmental Problems (IWEP) in Barnaul (Fig. 6) and the ISSA in Novosibirsk (all Siberian branches of the RAS). The Geo-Network Department of the Pryanishnikov All-Russian Institute of Agrochemistry (VNIIA) in Moscow is going to establish a new generation of an agri-environmental monitoring system for Russia and to rebuild monitoring stations in Siberia and the Far East. Scientists at the above institutes are involved in highly advanced research technologies and methods and high-ranking scientific publications. They are closely linked with the international scientific community and active in several national and international projects.

The Russian Academy of Sciences (*RAS*): RAS was the highest scientific institution in Russia, conducting basic research about the technological, economic, social and spiritual development of the country (IAP 2013). RAS incorporated 410 scientific institutes belonging to 9 specialized scientific departments, 3 regional divisions and 14 regional centres, RAS employed almost 100,000 people. In the period of 2001–2007 the total publishing output of RAS amounted to 60,000 book and journal titles, most of them edited by the RAS "NAUKA" Publishing House. Its scientists participate in international scientific cooperation and are active in various international organizations such as the United Nations, UNESCO, UNEP, IAEA,

Fig. 6 Research vessel of the Institute for Water and Environmental Problems (IWEP) in Barnaul. It is located on Lake Telozkoye in the Altai Mountains to monitor water quality and sediment

WHO and WMO. RAS researchers are in demand as top scientific experts by the industry and the business community. RAS has full membership relations with about 50 international non-governmental organizations (IAP 2013). Akademgorodok at Novosibirsk is the scientific centre of Siberia (Fig. 7).

The Russian Academy of Agricultural Sciences (*RAAS*): RAAS was the highest self-governed scientific organization of Russia, and responsible for agricultural research (UEAA 2006). It comprised 199 research institutes (year 2000) and 24 agricultural pilot stations, more than 400 pilot farms, 46 semi-industrial enterprises, and 47 breeding and biotechnological centres. About 14,200 scientists worked in research. RAAS owned 5.8 million ha of land for experimental farming including 1.7 million ha of cropping land. Livestock was more than 360 thousand cattle, 130,000 pigs, 65,000 sheep and 1.6 million heads of poultry. RAAS was structured into eight branch departments; one regional department, three scientific and methodological centres, a Siberian branch and regional centres in Siberia and

Fig. 7 Sculpture in Akademgorodok near Novosibirsk, the scientific center of Siberia and headquarter of the Siberian branch of the RAS. *Photo* Ralf Dannowski

the Far East. RAAS had an extended internal structure including administration departments for planning and coordination, a science organization section, an international relations department, pilot plants, a land and estate registration department, a construction and material provisions department, libraries and a printing press (UEAA 2006).

2.3 International Progress in Landscape Sciences

Internationally, research about the functioning of landscapes has experienced a rapid evolvement over recent years. Landscape research is based on a transdisciplinary view (Wiggering et al. 2006; Van Huylenbrok et al. 2007; Mander et al. 2007; Helming et al. 2011; Hermann et al. 2011). Landscape hydrology, soil hydrology and soil science are basic disciplinary compartments of landscape research studying water and soil altering processes in landscapes (Schindler et al. 2010; Lischeid and Natkhin 2011; Lischeid 2014a). As Western Europe is one of the leading agricultural regions worldwide and intensive agriculture significantly impacts on resources and ecosystems (Mueller et al. 2014a; Eulenstein et al. 2014), agri-environmental monitoring has to deliver data for detailed studies and decisions. Modern monitoring and evaluation tools have been developed and applied for research into landscape processes everywhere. Examples of those methodologies include measurement systems for gas, water and solute fluxes (Juszczak et al. 2013; Funk et al. 2014; Meissner et al. 2014; Schindler and Mueller 2010; Schindler 2014a, b). Siewert and Kučerik (2014) further developed thermogravimetry in soil science and identified the thermographic fingerprints of Siberian soils. Statistical methods for process identification and quantification from monitoring data (Lischeid 2014b), and sophisticated ecosystem models (Schaldach and Priess 2008; Wenkel et al. 2013; Nendel 2014) have provided quantifications and forecasts of ecosystems' responses. Improved sensor and data-processing technologies in remote sensing enable the spatio-temporal monitoring of large regions (Hese and Schmullius 2008; Gessner et al. 2012; Klein et al. 2014). Soil science has developed new assessment frameworks as a basis for assessing soil processes (WRB 2006, 2014; Zech et al. 2014) and functional performance (Mueller et al. 2010; 2014c). Drought has been identified as a main crop yield limiting process worldwide (Brown 2012; Mueller et al. 2012). Raising water productivity is thus a crucial item in the global struggle for food security (Brown 2012). Technologies for managing agro-ecosystems have to be based on principles of conservation agriculture (Kassam et al. 2011; Meinel et al. 2014; Suleimenov et al. 2014). Water savings in irrigated agriculture can be provided by sophisticated computer programs and new technological developments (Djanibekov and Sommer 2014; Michel and Dannowski 2014). Advanced irrigation technologies (Evans 2014) enable site-specific water distribution over fields and further improvements to water and fertilizer efficiency. Methods have been developed for evaluating grassland quality, as well as new grazing methods for protecting soils and vegetation from degradation (Behrendt et al. 2014; Mueller et al. 2014b, c). In order to prevent groundwater pollution from agriculture, methods for groundwater monitoring and risk assessment are now available (Dannowski et al. 2014; Eulenstein et al. 2014; Godbersen et al. 2014).

Not only the contents of research but also its organization and structure have changed in West European countries. Some former research fields such as soil mapping, environmental monitoring, construction of research equipment, or plant and animal breeding are no longer part of public research but instead in the hands of other state or private institutions. Transferring this system to research institutes located in the former GDR at the beginning of the 1990s enabled a better division of work, more creative research and a faster transfer of new methods into practice. Public State or Federal institutions provide soil functional mapping and act as connectors between research and practice (Hennings 2013). Companies such as Eijkelkamp Agrisearch Equipment, UGT Müncheberg, UMS Munich and others have developed leading technologies for environmental and agri-environmental monitoring jointly with leading researchers (Eijkelkamp 2014; Hertel and von Unold 2014; Meissner et al. 2014).

Since the 1990s some Russian scientists have benefited from the achievements of the international scientific community. Novel research methods have become available to those scientists who were able to integrate into international research networks and projects (Kerzencev and Meissner 2006). Knowledge of the English language was a main precondition. Lately, a new generation of Russian scientists has emerged, characterized by some typical features:

- Good disciplinary education based on great knowledge and traditions of Russian geography, soil science and hydrology as taught at the leading universities of Moscow, Tomsk, Barnaul and a few other cities
- Well linked with the international community, working in international trans-disciplinary cooperation projects and organizations (for example, the European Union programme "International Association for the promotion of cooperation with scientists from the independent states of the former Soviet Union—INTAS")
- Applying and developing modern research technology and methodology
- Aware that landscape research is a scale issue on both the temporal and spatial scales
- Able to apply both bottom-up and top-down processes and to handle large amounts of data to quantify them
- Publishing in international journals.

These characteristics are same as those of leading scientists in the international community. Consequently, some of these scientists work on international projects and/or abroad. It is a challenge for Russian science to integrate them into their new academic system.

2.4 *The Reform of the Russian Academic System*

The Russian research landscape is in a difficult transition phase. The Academy of Sciences, Academy of Medical Sciences and the Academy of Agricultural Sciences have merged to form a new Academy of Sciences under a Federal Law from 2013: "On the Russian Academy of Sciences, the reorganisation of the state academies of sciences and amendments to certain legislative acts of the Russian Federation". A new governmental body was founded, the Federal Agency for Scientific Organisations (FASO, Russian term ФАНО) responsible for all academic institutions by the decree "On the Federal agency for research organisations" (Schiermeier 2013). The reason for this reform was to create more effectiveness in science (Polterovich 2013; Voswinkel 2014). The current scientific system faces several challenges such as ageing of the academic staff, lack of publications in ranked international journals, bad presentation of institutes in the Internet or suboptimum international cooperation. These deficits have been recognized for some years and target settings have been developed in the Siberian Federal District (Table 1).

Table 1 Target settings for the development of innovations in Siberia (Strategy 2010, excerpt)

Indicator	2015	2020
Number of persons working in research and education	59,000	61,000
Proportion of young researchers (<40 years old) %	22	27
Number of international research centres	15	20–23
Proportion of internationally rated scientific journals %	5	7
Number of patents	3600	4200

Are those target settings still true and realistic? The reform was needed, but the way to operationalize it could become a problem. The new RAS will be an academic club without administrative power and service units. In the case of the former RAAS this means expropriation from all research stations and lands, and could have unforeseeable consequences for all running long-term trials. Overall and generally, freedom and independency of research could be threatened by administered reforms that are not based on the expertise of main stakeholders. Insecure basic financing for scientific institutes bears risks for high-quality research due to brain drain and more bureaucracy (Yablokov 2014).

New initiatives of international scientific–technical cooperation in landscape research would be very important for stabilizing the situation at this critical stage. It must be based on genuine partnership at eye level. This is currently still possible.

2.5 *Effects of the Market Economy on Ecosystems and Research Efficiency*

The Russian market economy capitalizes Siberian resources of gas, oil and minerals by wasteful mining and other industries with benefits for urban areas and with damage to the environment (Newell 2004). Rural regions of solely less profitable branches such as forestry, agriculture and fishery suffer from recession and the breakdown of their infrastructure. Farmers' acceptance of abandoning soil-destroying agriculture and following scientists' advice is low under these conditions.

The indigenous peoples who inhabit most of Siberia are seriously threatened by mining industries and the breakdown of the infrastructure in remote areas (Osherenko 1995; Vakhtin 1998; Semenova 2007; Yakovleva 2011). Programmes are urgently needed to help the small numbers of peoples in the north support themselves (Isachenko 2013).

Framework conditions for the development of cultural landscapes are important (Ragulina 2013). They are largely lacking in Siberia and there seem to be none at all in the Far East. Poverty is not a suitable framework for the development of environmental consciousness but instead promotes unsustainable management or illegal behaviour (poaching, logging). Overall, the introduction of the market economy has not diminished but exacerbated negative consequences for the environment (Shaw and Oldfield 2007) and for the culture of indigenous peoples (Isachenko 2013).

3 Deficits of Research in Land Management and Agronomy

The analysis of publications in the area of landscape and agri-environmental research referring to Siberia over the past 25 years shows interesting results and tendencies. Great progress has been made in analysing, understanding, predicting and coping with processes and problems of ecosystem functioning under the pressure of the market economy and climate change. The international community has invested a great deal in environmental research and monitoring programmes. The focus has been on Arctic research and monitoring, climate change issues such as carbon cycling, GHG emissions and water pollution.

Research about land- and water-resource-based industries such as agriculture, forestry and fishery has been a national issue and cannot meet international standards. Agricultural research has a well-developed experimental basis in the country but poor resources, obsolete analytics and a lack of access to leading technologies worldwide. Research in agronomy is particularly based on traditions and there is a lack of modern research technology. Numerous experiments have been conducted, and valuable data generated, but this work is largely de-central. Many Russian scientists working in the field of agricultural soil and water management are also largely isolated from the international research community. No Russian researcher is active in the International Soil and Tillage Research Organization (ISTRO), the world's leading scientific organization in this field of research, producing the well-respected scientific journal "Soil and Tillage Research". Russian researchers publish mainly in Russian. Many publications in Russian are freely available online and could promote the fast dissemination of results over the country. Because the methodology of trials is classical, the equipment for trial conduction is obsolete and biophysical conditions of plant growth and results are very site-specific; there is not much to transfer to other regions. Dissemination of those local results is mainly directed to advanced farmers and authorities in the same region. However, as agriculture is in a recession it cannot benefit from this knowledge. Experimental results are potentially important for the calibration of agro-ecosystem models. However, agri-environmental modelling is still underdeveloped.

Research about land and water resources, landscape processes and environmental monitoring in the Far East region of Russia is a white spot in ground-based research and monitoring land and water resources and processes.

4 Initiatives of International Cooperation and Communication in Landscape Research

International cooperation in research and the transfer of knowledge and education has been recognized as key factor for better interlinking Russian scientists with the international scientific community. International, multidisciplinary Arctic research is a flagship of cooperation in environmental research leading to better understanding of climate changing processes and their consequences for the development of landscapes (Polyakov et al. 2014).

During recent years some progress has also been achieved regarding cooperation in agri-environmental and landscape research. For example, the German Federal Ministry of Food and Agriculture and the former Russian Academy of Agricultural Sciences created a list of mutual cooperation projects between German and Russian researchers. The German-Russian project 05/07 “Indicators of fertility and function of agricultural soils” of this list formed the basis of cooperation between several Russian and German soil scientists and agronomists. The main partners were ISSA Novosibirsk and VNIIA Moscow from Russia, and ZALF Müncheberg from Germany. This enabled joint fieldwork about soil classification and evaluation on numerous agricultural sites in both countries. Another small-budget project, “Effect of climate change in boreal and sub-Arctic ecosystems on water quality and soil functions, code 01DJ12058” was supported by the German Federal Ministry of Education and Research (BMBF) and enabled this cooperation between ZALF Müncheberg and ISSA Novosibirsk to be deepened along with IWEP Barnaul. Overall, more than 15 joint publications appeared as outcome of both projects. Unfortunately, the current reform of the Russian academic system has disrupted those effective activities.

Currently, the above-mentioned better-funded projects “KULUNDA” and “SASCHA” (KULUNDA project 2014) focus on sustainable land management as a main topic of agri-environmental research. Some progress has been achieved in landscape planning due to Russian–German cooperation (Koroshev et al. 2014).

Another type of initiative was started by Russian and German soil scientists almost 20 years ago. Soil scientific and ecological summer schools and field excursions through the main landscapes of West Siberia have become established as a permanent educational institution for students and scientists from abroad (Siewert et al. 2014; Fig. 9). They are also a source of inspiration for new research activities. Scientists from Irkutsk have established further summer schools and excursions through the Baikal region and exciting regions of the Far East (Chepinoga et al. 2004). The German Academic Exchange Service (DAAD) supported the participation students in those summer schools in the framework of the “go east” submission (DAAD 2014).

5 Conclusions: Knowledge Gaps and Research Needs

A lot of progress has been achieved in understanding landscape processes in Siberia and in evaluating land and water resources. However, the dynamics of alterations of ecosystems due to climate change and unsustainable human impacts requires faster progress. A lack of knowledge and absence of reliable data have contributed to the harmful treatment of land and water resources. Scientists, decision-makers and other responsible people need more reliable data. Some gaps of knowledge and deficits of agri-environmental research and monitoring are:

- A lack of measurement and monitoring technologies for net primary production (NPP), heterotroph respiration and gas exchange between ecosystems and atmosphere. Reliable and automatic high-resolution measurement of the functional performance of the Geo–Bio system by exact balancing of water, carbon, sediments and other matter fluxes (lysimeters, GHG flux measurement systems, dust measurement systems).
- Soil physics, soil hydrology and modelling the soil–plant–atmosphere system are underdeveloped. Better ecosystem models and decision support systems that consider feedback from and the complexity of processes in landscapes are needed to make climate change scenarios more reliable.
- Many remote areas in the Tundra and Taiga of East Siberia and the Far East are not sufficiently covered with monitoring technology. Remote sensing approaches have emerged but must be better linked with automatically operating ground-level monitoring stations. Airborne data also need to be combined better with process models.
- Data analysis is largely based on traditional statistical approaches which do not consider autocorrelations of processes and landscape structures. Methods of modern explanatory statistical data analysis, which allow hidden structures and processes to be detected, should be part of landscape experiments (including agricultural trials) and data acquisition.
- Reliable but simple approaches for the assessment and monitoring of the functional status of land (including crop yield potentials) consistently over different zones and regions are lacking for Russia. They should be better compatible with recommendations and standards of the EU and the United Nations Food and Agriculture Organization (FAO).
- Chemical analytics of soil, water and plants has a great tradition in Russia but needs to be harmonized with international standards and leading developments. In the case of soil analyses, sample preparation methods are often difficult to compare with EU approaches.
- Analytical methods and evaluation frameworks for the functional status of aquatic ecosystems are also in need of harmonization.
- The environmental side effects of agriculture, their risks and real extent, are still not well researched. Environmental impacts on aquatic ecosystems, water resources and air quality should be monitored, evaluated and controlled. Those systems should meet international standards.

- Principles of Conservation Agriculture (CA) have been tested in several crop trials but their complex meaning for avoiding soil degradation is still not understood by researchers and decision-makers. Also, CA is still neglected in practice due to poor resources and obsolete machinery.

Progress can be achieved by the adoption and application of new methods for measuring, assessing, modelling, monitoring and controlling landscape processes. This includes the need for both methods and technologies for basic research to understand the ecosystem, and applied agri-environmental research for a fast transfer into practice. Most of these technologies are internationally available and could help to get better data for recognizing, understanding and possibly controlling land and water resources and processes. Some of them are presented in the following chapters. Environmental data and knowledge may contribute to the formation of a higher stage of public awareness about environmental problems and a basis for impact-assessment procedures in order to find optimal site-specific solutions for science-based landscape planning. We encourage decision-makers to install a sustainable platform for scientific technical cooperation in landscape research between Russia and the EU. This would be very important to maintaining high-level research and agri-environmental monitoring of the terrestrial and aquatic ecosystems of Siberia.

Appendix: Environmental Education Abilities in Soil-Ecological Summer Schools in West Siberia

Since 1995 soil-ecological summer schools have been taking place across bioclimatic zones in West Siberia plain and altitudinal belts in the Altai Mountains. They are organized annually by a group of scientists from Russia and Germany. The main goal of the excursions was to answer growing demands for better education and research on mutual interdependencies between climate, geological substrates, vegetation cover and other factors as a tool for practical land use improvement.

The sites of the summer school are selected as a logical sequence of changing climate conditions from north to south in West Siberia including plains (horizontal climate zones) as well as mountains (vertical climate zones) (Fig. 8). They focus on landscapes of exceptional beauty and highly interesting natural objects, both with extreme emotional impact on the participants' learning abilities as a tool of long-term motivation for sustainable land use. Virgin ecosystems and sites untouched by human activity are included for simplified teaching of the complex interrelations between factors of landscape formation unconcealed by the artefacts usually caused by a long history of land use or by the wide-scale pollution in developed countries.

The introduction lectures at every site provide information about their geology, relief, climate, vegetation cover, soil formation and history of human exploration together with the main aspects of local culture and challenges of social life

Fig. 8 Typical landscapes used by soil ecological summer schools **a** Mountain desert, **b** Mountain tundra, **c** Mountain forest tundra, **d** Southern Taiga, **e** Forest steppe. The floristic diversity of forest steppe grasslands in West Siberia is still very high due to extremely low human impacts. About 60 species of vascular plants per 100 m^2 have been found at some locations. *Photos* Christian Siewert

development. They try to mediate a simple access to understanding long-term needs in local productive land use taking into account global trends.

The field lectures which follow are the main method of education at the summer school. They are dedicated to illustrating the most important local features, which are sometimes incredible to foreign participants, using easily available natural materials from the surroundings. Teaching methods include experiences by means of hearing, touching, feeling or even tasting (Fig. 9). Short walks (around one hour)

(a)

(b)

Fig. 9 Teaching soil science and soil–vegetation interactions. **a** The profile shows a Chernozem in the forest steppe. From their inherent properties Chernozems are the most fertile soils of the globe. **b** Minutes and examinations are part of the open air summer school. *Photos* Pavel Barsukov

across the countryside without trails or prepared adventures allow participants to feel the landscapes under their feet, to catch its colours, sounds and smells. The personal experience obtained this way empowers the participants to gain their own insight into the complexity of natural conditions in a most unforgettable way. This supports open discussions providing a better understanding of the details of both local and global consequences of human land use.

The teaching goals and needs determine the organizational features which support a deep personal perception of the environment. Almost the entire route of the summer school, all accommodation and meals are held outdoors. The summer school participants must adapt to the weather conditions; they have to walk, live in tents, collect their own experience and "sense of places", and they have to accept everyday life in nature as a source of ecological knowledge. A specialized service team manages daily challenges including the completion of bureaucratic demands (e.g. visa formalities and registration of foreigners), reliable transport, a supply of tasty food, most possible accommodation and the organization of cultural events. It consists of drivers, a cooking team and assistants under the supervision of Russian scientists from leading research institutes and universities.

The results are reflected in excellent evaluation results, several multiplier effects, research projects, some generous funding by different organizations, and more. The participants especially appreciate the experience they gain, which supports the long-term mitigation of global change in land use and connected job opportunities. The following photographs provide some visual impressions about these courses. More information is given in the publication by Siewert et al. (2014).

References

Abramov NV, Salova EV (1998) The optimum parameters of factors determining the fertility of leached chernozem in the northern part of the forest-steppe zone in West Siberia. Eurasian Soil Sci 31(10):1131–1136

Afonin AN, Greene SL, Dzyubenko NI, Frolov AN (eds) (2008) Interactive agricultural ecological atlas of Russia and neighboring countries. Econ Plants Diseases Pests Weeds. Available at: http://www.agroatlas.ru. Accessed 28 Dec 2014

Anisimov O, Reneva S (2006) Permafrost and changing climate: the Russian perspective. Ambio 35(4):169–175. Royal Swedish Academy of Sciences 2006. http://permafrost.su/sites/default/files/Ambio.pdf. Accessed 30 Jan 2015

Ananicheva MD, Krenke AN, Gregory A. Kapustin GA (2011) Recent and forecasted change of glaciers of the Northeastern Asia. Geogr Environ Sustain 3(4):19–33, http://int.rgo.ru/wp-content/uploads/2011/12/GES3-11.pdf. Accessed 30 Jan 2015

Antipov AN, Korytny LM, Plyusnin VM (2006) Siberia and its geographical research, pp 3–36. In: Antipov AN (ed) Geography of Siberia. Research India Publications, New Delhi, p 235

Arzhanov MM, Demchenko PF, Eliseev AV, Mokhov II (2008) Simulation of characteristics of thermal and hydrologic soil regimes in equilibrium numerical experiments with a climate model of intermediate complexity. Izvestiya Atmos Oceanic Phys October 2008, 44(5):548–566 http://link.springer.com/article/10.1134/S0001433808050022. Accessed 30 Jan 2015

Bazhenova OI, Kobylkin DV (2013) The dynamics of soil degradation processes within the Selenga basin at the agricultural period. Geogr Nat Res 34(3):221–227

Beer C, Fedorov AN, Torgovkin Y (2013) Permafrost temperature and active-layer thickness of Yakutia with 0.5-degree spatial resolution for model evaluation. Earth Syst Sci Data 5:305–310. http://www.earth-syst-sci-data.net/5/305/2013/essd-5-305-2013.pdf. Accessed 30 Jan 2015

Behrendt A, Fischer A, Kaiser T, Eulenstein F, Ortmann S, Berger A, Mueller L (2014) Multi-Species Grazing on Deer Farms. In: Mueller L, Saparov A, Lischeid G. (eds), Novel measurement and assessment tools for monitoring and management of land and water resources in agricultural landscapes of Central Asia. Springer International Publishing, pp 491–501 http://link.springer.com/chapter/10.1007/978-3-319-01017-5_30. Accessed 27 Feb 2015

Bergen KM, Hitztaler SK, Kharuk VyI, Krankina ON, Loboda TV, Zhao T, Shugart HH, Sun G (2013) Human dimensions of environmental change in Siberia, In: Groisman PJ, Gutman G (eds) Regional environmental changes in Siberia and their global consequences (Chap. 7),

Springer Environmental Science and Engineering 2013, ISBN: 978-94-007-4568-1 (Print) 978-94-007-4569-8 (Online). http://link.springer.com/chapter/10.1007%2F978-94-007-4569-8_7. Accessed 30 Jan 2015

Blome T (2014) Influence of different permafrost processes on the large-scale energy and water cycles over Siberia reports on earth system science 150, p 159. http://pubman.mpdl.mpg.de/pubman/item/escidoc:2039431/component/escidoc:2039432/WEB_BzE_150.pdf. Accessed 30 Jan 2015

Boike J, Kattenstroth B, Abramova K, Bornemann N, Chetverova A, Fedorova I, Fröb K, Grigoriev M, Grüber M, Kutzbach L, Langer M, Minke M, Muster S, Piel K, Pfeiffer E-M, Stoof G, Westermann S, Wischnewski K, Wille C, Hubberten H-W (2012) Baseline characteristics of climate, permafrost, and land cover from a new permafrost observatory in the Lena River Delta, Siberia (1998 – 2011). Biogeosci Discuss 9:13627–13684. doi:10.5194/bgd-9-13627-2012

Brown LR (2012) Full planet, empty plates: the new geopolitics of food scarcity. Earth Policy Institute, p 160. ISBN: 978-0-393-34415-8, http://www.earth-policy.org/books/fpep. Accessed 30 Jan 2015

Callaghan TV, Velichko AA, Borisova OK (2011), Tundra in a changing climate, geography environment sustainability 3(4):4–18, http://int.rgo.ru/wp-content/uploads/2011/12/GES3-11.pdf. Accessed 30 Jan 2015

Chepinoga VV (2012) Wetland vegetation database of Baikal Siberia (WETBS). Biodiv Ecol. Vegetation databases for the 21st century 4:311. http://www.biodiversity-plants.de/biodivers_ecol/publishing/b-e.00107.pdf. Accessed 30 Jan 2015

Chepinoga V, Lopatovskaja O, Mehring M, Pudwill R (2004) Baikalsee-Exkursion 2004 Exkursionsbericht, http://exkursionen.pr-naturetours.de/Exkursionsbericht_2004.pdf. Accessed 28 Dec 2014

Chuprova IL (2006) Optimization of technogene landscapes in the high North (Оптимизация техногенных ландшафтов крайнего севера, Норильский промышленный район, п-ов Таймыр) Diss. Petrozavodsk 2006, http://earthpapers.net/optimizatsiya-tehnogennyh-landshaftov-kraynego-severa. Accessed 30 Jan 2015

DAAD (2014) Go East. https://goeast.daad.de/de/22624/index.html. Accessed 28 Dec 2014

Dannowski R, Schindler R, Cremer N, Eulenstein F (2014) Methods of In Situ Groundwater Quality Monitoring: Basis for the Efficiency Survey of Agricultural Groundwater Conservation Measures. In: Mueller L, Saparov A, Lischeid G (eds) Novel measurement and assessment tools for monitoring and management of land and water resources in agricultural landscapes of Central Asia. Springer International Publishing, pp 275–288. http://link.springer.com/chapter/10.1007/978-3-319-01017-5_16. Accessed 28 Dec 30 Jan 2015

Djanibekov N, Sommer R (2014) Integrated decision support for sustainable and profitable land management in the lowlands of Central Asia. In: Mueller L, Saparov A, Lischeid G (eds) Novel measurement and assessment tools for monitoring and management of land and water resources in agricultural landscapes of Central Asia. Springer International Publishing, pp 407–422. http://link.springer.com/chapter/10.1007/978-3-319-01017-5_24. Accessed 30 Jan 2015

Dronin N, Kirilenko A (2011) Climate change, food stress, and security in Russia. Reg Environ Change 11(Suppl 1):S167–S178. doi:10.1007/s10113-010-0165-x

Eberle J, Clausnitzer S, Hüttich C, Schmullius C (2013) Multi-source data processing middleware for land monitoring within a web-based spatial data infrastructure for Siberia. ISPRS Int J Geo-Inf 2:553–576. doi:10.3390/ijgi2030553

Eijkelkamp (2014) Eijkelkamp Agrisearch Equipment. http://en.eijkelkamp.com/. Accessed 30 Jan 2015

Eremina IG (2009) Alteration of Chernozem soil properties in Khakassia under long-term agricultural use (Изменение свойств черноземов хакасии при длительном ельскохозяйственном использовании) Dissertation Ulan-Ude 2009. http://www.dissercat.com/content/izmenenie-svoistv-chernozemov-khakasii-pri-dlitelnom-selskokhozyaistvennom-ispolzovanii. Accessed 11 Dec 2014

Ermolov YV, Makhatkov ID, Milyaeva EV (2011) Application of conductance measuring method for studying of self-cleaning of a salted peatbog, pp 168–170. In: Vompersky SE (ed) West Siberian peatlands and carbon cycle: past and present. Proceedings of the third international field symposium (Khanty-Mansyisk June 27–July 5, 2011), Novosibirsk, p 220. ISBN 978-5-9611-066-2

Eulenstein F, Tauschke M, Lana M, Sheudshen AK, Dannowski R, Schindler R, Drechsler H (2014) Nutrient balances in agriculture: a basis for the efficiency survey of agricultural groundwater conservation measures. In: Mueller L, Saparov A, Lischeid G (eds) Novel measurement and assessment tools for monitoring and management of land and water resources in agricultural landscapes of Central Asia. Springer International Publishing, pp 263–273. http://link.springer.com/chapter/10.1007/978-3-319-01017-5_15. Accessed 30 Jan 2015

Evans RG (2014) Advanced technologies for irrigated cropping systems. In: Mueller L, Saparov A, Lischeid G (eds) Novel measurement and assessment tools for monitoring and management of land and water resources in agricultural landscapes of Central Asia. Springer International Publishing, pp 475–489. http://link.springer.com/chapter/10.1007/978-3-319-01017-5_29. Accessed 30 Jan 2015

Fedorov A, Konstantinov P (2003) Observations of surface dynamics with thermokarst initiation, Yukechi site, Central Yakutia. In: Permafrost. Phillips, Springman & Arenson (eds) pp 239–243, Swets & Zeitlinger, Lisse. ISBN 90 5809 582 7. http://www.arlis.org/docs/vol1/ICOP/55700698/Pdf/Chapter_043.pdf. Accessed 30 Jan 2015

Funk R, Hoffmann C, Reiche M (2014) Methods for quantifying wind erosion in steppe regions. In: Mueller L, Saparov A, Lischeid G (eds) Novel measurement and assessment tools for monitoring and management of land and water resources in agricultural landscapes of Central Asia. Springer International Publishing, pp 315–327. http://link.springer.com/chapter/10.1007/978-3-319-01017-5_18. Accessed 30 Jan 2015

Gamzikov GP, Nozov V (2010) Role of crop nutrition in narrowing the yield gap for spring wheat in Siberia. Better crops with plant food. A Publication of the International Plant Nutrition Institute (IPNI) XCIV (94), No. 4, p. 9–11

Gessner U, Naeimi V, Klein I, Kuenzer C, Klein D, Dech S (2012) The relationship between precipitation anomalies and satellite-derived vegetation activity in Central Asia. Global and Planetary Change. http://dx.doi.org/10.1016/j.gloplacha.2012.09.007. Accessed 30 Jan 2015

Glagolev M, Kleptsova I, Filippov I, Maksyutov S, Machida T (2011) Regional methane emission from West Siberia mire landscapes. Environ Res Lett 6(4):7. doi:10.1088/1748-9326/6/4/045214. http://iopscience.iop.org/1748-9326/6/4/045214/pdf/1748-9326_6_4_045214.pdf. Accessed 30 Jan 2015

Godbersen L, Utermann J, Duijnisveld HM (2014) Methods in the exploratory risk assessment of trace elements in the soil-groundwater pathway. In: Mueller L, Saparov A, Lischeid G (eds) Novel measurement and assessment tools for monitoring and management of land and water resources in agricultural landscapes of Central Asia. Springer International Publishing, pp 289–314. http://link.springer.com/chapter/10.1007/978-3-319-01017-5_17. Accessed 30 Jan 2015

Golubeva EI, Plyushkyavichyute YuA, Rees WG, Tutubalina OV (2010) Remote sensing methods for phytomass estimation and mapping of tundra vegetation. Geogr Environ Sustain 3(3):4–13. http://www.geogr.msu.ru/GESJournal/contents.php?iid=5&menupos=5. Accessed 30 Jan 2015

Gordov EP, Bryant K, Bulygina ON, Csiszar I, Eberle J, Fritz S, Gerasimov I, Gerlach R, Hese S, Kraxner F, Lammers RB, Leptoukh G, Loboda TV, McCallum I, Obersteiner M, Okladnikov IG, Pan J, Prusevich AA, Razuvaev VyN, Romanov P, Rui H, Schepaschenko D, Schmullius CC, Shen S, Shiklomanov AI, Shulgina TM, Shvidenko AZ, Titov AG (2013) Development of information-computational infrastructure for environmental research in Siberia as a baseline component of the Northern Eurasia earth science partnership initiative (NEESPI) studies. In: Groisman PJ, Gutman G (eds) Regional environmental changes in Siberia and their global consequences (Chap. 2), Springer Environmental Science and Engineering. ISBN: 978-94-007-4568-1 (Print) 978-94-007-4569-8 (Online)

Guggenberger G, Rodionov A, Shibistova O, Grabe M, Kasansky OA, Fuchs H, Mikheyeva N, Zrazhevskaya G, Flessa H (2008) Storage and mobility of black carbon in permafrost soils of the forest tundra ecotone in Northern Siberia. Glob Change Biol 14:1367–1381. doi:10.1111/j.1365-2486.2008.01568.x

Gutman G, Groisman PYa, Gordov EP, Shiklomanov AI, Shiklomanov NI, Shvidenko AZ, Bergen KM, Baklanov AA (2013) Summary and outstanding scientific challenges for research of environmental changes in Siberia. In: Groisman PJ, Gutman G (eds) Regional environmental changes in Siberia and their global consequences (Chap. 9), Springer Environmental Science and Engineering 2013. ISBN: 978-94-007-4568-1 (Print) 978-94-007-4569-8 (Online). http://link.springer.com/chapter/10.1007/978-94-007-4569-8_9. Accessed Jan 30 2015

Helming K, Diehl K, Bach H, Dilly O, König B, Kuhlman T, Perez-Soba M, Sieber S, Tabbush P, Tscherning K, Wascher D, Wiggering H (2011) Ex ante impact assessment of policies affecting land use, Part A: analytical framework. Ecol Soc 16(1):1–17 Art.27

Hennings V (2013) Ackerbauliches Ertragspotential der Böden in Deutschland. http://www.bgr.bund.de/DE/Themen/Boden/Ressourcenbewertung-management/Ertragspotential/Ertragspotential_node.html. Accessed 30 Jan 2015

Hermann A, Schleifer S, Wrbka T (2011) The concept of ecosystem services regarding landscape research: a review. Living Rev Landscape Res 5(1). http://landscaperesearch.livingreviews.org/Articles/lrlr-2011-1/. Accessed 30 Jan 2015

Hertel C, von Unold G (2014) Third-generation lysimeters: scientific engineered monitoring systems. In: Mueller L, Saparov A, Lischeid G (eds) Novel measurement and assessment tools for monitoring and management of land and water resources in agricultural landscapes of Central Asia. Springer International Publishing, pp 175–184. http://link.springer.com/chapter/10.1007/978-3-319-01017-5_9. Accessed 30 Jan 2015

Hese S, Schmullius C (2008) Object oriented oil spill contamination mapping in West Siberia with Quickbird data. Object-based image analysis. Lecture notes in geoinformation and cartography, pp 367–382. http://link.springer.com/chapter/10.1007%2F978-3-540-77058-9_20. Accessed 30 Jan 2015

IAP (2013) The global network of science academies. Russian Academy of Sciences. http://www.interacademies.net/Academies/ByRegion/CentralEasternEurope/13383/13385.aspx. Accessed 30 Jan 2015

Illiger P, Frühauf M, Schmidt G, Meinel T, Belaev VI, Sianteva MM, Kasarjyan M (2014) Ökosystemkonversion und ihre Folgen bezüglich der Kohlenstoffsenken-Funktion in der westsibirischen Kulundasteppe. In: Grunewald K, Bastian O, Drozdov A, Grabovsky V (eds) Erfassung und Bewertung von Ökosystemdienstleistungen (ÖSD): Erfahrungen, insbesondere aus Deutschland und Russland. Bonn: Bundesamt für Naturschutz, (BfN-Skripten; 373), pp 300–319. http://www.bfn.de/fileadmin/MDB/documents/service/Skript_373.pdf. Accessed 23 Dec 2014

Isachenko AG (2013) Geographical aspects of life sustenance of small-numbered peoples of the north. Reg Res Russia 3(2):187–203

Ivanova GA, Conard SG, Kukavskaya EA, McRae DJ (2011) Fire impact on carbon storage in light conifer forests of the Lower Angara region, Siberia. Env Res Lett (6):045203. doi:10.1088/1748-9326/6/4/045203. http://iopscience.iop.org/1748-9326/6/4/045203/. Accessed 30 Jan 2015

Jacob D, Podzun R (1997) Sensitivity studies with the regional climate model REMO. Meteorology and atmospheric physics 63(1–2). International workshop on limited-area and variable resolution models, Beijing, PR China, pp 119–129. ISSN: 0177-7971. doi:10.1007/BF01025368

Juszczak R, Humphreys E, Acosta M, Michalak-Galczewska M, Kayzer D, Olejnik J (2013) Ecosystem respiration in a heterogeneous temperate peatland and its sensitivity to peat temperature and water table depth. Plant Soil 366(1–2):505–520

Kassam AH, Friedrich T, Shaxson F, Reeves T, Pretty J, de Moraes Sa JC (2011). Production systems for sustainable intensification—integrating productivity with ecosystem services.

Technology assessment—theory and praxis, special issue on feeding the world. http://www.fao.org/ag/ca/CA-Publications/TATuP_2011.pdf. Accessed 30 Jan 2015
Kerzencev AS, Meissner R (Eds.) (2006) Modelirovanie erozionnych processov na territorii malogo vodosbornogo bassejna, Nauka Moscow 2006, p 224. (in Russian) ISBN 5-02-03425-X
Kharuk VI, Ranson KJ, Dvinskaya ML, Im ST (2011) Wildfires in northern Siberian larch dominated communities. Env Res Lett (6):045208. doi:10.1088/1748-9326/6/4/045208. http://iopscience.iop.org/1748-9326/6/4/045208. Accessed 30 Jan 2015
Khmelev VA, Tanasienko AA (2009) Chernozem soils of Novosibirsk oblast: problems of their rational use and preservation. Contemp Prob Ecol 2(6):631–641
Khomutova MIu, Beer C, Lucht W, Gerten D, Thonicke K, Schmullius C (2007) Effects of soil freezing and thawing on vegetation carbon density in Siberia: a modeling analysis with the Lund-Potsdam-Jena dynamic global vegetation model (LPJ-DGVM). Global Biogeochem Cycles 21, GB1012, doi:10.1029/2006GB002760
Khvorostyanov DV, Krinner G, Ciais P, Heimann M, Zimov SA (2008) Vulnerability of permafrost carbon to global warming. Part I: model description and role of heat generated by organic matter decomposition. Tellus B 60:250–264. doi:10.1111/j.1600-0889.2007.00333.x
Kirsta YuB (2006) System-analytical modelling-Part II wheat biotime run and yield formation. Agroclimatic potential, the Le Chatelier principle, and changes in agroclimatic potential and climate in Russia and the U.S. Ecol Model 191(3–4):331–345, 5 February 2006. http://www.sciencedirect.com/science/article/pii/S0304380005002802. Accessed 30 Jan 2015
Klein I, Gessner U, Künzer C (2014) Generation of up to date land cover maps for Central Asia. In: Mueller L, Saparov A, Lischeid G (eds) Novel measurement and assessment tools for monitoring and management of land and water resources in agricultural landscapes of Central Asia. Springer International Publishing, pp 329–346. http://link.springer.com/chapter/10.1007/978-3-319-01017-5_19. Accessed 30 Jan 2015
Koneva IV, Batuev AR (2014) The biogeographical map series for Asian Russia. Geogr Nat Res 35(1):102–108. http://link.springer.com/article/10.1134/S1875372814010156. Accessed 30 Jan 2015
Khoroshev AW, Bastian O, Grunewald K (2014) Landschaftsökologische Grundlagen und Prinzipien der Bewertung von Ökosystemdienstleistungen. In: Grunewald K, Bastian O, Drozdov A, Grabovsky V (eds) Erfassung und Bewertung von Ökosystemdienstleistungen (ÖSD): Erfahrungen, insbesondere aus Deutschland und Russland. Bonn: Bundesamt für Naturschutz, (BfN-Skripten; 373), Sp. 64–87. http://www.bfn.de/fileadmin/MDB/documents/service/Skript_373.pdf. Accessed 23 Dec 2014
Korytny LM, Kichigina NV, Gartsman BI, Gubareva TS (2007) Rain floods of the far east and East Siberia extreme hydrological events: new concepts for security NATO science series, vol 78, pp 125–135
Krasnoshchekov YN, Evdokimenko MD, Cherednikova YS (2013) Effect of fires on the ecosystems of subTaiga forest-steppe forests in the southwestern Baikal Region. Contemp Prob Ecol 6(5):477–485. http://link.springer.com/article/10.1134%2FS1995425513050053. Accessed 30 Jan 2015
KULUNDA project (2014) Wie verhindert man die nächste “Global Dust Bowl”?—Ökologische und Ökonomische Strategien zur nachhaltigen Landnutzung in Russischen Steppen: Ein Beitrag zur Anpassung an den Klimawandel. http://www.kulunda.eu/. Accessed 28 Dec 2014
Lantuit H, Atkinson D, Overduin PP, Grigoriev M, Rachold V, Hubberten H-W, Grosse G (2011) Coastal erosion dynamics on the permafrost-dominated Bykovsky Peninsula, north Siberia, 1951–2006. Polar Res 30:7341. doi:10.3402/polar.v30i0.7341
Lapenis A, Shvidenko A, Schepaschenko D et al (2005) Acclimation of Russian forests to recent changes in climate. Global Change Biol 11:2090–2102. doi:10.1111/j.1365-2486.2005.001069.x
Lebedeva MP, Lopukhina OV (2006) Specificity of chemical and mineralogical composition of salts in Sor Solonchaks and Salt Lakes of the Kulunda Steppe, Western Siberia. In: 18th world congress of soil science—Philadelphia, Pennsylvania, USA, 9–15 July 2006. https://a-c-s.confex.com/crops/wc2006/techprogram/P19092.HTM. Accessed 30 Jan 2015

Lischeid G (2014a) Landscape hydrology of rural areas: Challenges and tools. In: Mueller L, Saparov A, Lischeid G (eds) Novel measurement and assessment tools for monitoring and management of land and water resources in agricultural landscapes of Central Asia. Springer International Publishing, pp 107–113. http://link.springer.com/chapter/10.1007/978-3-319-01017-5_5. Accessed 30 Jan 2015

Lischeid G (2014b) Non-linear approaches to assess water and soil quality. In: Mueller L, Saparov A, Lischeid G (eds) Novel measurement and assessment tools for monitoring and management of land and water resources in agricultural landscapes of Central Asia. Springer International Publishing, pp 365–378, http://link.springer.com/chapter/10.1007/978-3-319-01017-5_21. Accessed 30 Jan 2015

Lischeid G, Natkhin M (2011) The potential of land-use change to mitigate water scarcity in Northeast Germany: a review. Erde 142(1–2):97–113

Lopez CML, Brouchkov A, Nakayama H, Takakai F, Fedorov AN, Fukuda M (2007) Epigenetic salt accumulation and water movement in the active layer of central Yakutia in eastern Siberia. Hydrol Process 21:103–109. doi:10.1002/hyp.6224

Mack MC, Finlay JC, Demarco J, Chapin F, Schuur EA, Neff JC, Zimov SA (2010) Nitrogen and phosphorus in Yedoma soils of Northeast Siberia: stocks, fluxes and the ecosystem consequences of nutrient release from permafrost thaw. Am Geophys Union Fall Meet, abstract #GC52A-05. http://adsabs.harvard.edu/abs/2010AGUFMGC52A..05M. Accessed 30 Jan 2015

Mander Ü, Helming, K, Wiggering H (2007) Multifunctional land use: meeting future demands for landscape goods and services. Multifunctional Land Use, pp 1–13. http://link.springer.com/chapter/10.1007/978-3-540-36763-5_1. Accessed 30 Jan 2015

Meinel T, Grunwald L-C, Akshalov K (2014) Modern technologies for soil management and conservation in Northern Kazakhstan. In: Mueller L, Saparov A, Lischeid G (eds) Novel measurement and assessment tools for monitoring and management of land and water resources in agricultural landscapes of Central Asia. Springer International Publishing, pp. 455–464. http://link.springer.com/chapter/10.1007/978-3-319-01017-5_27. Accessed 30 Jan 2015

Meissner R, Rupp H, Seyfarth M (2014) Advanced technologies in lysimetry. In: Mueller L, Saparov A, Lischeid G (eds) Novel measurement and assessment tools for monitoring and management of land and water resources in agricultural landscapes of Central Asia. Springer International Publishing, pp 159–173. http://link.springer.com/chapter/10.1007/978-3-319-01017-5_8. Accessed 30 Jan 2015

Mi Y, van Huissteden J, Parmentier FJW, Gallagher A, Budishchev A, Berridge CT, Dolman AJ (2014) Improving a plot-scale methane emission model and its performance at a northeastern Siberian tundra site. Biogeosciences 11:3985–3999. doi:10.5194/bg-11-3985-2014

Michel R, Dannowski R (2014) Using soil–water–plant models to improve the efficiency of irrigation. In: Mueller L, Saparov A, Lischeid G (eds) Novel measurement and assessment tools for monitoring and management of land and water resources in agricultural landscapes of Central Asia. Springer International Publishing, pp 379–387. http://link.springer.com/chapter/10.1007/978-3-319-01017-5_22. Accessed 30 Jan 2015

Mikheeva I (2013) Use of soil survey database for the probabilistic evaluation of soil cover transformation in the semiarid zone of western Siberia developments in soil classification, land use planning and policy implications, pp 305–313. http://link.springer.com/chapter/10.1007/978-94-007-5332-7_15. Accessed 30 Jan 2015

Morgenstern A, Ulrich M, Günther F, Roessler S, Fedorova IV, Rudaya NA, Wetterich S, Boike J, Schirrmeister L (2013) Evolution of thermokarst in East Siberian ice-rich permafrost: a case study. Geomorphology 201: 363–379. http://dx.doi.org/10.1016/j.geomorph.2013.07.011. Accessed 30 Jan 2015

Moskalenko NG (2013) Impact of climate warming on vegetation cover and permafrost in West Siberia northern Taiga. Nat Sci 5(1A):144–148. http://dx.doi.org/10.4236/ns.2013.51A022. Accessed 11 Dec 2014

Mueller L, Schindler U, Mirschel W, Shepherd TG, Ball B, Helming K, Rogasik J, Eulenstein F, Wiggering H (2010) Assessing the productivity function of soils: a review. Agronomy for Sustainable Development 30(3):601–614 http://dx.doi.org/10.1051/agro/2009057

Mueller L, Schindler U, Shepherd TG, Ball BC, Smolentseva E, Hu C, Hennings V, Schad P, Rogasik J, Zeitz J, Schlindwein SL, Behrendt A, Helming K, Eulenstein F (2012) A framework for assessing agricultural soil quality on a global scale. Arch Agro Soil Sci 58(1):S76–S82

Mueller L, Schindler U, Ball BC, Smolentseva E, Sychev VG, Shepherd TG, Qadir M, Helming K, Behrendt A, Eulenstein F (2014a) Productivity potentials of the global land resource for cropping and grazing. In: Mueller L, Saparov A, Lischeid G (eds) Novel measurement and assessment tools for monitoring and management of land and water resources in agricultural landscapes of Central Asia. Springer International Publishing, pp 115–142. http://link.springer.com/chapter/10.1007/978-3-319-01017-5_6. Accessed 30 Jan 2015

Mueller L, Behrendt A, Shepherd TG, Schindler U, Ball BC, Khudyaev S, Kaiser T, Dannowski R, Eulenstein F (2014b) Simple field methods for measurement and evaluation of grassland quality. In: Mueller L, Saparov A, Lischeid G (eds) Novel measurement and assessment tools for monitoring and management of land and water resources in agricultural landscapes of Central Asia. Springer International Publishing, pp 199–222. http://link.springer.com/chapter/10.1007/978-3-319-01017-5_11. Accessed 30 Jan 2015

Mueller L, Schindler U, Shepherd TG, Ball BC, Smolentseva E, Pachikin K, Hu C, Hennings V, Sheudshen AK, Behrendt A, Eulenstein F, Dannowski R (2014c) The Muencheberg Soil Quality Rating for assessing the quality of global farmland. In: Mueller L, Saparov A, Lischeid G (eds) Novel measurement and assessment tools for monitoring and management of land and water resources in agricultural landscapes of Central Asia. Springer International Publishing, pp 235–248. http://link.springer.com/chapter/10.1007/978-3-319-01017-5_13. Accessed 30 Jan 2015

Nakano T, Wataru T, Inoue G, Fukuda M, Yasuoka Y (2006) Temporal variations in soil–atmosphere methane exchange after fire in a peat swamp forest in West Siberia. Soil Sci Plant Nutr 52:77–88. doi:10.1111/j.1747-0765. 2006.00004.x, http://www.agr.hokudai.ac.jp/env/ctc_siberia/bibliography/pdf/Nakano2006.pdf. Accessed 20 Dec 2014

Nechaeva EG, Belozerntseva IA, Naprasnikova EV, Vorobyeva IB, Davydova ND, Dubynina SS, Vlasova NV (2010) Monitoring and forecasting of the substance-dynamical state of geosystems in the Siberian regions (Мониторинг и прогнозирование вещественно-динамического состояния геосистем сибирских регионов). Novosibirsk, Nauka, p 360. ISBN 978-5-02-023315-7. http://irigs.irk.ru/docs/Nechaeva/Nechaeva_et_al_2010.pdf. Accessed 30 Jan 2015

Nendel C (2014) MONICA: a simulation model for nitrogen and carbon dynamics in agro-ecosystems. In: Mueller L, Saparov A, Lischeid G (eds) Novel measurement and assessment tools for monitoring and management of land and water resources in agricultural landscapes of Central Asia. Springer International Publishing, pp 389–405. http://link.springer.com/chapter/10.1007/978-3-319-01017-5_23. Accessed 30 Jan 2015

Newell J (2004) The Russian far east: a reference guide for conservation and development. McKinleyville, CA: Daniel & Daniel. p 466. http://urbansustainability.snre.umich.edu/wp-content/uploads/2011/04/RFE.04.pdf. Accessed 30 Jan 2015

Ollesch G, Demidov V, Volokitin M, Voskamp M, Abbt-Braun G, Meissner R (2008) Sediment and nutrient dynamics during snowmelt runoff generation in a southern Taiga catchment of Russia. Agric Ecosyst Environ 126(2008):229–242

Olonova MV, Zhang YM (2013) Alien invasive species in Siberia: current status and problem. J Arid Land 5(4):428–433

Osherenko G (1995) Indigenous political and property rights and economic/environmental reform in Northwest Siberia. Post-Soviet Geogr 36(4):225–237

Parham LM, Prokushkin AS, Pokrovsky OS, Titov SV, Grekova E, Shirokova LS, McDowell WH (2013) Permafrost and fire as regulators of stream chemistry in basins of the Central Siberian Plateau. Biogeochemistry 116:55–68. doi:10.1007/s10533-013-9922-5

Pavlov AV (1994) Current changes of climate and permafrost in the Arctic and sub-Arctic of Russia. Permafrost Periglac Process 5:101–110. doi:10.1002/ppp.3430050204

Pavlov AV, Moskalenko NG (2002) The thermal regime of soils in the north of Western Siberia. Permafrost Periglac Process 13:43–51. doi:10.1002/ppp.409

Pokrovsky OS, Viers J, Dupre B, Chabaux F, Gaillardet J, Audry S, Prokushkin AS, Shirokova LS, Kirpotin SN, Lapitsky SA, Shevchenko VP (2012) R. Geoscience 344:663–677

Polterovich V (2013) Reform of the Russian academy of sciences: an expert analysis: part I. Reform of the Russian academy of sciences: a project of the ministry of education and science CEMI RAS. 2013. MPRA Paper No. 49291. Online at: http://mpra.ub.uni-muenchen.de/49291/1/MPRA_paper_49291.pdf/. Accessed 30 Jan 2015

Polyakov IV, Bolton R, Greve R, Hutchings J, Kim SJ (2014) Promoting international, multidisciplinary efforts in detecting and understanding high-latitude changes, and searching for their global impacts. Polar Sci 8(2):53–56. http://www.sciencedirect.com/science/article/pii/S187396521400022X. Accessed 30 Jan 2015

Pomazkina LV, Lubnina EV, Zorina SYu, Kotova LG (1996) Dynamics of CO_2 evolution in grey forest soil of the Baikal forest-steppe. Biol Fertil Soils 23(3):327–331

Prokushkin AS, Pokrovskiy OS, Shirokova LS, Korets MA, Viers J, Prokushkin SG, Amon RMW, Guggenberger G, McDowell WH (2011) Sources and the flux pattern of dissolved carbon in rivers of the Yenisey basin draining the Central Siberian Plateau. Environ Res Let 6, Art. No. 045212

Pugacheva IYu, Sid'ko AF, Shevyrnogov AP (2010) A study of backscattered spectra dynamics of agricultural crops during growth period on the territory of the Krasnoyarskii Krai (Russia). Adv Space Res 45(10):1224–1230. http://dx.doi.org/10.1016/j.asr.2009.11.020. Accessed 30 Jan 2015

Puzyreva AYu (2013) Influence of cultivars and seeding time on yield and grain quality of barley in different agroclimatic zones of the Irkutsk Oblast. (Влияние сорта и сроков посева на урожайность и качество зерна ячменя в разных агроклиматических зонах Иркутской Области) dissertation Ulan-Ude 2013 http://www.bgsha.ru/files/docs/aspirantura/avtoref_puzirevoi.pdf. Accessed 30 Jan 2015

Quegan S, Beer C, Shvidenko A, McCallum I, Handoh IC, Peylin P, Roedenbeck C, Lucht W, Nilson S, Schmullius C (2011) Estimating the carbon balance of central Siberia using a landscape–ecosystem approach, atmospheric inversion and Dynamic Global Vegetation Models. Glob Change Biol 17:351–365. doi:10.1111/j.1365-2486.2010.02275.x

Ragulina MV (2013) Local community and cultural landscape an contemporary world (Локальное сообщество и культурный ландшафт в современном мире). Modern Prob Sci Edu (Современные проблемы науки и образования) 6(50) (Electronic Journal). URL: http://www.science-education.ru/pdf/2013/6/192.pdf. Accessed 30 Jan 2015

Romanenkov V, Rukhovich D, Koroleva P, McCarty JL (2014) Estimating black carbon emissions from agricultural burning. In: Mueller L, Saparov A, Lischeid G (eds) Novel measurement and assessment tools for monitoring and management of land and water resources in agricultural landscapes of Central Asia. Springer International Publishing, pp 347–364. http://link.springer.com/chapter/10.1007/978-3-319-01017-5_20. Accessed 30 Jan 2015

Romanov AN, Khvostov IV, Pavlov VE, Vinokurov YuI (2014) Remote monitoring of wetland areas of Western Siberia using SMOS (ESA) data. J Atmos Oceanic Opt 27(4):313–316. http://link.springer.com/article/10.1134/S1024856014040150. Accessed 30 Jan 2015

Romanovsky VE, Sazonova TS, Balobaev VT, Shender NI, Sergueev DO (2007) Past and recent changes in air and permafrost temperatures in Eastern Siberia. Global Planet Change 56(3–4): 399–413. http://www.sciencedirect.com/science/article/pii/S0921818106001974. Accessed 30 Jan 2015

Rozhkov VA, Alyabina IO, Kolesnikova VM, Molchanov EN, Stolbovoi VS, Shoba SA (2010) Soil-geographical database of Russia Eurasian Soil Sci Vol 43, Issue 1, pp 1–4. http://link.springer.com/article/10.1134/S1064229310010011. Accessed 11 Dec 2014

Rusalimova O, Barsukov P (2006) Decomposition of labile and recalcitrant soil organic matter of Gleyic Cryosols in permafrost region of Siberia. In: Hatano R, Guggenberger G (eds) Symptom of environmental change in Siberian permafrost region, pp 93–102, Hokkaido University Press, Sapporo. http://www.agr.hokudai.ac.jp/env/ctc_siberia/pdf_book/09_Barsukov.pdf. Accessed 28 Dec 2014

Rzaeva VV, Fedotkin VA (2013) Crop yield of a cereal-fallow rotation at different primary tillage (Урожайность культур зернопарового севооборота при возделывании по основной обработке почвы). http://sibac.info/sibac.info/2009-07-01-10-21-16/10112. Accessed 30 Jan 2015

Schaldach R, Priess JA (2008) Integrated Models of the Land System: A review of modelling approaches on the regional to global scale. Living Rev. Landscape Res 2(1). http://landscaperesearch.livingreviews.org/Articles/lrlr-2008-1/. Accessed 30 Jan 2015

Schepaschenko DG, Mukhortova LV, Shvidenko AZ, Vedrova EF (2013) The pool of organic carbon in the soils of Russia. Eurasian Soil Sci 46(2):107–116

Schepaschenko D, McCallum I, Shvidenko A, Fritz S, Kraxner F, Obersteiner M (2011) A new hybrid land cover dataset for Russia: a methodology for integrating statistics, remote sensing and in situ information. J Land Use Sci 6(4):245–259

Schiermeier Q (2013) Russian academy leader speaks out on reforms. Vladimir Fortov highlights the need for a clear separation between science and administration NATURE | NEWS 20. http://www.nature.com/news/russian-academy-leader-speaks-out-on-reforms-1.13791. Accessed 30 Jan 2015

Schindler U (2014a) A novel method for quantifying soil hydraulic properties. In: Mueller L, Saparov A, Lischeid G (eds) Novel measurement and assessment tools for monitoring and management of land and water resources in agricultural landscapes of Central Asia. Springer International Publishing, pp 145–158. http://link.springer.com/chapter/10.1007/978-3-319-01017-5_7. Accessed 30 Jan 2015

Schindler U (2014b) A field method for quantifying deep seepage and solute leaching. In: Mueller L, Saparov A, Lischeid G (eds) Novel measurement and assessment tools for monitoring and management of land and water resources in agricultural landscapes of Central Asia. Springer International Publishing, pp 185–198. http://link.springer.com/chapter/10.1007/978-3-319-01017-5_10. Accessed 30 Jan 2015

Schindler U, Mueller L (2010) Data of hydraulic properties of North East and North Central German soils. Earth Syst Sci Data 2(2):189–194. http://dx.doi.org/doi:10.5194/essd-2-189-2010. http://www.earth-syst-sci-data.net/2/189/2010/essd-2-189-2010.pdf, http://publ.ext.zalf.de/publications/3e0c8223-b0d3-44c9-9477-a6ec2d80f6a7.pdf. Accessed 30 Jan 2015

Schindler U, Müller L, Dannowski R, Barkusky D, Francis G (2010) Long-term measurements to quantify the impact of arable management practices on deep seepage and nitrate leaching. Long-term ecological research: between theory and application. Springer Science + Business Media B.V, Dordrecht, pp 243–252

Schirrmeister L, Froese D, Tumskoy V, Grosse G, Wetterich S (2013) Yedoma: late pleistocene ice-rich syngenetic permafrost of Beringia. In: Elias SA (ed) The encyclopedia of quaternary science, vol 3. Elsevier, Amsterdam, pp 542–552

Semenova T (2007) Political mobilisation of northern indigenous peoples in Russia. Polar Record 43(224):23–32

Semiletov IP, Shakhova NE, Sergienko VI, Pipko II, Dudarev OV (2012) On carbon transport and fate in the East Siberian Arctic land–shelf–atmosphere system. Focus on environmental, socio-economic and climatic changes in northern eurasia and their feedbacks to the global earth system. Environ Res Lett 7:015201. doi:10.1088/1748-9326/7/1/015201

Shaw DJB, Oldfield J (2007) Landscape science: a Russian geographical tradition. Ann Assoc Am Geogr 97(1):111–126. ISSN 0004-5608. http://eprints.gla.ac.uk/6812/. Accessed 30 Jan 2015

Shiklomanov NI, Nelson FE, Streletskiy DA, Hinkel KM, Brown J (2008) The circumpolar active layer monitoring (CALM) program: data collection, management, and dissemination strategies. In: Proceedings of the ninth international conference on permafrost. vol 29. http://permafrost.su/sites/default/files/ShiklomanovNik_final.pdf. Accessed 30 Jan 2015

Shirokova LS, Pokrovsky OS, Kirpotin SN, Desmukh C, Pokrovsky BG, Audry S, Viers J (2013) Biogeochemistry of organic carbon, CO2, CH4, and trace elements in thermokarst water bodies in discontinuous permafrost zones of Western Siberia. Biogeochemistry 113(1–3):573–593. http://link.springer.com/article/10.1007%2Fs10533-012-9790-4. Accessed 30 Jan 2015

Shoba SA, Stolbovoi VS, Alyabina IO, Molchanov EN (2008) Soil-geographic database of Russia. Eurasian Soil Sci 41(9):907–913

Shtro EV (2013) Improving the agrotechnology for brewing barley in the southern forest steppe of West Siberia (Совершенствование агротехнологии пивоваренного ячменя в южной лесостепи Западной Сибири) Dissertation Omsk 2013. http://vak2.ed.gov.ru/catalogue/details/119139?print=true. Accessed 30 Jan 2015

Shvidenko A, Schepaschenko D, McCallum I, Nilsson S (2011) Can the uncertainty of full carbon accounting of forest ecosystems be made acceptable to policymakers? Climatic change, vol 103, pp 137–157. In: White T, Jonas M, Nahorski Z, Nilsson S (eds) Greenhouse gas inventories: dealing with uncertainty. Springer, Netherlands, pp 137–157. doi:10.1007/978-94-007-1670-4_9. http://link.springer.com/chapter/10.1007%2F978-94-007-1670-4_9. Accessed 30 Jan 2015

Siewert C, Kucˇerik J (2014) Practical applications of thermogravimetry in soil science. Part 3: I interrelations between soil components and unifying principles of pedogenesis. J Therm Anal Calorim. doi:10.1007/s10973-014-4256-7

Siewert C, Barsukov P, Demyan S, Babenko A, Lashchinsky N, Smolentseva EN (2014) Teaching soil science and ecology in West Siberia: 17 years of field courses. Environ Edu Res 20 (6):858–876. http://dx.doi.org/10.1080/13504622.2013.839778. Accessed 28 Dec 2014

Skomoroshchenko OV, Korchagina IA, Shchitov AG, Yushkevich LV (2013) Effect of tillage systems and intensification factors on spring wheat yielding capacity in southern forest-steppe of West Siberia, Russian text. http://www.asau.ru/files/vestnik/2013/1/Agronomy_Jushkevich.pdf. Accessed 30 Jan 2015

Smolentseva EN, Smolentsev B, Pachikin K, Mueller L (2014) Assessing the soil quality and crop yield potentials of some soils of Eurasia. In: Mueller L, Saparov A, Lischeid G (eds) Novel measurement and assessment tools for monitoring and management of land and water resources in agricultural landscapes of Asian Russia. Springer International Publishing, Cham, pp 505–517. http://link.springer.com/chapter/10.1007%2F978-3-319-01017-5_31. Accessed 30 Jan 2015

Solntseva NP, Sadov AP (2000) Technogenic halogenesis in the soils of forest-tundra and northern Taiga ecosystems in Western Siberia. J Pochvovedenie (9):1127–1141. ISSN0032-180X Record Number 20013007536 http://www.cabdirect.org/abstracts/20013007536.html. Accessed 30 Jan 2015

Sorokin IB (2011) Bioresources in the intensification of agriculture in Siberia. http://agris.fao.org/agris-search/search.do?f=2011/RU/RU1103.xml;RU2011000177. Accessed 30 Jan 2015

Sorokin ND (2009) Microbiological monitoring of disturbed ground ecosystems of Siberia. Biol Bull 36(6):619–623. http://link.springer.com/article/10.1134/S1062359009060119. Accessed 30 Jan 2015

Stolboboi V (2000) Soils of Russia: Correlated with the Revised Legend of the FAO Soil Map of the World and World Reference Base for Soil Resources. Research Report, RR-00–13 IIASA, Laxenburg, Austria, pp 112

Stolbovoi V (2006) Soil carbon in the forests of Russia. Mitig Adapt Strat Glob Change 11 (1):203–222. http://link.springer.com/article/10.1007%2Fs11027-006-1021-7. Accessed 30 Jan 2015

Stolbovoi V, Fischer G (1997) A new digital georeferenced database of soil degradation in Russia. IIASA Laxenburg, INTERIM REPORT IR-97-084/November. http://citeseerx.ist.psu.edu/viewdoc/download?doi=10.1.1.56.689&rep=rep1&type=pdf. Accessed 30 Jan 2015

Strategy (2010) Strategy of the socio-economical development of Siberia until 2020 (Стратегия социально-экономического развития Сибири до 2020 года. Распоряжение от 5 июля 2010 г. №1120-р. http://www.sibfo.ru/strategia/strdoc.php#strategia. Accessed 30 Jan 2015

Suleimenov M, Kaskarbayev Z, Akshalov K, Yushchenko N (2014) Conservation agriculture for long-term soil productivity. In: Mueller L, Saparov A, Lischeid G (eds) Novel measurement and assessment tools for monitoring and management of land and water resources in agricultural landscapes of Central Asia. Springer International Publishing, pp 441–454. http://link.springer.com/chapter/10.1007/978-3-319-01017-5_26. Accessed 30 Jan 2015

Suzuki K, Kubota J, Ohata T, Vuglinsky V (2006) Influence of snow ablation and frozen ground on spring runoff generation in the Mogot experimental watershed, southern mountainous Taiga of Eastern Siberia. Nordic Hydrol 37(1):21–29 q IWA Publishing

Takakai F, Desyatkin AR, Lopez CML, Fedorov AN, Desyatkin RV, Hatano R (2008) CH_4 and N_2O emissions from a forest-alas ecosystem in the permafrost Taiga forest region, eastern Siberia, Russia. J Geophys Res: Biogeosci 113(G2). doi:10.1029/2007JG000521. http://onlinelibrary.wiley.com/doi/10.1029/2007JG000521/abstract. Accessed 30 Jan 2015

Tanasienko AA, Chumbaev AS (2010) Conditions of the formation of ice barriers in eroded chernozems of Western Siberia. Eurasian Soil Sci 43(4):417–426

Tandelov YuP (2012) Fertility of acid soils on agricultural lands of the Krasnoyarsk Krai (Плодородие кислых почв земледельческой территории Красноярского края) Красноярск 2012 http://tandelov-agro.narod.ru/olderfiles/1/JU.P.Tandelov_Plodorodie_kislyh_po-89447.pdf. Accessed 30 Jan 2015

Tarnocai C, Canadell JG, Schuur, EAG, Kuhry P, Mazhitova G, Zimov S (2009) Soil organic carbon pools in the northern circumpolar permafrost region. Global Biogeochem Cycles 23, GB2023. doi:10.1029/2008GB003327

Tchebakova NM, Rehfeldt GE, Parfenova EI (2006) Impacts of climate change on the distribution of *Larix spp.* and *Pinus sylvestris*. Mitig Adapt Strat Global Change 11(4):861–882. http://link.springer.com/article/10.1007%2Fs11027-005-9019-0. Accessed 30 Jan 2015

Tchebakova NM, Parfenova EI, Lysanova GI, Soja AJ (2011) Agroclimatic potential across central Siberia in an altered twenty-first century. Environ Res Lett 6 (Oct–Dec 2011) 045207. doi:10.1088/1748-9326/6/4/045207

Titlyanova AA (2008) Functioning and evolution of biogeocenoses in the Baraba lowland, West Siberia. Contemp Prob Ecol 1(6):629–632. http://link.springer.com/article/10.1134/S1995425508060027. Accessed 30 30

Titlyanova AA, Kudryashova SYa, Yakutin MV, Bulavko GI, Mironycheva-Tokareva NP (2001) The reserves of carbon in vegetative and microbial biomass of siberian ecosystems. Eurasian Soil Sci 34(8):837–849

Trubacheva NV, Rosseeva LP, Belan IA, Osadchaia TS, Kravtsova LA, Kolmakov IuV, Blokhina NP, Pershina LA (2011) Characteristics of common wheat cultivars of West Siberia carrying the wheat-rye 1RS.1BL translocation. Genetika 47(1):18–24 Engl. abstract at http://www.ncbi.nlm.nih.gov/pubmed/21446180. Accessed 30 Jan 2015

UEAA (2006) Union of european academies for science applied to agriculture. Food Nat http://www.euracadagri.com/eng/members/russia.php. Accessed 30 Jan 2015

Urban M, Hese S, Herold M, Pöcking S, Schmullius C (2010) Pan-Arctic land cover mapping and fire assessment for the ESA data user element, PFG Photogrammetrie, Fernerkundung und Geoinformation, Special Issue, PFG 2010 /4, 283–293

Vakhtin NB (1998) Indigenous people of the Russian Far North: land rights and the environment. Polar Geogr 22(2):79–104

Vaks A, Gutareva OS, Breitenbach SFM, Avirmed E, Mason A, Thomas AL, Osinzev AV, Kononov AM, Henderson GM (2013) Speleothems reveal 500,000-year history of siberian permafrost, Published Online February 21 2013. Science 12 340(6129):183–186. doi:10.1126/science.1228729

Van Huylenbroeck G, Vandermeulen V, Mettepenningen E, Verspecht (2007) Multifunctionality of agriculture: a review of definitions, evidence and instruments. Living Rev Landscape Res http://landscaperesearch.livingreviews.org/Articles/lrlr-2007-3/. Accessed 30 Jan 2015

Vasileva A, Moiseenko K (2013) Methane emissions from 2000 to 2011 wildfires in Northeast Eurasia estimated with MODIS burned area data. Atmos Environ 71:115–121

Vasil'evskaya VD, Grigor'ev VY, Pogozhev EY, Pogozhev EA (2012) Procedure for calculation of chemical indicators of ecological danger of tundra soil degradation and their validation (Review). Contemp Prob Ecol 5(5):481–496

Venevskaia I, Venevsky S, Thomas CD (2013) Projected latitudinal and regional changes in vascular plant diversity through climate change: short-term gains and longer-term losses. Biodiv Conser 22(6–7):1467–1483

Vorobyeva IB (2012) Changes in the Southern Siberian Forest-Steppes. Eurasian Steppes. Ecological Problems and Livelihoods in a Changing World. Plant and Vegetation vol 6, pp 425–443

Voswinkel J (2014) Das Ziel ist die Weltspitze. http://www.helmholtz.de/artikel/das-ziel-ist-die-weltspitze-2512/. Accessed 30 Jan 2015

Wenkel K-O, Berg M, Mirschel W, Wieland R, Nendel C, Köstner B (2013) LandCaRe DSS: an interactive decision support system for climate change impact assessment and the analysis of potential agricultural land use adaptation strategies. J Environ Manage 127(Supplement):S168–S183

Wiggering H, Dalchow C, Glemnitz M, Helming K, Müller K, Schultz A, Stachow U, Zander P (2006) Indicators for multifunctional land use: linking socio-economic requirements with landscape potentials. Ecol Ind 6(1):238–249

WRB (2006) World reference base for soil resources 2006, a framework for international classification, correlation and communication, FAO Rome, World soil resources reports 103, p 145

WRB (2014) World reference base for soil resources 2014, world soil resources report no 106, p 181. ISBN 978-92-5-108369-7, http://www.fao.org/3/a-i3794e.pdf. Accessed 14 Nov 2014

Yablokov A (2014) Academy 'reform' is stifling Russian science. Nature 511, 7 (03 July 2014) doi:10.1038/511007a. http://www.nature.com/news/academy-reform-is-stifling-russian-science-1.15486. Accessed 30 Jan 2015

Yakovleva N (2011) Oil pipeline construction in Eastern Siberia: Implications for indigenous people. Geoforum 42(2011):708–719

Zakharova EA, Kouraev AV, Rémy F, Zemtsov VA, Sergey N. Kirpotin SN (2014) Seasonal variability of the Western Siberia wetlands from satellite radar altimetry. J Hydrol 512(6):366–378. http://www.sciencedirect.com/science/article/pii/S0022169414001747. Accessed 30 Jan 2015

Zech R, Huang Y, Zech M, Tarozo RA, Zech W (2011) High carbon sequestration in Siberian permafrost loess-paleosols during glacials. Clim Past 7(2):501–509

Zech W, Schad P, Hintermaier-Erhard G (2014) Böden der Welt. 2nd edit. Springer 2014, p 164. http://www.springer.com/springer+spektrum/geowissenschaften/bodenkunde/book/978-3-642-36574-4. Accessed 30 Jan 2015

Zimov SA, Chuprynin VI, Oreshko AP, Chapin FS III, Reynolds JF, Chapin MC (1995) Steppe-Tundra transition: a herbivore-driven biome shift at the end of the Pleistocene. Am Nat 146(5):765–794

Zolnikov ID, Glushkova NV, Lyamina VA, Smolentseva EN, Korolyuk AYu, Bezuglova NN, Zinchenko GS, Puzanov AV (2011) Indication of the dynamics of natural-territorial complexes in southern Western Siberia in connection with climate changes. Geogr Nat Res 2:155–160

Part II
Methods and Case Studies for Understanding and Monitoring the Landscapes of Siberia

Chapter 3
Methods for Monitoring the Chemical Composition of Lake Baikal Water

Tamara Khodzher, Valentina M. Domysheva, Larisa M. Sorokovikova and Liudmila P. Golobokova

Abstract In the early 1990s, a system of comprehensive monitoring, including hydrophysical, chemical and biological investigations, was developed at the Limnological Institute of the Siberian Branch of the Russian Academy of Sciences to assess the current environmental state of Lake Baikal. Chemical monitoring of the Baikal water includes checking the dynamics of chemical components in the pelagic and littoral areas of the lake, as well as their input from the atmosphere and water from the lake's tributaries. The monitoring system allows scientists to assess historical trends of the chemical components in the lake and forecast possible changes in the biota habitat. The low concentrations of most components in the Baikal water initiated the development of more sensitive methods. New methods for analysing anions and persistent organic pollutants (PAHs and PCBs) were elaborated at the Institute and certified by the State Standard Committee of the Russian Federation. These methods feature high sensitivity, selectivity and fast analysis and are widely applied when monitoring the chemical composition of different environments in the Baikal region. The reliability of methods and quality of analytical analyses are checked annually according to international and Russian programmes on inter-laboratory calibration. The results do not deviate from the reference standard samples by more than 10 %, which attests to the reliability of new methods for chemically monitoring the Baikal water. Based on long-term data with the application of high-precision methods, we were able to assess the current chemical composition of the Baikal water. Present concentrations of pollutants in Lake Baikal, such as persistent organic pollutants (POPs) and heavy metals, are low and

T. Khodzher (✉) · V.M. Domysheva · L.M. Sorokovikova · L.P. Golobokova
Siberian Branch, Limnological Institute (LIN), Russian Academy of Sciences, 3, Ulan-Batorskaya Street, Irkutsk 664033, Russia
e-mail: khodzher@lin.irk.ru

V.M. Domysheva
e-mail: hydrochem@lin.irk.ru

L.M. Sorokovikova
e-mail: lara@lin.irk.ru

L.P. Golobokova
e-mail: lg@lin.irk.ru

L. Mueller et al. (eds.), *Novel Methods for Monitoring and Managing Land and Water Resources in Siberia*, Springer Water,
DOI 10.1007/978-3-319-24409-9_3

do not directly affect the composition of the water and biota. Moreover, the ecosystem of Lake Baikal is self-purifying. The water of Lake Baikal in its deep area is one of the purest natural waters in the world and can be used for drinking and other purposes. It is recommended to include water quality tests in the regular monitoring system, in particular checking the littoral zone where the coast is highly developed and near the mouths of the lake large tributaries, analysing such parameters as sanitary and microbiological characteristics, nutrients and biota. It is also necessary to regularly monitor the concentration of persistent organic compounds in the air, and in the lake water, bottom sediments and biota. Many of these compounds are mutagenic and carcinogenic, and they are also able to accumulate in the food chains and transfer from one organism to another.

Keywords Monitoring · Lake Baikal · Methods of analysis · Chemical composition · Water · Atmosphere · Trace gases

1 Introduction

The aim of this work is to estimate the modern chemical water composition of Lake Baikal and its tributaries based on long-term data using high-precision methods for analysis, as well as to assess possible changes associated with climate warming and the anthropogenic effect.

Lake Baikal is the world's largest freshwater reservoir (23,000 km^3), with unique flora and fauna, 60 % of them being endemic species. It is a UNESCO World Heritage Site and its significance has been set down in the Federal Law of the Russian Federation "On the Protection of Lake Baikal". Lake Baikal consists of three basins differing in depth, temperature conditions, water exchange and volume of water masses (Shimaraev et al. 1996; Hohmann et al. 1997).

Hydrochemical investigations of Lake Baikal and its tributaries started in 1925 by researchers of Baikal Limnological Station (Vereshchagin 1927). At the IV International Limnological Congress in Rome in 1927, Vereshchagin (1933) reported the results of studies on the hydrochemical characteristics of the water in the deep part of Southern Baikal.

Votintsev continued to study the hydrochemical composition of Baikal water in 1948 (Votintsev 1961; Votintsev and Glazunov 1963; Votintsev et al. 1975). As a result of these long-term investigations, the chemical balance of the lake was estimated.

A system of comprehensive monitoring, including hydrophysical, chemical and biological studies, was developed in the early 1990s by researchers at the Limnological Institute to assess the current ecological state of Lake Baikal. Complex chemical monitoring of Lake Baikal is based on the long-term experience of researchers from other countries in the field of water resource conservation (Baig et al. 2009; Najar and Khan 2012; O'Neil et al. 2012).

The chemical composition of trace gases, aerosol, precipitation and snow cover is monitored in the atmosphere above the lake and adjacent territory. Monitoring observations are performed in the mouth areas of the lake tributaries. In addition, processes of gas exchange and geochemical processes are studied at the "water–atmosphere" and "water–bottom" interfaces, respectively. The chemical composition of pore waters is also analysed.

Chemical monitoring is based on the following main principles:

- using high-precision methods for chemical analysis to measure major ions, nutrients and organic compounds;
- analysing long-term trends in changes to the chemical composition of surface and deep waters in the open part of Lake Baikal, atmospheric precipitation and waters of the lake tributaries;
- forecasting water quality under conditions of climate change and anthropogenic effects.

The monitoring system allows scientists to assess historical trends of the chemical components in the lake and forecast possible changes in the habitat of biota.

2 Materials and Methods

The chemical composition of lacustrine and riverine waters is monitored at the standard sites (Figs. 1 and 2).

The water and atmospheric air is sampled according to State Standard Specification (SSS) 17.1.5.05-85, RD 52.04.186-89, SSS R 51945-2002, SSS R ISO 12884-2007 and SSS 31861-2012. Baikal water from different depths is sampled on board a scientific research vessel with a special Rosette water sampler containing 24 water bottles. In the rivers, water is sampled from the surface with Nansen bottles. The following parameters are measured: pH, dissolved oxygen, conductivity, sulphate, chloride, bicarbonate, magnesium, calcium, sodium, potassium, nitrogen (nitrate, nitrite and ammonium), phosphorus (phosphate and total), silicon and organic matter. Trace elements, oil products and persistent organic pollutants (POPs) are analysed every 3–5 years. Those components whose retention time is several hours are analysed at the sampling site in the field. A portion of the sample is conserved following the analysis technique and transported to the Institute laboratory in a cooling box.

The low concentration of some components in the water of Lake Baikal requires more sensitive methods for analysis. Since the 1990s, the analytical laboratory at the Limnological Institute has elaborated and certified novel methods for determining ionic composition and POPs. These novel methods have been introduced into the monitoring system after their comparison with the traditional classical methods.

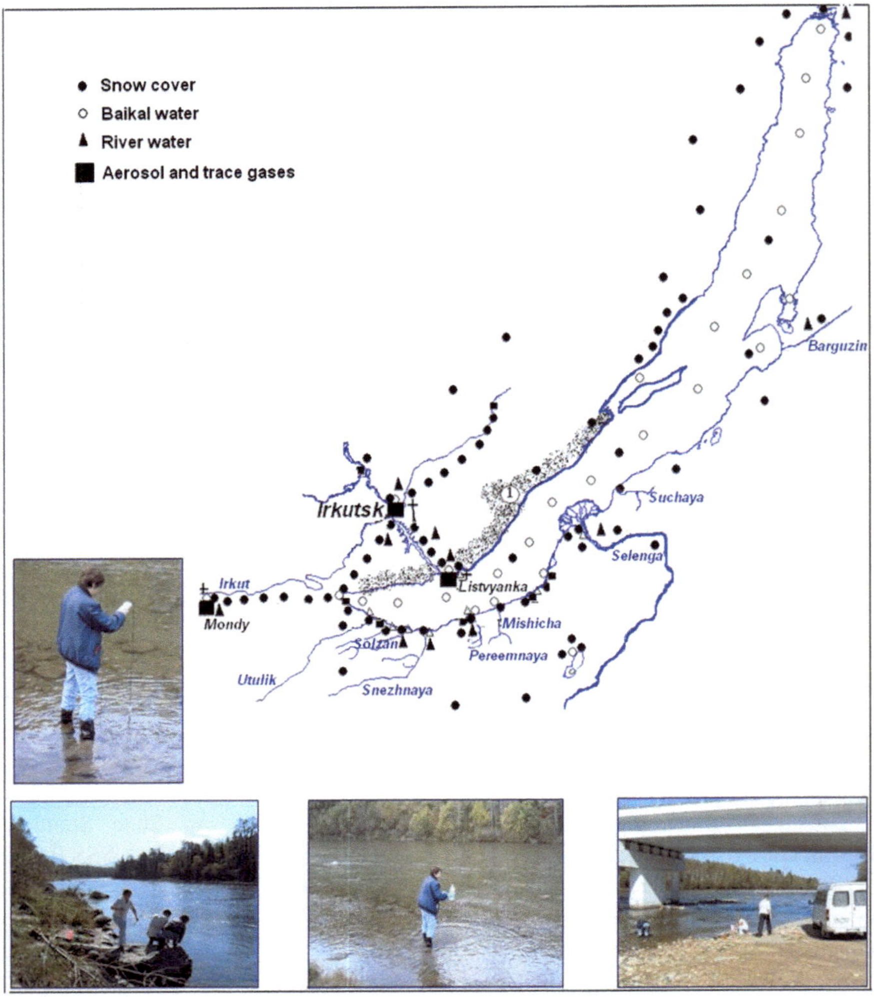

Fig. 1 Sampling scheme of water, aerosol, trace gases, precipitation and snow cover in Lake Baikal and in the Baikal region

2.1 *Methods for Determining Ionic Composition Using High-Performance Liquid Chromatography (HPLC)*

To perform traditional analytical methods, a large volume of a sample ($\sim$300 ml) is needed to determine the anion concentrations. The sensitivity of the method has high lower limits: the determination error is 20 % for SO_4^{2-}, 25–50 % for NO_{3-} and 50 % for Cl^- (Alekin et al. 1973; Fomin 2010). Small amounts of the substance in some objects of studies (precipitation, aerosol and pore waters) and low concentrations of certain components demand strict requirements regarding analytical

Fig. 2 Scientific research vessel "Vereshchagin", water sampling with a "Rosette" sampler

precision. Taking into consideration these peculiarities, the researchers at the Limnological Institute elaborated novel methods for analysing anion composition using HPLC. These methods were certified (Baram et al. 1999). The principle of this method is that a chromatographic column filled with inverse phase sorbent is saturated with molecules of quaternary amine salt containing a long-chain alkyl radical. The lipophilic part of the molecule of quaternary amine allows the modifier to firmly remain on the inverse phase when interacting with anions. As the analysed sample is injected into the prepared column, a counterion of the quaternary amine is substituted for anions of the sample, which are bound with the stationary phase. Anions deposited on the modified sorbent are eluted using a solvent that contains an absorbing anion in the UV area. The eluent, which contains a decreased concentration of the UV-absorptive addition substituted for a certain anion, is then recorded as it moves along the chromatographic column. The detection of the outgoing eluent in the UV area causes the emergence of a negative signal during the appearance of such zones. Chromatograms are recorded on a "Milichrom A-02" microcolumn liquid chromatograph (EcoNova, Novosibirsk, Russia) with a short, low-capacity inverse-phase column (2 × 75 mm, $V \approx 0.2$ ml) and UV photometric detection (Fig. 3).

This method is used to analyse the mass concentration of sulphate, chloride and bicarbonate within the range of 5–100 mg/l. The mass concentrations of anions are measured from peaks on chromatograms of the analysed samples. The extended relative measurement uncertainty ($k = 2$) is 7–10 %. The technique elaborated

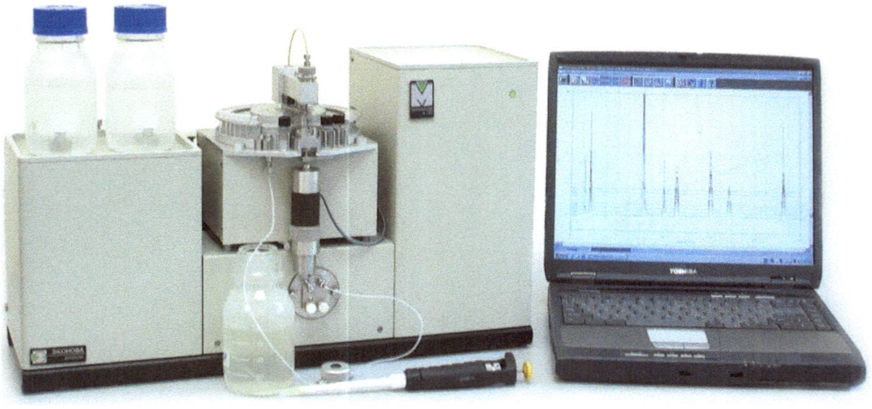

Fig. 3 "Milichrom A-02" chromatograph (Novosibirsk, Russia)

was modified and certified for solutions with low concentrations of Cl^- and SO_4^{2-} (Fig. 4).

The method's wide use in analysing different objects was due to the possibility of identifying a large number of inorganic and organic ions in a small amount of a sample, high sensitivity without pre-concentration, selectivity and fast analysis. This method has been successfully applied for the last decade to the analysis of ionic compositions of aerosol and natural waters of the Baikal region (Khodzher et al. 1999; Fukuzaki et al. 2002; Grachev et al. 2004; Domysheva et al. 2010a, b; Golobokova et al. 2011; Khodzher et al. 2011).

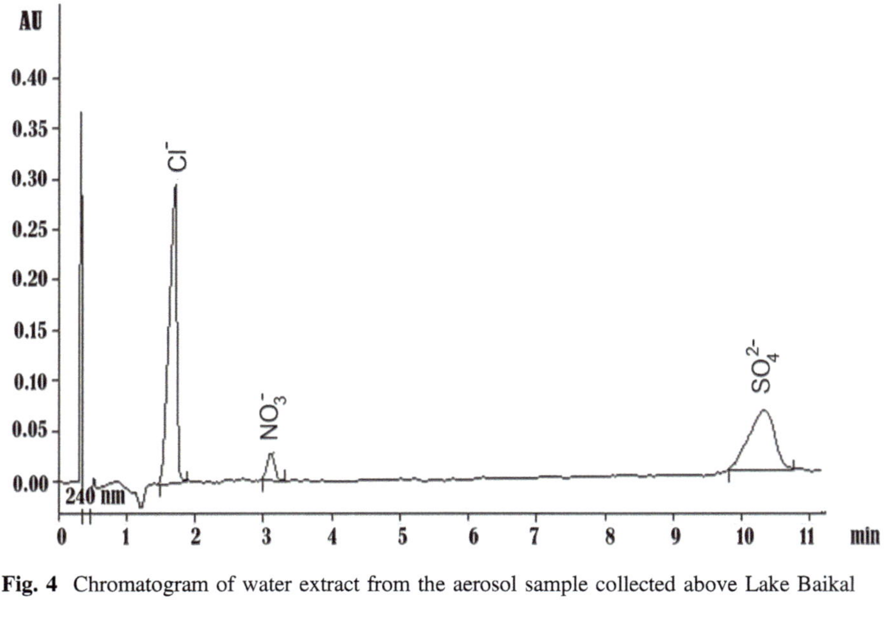

Fig. 4 Chromatogram of water extract from the aerosol sample collected above Lake Baikal

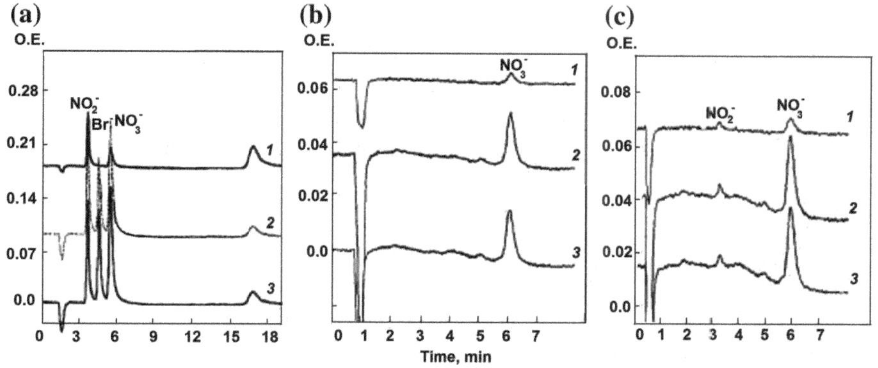

Fig. 5 Chromatograms of model mixture of anions (**a**), Baikal water sample (**b**) and extract from aerosol sample (**c**) at 226 (1), 204 (2) and 202 (3) nm. The sample volume is 10 μl. The anion concentration in the model mixture is 2.5 mg/l

Concentrations of nitrate and nitrite ions in the Baikal water and atmospheric precipitation range from a thousandth to a tenth of a milligram per litre. The traditional methods only allow nitrate and nitrite ions to be measured up to 0.2 and 0.01 mg/l, and are associated with different interference factors (Alekin et al. 1973; Fomin 2010).

To determine small concentrations of NO_{3-} (lower than 50 μg/l), a technique for measuring the mass concentrations of nitrate, nitrite, bromide and iodide anions within the range of 0.005–10 mg/l was developed based on HPLC with UV detection on an inverse-phase column modified with octadecyltrimethylammonium bromide using direct detection to register anions absorbing UV radiation (Vereshchagin et al. 2000). Mass concentrations of anions are measured from the peaks on the chromatograms of the analysed samples. The extended relative uncertainty of measurements ($k = 2$) is 15–25 % (FR.1.31.2008.04415) (Fig. 5).

This technique was successfully tested when analysing the anion composition of a wide spectrum of the environmental objects and atmospheric aerosol (Khodzher et al. 1999, 2004; Sorokovikova et al. 2000, 2004; Grachev et al. 2004; Netsvetaeva et al. 2004).

2.2 *Methods for Determining Persistent Organic Pollutants in Different Environments of the Baikal Region*

Methods for measuring mass concentrations of benzapyrene within the range of 10–1000 ng/dm^3 in water and atmospheric aerosol were developed using HPLC. The principle of this method is that the concentrate is divided on a "Milichrom A-02" chromatograph after polycyclic aromatic hydrocarbons (PAHs) with hexane are extracted from the sample and the extract is pre-concentrated on a rotor evaporator.

Using these methods, researchers analysed the distribution of benzapyrene and other PAH compounds in the atmospheric air and snow cover in the industrial and background areas of Pribaikalye, and in the lake water and biota (Gorshkov et al. 1998; Koroleva et al. 1998; Gorshkov and Marinayte 2000; Marinayte and Gorshkov 2002; Gorshkov et al. 2003; Belykh et al. 2006).

A method for measuring mass concentrations of polychlorinated biphenyls (PCBs) was developed using high-speed capillary gas chromatography with mass spectrometric detection of selected ions (Nikonova and Gorshkov 2012). PCBs are analysed in such natural objects as soil, suspended particles of snow water, Lake Baikal particulates and bottom sediments, tissues of the Baikal omul (*Coregonus migratorius,* Georgi 1775) and the fat of Baikal seals (*Phoca sibirica* Gm.). PCBs are extracted from natural samples and hydrolysates of the biological material, then extracts are purified on compact cartridges with silica gel or florisil (0.5 g of sorbent). The intra-laboratory precision when determining all the PCBs and groups with equal level of chlorination does not exceed 10 %, whereas in indicator compounds (PCBs 28, 52, 101, 118, 138, 153 and 180) it is 15 %.

2.3 Classical Methods of Analysis

Classical conventional methods of analysis are used to measure the pH, dissolved oxygen, conductivity, magnesium, calcium, sodium, potassium, ammonium nitrogen, phosphate, total phosphorus, silicon and organic matter (Wetzel and Likens 1991).

Since the mid-1990s, researchers at the Limnological Institute have analysed trace elements in the water of Lake Baikal and its tributaries, bottom sediments and atmospheric aerosol using inductively coupled plasma mass spectrometry (ICP-MS) on an Agilent 7500ce and standard samples.

2.4 Checking the Quality of Analyses Performed at Lake Baikal Within the Framework of the Monitoring Programme

The high quality of the quantitative chemical analysis of natural objects is an integral part of chemically monitoring the state of the environment and its pollution.

To perform monitoring studies at Lake Baikal, its tributaries and atmosphere, the certified Laboratory of Hydrochemistry and Atmosphere Chemistry (Certificate ROSS RU.0001.513855) developed a system of quality management based on international (ISO 9001:2000, ISO/IEG 17025:2005) and Russian (SSS R 17025-2009, SSS R ISO 9001-2008, SSS R ISO 9000-2008) standards. Since 2000, the quality of the analysis results obtained by the laboratory has been regularly controlled within the framework of two international programmes for testing "reference standard samples", artificial "acid rains" and "surface waters" (Khodzher et al. 2002, 2004). One of the programmes is carried out within the framework of Global Atmospheric Watch (GAW) under the aegis of the World Meteorological Organisation (WMO). Over 70 chemical laboratories from different countries participate in this programme (http://www.qasac-americas.org). Standard samples are sent for analysis to the Limnological Institute laboratory twice a year. The major ions, ammonium, pH value and conductivity are analysed. Another test is performed within the framework of the international programme "Acid Deposition Monitoring Network in East Asia—EANET" (http://www.acap.asia/~interlab/os/). The participants of this programme represent 23 laboratories from 13 countries of East Asia. The analysis quality is checked once a year with measurements of the major ions, nutrients, pH value and conductivity. Artificial atmospheric precipitation, surface water, soil and aerosol are used as standard samples. To confirm the quality of the results obtained, the laboratory was involved into two international quality control programmes (QA/QC Programmes). The International European Programme of Estimation and Monitoring of Acidification of Rivers and Lakes (EMEP) (http://www.nilu.no) estimates the quality of measurements of pH, specific conductivity, bicarbonate, nitrate, chloride, sulphate, calcium, magnesium, sodium, and potassium in atmospheric precipitation. Another international programme assesses the quality of measurements of pH, specific conductivity, bicarbonate, nitrate, chloride, sulphate, calcium, magnesium, sodium, potassium, total organic matter and trace elements in natural waters (http://kvina.niva.no/intercomparison2).

The analysis results obtained in the laboratory do not deviate from the "reference standard samples" within these programmes by more than 10 %, which attests to the confidence level of the results.

3 Results and Discussion

3.1 Chemical Composition of Baikal Water: Major Ions, Nutrients and Gas Composition

Lake Baikal is an inertial system: it needs approximately 400 years to completely substitute its waters for waters from its tributaries. The integral influence of certain mechanisms of exchange causes the annual partial renewal of deep Baikal waters. The age of the water at a depth of over 250 m changes from 7.2 to 11.1 years in different years and in different parts of the lake. The age is determined with the help of chemical tracers, such as dissolved atmospheric freon and tritium–helium (Weiss et al. 1991; Killworth et al. 1996; Peeters et al. 1997; Hohmann et al. 1997; Kodenev et al. 1998).

Lake Baikal is divided into southern, central and northern basins by the bottom elevations with depths of 1400, 1600 and 800 m, respectively. The littoral area occupies about 7 % of the surface water area (Fialkov 1983). However, it greatly affects the functioning of the lake ecosystem. To estimate the water quality in Lake Baikal under conditions of climatic change and active exploration of the shores, we started monitoring the chemical composition of the water in the pelagic area in 1993 and in the littoral zone in 2003.

Fundamental scientific investigations of the currents for the last 20 years (more than 250 expeditions with the participation of foreign scientists) have demonstrated the stability of the chemical composition of the water in the pelagic area of Lake Baikal due to the huge volume of water masses compared with the annual water flow (approx. 60 km^3) and intensive water exchange in the lake (Falkner et al. 1991; Grachev 2002; Grachev et al. 2004).

The water in the pelagic area of the lake has not undergone significant changes in comparison with the pre-industrial period (the 1950s–1960s) except the water in bays, shallow areas adjacent to the settlements and at the mouths of the lake's large tributaries. The water in the lake is low-mineralised (with total ions of approximately 96 mg/l), belonging to the bicarbonate class of the calcium group (Table 1).

High concentrations of oxygen (9–14 mg/l) at all depths including the maximum depth (1637 m) attest to the unique renewal mechanisms of the deep waters in spring and autumn (Shimaraev et al. 1996; Shimaraev and Domysheva 2013).

Concentrations of nutrients (nitrogen, phosphorus and silicon) are subject to seasonal fluctuations in the surface and near-bottom water layers of Lake Baikal due to their consumption and regeneration by biota (Weiss et al. 1991; Domysheva 2009).

The concentration of nutrients, which serves as one of the markers of water quality, is not high. It increases from the surface to the near-bottom layer and does not exceed 1.9 mg Si/l for silicon, 0.14 mg N/l for nitrate and 0.020 μg P/l for phosphorus. The results of studies show that since the 1950s no significant changes in the content and annual dynamics of nutrients have been recorded in the pelagic area of the lake. Fluctuations in phosphate and nitrate concentrations do not exceed the error of the method at certain horizons in the water column (Fig. 6).

Table 1 Mean concentrations of major ions in the water of Lake Baikal (mg/l, 2005–2012)

Ion	Southern Baikal	Central Baikal	Northern Baikal	Entire lake
HCO_3^-	66.3±1.8	66.7±1.5	65.7±1.4	66.3±1.6
SO_4^{2-}	5.2±0.1	5.2±0.2	5.2±0.1	5.2±0.1
Cl^-	0.4±0.02	0.4±0.03	0.4±0.02	0.4±0.03
Ca^{2+}	16.4±0.4	16.5±0.4	16.3±0.3	16.4±0.4
Mg^{2+}	3.0±0.1	3.0±0.1	3.0±0.1	3.0±0.1
Na^+	3.3±0.1	3.4±0.14	3.31±0.11	3.3±0.1
K^+	1.0±0.1	1.0±0.1	1.0±0.1	1.0±0.1

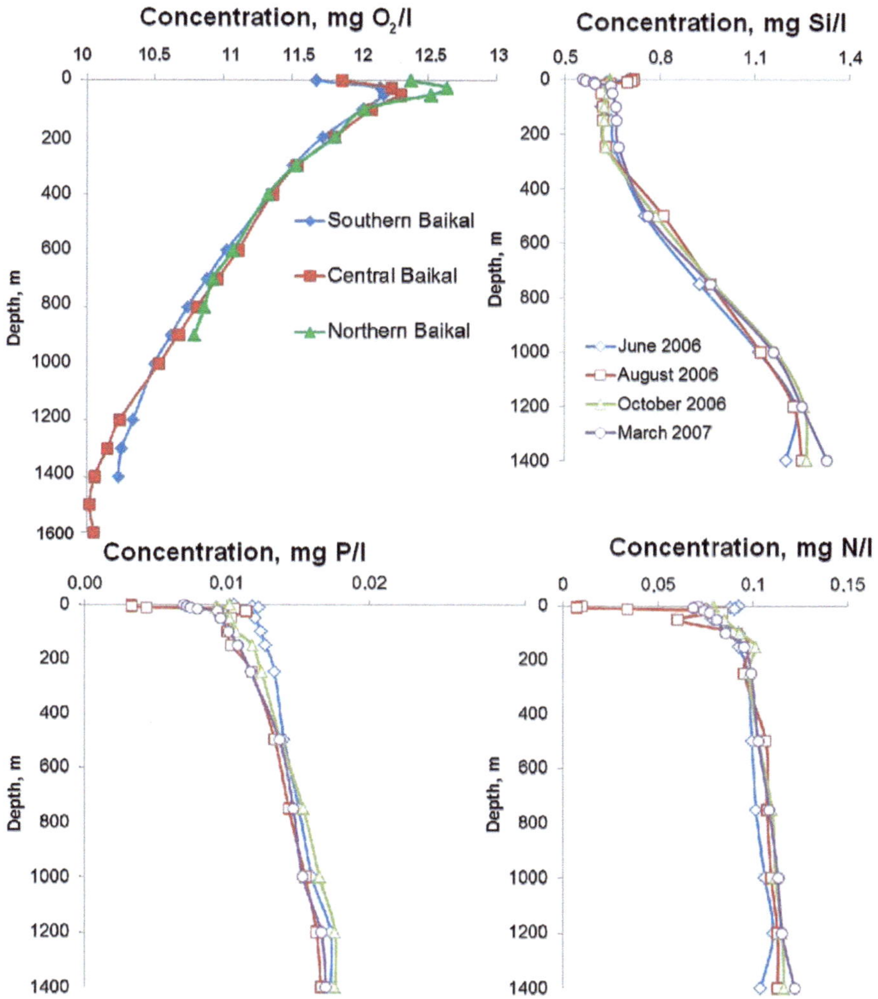

Fig. 6 Vertical profiles of nitrate, phosphate and oxygen concentration in the water of Southern Baikal in different years (1957–1958—data by Votintsev (1961); 1967–1968 and 1986–1989—data by Tarasova and Meshcheryakova (1992); 2000–2010—data by Domysheva (2009))

The analysis of phytoplankton, as one of the important chains of nutrient consumption and a means of preserving the stability of the lake ecosystem, shows that there is no eutrophication effect in the pelagic part of the lake except some areas of its littoral zone (Popovskaya 1991; Pomazkina et al. 2010).

The problem of the coastal zone pollution with litter in Lake Baikal is getting more urgent due to the development of tourism and the lake shores. Local pollution of the surface waters near the settlements, ports, in bays and the mouths of large tributaries was reported from the concentrations of nutrients (nitrogen and phosphorus), sanitary microbiological parameters and the development of certain groups of algae unusual for Lake Baikal. In July–August of 2011, high concentration of nutrients (phosphates up to 0.42 μg P/l and nitrate up to 0.20 mg N/l) were recorded in the popular tourist area (Listvyanka, Southern Baikal) along the shoreline for 4 km, whereas the background values for these nutrients are 0.007 μg P/l and 0.01 mg N/l. The belt structure of phytobenthos was disturbed: the mass development of the filamentous algae *Spirogya* within the depth range of 2.5–10 m and the disappearance of *Didymospheniageminate* colonies, which are common for the littoral area of the lake, were recorded (Kravtsova et al. 2012, 2014).

3.2 Persistent Organic Pollutants (POPs) in the Water and Biota of Lake Baikal

3.2.1 Polychlorinated Biphenyls (PCBs)

The first results on PCB determination in the surface waters and bottom sediments of Lake Baikal and in the soils of the Baikal region were obtained during international expeditions in the 1990s (Iwata et al. 1995; Kucklick et al. 1996) and were systematised in the book by Grachev (2002). Concentrations of PCBs in the waters of Lake Baikal are 130–1900 pg/l (Nikonova and Groshkov 2010). These values are comparable with those in Lakes Superior and Huron (Great Lakes of North America) and in the water bodies of some countries in South-East Asia (Grachev 2002).

The first data on the concentration of PCBs in living organisms of Lake Baikal were presented by Bobovnikov et al. (1985), Malakhov et al. (1986): 35 ng/g in zooplankton; 25 ng/g of wet weight in Baikal omul and 6600 ng/g of lipids in Baikal seals. Based on these data, the PCB content was regarded as a background value in the lake ecosystem for that period.

According to recent data, the levels of PCB accumulation in the Baikal zooplankton were as follows: 4–8 ng/g of wet mass (300–450 ng/g of lipids) in *Epischura baicalensis* and 8–14 ng/g of wet mass (300–700 ng/g of lipids) in *Macrohectopus branickii*. These values correspond to those of PCB accumulation levels in the background areas of the world (Nikonova and Gorshkov 2011). A Baikal seal *Phoca sibirica* accumulates PCBs in its fat from 7000 to 20000 ng/g of wet weight for males and from 7000 to 22000 ng/g of wet weight for females. These data are consistent with the data obtained earlier. The mean value of PCB accumulation in the Baikal seal is higher than in seals from the world's background areas (the Arctic). However, this value is 20 times less than that in the Baltic seals. The minimal value (7000 ng/g of lipids) is comparable with that in the Greenland seal *Phoca groenlandica*, which inhabits the Gulf of Saint Laurence, and three times less than in the grey seal *Halichoerus grypus* (a West-Atlantic population). The Baikal omul (*Coregonus migratorius*, Georgi, 1775) of 1–7 years old, a member of the benthic-deepwater morpho-ecological group, accumulates PCBs from 20 to 40 ng/g of muscle or 800–2500 ng/g of lipids during the feeding season.

The PCB concentrations recorded in omul are 10 times less than those in the representatives of herring *(Clupeidae)* in the Baltic Sea and Atlantic Ocean and three times less than the maximum admissible levels established for commercial fish (Nikonova and Gorshkov 2010).

3.2.2 Oil Products in the Water of Lake Baikal

Natural oil seepage is observed in Central and Southern Baikal in the form of bitumen in the coastal cliffs or oil patches on the water surface. The discharge of biodegraded oil from the lake bottom amounts to 2 tonnes a year at Cape Tolsty and 4 tonnes a year at Cape Gorevoy Utes (Kontorovich et al. 2007). Fractionation takes place at the deep sites: separation of heavy fractions forming ozokerite-like bitumen mounds at the lake bottom and oil enrichment with *n*-alkanes. The biodegradation of light oil in the water column and on the lake surface provides the purity of waters at the sites of oil seepage.

Emissions from ship engines are one of the main anthropogenic sources of oil products at Lake Baikal. Analysis of distribution of oil and oil products in the lake shows that the oil content in the pelagic area does not exceed 10 μg/l, *n*-alkanes—0.15 μg/l and PAH—0.012 μg/l. Extreme concentrations of oil products (up to 1.3×10^4 μg/l), *n*-alkanes (up to 500 μg/l) and PAHs (up to 20 μg/l) were recorded in the surface water samples in the area of natural oil seepage. However, the concentration of oil products decreases abruptly further away from the oil seepage (Gorshkov et al. 2010).

3.3 *Atmospheric Composition Above Lake Baikal*

The composition of the atmosphere above Lake Baikal has been monitored for many years. A large industrial Irkutsk–Shelekhov complex, comprising the cities of Shelekhov, Irkutsk, Angarsk and Usolye-Sibirskoye, with enterprises from the heat power engineering, chemical, oil-refining and aluminium industries and the production of construction materials, is located in the valley of the Angara River at a distance of 70–150 km from the western shore of Southern Baikal. Numerous instrumental measurements of the chemical composition of trace gases and aerosol above the water area of the lake, analysis of the snow cover and the application of mathematical models show that there is an insignificant input of anthropogenic impurities (compounds of sulphur, nitrogen and heavy metals) from the atmosphere into the southern basin. The data on the accumulation of persistent organic substances (PAHs and PCBs) in the snow cover demonstrate that the atmospheric pollution above Lake Baikal is of a local character and the contribution of regional transfer of these classes of organic pollutants is insignificant at the shore and water area of Southern Baikal. The accumulation of PAHs in the snow collected from the ice on the lake was 100 times less than in the industrial centres of the Irkutsk Oblast.

The ratio of the mass of sulphur, nitrogen and suspended particles deposited from the atmosphere onto the water surface of Southern Baikal to the total mass of sulphur and nitrogen emitted by industrial enterprises in Pribaikalye shows that approximately 10 % of sulphur, 6 % of nitrogen and 7 % of suspended particles of total volume of industrial emissions are deposited onto the lake when there is a prevailing north-western wind.

During summer expeditions, trace gases and atmospheric aerosol are analysed on board the ship in the surface layer above the entire water area of Lake Baikal. The majority of chemical impurities in the air above the lake water area are of fine submicron fraction with a particle size of less than 0.7 μm and a mass concentration of 1–2 $\mu g/m^3$. Up to 60 % of the dissolved components consist of calcium, bicarbonate, sulphate and nitrate ions. Elevated concentrations of ions are recorded in large particles (a coarse-dispersed fraction) within the size range of 1.3–2.1 and 4.2 μm. A quantitative analysis of the chemical composition of particulates shows that the carbon constituent prevails in the rounded particles, whereas alumosilicates, gypsum particles and other inclusions dominate in the particles of irregular shape and faceted particles (Golobokova et al. 2011).

The main dynamics of the variability of the diurnal and seasonal trend of the CO_2 content are studied in the surface layer of the atmosphere above Lake Baikal. The concentration of CO_2 decreases from March–April. Its annual minimum is registered in July–August, when photosynthesis reaches its maximum. Beginning from September, the CO_2 content increases in the atmosphere. The maximum amplitudes of the diurnal trend are observed in August, whereas minimal values are characteristic of winter periods (Sakirko et al. 2008).

The CO_2 gas exchange is studied at the "water–atmosphere" interface, revealing its diurnal and seasonal differences. The CO_2 flux changes its direction at this interface in the pelagic area from spring to summer. In spring, the carbon dioxide flux is directed into the atmosphere and in summer onto the water surface of the lake. The main factor behind the change in the flux direction is the seasonal dynamics of species composition of phytoplankton and the level of its maturity (Domysheva et al. 2010a, b, 2011, 2012). In summer, the flux of carbon dioxide from the atmosphere is three times higher in the pelagic area than in the littoral zone (Fig. 7) (Domysheva et al. 2010a, b).

3.4 *Tributaries of Lake Baikal*

The most important constituent of the Baikal monitoring is the monitoring of the chemical water composition in the tributaries. Long-term data on the tributaries' chemical water composition collected for many years allowed its intra- and inter-annual changes to be determined under the influence of natural conditions and anthropogenic factors (Votintsev et al. 1965; Drucker et al. 1997; Sorokovikova et al. 2004). Of special attention is the Selenga River, which brings approximately 50 % of the water and over 50 % of the chemical components, as well as small

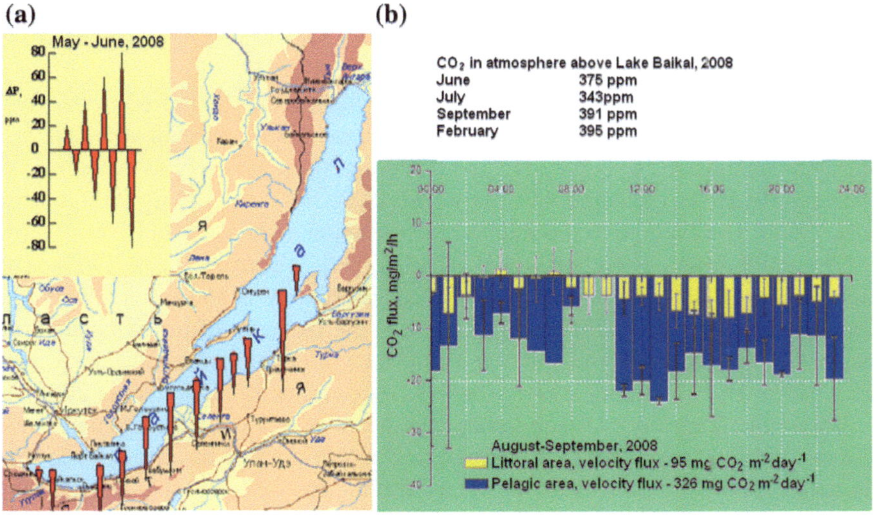

Fig. 7 Difference in the partial pressures of CO_2 in the water and atmosphere in the pelagic area (**a**), CO_2 flux in the "water–atmosphere" system (**b**) in the littoral and pelagic areas of Lake Baikal

rivers, whose basins are located in the industrial emissions zone (Obozhin et al. 1984; Sorokovikova et al. 2000, 2008, 2009, 2013; Sinyukovich et al. 2010; Bogdanov 2006; Tomberg et al. 2010). Table 2 shows the dynamics of the concentration of major ions in the lower course of the Selenga River since the mid-20th century. The rise of mineralisation and especially the increase in sulphate and chloride concentrations in the 1970s compared to the 1950s–1960s were caused by anthropogenic factors (Obozhin et al. 1984). The decrease in the concentrations of chemical components in the 1990s was caused by a production decline and the introduction of a closed water cycle at the Selenginsk Pulp and Cardboard Plant.

Recently, an increase in the concentrations of sulphate, calcium and other ions has been recorded in the Selenga River water, mainly due to low water in the river and an increase in the ground water inflow (Sinyukovich et al. 2010). An analysis of

Table 2 Mean concentrations of ions in the Selenga River water in different observation periods

Year	HCO_3^- (mg/l)	SO_4^{2-} (mg/l)	Cl^- (mg/l)	Ca^{2+} (mg/l)	Mg^{2+} (mg/l)	Na^+ (mg/l)	K^+ (mg/l)	Total ions (mg/l)
1950–1962[a]	88.7	6.5	1.2	20.7	4.6	4.1	1.1	127
1971–1974[b]	113.8	12.4	2.4	27.6	5.2	7.3	2.0	171
1995–1997[c]	97.7	8.4	1.9	20.9	5.8	6.5	1.4	143
2001–2004[c]	94.3	9.2	1.2	20.8	4.7	5.9	1.3	138
2010[c]	92.3	11.1	1.7	22.0	5.2	5.6	1.4	139

[a]Votintsev et al. (1965)
[b]Bogdanov (2006)
[c]Sorokovikova et al. (2013)

the long-term data shows that the input of sulphate from the Selenga River into Lake Baikal has increased by 25–30 % depending on the water level (Sorokovikova et al. 2008).

Elevated concentrations of nitrogen and phosphorus are registered in the river water. In the 1950s, the mean annual concentrations of mineral nitrogen and phosphorus amounted to 0.9 mg N/l and 0.078 μg P/l, respectively (Votintsev et al. 1965); in the 1970–1980s these values were 3.2 mg N/l and 0.110 μg P/l (Obozhin et al. 1984). The average concentration of phosphorus in the Selenga River water is 0.021 μg P/l, whereas earlier it was 0.013 μg P/l (Votintsev et al. 1965). The increase in the nitrogen and phosphorus concentrations in the water affects the functioning of the river ecosystem: changes in the structure of phytoplankton, an increase in small centric forms and an abundance of phytoplankton, whose values are characteristic of eutrophic water bodies (Popovskaya and Tashlikova 2008).

Within the framework of monitoring, significant attention is paid to small rivers located on the eastern shore of Lake Baikal, which are fed by atmospheric precipitation. The pH value of snow waters feeding the rivers is 4.6–5.5, and the minimal ratio of equivalent concentrations of cations to anions (0.60–0.77) is characteristic of them. This attests to the lack of alkaline and alkaline-earth ions to neutralise the anion acidity of strong acids (Netsvetaeva et al. 2013). The catchment areas of these rivers were subject to emissions from the Baikalsk Pulp and Paper Plant for many years (in December of 2013, this plant was shut down). As the rivers are fed with acidified precipitation while their waters feature low mineralisation (30 mg/l) this causes changes in the percentage composition of major ions in the river waters: a rise in the sulphate content and a decrease in bicarbonate and calcium (Sorokovikova et al. 2004). A tendency for the total ion content to increase (by 10–15 %) has been revealed in the water of the rivers.

4 Conclusions

Investigations of the chemical composition of the water of Lake Baikal and its basin using modern, sophisticated analytical equipment and high-precision methods allow researchers to obtain reliable results and to make long-term forecasts on the state of the lake ecosystem. To date, these studies are extremely important in view of global climate changes and intensifying anthropogenic effects in the region, and have to be performed under strict state control.

When it comes to the improvement of the modern monitoring system at Lake Baikal, it is necessary to develop novel approaches to its management. More attention should be paid to problematic zones which have appeared for the last decade in the Baikal region. First of all, detailed water monitoring should be conducted in the littoral area of Lake Baikal, where the coast is being intensively developed, and near the mouths of large tributaries of the lake, and this system should include such parameters as monitoring sanitary microbiological characteristics, nutrients and biota. It is necessary to regularly monitor the dynamics of concentration of persistent organic compounds in the air, water, bottom sediments and biota of the lake. Many of them are mutagenic and carcinogenic, and they are able to accumulate in the food chains during transfer from one organism to another. Due to the increase in water transport at the lake and the occurrence of natural oil seepage in the Baikal ecosystem, it is necessary to monitor oil products and volatile phenols, especially in the ports and settlements.

To conclude, it is necessary to stress again that at present the concentrations of pollutants (POPs and heavy metals) in Lake Baikal are low. All these substances enter the lake in small concentrations which do not directly affect the composition of the water or biota. Moreover, there is a powerful self-purification process in the ecosystem of Lake Baikal. The water in the deep part of Lake Baikal remains one of the cleanest natural waters in the world and can be used for drinking and other purposes.

References

Alekin OA, Semenov AD, Skopentsov BA (1973) Rukovodstvo po khimicheskomu analizu vod sushi [Manual on chemical analysis of inland waters]. Leningrad: Gidrometeoizdat, p 262

Baig JA, Kazi TG, Arain MB, Afridi HI, Kandhro GA, Sarfraz RA, Jamal MK, Shah AQ (2009) Evaluation of arsenic and other physico-chemical parameters of surface and ground water of Jamshoro, Pakistan. J Hazard Mater 166:662–669

Baram GI, Vereshchagin AL, Golobokova LP (1999) Microcolumn high-performance liquid chromatography with UV detection for the determination of anions in the environmental materials. J Anal Chem 54(9):962–965

Belykh OI, Gorshkov AG, Ryabchikova IA, Seryshev VA, Marinayte II (2006) Distribution and biological activity of polycyclic aromatic hydrocarbons in the ecosystem "snow source cover-soil-plant". Siberian Ecol J 11(6):93–802

Bobovnikova TI, Virchenko EP, Dibtseva AV, Yablokov AV, Pastukhov VD (1985) Marine mammals as indicators of pollution of background areas with organochlorine pesticides and polychlorinated biphenyls//Improvement of regional monitoring of the Lake Baikal environmental state L: Hydrometeoizdat 49–54

Bogdanov VT (2006) LIN archives, Limnological Institute of the Siberian Branch of the Russian Academy of Sciences, Irkutsk, Unpublished data

Domysheva VM (2009) Hydrochemistry of Lake Baikal: in encyclopedic reference book "Nature and People" part II. Habitat, Biota and Ecosystem, Lake Baikal, pp 71–72

Domysheva VM, Sakirko MV, Pestunov DA, Panchenko MV (2010a) Seasonal trend of CO_2 gas exchange in the system "atmosphere-water" in the littoral area of Lake Baikal. 1. Hydrological spring. Opt Atmos Ocean 23(12):1067–1074

Domysheva VM, Sakirko MV, Pestunov DA, Panchenko MV (2010b) Experimental estimation of carbon dioxide flux in the system "atmosphere-water" in the littoral and pelagic areas of Lake Baikal. DAN 431(6):822–826

Domysheva VM, Sakirko MV, Pestunov DA, Panchenko MV (2011) Seasonal trend of CO_2 gas exchange in the system "atmosphere-water" in the littoral area of Lake Baikal. 2. Hydrological summer. Opt Atmos Ocean 24(9):737–742

Domysheva VM, Sakirko MV, Pestunov DA, Panchenko MV (2012) Seasonal trend of CO_2 gas exchange in the system "atmosphere-water" in the littoral area of Lake Baikal. 3. Autumn. Opt Atmos Ocean 25(2):826–832

Drucker VV, Sorokovikova LM, Sinyukovich VN, Potemkina TG, Nikulina IG, Molozhavaya OA, Korovyakova IV (1997) Water quality of the Barguzin River under modern conditions. Geogr Nat Resour 4:72–78
Falkner KK, Measures CI, Herbelin SE, Edmond JM, Weiss RF (1991) The major and minor element geochemistry of Lake Baikal. Limnol Oceanogr 36:413–423
Fialkov VA (1983) Currents of the littoral area of Lake Baikal Novosibirsk, NAUKA publisher, p 192
Fomin G (2010) Water, control of chemical, bacterial and radiation safety according to international standards, 4th edn. Protector Moscow, Moscow, 1008 pp
Fukuzaki N, Khodzher T, Hara H (2002) Quality control of chemical analysis in acid deposition monitoring—improvement in ion balance by hydrogencarbonate determination. J Jpn Soc Atmos Environ 37(6):393–401
Golobokova LP, Filippova UG, Marinayte II, Belozerova OY, Gorshkov AG, Obolkin VA, Potemkin VL, Khodzher TV (2011) Chemical composition of atmospheric aerosol over the water area of Lake Baikal. Opt Atmos Ocean 24(3):236–241
Gorshkov AG, Marinayte II (2000) Monitoring of eco-toxicants in the environmental objects of Pribaikalye. Determination of polycyclic aromatic hydrocarbons in aerosols of industrial centres (exemplified by the city of Irkutsk). Opt Atmos Ocean 13(10):967–970
Gorshkov AG, Marinayte II, Obolkin VA, Baram GI, Khodzher TV (1998) Polycyclic aromatic hydrocarbons in the snow cover of the southern shore of Lake Baikal. Opt Atmos Ocean 11 (8):913–918
Gorshkov AG, Marinayte II, Baram GI, Sokov IA (2003) Application of high-performance liquid chromatography on short columns of a small diameter for determination of priority of polycyclic aromatic hydrocarbons in the environmental objects. J Anal Chem 58(8):862–869
Gorshkov AG, Marinayte II, Zemskaya TI, Khodzher TV (2010) Recent level of oil products in the water of Lake Baikal and its tributaries. Chem View Sustain Dev 18:711–718
Grachev MA (2002) On recent state of ecological system of Lake Baikal. SB RAS, Novosibirsk, 156 pp
Grachev MA, Domysheva VM, Khodzher TV, Korovyakova NV, Golobokova LP, Vereshchagin AL, Granin NG, Gnatovsky RY, Kostornova TY (2004) Deep water of Lake Baikal as a standard of fresh water. Chem View Sustain Dev 12:417–429
Hohmann R, Kipfer R, Peeters F, Piepke G, Imboden DM, Shimaraev MN (1997) Processes of deep-water renewal in Lake Baikal. Limnol Oceanogr 42(5):841–855
Iwata H, Tanabe S, Ueda K, Tatsukava R (1995) Persistent organochlorine residues in air, water, sediments and soils from the Lake Baikal region. Russia Environ Sci Technol 29:792–801
Khodzher TV, Gorshkov AG, Potemkin VL, Obolkin VA (1999) The composition of aerosol over the East Siberia. J Aerosol Sci 30(1):271–272
Khodzher TV, Semenov MY, Obolkin VA, Domysheva VM, Golobokova LP, Kobeleva NA, Netsvetaeva OG, Potemkin VL, Sergeeva MV (2002) Monitoring of acidic precipitation in the Baikal region. Chem View Sustain Dev 10(5):699–705
Khodzher TV, Golobokova LP, Netsvetaeva OG, Domysheva VM, Pogodaeva TV, Tomberg IV (2004) Results of testing of chemical parameters of artificial standard samples of rain and surface fresh waters. Opt Atmos Ocean 17(5–6):478–482
Khodzher TV, Obolkin VA, Potemkin VL, Golobokova LP, Netsvetaeva OG, Sorokovikova LM, Tomberg IV, Gromov SA (2011) National assessment of the state of acid deposition monitoring of Russia. Second Periodic Report on the State of Acid Deposition in East Asia. Part II National Assessments. Chap. 11: Russia. Jesada Luangjame. ACAP, Japan, pp 205–248
Killworth PD, Carmack EC, Weiss RF, Matear R (1996) Modeling deep-water renewal in Lake Baikal. Limnol Oceanogr 41(7):521–1538
Kodenev GG, Shimaraev MN, Shishmarev AT (1998) Determination of temporal renewal of deep Baikal waters using chemical tracers. Geol Geophys 39(6):842–850
Kontorovich AE, Kashirtsev VA, Moskvin VI, Burshtein LM, Zemskaya TI, Kostyreva EA, Kalmychkov GV, Khlystov OM (2007) Petroleum potential of Baikal deposits. Geol Geophys 48(12):1346–1356

Koroleva GP, Groshkov AG, Vinogradova TP, Butakov EV, Marinayte II, Khodzher TV (1998) Pollution of snow cover as a deposit environment (Southern Pribaikalye). Chem View Sustain Dev 6(4):327–337
Kravtsova LS, Izhboldina LA, Khanaev IV, Pomazkina GV, Domysheva VM, Kravchenko OS, Grachev MA (2012) Disturbances of vertical zoning of green algae in the littoral area of Listvennichnoye Bay, Lake Baikal, as a result of local anthropogenic effect. DAN 447 (2):227–229
Kravtsova LC, Izhboldina LA, Khanaev IV, Pomazkina GV, Rodionova EV, Domysheva VM, Sakirko MV, Tomberg IV, Kostornova TY, Kravchenko OS, Kupchinsky AB (2014) Nearshore benthic blooms of filamentous green algae in Lake Baikal. J Great Lakes Res 40:441–448
Kucklick JR, Harvey HR, Ostrom PH, Ostrom NE, Baker JE (1996) Organochlorines in the water and biota of Lake Baikal. Environ Toxicol Chem 15(8):1388–1400
Malakhov SG, Bobovnikova TI, Dibtseva AV (1986) Comprehensive global monitoring of the biosphere state. L. Hydrometeoizdat 2:113–122
Marinayte II, Gorshkov AG (2002) Monitoring of eco-toxicants in the environmental objects of Pribaikalye. Determination of polycyclic aromatic hydrocarbons in the snow cover of industrial centres. Opt Atmos Ocean 15(5–6):450–455
Najar IA, Khan AB (2012) Assessment of water quality and identification of pollution sources of three lakes in Kashmir, India, using multivariate analysis. Environmental Earth Sciences 66 (8):2367–2378
Netsvetaeva OG, Khodzher TV, Golobokova LP, Kobeleva NA, Pogodaeva TV (2004) Chemical composition of snow cover in natural reserves of Pribaikalye. Geogr Nat Resour 1:66–72
Netsvetaeva OG, Chipanina EV, Obolkin VA, Zimnik EA, Sezko NP, Lopatina IN, Khodzher TV (2013) Peculiarities of chemistry of atmospheric precipitation at stations Listvyanka (Irkutsk region) and Primorskaya (Primorye territory). Opt Atmos Ocean 6:466–471
Nikonova AA, Gorshkov AG (2011) Rapid chromatography for the determination of polychlorinated biphenyls by GS-MS in environmental monitoring. Anal Lett 44:1290–1300
Nikonova AA, Gorshkov AG (2012) Determination of polychlorinated biphenyls by fast chromatography mass-spectrometry in environmental and biological samples. J Anal Chem 67 (1):72–80
Nikonova AA, Groshkov AG (2010) Polychlorinated biphenyls in natural environments and biological objects of the Baikal natural territory. News of the National Academy of Sciences of Belarus. Part 1. Ser Chem 1:60–63
Obozhin VN, Bogdanov VT, Klikunova OF (1984) Hydrochemistry of rivers and lakes of Buryatia. Nauka, Novosibirsk 152
O'Neil JM, Davis TW, Burford MA, Gobler CJ (2012) The rise of harmful cyanobacteria blooms: The potential roles of eutrophication and climate change. Harmful Algae 14:313–334, doi:10. 1016/j.hal.2011.10.027
Peeters F, Kipfer R, Hohmann R, Hofer M, Imboden DM, Kodenev GG, Khozder TV (1997) Modeling transport rates in Lake Baikal: gas exchange and deep water renewal. Environ Sci Technol 31(10):2973–2982
Pomazkina GV, Belykh OI, Domysheva VM, Sakirko MV, Gnatovsky RY (2010) Structure and dynamics of phytoplankton in Southern Baikal (Russia). Algology 20(1):56–72
Popovskaya GI (1991) Phytoplankton of Lake Baikal and its long-term changes (1958–1990). Doctoral thesis, Novosibirsk 32
Popovskaya GI, Tashlikova NA (2008) Phytoplankton. The Selenga River delta as a natural biofilter and indicator of the condition of Lake Baikal 15:88–101
Sakirko MV, Panchenko MV, Domysheva VM, Pestunov DA (2008) Diurnal rhythms of carbon dioxide concentrations in the surface air layer and surface water of Lake Baikal in different hydrological seasons. Meteorol Hydrol 2:79–86
Shimaraev MN, Domysheva VM (2013) Trends in hydrological and hydrochemical processes in Lake Baikal under conditions of modern climate change. In: Goldman CR, Kumagai M,

Robarts RD (eds) Climate change and inland waters impacts and mitigation approaches for ecosystems and society. Wiley, New York, pp 43–66
Shimaraev MN, Domysheva VM, Gorbunova LA (1996) On dynamics of oxygen in Lake Baikal during spring mixing. DAN 347(6):814–817
Sinyukovich VN, Sorokovikova LM, Tomberg IV, Tulokhonov AK (2010) Changes of climate and chemical flow of the Selenga River. DAN 6:817–821
Sorokovikova LM, Sinyukovich VN, Golobokova LP (2000) Formation of ionic flow of the Selenga River under modern conditions. Water Resour 5:560–565
Sorokovikova LM, Netsvetaeva OG, Tomberg IV, Khodzher TV, Pogodaeva TV (2004) Effect of atmospheric precipitation on the chemical composition of river waters in the south Baikal region. Opt Atmos Ocean 17(5–6):423–427
Sorokovikova LM, Tomberg IV, Bashenkhaeva NV (2008) Chemical composition of waters in the Selenga River and its distributaries. Selenga River Delta Nat Biofilter Indic Condition Lake Baikal 15:88–101
Sorokovikova LM, Popovskaya GI, Tomberg IV, Bashenkhaeva NV (2009) Spatial-temporal variability in the content of nutrients and organic matter and phytoplankton of the Selenga River and its tributaries. Water Resour 36(4):465–474
Sorokovikova LM, Popovskaya GI, Tomberg IV, Sinyukovich VN, Kravchenko OS, Marinayte II, Bashenkhaeva NV, Khodzher TV (2013) The Selenga River water quality on the border with Mongolia at the beginning of the 21st century. Meteorol Hydrol 2:93–103
Tarasova EN, Meshcheryakova AI (1992) Modern state of hydrochemical regime of Lake Baikal. Nauka, Novosibirsk 141
Tomberg IV, Sorokovikova IV, Sinyukovich VN, Pogodaeva TV (2010) Flow of ions in the Selenga River delta under conditions of low water. Meteorol Hydrol 12:87–95
Vereshchagin GY (1927) Hydrochemical observations in Southern Baikal in the summer of 1926. In: Proceedings of the USSR AS, series A, pp 20–28
Vereshchagin GY (1933) Methods for field hydrochemical investigations and their use in hydrological practice. M. Hydrometeoizdat 33
Vereshchagin AL, Dudinskii VF, Golobokova LP, Baram GI, Grachev MA (2000) Determination of UV-absorbing anions in environmental samples by microcolumn high-performance liquid chromatography. J Anal Chem 55(10):1000–1002
Votintsev KK (1961) Hydrochemistry of Lake Baikal. Proc Limnological Inst SB RAS 3(23):310
Votintsev KK, Glazunov IV (1963) Hydrochemical regime of Lake Baikal near the settlement of Listvennichnoye. Proc Limnological Inst SB RAS 3(23):42–56
Votintsev KK, Glazunov VI, Tolmacheva AP (1965) Hydrochemistry of rivers in Lake Baikal basin. Nauka 491
Votintsev KK, Meshcheryakova AI, Popovskaya GI (1975) Cycle of organic matter in Lake Baikal. Nauka, Novosibirsk 189
Weiss RF, Carmack EC, Koropalov VM (1991) Deep-water renewal and biological production in Lake Baikal. Nature 349:665–669
Wetzel RG, Likens GE (1991) Limnological analyses. Springer, New York, 391 pp

Chapter 4
Microbiological Monitoring of Lake Baikal

Valentina V. Parfenova, Natalia L. Belkova, Olga S. Pestunova, Mariya Yu. Suslova and Olga N. Pavlova

Abstract The microbial community is not only a natural component of the ecosystem of rivers, reservoirs, and lakes, but also one of the main indicators of their ecological situation. For the public water supply, water reservoirs are often used which are affected by regular or accidental contamination, which greatly influences the water quality. Microbial indicators, limiting values of which are set by the relevant regulatory documents, are defined both for the water supply and for drinking water resources. We provide an overview of the main regulatory documents used in the Russian Federation to assess the microbiological quality of water resources and of methods and results about monitoring of Lake Baikal. Lake Baikal is a well-known example of an oligotrophic deep-water lake which serves as a source of drinking water. The microbial communities of Lake Baikal are formed in the unique extreme environmental conditions determining their metabolism: a low nutrient content and low average annual temperature in the water column. The microbiological indicators for the pelagic zone of the lake are quite constant, as was shown by long-term observations. However, in these ecological conditions the possibility must be taken into account that the water contains microorganisms potentially hazardous to human health. Systematic annual results have been

V.V. Parfenova · N.L. Belkova (✉) · O.S. Pestunova · M.Yu. Suslova · O.N. Pavlova
Siberian Branch, Limnological Institute (LIN), Russian Academy of Sciences,
Ulan-Batorskaya Str. 3, Irkutsk 664033, Russia
e-mail: nlbelkova@gmail.com

V.V. Parfenova
e-mail: parf@lin.irk.ru

O.S. Pestunova
e-mail: ok.c@mail.ru

M.Yu. Suslova
e-mail: suslova@lin.irk.ru

O.N. Pavlova
e-mail: pavlova@lin.irk.ru

N.L. Belkova
Irkutsk State University (ISU), Karl Marks Str. 1, Irkutsk 664003, Russia

L. Mueller et al. (eds.), *Novel Methods for Monitoring and Managing Land and Water Resources in Siberia*, Springer Water,
DOI 10.1007/978-3-319-24409-9_4

presented (since 2005) of the microbiological monitoring of coliform bacteria, thermotolerant coliforms (TTC), coliphages, as well as *Pseudomonas, Clostridium* and *Enterococcus* in Lake Baikal. It is shown that the spatial distribution of allochthonous organotrophic and opportunistic bacteria is associated with the local anthropogenic impact: settlements, deltas of the main tributaries or domestic wastewater discharge. In the deep layers of the pelagic zone of the lake, no groups of opportunistic bacteria were found. Moreover, as the water depth increased, changes were found in the structure of the microbial community. Oligotrophic and psychrotrophic microorganisms were predominant there, while the amount of organotrophic microorganisms did not exceed the background level. The widespread prevalence of uncultivable bacterial forms in the natural environment and the ability of pathogenic and opportunistic bacteria to move into an uncultivable state make them of sanitary significance. It has been demonstrated that new species of heterotrophic microorganisms can be cultivated when the culture media and cultivation conditions are adapted. The experiments show the possible mechanisms of survival for opportunistic bacteria. During the cultivation of *Enterococcus faecalis* in the Baikal water at a low temperature, cells moved into a viable, but uncultivable state and restored their ability to reproduce after the addition of the nutrient.

Keywords Microbiological monitoring · Lake Baikal · Sanitary rules · Enterococcus · Opportunistic bacteria · Heterotrophic bacteria · Cultivation

1 Bacteriological Criteria of Water Quality Accepted in the Russian Federation

The hygienic regulations on drinking water quality and the sanitary protection of water reservoirs in Russia propose strict controls on the direct and indirect influences of anthropogenic activity, public supply water use, and the discharge of various kinds of sewage into ponds, reservoirs, and streams. It is regulated by the Federal Law "On the sanitary-epidemiological welfare of the population" dated March 30, 1999 № 52-FZ (Federal Law 1999). On the basis of the federal law and the Statement on State Sanitary-epidemiological Regulations, approved by the Government of the Russian Federation and dated July 24, 2000 № 554, the following sanitary-epidemiological rules and regulations (SANPIN) have been constituted:

- "Hygienic Requirements for Surface Water Protection" SANPIN 2.1.5.980-00 (2000);
- "Drinking Water. Hygienic requirements for water quality of centralized drinking water supply. Quality control" SANPIN 2.1.4.1074-01 (2002);
- "Hygienic Requirements for Water Quality in Noncentralized Water Supply Systems. Sanitary Safeguarding of Sources" SANPIN 2.1.4.1175-02 (2003).

The SANPIN 2.1.5.980-00 sanitary rules (2000) set out hygienic standards for the composition and properties of water in water reservoirs for two categories of water consumption. The first category of water consumption includes the water reservoirs or parts thereof which could be used as a source of drinking water and household water use, as well as to supply the food industry with water. The second category includes the use of water reservoirs or parts thereof for recreational water consumers. Water quality requirements established for the second category of water consumption also apply to all areas of water reservoirs located within the boundaries of settlements. In SANPIN 2.1.4.1175-02 (2003), water quality requirements are established regarding the quality and properties of the noncentralized water supply (ground water, the capture of which is carried out by special equipment for general and individual use). The norms on controlled water quality indicators such as total coliform bacteria (TCB), thermotolerant coliforms (TTC) and coliphages are given in this SANPIN.

The discharge of domestic and industrial waste water and the development of recreational and navigational water use led to the need to increase the number of hygienic and microbiological parameters that characterize the quality of water, in order to evaluate it adequately. SANPIN 2.1.4.1074-01 (2002) states that the microbiological evaluation of the quality of drinking water should be carried out according to the following sanitary and bacteriological parameters: TTC, TCB, total bacterial number (TBN), coliphages, and spores of sulfite-reducing clostridia.

Currently, there are several organizations dealing with the standardization of water: the World Health Organization (WHO), the United States Environmental Protection Agency (U.S. EPA), the European Community (EC) and the Sanitary and Epidemiological Service of the Russian Federation. The latter is guided by the Sanitary-epidemiological Rules and Regulations (SANPIN). Table 1 summarizes the indicators used by international organizations as criteria for assessing the quality of water.

The acceptance of SANPIN was a major breakthrough in Russian water quality control. The regulations were based on the latest findings by Russian scientists and were in line with the WHO recommendations. In addition to SANPIN, in the Russian Federation there are guidelines according to which natural water sources are analyzed. Thus, the rules MUK 4.2.1884-04 "Sanitary-microbiological and

Table 1 Bacteriological indicators of water quality

Indicator	WHO	U.S. EPA	EC	SANPIN
Total bacterial number (TBN)	–	+	+	+
Total coliform bacteria (TCB)	+	+	+	+
Thermotolerant coliforms (TTC)	+	–	+	+
Fecal streptococci—*Enterococcus* cultivation	–	–	+	–
Fecal streptococci—*Enterococcus* PCR detection	–	+	+	–
Coliphages	–	–	–	+
Spores of clostridia	–	–	+	+

sanitary-parasitological analysis of water from surface water reservoirs" (2004) define enterococci as an additional indicator when a new source of centralized water supply has been chosen. This indicator is used to confirm the origin of fecal contamination in water at water sources and places of recreation. According to the methodological guidelines RD 52.24.633-2002 "Methodological basis for the creation and operation of environmental monitoring subsystem recourse freshwater ecosystems" (2002), the TBN has been proposed as a criteria for assessing anthropogenic impact on freshwater ecosystems.

2 Microbiological Monitoring of the Water of Lake Baikal

The microorganisms of Lake Baikal have been studied intensively in various fields of interest: physiological groups of bacteria, their total number and biomass, seasonal dynamics and vertical distribution in the water and bottom sediments (Kuznetsov 1957; Romanova 1958; Mladova 1971; Maksimova and Maksimov 1989; Lapteva 1990; Namsaraev and Zemskaya 2000). Numerous works have been carried out to determine the phylogenetic diversity of the microbial communities of Lake Baikal (Bel'kova et al. 1996; Denisova et al. 1999; Belkova et al. 2003; Belkova 2004). The investigation of the microorganisms and microbial communities in the water and bottom sediments of the lake is continuing, with the involvement of new, modern research methods (Gladkikh et al. 2011; Parfenova et al. 2013), because the microbial community is one of the key ecosystem indicators in environmental monitoring (Parfenova et al. 2009a).

Lake Baikal are located in Eastern Siberia and is the biggest and deepest freshwater reservoir in the world. The peculiarities of its physical and chemical parameters create a unique environment for all living organisms inhabiting it. There is a low water temperature throughout all the seasons and for almost the entire breadth of the lake, a low nutrient content and low total mineralization, and oxygen saturation up to maximum depths (Grachev 2002). In this regard, the lake is of particular interest from both the scientific and practical sides (in recent years the lake water has been widely used for bottling drinking water). One of the reasons for the detailed study of the composition and functioning of microbial communities in aquatic ecosystems is to study the microorganisms which determine or affect the quality of water, as freshwater reservoirs are the main sources of drinking water. Lake Baikal's microbiological indicators are monitored regularly, with up to 50 sampling stations in different parts of the lake (Fig. 1). One of the most important indicators of the quality of the Baikal water is the total bacteria number (TBN) in native samples, found by staining fixed samples with fluorescent dye 4′,6-diamidino-2-phenylindole (DAPI) (Fig. 2a, b). It was shown that a seasonal increase in TBN was observed in surface waters and the littoral area, where the values increased to 10^7 cells/mL. In the water column at high depths these values are more stable and do not exceed 10^6 cells/mL. In the near-bottom water the TBN

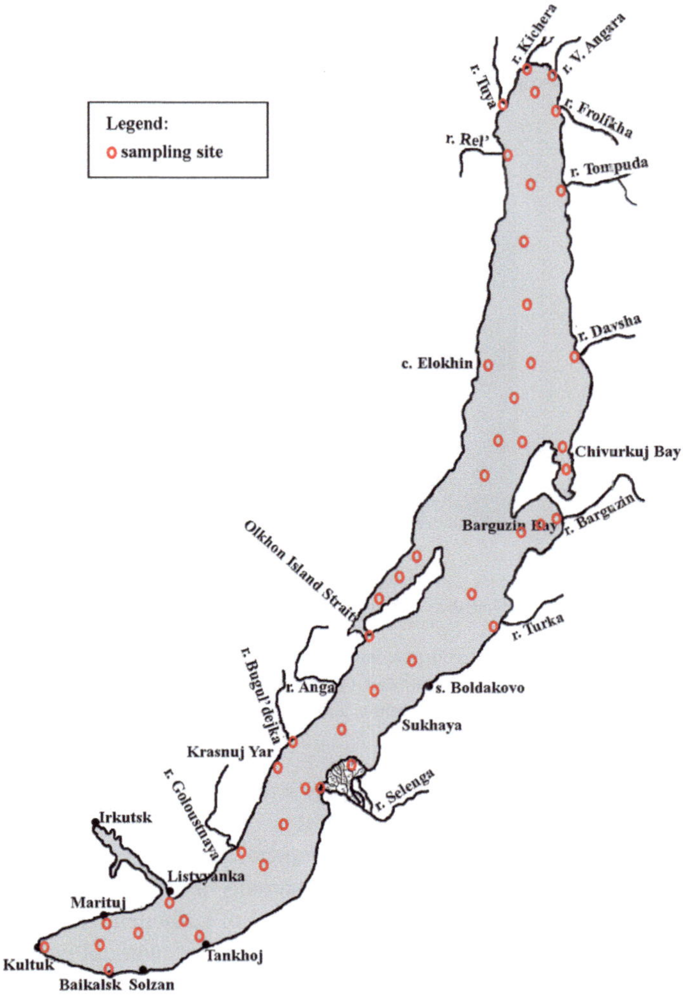

Fig. 1 Scheme of sampling sites for microbiological monitoring of Lake Baikal water quality

increases insignificantly and is associated with the influence of exchange processes occurring at the interface of water sediments.

Microorganisms in the aquatic environment are known to play a leading role in the destruction of organic compounds of different origin. One of the standard indicators of microbiological monitoring of Lake Baikal water is the enumeration of organotrophic bacteria. These bacteria react quickly to incoming organic matter. The temperature in the water column of the lake varies from 2 to 10 °C, with an average of 3.2 °C (Shimaraev 1978). Taking into account the low water temperature, it is assumed that there are a large number of psychrotrophic microorganisms among the autochthonous microbial community of the lake. In addition, low organic matter content and low mineralization determine the presence of

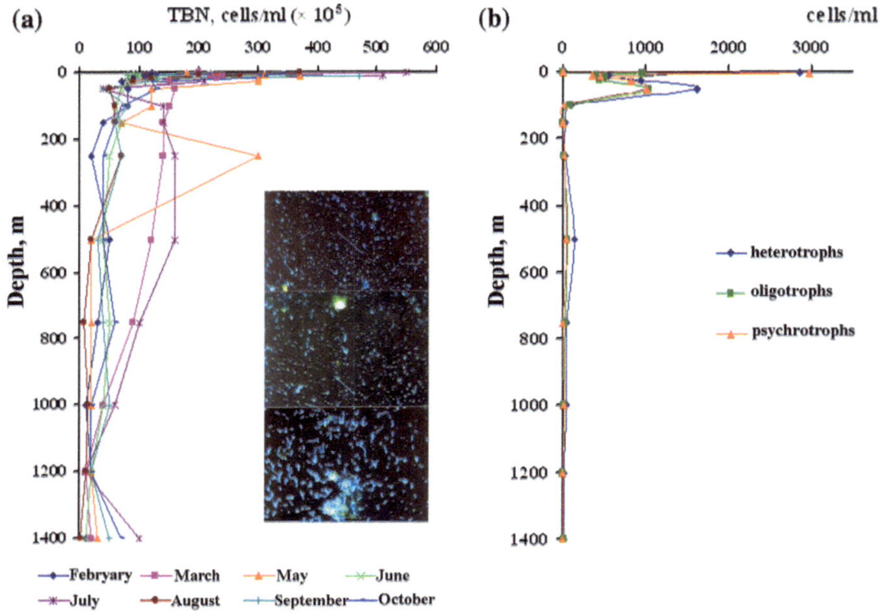

Fig. 2 Vertical distribution of microorganisms in the water column of the South Basin of Lake Baikal (2004–2007): **a** total bacterial number (*TBN*), insert—micrographs of microbial scenery, **b** total counts of heterotrophic, psychrotrophic and oligotrophic bacteria

oligotrophic microorganisms. Thus, to characterize the organotrophic microbial community of the lake water, physiological groups such as heterotrophic, psychrotrophic and oligotrophic bacteria are enumerated (Fig. 2b).

The temperature regime of the lake substantially affects the total number and variety of microorganisms as well as the survival of pathogenic and opportunistic bacteria. Molecular methods reveal the dominance of non-cultivable forms of bacteria in the water column of the lake. One of the dominant groups of Baikalian microorganisms consists of bacteria which are able to switch into an uncultivable state when adapting to the extreme conditions of the lake (Belkova 2004). Using alternative approaches for cultivation, new species have been identified for Lake Baikal. By filtration-acclimatization (Hahn et al. 2004) *Kaistobacter* sp., *Rhodopseudomonas* sp., *Actinotelluria* sp. and *Arsenicicoccus* sp. were isolated from the deep-water microbial communities. On the other hand, it is known that both pathogenic and nonpathogenic microorganisms can pass into an uncultivable state in which the cells do not form colonies on a nutrient agar medium, but remain viable (Kell et al. 1998; Oliver 2005). Perhaps this is the mechanism which allows different groups of microorganisms, including opportunistic ones, to survive and stay alive in Lake Baikal water.

In addition to its essential importance for cultivation, the temperature regime can be considered as a factor characterizing the possibility of manifestation of the pathogenic properties of microorganisms. Obviously, the ability of bacteria to survive and grow in laboratory conditions at a temperature of 37 °C suggests that they can survive and remain active in the human body. Investigations of the optimum temperature for heterotrophic bacteria growth showed that some of them could be effectively cultivated at 37 °C as well as at low temperatures varying between 4 and 8 °C. Growth tests at temperatures ranging from 4 to 37 °C were carried out for 385 strains isolated during different seasons from the Lake Baikal water on a diluted fish-peptone agar (FPA-10) medium (Table 2).

Table 2 shows that the strains isolated in the summer did not grow at low temperatures of 4 or 8 °C, demonstrating that mainly mesophilic bacteria grew from the samples collected during this season. In fall, the fraction of strains with a capacity to grow at 8 °C increased significantly, while the proportion of strains growing at 37 °C increased and remained relatively high when the incubation temperature was varied between 18 and 20 °C. In winter and spring, psychrotolerant microorganisms with growth at temperatures ranging from 4 to 20 °C grew in vitro more actively. At this time, the fraction of strains growing at temperature of 28 and 37 °C decreased 2–3-fold. It was found that in the deep-water column of Lake Baikal there are bacteria with a capacity to grow at temperatures varying over a wide range, from 2 to 37 °C. Clearly, it should be noted that this fact has to be taken into account during integrated environmental monitoring of the lake ecosystem to assess the water quality of Lake Baikal and define the role of these microorganisms in the epidemiological situation.

The microbiological contamination of natural waters mainly occurs due to fecally contaminated wastewater. According to published data, *Escherichia coli* is not a good indicator for many characteristics (Carrillo et al. 1985; Meier et al. 1997). Therefore, when assessing the quality of water intended for human consumption, *E. coli* should not be the only indicator organism determined, but also fecal enterococci (Fomin 2002). Enterococci are recommended to determine and confirm the nature of fecal contamination (Facklam et al. 1999). It is the reason for increasing interest in studying the ecological potential of opportunistic bacteria detected in the water of Lake Baikal. In sanitary and biological aspects, the spatial distribution in Lake Baikal and biodiversity of the representatives of the Enterobacteriaceae family and non-fermented group of bacteria have been studied (Pavlova et al. 2003; Drucker and Panasyuk 2006). For many years, these microorganisms were considered clinically insignificant. A reassessment of the pathogenic role of enterococci contributed to the detection of their virulence factors (Kayaoglu and Østavik 2004). In recent years, the role of microorganisms belonging to the genus *Enterococcus* increased as they acquired resistance to the vast majority of available antibiotics and extreme factors such as dryness, UV, low temperature, more than due to their wide distribution (Gilmore 2002). Moreover, it was found that they retain their viability and virulence even after the chlorination of drinking water. At present, interest in these bacteria has increased significantly due to their role in epidemiological terms. Therefore, the study of the possible permanent inflow of opportunistic bacteria into the lake, the species composition of allochthonous bacteria of the genus *Enterococcus*, the preservation of their viability

Table 2 Growth of heterotrophic bacteria isolated from Lake Baikal water at different temperatures in the laboratory experiments

Season, data and place of sampling	Total number of strains tested	Temperature, °C									
		4		8		18–22		28		37	
		Total	%	Total	%	Total	%	Total	%	Total	%
Winter, 1997, South Baikal	51	39	76.5	47	92.2	48	94.1	16	31.4	16	31.4
Spring, 1998, South Baikal	49	26	53.1	30	61.2	32	65.3	11	22.4	11	22.4
Summer, 1998, South and Middle Baikal	58	0	0	0	0	44	75.0	39	67.2	36	62.1
Fall, 1998, South and Middle Baikal	38	0	0	33	86.8	34	89.5	33	86.8	19	50.0
Winter, 1999, South Baikal	101	96	95.0	98	97.0	101	100	63	62.4	60	59.4
Spring, 1999, South Baikal	11	10	90.9	11	100	7	63.6	7	63.6	5	45.5
Summer, 1999, South and Middle Baikal	40	0	0	1	2.5	32	80.0	38	95.0	32	80.0
Fall, 1999, South and Middle Baikal	37	0	0	31	83.8	33	89.2	27	73.0	14	37.8

in the water, and their resistance to antibiotic and chlorination are important to the Lake Baikal monitoring system.

Studies of the distribution of opportunistic bacteria in Lake Baikal were carried out between 2004 and 2011. Is was shown that both allochthonous organotrophs and opportunistic bacteria, including those related to the genera *Pseudomonas* and *Enterococcus*, were confined to settlements, deltas of major tributaries, and the discharge point of undertreated domestic wastewater, where local anthropogenic influences were noted (Figs. 3, 4 and 5). A significant number of TCB and TTB near the shoreline of the south basin of Lake Baikal (Fig. 3) could be associated with the areas of local significant sources of pollution: the Sludyanka settlement (railway hub), the city of Baikalsk and the village of Kultuk (Grachev 2002; Parfenova et al. 2009b). Contamination of the Selenga shallow water and Barguzinsk Bay, located in the middle basin of Lake Baikal (Figs. 4 and 5), might be due to the influence of the two main tributaries of the lake—the rivers Selenga and Barguzin. Because of intensive use of water resources in the basin of these rivers, there was a change in the composition trend of natural waters and a reduction in their quality (Drucker et al. 1997).

The same situation was found in the north basin of the lake, where two main rivers Tuya and Verkhnyaya Angara affected the water quality (Figs. 4 and 5). Opportunistic bacteria were not found in the pelagic zone of the lake, in the samples collected from the deep water column, where the predominance of oligotrophic and

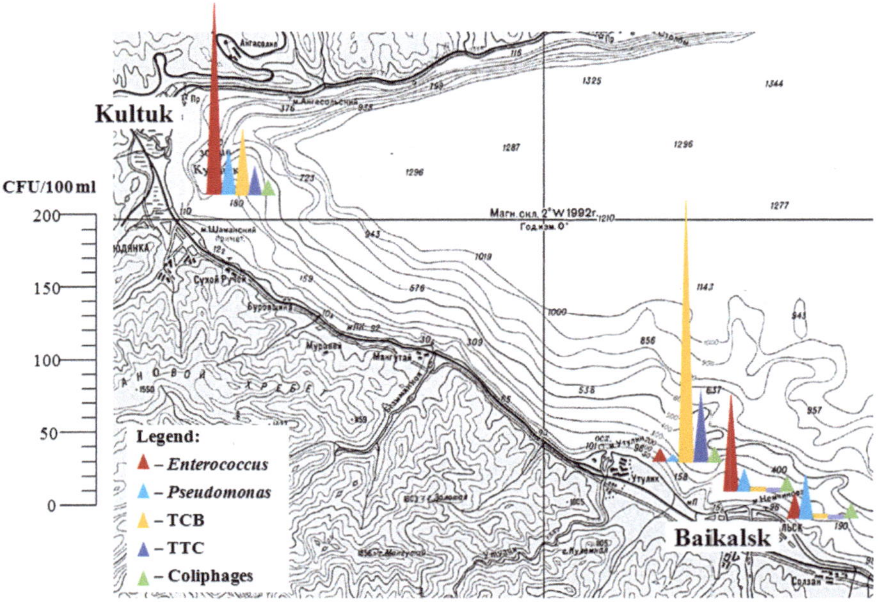

Fig. 3 Distribution of opportunistic bacteria in the surface water in the south basin of Lake Baikal near the settlement of Kultuk and the city of Baikalsk (data were collected in 2005)

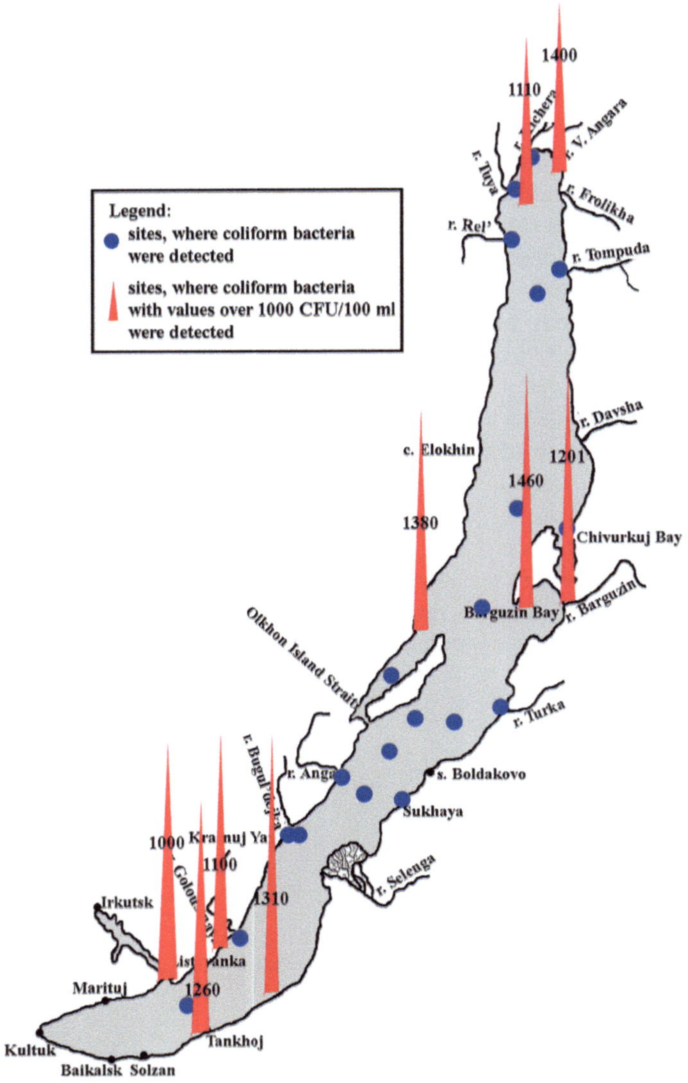

Fig. 4 Detection of TCB in the surface waters of Lake Baikal (data were collected in 2011)

psychrotrophic microorganisms was determined, while the content of organotrophic bacteria did not exceed the natural background (Table 3).

In many cases, there is no need to identify species of enterococci. However, when assessing the potential risk to humans for a potential source of water consumption, information on the distribution of this type of bacteria in the water column, as well as their viability properties at different temperatures and resistance to antibiotic and chlorination, is very important. Different species of enterococci have been identified in humans, animals and waters, and on plant debris. Most

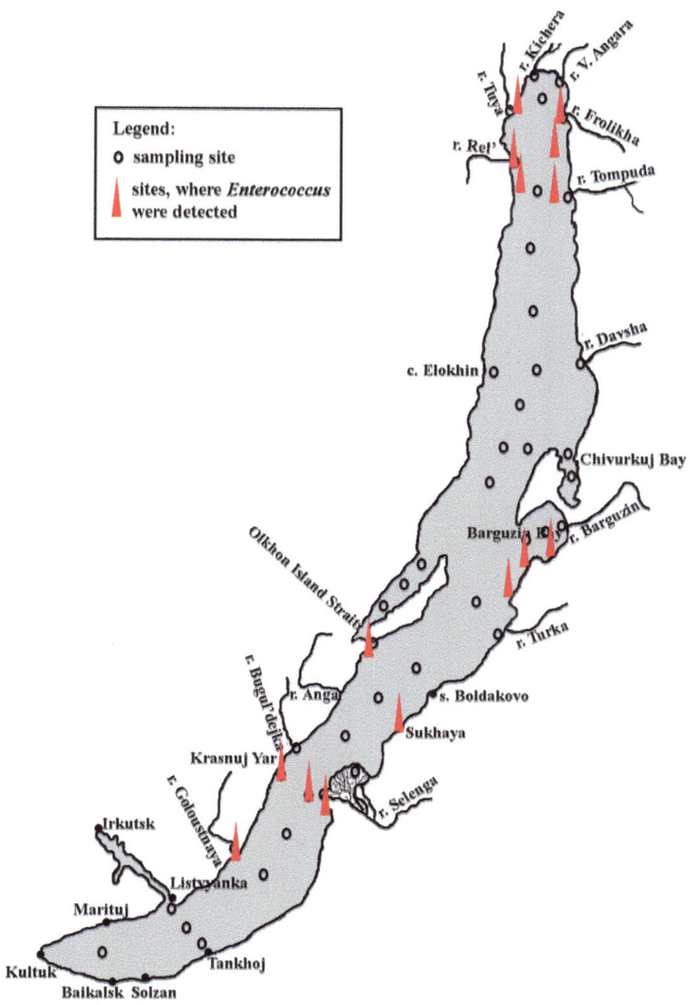

Fig. 5 Distribution of bacteria belonging to the genus *Enterococcus* in the surface water of Lake Baikal (data were collected in 2007)

human and animal waste is distributed in the environment and, therefore, the environment and water samples often contain enterococci. The presence in water of the main species *E. faecium* and *E. faecalis* is considered an indicator of fecal contamination (Gilmore 2002), but water samples may also contain enterococci, whose natural habitats are plants or animals, so the differentiation of the *Enterococcus* species is extremely important to monitor water quality. Thus, identifying the species composition of bacteria of the genus *Enterococcus* according to modern taxonomy and by standard microbiological methods is part of environmental microbiological monitoring of the state of the Lake Baikal

Table 3 Distribution of microorganisms in water samples collected in 2005 at the Kharauz–Krasnuy Yar transection (delta of Selenga River)

Distance from the shoreline, km	Depth, m	Heterotrophs, CFU/mL	Oligotrophs, CFU/mL	Psychrotrophs, CFU/mL	TNCM, CFU/mL	TCB, CFU/100 mL	TTB, CFU/100 mL	*Pseudomonas*, CFU/100 mL	*Enterococcus*, CFU/100 mL
14.5	0	6759	202	2560	1080	0	1	10	0
	5	1500	746	67	10	0	0	4	0
	270	23	61	87	0	0	0	3	0
10	0	121	109	117	3	0	0	3	0
	5	244	232	88	0	0	0	5	0
	100	821	980	63	24	0	0	6	0
7	0	162	36	42	12	166	133	4	0
	5	223	176	8	1	0	0	3	0
	28	92	55	32	2	0	0	3	0
5	0	294	114	72	1	0	0	20	0
	25	276	169	15	0	0	0	40	0
3	0	379	67	17	0	0	0	65	0
	17	332	225	23	3	0	0	33	0
1	0	1664	1280	726	78	53	67	80	22
	5	226	114	34	9	0	0	53	0

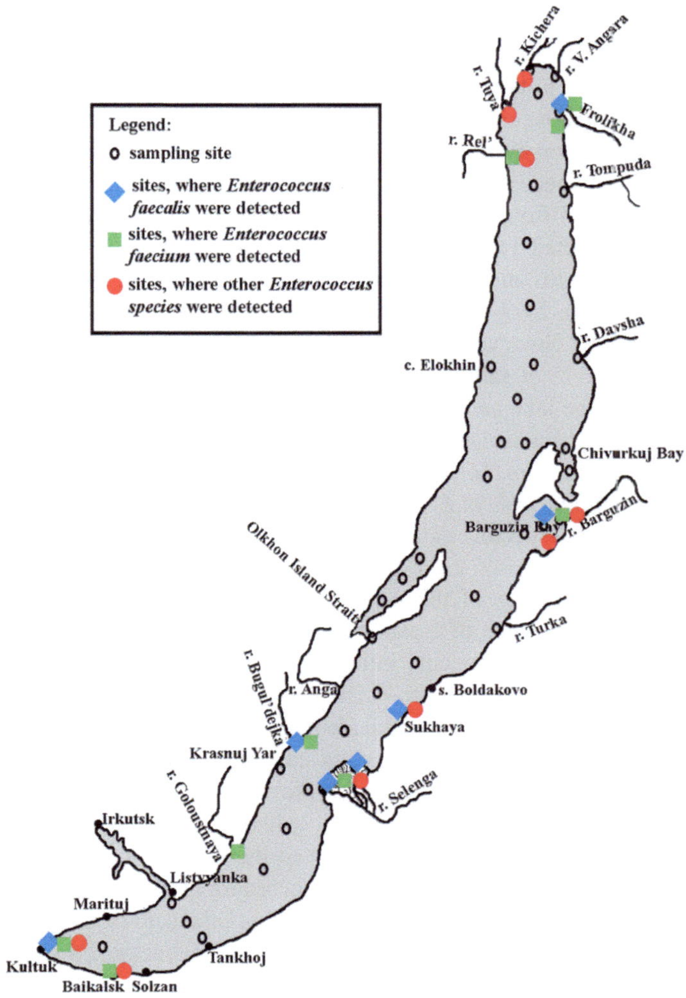

Fig. 6 Distribution of dominant and epidemiologically important species of genus *Enterococcus* in the water of Lake Baikal

ecosystem. During the studied period from 2005 to 2008, 306 strains belonging to the genus *Enterococcus* were isolated from Baikal water, 120 of which were identified at the species level (Fig. 6).

Studies have shown an association of *Enterococcus* bacteria detection with the littoral regions of the lake, where there is a local anthropogenic influence (Parfenova et al. 2010). Significant amounts of the bacteria tested in water samples collected in the South Baikal might be due to the influence of settlements (Kultuk and Baikalsk). In the Middle Baikal, the water quality and content of *Enterococcus* bacteria is influenced by the rivers Selenga and Barguzin—the largest tributaries of Lake Baikal. *E. faecium, E. avium, E. faecalis, E. mundi, E. hirae, E. durans* and

E. gallinarum were isolated from the areas studied (Fig. 6). Biodiversity studies of opportunistic bacteria of the genus *Enterococcus* have shown their higher numbers and species diversity in the South and Middle Baikal. Seven species were detected in these areas in contrast to the North Baikal, where only four species of enterococci were found. The diversity of the bacteria studied in the water of North Baikal is represented by the species *E. faecium, E. faecalis, E. mundtii* and *E. avium*. They were mainly isolated from the deltas of the northern tributaries of the lake, while they were not detected in the open waters of Lake Baikal. The most common species were *E. faecalis* and *E. faecium*, isolated from different areas of the lake. *E. faecalis* was found in 20 % of the water samples collected in the South Baikal, and 12 % in the Middle. However, the dominant species in these areas was *E. faecium*, found in 47 and 26 % of samples, respectively. Therefore, we can assume that the northern areas of the lake are subject to the lowest anthropogenic influence. Thus, among all of the strains isolated from the lake waters, 33 % were identified as *E. faecium*, 17 % as *E. avium*, 16 % as *E. faecalis*, 15 % as *E. mundtii*, 8 % as *E. hirae*, 6 % as *E. durans* and 5 % as *E. gallinarum*. Mainly they are of fecal origin, and might be considered specific indicators of water contamination by the feces of humans, animals and birds.

Currently, one estimate of the degree of negative human impact on aquatic ecosystems is the resistance of isolated bacteria to antibiotics, and this feature is used as a marker of anthropogenic influence. In terms of microbial ecology, this problem is also relevant, since the co-existence of allochthonous and autochthonous bacteria can cause horizontal gene transference, especially for antibiotic-resistant genes transferred from allochthonous resistant bacteria to autochthonous bacteria and vice versa (Moiseenko 1994). As a result, the autochthonous bacteria can acquire resistance or virulence properties, whereas allochthonous bacteria can acquire genes which help the bacteria adapt to the extreme environmental conditions of aquatic ecosystems (Lobova 2003). According to MUK 4.2.1890-04 (2004) particular attention should be paid to the definition of sensitivity and resistance of microorganisms belonging to the taxonomic groups characterized by a high frequency of dissemination of acquired resistance.

Resistance to different antibiotics was studied for 120 strains isolated from different areas of Lake Baikal and belonging to the genus *Enterococcus*: inhibitors of cell wall synthesis (vancomycin benzylpenicillin), inhibitors of protein and nucleic acids synthesis (erythromycin, tetracycline, streptomycin, gentamycin), inhibitors of RNA polymerase (rifampicin) and inhibitors of DNA synthesis of microbial cells (ciprofloxacin) (Kravchenko et al. 2008). The experiments characterized these microorganisms by antibiotic sensitivity. Thus, among 120 strains, microorganisms were characterized by high sensitivity to gentamycin (110 strains—92 %), benzylpenicillin (99 strains—83 %), and vancomycin (100 strains—83 %). The microorganisms showed an intermediate level of resistance to erythromycin (60 strains—50 %) and ciprofloxacin (41 strains—34 %). The highest resistance was determined to rifampicin and streptomycin (23 and 22 %, respectively). According to the study, the dominant species *E. faecium* was characterized as sensitive to antibiotics, but some of the strains were resistant to aminoglycoside antibiotics (streptomycin, gentamicin), benzylpenicillin, and rifampicin. The highest number of strains resistant to antibiotics was determined among the species *E. faecium*, especially isolated from water samples of the South and Middle Baikal, in comparison with microorganisms isolated from the North Baikal. They were resistant to seven antibiotics studied, with the exception of ciprofloxacin.

3 Peculiarities of Survival of Microorganisms in Lake Baikal

The problem of metabolically active but non-culturable asporogenous bacteria was discussed in many studies (Panasovets 2007; Roszak and Colwell 1987; Oliver 2005). Uncultivable forms of bacteria are not included in sanitary and epidemiological studies of water, soil and food, as they cannot be detected by standard microbiological methods (Sokolenko 2006). The widespread occurrence of these bacterial forms in nature and the ability of pathogenic and opportunistic bacteria to move into an uncultivable state leads to their social and hygienic significance.

The influence of temperature and the concentration of nutrients are two of the main factors that determine the survival and the presence of bacteria in aquatic ecosystems. According to previous studies (Whitesides and Oliver 1997; Oliver 2005), oligotrophy is the main abiotic factor which causes the transition of microorganisms into the uncultivable state. In this state, bacteria lose the ability to be cultured on a nutrient medium, but remain viable and exhibit a low level of metabolic activity (Byrd et al. 1991; Roszak et al. 1984). This strategy to survive has been described in many gram-negative bacteria (Kell et al. 1998; Oliver 2000a, b, c), but it has only recently been demonstrated that gram-positive species, including *E. faecalis*, may enter an uncultivable state. In this state, cells of enterococci are metabolically active and can resume active growth when normal growth conditions have been restored (Lleo et al. 1998; Signoretto et al. 2000).

For the experiment involving the conversion of heterotrophic microorganisms into an uncultivable state, strains of asporogenous bacteria were used. They were obtained by the direct cultivation of aliquots of natural water of Lake Baikal (*Pseudomonas, Acinetobacter*), after the acclimatization experiment on viable cells (*Pseudomonas, Arsenicicoccus*). It was shown that the low incubation temperature (4 °C), as well as cultivation without nutrients (in distilled water) leads the cells to transfer into an uncultivable state. The cells did not form colonies on solid media after 36 h of incubation. Adding nutrients to the media led to the restoration of the cells' ability to grow even after 60 days of incubation in adverse conditions (Belkova, unpublished data).

Experiments were carried out to study the possible mechanisms by which an enterococci species—*E. faecium*—adapts to Baikalian water. It was show that there was a slight increase in the number of bacterial cells cultured at 4 and 24 °C within 36 h of the cells' incubation in Baikalian water, after a sharp decline of the cells was

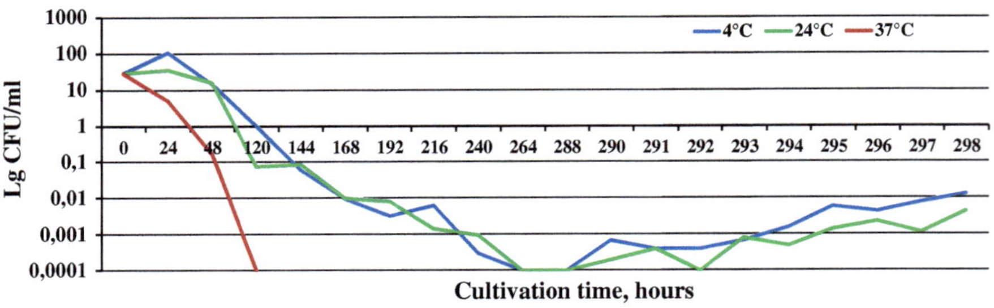

Fig. 7 Cultivation of the strain *E. faecium* No.8 in Baikalian water at different temperatures, read commas as points

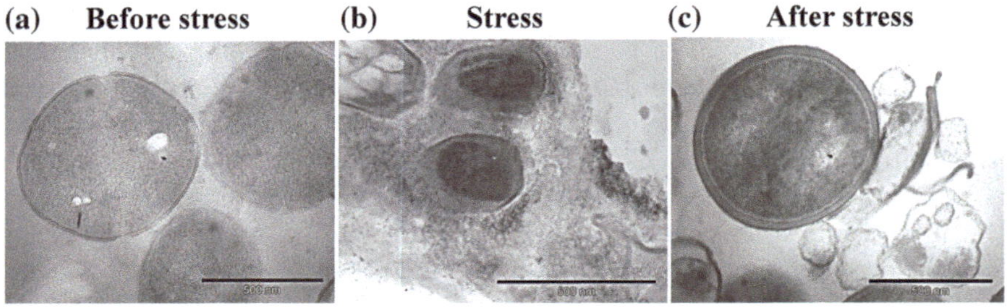

Fig. 8 Electron micrographs (TEM) of enterococci cells in an experiment on their survival in Baikalian water: **a** cultivable cells; **b** under stress conditions; **c** cell morphology after revival

detected (Fig. 7). After 12 days' incubation the cells of *E. faecium* are no longer cultivated in the Baikalian water, but the addition of nutrients (medium BHI) revived bacteria, which passed from an uncultivable state to a cultured one. The exceptions were the cells of *E. faecium* cultured at 37 °C, where the reduction in the cell number was determined immediately after inoculation and their death was observed during cultivation. The growth of these cells was not renewed even after nutrients were inoculated into the growth media.

Thus, the important peculiarities of Baikalian microorganisms are their ability to adapt and exist for a long time at a low temperature with minimum nutrient content, to fall into an uncultivable state in extreme conditions, and to take on cultivated forms upon the appearance of nutrients.

Further studies of cell adaptation in an uncultivable state were carried out using electron microscopy. It was shown that the cell size and the depth of the cell wall changed under stress conditions. This apparently promotes the survival of the bacteria at low temperatures and with an insufficient amount of organic matter for their growth. Electron microscopy has shown that at the beginning of the experiment, enterococci cells were in an intensively dividing state and had a thin cell wall and coccoid form (Fig. 8a). Under stress conditions, during cell cultivation in Baikalian water, cell maintenance occurs through the formation of a more dense and thick cell wall (Fig. 8b, c). This probably gives the bacterial wall extra strength, whereby the cells become more resistant to external stress factors. Epifluorescent microscopy of enterococci in an uncultivable state showed aggregates and cell clusters. From the literature it is known that cells of gram-negative bacteria in the transition to an uncultivable state have a coccoid form and vary in size (Catenrich and Makin 1991; Donelli et al. 1998), and are able to aggregate into clusters (Donelli et al. 1998).

Microbiological indicators of the pelagic zone of Lake Baikal have been fairly constant for a long time, according to long-term observations. Nevertheless, we cannot underestimate the influence of anthropogenic factors on the structure and composition of the microbial community. The entrance of allochthonous microorganisms, both pathogenic and opportunistic, into the lake ecosystem affects autochthonous microorganisms. This factor is of particular importance nowadays, when the active development of tourism and tourist and recreational zones on Lake Baikal has been going on.

4 Conclusions

The deterioration of the freshwater ecosystems has drawn increased attention to the sanitary requirements for the quality of drinking water and drinking water resources. The hygienic standardization of the quality of drinking water and sanitary protection of water reservoirs in the Russian Federation requires strict control of the direct and indirect impact of anthropogenic activities, domestic use and the discharge of the wastewaters into the ponds and streams, and is regulated by sanitary and epidemiological rules, norms, regulations, and guidelines.

Monitoring studies on microbiological indicators of Baikal water are conducted regularly in line with SANPIN standardized indicators and the general characteristics of aquatic heterotrophic microbial communities. They include enumerations of the TBN, the counts of heterotrophic, psychrotrophic, and oligotrophic bacteria and the spatial distribution of opportunistic bacteria, including those belonging to the genera *Pseudomonas, Clostridium* and *Enterococcus*.

Enterococci isolated from different sampling sites of Lake Baikal have been identified and their resistance to antibiotics tested to evaluate the potential danger of the water source to human health. Studies have shown the association of bacteria of the genus *Enterococcus* with the littoral areas of Lake Baikal, where there is a local anthropogenic influence. Enterococci of fecal origin have mainly been identified; most of them have shown antibiotic susceptibility. In general they could be considered specific indicators of water contamination by the feces of humans, animals and birds.

The temperature conditions of the lake substantially affect the total number and variety of microorganisms as well as the survival of pathogenic and opportunistic bacteria. Adaptation of the cultivation methods allows new strains to be isolated from the water of Lake Baikal. Experiments to transfer cells into an uncultivable state showed that both autochthonous and opportunistic heterotrophic microorganisms (as an example, *E. faecium* was used) cultured in the Baikal water and at low temperature fell into an uncultivable state, but their ability to divide was restored with the addition of nutrients. This might be a protective mechanism contributing to the survival of a variety of bacteria in the Baikal water.

Thus, microbiological monitoring of freshwater reservoirs, which are the main sources of drinking water, should include comprehensive information on indigenous heterotrophic and opportunistic microorganisms: a quantitative assessment of their diversity and a qualitative description of the composition, structure and dynamics of their distribution.

Appendix: Sample Preparation and Laboratory Methods

The determination of the total number of cultivable microorganisms (TNCM), total coliform bacteria (TCB), TTC, coliphages and spores of sulfite-reducing clostridia was carried out according to the methodological instructions MUK 4.2.1018-01 (2002).

Heterotrophic bacteria isolation from the water of Lake Baikal was performed in all seasons on the standard hydrological sections (Fig. 1). Water sampling was conducted using Nansen water samplers and sterile bottles. To account for the total psychrotrophic, heterotrophic and oligotrophic bacteria counts, cultivation were carried out on R2A (Sigma-Aldrich, USA), FPA/10 (g/L: nutrient broth—2.0, bacteriological agar—15.0) and PYA (g/L: peptone—1.0, yeast extract—1.0, bacteriological agar—15.0) media, respectively (Parfenova et al. 2006). Petri dishes with R2A medium were incubated at 4 °C, the others at 20–25 °C. Enumerations were made for heterotrophic and oligotrophic bacteria after 8–10 days of incubation, and after 12–14 days for psychrotrophic bacteria.

Morphological and biochemical studies were performed for all isolated strains using gram staining, and studying their morphological, physiological and biochemical characteristics. Further identification of the cultures was carried out using test systems MMTE-1 MMTE-2, a set of NIB indicators and test systems for the non-fermenting bacteria Lachema.

For total bacteria number (TBN) enumeration, epifluorescence microscopy was used. Water samples were filtered through 0.22 μm filters, stained with 4,6-diamino-2-phenylindole (DAPI) and visualized with the Olympus epifluorescent microscope (×100).

Analyses of bacteria of the genera *Enterococcus* and *Pseudomonas* were carried out by means of membrane filtration: a volume of water (50, 100 mL) was passed through a filter with a pore size of 0.45 μm, then plated on selective media. For enterococci detection, a selected nutrient agar was used containing 2,3,5-trifeniltetrazoliyhlorid and sodium azide (NaN_3), which inhibited the growth of gram-negative bacteria. Bacteria of the genus *Enterococcus* form red or pink colonies on this medium. Plates were incubated at 37 °C. For more accurate results, filters with grown colonies were transferred to a medium containing bile, esculin and ammonium iron(III) citrate. Plates were additionally incubated at 44 °C for 2 h.

Fecal enterococci have the ability to hydrolyze esculin and formed black-colored colonies. The results of enumeration express the number of colony-forming units (CFU) per 100 mL of water.

The antibiotic resistance of the strains was studied using the disk diffusion method for 8 antibiotics: streptomycin, tetracycline, vancomycin, benzylpenicillin, ciprofloxacin, erythromycin, gentamicin, and rifampicin. Analysis was conducted according to three groups: sensitive, intermediate and resistant (MUK 4.2.1890-04 2004).

A study of the cell morphology was carried out using transmission (TEM) microscopy. For TEM, visualization cells were fixed by glutaraldehyde at a final concentration of 2.5 % (2 h), washed with a phosphate buffer (pH 7.0) and postfixed with 1 % osmium oxide in a sodium phosphate buffer with potassium ferricyanide (0.2 %) for 2 h. Dehydration was performed in ethanol series: 30, 40, 50, 70 and 96 %, each for 10 min, then twice for 30 min in anhydrous acetone. The dehydrated material was impregnated into a mixture of epoxy resin (Araldite 502 Kit SPI USA) and acetone at a ratio of 1:1, then transferred to a Beem™ polypropylene capsule in a fresh mixture of epoxy resins with a catalyzer (250 μl of DMP-30 to 20 ml of resin) and polymerized at 37 °C (12 h) and at 60 °C (48 h). The thin sections were obtained on the microtome Ultracut R LEICA (Austria), and mounted on palladium grids. Samples were contrasted with lead citrate (Reynolds 1963) and washed in 0.02 M NaOH, and in distilled water. The thin sections were analyzed using a Leo 906E transmission electron microscope (Zeiss, Germany).

For the experiments on the survival of enterococci cells under extreme conditions, two strains of *E. faecium* were used. The first, *E. faecium* N8, was isolated in 2005 from the water near the Kultuk settlement, and the second, *E. faecium* S11, was isolated in 2006 from the water of the Selenga River delta. Laboratory experiments were conducted with enriched cultures. Cells were grown in a 50 mL flask containing 30 mL of liquid selective media BHI (Difco) at 37 °C for one day. After biomass growth, cells were centrifuged and washed with a saline buffer (0.9 % NaCl). Then, aliquots containing about 1.2×10^7 cells were inoculated into flasks with 100 mL of the non-sterile Baikalian water and incubated at different temperatures: 4, 20, and 37 °C. The control flask contained no bacteria. Enumerations of bacteria were performed daily by sampling an aliquot of 1 ml of water, which was plated on selective esculin agar with azide and kanamycin (HiMedia, M510). Experiments were carried out in triplicate and cells were cultured overnight at 37 °C.

References

Belkova NL (2004) Taxonomic diversity of deep water microbial community of Lake Baikal. (Таксономическое разнообразие микробного сообщества водной толщи озера Байкал) Summary of Ph.D thesis. Vladivostok: Far East University, p 20 (In Russian)

Belkova NL, Denisova LIa, Manakova EN, Zaichikov EF, Grachev MA (1996) Species diversity of deep water microorganisms in Lake Baikal, detected by 16S rRNA sequences. (Видовое разнообразие глубоководных микроорганизмов озера Байкал, выявленное по последовательностям 16S рРНК) Dokl Akad Nauk 348(5):692–695 (In Russian)

Belkova NL, Parfenova VV, Kostopnova TIa, Denisova LIa, Zaichikov EF (2003) Microbial biodiversity in the Lake Baikal water. Translated Microbiologiya 72(2):203–212

Byrd JJ, Xu H, Colwell RR (1991) Viable but nonculturable bacteria in drinking water. Appl Environ Microbiol 57:875–878

Carrillo M, Estrada E, Hazen TC (1985) Survival and enumeration of faecal indicators Bifidobacterium adolescentis and Escherichia coli in a tropical rain forest watershed. Appl Microbiol 50:468–476

Catenrich CE, Makin KM (1991) Characterization of the morphologic conversion of Helicobacter pylori from bacillary to coccoid forms. Scand J Gastroenterol 26:58–64

Denisova LYa, Bel'kova NL, Tulokhonov II, Zaichikov EF (1999) Diversity of bacteria at various depths in the southern part of lake Baikal as detected by 16S rRNA sequencing. Translated Microbiologiya 68(4):475–483

Donelli G, Matarrese P, Fiorentini C, Dainelli B, Taraborelli T, Di Campli E, Di Bartolomeo S, Cellini L (1998) The effect of oxygen on the growth and cell morphology of Helicobacter pylori. FEMS Microbiol Lett 168:9–15

Drucker VV, Panasyuk EYu (2006) Potentially pathogenic bacteria in a microbial community of Lake Baikal. Hydrobiologia 568(1):267–271

Drucker VV, Sorokovikova LM, Sinyukovich VN, Potemkina TG, Nikulina IG, Molozhavaya OA, Korovyakova IV (1997) Quality of water of river Barguzin in modern period. (Качество воды реки Баргузин в современных условиях) Geografiya i prirodnue resursy 4:74–80 (In Russian)

Facklam RR, Shaw DF, Teixeira LM (1999) Enterococcus. In: Murray PR, Baron FJ, Pfaller MA, Tenover FC, Yolken RH (eds) Manual of clinical microbiology, 7th edn. American Society for Microbiology, Washington, D.C., pp 297–305

Federal Law (1999) Federal Law « On the sanitary-epidemiological welfare of the population » (О санитарно-эпидемиологическом благополучии населения), March 30, 1999 № 52-FZ http://www.consultant.ru/document/cons_doc_LAW_164608/. Accessed on 16 Dec 2014

Fomin GS (2002) Water. Control of chemical, biological and radioactive safety according to international standards. (Вода. Контроль химической, бактериальной и радиационной безопасности по международным стандартам) Moscow: Publisher « Protector », pp 70–77 (In Russian)

Gilmore (2002) The Enterococci: pathogenesis, molecular biology, and antibiotic resistance. In: Gilmore MS (ed) American society for microbiology, Washington, p 439

Gladkikh AS, Belykh OI, Levina OV, Parfenova VV (2011) Metagenomic analysis of the bacterial communities from Lake Baikal. (Метагеномный анализ сообществ микроорганизмов из озера Байкал) Vestnik Ural'skoj meditsinskoj akademicheskoj nauki 4(1):76–77 (In Russian)

Grachev MA (2002) About modern situation of ecological system of Lake Baikal. (О современном состоянии экологической системы озера Байкал) Publisher of SD RAS, Novosibirsk, p 156 (In Russian)

Hahn MW, Stadler P, Qinglong LW (2004) The filtration-acclimatization method for isolation of an important fraction of the not readily cultivable bacteria. Microbiol Methods 57:379–390

Kayaoglu G, Østavik D (2004) Virulence factors of Enterococcus faecalis: relationship to endodontic disease. Crit Rev Oral Biol Med 15(5):308–320

Kell DB, Kapreylants AS, Weichart DH, Harwood CL, Barer MR (1998) Viability and activity in readily culturable bacteria: a review and discussion of the practical issues. Ant Van Leeuvenhoek 73:169–187

Kravchenko OS, Parfenova VV, Pavlova ON (2008) Antibiotikoresistance of bacteria genus Enterococcus isolated from water of Lake Baikal. (Антибиотикорезистентность бактерий рода *Enterococcus* изолированных из воды Байкала) Bull VSNTs SB RAMS (2):67 (In Russian)

Kuznetsov SI (1957) Microbiological characteristic of water and sediments of Baikal. (Микробиологическая характеристика вод и грунтов Байкала) In: Proceedings of Baikal Limnological Station AS USSR 15:45–50 (In Russian)

Lapteva NA (1990) Species characteristic of oligotrophic bacteria in Lake Baikal. (Видовая характеристика олиготрофных бактерий в озере Байкал) Microbiologiya 59(3):499–506 (In Russian)

Lleo MM, Tafi MC, Canepari P (1998) Nonculturable Enterococcus faecalis cells are metabolically active and capable of resuming active growth. Syst Appl Microbiol 21:333–339

Lobova TI (2003) Biological peculiarities of autochtonous and allochtonous bacteria from Lake Shira and markers for anthropogenic influence on aquatic ecosystem. (Биологические особенности автохтонных и аллохтонных бактерий из оз. Шира и маркеры антропогенного воздействия на водную экосистему) Summary of Ph.D. Thesis. Perm: Institute of ecology and genetic of microorganisms, p 24 (In Russian)

Maksimova EA, Maksimov VI (1989) Microbiology of water of Baikal. (Микробиология вод Байкала) Irkutsk: Publisher of Irkutsk State University, p 167 (In Russian)

Meier H, Koob C, Ludwig W, Amann R, Frahm E, Hoffmann S, Obst U, Shleifer KH (1997) Detection of Enterococcci with rRNA targeted DNA probes and their use for hygienic drinking water control. Water Sci Technol 35:437–444

Mladova TA (1971) On qualitative composition of bacterioplankton. (О качественном составе бактериопланктона) In: Proceedings of Baikal Limnological Station AS USSR 12(32):196–201 (In Russian)

Moiseenko NN (1994) Antibiotic-resistant enterobacteria in sewage and the water of surface reservoirs. Antibiot Khimioter 39(3):64–68

MUK 4.2.1018-01 (2002) Sanitary-microbiological analysis of drinking water (Санитарно-микробиологический анализ питьевой воды), Federal Centre Gossanepidnadzor Minzdrav Russia, Moscow, p 24 (In Russian)

MUK 4.2.1884-04 (2004) Sanitary-microbiological and sanitary-parasitological analysis of water from surface water reservoirs (Санитарно-микробиологический и санитарно-паразитологический анализ воды поверхностных водных объектов), Federal Centre Gossanepidnadzor Minzdrav Russia, Moscow (In Russian)

MUK 4.2.1890-04 (2004) Determination of sensibility of microorganisms to antibacterial substances: methodological instructions (Определение чувствительности микроорганизмов к антибактериальным препаратам: Методические указания), Federal Centre Gossanepidnadzor Minzdrav Russia, Moscow, p 91 (In Russian)

Namsaraev BB, Zemskaya TI (2000) Microbiological processes of carbon cycling in bottom sediments of Lake Baikal. (Микробиологические процессы круговорота углерода в донных осадках озера Байкал) Publisher of SD RAS, GEO, Novosibirsk, p 160 (In Russian)

Oliver JD (2000a) Problems in detecting dormant (VBNC) cells and the role of DNA elements in this response. In: Jansson JK, van Elsas JD, Bailey MJ (eds) Tracking genetically-engineered microorganisms. Landes Biosciences, Georgetown, pp 1–15

Oliver JD (2000b) The public health significance of viable but nonculturable bacteria. In: Colwell RR, Grimes DJ (eds) Nonculturable microorganisms in the environment. American Society for Microbiology Press, Washington, D.C, pp 277–299

Oliver JD (2000c) The viable but nonculturable state and cellular resuscitation. In: Bell CR, Brylinsky M, Johnson-Green P (eds) Microbial biosystems: new frontiers. Atlantic Canada Soc Microb Ecol, Halifax, pp 723–730

Oliver JD (2005) The viable but nonculturable state in bacteria. J Microbiol 43:93–100

Panasovets OL (2007) Development of liquid enriched nutrient media for isolation of salmonella from aquatic systems. (Разработка жидкой питательной среды накопления для выделения сальмонелл из водных объектов) Summary of Ph.D. thesis, Moscow: Moscow State University, p 22

Parfenova VV, Bel'kova NL, Denisova LYa, Zaychikov EF, Marsimenko SYu, Zakharova YuR, Poddubnyak NYu, Molozhavaya OA, Nikulina IG (2006) Investigation of species composition of cultivable heterotrophic microorganisms from Lake Baikal. (Изучение видового состава культивируемых гетеротрофных микроорганизмов оз. Байкал) Inland Water Biol 1:8–15 (In Russian)

Parfenova VV, Kravchenko OS, Pavlova ON, Kostornova TYa, Nikulina IG (2009a) Quality of water from Lake Baikal, problems and perspectives its usage. (Качество воды озера Байкал, проблемы и перспективы ее использования) Proizvodstvenno-technicheskii zhurnal. Vodoochistka. Vodopodgotovka. Vodosnabzhenie. 1:48–54 (In Russian)

Parfenova VV, Pavlova ON, Kravchenko OS, Kostornova TYa, Nikulina IG, Tomberg IV, Ivanov VG, Meyer Y, Jayter D, Fickle Ya, Baksman T, Obst Y (2009b) Studies on the local anthropic influence on the horizontal and vertical distribution of microorganisms in the water of Lake Baikal. (Изучение локального антропогенного влияния на горизонтальное и вертикальное распределение микроорганизмов в воде оз. Байкал) Hydrobiol J 45(2):51–62 (In Russian)

Parfenova VV, Pavlova ON, Kravchenko OS, Tulupova YuR, Kostornova TYa (2010) Investigation of the distribution of the bacteria of Enterococcus genus in the water of Lake Baikal, their species composition and degree of stability to antibiotics. (Изучение распределения видового состава и степени устойчивости к антибиотикам бактерий рода Enterococcus в воде озера Байкал) Sibirskii Ecologicheskij Zhurnal 4:633–640 (In Russian)

Parfenova VV, Gladkikh AS, Belukh OI (2013) Comparative analysis of biodiversity of bacterial communities of plankton and biofilms in Lake Baikal. (Сравнительный анализ биоразнообразия бактериальных сообществ планктона и биопленки в озере Байкал) Microbiologiia 82(1):94–105 (In Russian)

Pavlova ON, Drucker VV, Kostornova TYa, Nikulina IG (2003) Peculiarities of distribution of bacteria genus Pseudomonas in Lake Baikal. (Особенности распространения бактерий рода Pseudomonas в озере Байкал) Sibirskij Ecologicheskij Zhurnal 3:267–272 (In Russian)

RD 52.24.633-2002 (2002) Methodological instructions. Methodological basis for the creation and operation of environmental monitoring subsystem recourse freshwater ecosystems (Методические указания. Методические основы создания и функционирования подсистемы мониторинга экологического регресса пресноводных экосистем), Moscow, Roshydromet, p 40 (In Russian)

Reynolds ES (1963) The use of lead citrate at high pH as an electron-opaque stain in electron microscopy. J Cell Biol 17:208–212

Romanova AP (1958) Season dynamics of bacterioplankton, its horizontal and vertical distribution in South basin of Lake Baikal. (Сезонная динамика бактериопланктона, его горизонтальное и вертикальное распределение в Южной части Байкала) Bull SD AS USSR 7:114–124 (In Russian)

Roszak DB, Colwell RR (1987) Survival strategies of bacteria in the natural environment. Microbiol Rev 51:365–379

Roszak DB, Grimes DJ, Colwell RR (1984) Viable but nonrecoverable stage of Salmonella enteritidis in aquatic systems. Can J Microbiol 30:334–338

SANPIN 2.1.4.1074-01 (2002) Drinking water. Hygienic requirements for water quality of centralized drinking water supply. Quality control: sanitary rules and requirements (Питьевая вода. Гигиенические требования к качеству воды централизованных систем питьевого водоснабжения. Контроль качества. Санитарные правила и нормы), Federal Centre Gossanepidnadzor Minzdrav Russia, Moscow, p 104 (In Russian)

SANPIN 2.1.4.1175-02 (2003) Hygienic requirements for water quality in noncentralized water supply systems. Sanitary safeguarding of sources (Гигиенические требования к качеству воды нецентрализованного водоснабжения. Санитарная охрана источников), Federal Centre Gossanepidnadzor Minzdrav Russia, Moscow, p 29 (In Russian)

SANPIN 2.1.5.980-00 (2000) Hygienic Requirements for Surface Water Protection. Quality control: Sanitary rules and requirements (Гигиенические требования к охране поверхностных вод. Контроль качества: Санитарные правила и нормы), Federal Centre Gossanepidnadzor Minzdrav Russia, Moscow, p 27 (In Russian)

Shimaraev MN (1978) Temperature regime (Тепловой режим) Proceeding of AS USSR, SD, Limnological Institute, Novosibirsk, 16(36):102–117 (In Russian)

Signoretto C, Lleò MM, Tafi MC, Canepari P (2000) Cell wall chemical composition of Enterococcus faecalis in the viable but nonculturable state. Appl Environ Microbiol 66 (5):1953–1959

Sokolenko AV (2006) Uncultivable forms of bacteria: distribution in nature, inductors of uncultivable state and reversion. (Некультивируемые формы бактерий: распространение в природе, индукторы некультивируемого состояния и реверсии) Sovremennue naukoemkie tekhnologii 2:11–15 (In Russian)

Whitesides MD, Oliver JD (1997) Resuscitation of Vibrio vulnificus frcm the viable but nonculturable state. Appl Environ Microbiol 64:3025–3028

Chapter 5
Developing the Regional Indicator Indexes of Zooplankton for Water Quality Class Determination of Water Bodies in Siberia

Nadezhda Yermolaeva and Serafima Dvurechenskaya

Abstract It is shown that the characteristics of individual species of zooplankton used to determine the water quality class in Western Siberia differ from those in Europe. To define a class of water quality in the saprobic indexes, the lists of indicator organisms were used based on the presence or absence of relevant taxa and their quantitative ratio. This allowed a water body or its area to be rated in an appropriate class of water quality. Usually the indicator significance and saprobic valence of indicator species were found in the tables developed through long-term research and literature data. These data were evaluated for European water bodies. However, the values of the indicator significance of individual species depend substantially on the number of regional factors, which can produce an essential error in the calculation of the index and saprobity and in the definition of the water quality class, respectively. In this paper the values of the indicator significance (s) and indicator weight (J) for 111 species of zooplankton were calculated taking into account the regional peculiarities of the South of Western Siberia. The examples of calculation according to Pantle and Buck's saprobity index using indexes from the literature on one hand and calculated for the specific region on the other hand were discussed. It is shown that, using indexes obtained from regional features of the hydrochemical background of reservoirs and rivers, one can define the water quality class more exactly. Thus, the use of regional indexes is appropriate because it provides a more objective assessment of the state of the ecosystem.

N. Yermolaeva (✉) · S. Dvurechenskaya
Novosibirsk Department, Institute for Water and Environmental Problems (IWEP), SB RAS, Morskoy Prospect 2, Novosibirsk 630090, Russia
e-mail: hope@iwep.nsc.ru

S. Dvurechenskaya
e-mail: dvur@iwep.nsc.ru

L. Mueller et al. (eds.), *Novel Methods for Monitoring and Managing Land and Water Resources in Siberia*, Springer Water,
DOI 10.1007/978-3-319-24409-9_5

1 Introduction

Assessing the class of water quality in water bodies is extremely important to establish the ecological status of the water body as a whole. This problem can be solved in different ways: by direct chemical analysis or by bioindication techniques. In certain cases, the methods of bioindication are preferable for a prompt assessment and determination of the state of the water bodies for a sufficiently long time interval, in contrast to chemical methods, which give the water quality for a specific moment. As is known, zooplankton organisms allow us to determine both the effects of instant (snapshot) and chronic water pollution. Because of the expanded life-cycle of these organisms, their presence in a reservoir characterizes a prolonged period of time. The biological method of water quality assessment differs in its efficiency from chemical and bacteriological methods, which identify water pollution only at the time of sampling.

Methods of bioindication have been developed for the various components of the biota of aquatic ecosystems: phytoplankton, zooplankton, zoobenthos, and macrophytes. Zooplankton is one of the most sensitive components of the aquatic ecosystem components, because of its longer life cycle than that of phytoplankton, but shorter than that of zoobenthos. The zooplankton community, like any ecosystem community, is characterized by a constant species composition, dynamic stability, and definite inherent organization. Changes in the conditions of organisms' existence affect the species composition, quantitative indicators, the ratio of individual taxonomic groups, and the structure of zooplankton populations. Thus, zooplankton may serve as a characteristic of the aquatic environment.

The purpose of this work is to determine the indicator properties of zooplanktonic organisms for water bodies of the South of Western Siberia and to make a correction to the definition of a class of water quality.

The most usable index for estimating the degree of water pollution in terms of zooplankton appears to be the index of Pantle and Buck (*S*) as modified by Sladečhek (1973). The aim of this method is to compare the results of studies of water bodies in different areas. To calculate the index, one needs to know the magnitude of the indicator value of each organism *s* (degree of confinement of species to water bodies and its parts with a certain level of organic pollution).

Mostly, the values of the indicator significance *s* of zooplankton are determined from published tables developed from European water bodies (Unified Methods 1977; Makrushin 1974a, b). However, this approach, as noted by many authors (Finogenova and Alimov 1976; Andrushaitis et al. 1981), should be applied with substantial corrections due to the climatic, hydrological, and hydrochemical conditions of different regions. One also needs to take into account the differences in the fauna and hydrochemical regimes in European and Siberian water bodies. When the indexes calculated for the zooplankton of water bodies in the South of Western Siberia are compared with the widely used tabulated data obtained for water bodies in Europe, one can find that a number of species of zooplankton differ significantly in terms of the indicator significance. Thus, the water bodies where bioindication methods are used fall in completely different water quality classes. The error in the class definition of water quality can vary widely, depending on the species composition of zooplankton and on the quantity of indicator species. However, more often, the saprobity index using the tables of indicators generally appears to be overevaluated compared to the real condition of the water body. Using indexes obtained from the regional features of reservoirs and rivers, one can define the water quality class more exactly. Thus, the use of regional indexes is appropriate because of their more objective assessment of the state of the ecosystem.

At present, much attention is devoted to developing standards for the permissible impact of different types of negative effects on aquatic ecosystems and schemes for the complex use and protection of water bodies. These schemes are developed in order to determine the permissible anthropogenic load on water bodies, to determine water needs in the future, to ensure protection of water bodies, to ensure the sustainability of ecological systems, and to identify safe levels of contaminants, as well as other indicators of the impact on water bodies. Guidelines for the development of both standards and schemes require the actual state of the water body to be determined. The assessment of this state must be based on calculations of a regional indicator's values of the various components of aquatic ecosystems. In this context, the definition of regional characteristic values of hydrochemical and hydrobiological parameters is a current problem. In the different countries of Europe, this approach is implemented in the Water Framework Directive (WFD, "Directive 2000/60/EC").

Water quality standards established in accordance with physical, chemical, biological, and other parameters are based on the results of a retrospective analysis of monitoring of hydrobiological and abiotic (hydrochemical, etc.) parameters. The definition of a range of regional background values of water bodies must meet the criteria of ecological welfare of a water body (normal procreation of the main links of the ecological system). The analysis of the initial information on the water body's state includes data on biotic and abiotic characteristics of the water body and its catchment area. The hydrobiological characteristics (species diversity, the number of indicator organisms, biomass, production, reproduction rates of aquatic organisms, the composition and number of specially protected species of aquatic plants and animals, etc.) are considered as biotic parameters.

Differences in the fauna of Central European and Siberian reservoirs should also be considered. Finogenova and Alimov (1976) suppose that special studies are required to develop methods for assessing the features of the fauna of different regions in accordance with the zoogeographical division of inland waters. Moreover, it is important to take into account not only zoogeographical but also geochemical zoning characteristics. The reason is that the background concentrations of a number of chemicals such as nutrients and toxic substances in different regions may vary significantly. Thus, the study of the elemental chemical composition of soils, plants, and water showed that Western Siberia is a complex region in terms of biogeochemistry (Il'in 1973; Il'in and Syso 2004). Thus, the average copper content in the soils of the central and southern regions of the former USSR is 4.5–10.0 mg/kg, and in the south of Western Siberia it is up to 30.6 mg/kg (Il'in 1973). Nechayeva (2001) noted the increased values of iron content in the forest and forest-steppe landscapes of the Priob region. For Upper Ob, the extremely high content of dissolved iron is confined mainly to the flood period. At the same time, variation during the year may be due to both natural factors and anthropogenic pressure (Balykin et al. 2011). The phosphorus content in the reservoirs of the South of Western Siberia is usually below detection limits, whereas in the European part of Russia, phosphorus is the defining element in the eutrophication of water bodies.

The eutrophication of water bodies in the area under consideration is determined by the accumulation of a number of other nutrients: nitrogen, iron, organic carbon, etc. In connection with the above-mentioned factors, it is necessary to review the list of indicator organisms as closely as possible before expanding it to include each region based on the hydrochemical features of aquatic ecosystems.

2 Materials and Methods

The saprobity valence, indicator weight, and significance (saprobity index) of zooplankton species calculations were made on the basis of data on 132 lakes in the Novosibirsk and Altai regions, Northern Kazakhstan, on 10 transects in the Novosibirsk reservoir, and several small rivers in the South of Western Siberia, which were collected between 1990 and 2010 (Fig. 1). Photographs given in the Appendix show some of these water bodies sampled. The results of processing 1236 zooplankton samples were included in the analysis.

Zooplankton samples were collected by straining 50–100 L of water from a certain depth across the Apshteyn's net (mesh width 0.065 mm). Each sample of zooplankton was immediately fixed with 4 % neutral formalin. Zooplankton and hydrochemical samples were collected at the same sites. Investigations covered all hydrological seasons: winter, spring, autumn, and summer.

The species composition of zooplankton was determined from fixed samples using a binocular microscope MBS-9 with a magnification factor of 12–100×. A more accurate determination was made under the microscope on a glass slide in an aqueous solution of glycerol at a magnification factor of 400×. The species were identified using an identification guide.

The quantitative assessment consisted in calculating the number of organisms for each species in the sample as far as possible by age stages or group size. The estimation was carried out in a Bogorov chamber. The abundance was then recalculated per 1 m^3.

Samples of water for hydrochemical analysis were taken simultaneously with hydrobiological samples, at the same sites and analyzed in the laboratory of the Federal State Agency "VerhneObregionvodhoz" of the Russian Ministry of Natural Resources using standard methods for analyzing freshwater [List of procedures included in the State Register of methods of quantitative chemical analysis, 01.02.2009. Part I. Quantitative chemical analysis of water (http://www.fcao.ru/metodiki-kkha/2-uncategorised/114-perechni-metodik-vklyuchennykh-v-reestrpnd-f.html)]. These parameters were biochemical oxygen demand BOD_5, chemical oxygen demand COD, dissolved oxygen, pH, and the contents of ammonium, nitrite, nitrate, and phosphate ions (GOST 17.1.2.04-77 1977).

The method developed by Pantle and Buck is the most convenient one to analyze the zooplankton community (Sladeček 1973; Makrushin 1974a). The advantage of this method is to compare results of studies of water bodies of different areas. Quantitative assessment using the Pantle and Buck method considers the relative frequency of occurrence of organisms *h* and the value indicator significance of each organism *s*—the degree of confinement of species to water bodies in whole and separate areas of water bodies with a certain level of organic pollution. Both of these values *h* and *s* are included in the formula for calculating the saprobity index (*S*).

$$S = \frac{\sum (sh)}{\sum h}$$

On the Pantle and Buck scale *h* = 5 means an extremely high frequency (dominant), *h* = 4—high frequency (subdominant), *h* = 3—not frequent; *h* = 2—occasional, and *h* = 1—very rare (single one).

On the basis of hydrochemical parameters, the water quality class was estimated on a six-grade scale (Oksijuk et al. 1993; GOST 17.1.2.04-77 1977), and each class was assigned to a certain category of water saprobity according to the value of the Pantle and Buck index (*S*) (Table 1).

Water bodies with water quality class 4b—"heavily polluted" and 5—"dirty" were not found and were not discussed in this paper.

There can be some cases when hydrochemical and hydrobiological samples taken at the same time do not occur in the same class of water quality. In these situations it is desirable to specify both parameters, as it should be kept in mind that zooplankton may reflect the maximum impact of volley pollution, whereas a chemical sample reflects only the quality of the water at the time of sampling. Restructuring the zooplankton community disturbed due to pollution requires quite a long time.

Indexes of the saprobic valence (*a*) of zooplanktonic organisms were calculated using the equation proposed by Shitikov et al. (2003) for benthic organisms. This equation takes into account the proportions (relative abundance) of species, rather than simply the fact of their presence or absence, which improves the sensitivity of the calculated saprobity index.

The sample formed as described above was used to calculate the values of a_{ik}, the saprobic valence of species *i* in class *k*. The saprobic valence of a species according to Zelinka and Marvan (1961) indicates the extent to which it is characteristic of waters with a certain level of saprobity. In the simplest case, this parameter may be considered proportional to the relative abundance of species *i* in samples from biotopes of class *k* (Shitikov et al. 2003):

$$a_{ik} = \frac{10 \sum_{j=1}^{m_k} x_{ijk}}{\sum_{k=1}^{L} \sum_{j=1}^{m_k} x_{ijk}},$$

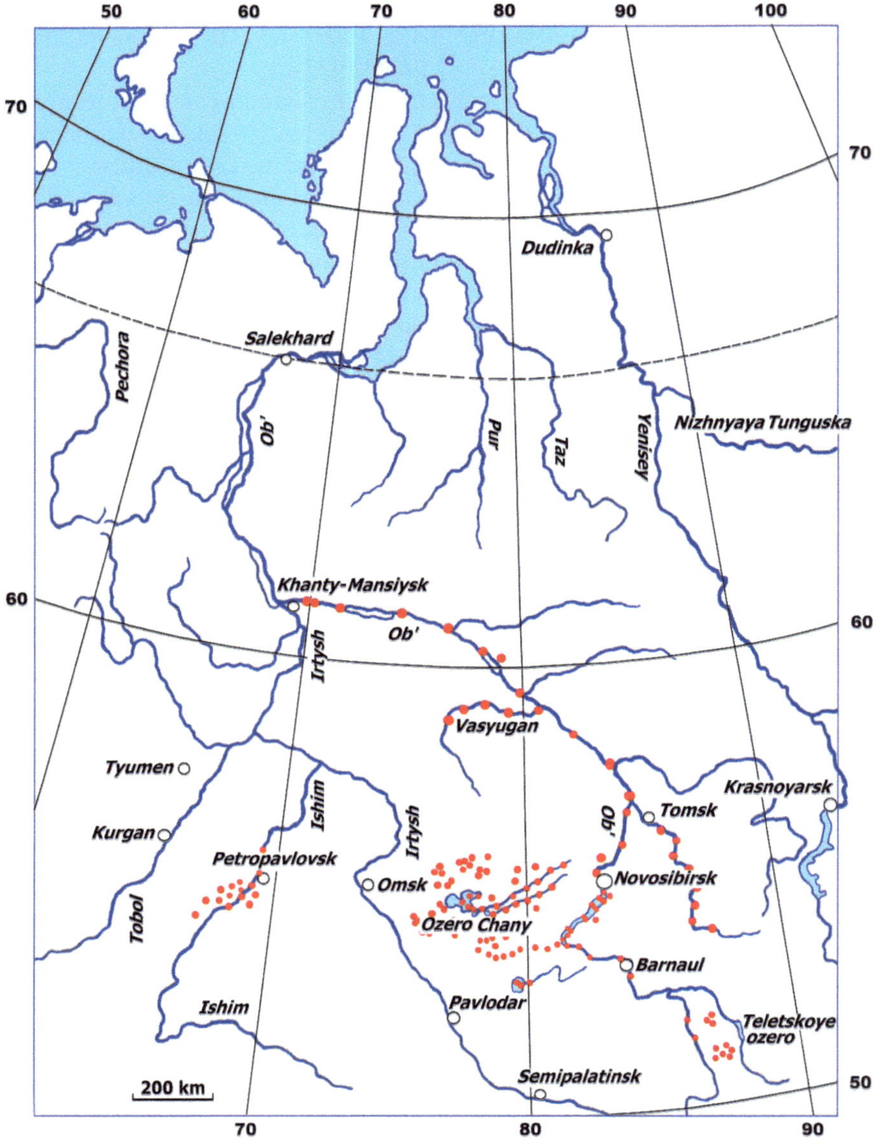

Fig. 1 Sampling locations

where L is the number of water quality classes (in our case, $L = 6$), and m_k is the number of measurements of k-th type. The resultant matrix with dimensionality $6n$ (with n being the total number of species per sample) makes it possible to estimate the role of each species and its occurrence frequency in a water body of a certain type (Table 2). The sum of valences calculated for each species in all six zones of water saprobity is 10 grades. We took x_{ik} to be equal to the proportion (%)

Table 1 The adequacy of hydrochemical classes and saprobic ranks of water quality on hydrobiological indicators (GOST 17.1.2.04-77 1977; Oksijuk et al. 1993)

Hydrochemical water quality classes						
	1	2 a	2 б	3 a	3 б	4 a
	Extremely clean	Very clean	Quite clean	Sufficiently clean	Slightly polluted	Moderately polluted
NH_4^+ mgN/l	<0.05	0.05–0.10	0.11–0.20	0.21–0.30	0.31–0.50	0.51–1.00
NO_2^- mgN/l	0	0.001–0.002	0.003–0.005	0.006–0.010	0.011–0.020	0.021–0.050
NO_3^- mgN/l	<0.05	0.05–0.20	0.21–0.30	0.31–0.50	0.51–0.70	0.71–1.00
PO_4^{3-} mgP/l	<0.005	0.005–0.015	0.016–0.030	0.031–0.050	0.051–0.100	0.101–0.200
$BOD_5 mgO_2$/l	<0.4	0.4–0.7	0.8–1.2	1.3–1.6	1.7–2.1	2.2–4.0
COD mgO/l	<8	8–12	13–18	19–25	26–30	31–40
Hydrobiological level of water quality index of Pantle and Buck (*S*)						
	$S < 0.5$	$0.5 \le S < 1.0$	$1.1 \le S \le 1.5$	$1.5 < S \le 2.0$	$2.0 < S \le 2.5$	$2.5 < S \le 3.0$
	xenosaprobic	oligosaprobic	oligo-β-mesosaprobic	β-mesosaprobic	β-α-mesosaprobic	α-mesosaprobic

of individuals of *i*th species from the total number of organisms in the sample. If the number of individuals (*N*) was used instead of the proportion ($x_{ik} = N_{ik}$), the equation acquired the form proposed by Tsimdin' (1979). In our case, this form of the equation was not suitable, because water objects included in the study differed markedly in terms of the quantitative parameters of zooplankton.

Proceeding from the distribution of saprobic valences a_{ik} in the above zones, we then calculated the values of indicator significance *s* and indicator weight *J* (Tsimdin' 1979) for some species and forms of zooplanktonic organisms, namely for Southern Siberia water bodies (Table 3).

In each of the saprobity zones (from oligosaprobic to polysaprobic) every species of zooplankton is characterized by some value of indicator significance: in oligosaprobic 1, in β-mesosaprobic 2, in α-mesosaprobic 3, in polysaprobic 4 (Sladečhek 1973).

The main zone is considered the one where the index of saprobic valence (a_{ik}) in the resulting matrix is the greatest. If the species is found in a more contaminated area, the indicator significance (*s*) increases proportionally to the value of saprobic valence (a_{ik}) for the same species in a contaminated area. For species occurring outside the main zone, in the less contaminated ones, the indicator significance (*s*) decreases (Table 2). For example, at a saprobity valence ratio of 9:1 the indicator significance changes to 0.1; at a ratio of 8:2 it changes to 0.2. When evaluating the transition zones (oligo-β-mesosaprobic, β-α-mesosaprobic, etc.) the decrease or increase in the indicator significance index is of 0.05. At the allocation of 1:8:1 or 2:6:2, the indicator significance does not change (Sladečhek 1973; Tsimdin' 1979).

The indicator weight (*J*), equal to 5, is a good indicator if all 10 points of saprobic valence (a_{ik}) are distributed in the same or neighboring saprobity areas at a ratio of 9:1; $J = 4$ if the saprobic valence is distributed in two adjacent areas at a ratio of 8:2; 7:3 or in three zones at a ratio of 1:8:1 (Sladečhek 1973). The indicator is bad if its saprobic valence is uniformly distributed in all the zones, and in the resulting matrix *J* is equal to 1.

A comparison of our data with those from the tables for European water bodies (Unified Methods 1977; Makrushin 1974b ; Tsimdin' 1979; Naberezhnyj 2010; Chertoprud and Chertoprud 2010) showed that, in some species such as *Alonella nana, Keratella hiemalis, Cyclops strenuus*, and others, they not only differ in value but also indicate a different class of water quality. This leads to serious errors in the calculated regional values of the Pantle and Buck index, which may vary widely depending on the species composition of a given zooplankton cenosis and the proportion of indicator species in it. In most cases, the saprobity index for Western Siberian water bodies proves to be overestimated.

3 Results and Discussion

The indexes of indicator significance were calculated exclusively for freshwater reservoirs [a total water mineralization (TDS) of no more than 1.0 g/dm^3]. Under the conditions of saltish and saline water the saprobity indexes have to be calculated

Table 2 Calculation of the values of indicator significance *s* from the distribution of saprobic valences a_{ik}

species	a_{ik}						s
	Xeno-saprobic (0.5)	Oligo-saprobic (1.0)	o-β-meso-saprobic (1.5)	β-meso-saprobic (2.0)	β-α-meso-saprobic (2.5)	α-meso-saprobic (3.0)	
Asplanchna priodonta Gosse	0.27	2.70	2.04	4.32	0.67	0.00	1.50
Asplanchna herricki Guerne	0.43	1.70	2.56	4.25	1.06	0.00	1.61
Brachionus angularis Gosse	0.00	1.37	2.88	3.84	0.68	1.23	1.74

Table 3 Indexes of indicator significance (s) and indicator weight (J) of some zooplankton species in water bodies of the South of Western Siberia

Indicator species	Indicator significance s					Indicator weight, J		
	South Western Siberia	European part of Russia and neighboring countries[a]	Small rivers of Latvia[c]	Moldova[d]	European part of Russia[e]	South Western Siberia	European part of Russia and neighboring countries[b]	Small rivers of Latvia[c]
Asplanchna priodonta Gosse	1.50	–	–	1.5	1.5	2	1	–
As. herricki Guerne	1.61	–	–	1.0	–	3	–	–
Brachionus angularis Gosse	1.74	2.5	–	2.5	2.5	3	3	–
Br. calyciflorus Pallas	1.70	–	2.6	2.5	2.5	3	3	3
Br. diversicornis (Daday)	1.69	–	–	2.0	–	4	–	–
Br. leydigiii Cohn	1.64	–	–	2.2	–	4	–	–
Br. quadridentatus Hermann	1.85	–	2.0	2.0	–	2	–	5
Br. q. melheni Barrois et Daday	1.38	–	–	2.0	–	3	–	–
Br. q. cluniorbicularis Skorikov	1.83	2.0	–	2.0	–	4	–	–
Br. plicatilis Müller	1.50	–	–	2.0	–	4	–	–
Br. urceus Linnaeus	2.00	–	2.5	2.2	–	4	4	3
Br. variabilis Hempel	1.86	–	–	2.0	–	4	–	–
Cephalodella gibba (Ehrenberg)	1.83	1.35	1.6	2.0	–	2	–	3

(continued)

Table 3 (continued)

Indicator species	Indicator significance *s*					Indicator weight, *J*		
	South Western Siberia	European part of Russia and neighboring countries[a]	Small rivers of Latvia[c]	Moldova[d]	European part of Russia[e]	South Western Siberia	European part of Russia and neighboring countries[b]	Small rivers of Latvia[c]
Colurella obtusa (Gosse)	1.65	–	1.4	0.8	–	3	–	3
Euchlanis deflexa Gosse	1.65	–	–	1.5	–	3	–	–
E. dilatata Ehrenberg	1.61	–	1.5	1.5	–	3	–	3
E. lyra Hudson	1.42	–	–	1.5	–	3	–	–
E. lyra larga Kutikova	1.50	–	–	1.5	–	4	–	–
Filinia longiseta (Ehrenberg)	1.25	2.35	1.2	2.5	2.3	4	2	4
F. terminalis Plate	1.69	–	–	1.5	–	4	–	–
F. major (Golditz)	1.70	–	–	2.0	–	3	–	–
Hexarthra mira (Hudson)	2.00	1.8	–	1.8	1.8	4	4	–
Keratella cochlearis (Gosse)	1.55	–	–	1.5	1.3	2	2	–
K. cochlearis tecta (Gosse)	1.50	–	–	1.2	–	2	–	–
K. hiemalis Carlin	1.92	–	–	1.15	–	3	–	–
K. quadrata (Müller)	1.65	–	1.8	1.5	1.3	3	2	3

(continued)

Table 3 (continued)

Indicator species	Indicator significance *s*					Indicator weight, *J*		
	South Western Siberia	European part of Russia and neighboring countries[a]	Small rivers of Latvia[c]	Moldova[d]	European part of Russia[e]	South Western Siberia	European part of Russia and neighboring countries[b]	Small rivers of Latvia[c]
K. q. longispina (Thiebaud)	1.77	–	–	–	–	4	–	–
Lecane luna (Müller)	1.48	1.55	–	1.5	–	2	–	–
L. ungulata (Gosse)	1.50	–	–	1.5	–	4	–	–
Lepadella obtusa Wang	1.67	–	–		–	3	–	–
L. ovalis (Müller)	1.70	–	1.0	1.5	–	3	–	5
Lophocharis oxysternon Gosse	1.71	–	–	1.2	–	5	–	–
Mytilina mucronata (Müller)	1.67	–	2.0	1.7	–	5	–	5
M. m. spinigera (Ehrenberg)	1.25	1.85	–	1.85	–	5	–	–
M. ventralis (Ehrenberg)	1.81	–	2.0	1.0	–	4	–	5
M. videns (Levander)	1.17	–	–	1.5	–	4	–	–
Notholca acuminata (Ehrenberg)	1.42	–	–	1.2	–	4	–	–
Platyias quadricornis (Ehrenberg)	1.44	–	–	1.8	–	3	–	–

(continued)

Table 3 (continued)

Indicator species	Indicator significance *s*					Indicator weight, *J*		
	South Western Siberia	European part of Russia and neighboring countries[a]	Small rivers of Latvia[c]	Moldova[d]	European part of Russia[e]	South Western Siberia	European part of Russia and neighboring countries[b]	Small rivers of Latvia[c]
Plationus polyacanthus (Ehrenberg)	1.33	1.8	–	–	–	4	–	–
Polyarthra dolichoptera Idelson	1.63	–	–	1.5	–	3	–	–
P. euryptera Wierzejski	1.67	–	–	1.2	–	3	–	–
P. major Burckhardt	1.36	–	–	1.2	–	3	–	–
P. minor Voigt	1.50	–	–	0.5	–	3	–	–
P. remata Skorikov	1.64	1.0	–	1.0	–	4	–	–
P. vulgaris Carlin	1.60	–	–	1.85	1.9	2	2	–
Proales similis de Beauchamp	1.38	–	–	–	–	3	–	–
Synchaeta oblonga Ehrenberg	1.83	–	–	1.75	–	3	–	–
S. stylata Wierzejski	1.50	–	1.6	1.0	–	3	–	3
Testudinella patina (Hermann)	1.50	1.85	1.7	1.85	–	3	–	4
Trichocerca cylindrica (Imhof)	1.50	–	2.7	1.0	–	3	–	4
Tr. capucina (Wierzejski et Zacharias)	1.33	–	–	1.0	–	4	–	–

(continued)

Table 3 (continued)

Indicator species	Indicator significance *s*					Indicator weight, *J*		
	South Western Siberia	European part of Russia and neighboring countries[a]	Small rivers of Latvia[c]	Moldova[d]	European part of Russia[e]	South Western Siberia	European part of Russia and neighboring countries[b]	Small rivers of Latvia[c]
Tr. elongata (Gosse)	1.60	–	–	1.5	–	2	–	–
Trichotria truncata (Whitelegge)	1.58	–	–	1.2	–	4	–	–
Acroperus harpae (Baird)	1.17	1.4	–	–	–	4	–	–
Alona affinis (Leydig)	1.44	1.1	–	–	–	2	–	–
A. guttata Sars	1.60	1.5	–	–	–	2	–	–
A. quadrangularis (O.F. Müller)	1.25	1.4	–	–	–	3	–	–
A. rectangula Sars	1.00	1.3	–	–	–	5	–	–
Alonella excisa (Fischer)	1.00	1.2	–	–	–	5	–	–
A. nana (Baird)	2.13	1.4	–	–	–	3	–	–
Bosmina longirostris (O.F. Müller)	1.53	1.55	–	–	1.5	1	1	–
Bythotrephes cederströmii Schödler	1.88	–	–	–	–	3	–	–
B. longimanus Leydig	1.10	1.0	–	–	1.0	4	–	–

(continued)

Table 3 (continued)

Indicator species	Indicator significance *s*					Indicator weight, *J*		
	South Western Siberia	European part of Russia and neighboring countries[a]	Small rivers of Latvia[c]	Moldova[d]	European part of Russia[e]	South Western Siberia	European part of Russia and neighboring countries[b]	Small rivers of Latvia[c]
Ceriodaphnia affinis Lilljeborg	1.68	1.5	–	–	–	3	–	–
C. quadrangula (O. F. Müller)	1.48	1.15	–	–	1.1	2	2	–
C. reticulata Jurine	1.69	1.7	–	–		4	–	–
Chydorus ovalis Kurz	1.63	1.2	–	–	–	2	–	–
Ch. sphaericus (O. F. Müller)	1.28	1.75	–	–	1.7	2	1	–
Ctenodaphnia magna Straus	1.58	3.4	–	–	3.4	4	3	–
Ct. similes Claus	1.88	–	–	–	–	4	–	–
Daphnia cucullata Sars	1.88	1.75	–	–	1.7	1	2	–
D. longispina O.F. Müller	1.58	2.05	–	–	1.9	2	1	–
D. pulex (De Geer)	1.72	2.5	–	–	2.8	3	4	–
Diaphanosoma brachyurum (Lievin)	1.52	1.4	–	–	1.4	4	3	–
Disparalona (Phrixura) rostrata (Koch)	1.13	1.3	–	–	–	4	–	–

(continued)

Table 3 (continued)

Indicator species	Indicator significance *s*					Indicator weight, *J*		
	South Western Siberia	European part of Russia and neighboring countries[a]	Small rivers of Latvia[c]	Moldova[d]	European part of Russia[e]	South Western Siberia	European part of Russia and neighboring countries[b]	Small rivers of Latvia[c]
Eubosmina longispina Leydig	1.50	–	–	–	–	2	–	–
Eu. coregoni Baird	1.54	–	–	–	0.9	3	3	–
Eurycercus lamellatus (O.F. Müller)	1.62	1.2	–	–	–	4	–	–
Graptoleberis testudinaria (Fischer)	1.21	1.5	–	–	–	4	–	–
Iliocryptus acutifrons Sars	1.89	–	–	–	–	5	–	
Leptodora kindtii (Focke)	1.75	1.65	–	–	1.7	2	2	–
Macrotrix hirsuticornis Norman and Brady	1.64	–	–	–	–	4	–	–
Moina brachiata (Jurine)	2.28	3.4	–	–	3.4	3	3	–
M. macrocopa (Straus)	2.15	2.75	–	–	–	3	–	–
Oxyurella tenuicaudis (Sars)	1.50	–	–	–	–	3	–	–

(continued)

Table 3 (continued)

Indicator species	Indicator significance *s*					Indicator weight, *J*		
	South Western Siberia	European part of Russia and neighboring countries[a]	Small rivers of Latvia[c]	Moldova[d]	European part of Russia[e]	South Western Siberia	European part of Russia and neighboring countries[b]	Small rivers of Latvia[c]
Peracantha truncata (O.F. Müller)	1.54	1.3	–	–	–	3	–	–
Picripleuroxus striatus Schoedler	1.64	1.5	–	–	–	2	–	–
Pleuroxus aduncus (Jurine)	1.52	–	–	–	–	4	–	–
Polyphemus pediculus (Linnaeus)	1.63	1.3	–	–	1.3	4	4	–
Pseudochydorus globosus (Baird)	1.77	1.2	–	–	–	4	–	–
Sida crystallina (O. F. Müller)	1.50	1.8	–	–	1.3	4	–	–
Simocephalus vetulus (O.F. Müller)	1.80	1.5	–	–	1.5	4	–	–
Scapholeberis mucronata (O.F. Müller)	1.79	2.0	–	–	–	3	–	–
Cyclops furcifer Claus	1.50	–	–	–	–	4	–	–

(continued)

Table 3 (continued)

Indicator species	Indicator significance *s*					Indicator weight, *J*		
	South Western Siberia	European part of Russia and neighboring countries[a]	Small rivers of Latvia[c]	Moldova[d]	European part of Russia[e]	South Western Siberia	European part of Russia and neighboring countries[b]	Small rivers of Latvia[c]
C. s. strenuus Fischer	1.14	2.25	–	–	2.2	2	2	–
C. v. vicinus Uljanin	1.38	–	–	–	2.1	3	3	–
Eucyclops serrulatus Fischer	1.65	1.85	–	–	–	4	–	–
Macrocyclops albidus (Jurine)	1.93	–	–	–	–	3	–	–
M. fuscus (Jurine)	2.00	–	–	–	–	5	–	–
Megacyclops gigas (Claus)	1.88	–	–	–	–	4	–	–
M. viridis (Jurine)	1.69	–	–	–	–	2	–	–
Mesocyclops leuckarti (Claus)	1.74	–	–	–	1.2	2	3	–
Paracyclops f. fimbriatus (Fischer)	1.65	–	–	–	–	3	–	–
Thermocyclops oithonoides (Sars)	2.33	–	–	–	–	4	–	–
Acanthodiaptomus denticornis Wierzejski	2.50	–	–	–	–	4	–	–
Eudiaptomus gracilis Sars	1.50	–	–	–	1.2	3	3	–

(continued)

Table 3 (continued)

Indicator species	Indicator significance s					Indicator weight, J		
	South Western Siberia	European part of Russia and neighboring countries[a]	Small rivers of Latvia[c]	Moldova[d]	European part of Russia[e]	South Western Siberia	European part of Russia and neighboring countries[b]	Small rivers of Latvia[c]
E. graciloides Lilljeborg	1.66	–	–	–	–	3	–	–
E. vulgaris (Schmeil)	1.55	–	–	–	–	4	–	–
Heterocope appendiculata Sars	1.67	–	–	–	–	4	–	–
Mixodiaptomus theeli (Lilljeborg)	1.94	–	–	–	–	3	–	–
Neutrodiaptomus incongruens (Poppe)	1.73	–	–	–	–	3	–	–

[a]Unified Methods (1977)
[b]Makrushin (1974b)
[c]Tsimdin (1979)
[d]Naberezhnyj (2010)
[e]Chertoprud and Chertoprud (2010)

specifically, since the growth of mineralization determines the species composition of zooplankton and indicator species characteristics. When the critical point of salinity is crossed, the indicator significance of individual species is not representative, and the water quality is determined only by integral parameters (transparency, quantitative indicators of individual groups of aquatic organisms, primary production, hydrochemical characteristics, etc.).

Here is an example of the index calculation for a hypothetical reservoir, based on types and the most common species in waters of the South of Western Siberia (Table 4). The table illustrates the indexes of indicator significance which we obtained for Western Siberia for a number of species and those indexes taken from the literature, calculated for European water bodies (Unified Methods 1977; Makrushin 1974b; Tsimdin' 1979; Naberezhnyj 1984; Chertoprud and Chertoprud 2010).

Let us try to analyze the results of the Pantle and Buck index (*S*) calculation using regional saprobity valence indexes obtained for specific regional changes in comparison with the indexes developed for European water bodies taken from the published tables (a common occurrence).

If we calculate the Pantle and Buck index based on the indicator significance values of zooplankton taken from European water bodies, we get S = 1.95. The area is shown to be β-mesosaprobic, gravitating toward β-α-mesosaprobic. If we calculate the same characteristics using the regional value *s*, we have S = 1.48. The area turns to oligo-β-mesosaprobic. The error appears to be very substantial. This error can vary widely, depending on the species composition of zooplankton, the species

Table 4 Comparison of the calculated index of indicator significance (*s*) and the Pantle and Buck index (*S*) with literature data

Indicator species of zooplankton	Frequency of occurrence, h	*s** Western Siberia	*s** Europe
Brachionus angularis Gosse	5	1.74	2.50
Filinia longiseta (Ehrenb.)	4	1.25	2.35
Brachionus quadridentatus cluniorbicularis Skorikov	2	1.83	2.0
Lecane luna (Müller)	1	1.48	1.55
Alona quadrangularis (Müller)	3	1.25	1.4
Alona rectangular Sars	2	1.00	1.3
Bosmina longirostris (Müller)	3	1.53	1.55
Leptodora kindtii (Focke)	1	1.75	1.65
Daphnia longispina Müller	4	1.58	2.05
Chydorus sphaericus (Müller)	4	1.28	1.75
Sida crystallina (Müller)	1	1.50	1.80
Moina macrocopa (Straus)	1	2.15	2.75
Scapholeberis mucronata (Müller)	1	1.79	2.00
Cyclops strenuus Fisch.	2	1.14	2.25
Eucyclops serrulatus Fisch.	2	1.65	1.85
Pantle and Buck Index (*S*)		**1.48**	**1.95**

**s*—index of species indicator significance

fraction in the abundance of indicator species, but in general most saprobity indexes using indicators from the Atlas of Saprobic Organisms (Unified Methods 1977) are overestimated compared to the real condition of the water body, or section thereof. When the classes of water quality are compared to those based on hydrochemical characteristics according to the above saprobity indexes, one can detect the inadequacy. According to hydrochemical indexes, the reservoir falls into one class of water quality and according to hydrobiological ones it falls into another, usually the worst one. Therefore, for the conditions of the South of Western Siberia and Altai Krai, reservoirs preferably need the regional values of the indicator species of zooplankton (Yermolaeva and Dvurechenskaya 2007, 2014; Ermolaeva and Dvurechenskaya 2013).

The authors have obtained indexes of indicator significance and indicator weight for 111 species of zooplankton that are recommended for use when calculating the Pantle and Buck index in aquatic ecosystems of Western Siberia (Table 3).

4 Conclusions

1. Monitoring the quality of freshwater bodies by biological indicators should lead to a reliable allocation of measurement data into water quality classes.
2. This task is challenging as biological objects may vary and have been adapted to different regions. Their indicator values may be different in other regions.
3. We developed and tested a procedure for calculating reliable regional saprobity indexes based on zooplankton analyses in Western Siberia using equations developed by Shitikov et al. (2003), Zelinka and Marvan (1961), and finally Pantle and Buck (1955), modified by Sladečhek (1973).
4. Our regional indicator values reflect the ecological status of open freshwater ecosystems better than classifications based on indicator values derived from tabulated literature data. Hydrochemical data confirmed the adequate quality of our classifications.
5. The methodology should be tested in other regions where those regional indicator values do not yet exist. It should be underpinned by new multifactorial statistical approaches (Lischeid 2014).
6. Using indexes obtained from regional features of reservoirs and rivers hydrochemical background of, one can define the water quality class more exactly. Thus, the use of regional indexes is appropriate because it provides a more objective assessment of the state of the ecosystem.

Appendix: Photographs of Some Water Bodies Sampled

(see that Figs. 2, 3, 4, 5, 6 and 7)

Fig. 2 Lake Teletskoye is located in the north-east of the Altai Mountains. It is a unique natural complex of our planet. It is one of the deepest lakes in Russia, ranking second in Russia's natural freshwater supply. In 1998 Lake Teletskoye was included in the list of UNESCO World Heritage sites. It is an oligosaprobic lake (Photo courtesy of Evgenia Zarubina)

Fig. 3 Katun' is the largest river of the Altai Mountains. The Katun' originates in the glacier on the Southern slopes of Belukha Mountain. Flowing together with Biya, the Katun' forms the Ob River, one of the largest in Siberia. In terms of water quality, the Katun' river is mainly o-β-mesosaprobic (Photo courtesy of Evgenia Zarubina)

Fig. 4 Abushkan (**a**) and Bolshoy Kurgan (**b**) are lakes of the Ob-Irtysh interfluve (Barabinskaya steppe). They are small saucer-shaped lakes with depths up to 3–4 m. In terms of water quality these lakes are from o-β-mesosaprobic to α-mesosaprobic

Fig. 5 Small lowland rivers: the Kargat river (**a**) and Chulym river (**b**). They are rivers of the closed flow region of the Ob-Irtysh interfluve. These rivers serve as the main source of Lake Chany's water supply. In terms of water quality these rivers are mainly β-mesosaprobic

Fig. 5 (continued)

Fig. 6 Fadikha is a running-water lake in the lower stream of the Chulym river. In terms of water quality it is β-mesosaprobic

Fig. 7 The Novosibirsk reservoir is located on the Ob River in the South of Western Siberia. Length of the reservoir: 220 km. Water area: 1082 km^2. Maximum depth: 25 m. Quality of the water: β-mesosaprobic

References

Andrushaitis GP, Tsimdin' PA, Parele EA, Daksh LV (1981) Ecological indication of water quality of small rivers. Scientific foundation of surface water quality monitoring on hydrobiological parameters. (Экологическая индикация качества вод малых рек // Научные основы контроля качества поверхностных вод по гидробиологическим показателям). Hydrometeoizdat, Leningrad, pp 59–65 (in Russian)

Balykin SN, Balykin DN, Saltykov AV, Puzanov AV (2011) The dynamics of soluble iron compounds in altai surface water. Basic and innovation aspects of biogeochemistry (Динамика содержания растворимых форм железа в поверхностных водах Алтая. / Фундаментальные и инновационные аспекты биогеохимии) Proceedings of Int. Biogeochem. School. GEOKHI RAS, Moscow, 365 p (in Russian)

Chertoprud MV, Chertoprud ES (2010) A short guide to freshwater invertebrates for Center of European Russia (Краткий определитель беспозвоночных пресных вод центра Европейской России). Moscow, 184 p (in Russian)

Ermolaeva NI, Dvurechenskaya SYa (2013) Regional indices of the indicator significance of zooplanktonic organisms in water bodies of southern Western Siberia. Russ J Ecol 44(6): 527–531

Finogenova NP, Alimov AF (1976) Assessment of the degree of pollution on the composition of benthic. Methods of biological analyses of fresh water (Оценка степени загрязнения вод по составу донных животных // Методы биологического анализа пресных вод) Zool Inst Akad. Nauk SSSR, Leningrad, pp 95–106 (in Russian)

GOST (State Standard) 17.1.2.04-77 (1977) Nature conservation. Hydrosphere. Indicators of status and rules for taxation of fishery water bodies. (ГОСТ 17.1.2.04-77 Охрана природы. Гидросфера. Показатели состояния и правила таксации рыбохозяйственных водных объектов) (in Russian)

Il'in VB (1973) Biochemistry and agrochemistry of microelements (Manganese, copper, molybdenum, boron) in Southern Part of Western Siberia. (Биогеохимия и агрохимия микроэлементов (марганец, медь, молибден, бор) в южной части Западной Сибири). Science, Novosibirsk, 392 p (in Russian)

Il'in VB, Syso AI (2004) Microelements and heavy metals in soils and plants of Novosibirsk Region (Микроэлементы и тяжелые металлы в почвах и растениях Новосибирской области) SB RAS, Novosibirsk, 229 p (in Russian)

Lischeid G (2014) Non-linear approaches to assess water and soil quality. In: Mueller L, Saparov A, Lischeid G (eds) Novel measurement and assessment tools for monitoring and management of land and water resources in agricultural landscapes of Central Asia. Springer International Publishing, Cham, pp 356–378

Makrushin AV (1974a) Biological analysis of water quality (Биологический анализ качества вод) Zool Inst Akad. Nauk SSSR, Leningrad, 60 p (in Russian)

Makrushin AV (1974b) Bibliographic index for biological analysis of waler quality, with a list of indicator organisms of water pollution (Библиографический указатель по теме «Биологический анализ качества вод» с приложением списка организмов-индикаторов загрязнения) Zool Inst Akad. Nauk SSSR, Leningrad, 53 p (in Russian)

Naberezhnyiy AI (2010) Ecological certificates of rotifers for reservoirs of Moldova (Экологические паспорта коловраток водоемов Молдовы). Herald Hydrobiology (in Russian). http://hydrobiologist.wordpress.com/. Accessed 10 Nov 2014

Nechaeva EG (2001) Landscape and geochemical system of Asian Russia. Geogr Nat Res (Ландшафтно-геохимическое районирование Азиатской России // География и природные ресурсы)(1): 12–18 (in Russian)

Oksiyuk OP, Zhukinskii VN, Braginskii LP (1993) A complex ecological classification of the quality of land surface waters (Комплексная экологическая классификация качества поверхностных вод суши) Gidrobiol Zhu. 29(3): 42–76 (in Russian)

Pantle R, Buck H (1955) Die biologische Ueberwachung der Gewaesser und die Darstellung der Ergebnisse. Gas–und Wasserfach 96(18):604–618

Shitikov VK, Zinchenko TD, Golovatyuk LV (2003) A method for assessing the quality of surface waters by macrobenthos based on species indicator valences (Метод оценки качества поверхностных вод по макробентосу на основе видовых индикаторных валентностей). In: Proceedings of the 2nd international conference on lake ecosystems: biological processes anthropogenic transformation and water quality, Minsk–Naroch, pp 537–540 (in Russian)

Sladeček V (1973) System of water quality from the biological point of view. Arch Hydrobiol Ergeb Limnol (3):218

Tsimdin' PA (1979) Rotifers as indicators ot saprobity (Коловратки как биоиндикаторы сапробности). Gidrobiol Zh 15(4):63–67 (in Russian)

Unified Methods (1977) Unified methods for water assessment, part 3 (Methods for biological analysis of waters), Appendix 2. (Atlas of saprobic organisms). (Унифицированные методы исследования качества вод (1977). Ч. Ш. Методы биологического анализа вод. Приложение 2. Атлас сапробных организмов). SEV 1977, Moscow, 227 p (in Russian)

Yermolaeva NI, Dvurechenskaya SYA (2007) Indicator significance of the various groups of zooplankton of limnological systems in Western Siberia (Индикаторное значение различных групп зоопланктона лимнических систем Западной Сибири). In: Proceedings of the international conference (selected papers) "bioindication in freshwater ecosystems monitoring". LEMA, St. Petersburg, pp 217–221 (in Russian)

Yermolaeva NI, Dvurechenskaya SYA (2014) Determination of indicative significance of zooplankton subject to regional distinctions reservoirs in the south of Western Siberia (Определение индикаторной значимости зоопланктона с учетом региональных особенностей водоемов юга Западной Сибири). Water Chem Ecol 5(71):60–67 (in Russian)

Zelinka M, Marvan P (1961) Zur Präzisierung der biologischen Klassifikation der Reinheit fließender Gewässer. Arch Hydrobiol 57:389–407

Chapter 6
Measuring and Estimating Fluxes of Carbon, Major and Trace Elements to the Arctic Ocean

Oleg S. Pokrovsky

Abstract This chapter describes the methods and case studies of element flux measurements in the Arctic and subarctic rivers, in the Russian boreal and subarctic zone, along the gradient of permafrost-free terrain to continuous permafrost settings, developed on various lithology and vegetation coverage. The majority of existing flux measurements is based on a combination of daily discharges from Russian Hydrological Survey gauging stations with grab samples or year-round sampling of dissolved and particulate load following the chemical analysis. In this chapter, a new, geochemical-based perspective on the functioning of aquatic boreal systems is described which takes into account the role of the following factors on riverine element fluxes: (1) the specificity of lithological substrate; (2) the importance of organic and organo-mineral colloidal forms, notably during the spring flood; (3) the role of permafrost presence within the small and large watersheds; and (4) the governing role of terrestrial vegetation in element mobilization from rock substrate to the rivers. This kind of multiple approach allows a first-order prediction of element fluxes in a scenario of progressive warming in high latitudes. Two novel dimensions added to the existing knowledge on element transport from the land to the Arctic Ocean by the Russian boreal and subarctic rivers are (i) evaluation of colloidal flux of dissolved substances and low molecular weight (LMW) fraction and (ii) assessing, for the first time, the isotopic signatures of Ca, Mg, Si, and Fe in several case watersheds of various lithology and permafrost coverage. The results of this study and available data from the literature demonstrate that, while climate warming will certainly affect the wintertime element fluxes and speciation, it is unlikely to change the nature and magnitude of the main fraction of trace elements

O.S. Pokrovsky (✉)
Geoscience and Environment Toulouse, UMR 5563 CNRS, Avenue Edouard Belin, 31400 Toulouse, France
e-mail: oleg@get.obs-mip.fr

O.S. Pokrovsky
Institute of Ecological Problems of the North, Ural Branch of Russian Academy of Sciences, Arkhangelsk, Russia

O.S. Pokrovsky
BIO-GEO-CLIM Laboratory, Tomsk State University, Tomsk, Russia

L. Mueller et al. (eds.), *Novel Methods for Monitoring and Managing Land and Water Resources in Siberia*, Springer Water,
DOI 10.1007/978-3-319-24409-9_6

TE flux to the ocean. This fraction of the flux occurs in colloidal form during several weeks of the spring flood. At the present time, it is not strongly affected by climate change, or this influence is within the uncertainty of the flux measurements. Overall, the major changes in the chemical and isotopic nature of riverine fluxes to the Arctic Ocean from Northern Eurasia in a climate warming scenario are likely to be linked to the change in the vegetation (species, biomass and geographical extension), rather than temperature and hydrology. The increase in the depth of the active layer has an influence of second-order importance on the riverine fluxes given that the majority of continental flux to the Arctic Ocean is formed on permafrost soils, highly homogeneously over the depth profile.

Keywords Arctics • River discharge • Carbon flux • Element flux • Climate change

1 Introduction

The importance of boreal and subarctic continental zones in the regulation of both organic and inorganic carbon cycle on the Earth follows from (1) the high storage of C_{org} in soils, (2) the dominance of humid climate supporting abundant vegetation and yielding high riverine fluxes of both carbon and nutrients from the land to the Arctic Ocean, (3) the presence of shallow soils and fresh rocks exposed to the surface and capable of taking up atmospheric CO_2 during weathering, and (4) increase in the physical degradation of rocks due to frost action and freezing cycles. Due to the high importance of the Arctic Ocean and permafrost-dominated subarctic continental zones in the carbon cycle on the Earth and the high vulnerability of circumpolar zones to climate warming, most works conducted have been devoted to the biogeochemistry of organic carbon and sediments in large rivers (Gordeev et al. 1996, 2004; Dittmar and Kattner 2003; Gordeev 2006; Magritsky 2010, Holmes et al. 2012), lakes (Pokrovsky et al. 2012a, b; Shirokova et al. 2013a, b; Moiseenko et al. 2013), and soils (Botch et al. 1995; Beilman et al. 2009) of the Russian boreal circumpolar zone. At the same time, the main factors controlling biogeochemical cycles of other major and trace elements in the watersheds of the Arctic Ocean and their response to climate change are only just beginning to be understood (e.g. see White et al. 2007). The main specificities of the boreal Eurasian high-latitude watersheds extended over cold and humid zone on the organic-rich soils on poorly weathered rock substrate, making them drastically different from temperate and tropical regions, are: (i) extremely high seasonal variability of element concentration in rivers producing the majority of annual organic carbon (OC) flux and related metals during 2–3 weeks of the spring flood; (ii) specific poorly humified, fulvic-like dissolved organic matter of LMW in summer and high molecular weight (HMW) in spring, and at the same time the presence of old (Pleistocene age) dissolved OC; (iii) high concentration of dissolved Fe-producing organo-ferric colloids as the main vectors of TE transport in surficial fluids; (iv) high abundance, up to 70–80 % of the watershed area, of glacial or thermokarst lakes capable of controlling the solute transport from the soil to the river, and (v) abundant coniferous forest of essentially deciduous *Larix* spp. throughout Northern Siberia.

Unlike many regions of the world, the Arctic and subarctic zones exhibit extreme variations in the discharge and chemical elements concentration (Gordeev and Sidorov 1993; Gordeev et al. 1996; Gordeev 2006; Gislason et al. 1996; Stefansson and Gislason 2001; Gaillardet et al. 2003; Rember and Trefry 2004; Zakharova et al. 2005; Guay et al. 2010; Bagard et al. 2011; Huser et al. 2011; Prokushkin et al. 2011; Stedmon et al. 2011; Guo et al. 2012). The quantitative description of these systems therefore requires an understanding of how weathering rates and riverine fluxes of major and trace elements as well as their main carrier (OC) vary seasonally. However, this high seasonality implies significant variations in the source of the elements in the river flow over the year, which is further accentuated by the high variability of the depth of the active layer and relevant contribution of mineral soil weathering processes compared to the leaching of the organic horizon of soils. As such, the important parameter controlling the chemistry of fluxes in seasonal resolution becomes the relative role of mineral dissolution versus plant litter dissolution in the rate of chemical denudation and element fluxes from the land to the ocean. Although, several recent studies have used isotopic techniques in an attempt to resolve the sources of elements in Eurasian rivers (Reynolds et al. 2006; Engstrom et al. 2010; Lambelet et al. 2013), the contribution of mineral versus plant litter remains poorly explained, particularly for boreal watersheds.

The purpose of this chapter is to present the examples of recent estimations of fluxes in the boreal and subarctic regions of Siberia and European Russia, describe the high seasonality of these fluxes, evaluate the role of colloidal transport of OC, major, and trace elements, and give a first-order estimation of the stable isotope signature of several major elements (Ca, Mg, Si and the main colloidal carrier—Fe) in boreal Russian rivers draining to the Arctic Ocean.

2 Assessing the River Fluxes in the Boreal and Subarctic Regions of Siberia and European Russia

The main source of information for calculating the fluxes of dissolved and suspended material from the land to the Arctic Ocean remains, the data collected by the Russian Hydrometeorological Survey (RHS) at key gauging stations in the freshwater zone of almost all major Arctic rivers and many other interland stations using a standard procedure (Gordeev et al. 1996; Nikanorov et al. 2010a, b; Magritsky 2010). These data comprise the daily discharge (water level) and several (typically from 5 to 15) measurements of dissolved load. Since, both the sample handling procedure and the chemical analysis adopted by the RHS differ from those used for river water sample processing in the academic science these days, a calibration of

RHS data with independent sampling, filtration, and analysis is often required. Examples include the assessment of nutrient transport to the Arctic Ocean by Siberian rivers (Gordeev et al. 1996; Holmes et al. 2000, 2001) and European rivers (Leonov and Chicherina 2004), and major elements and OC flux measurements in both permafrost-free (Pokrovsky et al. 2010) and permafrost-bearing regions (Pokrovsky et al. 2005a, b; Cooper et al. 2005; Frey et al. 2007a, b). While, for the nutrients, a rigorous procedure of sampling, filtration, storage, and analysis is mandatory, the dissolved inorganic load (Ca, Mg, Si, Cl, SO_4, HCO_3) and the total organic carbon can be adequately approximated (i.e., within 10–30 %) using available RHS data.

In this section, we present the methodology of flux calculation and case studies in both permafrost-bearing and permafrost-free zones of Northern Eurasia. The study sites are shown on the map of permafrost distribution in the Russian Federation (Fig. 1). We investigated monolithological (granites, gneisses, carbonates, basalts) and mixed lithological (sedimentary rocks, carbonates, sandstones, claystones) river basins of both permafrost-free and continuous permafrost zones (European part of NW Russia, Western and Central Siberia, respectively). For comparison, discontinuous permafrost sites of the Aldan shield in the Transbaikal region (granite-gneisses) and the Kamchatka Peninsula (basalts, tuffs and sedimentary rocks) will also be considered. Boreal and subarctic permafrost-free zones include the Karelia region, located in North-West Russia on the Eastern Fennoscandian Shield, and the Severnaya Dvina River basin of the Arkhangelsk region on sedimentary rocks (carbonates, clays). All sites can be considered as pristine with minimal impact from human activity except long-range atmospheric

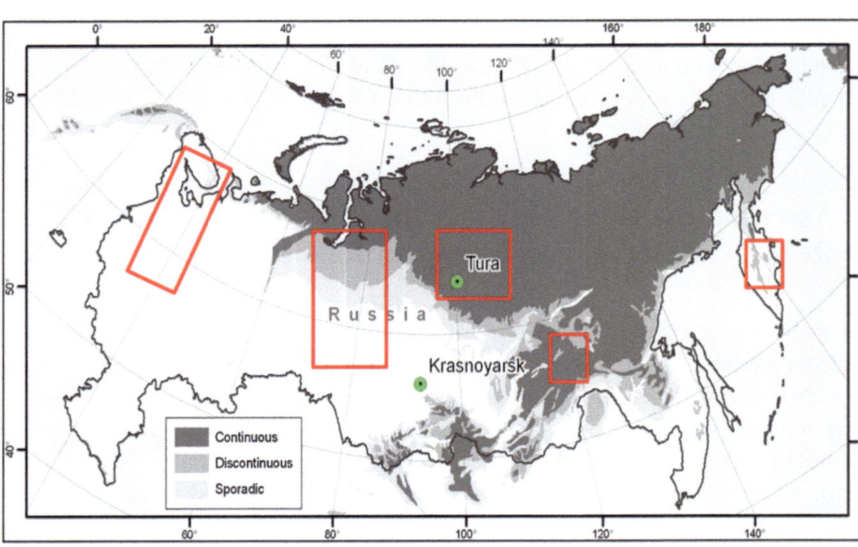

Fig. 1 A map of permafrost distribution in Russia (from Brown et al. 2002) and selected study sites. The main study sites described in this chapter are outlined in *red*

deposition. In the continuous permafrost zone, we selected a basaltic province located within the Central Siberian Plateau which includes a considerable part of the basins of the N. Tunguska and Podkamennaya Tunguska rivers, the tributaries of the Yenisey River.

Available data from systematic surveys by the Hydrometeorological State Committee of the former USSR Goskomgidromet and later Roskomgidromet, which were used to estimate the annual water and suspended and dissolved elements discharge for rivers of the Central Siberia region, are published in the annual issues of the State Water Cadastre (Hydrological Yearbooks of Yenisey Basins 1954–1975) and further generalized in "Resources of surface waters of the USSR 1973." The river watershed size varies from 9000–447,000 km^2. The data from the Hydrological Survey include, for each 5 hydrological stations in Central Siberia (N. Tunguska at Tura, N. Tunguska at Bolshoi Porog, Tembenchi, Eratchimo and Taimura), the daily water discharge, from 4–11 measurements per year of major cations, anions, silica, iron, and OC and 15–30 measurements per year of total suspended matter (SM). Daily discharge values for all rivers studied were obtained from the stage-discharge rating curve established by the Hydrological Survey for each gauging station according to International Standards (ISO 1983). Analyses of solutes and SM performed by the Hydrological Survey are described elsewhere (Semenov 1977; Soyer and Semenov 1971; Gordeev and Sidorov 1993; Zakharova et al. 2005).

The concentrations of calcium, magnesium, sodium and potassium, sulfate, bicarbonate, chloride, and total mineralization demonstrate a strong power dependence on water discharge. Because this dependence is valid over the entire observation period (we verified it for the decade of 1965–1975), it was used to estimate the mean multi-annual concentrations. For this purpose, all chemical analysis data for a given period of observation were used to generate the coefficients k and n in Eq. 1

$$C_i = k \cdot Q_i^n \tag{1}$$

where C_i represents the measured ion concentration for a given day of the year, Q_i is the water discharge for this day, and k and n are empirical constants for each river. For each gauging station, the value of the mean annual water discharge available for the period of observations (Resources 1973) was used to calculate the mean multi-year concentration of component (Table A1) using Eq. 1. A similar method was used to estimate the mean monthly and annual concentrations of elements in rivers of the Lena Basin (Gordeev and Sidorov 1993) and Transbaikal region (Zakharova et al. 2005).

The mean multi-year flux of element i (R_i) is calculated as

$$R_i = C_i^* \cdot W/A \tag{2}$$

where C_i* is the mean multi-year concentration of element i calculated as described above and corrected for atmospheric input; A stands for the watershed area (km^2)

and W is the multi-year average water discharge (km^3/y) for the period considered, based on the data of the Hydrological Survey Database. The mean multi-year concentrations were corrected for atmospheric input by subtracting the average multiannual concentration of elements in the rain measured at the nearest meteorological station in the region. The sea-salt normalized corrections, which use only Cl^- concentration in the precipitation according to Négrel et al. (1993), gave similar results within ±15 % uncertainty. Note that the seasonal variations of atmospheric precipitates (by an order of magnitude) yield significant uncertainty on the C_i* value (i.e., 15 %).

The annual SM fluxes are also available from the USSR Hydrological Survey database (Resources 1973). The method of SM flux estimation used by the Hydrological Survey is based on the time interpolation between the values measured at the main phases of hydrological regime and, more frequently, when the water level rises (15–30 times per year). For several watersheds, this technique was compared with the more accurate combined method of chemical flux calculation. In this method, we used the exponential relation between C and Q (Eq. 1) to calculate the daily concentration from the daily discharge exclusively for the high-water periods (spring melt and summer storms). For the low water level period, when both the discharge and element concentration are more or less constant, the daily concentrations were calculated by means of time linear interpolation (Zaslavskaya and Tikhotskaya 1978). This allowed us to calculate the mean annual discharge-weighted concentration for each element. Because the water flow during the low water period on the rivers studied is negligible, the difference in the mean annual concentration estimation between the two methods does not exceed 20 % for cations (Ca, Mg) or 8 % for TDS and TDS_c defined as $[Ca^{2+}] + [Mg^{2+}] + [Na^+] + [K^+] + [SiO_2] + [SO_4^{2-}] + [Fe]$ and $[Ca^{2+}] + [Mg^{2+}] + [Na^+] + [K^+]$, respectively.

As a result of the difficulties in conducting high-resolution geochemical studies in space and time in these remote and hostile regions, the majority of the available geochemical information was collected during the easiest time of sampling, the summer/fall low flow period (Guieu et al. 1996; Huh et al. 1998; Huh and Edmond 1999; Gebhardt et al. 2004). Recently, the fluxes of suspended and conventionally dissolved (<0.22 μm) major components of the Arctic rivers became available, thanks to the thorough PARTNERS program (Cooper et al. 2008; Holms et al. 2012 and references therein). In each given watershed, the main method of flux calculation from year-round chemical measurements is based on regularly sampling filtered river water and calculating the discharges either from direct gauging station measurements or from interpolation/extrapolation to/from adjacent gauging stations. An example of this kind of approach is described below.

Daily discharges for three large Siberian rivers, the N. Tunguska and Kochechum at Tura, and the Podkamennaya Tunguska at Baikit, were obtained from the ROSHYDROMET for the period of 2006–2010. Long-term (1939–1995) monthly discharges of these rivers were obtained from the R-ArcticNet Database (www.r-arcticnet.sr.unh.edu/v4.0/main.html). For another large Siberian river that does not have a gauging station (the Kochechum River, KO), the daily and monthly discharges were estimated using data from the TE, assumed to be analog with the river with a basin with similar geomorphological characteristics and long-term records of individual flow measurements. In this data set, the KO to TE long-term annual discharges (29.96 and 7.97 km^3, respectively), yielded the ratio 3.76, which was used to estimate the KO daily discharge from TE data (see Prokushkin et al. 2011 for details).

Historic data for discharges, DIC, and total organic carbon (TOC, the sum of dissolved and particulate organic carbon) concentrations in NT and TE were collected by the State Service of Hydrometeorology and Environment Monitoring (USSR) for the period of 1966–1975 (Hydrological Yearbooks of Russian Hydrological Survey 1954–1975). First, for a comparative analysis of all five rivers, the monthly, seasonal, and annual concentrations of DOC and DIC were calculated as simple averages. Additionally, monthly, seasonal, and annual concentrations of DOC and DIC have been calculated for the TE, KO (estimated using Tembenchi river discharge), and NT as flow-weighted means. The calculation of mean annual concentrations for the Podkamennaya Tunguska River is based on the contribution of each month to the annual runoff. Then, seasonal and yearly river C loads (mass C $period^{-1}$) were estimated using the daily load (mass C day^{-1}) obtained from a time series of paired river flow and constituent concentration data used to construct a calibration curve. Finally, the area-normalized DOC and DIC exports with rivers were expressed per watershed area (g C m^{-2} a^{-1}).

3 Contribution of Different Sources to the Element Fluxes

Apart from the atmospheric depositions, the majority of solutes in the river water originate from the ground (bedrock dissolution), mostly via underground feeding, during the winter base flow. In the permafrost region, the ground ice thawing during the summer (active) period adds solutes to the two main sources of the river's dissolved load: mineral weathering in the soil column and plant litter degradation at the surface of the soil. An example of the annual flux separation into several main sources is presented in Fig. 2. Precise knowledge of these sources is essential to predict the response of Arctic ecosystems to global warming as the plant litter production and degradation respond more rapidly to environmental change than mineral dissolution.

The crucial parameter for estimating litter impact is the ratio between the export flux (*J*) of the elements from degrading litter and the primary production (NPP). The *J*/NPP ratio is extremely low for biogenic elements such as P, K, N, C (0.01–0.02) but may be as high as 0.88 for mineral components in the taiga region (Bazilevitch 1976). Thus, for TDS_c = Na + K+ Ca + Mg we will consider the range of *J*/NPP between 0.1 as for overall detritus production (Vogt et al. 1982; Schlesinger 1997) and 0.88. Assuming steady-state conditions of forest biomass and soil thickness and converting to element content in the dry biomass using available literature data (Pokrovsky et al. 2005a, b; Viers et al. 2013), this yields for Si, Ca,

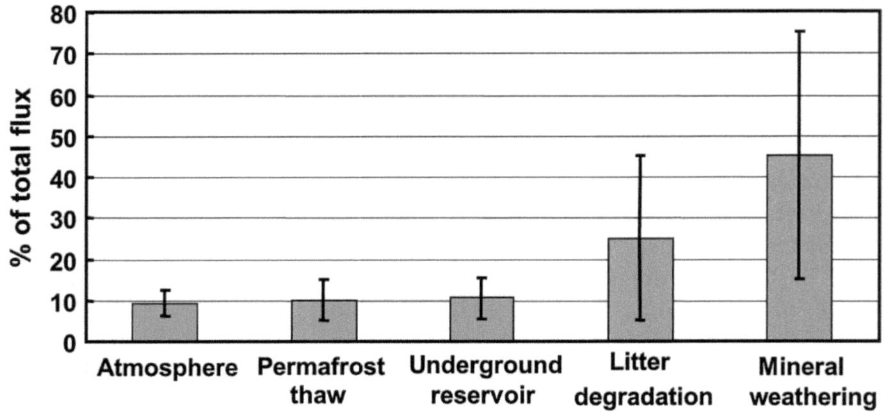

Fig. 2 The relative contribution of atmospheric deposition, permafrost melting, underground feeding, plant litter degradation, and rocks and soil mineral weathering to the total dissolved cation (Ca + Mg + Na + K) flux of the N. Tunguska River, the largest tributary of the Yenisey River (Pokrovsky et al. 2005a, b). Note that large uncertainties are associated with flux evaluation, linked to variations in the element turnover coefficient in the vegetation

Mg, Na, Fe, and Al minimal values of annual fluxes issued from litter degradation equal to 0.34, 0.17, 0.06, 0.016, 0.09, and 0.1 t/km^2/y, respectively. The maximal values may be $\sim$9 times higher.

Another potential source of elements in rivers during the warm high-water season is permafrost thawing. The concentrations of all major elements and DOC in permafrost ice veins and in ground ice are comparable to or higher than those in soil porewaters, groundwaters, and rivers (Pokrovsky et al. 2005a, b, 2006). To approximate the potential impact of permafrost ice thawing on river chemistry, we have hypothesized that 40 $\pm$ 20 cm of the soil active layer, containing $\sim$5 % of ground ice, as follows from available field data on the water content of frozen samples of deep soil horizons, can thaw during the summer. This gives a value of TDS_c flux issuing from permafrost melting of 0.5 $\pm$ 0.15 t/km^2/y.

Based on these data, we estimated the relative proportion of various sources making up the river cationic composition, both on an annual scale and during the summer/fall (July to October) (Fig. 2). For this, we used multi-annual data of atmospheric precipitates (75 % occur in August/September) and total riverine fluxes (F_{tot}) as described previously. We also assumed that the underground input ($F_{undeground}$) is constant over the year, but only dominates the river cationic discharge in the winter (November to April). The contribution of salts to summertime river composition can be approximated by comparing the Cl concentration in rivers in the summer (i.e., 8.25 mg/L for the Tembenchi River of Central Siberia) and winter (400–1300 mg/L), which yields a contribution of groundwater to the annual flux equal to $\sim$1.3 $\pm$ 0.7 %. Finally, we assumed that the permafrost thawing

($F_{permafrost}$) and the litter degradation (F_{litter}) occur during the summer/fall (July to October). Therefore, the overall flux is given as

$$F_{tot} = F_{atmosphere} + F_{undeground} + F_{litter} + F_{permafrost} + F_{RW} \quad (3)$$

where F_{RW} is direct basalt and soil mineral chemical weathering via rock–water interaction with soil porewater fluids. It can be seen from Fig. 2 that both during the active period and on the annual scale, the largest uncertainty stems from litter degradation processes; more precisely, the *J*/NPP value. The exact contribution of direct mobilization of elements from the rocks cannot be assessed, but ranges between 20 and 70 % of the total river TDS_c flux.

4 Seasonality of These Fluxes: Specificity of Boreal and Subarctic Zone

For practical reasons, it is much easier to initiate year-round sampling for chemical analysis at an already operating site then to construct and maintain a new gauging station for daily discharges. However, many former Russian Hydrometerological Stations do not operate any more, and only a small fraction (from 10 to 20 %) of them possesses any chemical sampling. Ideally, the best strategy for rigorous flux assessment would be to organize collaboration with the existing Russian Hydrological Survey system, organizing regular chemical sampling of dissolved components (via on-site filtration with disposable filter holders or filters and syringes), and SM via the sedimentation of large volumes of the river water, decantation, and bottom suspension freezing in the laboratory. When the sampling analysis is carried out, chemical fluxes are calculated based on hydrological fluxes measured in parallel, by interpolating between neighboring data points.

An example of the strategy for seasonal flux calculation is described below. The Central Siberian basaltic plateau, underlain by continuous permafrost and covered by abundant larch forests, offers a unique natural site to assess the contribution of different sources of elements to the rivers (Pokrovsky et al. 2005a, b, 2006; Bagard et al. 2011; Prokushkin et al. 2011). Similar to the non-permafrost zone, the majority of the OC flux occurs during the freshet, whereas the HCO_3^- flux, reflecting inorganic CO_2 consumption during weathering, is similar for the spring flood and summer base flow, but negligibly small during the winter period (Fig. 3).

The majority of the metal flux (of Fe, Al, other low mobile trace elements) occurs during the spring flood, which lasts less than a month and contributes up to 60–80 % of the total water discharge to the ocean (Fig. 4). In contrast, the winter base flow, lasting from October to May and delivering only 10–15 % of the annual water discharge, still contributes to approximately 35 % of the annual Ca flux. This is certainly linked to high concentration of Ca and other major components in the river water during the winter, when Siberian rivers are fed by taliks connected to groundwater reservoirs of fluids. During this period, the TDS increases to ≥ 1 g L^{-1}.

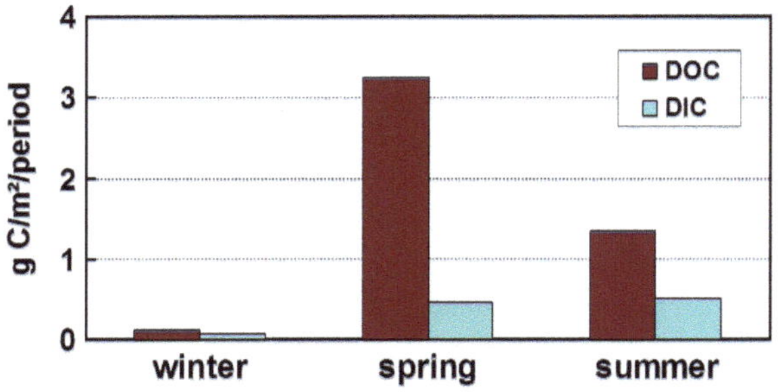

Fig. 3 Winter, spring, and summer fluxes of dissolved organic and inorganic carbon (*DOC* and *DIC*, respectively) of the largest tributary of the Yenisey River, 1966–1975 and 2006–2010. Note that most carbon exported from the watershed is in organic, not inorganic form. This is a major, typical feature of all Eurasian Arctic rivers, contrasting with the Yukon and Mackenzie (see Prokushkin et al. 2011). As a result, many features of carbon transport and element fluxes at the watershed established for North America might not be applicable to Northern Eurasia

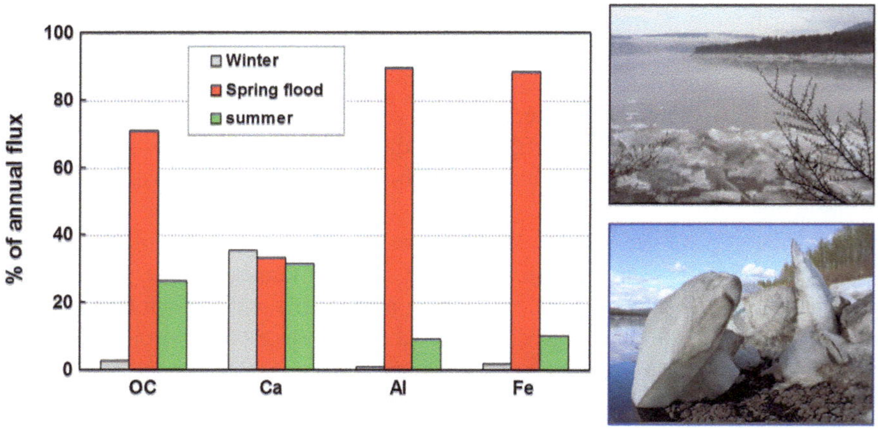

Fig. 4 Annual breakdown of dissolved organic carbon, Ca, Al, and Fe fluxes on the Kochechumo River, the largest tributary of the river N. Tunguska (Bagard et al. 2011), courtesy of Marie-Laure Bagard. The spring flood (photos on the *right*) is the most important period in terms of dissolved component transport

The contribution of the spring flood to the annual $Ca(HCO_3)_2$ flux is around 30 %. During these 2–3 weeks of the freshet, the mineral soil is still frozen and the sources of solutes in rivers are mostly upper organic-rich soil horizon and degrading plant litter. The intensity of HCO_3^- leaching in the form of CO_2 emission from the plant litter depends on the nature of the organic substrate and the activity of soil microorganisms. As such, a significant proportion of annual CO_2 consumption during weathering should be linked to the plant litter degradation in topsoil organic layers rather than to direct dissolution of parent rocks in subsoil interstitial solutions. As a result, the main response of the system to climate warming seems to be in the change of element uptake flux by roots, which is linked merely to the plant biomass rather than to the changing soil hydrological or chemical regimes.

Moreover, the annual dissolved carbon flux from the watershed, reflecting the total CO_2 consumption by vegetation and chemical weathering, is largely dominated by OC, constituting 80 % of the total C flux on average (Fig. 3). It means that only 20 % of atmospheric carbon is directly linked to basalt mineral dissolution, whereas most carbon taken up from the atmosphere is transported by the river to the ocean in the form of plant litter degradation products. The distinctive feature of Siberian larch forest, capable of providing an essential part of riverine element fluxes, is that it is developed on permafrost terrain. Coniferous larch species shed needles in September, when the element uptake by roots is limited as snow cover appears and the soil starts to freeze. Since, the aqueous reactivity of plant biomass is the highest at the very beginning of the reaction (e.g. see Fraysse et al. 2010), the litter degrades very fast in the late fall and in the next spring during the massive snow melt. In this period, both the mineral and organic layers of soil, especially the rooting zone, are still fully frozen. The release of elements from larch needles happens directly at the site of deposition, under the tree canopy, and continues further in the river water if, during the high-water level period, the needles are transported by the meltwater. As a result, compared to temperate or non-permafrost forests, the amount of recycling of elements from the degrading litter back to the plants is lower, and, consequently, the fraction of element that can be released into the river is significantly higher.

It is important to note that most (approx. >85 %) annual trace metal flux in Central Siberian rivers occurs during freshet, following the OC flux. The transport of insoluble trace metals in Siberian rivers is linked to Fe and Al as the major constituents of river organo-mineral colloids (Pokrovsky et al. 2006; Bagard et al. 2011). Given that this flux is mainly formed when the mineral soil is frozen, the importance of plant litter in providing the majority of dissolved trace metals is incontestable, as is also supported by mass balance calculations (Pokrovsky et al. 2006). As such, while climate warming will certainly affect the wintertime element fluxes and speciation, it is unlikely to change the nature and magnitude of the TE fluxes to the ocean which occurs in colloidal form during several weeks of the spring flood.

5 Role of Colloidal Transport of Organic Carbon, Major, and Trace Elements

The boreal aquatic zone is dominated by colloids, the main vectors of the transport of most trace insoluble and even major elements such as Ca. Compared to the tropical and temperate zones, colloids in the boreal zone (1) are richer in Fe(III) and (2) contain a higher proportion of low-molecular weight ($LMW_{<1\ kDa}$) labile organic matter.

All boreal and circumpolar regions of both European Russia and Siberia contain significant amount of OC in the peat soil of the bog zones and as the forest litter of the highly productive taiga zone. Even in the tundra zone, the river valleys are often forested; as a result, plant litter decomposition at the watershed, together with bog zone feeding, is a very important source of dissolved organic carbon in the rivers. However, depending on the lithology of the underlying rocks and mineral soil composition, plant-derived contemporary and ancient organic carbon may be delivered to the rivers in the form of colloids or be retained in soils as thick horizons of Ca humates and fulvates. Historically, most boreal and subarctic rivers were investigated in the Scandinavia and Russian European zone. The influence of glacial deposits in this region on the water–rock interaction in the weathering zone is very high. The quaternary moraine and till deposits formed after the glacial erosion of Precambrian granites and granito-gneisses provide a high concentration of Al (and Fe) in essentially podzol soils. The mobilization of Fe–Al-OM compounds bearing various trace metals from the soils to the river is therefore one of the major processes determining the colloidal status of boreal waters. Interestingly, rivers draining the carbonate rocks in the boreal zone rarely have more than 5–7 mg/L of DOC, which is a factor of 2–3 lower than that in silicate-bearing watersheds. An example of the influence of carbonate (and gypsum) lithology on colloids and OC fluxes can be found in the analysis of the Severnaya Dvina River, draining mostly clays and silicate moraine, and its major northern tributary, the Pinega River, draining carbonate and gypsum rocks (Pokrovsky et al. 2010).

To assess the proportion of colloids, entities between 1 nm ($\sim$1 kDa) and 0.22 μm in size, we consider an example from the Severnaya Dvina River. For this largest European Arctic river, the relative proportion of colloidal (1 kDa–0.22 μm), conventionally dissolved (<1 kDa) and suspended (>0.22 μm) element fluxes from the land to the Arctic Ocean has become available (Pokrovsky et al. 2010; Shevchenko et al. 2010). Ultrafiltration and dialysis, systematically conducted over all seasons of the year, allowed an integrated assessment of three main pools of elements delivered to the ocean by this boreal river (Fig. 5).

Note that the majority of insoluble elements (Fe, Al, REEs, Zr, Th) are transported both as colloidal and suspended fraction, whereas U and OC are delivered essentially in colloidal form. Divalent metals are equally distributed among <1 kDa, colloidal, and suspended fractions. The last result is important in view of the micronutrient supply to the Arctic Ocean, given that the <1 kDa fraction, comparable with the pore size of cellular membranes, may be potentially bioavailable. A similar distribution of colloidal and dissolved annual fluxes has been observed for a large tributary of the Yenisey River (Bagard et al. 2011; Prokushkin et al. 2011). The relative proportion of HMW (colloidal, 1 kDa–0.22 μm) DOC in Central Siberian rivers exhibited both seasonal and spatial variations. The highest proportion of HMW DOC (60–70 %) is observed during the spring flood period, in accordance with recent data from another subarctic, but permafrost-free river, the Severnaya Dvina (Pokrovsky et al. 2010). During the summer base flow, the LMW (<1 kDa) fraction of the largest Yenisey tributaries constitutes from 60–90 % of DOC (Prokushkin et al. 2011). To our knowledge, this kind of information covering

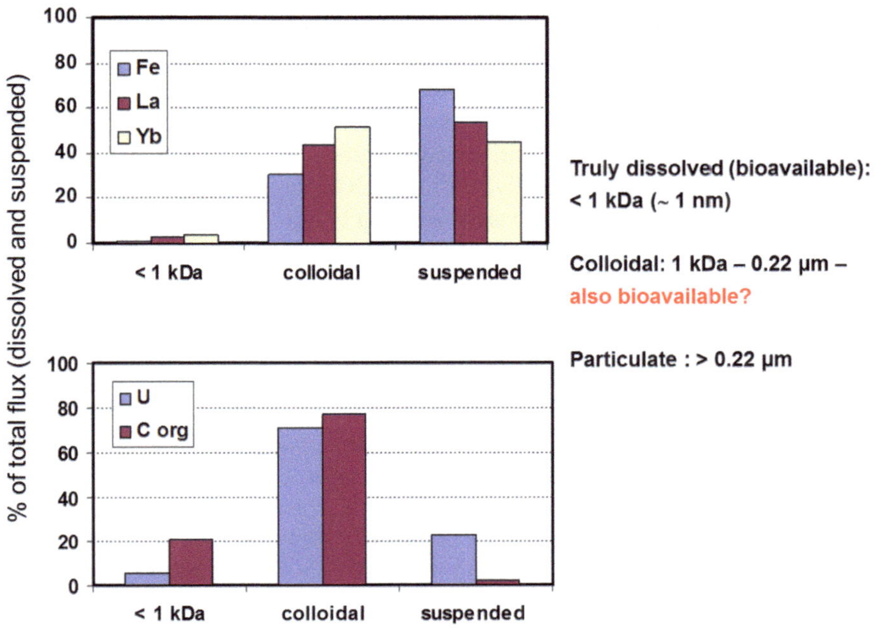

Fig. 5 Histogram of three pools of fluxes for the Severnaya Dvina River (Pokrovsky et al. 2010)

the entire year is not available for other large temperate and tropical rivers. Therefore, a rigorous comparison of the degree of colloidal transport to the Arctic Ocean and to the other oceans is at present impossible. However, it can be speculated that, normalized to its relatively small volume and surface area, the Arctic Ocean receives by far the highest colloidal element flux from the land, and this colloidal input occurs during 1–2 months of the spring flood.

The lithology of rocks exposed at the watershed determines to an essential degree, the colloidal status of TE in surface waters. The commonly established opinion that the boreal waters are always organic-rich corresponds to the majority of cases where the watershed lithology is essentially igneous silicate granitic and basaltic rocks or peatlands over the clastic (silicate) sediment. In these settings, the retention of Al and Fe humates in soils is relatively small, and the majority of dissolved OC is indeed transported from the soils to the river. In contrast, in case of carbonate (Ca-rich) rocks, significant retention of plant-derived OM occurs in the soil and the waters are much poorer in OC and, consequently, organic colloids. Carbonate-dominated watersheds are rare in the historically mostly studied boreal zone (European subarctic) but quite common further to the east (Kyloi river, Anabar, tributaries of the Lena River draining Neoproterozoic and Cambrian carbonates and low reaches of the Lena; some right tributaries of the Yenisey river). The degree of colloidal transport in a river should increase with an increase in (i) the proportion of silicate versus carbonate rocks on the watershed, (ii) bog coverage, (iii) coniferous versus deciduous tree fraction in the vegetation. It is likely to decrease with an increase in (i) the proportion of deciduous trees, (ii) permafrost coverage and thickness of the watershed, (iii) the proportion of lakes, and (iv) runoff. As a result of the combination of several factors, the net value of the colloidal flux is a complex function of the watershed parameters. This flux is also seasonally dependent. For example, forested carbonate watersheds developed on permafrost-bearing terrain may deliver as much DOC (and as many colloids) during the spring flood as do the silicate watersheds on a bog-rich watershed on the annual scale.

Tentatively, all Russian boreal/subarctic rivers can be classified in the order corresponding to the degree of colloidal transport, from the west to the east: Onega ≥ Severnaya Dvina > Kem ≥ Mezen ~ Pechora > Kuloi in the European Arctic Zone. In Western Siberia, this order is likely to be: Ob ~ Taz ~ Pur ~ Nadym > Pyasina > Nizhnyaya Taimura. Among the largest rivers of Central and Eastern Siberia, the colloidal proportion is likely to follow the order Yenisey > Lena > Olenek > Khatanga ~ Indigirka > Kolyma > > Anabar ≥ Olenek > Yana. This order does not necessarily follow the average annual DOC concentration and fluxes, as the colloidal transport of TE depends on the presence of Al, Fe, and the molecular size of dissolved organics. Note that the majority of the above landscape-based factors remain to be quantified.

6 Fluxes of Non-traditional Stable Isotopes (Ca, Mg, Si, Fe) Carried by Boreal Russian Rivers to the Arctic Ocean

The study of non-traditional stable isotopes, which has become possible only over the past decade thanks to the progress in ICP MS multi-collector instruments, which can provide a conceptually new view of Siberian and, more generally, boreal riverine fluxes in terms of (1) their source, (2) the origin of their seasonal variations, and (3) their impact on the Arctic Ocean chemical and isotopic composition. Four case studies are considered below regarding four major contrasting components of dissolved (<0.22 μm) river load: Ca, Mg, Si, and Fe. Here, Fe is considered as major component since it is the main constituent of organo-ferric colloids, the major carriers of many trace elements via the rivers to the ocean (Pokrovsky et al. 2012a, b).

6.1 Sr and Ca Isotopes

First, we consider the radiogenic Sr isotopes, allowing us to trace the source of alkaline earth elements in the river during different seasons. At the large scale, the radiogenic Sr isotopic variations observed in the Kochechum river water over the hydrological year (Fig. 6) are explained by variations in the origin of the water contributing to the river flux because the river is predominantly fed by deep

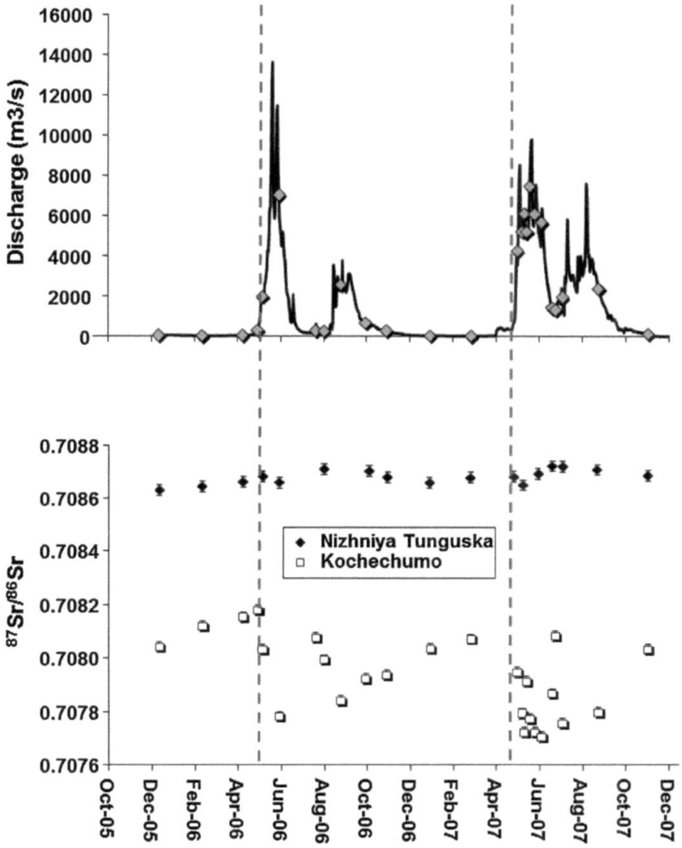

Fig. 6 The radiogenic Sr isotopic variations observed in the Kochechum and the N. Tunguska (Yenisey tributary) water observed over the hydrological year (Bagard et al. 2011), courtesy of Marie-Laure Bagard

brine-related underground waters during the winter low-water period. These underground waters reach the river through unfrozen paths in the permafrost created by the thermal effect of the large water body (Bagard et al. 2011).

The Ca isotopes exhibit quite small variations (±0.3 ‰) over the hydrological year. The winter waters appear to be slightly enriched in ^{40}Ca compared with spring/summer waters ($\delta^{44/40}Ca = 0.60 \pm 0.06$ ‰), and ($\delta^{44/40}Ca = 0.77 \pm 0.04$ ‰, respectively: Fig. 7). This difference may result either from (i) secondary inorganic or organic fractionation processes affecting riverine Ca or (ii) differences in the Ca isotopic composition of the primary source of the dissolved Ca.

As a result, although a significant portion of the riverine Ca could have transited through the vegetation (Pokrovsky et al. 2005a, b), this process does not appear to significantly impact the Ca isotopic composition in the Central Siberian rivers. This conclusion is in agreement with the conclusions by Schmitt et al. (2003) and Tipper et al. (2008, 2010) for major world rivers.

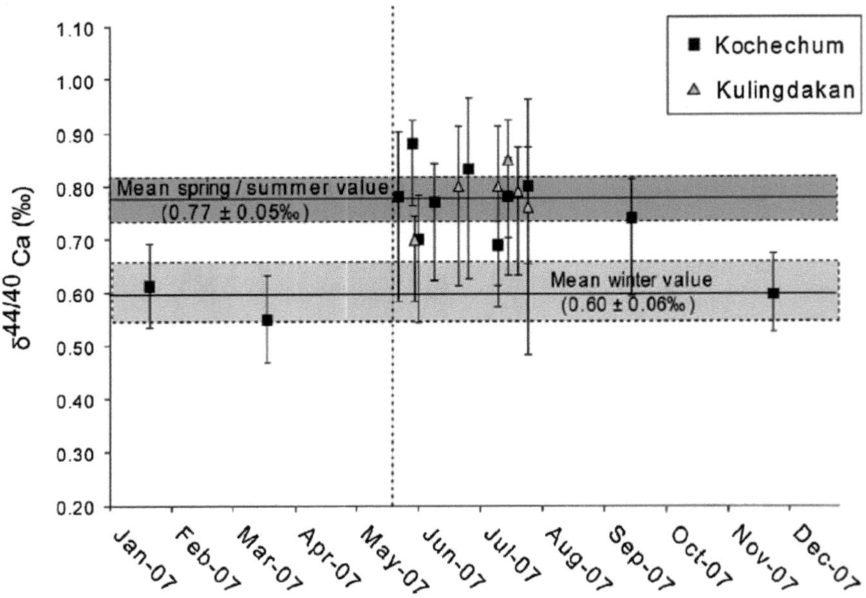

Fig. 7 Ca isotopic compositions of Kochechum river waters (the largest tributary of the N. Tunguska River, the Yenisey tributary) and Kulingdakan stream (a small watershed of ~30,000 km^2, a tributary of the Kochechumo river) waters during the year 2007 (Bagard et al. 2013), courtesy of Marie-Laure Bagard

A recent study (Bagard et al. 2013) demonstrated that, during the winter, 80–90 % of the riverine Ca in the Kochechum waters originates from deep brine inputs from the underlying sedimentary series, whereas during the summer, most of the Ca originates from the surface and subsurface soil pools. The varying proportions of these two major sources of Ca might explain the riverine Ca isotopic composition, which shows similar variations over the year. The summer waters show isotopic compositions similar to the basalts and to the small (~25,000 km^2 watershed) stream waters, which drain only these basalts, indicating that high spring/summer $\delta^{44/40}$Ca values in the Kochechum river waters reflect the weathering of this lithology. This fully confirms the observation made by Tipper et al. (2010), who showed that rivers draining silicate rocks usually carry the isotopic signature of the latter. The Kochechum winter waters with low $\delta^{44/40}$Ca values could subsequently carry the imprint of the sedimentary rocks seated underneath the basaltic floods (Alexeev et al. 2007). This possibility suggests that abiotic sources predominantly influence the $\delta^{44/40}$Ca of Kochechum river waters. It follows that these source-induced variations remain severely limited (0.17 ‰) compared with the intrinsic $\delta^{44/40}$Ca lithological variations, confirming that, in most settings, signals of this kind may be buffered, similar to the biological signal, and that the river dissolved $\delta^{44/40}$Ca is mainly controlled by the heterogeneity of the lithological sources, as also suggested by Tipper et al. (2010).

6.2 Mg Isotopes in Siberian Rivers

The Mg isotopic composition of large rivers (the Kochechum and Nizhnaya Tunguska rivers, tributaries of the Yenisey River) was recently measured in order to unravel the different sources of Mg generated by basalt weathering in Central Siberia under conditions of permafrost and deciduous larch forest (Mavromatis et al. 2014). During the winter base flow, the dissolved Mg isotope composition of large rivers is significantly lighter compared to the source basaltic rocks and atmospheric deposition, suggesting a deep underground source such as sedimentary carbonate rocks. During spring flood and in the summer-fall season, δ^{26}Mg increases by 0.2–0.3 ‰ and approaches the Mg isotope composition of ground vegetation (dwarf shrubs, mosses) and the soil organic horizon. Overall, riverine waters are 0.6–1.0 ‰ lighter than the unaltered bedrock and deep minerals soil horizon.

In winter (October–May), only large rivers could be sampled and, as can be seen in Fig. 8, they both exhibit significant depletion in ^{26}Mg. At the maximum discharges, during the spring flood (beginning of June), the heavier Mg isotopes dominate in the dissolved fraction. Considering that the isotopic composition of the Kuligdakan stream is similar to that of the Kochechum river during the spring flood (Fig. 8), it can be assumed that the isotopic composition of the dissolved fraction reflects the Mg-isotope composition of the host rock. The high discharge during this time of the year almost excludes the formation of secondary phases that alter the Mg isotope composition of the aqueous phase. Similar processes affect the isotopic composition of the dissolved fraction in the N. Tunguska river. The observed difference between the high- and low-flood season is likely the result of deep underground waters having a longer residence time within the watershed, allowing

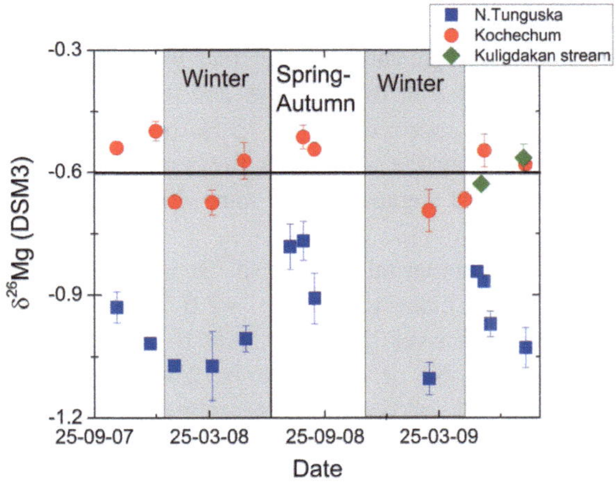

Fig. 8 Evolution of Mg isotopic composition in the two large rivers (Kochechum and N. Tunguska) and Kulingdakan stream over the course of the hydrological year between September 2007 and August 2009. The solid line represents atmospheric precipitation. Modified after Mavromatis et al. (2014), courtesy of Vasileios Mavromatis

a higher degree of interaction between the reacting fluids and the surrounding sedimentary (carbonate) rocks and, as a result, the precipitation of secondary clay silicate. Both clay silicates and clays were earlier shown to preferentially take up the light Mg isotopes, enriching the remaining aqueous phase in ^{26}Mg. This scheme allows us to explain the seasonal behavior of both large rivers examined in this study.

On the scale of other large Siberian rivers, one may anticipate quite low δ^{26}Mg values, close to −1 ‰ (light Mg) in rivers of the Eastern Siberia (such as the Lena River) draining Early Phanerozoic and Precambrian carbonate rocks or the rivers of the European Russian boreal zone (such as the Severnaya Dvina or Pechora) draining sedimentary rocks. In contrast, rivers draining igneous and effusive rocks including granites (Aldan, upper reaches of Anabar) and basalts (Kotyi, Kheta) are likely to bear less negative δ^{26}Mg. However, given that the majority of the flux occurs during the spring high flow, the global value of δ^{26}Mg from Russian boreal rivers to the Arctic Ocean between −0.5 and −1.5 ‰ seems to be a reasonable approximation. Climate warming is unlikely to significantly impact the isotopic signature of Mg flux in Siberian rivers.

6.3 Features of Si Isotopic Composition

Si isotopic composition has been studied in two large Siberian rivers and a small permafrost-dominated watershed. In winter (October–May), only large rivers could be sampled. They exhibited significant enrichment in heavy isotope with δ^{30}Si equal to 2.0–2.5 ‰ and 1.0–1.2 ‰ for the Kochechum and N. Tunguska rivers, respectively. At the maximum of discharges, during the spring flood (beginning of June), light Si isotopes dominated in the dissolved fraction (δ^{30}Si = 0.8 ± 0.2 ‰ for both large rivers) and the δ^{30}Si value of the river water approached that of particulate SM and weathered basalt (0.25 ± 0.16 and 0.32 ± 0.13 ‰, respectively). Note that the difference of approximately 1.75 ‰ between the Kochechum and N. Tunguska measured during the winter low flows almost completely disappears during the spring flood, thus suggesting the same dominant source of Si in both systems.

During the frost-free summer/fall season, δ^{30}Si in large rivers systematically increases, achieving values close to +1.5 … +2.0 ‰ in August. These highly positive values of dissolved river load are closer to the δ^{30}Si of the larch litter (1.47 ± 0.14 ‰), minerotrophic peat bog waters (2.51 ± 0.23 ‰), and interstitial waters of the organo-mineral soil horizon representing the suprapermafrost flow of the Kochechum river watershed (+2.0 … +2.2 ‰).

Continuous year-round measurements of δ^{30}Si in large rivers' waters allow us to assess the average discharge-weighted isotopic composition for different seasons and the integral annual δ^{30}Si value. The spring melt period exhibits the lightest average isotopic composition (1.11 and 0.85 ‰ for the Kochechum and Nizhnyaya Tunguska rivers, respectively), whereas the heaviest silicon is observed in the winter and summer low flow in the Kochechum River (2.18 and 1.97 ‰, respectively). Finally, the annual average discharge-weighted isotopic signature is equal to 1.67 ± 0.15 ‰ and 1.08 ± 0.10 ‰ for the Kochechum and N. Tunguska rivers, respectively.

Using the approach of monthly and seasonal discharge-weighted flux calculation elaborated for organic and inorganic carbon as described above (Prokushkin et al. 2011), seasonal fluxes of dissolved Si were calculated to assess the average annual value of Si isotopic composition. The average annual discharge-weighted δ^{30}Si values of 1.67 and 1.08 ‰ found for the Kochechum and N. Tunguska rivers are comparable with summertime results of grab samples on other larger Arctic rivers, such as the Lena, Yenisey, Ob, and MacKenzie, which exhibit δ^{30}Si equal to 1.17 ± 0.13, 2.30 ± 0.09, 1.78 ± 0.14, and 1.36 ± 0.19 ‰, respectively (Reynolds et al., unpublished). They are also close to δ^{30}Si measured in the boreal permafrost-free Kalix River (+0.7 to +1.5 ‰, Engström et al. 2010) and in large tropical rivers (+0.91 ± 0.09 ‰ in the Congo, Hughes et al. 2011; 1.51 ± 0.19 ‰ in the Ganges-Brahmaputra system, Georg et al. 2009). At the same time, the values of the Central Siberian rivers are higher than the average of small Icelandic catchments (0.63 ± 0.38 ‰, Georg et al. 2007), and Swiss rivers (0.84 ± 0.19 ‰, Georg et al. 2006). This difference can be explained by the significantly lower amount of suspended material in Siberian rivers compared to the mountainous Icelandic and Swiss rivers of high runoff, where enhanced water and sediments discharge inhibited clay formation which, together with scarcer vegetation, results in lower δ^{30}Si in the waters.

Results obtained on the rivers Kochechum and N. Tunguska point to the importance of studying riverine isotopic composition during high-water events such as the spring flood in the boreal zone. For example, considering the summer/fall period alone, corresponding to the period of “easiest” sampling, may yield up to 1 ‰ higher δ^{30}Si than the actual annual discharge-weighted isotopic signature of dissolved Si flux to the Arctic Ocean.

The environmental and biogeochemical contexts of large Siberian rivers making them different from the other rivers of the world are as follows: (i) the presence of very deep underground waters bearing the signature of mineral transformation reactions occurring under high mineral/fluid ratio; (ii) different depth of the active layer and significant uptake of light Si isotopes during short vegetative seasons; (iii) the deciduous character of *Larix,* producing high amounts of litterfall in a short period of time; (iv) the significant enrichment of Siberian larches in ^{30}Si (+1.2 … +1.5 ‰) and thus significant contrasts between isotopically light mineral and isotopically heavy organic pool of solid Si; (v) the seasonal change in the depth of the active layer and the respective involvement of different soil horizons in Si supply to the river during the frost-free season, and (vi) pulses of RSM and its interaction with dissolved Si load during high-flux period in large Siberian rivers. All these aforementioned factors will exert a tight control on the evolution of the aqueous Si isotopic composition delivered by Siberian rivers to the Arctic Ocean in a climate warming scenario. The climate warming in boreal and subarctic regions in the next 20–100 years will result, notably, in (i) an increase in the thaw layer thickness and an elongated growing season, (ii) vegetation changes and increased plant productivity, (iii) amplified winter discharges and annual river runoff, (iv) an intensification of weathering rates in terrestrial ecosystems, and (v) an increased autochthonous biological activity. These processes are capable of shifting the overall dissolved silicon isotope ratio toward both positive and negative values as discussed below.

The increase in the biomass production and the active layer depth during climate warming should enhance light isotope removal from interstitial soil solutions via root uptake and secondary mineral formation. Both processes are likely to increase the concentration of heavy Si isotope in the river water. Similarly, the rise in the water temperature and the increase in the length of the frost-free period should provoke an increase in the aquatic macrophyte biomass, and the productivity of planktonic and periphytic diatoms in the river channel. This will again increase the concentration of heavy isotopes in large rivers during the summer base flow.

The factors capable of shifting the Si isotopic composition of the river water toward more negative values than at present are the increase in chemical weathering linked to permafrost thawing (Tank et al. 2012), including mineral and litterfall dissolution, and the increase in the river suspended flux. At present, quantitative estimations of the expected magnitude of each of those isotope-controlling processes are lacking and the reconstruction of the exact evolution of the isotopic composition of the dissolved Si-based solely on the analysis of governing factors is impossible.

6.4 Fe Isotope Fluxes

There are very few data available on Fe isotope composition in boreal and subarctic rivers. In a recent study by Ilina et al. (2013), small subarctic streams are reported to have a significantly heavier isotopic composition compared to previously reported large temperate and tropical rivers (Beard et al. 2003; Fantle and DePaolo 2004; Bergquist and Boyle 2006; Escoube et al. 2009). This suggests that the dissolved load of previously studied large rivers cannot serve as a proxy for Fe isotopic composition in the river discharge from the land to the Arctic Ocean, and direct measurements of $\delta^{57}Fe$ in all large Arctic rivers are necessary.

The difference of $\delta^{57}Fe$ in the dissolved fraction (<0.22 μm) between different types of surface waters of Russian Arctic and subarctic rivers reaches 3 ‰. This range is larger than that reported in other natural settings from lower latitudes, such as lake and river SM for particles in the Aha lake, China (2.0 ‰, Song et al. 2011); the dissolved fraction of Fe-rich Lake Nyos, Cameroon (2.7 ‰, Teutsch et al. 2009), and Fe isotopic variation in the Amazon basin (2.25 ‰, Bergquist and Boyle 2006). Moreover, the range of isotopic variations in subarctic surface waters of the Arctic Ocean watershed is significantly larger than that of boreal river and Baltic Sea SM (0.03, 0.66, and 0.3 ‰, Beard et al. 2003, Ingri et al. 2006, and Gelting et al. 2010, respectively) and a temperate river (from −1.7 to −0.3 ‰ and 0.55–0.78 ‰, Fantle and DePaolo 2004, and Escoube et al. 2009, respectively).

The trend of isotopic composition change among colloids and particles is similar between subarctic and temperate rivers and streams (Fig. 9). However, at otherwise close Fe/C ratios, the absolute values of δ^{57}Fe are distinctly shifted by −1 ‰ in the temperate Senga River (draining podzol soils developed on quaternary sand alluvial deposits) compared to the subarctic Palojoki River (draining Archean and Proterozoic acidic and mafic magmatic rocks overlaid by thin podzol soils). It could be that the difference in lithological substrate and the regime of groundwaters control the initial isotopic composition of surface waters (<0.2–0.4 μm). A systematic enrichment of filtrates in the heavy isotope with the decrease in the size fraction and Fe/C_{org} ratio strongly suggests that the small size organic-rich colloids are isotopically heavy and the large size mineral-rich colloids are lighter. The degree to which this observation can be extended to other boreal rivers and streams still remains unknown.

Among the various processes responsible for isotopically heavy Fe in high-latitude water bodies would be the production of Fe isotopically heavy LMW organic ligands via photoreduction; heterotrophic bacterioplankton respiration; and phytoplankton metabolism may be different between small subarctic streams and large temperate rivers. At present, we cannot link these differences to distinct features of landscape characteristics, stream chemistry, or microbial activity. However, one can expect that the photoreduction of the subarctic waters should be more pronounced due to significantly longer periods of solar irradiation at high latitudes compared to temperate zones.

Moreover, high enrichment by the heavy isotope of LMW fractions (<1–10 kDa) may add very important constraints on the terrestrial isotopic signature of Fe

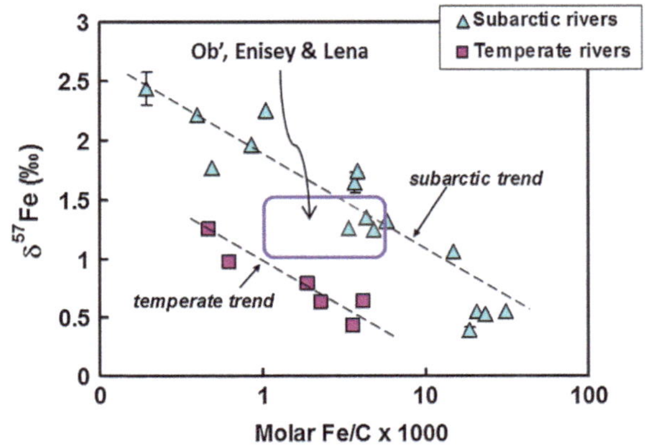

Fig. 9 δ^{57} Fe as a function of molar Fe/C_{org} ratio (log scale) in filtrates and ultrafiltrates of small temperate and subarctic rivers (Ilina et al. 2013), courtesy of Svetlana M Ilina. The rectangle depicting the hypothetical position of the three largest Siberian rivers is based on 0.22 μm-filtered data on the Ob, Lena, and Yenisey from the PARTNERS database (http://ecosystems.mbl.edu/partners), all year-round. Fe isotopic measurements are not available in the PARTNERS database, but extrapolated from the range of Fe/C_{org} in dissolved fraction and geographic latitudes of the river basin

transport to the Arctic Ocean. This small, previously overlooked soluble fraction is: (i) less likely to coagulate in the estuarine zone and (ii) potentially more bioavailable for aquatic microorganisms compared to the <0.22 μm fraction usually considered. Indeed, it has recently been shown that the isotopic fractionation of Fe during estuarine mixing is negligible (de Jong et al. 2007; Escoube et al. 2009), even when there is a coagulation of Fe oxy(hydr)oxide (Bergquist and Boyle 2006). In estuarine mixing experiments, it was also shown that the iron oxyhydroxide colloids of the river water aggregate into much larger particles, while the organic colloidal phase remained virtually unaffected (Stolpe and Hassellöv 2010). Finally, it has been recently demonstrated that, along the salinity gradient from 0–15 ‰ in the White Sea, the most labile, LMW fraction of Fe increases its concentration by a factor of 5, whereas the concentration of <0.22 μm fraction decreases followed by the "classic" coagulation process (Pokrovsky et al. 2014). Therefore, the possibility is not excluded, the isotopic signature of the most labile and potentially bioavailable Fe fraction delivered by small coastal rivers and streams to the Arctic Ocean might be up to +4 ‰ heavier compared to (1) the silicate terrestrial source material transported in the form of river SM and (2) conventionally dissolved (<0.22 μm) Fe pool. The trend shown in Fig. 9 is also consistent with increasingly heavier δ^{57}Fe delivered by rivers to the ocean as the latitude increases, as inferred from soil studies (Poitrasson et al. 2008). The climate warming at high latitudes should intensify both photoreduction and plankton exometabolite production due to the water temperature increase; thus, the relative input of isotopically heavy Fe to the Arctic ocean will further increase in the next decades, thereby providing another, new tracer of ongoing environmental changes.

The Fe isotopic signatures in other subarctic rivers have not yet been investigated. Unpublished data collected by R. Escoube on the Severnaya Dvina River and its tributaries in 2007–2008 suggest that δ^{57}Fe is close to +0.8 … +0.9 ‰ during the spring flood. This is generally consistent with the position of 0.22 μm-filtered fraction of small subarctic rivers reported by Ilina et al. (2013). The isotopic signature of Fe flux in other subarctic rivers is at present unknown as no single measurement of δ^{57}Fe in the rivers of the Arctic ocean basin is available. However, considering the dependence between the Fe isotopic signature and the Fe/C ratio in the dissolved fraction, the largest Siberian rivers—the Ob, Yenisey, and Lena—are likely to have a year average, discharge-weighted δ^{57}Fe of Fe flux between +1 and +1.5 ‰. This high ratio reflects the relatively low Fe/C ratio in the dissolved fraction of the largest rivers (compared to smaller, underground-fed rivers) and the generally temperate to subarctic context of the watershed. The isotopic signature is likely to progressively increase in the order winter base flow < summer base flow ≤ spring flood, reflecting the decrease in the Fe/C ratio and the increase in small-size colloids in this order.

7 Conclusions

Seasonal aspects of fluxes of carbon and related transport by the rivers to the Arctic Ocean can only be assessed via systematic, year-round studies. Although the Russian Hydrological Survey gauging stations still in operation are perfectly suitable for such an assessment, they require careful sampling and on-site filtration. It is important to note that the majority (approx. >85 %) of the annual trace metal flux in Central Siberian rivers occurs during freshet following the OC flux. The transport of insoluble trace metals in Siberian rivers is linked to Fe and Al as the major constituents of river organo-mineral colloids (Pokrovsky et al. 2006; Bagard et al. 2011). Given that this flux is mainly formed when the mineral soil is frozen, the importance of plant litter in providing the majority of dissolved trace metals is incontestable, as also supported by mass balance calculations (Pokrovsky et al. 2006). As such, while climate warming will certainly affect the wintertime element fluxes and speciation, it is unlikely to change the nature and magnitude of the TE fluxes to the ocean which occurs in colloidal form during several weeks of the spring flood.

During the winter base flow, the dissolved Mg isotope composition of large rivers is significantly lighter compared to the source basaltic rocks and atmospheric deposition, suggesting a deep underground source such as sedimentary carbonate rocks. During the spring flood and in the summer/fall season, δ^{26}Mg increases by 0.2–0.3 ‰ and approaches the Mg isotope composition of ground vegetation (dwarf shrubs, mosses) and the soil organic horizon. Overall riverine waters are 0.6–1.0 ‰ lighter than the unaltered bedrock and deep minerals soil horizon.

The majority of climate-affected environmental processes controlling Si transport from the land to the sea may induce an increase in the water-rock ratio in the watersheds and plant uptake of Si from soil solution. This may decrease the average discharge of weighted aqueous silica concentration and the δ^{30}Si value in soil porewaters and rivers, therefore modifying the isotopic signature of Si fluxes to the Arctic Ocean.

The importance of the LMW isotopically heavy fraction for Fe transport from the land to the Arctic Ocean stems from its higher mobility through the estuarine mixing zone and its potentially high bioavailability. As a result, the small-size, Fe- and organic carbon-rich coastal rivers and streams may play a very important, yet underestimated role in the overall Fe chemical and isotopic budget of the Arctic Ocean.

Acknowledgments I would like to thank numerous colleagues who contributed to various parts of this manuscript: Bernard Dupré, Jacques Schott, Jerome Viers, Francois Chabaux, Anatoly Prokushkin, Elena Zakharova, Vladimir Shevchenko, Liudmila Shirokova, Vasileios Mavromatis, Sergey Lapitsky, Svetlana Ilina, Marie-Laure Bagard, Ekaterina Vasyukova, Sergey Kirpotin, Sergey Vorobiev, and Sergey Kulizhsky. Supports from the grant BIO-GEO-CLIM No 14. B25.31.0001 of the Russian Ministry and Education at Tomsk State University and RFFI No 13-05-00890, No 12-05-91055-CNRS_a No 14-05-98815-sever are greatly acknowledged.

References

Alexeev S, Alexeeva L, Borisov V, Shouakar-Stash O, Frape S, Chabaux F, Kononov A (2007) Isotopic composition (H, O, Cl, Sr) of ground brines of the Siberian platform. Russ Geol Geophys 48:225–236

Bagard ML, Chabaux F, Pokrovsky OS, Prokushkin AS, Viers J, Dupré B, Stille P (2011) Seasonal variability of element fluxes in two Central Siberian rivers draining high latitude permafrost dominated areas. Geochim Cosmochim Acta 75:3335–3357

Bagard ML, Schmitt AD, Chabaux F, Pokrovsky OS, Viers J, Stille P, Labolle F, Prokushkin AS (2013) Biogeochemistry of stable Ca and radiogenic Sr isotopes in larch-covered permafrost-dominated watersheds of Central Siberia. Geochim Cosmochim Acta 114:169–187

Bazilevitch NI (1976) Biogenic and abiogenic processes in forest, step, and desert ecosystems. Int Geogr 76:58–62 (Section 4: Biogeography and Geography of Soils)

Beard BL, Johnson CM, Von Damm KL, Poulson RL (2003) Iron isotope constraints on Fe cycling and mass balance in oxygenated earth oceans. Geology 31:629–632

Beilman DW, MacDonald GM, Smith LC, Reimer PJ (2009) Carbon accumulation in peatlands of West Siberia over the last 2000 years. Global Biogeochem Cycles 23, GB1012, doi:10.1029/2007GB003112

Bergquist BA, Boyle EA (2006) Iron isotopes in the Amazon River system: weathering and transport signatures. Earth Planet Sci Lett 248(1–2):54–68

Botch MS, Kobak KI, Vinson TS, Kolchugina TP (1995) Carbon pools and accumulation in peatlands of the former Soviet Union. Global Biogeochem Cycles 9:37–46. doi:10.1029/94GB03156

Brown J, Ferrians Jr OJ, Heginbottom JA, Melnikov ES (2002) Circum-arctic map of permafrost and ground ice conditions. National snow and ice data center/World data center for glaciology, Boulder, Digital media. http://nsidc.org/data/docs/fgdc/ggd318_map_circumarctic/. Accessed 22 Feb 2015

Cooper LW, Benner R, McClelland JW, Peterson BJ, Holmes RM, Raymond R, Hansell D, Grebmeier J, Codispotti L (2005) Linkages among runoff, dissolved organic carbon, and the stable oxygen isotope composition of seawater and other water mass indicators in the Arctic Ocean. J Geophys Res 110:G02013. doi:10.1029/20058JG000031

Cooper LW, McClelland JW, Holmes RM, Raymond PA, Gibson JJ, Guay CK, Peterson BJ (2008) Flow-weighted values of runoff tracers ($\delta^{18}O$, DOC, Ba, alkalinity) from the six largest Arctic rivers. Geophys Res Lett 35:L18606. doi:10.1029/2008GL035007

de Jong J, Schoemann V, Tison J-L, Becquevort S, Masson F, Lannuzel D, Petit J, Chou L, Weis D, Mattielli N (2007) Precise measurement of Fe isotopes in marine samples by multi-collector inductively coupled plasma mass spectrometry (MCICP-MS). Anal Chim Acta 589:105–119

Dittmar Th, Kattner G (2003) The biogeochemistry of the river and shelf ecosystem of the Arctic Ocean: a review. Mar Chem 83:103–120

Engström E, Rodushkin I, Ingri J, Baxter DC, Ecke F, Osterlund H, Ohlander B (2010) Temporal isotopic variations of dissolved silicon in a pristine boreal river. Chem Geol 271:142–152

Escoube R, Rouxel OJ, Sholkovitz E, Donard OFX (2009) Iron isotope systematics in estuaries: the case of North River, Massachusetts (USA). Geochim Cosmochim Acta 73(14):4045–4059

Fantle MS, DePaolo DJ (2004) Iron isotopic fractionation during continental weathering. Earth Planet Sci Lett 228:547–562

Fraysse F, Pokrovsky OS, Meunier J-D (2010) Experimental study of terrestrial plant litter interaction with aqueous solutions. Geochim Cosmochim Acta 74:70–84

Frey KE, McClelland JW, Holmes RM, Smith LC (2007a) Impacts of climate warming and permafrost thaw on the riverine transport of nitrogen and phosphorus to the Kara Sea. J Geophys Res 112: G04S58. doi:10.1029/2006JG000369

Frey KE, Siegel DI, Smith LC (2007b) Geochemistry of west Siberian streams and their potential response to permafrost degradation. Water Resour Res 43:W03406. doi:10.1029/2006WR004902

Gaillardet J, Millot R, Dupré B (2003) Chemical denudation rates of the western Canadian orogenic belt: the Stikine terrane. Chem Geol 201:257–279

Gebhardt AC, Gaye-Haake B, Unger D, Lahajnar N, Ittekkot V (2004) Racent particulate organic carbon and total suspended matter fluxes from the Ob and Yenisei Rivers into the Kara Sea (Siberia). Mar Geol. 207:225–245

Gelting J, Breitbarth E, Stolpe B, Hassellöv M, Ingri J (2010) Fractionation of iron species and iron isotopes in the Baltic Sea euphotic zone. Biogeosciences 7(8):2489–2508

Georg RB, Reynold BC, Frank M, Halliday AN (2006) Mechanisms controlling the silicon isotopic compositions of river waters. Earth Planet Sci Lett 249:290–306

Georg RB, Reynolds BC, West AJ, Burton KW, Halliday AN (2007) Silicon isotope variations accompanying basalt weathering in Iceland. Earth Planet Sci Lett 261:476–490

Georg RB, West AJ, Basu AR, Halliday AN (2009) Silicon fluxes and isotope composition of direct groundwater discharge into the Bay of Bengal and the effect on the global ocean silicon isotope budget. Earth Planet Sci Lett 283:67–74

Gislason SR, Arnorsson S, Armannsson H (1996) Chemical weathering of basalt as deduced from the composition of precipitation, rivers and rocks in Southwest Iceland: effect of runoff, age of rocks and vegetative/glacial cover. Am J Sci 296:837–907

Gordeev VV (2006) Fluvial sediment flux to the Arctic Ocean, Geomorphology 80:94–104

Gordeev VV, Sidorov IS (1993) Concentration of major elements and their outflow into the Laptev Sea by the Lena River. Mar Chem 43:33–45

Gordeev VV, Martin JM, Sidorov IS, Sidorova MV (1996) A reassessment of the Eurasian river input of water, sediment, major elements, and nutrients to the Arctic ocean. Am J Sci 296:664–691

Gordeev VV, Rachold V, Vlasova IE (2004) Geochemical behavior of major and trace elements in suspended particulate material of the Irtysh river, the main tributary of the Ob river. Siberia Appl Geochem 19:593–610

Guay CKH, Zhulidov AV, Robarts RD, Zhulidov DA, Gurtovaya TY, Holmes RM, Headley JV (2010) Measurements of Cd, Cu, Pb and Zn in the lower reaches of major Eurasian arctic rivers using trace metal clean techniques. Environ Pollut 158:624–630

Guieu C, Huang WW, Martin J-M, Yong YY (1996) Outflow of trace metals into the Laptev Sea by the Lena River. Mar Chem 53:255–267

Guo L, Cai Y, Belzile C, Macdonald R (2012) Sources and export fluxes of inorganic and organic carbon and nutrient species from the seasonally ice-covered Yukon River. Biogeochemistry 107(1-3):187–206

Holmes RM, Peterson BJ, Gordeev VV, Zhulidov AV, Meybeck M, Lammers RB, Vörösmarty CJ (2000) Flux of nutrients from Russian rivers to the Arctic Ocean: can we establish a baseline against which to judge future changes? Water Resour Res 36(8):2309–2320

Holmes RM, Peterson BJ, Zhulidov AV, Gordeev VV, Makkaveev PN, Stunzhas PA, Kosmenko LS, Köhler GH, Shiklomanov AI (2001) Nutrient chemistry of the Ob' and Yenisey Rivers, Siberia: results from June 2000 expedition and evaluation of long-term data sets. Mar Chem 75:219–227

Holmes RM, McClelland JW, Peterson BJ, Tank SE, Bulygina E, Eglinton TI, Gordeev VV, Gurtovaya TY, Raymond PA, Repeta DJ, Staples R, Striegl RG, Zhulidov AV, Zimov SA (2012) Seasonal and annual fluxes of nutrients and organic matter from large rivers to the Arctic Ocean and surrounding seas. Estuaries Coasts 35:369–382. doi:10.1007/s12237-011-9386-6

Hughes HJ, Sondag F, Cocquyt C, Laraque A, Pandi A, André L, Cardinal D (2011) Effect of seasonal biogenic silica variations on dissolved silicon fluxes and isotopic signatures in the Congo River. Limnol Oceanogr 56(2):551–561

Huh Y, Edmond JM (1999) The fluvial geochemistry of the rivers of Eastern Siberia: III. Tributaries of the Lena and Anabar draining the basement terrain of the Siberian Craton and the Trans-Baikal Highlands. Geochim Cosmochim Acta 63:967–987

Huh Y, Panteleyev G, Babich D, Zaitsev A, Edmond JM (1998) The fluvial geochemistry of the rivers of Eastern Siberia: II. Tributaries of the Lena, Omoloy, Yana, Indigirka, Kolyma, and Anadyr draining collisional/accretionary zone of the Verkhoyansk and Cherskiy ranges. Geochim Cosmochim Acta 62:2053–2075

Huser BJ, Kohler SJ, Wilander A, Johansson K, Folster J (2011) Temporal and spatial trends for trace metals in streams and rivers across Sweden (1996–2009). Biogeosciences 8:1813–1823

Hydrological Yearbooks, State Gidromet (1954–1975) Yenissey basin. vol 7, parts 2–4

Ilina SM, Poitrasson F, Lapitsky SA, Alekhin YuV, Viers J, Pokrovsky OS (2013) Extreme iron isotope fractionation between colloids and particles of boreal and temperate organic-rich waters. Geochim Cosmochim Acta 101:96–111

Ingri J, Malinovsky D, Rodushkin I, Baxter DC, Widerlund A, Andersson P, Gustafsson Ö, Forsling W, Öhlander B (2006) Iron isotope fractionation in river colloidal matter. Earth Planet Sci Lett 245(3–4):792–798

ISO (1983) Liquid flow measurement in open channels. Handbook 16, International Standards Organization

Lambelet M, Rehkämper M, van de Flierdt T, Xue Z, Kreissig K, Coles B, Porcelli D, Andersson P (2013) Isotopic analysis of Cd in the mixing zone of Siberian rivers with the Arctic Ocean-new constraints on marine Cd cycling and the isotope composition of riverine Cd. Earth Planet Sci Lett 361:64–73

Leonov AB, Chicherina OV (2004) Export of biogenic components of the riverine flux to the White Sea. Water Resour 31(2):170–192

Magritsky DV (2010) Annual suspended matter flow of the Russian rivers belonging to the Arctic Ocean basin and its anthropogenic transformation. Vestnik Moscow State Unive Ser Geogr 5(6):17–24

Mavromatis V, Prokushkin AS, Pokrovsky OS, Viers J, Koretz M (2014) Mg isotopes in permafrost-dominated Central Siberian larch forest watersheds. Geochim Cosmochim Acta 147:76–89

Moiseenko TI, Skjelkvale BL, Gashkina NA, Shalabodoc AD, Khoroshavin VYu (2013) Water chemistry in small lakes along a transect from boreal to arid ecoregions in European Russia: effects of air pollution and climate change. Appl Geochem 28:69–79

Négrel P, Allègre CJ, Dupré B, Lewin E (1993) Erosion sources determined by inversion of major and trace element ratios and Sr isotopic ratios in river water. The Congo basin case. Earth Planet Sci Lett 120:59–76

Nikanorov AM, Bryzgalo VA, Kosmenko LS, Reshetnyak OS (2010a) The role of chemical river runoff in the anthropogenic transformation of the state of the aquatic environment in the Enisei mouth area. Water Resour 37(4):471–480

Nikanorov AM, Smirnov MP, Klimenko OA (2010b) Long-term trends in total and anthropogenic discharge of organic and biogenic substances by Russian rivers into the Arctic and Pacific Seas. Water Resour 37(3):361–371

Poitrasson F, Viers J, Martin F, Braun JJ (2008) Limited iron isotope variations in recent lateritic soils from Nsimi, Cameroon: implications for the global Fe geochemical cycle. Chem Geol 253:54–63

Pokrovsky OS, Schott J, Kudryavtzev DI, Dupré B (2005a) Basalts weathering in Central Siberia under permafrost conditions. Geochim Cosmochim Acta 69:5659–5680

Pokrovsky OS, Dupré B, Schott J (2005b) Fe-Al-organic colloids control the speciation of trace elements in peat soil solutions: results of ultrafiltration and dialysis. Aquat Geochem 11:241–278

Pokrovsky OS, Schott J, Dupré B (2006) Trace element fractionation and transport in boreal rivers and soil porewaters of permafrost-dominated basic terrain in Central Siberia. Geochim Cosmochim Acta 70:3239–3260

Pokrovsky OS, Viers J, Shirokova LS, Shevchenko VP, Filipov AS, Dupré B (2010) Dissolved, suspended, and colloidal fluxes of organic carbon, major and trace elements in Severnaya Dvina River and its tributary. Chem Geol 273:136–149

Pokrovsky OS, Shirokova LS, Zabelina SA, Vorobieva TY, Moreva OY, Klimov SI, Chupakov AV, Shorina NV, Kokryatskaya NM, Audry S, Viers J, Zouiten C, Freydier R (2012a) Size fractionation of trace elements in a seasonally stratified boreal lake: control of organic matter and iron colloids. Aquat Geochem 18:115–139

Pokrovsky OS, Viers J, Dupré B, Chabaux F, Gaillardet J, Audry S, Prokushkin AS, Shirokova LS, Kirpotin SN, Lapitsky SA, Shevchenko VP (2012b) Biogeochemistry of carbon, major and trace elements in watersheds of Northern Eurasia drained to the Arctic Ocean: the change of fluxes, sources and mechanisms under the climate warming prospective. C.R Geoscience 344:663–677

Pokrovsky OS, Shirokova LS, Kirpotin SN (2014) Biogeochemistry of Thermokarst Lakes of Western Siberia, Nova Science Publ. Inc., NY, 163 pp. 2014

Prokushkin AS, Pokrovsky OS, Shirokova LS, Korets MA, Viers J, Prokushkin SG, Amon R, Guggenberger G, McDowell WH (2011) Sources and export fluxes of dissolved carbon in rivers draining larch-dominated basins of the Central Siberian Plateau. Environ Res Lett 6: 045212 (14 pp). doi:10.1088/1748-9326/6/4/045212

Rember RD, Trefry JH (2004) Increased concentrations of dissolved trace metals and organic carbon during snowmelt in rivers of the Alaskan Arctic. Geochim Cosmochim Acta 68(3):477–489

Resources of Surface Waters of the USSR (1973) The main hydrological characteristics. In: Zilbershtein IA, Stolyarchuk IS (eds) Yenisseisky region, vol 17, Gidrometeoizdat, Leningrad

Reynolds BC, Frank M, Halliday AN (2006) Silicon isotope fractionation during nutrient utilization in the North Pacific. Earth Planet Sci Lett 244:431–443

Schlesinger WH (1997) Biogeochemistry. An analysis of global change. Academic Press, Massachusetts, 588 pp

Schmitt A-D, Chabaux F, Stille P (2003) The calcium riverine and hydrothermal isotopic fluxes and the oceanic calcium mass balance. Earth Planet Sci Lett 6731:1–16

Semenov AD (ed) (1977) Guide on the chemical analysis of continental surface waters, (Руководство по химическим анализам пресных вод) Gidrometeoizdat, Leningrad, 344 pp (In Russian)

Shirokova LS, Pokrovsky OS, Viers J, Klimov SI, Moreva OYu, Zabelina SA, Vorobieva TYa, Dupré B (2010) Diurnal variations of trace elements and heterotrophic bacterioplankton concentration in a small boreal lake of the White Sea basin. Ann Limnol Int J Lim 46:67–75. doi:10.1051/limn/2010011

Shirokova LS, Pokrovsky OS, Kirpotin SN, Desmukh C, Pokrovsky BG, Audry S, Viers J (2013a) Biogeochemistry of organic carbon, CO_2, CH_4, and trace elements in thermokarst water bodies in discontinuous permafrost zones of Western Siberia. Biogeochemistry 113:573–593

Shirokova LS, Pokrovsky OS, Moreva OY, Chupakov AV, Zabelina SA, Klimov SI, Shorina NV, Vorobieva TY (2013b) Decrease of concentration and colloidal fraction of organic carbon and trace elements in response to the anomalously hot summer 2010 in a humic boreal lake. Sci Total Environ 463–464:78–90

Song L, Liu C-Q, Wang Z-L, Zhu X, Teng Y, Liang L, Tang S, Li J (2011) Iron isotope fractionation during biogeochemical cycle: information from suspended particulate matter (SPM) in Aha Lake and its tributaries, Guizhou. China Chem Geol 280(1–2):170–179

Soyer VG, Semenov AD (1971) Photochemical method of organic carbon determination (Фотохимические методы анализа органического углерода). Gidrokhim Materialy 56:111–120 (in Russian)

Stedmon CA, Amon RMW, Rinehart AJ, Walker SA (2011) The supply and characteristics of colored dissolved organic matter (CDOM) in the Arctic Ocean: Pan Arctic trends and differences. Mar Chem 124:108–118

Stefansson A, Gislason SR (2001) Chemical weathering of basalts, Southwest Iceland: effect of rock crystallinity and secondary minerals on chemical fluxes to the oceans. Am J Sci 301:513–556

Stolpe B, Hassellöv M (2010) Nanofibrils and other colloidal biopolymers binding trace elements in coastal seawater: significance for variations in element size distributions. Limnol Oceanogr 55(1):187–202

Tank SE, Frey KE, Striegl RG, Raymond PA, Holmes RM, McClelland JW, Peterson BJ (2012). Landscape-level controls on dissolved carbon flux from diverse catchments of the circumboreal. Glob Biogeochem Cycles 26: GB0E02. doi:10.1029/2012GB004299

Teutsch N, Schmid M, Müller B, Halliday AN, Bürgmann H, Wehrli B (2009) Large iron isotope fractionation at the oxic-anoxic boundary in Lake Nyos. Earth Planet Sci Lett 285(1–2):52–60

Tipper E, Galy A, Bickle M (2008) Calcium and magnesium isotope systematics in rivers draining the Himalaya-Tibetan-Plateau region: lithological or fractionation control? Geochim Cosmochim Acta 72:1057–1075

Tipper E, Gaillardet J, Galy A, Louvat P, Bickle M, Capmas F (2010) Calcium isotope ratios in the world largest rivers: a constraint on the maximum imbalance of oceanic calcium fluxes. Glob Biogeochem Cycles 24: GB3019

Viers J, Prokushkin AS, Pokrovsky OS, Auda Y, Kirdyanov AV, Beaulieu E, Zouiten C, Oliva P, Dupré B (2013) Seasonal and spatial variability of elemental concentrations in boreal forest larch foliage of Central Siberia on continuous permafrost. Biogeochemistry 113:435–449

Vogt KA, Grier CC, Meier CE, Edmonds RL (1982) Mycorrhizal role in net primary production and nutrient cycling in *Abies amabilis* ecosystems in western Washington. Ecology 63:370–380

White D, Hinzman L, Alessa L, Cassano J, Chambers M, Falkner K, Francis J, Gutowski Jr WJ, Holland M, Holmes RM, Huntington H, Kane D, Kliskey A, Lee C, McClelland J, Peterson B, Rupp TS, Straneo F, Steele M, Woodgate R, Yang D, Yoshikawa K, Zhang T (2007) The arctic freshwater system: changes and impacts. J Geophys Res 112: G04S54. doi: 10.1029/2006JG000353

Zakharova EA, Pokrovsky OS, Dupré B, Zaslavskaya MB (2005) Chemical weathering of silicate rocks in Aldan Shield and Baikal Uplift: insights from long-term seasonal measurements of solute fluxes in rivers. Chem Geol 214:223–248

Zaslavskaya MB, Tikhotskaya E (1978) Characteristics of the chemical fluxes in the Upper Amur Basin. In: Chemical composition of natural waters in Amur and Transbaikal Region. Vladivostok, pp 46–56 (in Russian)

Chapter 7
Measuring Snowmelt in Siberia: Causes, Process, and Consequences

Alexander S. Chumbaev and Anatoly A. Tanasienko

Abstract The soil erosion on arable lands of Siberia is widespread. More than 50 % of all farmlands are subject to erosion to various degrees. Erosion is the main process of soil degradation in West Siberia which can lead to a catastrophic decrease in the fertility of soils, and pose a threat to food security in the region. Studying the causes, the process of soil erosion and its consequences is an important question both for science and for farm production. The purpose of this work is to show the main methods and devices used to define the quantity and quality of surface snowmelt water runoff, and also the damage caused by this in the form of soil erosion. To quantify the overall snowmelt erosion process, the following parameters need to be measured: the total pre-winter water reserve of soil, snow depth, snow water equivalent, depth of soil frost penetration, volume of snowmelt water runoff, runoff coefficient, water stream temperature, and soil loss with surface snowmelt water runoff. Research takes place in 3 stages: (1) preparatory stage, during which the late fall period soil water supply is defined and the runoff and thermometric plots are constructed; (2) studying the process of accumulation of solid atmospheric precipitation, the nature of its distribution over the territory, and also the influences of snow depth on the frost penetration in soils; and (3) monitoring the snowmelt process in spring, during which the intensity of snowmelt, the volumes of a superficial drain of snowmelt waters, and the damage caused by them to a soil cover are defined. One special feature of the Siberian soils during the cold period of the year is the intra-soil ice sheet, which is largely impenetrable to melting water and positive temperatures. This ice sheet in Siberian soils is one of the reasons for snowmelt water runoff forming. Over a period of 45 years we measured a mean annual soil loss of 6 t/ha by snowmelt erosion on arable land in West Siberia.

Keywords Snowmelt · Surface runoff · Soil erosion · Siberia · Measurement

A.S. Chumbaev (✉) · A.A. Tanasienko
Siberian Branch, Institute of Soil Science and Agrochemistry (ISSA),
Russian Academy of Sciences, Ac. Lavrentieva av. 8/2, Novosibirsk 630090, Russia
e-mail: chas3079@gmail.com

A.A. Tanasienko
e-mail: tanas40@ngs.ru

L. Mueller et al. (eds.), *Novel Methods for Monitoring and Managing Land and Water Resources in Siberia*, Springer Water,
DOI 10.1007/978-3-319-24409-9_7

1 Introduction

About 52 million hectares of the lands of Siberia (West Siberia, Middle Siberia and the Far East) are involved in the sphere of agricultural use. About a half of these lands are used as arable lands (Tanasienko 2007). The most valuable arable soils are Phaeozems and Chernozems, which provide a significant amount of grain owing to their high initial fertility. It should be noted that about 60 % of all agricultural lands of Russia, including 53 % of arable lands, are located on slopes with a grade of up to 10° (17.6 %) (Shurikova 1981). The ploughing of steep slopes on the farmlands of Siberia has led to significant water erosion. About 15 million hectares of agricultural lands are prone to erosion, of which 9 million hectares have already been damaged (Tanasienko 2003).

Slopes of more than 3° (5.2 %) are most at risk of erosive processes. This holds for about one-third to half of tilled lands in West Siberia. About 10 % of all arable lands are located on slopes of 6°–9° (10.5–15.8 %) and about 5 % have slopes of more than 9° (15.8 %). During land reclamation in West and East Siberia during the 1950s and 1960s, lands at grades of 8°–10° (14.1–17.6 %), and—in the foothills—even up to 20° (36.4 %) were ploughed up. This was more than 5 % of the arable land area. Soils on such extreme slopes have been the object of very strong sheet erosion. The greatest soil erosion occurs on convex sites of slopes where, at the same time as the increase in the incline, both the area of a watershed and the rate of a slope flow increase. On concave sloping sites there is the least destruction of soil cover, though here the rate of slope flow increases but, as the incline is reduced, its transporting ability decreases and there is an accumulation of soil particles.

On direct slopes the rate of a stream increases downhill, and sharply expressed soil loss is shown approximately from the middle of the slope up to the beginning of its bottom third.

Besides topographic conditions, further natural factors increase the water erosion risk of arable lands. These are irregular precipitation with a maximum in the second part of the summer, and strong and deep freezing of the well-moistened soil, preventing the infiltration of snowmelt water in the spring. The proportions of solid and liquid atmospheric precipitation are important in particular (Tanasienko and Chumbaev 2010).

Water erosion is among the worst soil degrading processes on a global scale because of its irreversible negative effects on soil productivity potentials and significant damage for adjacent media in the hydrosphere in particular (Mueller et al. 2014). However, processes and patterns differ between climates and landscapes and require quantitative data to be understood and monitored (Tanasienko et al. 2011, Shokparova et al. 2014). Quantifying snowmelt erosion as a specific kind of water erosion also requires a specific methodology of measurements.

The aim of this chapter is to present a field methodology for defining the quantity and quality of the liquid, and solid components of surface snowmelt water runoff in hydrological years with various levels of snowiness in the south-east part of West Siberia.

2 Methods of Measuring Snowmelt Erosion

2.1 Process Analysis

In West Siberia, the destruction of arable soils on sloping surfaces generally occurs at the expense of the surface drainage of waters of various genesis: storm rainfall and snowmelt. The annual occurrence of storm rainfall on this territory which leads to the development of water erosion on agricultural fields with an intensity of more than 0.5 mm/min (Mirtskhulava 1970) is very small (less than 10 %), therefore the leading role in the destruction of arable soils on sloping surfaces is played by snowmelt waters, instead of storm rainfall as in the European part of Russia.

The development of the erosive processes initiated by thawed snow on soils on sloping surfaces is possible only in the case of the simultaneous existence of various natural factors. Studying these factors allows us to understand the factors, process, and consequences of the development of snowmelt water soil erosion in Western Siberia. The most important parameters which need to be studied for a definition of soil erosion are presented in Table 1.

As shown by Orlov (1983), the intensity of snowmelting and related processes in the spring is a very critical phase of high-erosion risk. Snowmelt erosion can be considered as a specific type of water erosion. However, the specificity of underlying processes has yet to be understood. We suppose that the specific features of the spring snowmelting in West Siberia are predetermined in the fall, when the soil humus horizon is saturated with moisture up to the field capacity (FC). In the winter, an ice-rich impermeable barrier is annually formed in the well-moistened humus horizon under the impact of the low temperatures. This barrier favors the intensive snowmelt runoff in the spring (Tanasienko and Chumbaev 2010).

In December, January, February, and March there is a further frost penetration in the soil. In this time, various quantities of solid atmospheric precipitation falls and by the end of March or the beginning of April the snow cover depth reaches a maximum. The thawing of the snow cover, as a rule, begins in the first half of April. Regarding the character of spring weathers and their synoptic conditions, and also on the basis of the prevalence of one of two main factors of snowmelting, the spring

Table 1 Relevant parameters of the snowmelt erosion process

Number	Parameters	Unit	Measurement period
1	The total pre-winter water reserve of soil	mm	The late fall
2	Snow depth	cm	End of winter period
3	Snow water equivalent (SWE)	mm	
4	Depth of soil frost penetration	cm	Before snowmelt period
5	Volume snowmelt water runoff	m^3/ha	During snowmelt period
6	Water stream temperature	°C	
7	Runoff coefficient (Cr)		After snowmelt period
8	Soil loss with surface snowmelt water runoff	t/ha	

weathers of the southeast of Western Siberia are subdivided into three types by Rutkovskaya (1962): *radiation*, *advection*, and *mixed* with *radiation-advection* and *advection-radiation* subtypes. The snowmelt period duration, intensity of a runoff, and removal of soil material on slope surfaces is controlled by the type of spring weathers.

The radiation type is characterized by sunny, calm weather with a fast increase in positive daytime air temperatures and preservation of negative temperatures at night, and also, as a rule, the increase during snowmelt of an intra-soil ice layer. Spring weathers of this type occur in 30 % of cases. With snowmelt of this type, high modules and the greatest volumes of superficial drainage and washout of soils for a fairly short period (4–7 days) are typical.

A gradual increase in positive daytime air temperatures and the preservation of negative night air temperatures is peculiar to the advection type of spring weather. A prevalence of low and high cloudiness during daylight hours, as Putilin and Tanasienko (1998) suppose, defines the long period of snowmelt (18–27 days). With such a long snowmelt the intra-soil ice barrier manages to collapse and by the end of snowmelt the temperature of a soil profile becomes positive. Surface runoff in such conditions is low. Spring weathers of this type occur in 15–20 % of cases.

Radiation-advection and advection-radiation subtypes of spring weathers are characterized by a gradual increase in positive daytime air temperatures and low negative nighttime air temperatures. Also, the return of negative daytime air temperatures, which can last 1–5 days, is possible and may be repeated several times in one spring period. In these subtypes of spring snowmelt, cloudy, windy weather with clearings is usually observed. The repeatability of such weathers is 30–35 %. The period of snow cover thawing with such subtypes of snowmelt is 10–14 days, with the formation of a high runoff coefficient of 0.55–0.65.

Various weather conditions during snowmelt directly influence two main characteristics of spring snowmelt: the intensity of the runoff and turbidity of snowmelt waters which the researcher measures directly in a field. Then, using these data it is possible to calculate the runoff depth and volume of snowmelt runoff, and the coefficient of drainage and soil washout as a whole during the snowmelt period. Having processed snowmelt water samples in vitro, the researcher can obtain data on the texture of the soil material which was in the snowmelt water and the quantity of humus and micro- and macroelements in the thawed snow taken out with a runoff.

A surface snowmelt runoff is divided into three phases: the beginning, the middle (the most intensive runoff), and the end. In the initial and final phases, the intensity of surface snowmelt runoff drainage on slopes on arable land does not exceed 2.5 l/s/ha (about 1 mm/h/ha). These phases are characterized by the maximum duration of water contact with the soil surface, but streams with low kinetic energy that positively affects the amount of solid components carried out of the soil. In the initial and final phases, the snowmelt waters are absolutely transparent. The snowmelt water turbidity in these phases is minimal and varies generally within 1–3 g/l.

The most intensive phase of snowmelt runoff is noted when the intensity of the surface runoff reaches more than 2.5 l/s/ha. On a slope with a southerly aspect its duration lasts 2–4 hours per day (from 3 to 5 p.m.) and, on a northerly slope where the snowmelt does not even stop at night, it reaches 8–10 hours per day. The kinetic energy of a stream in this phase of a runoff is at a maximum. Therefore, depending on the texture and structure of soils, the snowiness of the winter and the type of spring weather, the turbidity of snowmelt water varies from 5–90 g/l.

The water turbidity is the indicator characterizing the reduction in water transparency because of the existence of inorganic and organic fine suspensions. Soil particles of various size are found as inorganic suspensions in snowmelt water streams. In the initial and final phases of a runoff, streams of snowmelt water are sated with water-soluble humus that gives them a light yellow color.

At the same time, with snowmelt water sampling it is necessary to measure the water stream temperature: the higher the snowmelt water temperature, the higher the turbidity of a water stream. In the first days of a runoff, the temperature varies between 0.4 and 0.8 °C, therefore the turbidity, as a rule, is minimal (0.5–2 g/l), even on the ploughed land. The maximum turbidity can amount to 100 g/l by the last days of intensive snowmelt runoff, when the top soil layer is oversaturated. The water stream temperature can reach 10–12 °C which increases the suspension ability and solubility of humus and other substances.

The texture of the suspended soil material depends on the runoff intensity. In the initial and final stages of snowmelt, the removal of particles <0.01 mm was most dominant. The suspended sediment consisted of up to 60 % fine particles, most of which were represented by clay (<0.01 mm). The total clay removal by runoff from the ploughed layer in high-snow and very high-snow years varied from 3300–4200 kg/ha of which 1260–1500 kg/ha of losses occurred during the initial and final stages of snowmelt.

2.2 *Field and Laboratory Measurements*

The calculation of discharged water and eroded soil material from a given area is the desired result of snowmelt erosion studies. It requires the following measurements:

- define representative microcatchments and create short-term runoff plots (Fig. 1);
- determine the soil total reserves of water in the late fall before the soil freezing in winter and after the snowmelt period;
- observe the air temperature regime in the late fall, in winter, and after snowmelt period;
- determine the soil freezing depth;
- calculate the velocity of frost penetration and thawing of soils;
- record existence of the icy barrier both visually and by temperature loggers;
- determine the water reserve in the snow before snowmelting on differently eroded soils in years with different amounts of snow;
- quantify the discharged amount of snowmelt water;

Fig. 1 Short-term runoff plot with a minimum size of 30 × 5 m

- measure the temperature of snowmelt waters and the content of suspended soil particles;
- estimate losses of soil material at soil profiles in flow pathways;
- analyze pH and various chemical elements (C, N, Ca, Mg, K, Na) in the snowmelt water and in the sedimented soil material;
- estimate ecological consequences of erosive processes.

Research on the snowmelt erosion process can be divided into stages.

(I) **Preparatory stage and field work before the winter**
The freezing and snowfall period in the Forest Steppe zone of West Siberia begins in November and ends in April. Field spadework includes:

1. We used short-term runoff plots (Fig. 1) to study the snowmelt velocity and volume of snowmelt water surface drainage. These plots must be built in fall, while soils are not frozen.
   Short-term runoff plots should have a size of 30–60 m in length by 5–10 m in width. The runoff plots are normally oriented from the upper to the lower part of the slope. Each runoff plot is bordered with 20–25-cm-high walls made of the upper part of the soil. This border stops external surface snowmelt waters from flowing inside a plot. A gutter made of galvanized sheet metal is attached at the lower end of each short-term runoff plot. Usually, short-term runoff plots are constructed in the lower part of a slope where the speed of the water streams is maximal. Depending on research objectives, short-term runoff plots can be constructed on slopes of varying aspect, and also on a different position on the catena.

2. During the same period, it is necessary to measure the representative pre-winter water content of different key plots on the catena. Our key plots are oriented toward areas of the catena with soils that are eroded to varying degrees: a virgin, non-eroded arable soils, slightly, moderately, and strongly eroded soils. Volumetric soil samples are taken at three positions on each plot in depth increments of 10 cm down to a depth of 1.5 m (Fig. 2). The water content of soil samples has to be analyzed in the laboratory by drying the samples at 105 °C until the weights are constant (minimum drying time 4 h) (Vadyunina and Korchagina 1961).
3. Installation of temperature sensors. Soil temperature monitoring during the cold period of the year is necessary to calculate the soil freezing and thawing rate and frost penetration into soils. One feature specific to the Siberian soils during the cold period of the year is the intra-soil ice sheet, which is impenetrable to melting water and positive temperatures. The Siberian soils become waterproof at moisture contents of 20 % and more and at a temperature of −2 °C and lower. The ice sheet can not only be

Fig. 2 Soil sampling for water content. Another measurement has to be carried out in spring before, during, and after the snowmelt. Volumetric samples from core cylinders are important to calculate the amount of water which infiltrates the soil during the snowmelt

recognized visually, but also estimated from data on soil temperature and moisture content (Tanasienko and Chumbaev 2010). Therefore, it is necessary to set up a thermometric plot near each runoff plot. We obtained reliable data using small (roughly coin-sized) electronic temperature recorders (Thermochron iButton Device 2013).

To study the temperature regime of the arable layer of soils (on the soil surface and at depths of 5, 10, 15 and 20 cm), temperature recorders must be put into individual protective caps and then placed in the soils. To monitor the temperature of subsurface layers of soil, a number of holes were drilled with a diameter of 5 cm and at depths of 40, 60, 80, 100, 120, 140, and 160 cm. Temperature recorders were fixed at the lower end of plastic pipes with a metallic cap, and thus placed at the bottom of the hole in close contact to the soil. A wooden bar was inserted into the plastic pipe, insulating the temperature recorders against water and heat flow from above (Fig. 3).

The temperature recorders were initially programmed for temperature measurement through equal time intervals. The velocity of frost penetration into the soils of West Siberia can range from 1–6 cm per day depending on weather conditions. It is very important to determine when the frost starts penetrating the soils. But it is more important to fix the moment when the soil starts to defrost in the spring and monitor the daily dynamics of soil temperatures at various depths during the snowmelt period. The daily amplitude of air temperatures during the snowmelt period in this region is great, ranging from +16 °C in the afternoon, to −15 °C at night (Tanasienko 2003). Therefore, temperature recorders are programmed with a measurement interval of 3 h.

Fig. 3 Thermometric plot

On a temperature platform we established a landmark (usually a 2.5 m stick), on which were fixed independent devices between 1 and 5 (depending on the purpose of the research) for recording the temperature of the air at various heights above the soil surface. In winter, all the temperature platforms are under snow cover and we can only find their location by means of these landmarks.

Round the thermometric platform, a protective shaft was constructed which prevents thawed snow from hitting the platform, protecting it from snowmelt runoff.

(II) **Research during the winter period**

Accurate data on the snow depth accumulated for the winter period and data on a snow water equivalent are necessary to study the snowmelt process during the spring period in Siberia. The snow water equivalent is the total amount of water in solid and liquid form stored in the snow cover. Measurements at the end of March before the snow melts provide the best estimate.

Snow survey stage

The snow survey is the measurement of the snow depth and the snow water equivalent of an elementary watershed. The snow depth and SWE are defined in snow surveys. We carried them out in Western Siberia at the end of March every year, because the SWE here reaches its maximum values during this period.

The snow depth was determined using a metal snow scale (Fig. 4). Not less than 25 measurements of snow depth per elementary unit under study (plot, watershed) provide statistically proven average values. The SWE was calculated using the BC-1 snow gauge three times at each snow depth measurement point (Fig. 5).

Fig. 4 Determining the snow depth

Fig. 5 Calculating the snow water equivalent. The mechanical balance is a very robust and reliable solution under severely frosty conditions

(III) **Research during the spring snowmelting period**

The runoff depth is the amount of water which flows down from an elementary watershed for any period, expressed in the form of the layer (in mm) which has been evenly distributed on the areas (Taratunin 2008). An area of 1 ha with a runoff depth of 1 mm holds 10,000 l of water. To determine the snowmelt runoff depth on a runoff plot, measure the intensity of runoff—the amount of water (mm) which is forming as the snow melts per unit of time. The intensity of runoff is expressed in liters per second from a 1-ha runoff area (l/s ha).

Using the measuring cylinder (1 l) and a stopwatch on a experimental runoff plot, we determined the snowmelt runoff depth at 1-hour intervals. At first, we defined the intensity of the runoff. For this purpose, we noted the time (in seconds) it took for the measuring cylinder to be filled with snowmelt water. The intensity of the runoff (l/s) was multiplied by the corresponding coefficients to calculate the runoff depth from 1 ha (Example 1).

The intensity of the snowmelt runoff varies widely within 1 day. Summarizing the drainage volume for each hour gives us the volume of snowmelt runoff in 1 day. As the snowmelt lasts some days, we summarized the daily volume of snowmelt runoff during the snowmelt.

The runoff coefficient is the ratio between the quantity of the runoff and the quantity of precipitation in a watershed. In the case of snowmelt erosion, the runoff coefficient (Cr) shows the proportion of water stored in the snow cover which underlies the surface runoff (Example 2).

On an elementary watershed of 10–15 ha the volume of snowmelt runoff was determined in the lower part of a narrow drain. Knowing the depth, width, and speed of a water stream per unit of time in a narrow open drain and the elementary watershed, it is possible to calculate the intensity of the snowmelt runoff. We determined the depth and width of the water stream in a narrow open drain using a ruler or a metal snow scale. Traditionally, the speed of a water stream can be determined by putting lightweight pieces of material (foam plastic for example) in a water stream and determining the time they take to pass a certain distance through the water stream. All measurements were carried out 3–5 times. It is possible and more accurate to measure the speed of the water stream using a hydrological rotator (Fig. 6). We used the speed of the water stream data to calculate the current drainage intensity of thawed snow (Example 3).

The turbidity of thawed snow was determined in the laboratory. For this, we took snowmelt water samples in the field each hour. On an experimental runoff plot for sampling we used the 1 l measuring cylinder. Snowmelt water from the measuring cylinder was merged in plastic jar and delivered to the laboratory. On the same day, the solid phase of the soil was separated from the liquid either by means of filtering, or centrifugation. The filtered water for further analysis was placed in the refrigerator. The filtrate was weighed and the quantity of solid phase in the soil was defined for 1 l of snowmelt water sample. The turbidity was evaluated in grams per 1 l of water (g/l).

On an elementary watershed of 10–15 ha the snowmelt water was also sampled hourly, but at once collected in 1 l plastic jars (Fig. 7). It should be noted that in the initial and final phases of drainage, snowmelt water runoff generally transports fine soil particles which are in a suspension and the height of particle lifting is commensurable with the stream depth (Kuznetsov and Glazunov 2004). In the phase of most intensive drainage (>1 mm/h), the snowmelt water stream is capable of transferring not only suspended fine (particles <0.01 mm), but also coarsely dispersed soil particles (>0.01 mm), (Fig. 8).

Fig. 6 Determining the speed of a water stream using a "GR-99" hydrological rotator

Fig. 7 Water sampling for the turbidity of the snowmelt waters

Fig. 8 Soil material that was taken out from agricultural lands by the surface drainage of snowmelt waters

Having obtained daily data on the volume of drained thawed snow and sediment carried out by a runoff, the researcher can sum up the result to show how much of the snowmelt waters migrated from the slope surfaces during a superficial drainage and what quantity of the solid phase of the soils was taken out from agricultural lands in a hydrographic network by that water for the snowmelt period (Example 4).

The products of solid runoff contain various quantities of the biogenic elements participating in the formation of the soil structure and providing cultivated crops with food compounds. Among these elements the major role is played by Organic Carbon (humus), Nitrogen, Phosphorus, Potassium, Natrium, Calcium, and Magnesium. The Carbon content is determined using Tyurin's method. The content of other elements is determined using standard methods (Arinushkina 1970). Tyurin's method is a modification of a volume method for defining the organic carbon of the soil by oxidizing it with potassium dichromate in a strongly acid medium until CO_2 forms according to the equations:

1. $2K_2Cr_2O_7 + 8H_2SO_4 = 2K_2SO_4 + 2Cr_2(SO_4)_3 + 8H_2O + 3O_2$;
2. $3C + 3O_2 = 3CO_2$.

Carbon is recalculated in the soil humus by multiplying the percentage of Carbon by the coefficient 1.724.

It should be noted that the same elements are in the liquid phase of a runoff. High contents are observed particularly clearly at the initial and final phase of the snowmelt period, when the duration of the contact of thawed snow with the surface of slightly thawed soils is at a maximum.

When data is available on the mass of the soil loss, the volume of snowmelt runoff after different hydrological years by snowiness, the maintenance of either or another biogenic element in the solid and liquid component of a runoff, it is possible to calculate the total losses of Humus, Nitrogen, Phosphorus, Potassium, Natrium, Calcium, Magnesium, and also other chemical elements.

3 Results and Discussion

On the basis of long-term measurements (more than 40 years) of snowmelt processes on arable lands in various regions of West Siberia (Predsalairye, the Kuznetsk Hollow, the Biye-Chumyshsky Hills and Priobye) our team found that the snow cover contained 51–187 mm of water (average 124 mm). We classified hydrological years as follows: low-snow, 51–90 mm; normal-snow, 91–130 mm; high-snow, 131–190 mm; (Table 2). It should be noted that the amount of precipitation in high-snow years is more than twice that in low-snow winters.

Table 2 Long-term data on snowmelt erosion in West Siberia

Characteristics of the hydrological year	Statistical parameters				
	n	Range	*M* ± *m*	*σ*	*V* (%)
	Solid atmospheric precipitations				
		Range (mm)	*M* ± *m* (mm)	*σ* (mm)	
1969–2013	45	51–187	124.4 ± 5.2	35.4	28
Low-snow	11	51–89	76.1 ± 4	13.2	17
Normal-snow	13	101–124	114 ± 2.3	8.5	2
High-snow	21	133–187	156 ± 3.1	14.6	9
	The volume of a snowmelt runoff				
1969–2013	45	2–118	58.9 ± 4.1	27.6	7
Low-snow	11	10–51	36 ± 4.1	13.7	38
Normal-snow	13	35–99	64 ± 5.2	18.8	29
High-snow	21	2–118	68 ± 6.8	31.4	46
	Soil frost penetration				
	n	Range (cm)	*M* ± *m* (cm)	*σ* (cm)	*V*
1969–2013	45	15–181	115 ± 6.7	45.2	39
Low-snow	11	47–181	140 ± 11.3	37.4	27
Normal-snow	13	15–163	119 ± 12.5	45	38
High-snow	21	17–175	99.1 ± 9.7	44.2	45
	Soil loss				
	n	Range (kg/ha)	*M* ± *m* (kg/ha)	*σ* (kg/ha)	*V*
1969–2013	45	1–28762	6036.4 ± 1060	7110.6	18
Low-snow	11	1–5994	2405 ± 598.3	1984.5	83
Normal-snow	13	580–13566	3889 ± 1178.5	4249	109
High-snow	21	1–28762	9267 ± 1917.6	8787.5	94.8
	Snowmelt period				
	n	Range (days)	*M* ± *m* (days)	*σ* (days)	*V*
1969–2013	45	4–23	8.7 ± 0.6	4	6.9
Low-snow	11	4–17	9 ± 1.4	4.7	55
Normal-snow	13	4–17	9 ± 1	3.6	39
High-snow	21	5–23	8.5 ± 0.9	4.1	48
	Runoff coefficient				
	n	Range	*M* ± *m*	*σ*	*V*
1969–2013	45	0.1–0.8	0.5 ± 0.03	0.2	5.8
Low-snow	11	0.2–0.7	0.5 ± 0.05	0.2	35
Normal-snow	13	0.3–0.8	0.6 ± 0.04	0.2	28
High-snow	21	0.1–0.8	0.4 ± 0.05	0.2	49.2

n number of measurements (years), *Range* minimum to maximum, referring to the whole data pool replications included, *M* arithmetic mean, *m* standard deviation, *σ* mean-square deviation, *V* coefficient of variation

In low-snow years, the volume of thawed snow drainage averages 26 mm while in high-snow years it is 2.4 times more. The runoff of 50–60 mm observable after normal-snow and high-snow winters leads to considerable soil loss in the arable Chernozems. Therefore, within the south-east part of West Siberia, the snowmelt water runoff has a volume varying 33 % of the water reserve in the snow after low-snow winters and 41 % after high-snow winters.

In low-snow winters, the temperature remains fixed at −2–4 °C in a 20 cm layer of arable soils because there is a small quantity of snow in November. Even in high-snow hydrological years, when the soil snow cover is sizeable, in November the temperature of an arable layer is established at −1 °C. Therefore, in different hydrological years in terms of snowiness the arable layer is in a frozen state at the end of November.

The maximum depth of frost penetration in Chernozems depending on the snowiness of the hydrological year is observed at the end of March. In low-snow winters, the zero isotherm on average falls to lower than 140 cm while in high-snow winters it is fixed at a depth of about 100 cm.

Depending on the snow depth, type of spring weather, depth of frost penetration in a soil profile, and also the type of tillage, various quantities of soil material are carried away by runoff from agricultural fields of West Siberia. Zaslavsky (1979) offered the following scale of annual soil loss: low-soil loss less 5 tonnes/ha; medium soil loss 5–10 tonnes/ha; strong soil loss 10–20 tonnes/ha; very strong soil loss 20–50 tonnes/ha; extremely strong soil loss more than 50 tonnes/ha. Thus, on Zaslavsky's scale, the annual washout in the territory studied can be diagnosed as a medium soil loss. The soil loss caused by snowmelt runoff depends on both spring weather and the type of land use. The maximum soil losses were observed when the snowmelt runoff was on ploughed land, and at least on grasslands and virgin lands (Tanasienko 2003; Tanasienko et al. 2013).

The duration of the snowmelt period depends on the weather in this period. In our studies, the snowmelt period lasted within a wide range: the shortest period was fixed as 4 days under the radiation type of spring weather, the longest period was fixed as 23 days under the advection type of spring weather.

4 Conclusions

Studies of water, sediment, and solute transport during the snowmelt in the conditions of West Siberia are subdivided into three periods:

1. The preparatory period during which key points with thermometric and runoff plots are built and late fall soil moisture is measured at key sites.
2. Research during the winter period. At this stage there is a monitoring of the accumulation of snow cover in the territory studied, and the speed and depth of frost penetration in soils is established.
3. Studying the spring snowmelt, soil thawing.

To quantify the overall snowmelt erosion process, the following parameters need to be measured: the total fall soil reserve of water, snow depth, snow water equivalent, depth of soil frost penetration, volume of snowmelt water runoff, runoff coefficient, water stream temperature, and soil loss with surface snowmelt water runoff.

The amount of solid precipitation, snow depths, and their water equivalents are basic parameters of snowmelt erosion in spring. Solid precipitation in West Siberia ranges from 51 mm in low-snow winters to 187 mm in high-snow winters.

The snowmelting process starts in the first and second 10 days of April forming superficial and intrasnow and soil runoffs of various volumes. Snow melts in the conditions of deep and strong frost penetration in soils at the waterproof ice screen created in winter period. Therefore the long-term average runoff reached 50 % of the precipitation of the cold period.

Average annual soil losses due to water erosion in West Siberia were about 6 tonnes/ha, so can be diagnosed as a medium soil loss (on Zaslavsky's scale).

Appendix 1: Calculation of Hydrological Data from Measurements

Example 1

The length of an experimental runoff plot is 50 m, the width is 10 m. The area of an experimental runoff plot is 500 m^2. In this case, to transfer the size of an experimental runoff plot to 1 ha, apply the coefficient **20** (500 × 20 = 10,000 m^2 = 1 ha). The measuring cylinder with a capacity of 1 l was filled with snowmelt water for 5 s. Thus, the intensity of snowmelt runoff from an experimental runoff plot in 500 m^2 during 1 s is:

1 l/5 s = 0.2 l/s.
During 1 h at this intensity, the flow from this runoff plot was:
0.2 l/s × 3600 s = 720 l/h
During 1 h, the flow from an area of 1 ha was:
720 l/h × 20 (coefficient) = 14,400 l/h/ha, or runoff depth = 1.44 mm/h.

Example 2

Data from a snow survey showed that for the cold period of the hydrological year (November–March), 140 mm of water were stored in the snow cover. The snowmelt runoff was 70 mm. The runoff coefficient in this case is 0.5 (70 mm/140 mm = 0.5).

Example 3
The depth of a water stream is 20 cm, the width is 25 cm. The speed of the water stream is 80 cm/s. The watershed area is 10 ha. The current drainage intensity of thawed snow of this watershed is:

$$\frac{20\,\text{cm} \times 25\,\text{cm} \times 80\,\text{cm/s}}{1000^{*} \times 10\,\text{ha}} = 4\ \text{l/s ha or } 1.44\ \text{mm/h}.$$

*1000—coefficient for transforming the volume of snowmelt runoff measured, in Example 3, in cm^3, into liters.

Example 4
Snow water equivalent (SWE)—150 mm. The type of spring weather is radiation, snowmelting period—5 days.

Day 1

Daily Rd1 (Runoff depth)—3 mm;
Daily *M* soil1 (mass of solid phase of the soils in the snowmelt waters)—250 kg/ha.

Day 2

Daily Rd2—7 mm;
Daily *M* soil2—600 kg/ha.

Day 3

Daily Rd3—56 mm;
Daily *M* soil3—1830 kg/ha.

Day 4

Daily Rd4—24 mm;
Daily *M* soil4—1220 kg/ha.

Day 5

Daily Rd5—4 mm;
Daily *M* soil5—370 kg/ha.

1.
$$\text{Rd} = \text{Rd1} + \text{Rd2} + \text{Rd3} + \cdots\ \text{Rd}$$

$$\text{The runoff coefficient (Cr)} = \frac{\text{Rd}}{\text{SWE}} = \frac{94}{150} = 0.6$$

2.
$$M\,\text{soil} = M\,\text{soil1} + M\,\text{soil2} + M\,\text{soil3} \cdots$$
$$M\,\text{soil} = 250 + 600 + 1830 + 1220 + 370 = 4270\ \text{kg/ha}$$

References

Arinushkina EV (1970) Soil chemical analysis instruction (Руководство по химическому анализу почв). Moscow State University Publishing, Moscow 488 pp (in Russian)

Kuznetsov MS, Glazunov GP (2004) Erosion and protection of soils (Эрозия и охрана почв). "Kolos" Moscow, 352 pp (in Russian)

Mirtskhulava CE (1970) Engineering methods of calculation and forecast of a water erosion (Инженерные методы расчета и прогноза водной эрозии). Moscow, 237 pp (in Russian)

Mueller L, Schindler U, Ball BC, Smolentseva E, Sychev VG, Shepherd TG, Qadir M, Helming K, Behrendt A, Eulenstein F (2014) Productivity potentials of the global land resource for cropping and grazing. In: Mueller L, Saparov A, Lischeid G (eds) Novel measurement and assessment tools for monitoring and management of land and water resources in agricultural landscapes of Central Asia. Environmental Science and Engineering 2014. Springer International Publishing, Berlin, pp 115–142. http://link.springer.com/chapter/10.1007/978-3-319-01017-5_6. Accessed on 27 May 2014

Orlov AD (1983) Erosion and erosion-dangerous lands of West Siberia (Эрозия и эрозионноопасные земли Западной Сибири). Nauka, Novosibirsk, 208 pp (in Russian)

Putilin AF, Tanasienko AA (1998) Geography of erosive processes of the south of Western Siberia. Topical issues of geology and geography of Siberia (География эрозионных процессов юга Западной Сибири // Актуальные вопросы геологии и географии Сибири). Tomsk 4:78–80 (in Russian)

Rutkovskaya NV (1962) Snowmelting in the southeastern part of West Siberian plain (Таяние и сход снежного покрова на юго-востоке Западно-Сибирской низменности). In: Rikhter (ed) Snow cover: distribution and role in national economy (Снежный покров, его распределение и роль в народном хозяйстве). USSR Acad Sci, Moscow, pp 104–113 (in Russian)

Shokparova DK, Kakimjanov EK, Meyer BC (2014) Erosion rates depending on slope and exposition of cropped chestnut soils. In: Mueller L, Saparov A, Lischeid G (eds) Novel measurement and assessment tools for monitoring and management of land and water resources in agricultural landscapes of Central Asia. Environmental Science and Engineering 2014. Springer International Publishing, Berlin, pp 549–561. http://link.springer.com/chapter/10.1007/978-3-319-01017-5_34. Accessed on 27 May 2014

Shurikova VI (1981) Eroded soil fertility and different methods its improvement (Плодородие почв, подверженных водной эрозии, и пути его повышения). VASKHNIL Publishing, Moscow, 42 pp (in Russian)

Tanasienko AA (2003) Specific features of soil erosion in Siberia (Специфика эрозии почв в Сибири). SB RAS Publishing, Novosibirsk, 173 pp (in Russian)

Tanasienko AA (2007) Soil erosion in Siberia (Эрозия почв Сибири). Soil erosion: scientific issues of Docuchaev's Soil Institute, Moscow, pp 34–62 (in Russian)

Tanasienko AA, Chumbaev AS (2010) Conditions of the formation of ice barriers in eroded Chernozems of Western Siberia. Eurasian Soil Sci 43(4):417–426

Tanasienko AA, Yakutina OP, Chumbaev AS (2011) Effect of snow amount on runoff, soil loss and suspended sediment during periods of snowmelt in southern West Siberia. Catena 87:45–51

Tanasienko AA, Yakutina OP, Chumbaev AS, Miller GF (2013) Conditions and intensity of erosive and accumulative processes in the forest-steppe of Predsalairye (Условия и интенсивность эрозионно-аккумулятивных процессов в лесостепи Предсалаирья). Eurasian Soil Sci 11:1397–1408 (in Russian)

Taratunin AA (2008) Inundations in Russian Federation (Наводнения на территории Российской Федерации). FGUP RosNIIVkH Publishing, Ekaterinburg, 432 pp (in Russian)

Thermochron iButton Device. Available from: http://www.maximintegrated.com/datasheet/index.mvp/id/4023. Accessed on 25 May 2014

Vadyunina AF, Korchagina ZA (1961) Methods of research of physical properties of soils and grounds (Методы исследования физических свойстве почв и грунтов). State publisher "The higher school", Moscow, 346 pp (in Russian)
Zaslavsky MN (1979) Soil erosion (Эрозия почв). "Mysl" Publisher, Moscow, 245 pp (in Russian)

Chapter 8
Estimation of Biomass and Net Primary Production (NPP) in West Siberian Boreal Ecosystems: In Situ and Remote Sensing Methods

Anna Peregon, Natalia P. Kosykh, Nina P. Mironycheva-Tokareva, Philippe Ciais and Yoshiki Yamagata

Abstract The goal of this study is to identify the current state of in situ observations and remote sensing data and methods used to assess biomass and net primary production (NPP) in West Siberian natural ecosystems, and consider perspectives for future developments. The natural ecosystems of the boreal region mainly consist of two classes: wetlands and forests, where one is very different from the other, requiring different methods for biomass assessment. Basically, two methods are available to estimate NPP and biomass in regional terrestrial ecosystems: (1) extrapolating the local field measurements up to a larger region, using the vegetation or land cover maps and (2) modeling productivity and plant biomass at regional and grid point scales, with or without the use of remote sensing data and techniques. The first method was predominantly used to estimate wetland biomass, having an extensive dataset of direct in situ measurements in both the above- and below-ground fractions of biomass. So far, no direct methods based on remote sensing data have been elaborated for biomass estimations in wetland ecosystems

A. Peregon (✉) · P. Ciais
CEA Centre de Saclay, Laboratoire des Sciences Du Climat et de l'Environnement (LSCE), Gif-sur-Yvette, France
e-mail: anna.peregon@lsce.ipsl.fr

P. Ciais
e-mail: philippe.ciais@lsce.ipsl.fr

N.P. Kosykh · N.P. Mironycheva-Tokareva
Siberian Branch, Institute of Soil Science and Agrochemistry (ISSA), Russian Academy of Sciences, Ac. Lavrentieva av. 8, Novosibirsk 630090, Russia
e-mail: kosykh@issa.nsc.ru

N.P. Mironycheva-Tokareva
e-mail: nina@issa.nsc.ru

Y. Yamagata
Center for Global Environmental Research (CGER), National Institute for Environmental Studies (NIES), Tsukuba, Japan
e-mail: yamagata@nies.go.jp

L. Mueller et al. (eds.), *Novel Methods for Monitoring and Managing Land and Water Resources in Siberia*, Springer Water,
DOI 10.1007/978-3-319-24409-9_8

and soil carbon inventories. In forest ecosystems, the biomass can be estimated by processing satellite data from high-resolution radiometers (AVHRRs). The radar or LIDAR remote sensing approaches hold great promise for direct observations of the three-dimensional structure (3D) of the above-ground vegetation that can be used for relatively straight-forward calculations of carbon storage, but the method works only in low to medium biomass ecosystems. The SAR-based biomass retrievals were found to be fairly uncertain in mature forests with high biomass values, as the Synthetic Aperture Radar (SAR) signal often saturates at ~70 tonnes/ha. The estimation errors in terms of RMSE are typically found at 25–30 % of the mean biomass. The methods should be further refined to reduce uncertainties and to make them operational over the vast region of Siberia.

Keywords Vegetation · NPP · Forest · Wetlands · Remote sensing · Field methods

1 Current State and Future Requirements of Biomass Observations in Siberia

1.1 *Objectives and Framework*

The boreal ecosystems, both the forest and wetlands/peatlands, play a critical role in the global uptake and storage of anthropogenic carbon, and their changes in response to the climate hold the key to improve the scientific understanding of the global carbon cycle and making accurate future projections. The build-up of CO_2 in the atmosphere has the potential to affect net primary production (NPP), the rates of biomass accumulation, and carbon storage, particularly in northern high latitudes (Ciais et al. 2013).

Uncertainties in inventories, observations, and analyses of biomass in natural ecosystems of Siberia and throughout Northern Eurasia need to be dramatically reduced to support effective policies and reporting such as under the United Nations Framework Convention on Climate Change (UNFCCC) (Shvidenko et al. 2010; Dolman et al. 2012).

In the global scale, one recent development is the NASA-driven MODIS GPP/NPP Project (MOD17), which provides continuous estimates of Gross/Net Primary Production (GPP/NPP) across Earth's entire land surface (available online at www.ntsg.umt.edu/project/mod17).

1.2 *Wetlands*

Earlier studies (collated in Campbell et al. 2000) have described the distribution of biomass within northern continental wetlands, primarily above ground. However, there have been relatively few measurements of below-ground biomass and production. Ignoring the below-ground component (which is the usual practice because the measurement of root production and mean root turnover rates in peatlands is difficult) excludes about 50 % of biomass and a large portion (from 40 to 90 %) of annual production, according to estimations by Coleman (1976), Aerts et al. (1992), Titlyanova et al. (1999), Moore et al. (2002).

Basically, two methods are available to estimate the NPP and biomass of regional terrestrial ecosystems: (1) extrapolating the local field measurements up to a larger region, using the vegetation or land cover maps and (2) modeling plant biomass and productivity at regional and grid point scales. Modern vegetation models (Woodward 1987; Prentice et al. 1992) have improved the representations of the global distribution of vegetation, but still have biases when reproducing observed vegetation distribution, due to their approximation about the climate envelope of different plant functional types. Consequently, Townshend et al. (1991) argued that data gained from remote sensing of the actual land cover needs to be used for more accurate classifications of global vegetation. The theory and rationale for using remote sensing in the model-based estimation of photosynthesis and NPP (so-called light use efficiency models), including wetland vegetation models, are described in Potter et al. (1993) and Ruimy et al. (1994). However, since it is still difficult to model NPP and wetland biomass with process-based models due to the lack of high-resolution regional land cover maps, a simple inventory-type approach to combine GIS data and point observation data can be a reliable, alternative to light use efficiency models (the method is further described in Sect. 2).

Over the past few decades, direct biomass measurements in West Siberian wetlands have been collected through various programs and projects, with efforts primarily led by the Institute of Soil Science and Agrochemistry (ISSA SB RAS, Novosibirsk, Russia) (Kosykh et al. 2008a, b, c; Peregon et al. 2008; Kosykh et al. 2009; Naumov and Kosykh 2010; Kopoteva and Kosykh 2011) and from the Institute of Monitoring of Climatic and Ecological Systems (IMCES SB RAS, Tomsk, Russia) (Golovatskaya 2009; Golovatskaya and Dyukarev 2009). The field measurements comprise both the above- and below-ground fractions of biomass (for details, see Sect. 2 of this chapter). However, the spatial coverage of these field measurements has either stagnated, with many regions (forested-steppe, tundra) remaining unsampled, or moderately increased through the establishment of in situ monitoring stations for different wetland types and fine-scale typological units that create land cover mosaics in the boreal region of Western Siberia.

To date, efforts to monitor and report wetland biomass in West Siberia at a regional scale have been based mostly on limited point-scale in situ observations at three test sites along a wide north-south climate (and latitudinal) gradient in WS, a total of 5500 plant samples collected between 1993 and 2001, with consequent upscaling to larger regions. The presence of topography, with such microscale features as hummocks, hollows, ridges, open water pools, and distance from the peatland margin (macrotopographical position) have a profound effect on the plant species distribution and productivity (Andrus et al. 1983; Moore et al. 2002). Existing global maps of wetland types, NPP/biomass, and CH_4 emissions (e.g.

Matthews and Fung 1987; Aselmann and Crutzen 1989; Loveland et al. 2000; Lehner and Döll 2004; Petrescu et al. 2010; Melton et al. 2013) suffer from the crude representation of the wetland typology for Western Siberia and point out the need for more detailed land cover information. One estimate of the area, carbon storage, and accumulation rates in mires of the former Soviet Union (FSU) was reported by Botch et al. (1995). In their study, a major source of uncertainty was attributed to poor reporting on shallow mires that were not represented in Russian peat inventories. The area of peatlands in the FSU might be twice as large as the area reported by Kivinen and Pakarinen (1981) or Tyuremnov (1976), who based their estimates on data from the Peat Fund of the USSR, which considers mostly commercial peatlands. Peat and carbon content data for west Siberian peatlands have been summarized recently by Efremov and Efremova (2001), Sheng et al. (2004), and Smith et al. (2004). A recent attempt to address spatial heterogeneity of natural ecosystems in the vast region of Western Siberia was made by Peregon et al. (2009) for biomass inventories only. There is (to our knowledge) no other data product providing the regional distribution of wetland biomass and net primary productivity (NPP) over the region of West Siberia based upon in situ data apart from the Peregon et al. (2008) data shown in Fig. 1.

For the carbon quantity in soils, the current situation is that there are neither any continuous, standardized, and georeferenced soil carbon inventories in the whole Siberian region, nor a consistent network of detailed soil carbon measurement plots for use in model parameterization, except for the distributed GIS to estimate the soil carbon stock in the boreal region of Western Siberia (Kudryashova et al. 2011). Particularly in carbon-rich soils such as frozen soil and peat, the uncertainties are very large (Tarnocai et al. 2009), as the estimates of carbon stocks. Recent studies estimated the Soil Organic Carbon (SOC) in soils and other consolidated deposits using a polygon-based digital database adapted for use in Geographic Information Systems (GIS) and an upscaling technique in the northern circumpolar permafrost region, while quantifying substantial uncertainty ranges and identifying remaining data gaps (Hugelius et al. 2013). There are no direct methods based on remote sensing data elaborated so far for NPP, biomass estimations, and soil carbon inventories.

1.3 Forests

The estimates of above-ground forest biomass (AGB) in Siberia and in Northern Eurasia in general are highly uncertain, despite the global importance of AGB for ecosystem services and its role as a carbon store. Until recently, remote sensing-based estimates of AGB and carbon storage have been limited to approximations based on (i) combining remotely sensed land cover type with representative carbon (biomass) values derived from in situ inventory samples, (ii) processing satellite data from high-resolution radiometers, and (iii) adapting the sensitivity of radar scattering to biomass in low to medium biomass ecosystems.

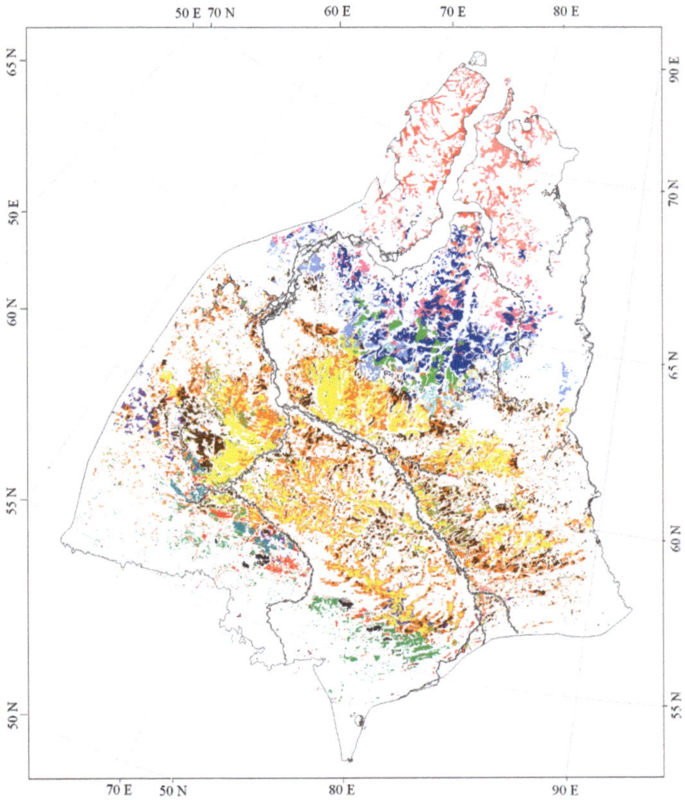

NPP	LB	
7.3	27.3	Polygonal-roller and polygonal-fissure mires
8.0	29.2	Polygonal mires combined with grass and grass-mosses mires
3.9	10.8	Patterned (hollow and hollow-pool) flat-palsa bogs
5.3	11.1	Flat-palsa and high palsa bogs
4.5	15.1	Shrub-dominated tussock mires
5.4	13.6	High palsa-hollow and pool-hollow patterned mires
5.3	10.9	High palsa and flat palsa mires
4.7	9.7	Sphagnum-dominated bogs with pools and open stand of trees
4.7	9.4	Ridge-hollow, ridge-hollow-pool and ridge-pool patterned bogs
11.2	20.3	Forested (treed) shrubs- and moss-dominated mires
6.7	15.3	Moss-dominated treed (pinus) mires
6.8	15.5	Ridge-hollow patterned bogs
20.1	27.4	Grass-mossy mires and oligotrophic pine-dwarf shrub-Sphagnum raised bogs
15.0	31.0	Grass- and grass-moss dominated mires
20.6	42.6	Grass-sedge, Sphagnum-sedge hardwood swamp
20.7	42.6	Reed and sedge-reed mires concerning oligotrophic raised bogs (ryam)
17.0	14.0	Reed-sedge and grassy mires
42.8	27.2	Grass-dominated mires on salty soils

Fig. 1 Live biomass (LB, t/ha) and net primary production (NPP, t/ha/year) in West Siberian wetlands (excluding woody layer). Map reproduced from Peregon et al. (2008) with permission from John Wiley and Sons and from AGU Publisher

Optical satellite images have a long history of being used to estimate forest parameters and assess wooden biomass with different results in terms of quality. Currently, there are several limitations on the use of remote sensing to measure biomass (Le Toan et al. 2011). In general, optical data are not physically related to biomass, although estimates of biomass have been obtained from the Leaf Area Index (LAI), derived from optical greenness indices. However, these are neither robust nor meaningful above a low LAI value. For instance, Myneni et al. (2001) used optical data from the Advanced Very High-Resolution Radiometer (AVHRR) sensors to infer biomass changes in northern forests over the period of 1981–1999, and concluded that Eurasia was a large sink. However, field data indicate that the Eurasian sink is much weaker (Beer et al. 2006).

The radar or LIDAR remote sensing approaches hold great promise for direct observations of the three-dimensional structure (3D) of the above-ground vegetation which can be used to make relatively straight-forward calculations of carbon storage in above-ground biomass in boreal ecosystems (Pan et al. 2011; Thurner et al. 2014), but their sensitivity to forest biomass depends on the radar frequency. When longer wavelength microwaves are used (>20 cm), the detected radiation is mostly due to backscattering from the branching elements and stems of the trees. As a result, long wavelength Synthetic Aperture Radar (SAR) has a stronger and more universal relationship than optical or short wavelength microwave sensors which are sensitive to leaf characteristics, where relationships with the woody component of vegetation are indirect and thus highly site- and season-specific (Mitchard et al. 2009). In addition, the weather conditions (cloudiness) and other atmospheric disturbances make the optical images prone to errors (Lillesand et al. 2004).

In recent years, several studies have demonstrated the potential of microwave remote sensing for forest stem volume (or biomass) estimations in the boreal forests. It has been proven at L- and P-band by means of backscatter analysis (Le Toan 1992; Ranson et al. 1997; Rauste 2005; Santoro et al. 2006), and using interferometric coherence technique (Askne et al. 2003; Eriksson et al. 2003; Gaveau et al. 2003; Neumann et al. 2012); at C-band using ERS-1/2 repeat-pass interferometry (Santoro et al. 2002, 2007), and combining ERS tandem interferometric coherence and JERS backscatter (Balzter et al. 2002; Tansey et al. 2004; Wagner et al. 2003). Some studies have considered SAR application for estimations of AGB in Central Siberia (Schmullius et al. 2001; Tansey et al. 2004; Wagner et al. 2003; Thiel et al. 2009), in Western Siberia (Peregon and Yamagata 2013), along with the map of carbon stock and density of northern boreal and temperate forests at 0.01° resolution (Thurner et al. 2014). Spatially, explicit datasets on forest carbon stocks in the Northern Hemisphere have been rare and inconsistent up to now. The availability of extensive observations by the Envisat Advanced Synthetic Aperture Radar (ASAR) has boosted the development of an algorithm to retrieve forest growing stock volume (GSV) (Santoro et al. 2011, 2013). Remote sensing techniques are, however, unable to measure below-ground biomass and carbon in the soil.

A general experience is that the long wavelength (L-band) SAR is adequate for biomass estimations (Rosenqvist et al. 2007; Thiel et al. 2006) with the accuracy of biomass estimates being up to 12.9 t/ha, $R^2 = 0.86$ in the Northeastern United States with ALOS/PALSAR dual-polarization L-band (Cartus et al. 2012). Radar sensitivity to canopy biomass decreases from moderate to dense canopies, where the signal no longer penetrates through the entire canopy. This saturation level depends on the frequency, the polarization mode, incidence angle, the type of forest, foliage structure, and moisture conditions (Rauste 2005). As a result, a wide range of sensitivities has been reported, but the threshold stem volume above which radar signals saturate rarely exceeds 100 m^3/ha for L-band polarimetric algorithms (Imhoff 1995; Gaveau et al. 2003; Rauste 2005; Santoro et al. 2006; Rauste et al. 2008). It makes the SAR data useful for accurate estimates of carbon stocks in relatively homogeneous, young, or sparse boreal forests, but less useful in complex canopies and mature, higher biomass forests (Le Toan et al. 2004).

The PALSAR instrument on ALOS, built using the JERS-1 L-band SAR technology, provided the first systematic observations collected across the Northern hemisphere and over the Globe. The data from ALOS/PALSAR have been used for generating forest change and derived biomass maps, but failed in 2011. A replacement satellite is planned.

Although promising results have been achieved, the estimation of forest biomass in Western Siberia remains problematic and goes far behind the operational stage, not only because of the effect of saturation of the radar signal at high-biomass levels, but also because of the spatial heterogeneity of natural ecosystems, leading to large uncertainties in biomass estimates. Moreover, there is still a gap in large areas of Siberia, where ground measurements are limited and no studies have been conducted based on remote sensing data in its Western part in the same way the Siberia II project (2002–2005, http://www.siberia2.uni-jena.de/index.php) focused on Central Siberia. There is no universally accepted methodology for assessing the AGB of woody boreal landscapes. Thus, there is a continued need for both new experimental data and further improvement to the existing models for biomass estimation from SAR data.

On the global scale, several research programs are underway to implement the use of SAR and airborne/spaceborne Lidar, to derive estimates of vegetation above-ground biomass (e.g. Saatchi et al. 2007). Satellite missions such as the ESA's BIOMASS P-band radar (Le Toan et al. 2011) or a concept based upon the CEOS strategy for carbon observations from space (available at ceos.org) are currently being considered for this purpose (Ciais et al. 2014).

While there is currently no spaceborne sensor operational to map global forest biomass, recent advances in active remote sensing technologies demonstrate the possibility of high-resolution, globally consistent estimates of the above-ground biomass and carbon stocks with significantly reduced uncertainties in the estimates. Remote sensing techniques integrating spaceborne imaging and airborne LIDAR with pattern recognition methods (e.g. CLASLITE; www.claslite.ciw.edu/) have demonstrated a strong capability for tracking and quantifying biomass and structural changes in forest undergoing deforestation at national and county scales (Asner et al. 2010). Forest height and canopy profile metrics have been derived from the Geoscience Laser Altimeter System (GLAS) on the ICESat satellite and used to estimate above-ground biomass (Lefsky et al. 2005). ICESat height samples and MODIS data have been merged to create the first global canopy height product (Lefsky 2010).

Regarding forest biomass and soil carbon stock inventories, it is critical to have available measurements at high-spatial resolution, since the scale of disturbance is often of tens of meters. A scale of ~1–4 ha defines the main spatial resolution required for satellite observations of biomass if the purpose of these measurements is to resolve fine-scale disturbances. A key challenge is to bring remote sensing measurements to a level of long-term consistency and accuracy so that they can be efficiently combined in models to reduce uncertainties, in synergy with ground-based data (Ciais et al. 2014).

2 Methods for Estimating NPP/Biomass in Wetlands

To obtain reasonable estimates of the spatial distribution of NPP and biomass for wetlands, one needs to measure these variables in the field at a representative set of locations along a wide north–south climate (i.e. latitudinal) gradient of WS (Peregon et al. 2008). Test sites for field observations must be selected in each of three subzones of the taiga region (northern, middle and southern taiga), and represent different wetland types and topography. Field observations would also be necessary in other geographical locations with extensive wetland massifs beyond the bounds of the taiga region (forested-steppe, tundra).

Step 1: Field observations

At each test site, detailed geobotanical descriptions are recorded and biomass sampling conducted at all major topographical features which are typical of the test site (Peregon et al. 2008). Field sampling should be repeated two or three times during the growing season at the same test sites for several consecutive years to obtain information on interannual variability and to collect representative data.

The total amount of biomass is divided into two fractions: live (phytomass) and dead biomass. Dead biomass contains partly decomposed plant residue, plant litter, dead mosses including the yellow-colored parts of *Sphagnum,* and dead roots of grasses and shrubs. Observed live biomass (LB) consists of above-ground (AG), land-surface (LS), and below-ground (BG) components: that is, LB = AG + LS + BG. The AG component is determined by clipping vegetation (herbs, shrubs and dwarf shrubs) at the top of the moss surface (i.e. the so-called "clipping line") over a 0.25-m^2 square, replicated 7–10 times at 50-cm intervals along the designated transect, which is about 5 m long. For further estimates of NPP, the above-ground phytomass is divided into several fractions: the green parts of herbs which are capable of photosynthesis; the shrub leaves of current and previous years; the shoots and woody parts (both perennial for shrubs growing on wetlands, and those that formed in the current year for shrubs and grasses); the upper part of shoots down to the dark line of brown mosses; and the photosynthetically active upper parts of lichen stems. Living bryophytes and lichens below the clipping line are considered as LS, whereas stems, stem bases, rhizomes, roots of vascular plants, grasses, shrubs, and dwarf shrubs which are located below the clipping line should be considered as BG. On the same plot, where the above-ground fraction has been clipped, one needs to sample 1 dm^3 cores at 10 cm increments down to a depth of 30 cm below the moss surface to estimate the LS- and BG-based on the acrotelm–catotelm model developed by Clymo (1984). Clipped (raw) material is then sorted by species, oven dried at 108 °C and weighed to express the mass in dry weight (see Peregon et al. 2008 for full experimental notes).

Like the biomass, the NPP also consists of above-ground (ANPP), land-surface (LNP), and below-ground (BNP) components. The ANPP of grasses can be defined as the maximum seasonal weight of the standing green phytomass. The ANPP of deciduous shrubs is determined as the maximum seasonal weight of leaves and shoots grown in the current year. The weight of green biomass grown in the current year, such as the shoots and foliage, and clipped at the end of the growing season corresponds to the ANPP of evergreen shrubs. The production of lichens is estimated as the difference between the maximum and minimum seasonal weights of live biomass. The production of *Sphagnum* mosses (LNP) is estimated using the measurements of the basal cover as a percentage, linear growth in mm, and shoot density in the number of shoots per dm^2 (Peregon et al. 2008; Naumov and Kosykh 2010). The linear growth of *Sphagnum* is measured using "individual tags," which allowed us to measure the NPP of plants growing obliquely in hollows. In this method, the stems are tagged with thin metal rings, just below their crowns, as shown in Fig. 2. About 70–100 tagged plants per community have usually been sampled one year after tagging. For each sample, the linear shoot growth represents the distance between the level a—just below the plant's crown and level b—where the ring was placed in the previous year. The dry weight of annual growth corresponds to the NPP of mosses. The total number of *Sphagnum* stems per unit area was additionally examined to express the LNP in the common unit of g/m^2/year. The sum of these estimated annual increments of different species of bryophytes, corrected for their basal cover, is the LNP.

Normally, BNP of vascular plants is conventionally assumed to be 50–80 % of NPP based on Wallén (1992). This method, however, allowed us to determine the below-ground components formed during the current year. Rhizomes and stem bases have to be divided into two groups: young (i.e. grown during the sampling year) and old (i.e. the rest), distinguished as follows. Young stem bases have many fine roots, which are lighter in color and originate directly from the stems. Newer growths of rhizomes are also lighter in color, and accompanied by vegetative shoots on the top. The roots of grasses and sedges can be divided into four groups according to their age and condition (i.e. whether they were alive or dead) indicated by a number of visible morphological features, such as length, color, diameter, turgor, degree of lateral ramification, position on below-ground rhizomes, etc. Dwarf shrub roots were divided into three groups based on the same criteria.

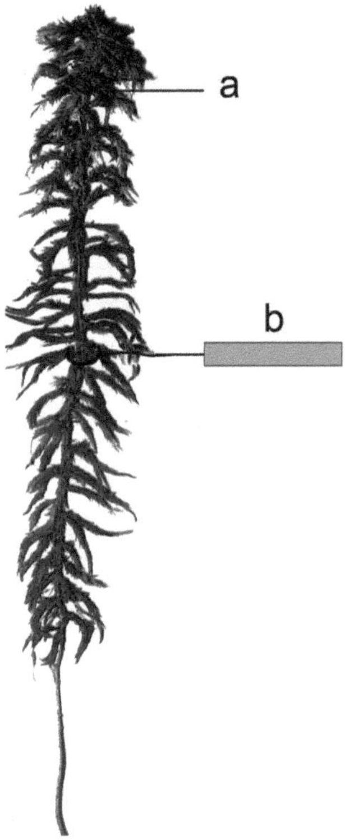

Fig. 2 Method of "individual tags" used to estimate the linear growth of *Sphagnum* mosses. Stems are tagged with thin metal rings at level *a* —just below the plant's crown and level *b*—where the ring was placed in the previous year.
Figure reproduced from Peregon et al. (2008) with permission from John Wiley and Sons and from AGU Publisher

Accordingly, the BNP is defined as the weight of below-ground biomass formed in the current year.

Step 2: Data synthesis

A "synthesis method" for combining data at several different spatial resolutions was developed to build a regional-scale NPP and live biomass inventory in Peregon et al. (2008). A multi-scale approach was applied to map the NPP and live biomass based on an existing wetland typology map at a scale of 1:2,500,000 with 20 wetland classes with obvious elements of heterogeneity (Romanova et al. 1977). The map was geometrically corrected and digitized. Obviously, the procedure is prone to error due to the use of hardcopy products and digitization, although the planimetric errors typically appeared to be at most 1.5–2 km, which seems reasonable at this (regional) scale. To address heterogeneity, average area fractions for each microlandscape/landscape type, composing the vegetation mosaic of those 20 classes, were derived from remote sensing and ground survey data. This process comprised of two steps.

First, the wetland typology map was refined by manual classifications based on satellite imagery to provide more detail on the spatial structure in boreal regions, over the most heterogeneous landscapes. The images with a spatial resolution of

30 m (LANDSAT TM, ETM+) from the Global Land Cover Facility (glcf.umiacs.umd.edu/) were processed via visual interpretation at a scale of 1:100,000–1:200,000 at each test site based on a classification system developed by Lapshina and Vasiliev (2001). Images were interpreted on a computer monitor for color, tone, and texture. About 30 classes of forest, paludified forest and wetland ecosystems were derived for the boreal region. At this scale, the satellite images were classified into 10 wetland classes, compared on a large-scale wetland typology map (Romanova et al. 1977), which used only 3 classes within the boreal region. However, there are also elements of a composite nature presented in this improved classification, such as open patterned wetlands (raised string bogs), that differ in the area fraction occupied by ridges, hollows, and small lakes.

Finally, we evaluated the fractional area coverage of microlandscape elements (ridges, hollows, and lakes) for each type of patterned wetland by interpreting the high-resolution satellite imagery (QuickBird, 0.6 m/pixel) (Peregon et al. 2009). Then, this information was applied to elucidate the structure of patterned wetlands on satellite image-based maps for the same climatic (or geographical) region, and the ground survey NPP, and biomass data were scaled up corresponding to wetland type on the regional wetland typology map.

3 Methods for Estimating NPP/Biomass in Forests

Direct in situ measurements of biomass in forest ecosystems imply labor-intensive measurements both in the field and in the laboratory. The usual practice of the Russian State Forest Account (SFA) consists of field measurements of major forest parameters in all types of forest over the test site, and further extrapolating the local measurements up to the larger regions using airborne or satellite imagery-based land cover maps, or so-called "stand boundary" maps. Stand boundary maps are created by displaying homogeneous forest cover (forest stands of same species composition, average height and diameter of tree stems, basal area, growing stock, and therefore the NPP and total biomass). Under the given definition of the stand as a homogeneous unit, the set of forest parameters is attributed to the whole area of the same "stand boundary" with a certain level of confidence. The SFA is available for about 2000 forest enterprises for Russia as a whole, as well as for all other countries of the former USSR for the period from 1961 to today, with an interval, as a rule, of 5 years (Shvidenko et al. 2007b). The in situ data on forest parameters is the property of SFA, has a certain commercial value, and is not freely available for scientific purposes. The standard methods of field measurements in the forestry sector (Sagreev 1975; Vorobyov et al. 1986; Sagreev et al. 1992, etc.) can be adopted for research purposes in order to conduct proper field campaigns.

Step 1: In situ data

The estimation of the whole stand biomass is based on the calculation of the forest stem volume using empirical allometric relations based on tree height, average age, diameter of tree stems, and number of trees per ha, using the growth (yield) tables (YTs) elaborated for major tree species in the region by Shvidenko et al. (2007a). Forest inventories provide important information such as age structure, allowing for model-based approaches to factor out the impacts of age-related changes in the assessment of additional sink, although the uncertainty of the initial YTs cannot be estimated in any formal way (Shvidenko et al. 2007b).

The forest biomass can be further assessed for a certain stem volume using a semi-empirical phytomass model developed by the International Institute for Applied Systems Analyses (IIASA) (Shvidenko et al. 2007b). The model was parameterized for dominant tree species, site indexes, and ecological regions based on an extensive set of sample plots collected in the forests of Northern Eurasia. The IIASA model provides estimates for not only commercial wooden biomass, but also the total above-ground biomass, including the fractions of stem wood, bark, branches, and foliage, but excluding shrubs, grasses, and necromass.

Assessing the actual accuracy of the models of phytomass using traditional statistical methods is difficult as it is not always possible to estimate the accuracy of forest inventories, particularly in mixed and relatively uneven-aged stands (Shvidenko et al. 2007b). The growing stock identified by the Russian forest inventory is biased for mature forests (Shvidenko and Nilsson 2002). However, it has been shown that the total phytomass of Russian forests as a whole and for large regions can be estimated with the "summarized" error (i.e. a function of random and systematic errors, which cannot be separated for a majority of cases) in the range of 4–7 % (a priori confidential probability of 0.9) under the assumption that the entire system of accounting does not have unrecognized biases (Shvidenko et al. 2001).

Step 2: Remote sensing techniques

Processing SAR signal/imagery offers ways to estimate the forest biomass over large regions (see Sect. 1.3).

Since, the boreal region is prone to great seasonal variability in temperature and snow cover, this could in turn contribute to seasonal variability in the SAR data (Pulliainen et al. 1999; Santoro et al. 2011). It is imperative to analyze the SAR data taken in the same season and under similar climatic conditions. The summer and early-autumn SAR acquisitions have shown a better relationship between the SAR data and biomass (Rauste 2005; Rauste et al. 2008; Santoro et al. 2006, 2009). Futhermore, to complete the procedures, the SAR data should be preprocessed before the regression analysis as follows: (i) geometrical correction, (ii) radiometrical normalization, and (iii) averaging of SAR signals to reduce the effect of speckle and spatial heterogeneity of the forest stands. The digital numbers of SAR signal amplitude must be converted to backscattering coefficients in dB based on equations specially elaborated for each SAR data set.

The next procedure is to test various single-variable and multivariate regression models to calibrate a relationship between the forest biomass derived from field measurements and the radar backscatter signal. In previous studies, the most reliable results were obtained using HV-polarized backscatter with the Water Cloud model developed by Attema and Ulaby in 1978 (Fransson and Israelsson 1999;

Kurvonen et al. 1999; Askne et al. 2003; Santoro et al. 2006). Based on the equation and regression coefficients in the forward model, an inverse relationship could be established, with the forest biomass as the dependent variable and the SAR backscattering coefficient as the independent variable. The main difficulty of biomass retrieval using the Water Cloud model is that the model produces uncertain results when inverted (Askne et al. 2003). Since the model is essentially exponential, it approaches an asymptotic limit, and in some cases the observation values fall outside the range covered by the model curve. In these cases, a strict inversion of the model would result in either infinite or negative estimates, neither of which represents the real biomass. Following Askne et al. (2003), these estimates should be arbitrarily set equal to either the highest biomass observed in the data set or to zero biomass. In the final step, the model performance for biomass retrieval should be tested against a validation subset of in situ biomass values to show an agreement between the observed forest biomass and the biomass retrieved from SAR backscatter, and to consider the uncertain results in the Water Cloud model (see Fig. 3 as an example).

Along the whole range of ecosystems, the major confusion in the results of the Water Cloud model was caused by biomass estimates in upland mature forests with high-biomass values, as the SAR signal often appeared above the saturation level (Fig. 4). Overall, the uncertainties could be related to the combined effect of the following factors: (1) spatial heterogeneity of forest stands (Rauste et al. 1994; Saatchi et al. 2012); (2) radar calibration and orthorectification (van Zyl 1993); and (3) field estimation errors propagating through the analysis of forest biomass (Balzter et al. 2002; Santoro et al. 2006; Shvidenko et al. 2007b). In addition, backscatter responds differently with respect to soil and vegetation moisture conditions and surface topography, adding to observed prediction errors (Mitchard et al. 2009; Sandberg et al. 2011).

While the SAR-based biomass retrievals involve high remaining uncertainty [the estimation errors in terms of RMSE typically found at 25–30 % of mean biomass in Peregon and Yamagata (2013)], the uncertainties were mainly attributed to

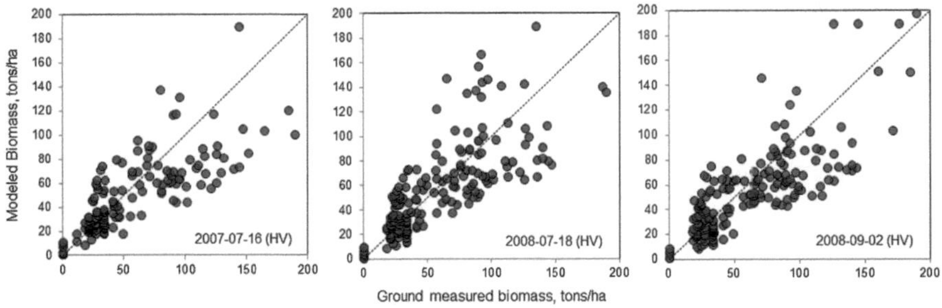

Fig. 3 Ground-measured above-ground forest biomass (AGB, metric ton/ha) versus AGB retrieved from SAR backscatter at test site in south taiga, Western Siberia (near 56° 51′N, 78° 53′ E) based on three scenes of ALOS/PALSAR acquired in HV polarization mode. Panels reproduced from Peregon and Yamagata (2013) with permission from Elsevier publisher

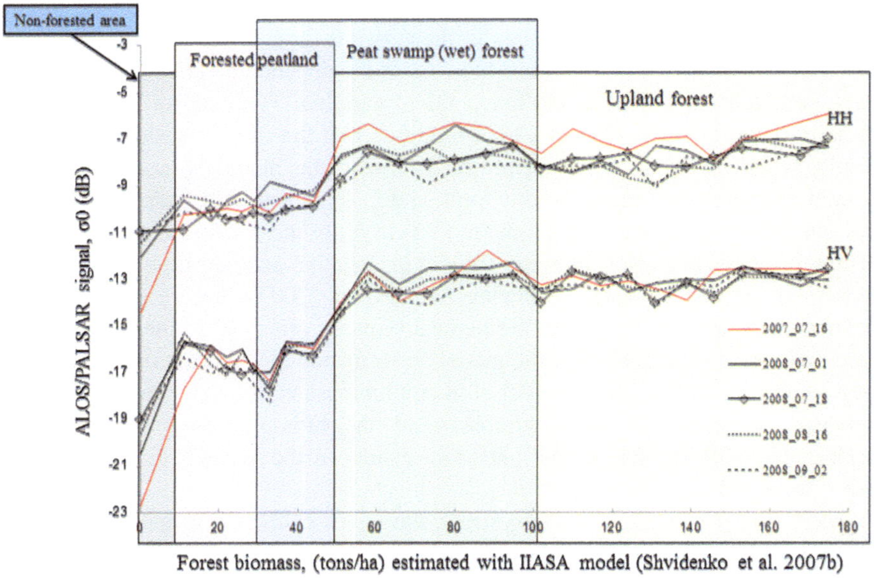

Fig. 4 The relationship between the average mean SAR backscatter (σ^0 averaged within each 10-tonne/ha interval of AGB) in *HH* and *HV* polarization modes and corresponding forest biomass (metric tons) for the whole range of natural ecosystems and environmental conditions. The ecosystem type is specified from field measurements. Figure reproduced from Peregon and Yamagata (2013) with permission from Elsevier publisher

estimates of biomass in mature forests when the threshold biomass of 60–70 tonnes/ha were exceeded. This approach is also challenging in sparse, low-productive, or young forests on shallow peat in near wetland areas. These natural ecosystems are often omitted in official forest inventories as improper for commercial use, although they have a large potential as sources/sinks of atmospheric carbon. Overall, the SAR applications would give more details to the total carbon accounting in Western Siberia where more accurate national or large-scale forest inventories hardly exist, and where the ground-based inventory is costly and time-consuming.

4 Conclusions

1. No unified method has yet been elaborated for estimating forest and wetland NPP/biomass in Siberia at once.
2. Direct in situ measurements of NPP/biomass at test sites and their further extrapolation to larger regions are an alternative to ecosystem modeling studies in wetland ecosystems.

3. In forest ecosystems, the NPP/biomass values can be accessed based on high-resolution radiometer data, along with radar- (SAR) or LIDAR-based remote sensing approaches.
4. All of the methods used are prone to uncertainties accounting for up to 30 % of the estimated NPP/biomass values.

References

Aerts R, De Caluwe H, Kanings H (1992) Seasonal allocation of biomass and nitrogen in four Carex species from mesotrophic and eutrophic fen as affected by nitrogen supply. J Ecol 80:653–664

Andrus R, Wagner DJ, Titus JE (1983) Vertical distribution of Sphagnum mosses along hummock-hollow gradients. Can J Bot 61:3128–3139

Aselmann I, Crutzen P (1989) Global distribution of natural freshwater wetlands and rice paddies, their net primary productivity, seasonality and possible methane emissions. J Atm Chem 8 (4):307–358

Askne J, Santoro M, Smith G, Fransson JES (2003) Multitemporal repeat-pass SAR interferometry of boreal forests. IEEE Trans Geosci Rem Sens 41:1540–1550

Asner GP, Powell GVN, Mascaro J, Knapp DE, Clark JK, Jacobson J, Kennedy-Bowdoin T, Balaji A, Paez-Acosta G, Victoria E, Secada L, Valqui M, Hughes RF (2010) High-resolution forest carbon stocks and emissions in the Amazon. Proc Natl Acad Sci USA 107:16738–16742

Attema EPW, Ulaby FT (1978) Vegetation modeled as a water cloud. Radio Sci 13:357–364

Balzter H, Talmon E, Wagner W, Gaveau D, Plummer S, Yu JJ, Yu JJ, Quegan S, Davidson M, Le Toan T, Gluck M, Shvidenko A, Nilsson S, Tansey K, Luckman A, Schmullius C (2002) Accuracy assessment of a large-scale forest cover map of central Siberia from synthetic aperture radar. Can J Rem Sens 28(6):719–737

Beer C, Lucht W, Schmullius C, Shvidenko A (2006) Small net uptake of carbon dioxide by Russian forests during 1981–1999. Geophys Res Lett 33:L15403. doi:10.1029/2006GL026919

Botch MS, Kobak K, Vinson TS, Kolchugina TP (1995) Carbon pools and accumulation in peatlands of the Former Soviet Union. Global Biogeochem Cycles 9:37–46

Campbell C, Vitt DH, Halsey LA, Campbell ID, Thormann MN, Bayley SE (2000) Net primary production and standing biomass in northern continental wetlands. Nat For Serv North For Cent, Edmonton, Alberta, Inf Rep NOR-X-369

Cartus O, Santoro M, Kellndorfer J (2012) Mapping forest aboveground biomass in the Northeastern United States with ALOS Palsar dual-polarization L-band. Rem Sens Env 124:466–478. doi:10.1016/j.rse.2012.05.029

Ciais P, Sabine C, Bala G, Bopp L, Brovkin V, Canadell J, Chhabra A, DeFries R, Galloway J, Heimann M, Jones C, Le Quéré C, Myneni RB, Piao S, Thornton P (2013) Carbon and other biogeochemical cycles. In: Stocker TF, Qin V, Plattner G-K, Tignor M, Allen SK, Boschung J, Nauels A, Xia Y, Bex V, Midgley PM (eds) Climate change 2013: the physical science basis. Contribution of working group I to the fifth assessment report of the intergovernmental panel on climate change. Cambridge University Press, Cambridge, United Kingdom, New York, NY, USA, pp 465–570

Ciais P, Dolman AJ, Bombelli A, Duren R, Peregon A, Rayner PJ, Miller C, Gobron N, Kinderman G, Marland G, Gruber N, Chevallier F, Andres RJ, Balsamo G, Bopp L, Bréon F-M, Broquet G, Dargaville R, Battin TJ, Borges A, Bovensmann H, Buchwitz M, Butler J, Canadell JG, Cook RB, DeFries R, Engelen R, Gurney KR, Heinze C, Heimann M, Held A, Henry M, Law B, Luyssaert S, Miller J, Moriyama T, Moulin C, Myneni RB, Nussli C, Obersteiner M, Ojima D, Pan Y, Paris J-D, Piao SL, Poulter B, Plummer S, Quegan S,

Raymond P, Reichstein M, Rivier L, Sabine C, Schimel D, Tarasova O, Valentini R, Wang R, van der Werf G, Wickland D, Williams M, Zehner C (2014) Current systematic carbon cycle observations and the need for implementing a policy-relevant carbon observing system. Biogeosci 11:3547–3602. doi:10.5194/bg-11-3547-2014

Clymo RS (1984) The limits to peat bog growth. Phylos Trans R Soc Lond Biol Sci 303:605–658

Coleman DC (1976) A review of root production processes and their influence on soil biota in terrestrial ecosystems. In: Anderson JM, Macfadyen A (eds) The role of terrestrial and aquatic organisms in decomposition processes. Blackwell Scientific, Oxford, pp 417–434

Dolman AJ, Shvidenko A, Schepaschenko D, Ciais P, Tchebakova N, Chen T, van der Molen MK, Belelli Marchesini L, Maximov TC, Maksyutov S, Schulze ED (2012) An estimate of the terrestrial carbon budget of Russia using inventory-based, eddy covariance and inversion methods. Biogeosc 9:5323–5340. doi:10.5194/bg-9-5323-2012

Efremov SP, Efremova TT (2001) Present stocks of peat and organic carbon in bog ecosystems of west Siberia. In: Bleuten W, Lapshina ED (eds) Carbon storage and atmospheric exchange by West Siberian Peatlands. Utrecht University, Utrecht, pp 73–78

Eriksson LEB, Santoro M, Wiesmann A, Schmullius C (2003) Multi-temporal JERS repeat-pass coherence for growing stock volume estimation of Siberian forest. IEEE Trans Geosci Rem Sens 41(7):1561–1570. doi:10.1109/TGRS.2003.814131

Fransson JES, Israelsson H (1999) Estimation of stem volume in boreal forests using ERS-1 C- and JERS-1 L-band SAR data. Int J Rem Sens 20(1):123–137

Gaveau DLA, Balzter H, Plummer S (2003) Forest woody biomass classification with satellite-based radar coherence over 900,000 km^2 in Central Siberia. For Ecol Manage 174:65–75

Golovatskaya EA (2009) Biological productivity of oligotrophic and eutrophic mires in the Southern Taiga of Western Siberia. J Siberian Fed Univ. Biology 2(1):38–53 (in Russian) (Головацкая ЕА (2009) Биологическая продуктивность олиготрофных и эвтрофных болот южнотаежной подзоны Западной Сибири. Журнал Сиб Фед Унив, Сер Биология, 2 (1):38–53)

Golovatskaya EA, Dyukarev EA (2009) Carbon budget of oligotrophic mire sites in the Southern Taiga of Western Siberia. Plant Soil 315(1–2):19–34

Hugelius G, Strauss J, Zubrzycki S, Harden JW, Schuur EAG, Ping CL, Schirrmeister L, Grosse G, Michaelson GJ, Koven CD, O'Donnell JA, Elberling B, Mishra U, Camill P, Yu Z, Palmtag J, Kuhry P (2014) Improved estimates show large circumpolar stocks of permafrost carbon while quantifying substantial uncertainty ranges and identifying remaining data gaps. Biogeosc Discuss 11:4771–4822. doi:10.5194/bgd-11-4771-2014

Imhoff M (1995) Radar backscatter and biomass saturation: ramifications for global biomass inventory. IEEE Trans Geosci Rem Sens 33(2):511–518

Kivinen E, Pakarinen P (1981) Geographical distribution of peat resources and major peatland complex types in the world. Ann Acad Scient Fenn Ser A 132:28

Kopoteva TA, Kosykh NP (2011) Comparative evaluation of the structure of phytomass and productivity of mesotrophic and dwarf shrub-Sphagnum Mires of the Taiga Zone. Contemp Prob Ecol 2:301–307

Kosykh NP, Mironycheva-Tokareva NP, Peregon AM, Parshina EK (2008a) Biological productivity of bogs in the middle Taiga subzone of Western Siberia. Russ J Ecol 39(7):8–16

Kosykh NP, Koronatova NG, Naumova NB, Titlyanova AA (2008b) Above- and below-ground phytomass and net primary production in boreal mire ecosystems. Wetlands Ecol Manage 16:139–153. doi:10.1007/s11273-007-9061-7

Kosykh NP, Mironycheva-Tokareva NP, Parshina EK (2008c) Biological productivity of mires in forest-tundra environments of Western Siberia, Vestnik TGPU, Seriya. Biologicheskie nauki 4 (78):53–57 (in Russian) (Косых НП, Миронычева-Токарева НП, Паршина ЕК (2008c) Биологическая продуктивность болот лесотундры Западной Сибири. Вестник ТГПУ, Сер: Биол науки, Вып. 4(78): 53–57)

Kosykh NP, Mironycheva-Tokareva NP, Parshina EK (2009) The carbon and macroelements budget in the bog ecosystems of the middle taiga in Western Siberia. Int J Environ Stud 66:485–493

Kudryashova SY, Baikov KS, Titlyanova AA, Dits LYu, Kosykh NP, Makhatkov ID, Shibareva SV (2011) Distributed GIS for estimation of soil carbon stock of West Siberia boreal zone. Contemp Prob Ecol 4(5):475–486

Kurvonen L, Pulliainen J, Hallikainen M (1999) Retrieval of biomass in boreal forests from multitemporal ERS-1 and JERS-1 SAR images. IEEE Trans Geosci Rem Sens 37:198–205

Lapshina ED, Vasiliev SV (2001) Carbon storage and atmospheric exchange by West Siberian peatlands. In: Bleuten W, Lapshina ED (eds) Appendix 1: Land unit maps. FGUU Scientific Reports 2001-1, Utrecht (NL), 165 p. ISBN 90-806594-1.X

Le Toan T, Beaudoin A, Riom J, Guyon D (1992) Relating forest biomass to SAR data. IEEE Trans Geosci Rem Sens 30(2):403–411

Le Toan T, Quegan S, Woodward I, Lomas M, Delbart N, Picard C (2004) Relating radar remote sensing of biomass to modeling of forest carbon budget. Clim Change 76:379–402

Le Toan T, Quegan S, Davidson M, Balzter H, Paillou P, Papathanassiou K, Plummer S, Saatchi S, Shugart H, Ulander L (2011) The BIOMASS mission: mapping global forest biomass to better understand the terrestrial carbon cycle. Rem Sens Env 115:2850–2860. doi:10.1016/j.rse.2011.03.020

Lefsky MA (2010) A global forest canopy height map from the moderate resolution imaging spectroradiometer and the geoscience laser altimeter system. Geophys Res Lett 37:L15401. doi:10.1029/2010GL043622

Lefsky MA, Harding DJ, Keller M, Cohen WB, Carabajal CC, Del Bom Espirito-Santo F, Hunter MO, de Oliveira R Jr (2005) Estimates of forest canopy height and aboveground biomass using ICESat. Geophys Res Lett 32:L22S02. doi:10.1029/2005GL023971

Lehner B, Döll P (2004) Development and validation of a global database of lakes, reservoirs and wetlands. J Hydrol 296(1–4):1–22. http://www.wwfus.org/science/data.cfm. Accessed on 15 Nov 2014

Lillesand TM, Kiefer RW, Chipman JW (2004) Remote sensing and image interpretation, 5th edn. Wiley, New Jersey, New York, p 763

Loveland TR, Reed BC, Brown JF, Ohlen DO, Zhu J, Yang L, Merchant JW (2000) Development of a global land cover characteristics database and IGBP DISCover from 1-km AVHRR data. Int J Rem Sens 21(6/7):1303–1330 (The global land cover characterisation (GLCC) dataset, Eurasia land cover characteristics, data base version 2.0 (available online at http://edc2.usgs.gov/glcc/eadoc2_0.php#down. Accessed on 15 Nov 2014)

Matthews E, Fung I (1987) Methane emission from natural wetlands: Global distribution, area, and environmental characteristics of sources. Global Biogeochem Cycles 1(1):61–86

Melton JR, Wania R, Hodson EL, Poulter B, Ringeval B, Spahni R, Bohn T, Avis C, Beerling DJ, Chen G, Eliseev AV, Denisov SN, Hopcroft PO, Lettenmaier DP, Riley WJ, Singarayer JS, Subin ZM, Tian H, Zürcher S, Brovkin V, van Bodegom PM, Kleinen T, Yu ZC, Kaplan J (2013) Present state of global wetlands extent and wetland methane modelling: conclusions from a model inter-comparison project (WETCHIMP). Biogeosciences 10:753–788. doi:10.5194/bg-10-753-2013

Mitchard ETA, Saatchi SS, Woodhouse IH, Nangendo G, Ribeiro NS, Williams M, Ryan CM, Lewis SL, Feldpausch TR, Meir P (2009) Using satellite radar backscatter to predict above-ground woody biomass: a consistent relationship across four different African landscapes. Geophys Res Lett 36:L23401. doi:10.1029/2009GL040692

Moore TR, Bubier JL, Frolking SE, Lafleur PM, Roulet NT (2002) Plant biomass and production and CO_2 exchange in an oligotrophic bog. J Ecol 90:25–36

Myneni RB, Dong J, Tucker CJ, Kaufmann RK, Kauppi PE, Liski J, Zhou L, Alexeyev V, Hughes MK (2001) A large carbon sink in the woody biomass of Northern forests. Proc Natl Acad Sci USA 98:14784–14789. doi:10.1073/pnas.261555198

Naumov AV, Kosykh NP (2010) The structure and functional features of sphagnum cover of the northern west siberian mires in connection with forecasting global environmental and climatic

changes–VI Sphagnum and Peatlands, pp 299– 317, In: Tuba Z, Slack NG, Stark LR (eds) Bryophyte ecology and climate change. Cambridge University Press, Cambridge, 491 p
Neumann M, Saatchi SS, Ulander LMH, Fransson JES (2012) Assessing performance of L- and P-Band polarimetric interferometric SAR data in estimating boreal forest above-ground biomass. IEEE Trans Geosci Rem Sens 50:714–726
Pan Y, Birdsey RA, Fang J, Houghton R, Kauppi PE, Kurz WA, Phillips OL, Shvidenko A, Lewis SL, Canadell JG, Ciais P, Jackson RB, Pacala SW, McGuire AD, Piao S, Rautiainen A, Sitch S, Hayes D (2011) A large and persistent carbon sink in the world's forest. Science 333:988–993. doi:10.1126/science.1201609
Peregon A, Yamagata Y (2013) The use of ALOS/PALSAR backscatter to estimate above-ground forest biomass: a case study in Western Siberia. Rem Sens Env 137:139–146. doi:10.1016/j.rse.2013.06.012
Peregon A, Maksyutov S, Kosykh NP, Mironycheva-Tokareva NP (2008) Map-based inventory of wetland biomass and net primary production in western Siberia. J Geophys Res Biogeosci 113: G01007. doi:10.1029/2007JG000441
Peregon A, Maksyutov S, Yamagata Y (2009) An image-based inventory of the spatial structure of West Siberian wetlands. Environ Res Lett 4:045014. doi:10.1088/1748-9326/4/4/045014
Petrescu, AMR, van Beek LPH, van Huissteden J, Prigent C, Sachs T, Corradi CAR, Parmentier FJW, Dolman AJ (2010) Modeling regional to global CH4 emissions of boreal and arctic wetlands. Global Biogeochem Cycles 24:GB400. doi:10.1029/2009GB003610
Potter CS, Randerson JT, Field CB (1993) Terrestrial ecosystem production: a process model based on global satellite data end surface data. Global Biogeochem Cycles 7:811–841
Prentice C, Cramer W, Harrison S, Leemans K, Monserub R, Solomon R (1992) A global biome model based on plant physiology and dominance, soil properties and climate. J Biogeogr 19:117–134
Pulliainen J, Kurvonen L, Hallikainen M (1999) Multitemporal behavior of L- and C-band SAR observations of boreal forests. IEEE Trans Geosci Rem Sens 37(2):927–937
Ranson KJ, Sun G, Lang RH, Chauhan NS, Cacciola RJ, Kilic O (1997) Mapping of boreal forest biomass from spaceborne synthetic aperture radar. J Geophys Res 102(D24):29599–29610
Rauste Y (2005) Multi-temporal JERS SAR data in boreal forest biomass mapping. Rem Sens Env 97(2):263–275
Rauste Y, Häme T, Pulliainen J, Heiska K, Hallikainen M (1994) Radar-based forest biomass estimation. Int J Rem Sens 15:2797–2808
Rauste Y, Lonnqvist A, Ahola H (2008) Mapping boreal forest biomass with imagery from polarimetric and semi-polarimetric SAR sensors. Ambiencia, Guarapuava, PR, 4, Edicao Especial 171–180
Romanova EA, Bybina RT, Golitsyna EF, Ivanova GM, Usova LI, Trushnikova LG (1977) Tipologicheskaya karta bolot Zapadno-Sibirskoi ravniny (Wetland typology map of west Siberian lowland). GUGK, Leningrad, Scale 1:2500000
Rosenqvist A, Shimada M, Ito N, Watanabe M (2007) ALOS PALSAR: a pathfinder mission for global-scale monitoring of the environment. IEEE Trans Geosci Rem Sens 45:3307–3316
Ruimy A, Sauguer B, Dedieu G (1994) Methodology for the estimation of terrestrial net primary production from remotely sensed data. J Geophys Res 99:5263–5283
Saatchi SS, Houghton RA, Alvala R, Soares JV, Yu Y (2007) Distribution of aboveground live biomass in the Amazon basin. Glob Change Biol 13:816–837
Saatchi S, Ulander L, Williams M, Quegan S, LeToan T, Shugart H, Chave J (2012) Forest biomass and the science of inventory from space. Nat Climate Change 2:826–827
Sagreev VV (1975) Basic principles of forest inventory. In: Sagreev VV, Vagin AB (eds) High school. Moscow, 264 pp (in Russian)
Sagreev VV, Sukhikh VI, Shvidenko AZ, Gusev NN, Moshkalev AG (1992) All-union standards for forest inventory, Urozhai. Moscow, 495 pp (in Russian). (Загреев ВВ, Сухих ВИ, Швиденко АЗ, Гусев НН, Мошкалев АГ (1992) Общесоюзные нормативы для таксации лесов. Справочник, Москва, Колос, 495с)

Sandberg G, Ulander LMH, Fransson JES, Holmgres J, Le Toan T (2011) L- and P-band backscatter intensity for biomass retrival in hemiboreal forest. Rem Sens Env 115:2874–2886

Santoro M, Askne J, Smith G, Fransson JES (2002) Stem volume retrieval in boreal forests with ERS-1/2 interferometry. Rem Sens Env 81(1):19–35

Santoro M, Eriksson L, Askne J, Schmullius C (2006) Assessment of standwise stem volume retrieval in boreal forest from JERS-1 L-band SAR backscatter. Int J Rem Sens 27(16):3425–3454. doi:10.1080/01431160600646037

Santoro M, Shvidenko A, McCallum I, Askne J, Schmullius C (2007) Properties of ERS-1/2 coherence in the Siberian boreal forest and implications for stem volume retrieval. Rem Sens Env 106:154–172

Santoro M, Fransson JES, Eriksson LEB, Magnusson M, Ulander LMH, Olsson H (2009) Signatures of ALOS PALSAR L-band backscatter in Swedish forest. IEEE Trans Geosci Rem Sens 47(12):4001–4019

Santoro M, Beer C, Cartus O, Schmullius C, Shvidenko A, McCallum I, Wegmüller U, Wiesmann A (2011) Retrieval of growing stock volume in boreal forest using hyper-temporal series of Envisat ASAR ScanSAR backscatter measurements. Rem Sens Env 115(2):490–507

Santoro M, Cartus O, Fransson JES, Shvidenko A, McCallum I, Hall RJ, Beaudoin A, Beer C, Schmullius C (2013) Estimates of forest growing stock volume for Sweden, Central Siberia and Québec using Envisat advanced synthetic aperture radar backscatter data. Rem Sens 5:4503–4532. doi:10.3390/rs5094503

Schmullius C, Baker J, Balzter H, Davidson M, Eriksson L, Gaveau D, Gluck M, Holz A, Luckman AJ, Marschalk U, McCallum I, Nilsson S, Oeskog A, Quegan S, Rauste Y, Roth A, Shvidenko A, Le Toan T, Tansey KJ, Vietmeier J, Wagner W, Wegmüller U, Wiesmann A, Yu JJ (2001) SAR imaging for boreal ecology and radar interferometry applications (SIBERIA). Centre of Earth Observation, 4th Framework Programme of the European Community. Final report Co. No. ENV4-CT97-0743-SIBERIA (Luxembourg: EC)

Sheng Y, Smith L, MacDonald GM, Kremenetski KV, Frey KE, Velichko AA, Lee M, Beilman DW, Dubinin P (2004) A high-resolution GIS-based inventory of the west Siberian peat carbon pool. Global Biogeochem Cycles 18:GB3004. doi:10.1029/2003GB002190

Shvidenko A, Nilsson S (2002) Dynamics of Russian forests and the carbon budget in 1961–1998: an assessment based on long-term forest inventory data. Clim Change 50:5–37

Shvidenko AZ, Schepaschenko DG, Nilsson S (2001) Aggregated models of phytomass of major forest forming species of Russia. Forest Inv Forest Plann 1(30):50–57 (in Russian) (Швиденко АЗ, Щепащенко ДГ, Нильссон С (2001) Агрегированные модели фитомассы основных лесообразующих пород России. Лесная таксация и лесоустройство 1:50–57)

Shvidenko A, Schepaschenko D, Nilsson S, Boului Y (2007a) Tables and models of growth and productivity of forests of major forming species of Northern Eurasia (standard and reference materials). Federal Agency of Forest Management of the Russian Federation and International Institute for Applied Systems Analysis, Moscow, 803 pp

Shvidenko A, Schepaschenko D, Nilsson S, Bouloui Y (2007b) Semi-empirical models for assessment biological productivity of Northern Eurasian forests. Ecol Modelling 204:163–179

Shvidenko A, Schepaschenko D, McCallum I (2010) Can the uncertainty of full carbon accounting of forest ecosystems be made acceptable to policymakers? Clim Change 103:137–157. doi:10.1007/s10584-010-9918-2

Smith LC, McDonald GM, Velichko AA, Beilman DW, Borisova OK, Frey KE, Kremenetski KV, Sheng Y (2004) Siberian peatlands a net carbon sink and global methane source since the early Holocene. Science 303:353–356

Tansey KJ, Luckman AJ, Skinner L, Balzter H, Strozzi T, Wagner W (2004) Classification of forest volume resources using ERS tandem coherence and JERS backscatter data. Int J Rem Sens 25(4):751–768

Tarnocai C, Canadell JG, Schuur EAG, Kuhry P, Mazhitova G, Zimov S (2009) Soil organic carbon pools in the northern circumpolar permafrost region. Global Biogeochem Cycles 23: Gb2023. doi: 10.1029/2008GB003327

Thiel C, Drezet P, Weise C, Quegan S, Schmullius C (2006) Radar remote sensing for the delineation of forest cover maps and the detection of deforestation. Forestry 79(5):589–597. doi:10.1093/forestry/cpl036

Thiel CJ, Thiel C, Schmullius C (2009) Operational large-area forest monitoring in Siberia using ALOS PALSAR summer intensities and winter coherence. IEEE Trans Geosci Rem Sens 47 (12):3993–4000. doi:10.1109/TGRS.2009.2021469

Thurner M, Beer C, Santoro M, Carvalhais N, Wutzler T, Schepaschenko D, Shvidenko A, Kompter E, Ahrens B, Levick SR, Schmullius C (2014) Carbon stock and density of northern boreal and temperate forests. Global Ecol Biogeogr 23:297–310. doi:10.1111/geb.12125

Titlyanova AA, Romanova IP, Kosykh NP, Mironycheva-Tokareva NP (1999) Pattern and process in above-ground and below-ground components of grassland ecosystems. J Veg Sci 10:307–320

Townshend J, Justice C, Li W, Gurney C, McManus J (1991) Global landcover classification remote sensing: present capabilities and future possibilities. Rem Sens Env 35:243–255

Tyuremnov SN (1976) Torfyanye mestorozdenya. Nedra, Moscow, 488 pp (in Russian) (Тюремнов СН (1976) Торфяные месторождения, Москва, Недра, 488 с)

van Zyl J (1993) The effect of topography on radar scattering from vegetated areas. IEEE Trans Geosci Rem Sens 31(1):153–160

Vorobyov GI (1986) Encyclopedia of forest. In: Anuchin NA, Atrokhin VG, Vinogradov VN et al (eds) Sov Encyclopedia. Moscow, 631 pp (in Russian) (Воробьев ГИ (1986) Лесная энциклопедия//Анучин НА, Атрохин ВГ, Виноградов ВН и др., Москва, Сов Энциклопедия, 631с)

Wagner W, Luckman A, Vietmeier J, Tansey K, Balzter H, Schmullius C et al (2003) Large-scale mapping of boreal forest in Siberia using ERS tandem coherence and JERS backscatter data. Rem Sens Env 85:125–144

Wallén B (1992) Methods for studying below-ground production in mire ecosystems. Suo 43(4–5):155–162

Woodward FI (1987) Climate and plant distribution. Cambridge University Press, Cambridge 174 pp

Chapter 9
GIS and Remote Sensing Data-Based Methods for Monitoring Water and Soil Objects in the Steppe Biome of Western Siberia

Ivan D. Zolnikov, Nadezhda V. Glushkova, Elena N. Smolentseva, Darya A. Chupina, Denis V. Pchelnikov and Viktoriya A. Lyamina

Abstract This article considers the original techniques based on GIS and remote sensing data (RSD) that permit the short-term and retrospective monitoring of lakes (water bodies) and Solonchaks (soils). These components of landscapes in the steppe biome of Western Siberia (WS) are indicators of ecosystem dynamics. The testing methods were consistently performed first at key sites, then over large areas of the steppe biome. Four key sites were selected for space-time monitoring of lakes and solonchaks of the south of the West Siberian Plain based on a series of multi-temporal satellite image. The plots are located submeridionally to reflect the change in climatic conditions from north to south, especially moisture, and their influence on the parameters of the objects investigated. The monitoring of lakes and Solonchaks was conducted on the basis of various satellite systems depending on the research scale. On the medium scale, Landsat and Spot imagery were used over

I.D. Zolnikov (✉) · N.V. Glushkova · D.A. Chupina · D.V. Pchelnikov · V.A. Lyamina
V.S. Sobolev Institute of Geology and Mineralogy, Siberian Branch of Russian Academy of Sciences, Koptyuga Ave. 3, Novosibirsk 630090, Russia
e-mail: zol@igm.nsc.ru

N.V. Glushkova
e-mail: hope@igm.nsc.ru

D.A. Chupina
e-mail: dasha.chupina@gmail.com

D.V. Pchelnikov
e-mail: denmail@ngs.ru

V.A. Lyamina
e-mail: balandis@igm.nsc.ru

I.D. Zolnikov · N.V. Glushkova · V.A. Lyamina
Novosibirsk State University, 2 Pirogova Str, Novosibirsk 630090, Russia

E.N. Smolentseva
Siberian Branch, Institute of Soil Science and Agrochemistry (ISSA), Russian Academy of Sciences, Ac. Lavrentieva av. 8/2, Novosibirsk 630090, Russia
e-mail: parabraunerde@rambler.ru

L. Mueller et al. (eds.), *Novel Methods for Monitoring and Managing Land and Water Resources in Siberia*, Springer Water,
DOI 10.1007/978-3-319-24409-9_9

a chronological interval of 20 years. The dynamics of indicators were studied on a small scale on 250 m Moderate Resolution Spectroradiometer images (MODIS). Special indices were developed to delineate the Solonchaks and water bodies on MODIS images. For the purposes of retrospective monitoring, ancient lake basins (50–60 thousand years old) were mapped using a complex method based on a combination of various data: geologic and topographic maps and RSD. The morphometric analysis was carried out on the basis of a Shuttle radar topographic mission (SRTM) digital model. Lakes of the mid-1970s and 1980s were also digitized from topographic maps. The results of the delineation of lakes on MODIS images were also used. The obtained monitoring results allow us to plot the increasing degradation of lakes and an increasing area of Solonchaks in the steppe biome of WS and, as a consequence, an increase in aridity. Combining paleogeographic data based on the analysis of SRTM, topographic maps of the last century, and more recent satellite imagery led to the conclusion that large bodies of water in arid became fragmented and then gradually dried out. Thus, the dynamics of ecosystems based on the selected indicator objects (lakes and Solonchaks) can be described not by a model of gradual change from south to north, but by a model of focal and discrete transformations throughout the area studied, regardless of their position on the submeridional transect. However, the analysis of the spatial distribution of objects-indicators in the region allows us to assess the situation more adequately and go from simplified models of the spatiotemporal dynamics of ecosystems due to the aridization of climate to more complex ones. The obtained results show that the methods based on GIS analysis of digital elevation models in conjunction with the processing of remote sensing data on all scales are effective for retrospective and current monitoring of ecosystem dynamics under the influence of global environmental changes. Proven technologies are promising to construct historical and current maps showing the status of individual objects (e.g., water and soil) of ecosystem by means of medium- and small-scale ecological mapping and environmental monitoring.

Keywords Monitoring · Aridization · Digital elevation models · Solonchaks · Arid ecosystems · Multispectral images · Remote sensing · GIS

1 Introduction

The steppe biome of Western Siberia (WS) is an ecosystem on a high hierarchical level. The composition of this ecosystem includes steppe and forest steppe zones. This territory is characterized by subhumid and subarid conditions. The area of such lands, as estimated by Zolotokrylin and Cherenkova (2009) in WS, is 640,000 km^2. The main regional hydrological feature of this site is the large number of shallow salt lakes. The reasons for this phenomenon are the flat terrain, the poor development of the river network, and the drainless nature of the area. The water supply of

Fig. 1 Drying lake and the formation in its place Solonchaks. *Left* the lake, after the evaporation of the water (the view from Landsat ETM+ 2010), *Right* Solonchaks with salt crusts in the place of the lake (view from the ground, 2008)

lakes is largely dependent on atmospheric humidity. Due to the tendency toward aridity and extreme climate in WS (Grusa et al. 2001, 2008) changes are occurring in the structure of the landscape of the steppe biome. Ecosystems are responding the most rapidly to climate change, as most of the biome is an area of insufficient moisture. For example, in the 2001–2008 period, many of these lakes dried up in connection with droughts. In place of the dry lake, highly saline soil formed: Solonchaks (Fig. 1). This was confirmed by the data from ground observations which we conducted in 2001–2008 at one of the key sites (Meyer et al. 2008). As a result, the area occupied by Solonchaks is increasing.

Thus, the change in the area of the water surface of lakes and Solonchaks and their spatial dynamics are indicators of the landscape level. To assess and forecast changes in the parameters of these objects, both short-term and retrospective monitoring is necessary. Solving this problem requires methods of accounting the area occupied by Solonchaks and lakes in the steppe biome of WS, as well as algorithms for studying spatiotemporal changes in their areas. In addition, these methods should provide regional and global monitoring.

Earth remote sensing data (RSD) and GIS are becoming increasingly important in monitoring the status of regional and local ecosystems. At the turn of the century, RSD of low-spatial resolution (NOAA, MODIS, etc.) were widely used for analysis in arid steppe biome and adjacent areas. Thus, for example, in 1992–1999 the relationship between LST and NDVI (LST/NDVI) was explored in desert and semidesert ecosystems of Mongolia to monitor drought parameters (Bayarzhagal and Karnely 2005). To study the dynamics of foci of desertification, taking the example of the arid land of North-Western Caspian, the average correlation coefficient between albedo and surface temperature was used (Zolotokrylin and Titkova 2011). To detect climate extremes (droughts and waterlogging after heavy rainfall in arid lands), the satellite index SCEI was proposed (Zolotokrylin and Titkova 2012).

However, there are also examples of the conjugate use of space multispectral images of low (MODIS), medium (Landsat, Aster, Spot), and high (Quick-Bird) spatial resolution to assess ecosystem changes in the south of WS (Zolnikov et al. 2010, 2011; Lyamina et al. 2010).

As a result, we set ourselves a task: to develop methods based on the processing of remote sensing (RS-data) and GIS to assess and monitor the dynamics of water bodies (lakes) and soils (Solonchaks) in the steppe biome of WS. This approach is the most promising for monitoring large areas.

2 Objects

The territory is situated in the southern part of the West Siberian Plain between the Ob and Irtysh Rivers (Fig. 2). The distance from north to south is about 600 km, and from west to east 200 km. It is characterized by complex topographical and climatic conditions. For example, there is a diversity of landforms: crests and crest-like rises, ridges, interridge dips, and huge flat plains. The climate is continental and subarid/arid, with major transient and long-term fluctuations in temperature and humidity. The average air temperature ranges between 0.6 and 2.9 °C. The average annual precipitation varies from 230 to 350 mm on the territory investigated. The average long-term Thornthwaite's Index of Moisture (TMI) varies

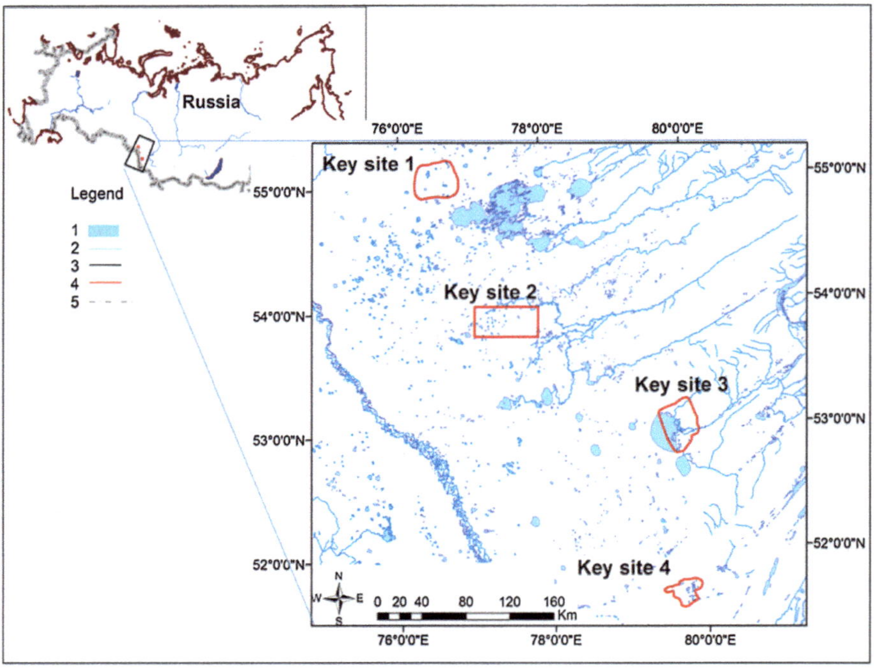

Fig. 2 The study area and the location of key sites. Legend: *1* lakes, *2* river, *3* image border, *4* key site borders, *5* administrative boundary

for the period 1936–2000 years between 0.20 and 0.65 (Zolotokrylin and Cherenkova 2009). The lithogenic basis of landscapes forms a thick layer of Quaternary sediments: subaerial loess and lake-alluvial deposits of various grain sizes. A characteristic geochemical feature of this part territory of the steppe biome of WS is the processes of continental salt accumulation in the soils, groundwater, and surface water of transit and accumulative landscapes. The natural vegetation in the zone is meadow, true, and dry steppes. However, in the modern era, such vegetation is preserved only in fragments on unplowed areas, since most of the area is under cultivation. The main background of intrazonal natural vegetation consists of salty steppe, halophytic vegetation, saline meadows, and reed marshes.

The soils studied are Solonchaks. They are the intrazonal soils of arid and semiarid regions that have a high concentration of soluble salts in solum at some time in the year. In the territory studied, Solonchaks are located in the lowest positions of a relief and form microbelts round lakes, and also arise in place of dried-up lakes. West Siberian Solonchaks are formed naturally. Salts precipitate from the ground water as a result of intense evaporation in hot summers, forming a salic horizon (IUSS 2014). In WS, the Solonchaks are most often found where the salt accumulation is greatest on the surface of the soil [external Solonchaks (IUSS 2014)]. A salt crust formation is very typical.

The importance of the selected objects (lakes and Solonchaks) for monitoring purposes is defined by more than just their landscape characteristics. We have also established the importance of their spectral characteristics for regional monitoring. In particular, it was previously found (Lamina et al. 2010), that at the transition from high- to low-spatial resolution RSD in the hierarchical sequence Quick-Bird—Landsat—MODIS, corresponding to the transition from the local to the regional and global ecosystem level, there is a generalization of ecosystems. This appears in the simplification of natural/territorial complexes from the complicated structure mosaic associations to spectrally homogeneous areas on satellite images. At the same time, lakes and salt marshes have spectral characteristics that vary with the smallest of all the generalizations.

Testing was consistently performed first at key sites, then over large areas of the steppe biome of WS. Four key sites were selected (Fig. 2) along the meridional transect between the rivers Ob and Irtysh for space-time monitoring of lakes and Solonchaks in the south of the West Siberian Plain using a series of multi-temporal satellite images (LANDSAT, SPOT). The sites are located submeridionally, to reflect the change in climatic conditions, especially moisture, from north to south. Each polygon has a database. It includes topographic and thematic raster maps (for example, data from the State Geological Survey), satellite imagery from various satellite systems, and the results of field studies. At each key site, a mixture of objects was studied: lakes and Solonchaks.

3 Methods: Land Cover Data and Its Processing

3.1 Method of Monitoring the Dynamics of the Lake and Solonchaks Area (Medium Scale)

We developed and tested various methods for estimating the spatial and temporal dynamics of soil and water facilities in the steppe biome of WS. Different satellite system data were used depending on the tasks and the scale of monitoring.

The method is based on the principle of mapping the selected objects on the basis of RSD through GIS. Initially, the method of mapping and monitoring the dynamics of the area of lakes and Solonchaks was worked out at Key Site 2 (Zolnikov et al. 2011). To evaluate change, the location and size of the water areas of lakes and Solonchaks areas were compared at different times and intervals. The technological scheme of the study consists of several stages (Fig. 3).

The paper used a series of multi-temporal images (Table 1) from the summer and autumn seasons: Landsat TM (1 August 1989), Landsat ETM+ (September 3, 2001), SPOT-2 (Aug. 28, 2008). When selecting the images, the time of shooting was taken into account, due to seasonal fluctuations in water level in the lakes. The lowest level, according to ground-based observations, is established in July, and remains until the middle of September.

Landsat sensor data has six channels in the visible, near, and short-wave infrared region with a spatial resolution of 30 m and a channel in the thermal range (resolution 120 m for TM and 60 m for the ETM+, respectively). The French satellite SPOT-2 has 3 channels in the visible and near-IR region with a spatial resolution of 20 m. Preliminary processing of satellite images included radiometric and

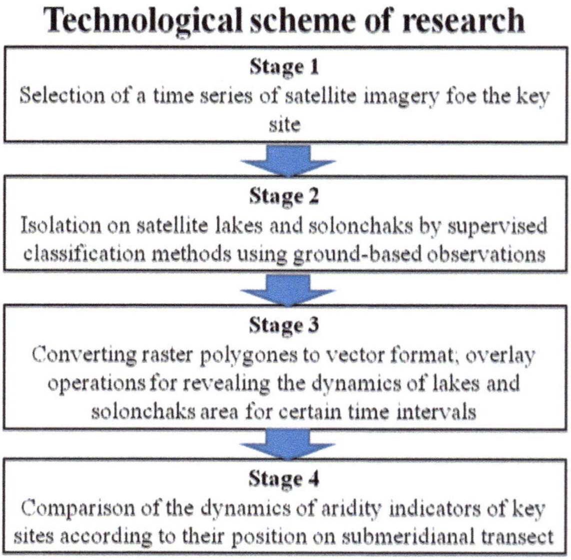

Fig. 3 Basic workflow of the proposed approach

Table 1 Key sites and Earth remote sensing data (RSD) for its monitoring

Key site		Coordinate of the key site center (latitude N/longitude E)	Area, km^2	RSD (day, month, year)
No.	Name			
1	Lake Chany	55.000/76.540	1115	Landsat TM (1.08.1989 and 7.09.2011), Landsat ETM+ (3.09.2001)
2	Lake Bagan	53.920/77.480	1566	Landsat TM (1.08.1989), Landsat ETM+ (3.09.2001), SPOT (28.08.2008)
3	Lake Kulundinskoe	53.043/79.780	1137	Landsat TM (4.09.1989 and 22.09.2010), Landsat ETM+ (4.08.2001)
4	Lake Malinovoe	51.696/79.332	468	Landsat TM (26.07.1989 and 22.09.2010), Landsat ETM+ (4.08.2001)

atmospheric correction. SPOT images were additionally linked to Landsat data. The satellite images were processed and classified out using the ENVI software package, while density schemes and mesh overlay operations with vector objects were calculated using the ArcGIS software package ArcGIS.

Solonchak areas and lakes were identified on Landsat images (5 channels) TM, ETM+, and SPOT (3 channels) using the maximum likelihood classification method at a reference site from pictures. The reference sites selected were Solonchaks, water bodies, arables, forests, natural herbaceous vegetation (hayfields and pastures). In total, 5 classes were allocated. The resulting image processing classes of Solonchaks and water bodies were converted from raster to vector layers (Fig. 4a). According to the vectors obtained, separate density grids were constructed for Solonchaks and lakes. It was necessary to compare images with different spatial resolution (Landsat: 28.5 m; SPOT: 20 m). Density grids were built with a 5 m cell size and a radius of 50 m. These parameters were chosen empirically so that the result reflected the most contrasting border vector objects obtained in the classification of images, and thus smoothed objects of the size of one pixel.

Next, we calculated the difference (Δ) of density grids of different years: (1) Landsat ETM+ 2001—Landsat TM 1989 (2) SPOT-February 2008—Landsat ETM+ 2001 (3) SPOT-February 2008—Landsat TM 1989. The resulting delta (Δ) was reclassified into grids with values of −1, 0, 1. The thresholds for reclassification were obtained by statistical analysis of the spatial distribution and density characteristics. The value 0 corresponds to the territory in which there was no change in the situation for the last period of time; a value of −1 corresponds to the area where mappable objects (lakes or salt flats) disappeared (the lake dried, or, for example, Solonchaks were flooded), and a value of 1 corresponds to the area where the mappable objects appeared (Fig. 4b). Further, the grids obtained were converted into vector format for further overlay operations. They are shown schematically in the flowchart in Fig. 3 (stages 1–3).

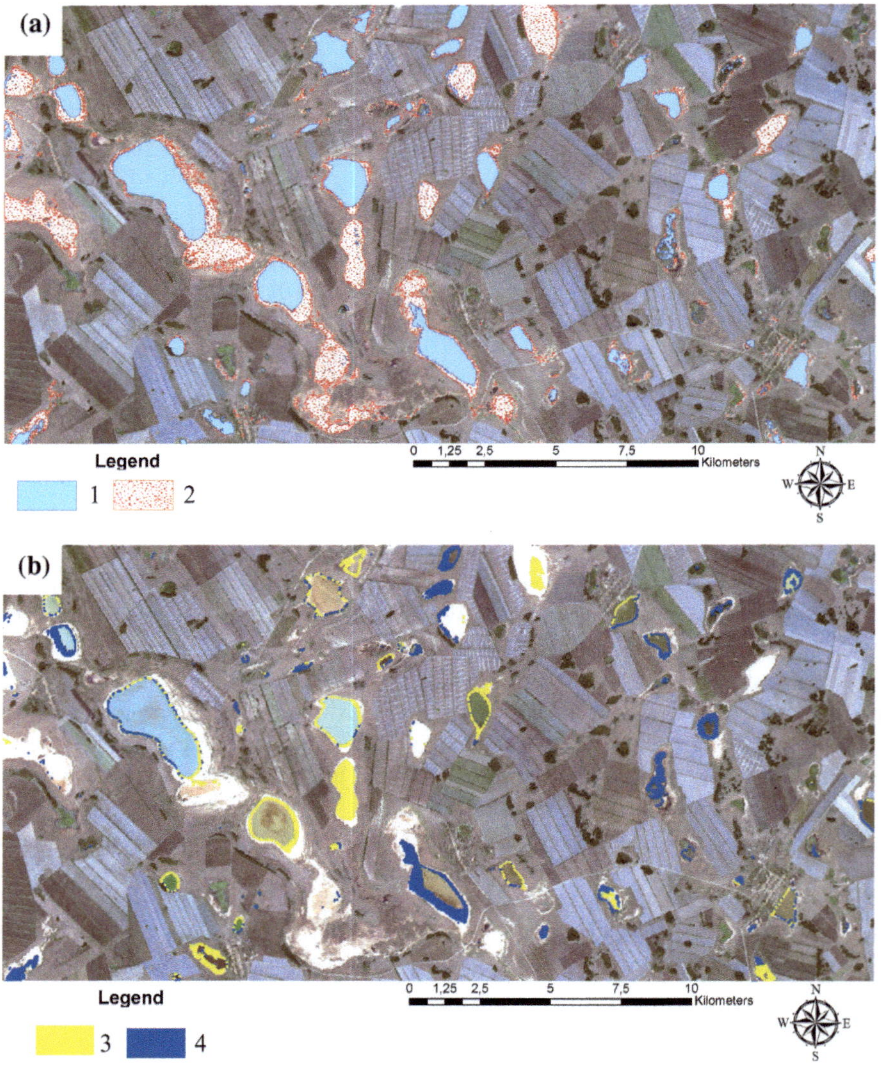

Fig. 4 The sequence of operations. **a** Receiving vector layers: *1* lakes, *2* Solonchaks; **b** Example of the dynamics of water bodies: *3* the disappearance of water, *4* the appearance of water (between 1989 and 2001)

Due to the need to investigate the spatiotemporal dynamics of the lake and Solonchaks area in the entire steppe biome of WS, the method described above was used in (extrapolated to) other key areas. It was also planned to establish the dependence of the dynamics on the location of the key area on the submeridional transect.

The paper used a series of multi-temporal images from the same season as well as key site No. 2 (Table 1). Then, using supervised classification methods (maximum likelihood method and the method of index images), lakes and Solonchaks were isolated on each of the images. The classification results are converted from raster to vector format. After that there were overlay operations with vector data. Furthermore, the area occupied by Solonchaks and lakes was calculated as a percentage of the total area of each polygon (for 100 % all key site areas were taken). This made it possible to compare the results with each other (Fig. 3, step 4) and to assess the dynamics of parameters depending on the position of the key site on the submeridional transect.

3.2 *Method for Monitoring Selected Indicators at the Regional Level*

For the transition from medium- to small-scale research, the distribution of the indicators on the territory of the steppe biome was analyzed on MODIS images at low-spatial resolution. This work used the products MODIS Level2G: MOD09GQ and MOD09GA. An analysis of the spectral responses revealed that the recognition of objects such as Solonchaks, soil, and water are best with red and near-infrared channels (channels 1 and 2 on the MODIS sensor with a spatial resolution of 250 m, spectral bands 620–670 nm and 841–876 nm, respectively). At the decomposition in two-dimensional (2D) feature space, the analyzed objects were most clearly differentiated, in comparison with other MODIS channels. The set of spectral values for the objects are represented by triangles (Fig. 5).

To highlight the Solonchaks soils, vegetation, and water, an application was written in IDL at the Novosibirsk GIS center. Applied to a 2D space (the algorithm works in the spectral domain on channels 1 and 2), a new axis was created

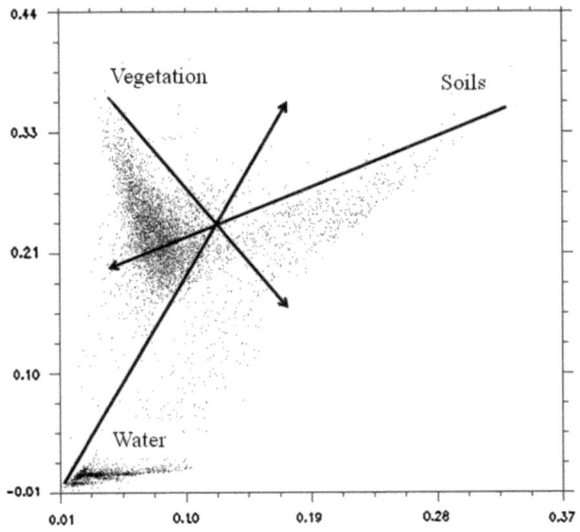

Fig. 5 The decomposition of MODIS images (2001) in two-dimensional feature space

responsible for conductivity, vegetation, and soil. One of the vertices of the triangle is taken as the origin. The medians of the vertices of the triangle are worked out. For each pixel in the image the projections on these medians are found. Distances from the vertices of the triangle to the points of projection represent the indexes reflecting the degree of similarity of the object on three different medians with either water, a saline soil or vegetation. The selection thresholds for these indices allowed us to identify the Solonchaks and lakes on MODIS images. Based on the selected objects, density grids were also built with the same cell size and window diameter as for the previous grids.

3.3 The Method of Retrospective Assessment of the Spatial Distribution of Lakes

The relief of the steppe biome of the West Siberian Plain between the rivers Ob and Irtysh keeps traces of the ancient eras of arid climate. Paleolake basins and lake-like expansions of the river valleys are widespread in the study area. As with modern lakes, paleolakes mainly filled due to the accumulation of surface water runoff, so the conditions of atmospheric moisture had a significant influence on their areas (Shnitnikov 1957).

The ratio of the areas of modern and ancient lakes surfaces allows us to indirectly evaluate the atmospheric moisture in the Paleozoic era retrospectively. The distribution of paleolakes can be investigated by the morphometric analysis of geological and geomorphological landscape basics, in this case the relief. The relief on the plain is relatively stable and is regarded by us as a static framework of modern ecosystems in the WS steppe biome. To identify the main characteristics of this framework, we used morphometric analysis of a digital elevation model (DEM). For studying reliefs this is currently a widely used DEM (Siart et al. 2009; Drăguţ and Eisank 2012; Bishop et al. 2003; Hengl and Reuter 2008).

For medium-scale mapping, it is advisable to use a DEM created based on Shuttle Radar Topographic Mission (SRTM). The SRTM carried out an interferometric radar survey of the globe in February 2000 from aboard the Space Shuttle Endeavor (Farr et al. 2000; Rabus et al. 2003). These data are available for free (http://srtm.csi.cgiar.org). SRTM DEM has good accuracy for medium-level research. However, there are methods to improve the quality of DEM and its accuracy (Tulu 2005). The original size of the cell for 55°–56° latitude is approximately 90 × 60 m. The cell size (or pixel size) is selected in each case individually, in accordance with the objectives of the study and the size of typical geomorphological sites. For the territory studied, DEM was aligned to the pixel size of 60 × 60 m.

Paleolake basins are clearly displayed on the DEM SRTM. For their allocation, the morphometric analysis of SRTM DEM method was used, as previously tested (Chupina et al. 2012; Chupina 2014). The technique can be described briefly as follows. The preliminary stage involves the study of geological maps, literary sources on the morphology, origin, and age of the study area topography. Based on the analysis of these data a set of landforms and morphocomplexes (typical combination of geomorphological objects) is allocated, most commonly within the study area. This is necessary for automated mapping, because the method is aimed at the recognition of already known forms and morphocomplexes in morphometric character space. Morphometric indices are constructed using DEM. Then an analysis is carried out to find which ones are more effective for the isolation of landforms or morphocomplexes.

The objectives of this study did not include mapping all the landforms and morphocomplexes prevalent in the area. We needed to allocate paleolake basins on DTM. To isolate them, a number of morphometric parameters were calculated using DEM: slope, aspect, the mean slope, the range of slope, relative elevation heights. Most effectively, paleolake basins were allocated on the basis of the statistical distribution of the values of the relative elevation of relief. First, the average height of the area and the overall elevation trend were calculated. Technically, this is implemented using the "Focal statistics" procedure in ArcGIS 10.2. For the trend to have a regional scale, the moving window must be large enough, for example, to eliminate the influence of macrorelief. In most cases, the radius of the moving window was 10 km. Then, the overall trend is subtracted from the digital terrain model using a Raster Calculator. This indicator clearly highlighted the positive and negative forms of the relief. According to this parameter, a range of values were selected corresponding to the paleolake basin. The resulting objects were converted to vector format.

In addition, to assess the dynamics of the lake area in retrospect, the lake was digitized from topographic maps of the mid-twentieth century (1970s, 1980s), on a scale of 1: 200,000. After the vector objects were obtained both from the mid-twentieth century lakes, and paleolake basins, the window size was selected to calculate the density distribution of these forms. The calculation of this parameter is implemented in the procedure "Density" in the Spatial Analyst module in ArcGIS 10.2. Due to the fact that there was zoning of large territories, and the paleolake surface can reach a diameter of 10 km, the radius of the moving window was chosen to be 25 km. In the window of the specified size, density maps were built showing the area of distribution of these forms. The indicators are normalized to 100% filling of the window and divided into intervals based on the analysis of histograms of the distribution of values.

For a comprehensive assessment of the situation, we compared the prevalence distribution of paleolake basins (for the last 50–60 thousand years), obtained according to DEM, with the distribution of modern lakes, digitized from topographic maps of the mid-twentieth century, and then with lakes obtained from MODIS data using the method described in Sect. 3.2. This sequential comparison allowed us to reconstruct moisture conditions as a short-term and long-term retrospective.

4 Results and Discussion

4.1 The Dynamics of the Area of Lakes and Solonchaks in Key Sites and for the Whole Submeridional Transect

The results of quantitative estimations of lake and Solonchaks areas are shown in Table 2.

Key Site 1 is characterized by the relatively small area of Solonchaks (<1 %). In Key Site 2 Solonchaks occupy from 3.00 to 4.60% in various years. The variation in the Solonchaks area in Key Site 3 at different years is 14.55–20.89 %, while it is 17.59–22.76 % in Key Site 4. These data show an increase in the relative area of Solonchaks in key sites from north to south. The relative areas of lakes in key sites differ slightly. The biggest area of lakes was observed in 1989 in Key Site 3. The southernmost Key Site 4 is characterized by a very high-specific area of Solonchaks (22.76 % in 2010) and the smallest area of lakes (0.96 %).

To assess the short- and long-term dynamics of the studied objects' area the base year (year of comparison) 1989 was chosen. The dynamics of the areas is expressed as the difference between the area of objects measured in certain years, where data from the subsequent (later) year is subtracted from the previous year, (earlier). For Key Area 1, the time interval from 1989–2001 was characterized by an increase in the area of Solonchaks and a decrease in the area of lakes (Fig. 6). The interval from 2001–2011 shows an increase in the area of the Solonchaks and lakes. In general, in Site 1 for the period of 22 years (1989–2011) we observed a trend of reduction in the lakes' area and an increase in the area of Solonchaks. Key Site 2 differs in terms of the dynamics of indicators for the year 2001 from Site 1, but is similar to Site 4. That is, at Sites 2 and 4 the year 2001 revealed the expansion of the lakes and thus a reduction in the area of Solonchaks. At Site 3, near Lake Kulundinskoye, a short peak of humidity observed in 2001 for Sites 2 and 4 is not detected. On the contrary, there is a constant tendency for lakes to degrade and a slight increase in

Table 2 Areas of Solonchaks and lakes in key areas of the submeridional transect by assessment on the basis of RSD

Year	Solonchak area		Lake area		Year	Solonchak area		Lake area	
	km^2	(%)	km^2	(%)		km^2	(%)	km^2	(%)
Key site 1					Key site 3				
1989	2.41	0.22	44.55	4.00	1989	165.40	14.55	87.20	7.67
2001	5.28	0.47	29.09	2.61	2001	237.50	20.89	37.30	3.28
2011	6.48	0.58	39.67	3.56	2010	172.00	15.13	22.30	1.96
Key site 2					Key site 4				
1989	71.99	4.60	38.38	2.45	1989	96.10	20.53	14.80	3.16
2001	47.03	3.00	45.13	2.88	2001	82.30	17.59	35.00	7.48
2008	63.70	4.07	36.40	2.32	2010	106.50	22.76	4.50	0.96

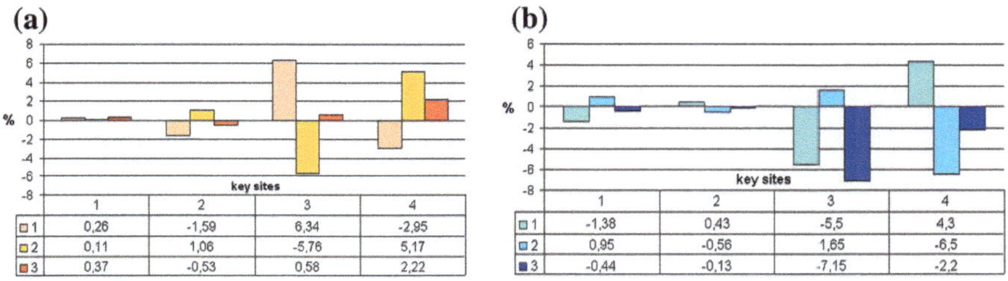

Fig. 6 The dynamics of the relative area of Solonchaks (**a**) and lakes (**b**) at different timesteps. Legend: *1–3* dates: *1* 2001 versus 1989, *2* 2011 versus 2001 (Key Site 1) or 2008 versus 2001 (Key Site 2) or 2010 versus 2001 (Key Site 3 and 4), *3* 2011 versus 1989 (Key Site 1) or 2008 versus 1989 (Key Site 2) or 2010 versus 1989 (Key Site 3 and 4). Read *commas* as points. The *minus* sign means a decrease in the area, a *plus* sign means that it increases

the area of Solonchaks. The overall dynamics of the studied parameters over a longer period (19–22 years) show a trend of reduction in the lakes' area in all key sites and an increase in the area of Solonchaks in three of the four sites. In addition, the amplitude of the specific area of lakes and Solonchaks increases several times from north to south. Therefore, the most contrasting reactions of local ecosystems to climate change have been recorded in the more southern areas.

However, this general trend is complicated by the short-period climatic cycles and there is a local specificity: breaking trend patterns of the aridization processes. In particular, asynchronous dynamics of aridity indicators were revealed in Key Site 3 with respect to the other three sites.

Thus, the study of four sites located from north to south led to the preliminary conclusion that the regularities in the aridization processes' appearance have a submeridional character. However, it is the largest lake, Kulundinskoye, which responds to climate change differently and violates the regularity on the transect. Apparently, this is due to the fact that due to its size (or geographic specificity) it is characterized especially by local climatic conditions.

4.2 *Results of Retrospective Assessment of the Spatial Distribution of Lakes*

As a result of the morphometric analysis, the relief map of paleolakes was drawn within the steppe biome of WS. This map showed that lakes occupy larger areas than today, and are unevenly distributed. This uneven distribution is caused by local specificity of the geological and geomorphological landscape basis and of local-regional atmospheric moisture transfer. The results of the comparative analysis of the lakes' distribution at different times in the steppe biome of WS generally confirmed that they were gradually drying up and disappearing.

Allocated on the basis of morphometric analysis, paleolake basins allow us to estimate the maximum lake distribution over the territory during periods of moisture.

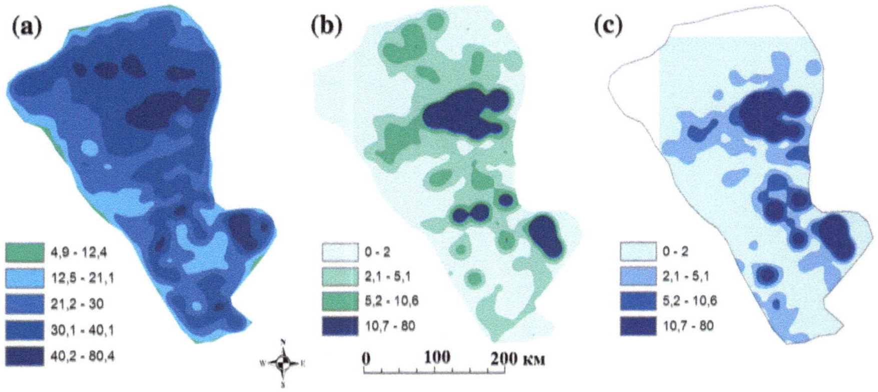

Fig. 7 Scheme of density distribution * of objects. **a** Lake surfaces, derived from morphometric analysis of DEM; **b** lakes, vectorized from topographic maps (1970s–1980s); **c** lakes, derived from MODIS images (2001); (*Legend is shown in the percentage of moving window area (of a 25 km radius), read *commas* as points.)

Ancient lake surfaces are most common in the northern and central parts of the study region (Fig. 7a), where they occupy 30–40 % of the territory, with some (Lake Chany) occupying 40–80 %. The distribution of lake surfaces in the southern part of the region is more complex. Here, there is a plain with a frame of ancient lake surfaces at its edges, and fewer lakes in its central part (12.5–21 %) (Fig. 6a). The resulting paleogeographic scheme agrees well with the maps of Quaternary deposits of the state geological mapping at a scale of 1: 200,000, where the paleolake surface is shown as lake and marsh sediments of the Late Pleistocene-Holocene.

Zoning in the degree of modern lake density revealed that paleolake areas exceed the area of modern lakes (Fig. 7b, c). The values of the density distribution are not comparable, because the basic range of modern lakes varies from 0 to 10 % (on the area of the moving window of 25 km in radius), while the old lake surfaces throughout occupy more than 12 %, and are concentrated mainly in the range from 20 to 40 %. Areas of increased modern lake density have formed only around the large lakes, such as Chany and Kulunda and some others. Thus, the submeridional trend was not observed for lake density. It has a mosaic distribution across the study area, and there is some unevenness in the direction from east to west, related to the ancient stream flow, and the continuation of their axes.

5 Conclusion

Comparative analysis of density maps revealed that the regional pattern of the dynamics of the lake and Solonchaks area in the steppe biome of WS is not limited to the gradual increase in land aridization from south to north in the framework of a uniform trend of meridional direction. However, the analysis of the spatial distribution of indicator objects in the region allows us to more adequately assess the situation and move from simplified to more complex models of spatiotemporal dynamics of ecosystems due to the climate aridization.

Combining paleogeographic data based on the SRTM analysis, topographic maps of the middle of the last century and more recent RSD led to the conclusion that during the aridization of territories, large bodies of water were fragmented, and then gradually dried out. Thus, the dynamics of ecosystems based on selected indicators can be described more adequately not by a model of gradual change from south to north, but a model of focal and discrete transformations over almost the entire territory, regardless of the position in the submeridional transect.

The studies conducted have shown the expediency of the integrated use of raster image processing methods and grid modeling using ArcGis for morphocomplex mapping. The use of relative morphometric parameters calculated for certain neighborhoods by moving window allows us to move from mapping landforms and their elements in favor of mapping terrain types that are a natural combination of typical geomorphological objects.

The results obtained show that the GIS analysis of digital elevation models in conjunction with the processing of RSD of all scales are effective methods of retrospective and current monitoring of ecosystem dynamics under the influence of global environmental and climate changes. Proven technology promises to build retrospective and current maps showing the status of individual objects (such as water and soils) in ecosystems with medium- and small-scale ecological mapping and environmental monitoring.

Acknowledgment This work was funded by project 4.13., subprogram "The problems of desertification in Central Asia" program RAS No. 4.6.

References

Bayarzhagal Yu, Karnely A (2005) Indexes NDVI and LST proposed NOAA-AVHRR for revealing drought in Mongolia. Arid Syst 11(6–27):73–782

Bishop MP, Shroder JF Jr, Colby JD (2003) Remote sensing and geomorphometry for studying relief production in high mountains. Geomorphology 55:345–361

Chupina DA (2014) Automatic selection of forms and complexes of the relief on the basis of morphometric GIS analysis (for example Vengerovsky district of Novosibirsk Region). Geomorphology 3:43–50

Chupina DA, Zolnikov ID, Lyamina VA (2012) GIS mapping of morfocomplexes of Chuya basin based on morphometric characteristics. Geol Min Resour Siberia 2:96–103

Drăguţ L, Eisank C (2012) Automated object-based classification of topography from SRTM data. Geomorphology 141:21–33

Farr TG, Hensley S, Rodriguez E, Martin J, Kobrick M (2000) The shuttle radar topography mission. In: Proceedings of the CEOS SAR workshop, 26–29 October 1999. ESA SP-450, CNES, Toulouse, pp 361–363

Grusa GV, Bardin MY, Rankova EY, Rocheva EV, Sokolov YY, Samokhina OF, Platov TV (2001) Of changes in air temperature and precipitation on the territory of Russia in the twentieth century. In: Condition and complex monitoring of the environment and climate.

Limits changes (Об изменениях температуры воздуха и атмосферных осадков на территории России в XX веке. Состояние и комплексный мониторинг природной среды и климата. Пределы изменений:). Nauka, Moscow, pp18–39 (in Russian)

Grusa GV, Zaitsev AS, Karol IL, Kattsov VM, Kobysheva NV, Myaleshka VP, Meshcherskaya AV, Mirvis VM, Resetnicov AI, Sporyshev PV et al (2008) Assessment report on climate change and its consequences on the territory of the Russian Federation. (Оценочный доклад об изменениях климата и их последствий на территории Российской Федерации). Climate change. vol 1 228 pp (in Russian)

Hengl T, Reuter HI (eds) (2008) Geomorphometry: concepts, software, applications. developments in soil science, vol. 33. Elsevier, Amsterdam, 772 pp

IUSS (2014) World reference base for soil resources (2014), World soil resources report No 106, 181 p. ISBN 978-92-5-108369-7, http://www.fao.org/3/a-i3794e.pdf, Accessed 14 Nov 2014

Lyamina VA, Koroljuk AY, Zolnikov ID, Smolentsev BA, Laschinsky NN (2010) Generalization of landscape conditions in the spectral characteristics of satellite images of different spatial resolution. Stud Earth Space 4:77–84

Meyer BC, Schreiner V, Smolentseva EN, Smolentsev BA (2008) Indicators of desertification in the Kulunda Steppe in the south of Western Siberia. Arch Agron Soil Sci 54(6):585–603

Rabus B, Eineder M, Roth A, Bamler R (2003) The shuttle radar topography mission—a new class of digital elevation models acquired by spaceborne radar. ISPRS (International Society for Photogrammetry and Remote Sensing) J Photogrammetry Remote Sens 57: 241–262

Shnitnikov AV (1957) Lakes of Western Siberia and Northern Kazakhstan and centuries of variability of steppes humidity (Озёра Западной Сибири и Северного Казахстана и многовековая изменчивость увлажнённости степей). Trudy Lab. ozerovedeniya Academy of Sciences USSR, vol 5. 112 p (in Russian)

Siart C, Bubenzer O, Eitel B (2009) Combining digital elevation data (SRTM/ASTER), high resolution satellite imagery (Quickbird) and GIS for geomorphological mapping: a multi-component case study on Mediterranean karst in Central Crete. Geomorphology 112:106–121

Tulu MD (2005) SRTM DEM suitability in runoff studies. ENSCHEDE, The Netherlands ITC. MSc., 87 p. http://www.itc.nl/library/papers_2005/msc/wrem/mesay.pdf, Accessed 01 Apr 2015

Zolnikov ID, Lyamina VA, Koroljuk AY (2010) Complex technology mapping and monitoring of heterogeneous vegetation cover. Geogr Nat Resour 2:126–131

Zolnikov ID, Glushkova NV, Lyamina VA, Smolentceva EN, Koroljuk AYu, Bezuglova NN, Zinchenko GS, Puzanov AV (2011) Indication of the dynamics of natural-territorial complexes of Western Siberia south in connection with climate change. Geogr Nat Resour 2:155–160

Zolotokrylin AN, Cherenkova EA (2009) The area of the dry plain lands of Russia. Arid Ecosyst 15 No. 1(37):5–12

Zolotokrylin AN, Titkova TB (2011) The trend of desertification of north-west Caspian by MODIS-data. Mod Probl Remote Sens Earth Space 8(2):217–225

Zolotokrylin AN, Titkova TB (2012) Satellite index of climate extremes of drylands. Arid Ecosyst 18 No. 4(53):5–12

Chapter 10
Significant Siberian Vegetation Change is Inevitably Brought on by the Changing Climate

Nadezhda M. Tchebakova, Elena I. Parfenova and Amber J. Soja

Abstract The redistribution of terrestrial ecosystems and individual species is predicted to be profound under Global Climate Model simulations. We modeled the progression of potential vegetation and forest types in Siberia by the end of the twenty-first century by coupling large-scale bioclimatic models of vegetation zones and major conifer species with climatic variables and permafrost using the B1 and A2 Hadley Centre HadCM3 climate change scenarios. In the projected warmer and dryer climate, Siberian taiga forests are predicted to dramatically decrease and shift to the northeast, and forest–steppe, steppe, and novel temperate broadleaf forests are predicted to dominate most of Siberia by 2090. The permafrost should not retreat sufficiently to provide favorable habitats for dark (*Pinus sibiric*, *Abies sibirica*, and *Picea obovata*) taiga, and the permafrost-tolerant *L. dahurica* taiga should remain the dominant forest type in many current permafrost-lain areas. Water stress and fire-tolerant tree species (*Pinus sylvestris* and *Larix* spp.) should have an increased advantage over moisture-loving tree species (*P. sibirica, A. sibirica*, and *P. obovata*) in a new climate. Accumulated surface fuel loads due to increased tree mortality from drought, insects, and other factors, especially at the southern forest border and in the Siberian interior (Yakutia), together with an increase in severe fire weather, should also lead to increases in large, high-severity fires that are expected to facilitate vegetation progression toward a new equilibrium with the climate. Adaptation of the forest types and tree species to climate change in the south may be based on the genetic means of individual species and human willingness to aid

N.M. Tchebakova (✉) · E.I. Parfenova
V.N. Sukachev Institute of Forest, Russian Academy of Sciences, Siberian Branch (SIF SB RAS), Akademgorodok No. 50/28, Krasnoyarsk 660036, Russia
e-mail: ncheby@ksc.krasn.ru

E.I. Parfenova
e-mail: lyeti@ksc.krasn.ru

A.J. Soja
NASA Langley Research Center, 21 Langley Boulevard, Mail Stop 420, Hampton, VA 23681-2199, USA
e-mail: Amber.J.Soja@nasa.gov

L. Mueller et al. (eds.), *Novel Methods for Monitoring and Managing Land and Water Resources in Siberia*, Springer Water,
DOI 10.1007/978-3-319-24409-9_10

migration, perhaps by seeding. Additionally, useful and viable crops could be established in agricultural lands instead of failing forests.

Keywords Vegetation · Siberia · Climate change · Scenario

1 Introduction

The climate has changed abruptly in the upper latitudes of the Northern Hemisphere over the recent decades, and this is where the largest increases in climate change are expected to occur (IPCC 2013). Observed temperature records across Siberia during the last decades of the twentieth century demonstrate the largest annual temperature anomalies of up to 3.5 °C for 100 years in the East Siberian interior: Transbaikalia and Yakutia (Gruza and Rankova 2004; Maximov 2007). In Western and Central Siberia, annual temperature increases have been noted up to 1.5–2 °C (Bazhenova and Martyanova 2003; Ponomarev et al. 2005; Tchebakova et al. 2011).

General Circulation Model (GCM) projections suggest that there will be a significant climate change in Siberia during the twenty-first century. Consequently, the vegetation is expected to shift, particularly at the margins (southern, northern and altitudinal ecosystem boundaries), and these shifts could be abrupt, particularly if they are driven by climatic extremes (droughts, storms, intense fire weather, and infestations), which are also predicted (IPCC 2007, 2013). Together, the amount and rate of change are expected to have deep impacts on Siberian forests, their composition, structure, and productivity.

Forest shifts northward will leave behind lands potentially suitable for agriculture. Agriculture in Siberia and Russia as a whole may benefit from climate change (Izrael and Sirotenko 2003; Tchebakova and Parfenova 2013). In European Russia, climate change over the last three decades, 1975–2005, has favored agriculture in 70 % of Russia's federal subjects (administrative units), producing 85 % of its trade grain. However, during this same time period, West Siberia and East Siberia experienced crop decreases due to climate drying across arable lands (Sirotenko et al. 2007).

Our goals are to (1) model the potential effect of two contrasting climate change scenarios on the progression of potential forest cover and major forest-forming tree species in Siberia during the twenty-first century; (2) predict locations ("hot spots") where possible changes in climate would create new habitats for novel forest; and (3) evaluate the potential agricultural change from pre-2010 to the end of the century using HadCM3 climate change projections. This work is important to understand where vegetation change is expected to occur, particularly agricultural lands that offer human sustenance, and also to determine potential ecosystem–biodiversity endangerment and extinction threats.

The study area is Siberia within the window, 60–140°E and 48–72°N, bounded by the longitudes 60°E along the Ural mountains in the west and by 140°E in the

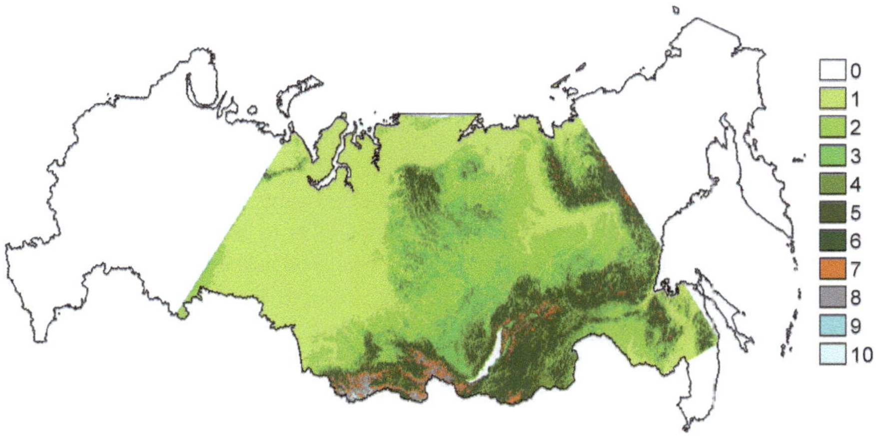

Fig. 1 Study area is in Siberia between 60–140°E and 48–72°N. In the center: Krasnoyarsk, Republics of Khakassia and Tyva; the latter two are in the south, north of the Mongolian border. Elevation: *1* <200 m; *2* 200–400 m; *3* 400–600 m; *4* 600–800 m; *5* 800–1000 m; *6* 1000–1500 m; *7* 1500–2000 m; *8* 2000–2500 m; *9* 2500–3000 m; *10* >3000 m

east, mainly beyond the monsoon influence and by the latitudes 72°N along the Arctic seashore in the north and by 48°N in the south, mainly along the Russian border (Fig. 1).

2 Methods

Climate change scenario simulations are used to initialize a bioclimatic vegetation model to assess potential climate-driven vegetation change. To provide a range of warming by the end of the twenty-first century, we apply recent HadCM3 GCM projections from the Hadley Center (IPCC 2007) that suggest a July temperature increase of 4–6 °C and a January temperature increase of up to 9 °C in the B1 Special Report on Emission Scenarios (SRES), and 6–7 °C and >9 °C in the A2 SRES scenarios, respectively. This amount of change is predicted to occur for 100 years, meaning that it would occur at rapid rate of up to 0.8 °C per decade.

In our forest change simulations, we used our large-scale Siberian BioClimatic zonal vegetation Model (SiBCliM, Tchebakova et al. 2009) and our tree species model (Tchebakova and Parfenova 2012), called TreeBCliM here, to predict vegetation zones (zonobiomes) and major forest-forming tree species, respectively. Both models are static "envelope-type" models (Box 1981) that determine unique climatic limits for a vegetation class and a tree species from three bioclimatic indices: growing degree days (GDDs) above 5 °C, representing plant requirements for warmth; negative degree days (NDDs) below 0 °C, characterizing plant cold tolerance, and an annual moisture index (AMI) characterizing plant drought resistance. The AMI is the ratio of GDDs above 5 °C to annual precipitation.

The borders between vegetation zones were shown to correspond well to specific values of ratios between heat and available water (Dokuchaev 1899). Different expressions of those ratios have been developed and used since then: deficit to potential evaporation (Thornthwaite 1948), precipitation to potential evaporation for a given period (Ivanov 1954), potential evaporation to annual precipitation (Budyko 1974), actual to potential evaporation (Prentice et al. 1992), the difference between annual precipitation and potential evapotranspiration (Hogg 1997), actual evaporation and deficit (Stephenson 1998), and GDDs to annual or summer precipitation (Rehfeldt et al. 1999). However, both potential and actual evaporation are not usually recorded in most weather stations. Instead, they are calculated from temperature, precipitation, air humidity, sunshine or cloudiness, albedo, etc. These empirical relationships are regional and may produce large estimation errors in specific habitats over a vast and complex terrain such as Siberia. Thus, we used directly measured temperature and precipitation, arriving at a straightforward AMI.

Additionally, the presence/absence of continuous permafrost was explicitly included in the models as limiting the forests and tree species distribution in Siberia. Permafrost occurs across 80 % of Siberia, and it is an important ecological factor controlling the vegetation distribution across Siberia. Forests develop in this region only because additional summer moisture from melting permafrost provides the moisture necessary to sustain tree growth where otherwise the vegetation would be steppe or semidesert (Shumilova 1962). Permafrost is not included in most models, and its absence can result in unrealistic and inaccurate predictions. Superimposing the tree species range (Pozdnyakov 1993) and an Active Layer Depth (ALD), the depth to which permafrost thaws (Malevsky-Malevich et al. 2001), we empirically defined ALD. Depths that are equal to $\sim$2 m and greater allow conifers to thrive, and only one conifer, *Larix dahurica* (*L. gmelini* + *L. cajanderii*), can survive under an ALD of < $\sim$2 m (Pozdnyakov 1993; Abaimov et al. 2002).

Siberian zonal (climax) vegetation was classified by geobotanists as follows (Fig. 2): 1. Tundra; 2. Forest–tundra; Northern Taiga: 3. light leaf (*Larix spp.* and *Pinus sylvestris*) and 4. dark leaf (*Pinus sibirica, Picea obovata,* and *Abies sibirica*); Middle Taiga: 5. light leaf and 6. dark leaf; Southern Taiga and Subtaiga: 7. light leaf and 8. dark leaf including highly productive lowland montane "chern" taiga, an analogue to European nemoral forests with tall herbs and ferns; 9. Forest–steppe; 10. Steppe; and 11. Semidesert (Shumilova 1962; Nazimova et al. 2006). These 11 contemporary boreal vegetation classes and 4 temperate classes (steppe, semidesert, broadleaf forest, and forest–steppe) are important in future Siberian climates and so were included in SiBCliM. These 14 classes were aggregated in 9 zonal classes analogous to zonobiomes (Walter 1985) and used in further analyses: 1. Tundra; 2. Forest–tundra; 3. Dark taiga; 4. Light taiga; 5. Mixed conifer broadleaf; 6. Broadleaf; 7. Steppe; 8. Semidesert; 9. Forest–steppe (Table 1 and Fig. 4a–b).

Envelope limits for each vegetation class in SiBCliM were delineated from the classification of 150 Siberian weather stations using the three climatic indices (Fig. 2). The cold parameter, negative degree days (NDD_0), equal to 3500–4000 °C, tends to separate dark- and light-leaf conifers, correspondingly, with milder and severe winters. Climate envelopes of GDD_5, NDD_0, and AMI for each conifer were

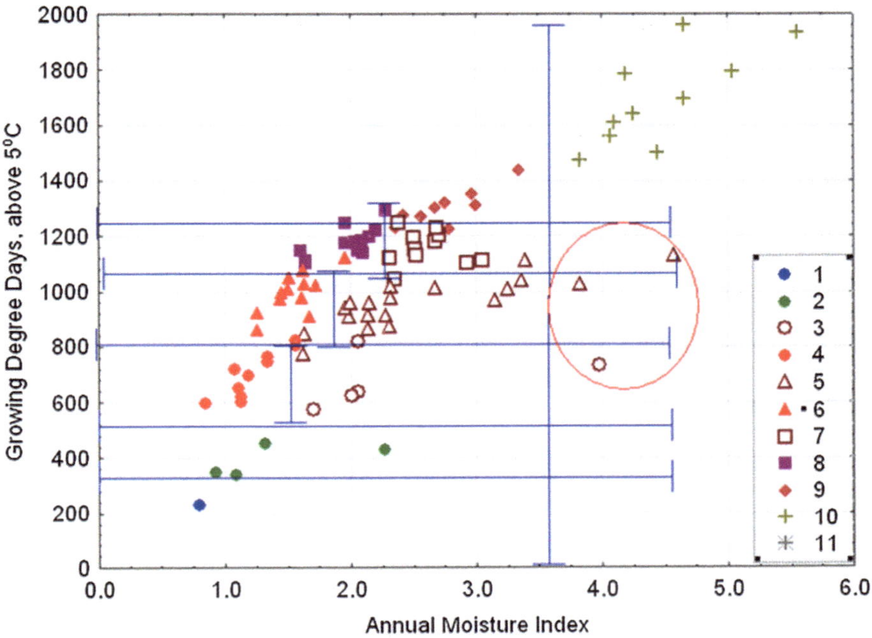

Fig. 2 Classification of Siberian biomes in heat (GDD_5) and moisture (AMI) space. Lines show envelope limits for each zonal climax biome. The red circle shows the permafrost zone in which the forest survives only due to additional moisture from thawing permafrost in the summer. Vegetation key: *1* Tundra; *2* Forest-tundra; Northern Taiga; *3* light leaf and *4* dark leaf; Middle Taiga; *5* light leaf and *6* dark leaf; Southern Taiga and Subtaiga; *7* light leaf and *8* dark leaf; *9* Forest-steppe; *10* Steppe; and *11* Semidesert

Table 1 Potential biome area (%) that shows the change in ecozones from the baseline 1960–1990 to the 2090 Hadley climate scenarios

Biome	1960–1990	B1 2090	A2 2090
Tundra	5.8	1.8	0.6
Forest-tundra	19.9	2.3	1.0
Dark taiga	15.0	7.6	1.3
Light taiga	46.8	32.4	10.6
Mixed conifer-broadleaf	0.6	15.3	5.0
Broadleaf	0.1	18.8	42.9
Forest-steppe	0.8	4.1	4.9
Steppe	4.5	7.5	19.4
Semidesert	6.3	10.2	14.4

found using gene–ecological studies (data from about 250 provenances for *P. sylvestris* and 250 for *Larix spp.*, Rehfeldt et al. 1999, 2002), vegetative climatic limits using species range maps, and various publications (Sokolov et al. 1977;

Popov 1980; Polikarpov et al. 1986; Shugart et al. 1991; Tchebakova et al. 2003; Tchebakova and Parfenova 2006).

Each forest zonobiome and forest-forming conifer tree distribution from 1960–1990 to 2090 was modeled by coupling SiBCliM and TreeBCliM with bioclimatic indices and the permafrost distributions calculated for the base period and HadCM3 2090 climates. The current permafrost border was approximated by the regression relating to this layer on the map by Malevsky-Malevich et al. (2001) and our three climatic indices (R^2 = 0.7). The future permafrost border was approximated from Stefan's theoretical formula based on temperature change (Dostavalov and Kudryavtsev 1967). Current climatic variables and indices were mapped using data taken from about 1200 Siberian weather stations across the study area using Hutchinson's (2000) thin plate splines on 1-km Digital Elevation Model (DEM) grids (Fig. 3).

Future bioclimatic indices were calculated using climatic anomalies for 2090 derived from two climate change scenarios: the HadCM3 A2 (harsh) and B1 (moderate), taken from the Hadley Centre in the United Kingdom and based on the SRES (IPCC 2007).

Kappa (K) statistics (Monserud and Leemans 1992) were used to evaluate both models' performance. The modeled contemporary climate zonobiome distribution was compared to the actual vegetation map (Isachenko et al. 1988), and the modeled tree species distribution was compared to the "Forests of the USSR" map

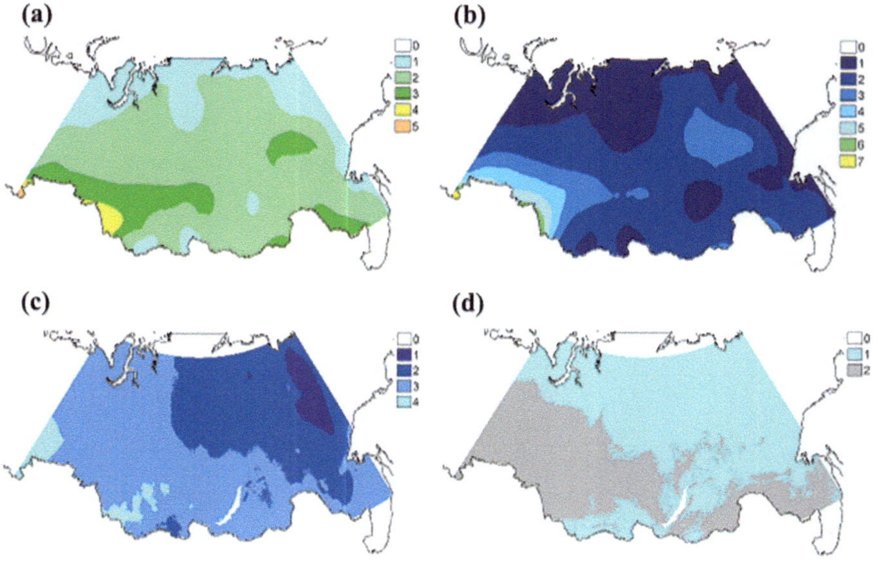

Fig. 3 Current climate layers. **a** Growing degree-days 5 °C (1. <500, 2. 500–1000, 3. 1000–1500, 4. 1500–2000, 5. 2000–2500); **b** AMI (1. <1, 2. 1.1–2, 3. 2.1–3, 4. 3.1–4, 5. 4.1–5, 6. 5.1–6, 7. >6); **c** Negative degree-days 0 °C (1. <(−6000), 2. (−6000)–(−4000), 3. (−4000)–(−2000), 4. > (−2000); **d** Permafrost (*light blue*)

(Isaev 1990) across Siberia (Tchebakova et al. 2009, 2012) and Russia (Shuman et al. 2014). The comparison between the actual vegetation maps (Fig. 4a) and our modeled (Fig. 4b) maps showed the overall agreement was "fair" to "good" (K = 0.53), and disparate vegetation class agreements showed matches across Siberia ranging from "excellent" (K > 0.7) to "poor" (K < 0.4) (Tchebakova et al. 2009). Thus, K statistics showed that SiBCliM performed at a fair level when modeling Siberian vegetation. A comparison of real and modeled tree species distributions in central Siberia showed a fair agreement. Thus, 73 % of the real *P. sibirica* range, 34 % of the *A. sibirica* range, 64 % of the *P. sylvestris* range, and 46 % of the *Larix spp.* range were within their climatic potential ranges (Tchebakova and Parfenova 2012).

3 Results

Vegetation change. The model simulations indicated that Siberian vegetation would be severely altered by 2090: A moderate change in vegetation is predicted from the B1 scenario, but dramatic changes are predicted from the A2 scenario (Table 1; Fig. 4).

Forest zones should shift far northward hundreds of kilometers, replacing previous biomes. According to the moderate scenario, habitats for northern vegetation classes (tundra and forest–tundra) should decrease from 10 to 1.5 %, and tundra should nearly disappear. Southern nonarboreal habitats (steppe and semidesert) should expand 1.5- and 3-fold, correspondingly, according to the moderate B1 and the harsh A2 scenario. The boreal taiga as a whole including both dark-leaf (*P. sibirica*, *A. sibirica*, and *P. obovata*) and light-leaf (*Larix spp.*, *P. sylvestris*) taiga should decrease by 2- to up to 5-fold in the B1 and A2 2090 climate scenarios. The light-leaf taiga component should decline by 5 times, in contrast with a 10-factor decrease in the dark-leaf component. Thus, the future climate is predicted to be much warmer and also dryer, and thus should be more suitable for water stress-resistant, light-leaf tree species. The SiBCliM also predicted temperate broadleaf forest (e.g. with *Tilia sibirica, Quercus robur*) and forest–steppe habitats in the south, which are nearly nonexistent today (~1 %). Temperate broadleaf and mixed conifer-broadleaf forests should expand up to 50 % in the warmer A2 2090 climate, while light taiga should still dominate in the less warmer B1 climate (Fig. 4, Table 1). A large portion of this light taiga should remain *L. dahurica* taiga. Despite the drastic increase in temperatures predicted at these high latitudes, permafrost should not thaw enough to support tree species other than Dahurian larch, which can withstand shallow permafrost. Both dark- and light-leaf taiga should substantially decrease, while grasslands should expand from the south, and permafrost should gradually retreat in the northeast. A notably dramatic picture of shrinking taiga is shown in Fig. 4 and Table 1 in the hot, dry A2 climate.

One of the major climate effects on vegetation in a changing climate would be "hot spots," where the first signs of vegetation change are expected to appear. Here,

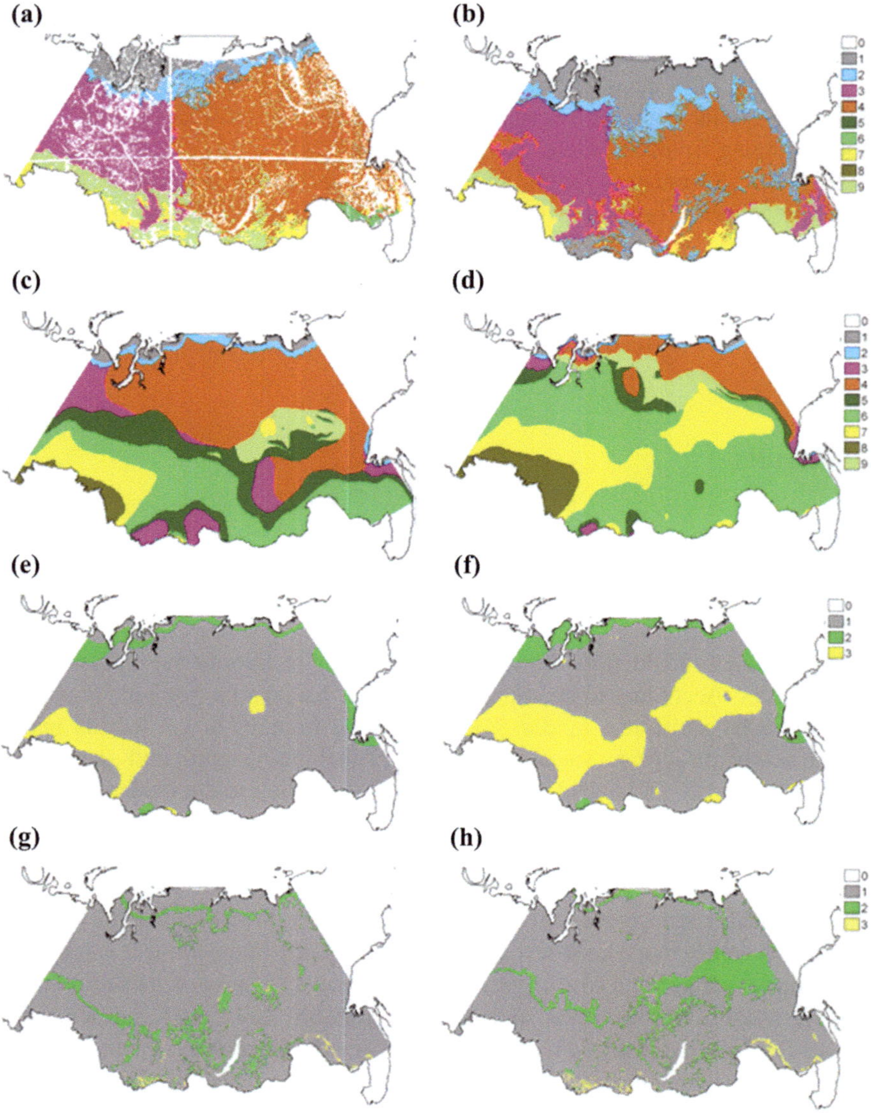

Fig. 4 Vegetation distribution: (**a**) actual (Isachenko et al. 1988); (**b**) SiBCliM contemporary climate simulation; 2090 Hadley climate scenarios for (**c**) B1 and (**d**) A2; "Hot spot" distribution (**e–f**): climate-induced tundra-to-forest transition (*green*), forest-to-steppe transitions (*yellow*), and no change (*gray*) in (**e**) B1 and (**f**) A2 2090 climate change scenarios; Vegetation-change-induced albedo feedback (**g–h**): potential forest return (*green*) and potential steppe return (*yellow*). Vegetation key: 1. Tundra; 2. Forest-tundra; 3. Dark taiga; 4. Light taiga; 5. Mixed conifer-broadleaf; 6. Broadleaf; 7. Steppe; 8. Semidesert; 9. Forest-steppe

we show major hot spots of potentially rapid tundra-to-forest transition in the north and highlands, and major forest-to-steppe transitions in the south, modeled using our SiBCliM as the difference in vegetation in 2090 and the current climate (Fig. 4e–f). The tundra-to-forest transitionary hot spots in the north were not as large as the forest-to-steppe hot spots in the south because these transitions are expected to be inhibited by the Arctic seas. Two forest-to-steppe hot spots modeled in southern Siberia and Yakutia (A2 2090 scenario) are notably large, covering about a quarter of the study area.

Estimated vegetation alteration-induced albedo change between now and 2090 across Siberia showed that albedo would increase over 44 % of the area in the southern and middle latitudes in Siberia due to the forest retreat. Albedo would decrease in 56 % of the territory, mainly in the northern latitudes and highlands, where tundra would be replaced by less reflective forest and in southern grasslands where the snow-free period would be extended (Tchebakova and Parfenova 2013). Resulting warming due to effects of albedo and snow cover change would be greatest at high latitudes and lesser warming or some cooling would occur at middle and low latitudes.

We modeled the potential albedo-induced effect on GDDs using the B1 and A2 climate change scenarios. With the additional albedo-corrected land cover change, simulations show additional warming in the north, promoting further forest advance into the tundra, and minimal cooling in the forest–steppe ecotone, promoting forest return in the south (Vygodskaya et al. 2007; Tchebakova and Parfenova 2013). Due to albedo feedback effects, the forest may advance further into the tundra and may retain more area in grasslands, gaining an additional 6–8.5 % (Fig. 4g–h).

Major forest-forming tree species. The boreal forests across Siberia consist largely of eight conifers (Pozdnyakov 1993), the largest of which comprises about 50 % of the species, *Larix spp.* (four species: *Larix sibirica*, *L. sukaczewii, L. gmelini and L. cajanderii*), the latter two were recently split from their parental species (*L. dahurica*). Remaining are 13 % *P. sylvestris*, 7 % *P. obovata*, 6 % *P. sibirica*, and 2 % *A. sibirica*. In Russian forest literature, larch and pine are light-leaf species that form light-colored, relatively reflective forests versus spruce, fir, and Siberian pine, dark-leaf species that establish dark forests.

During the century, the warming and drying climate should become increasingly more suitable for the so-called light-leaf conifers: *P. sylvestris* and *Larix spp.* are resistant to limited moisture and frequent fire events (Table 2, Fig. 5). The moderate B1 climate appears to be very favorable for *P. sylvestris*—75 % versus ~53 % in present day Siberia, while the harsh A2 climate does not favor its range expansion, *~59 % versus ~53 %*. Taking into account further permafrost retreat in the dryer A2 scenario, the steppe would expand at the expense of both pine and larche ranges (Fig. 5). Additionally, their potential climatic ranges reduced by ALD < 1.5–2 m.

For the *Larix* genera as a whole, the future climate should be about the same, favorable in the B1 climate and worse in the A2 climate, in which the lack of moisture limits the forest range and favors the steppe extension in Yakutia (Table 2 and Fig. 5). The larch biome across Siberia would still dominate Siberia by the end

Table 2 Major tree species potential range (%) from the 1960–1990 climate to the 2090 Hadley climates

Tree species	1960–1990	B1 2090	A2 2090
Pinus sibirica	15.9	16.2	5.4
Abies sibirica	8.4	5.5	1.8
Picea obovata	20.3	31.2	28.7
Larix spp.	76.3	78.9	60.0
Pinus sylvestris	53.4	75.7	58.8

Note that species ranges overlap and therefore do not add up to 100 %

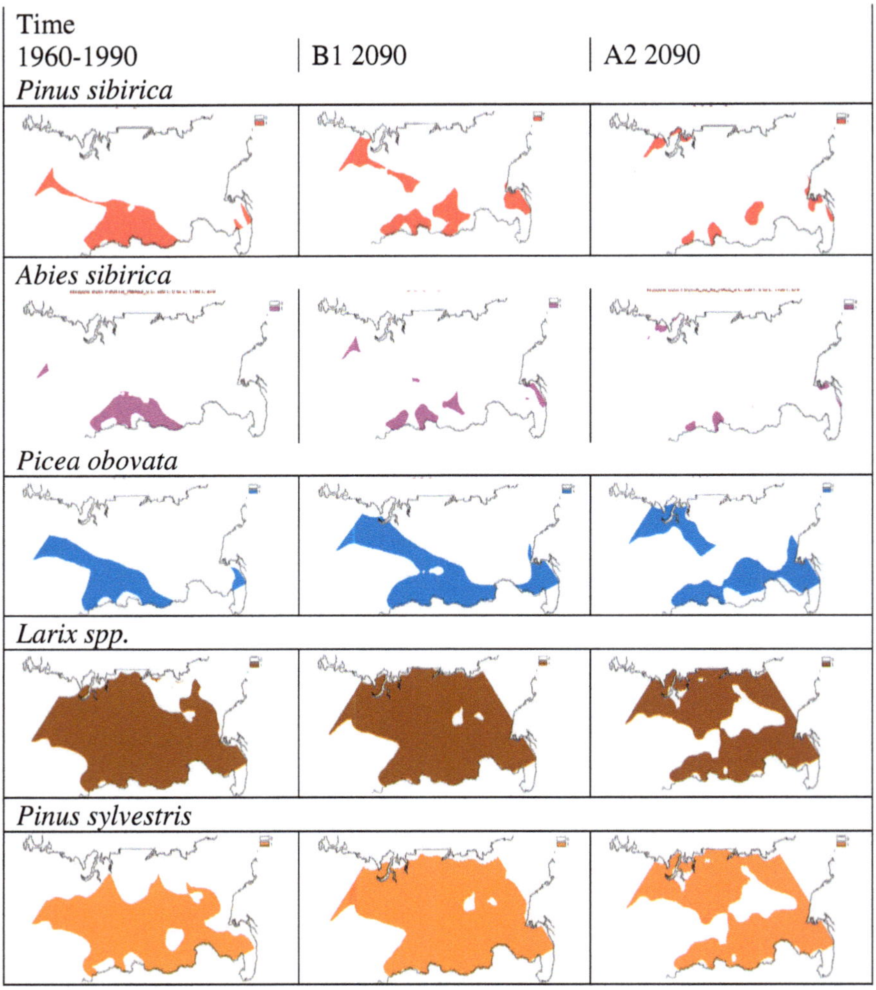

Fig. 5 Major tree species distribution in contemporary and in warmer and dryer Had B1 and A2 2090 climates

of century, but its range would change. The unique Dahurian larch, which can withstand shallow permafrost, would retreat eastward following the permafrost retreat.

The structure of the larch biome would change in a warmer climate: from the proportion 38:33 (*L. sibirica* and *Laris sukaczewii* vs. *L. dahurica*) by 1990 to 50:24 in favor of *L. sibirica* and *Larix sukaczewii*, which can outcompete Dahurian larch beyond permafrost (Tchebakova et al. 2010).

Dark-leaf tree species are not widely spread in Siberia because they are water loving and demand moisture. The habitat is sufficiently moist in West Siberia, and they are usually found growing on elevated tablelands in Central and East Siberia as well as on windward mountain macroslopes in the southern mountains. In the warmer B1 climate, Siberian pine range would not change in terms of area but would shift to the north on plains and upslope in the mountains to moister habitats. Siberian fir, a more warmth-demanding species, would decrease 1.5-fold in the area. The spruce range would increase 1.5- fold because the climate continentality would decrease, which is its preference, and it is less sensitive to predicted dryer climates than other Siberian dark-leaf conifers (Polikarpov et al. 1986).

In the harsh A2 climate, the Siberian pine range would decrease 2-fold, and fir would decrease 4-fold. In contrast, spruce range would increase 1.5-fold. The range of Siberian pine and fir would be very fragmented, and suitable habitats would remain primarily in the high-altitude mountains. The spruce range would not be fragmented because it would follow the gradually retreating permafrost. Additionally, the dark-coniferous "chern" taiga of fairy tales, which is dominated by the mix of Siberian pine, fir, and aspen, would be minimized in the dry future climate over the plain, especially in the A2 climate. The "chern" taiga would shift into higher, moister mountains from the mountain foothills.

4 Discussion

Our envelope-type vegetation and forest models are based on climate vegetation classifications. This approach is best known and the simplest to predict the equilibrium response of vegetation to climate change. However, the major disadvantage of this type of model is that vegetation/forest types will not change and shift as a whole under climate change in the future. The vegetation/forest is made up of a number of species which will individually respond to a changing climate and may comprise not only known but also unknown vegetation/forest types (Peng 2000).

Both zonal vegetation and tree species bioclimatic models accomplished good performance in modeling potential Siberian forests and showed fair to good matches between actual and modeled forests. Potential tree ranges are larger than real ones because they do not consider soil, competition, and disturbance factors. Those matches might be higher because historically part of the primary conifer forests were replaced by secondary birch and aspen forests after large disturbances (clear cuts and wildfire).

Our analyses on Siberian forest types and forest-forming species demonstrate the far-reaching effects of a changing climate on the ecologic distribution of future forests. Forest zones and tree species boundaries are expected to change at the same time. Because analogues to the future forests of Siberia exist contemporarily, one can assume that the vegetation is capable of adjusting to the predicted changes. However, the redistribution of forest zones and tree species will require long periods to adjust to the amount of change being predicted. From the ecological perspective, therefore, it is the speed of warming rather than the absolute amount of warming that is most foreboding.

The future climate is predicted to be dryer. Forest–steppe and steppe, rather than the boreal conifer forest, would dominate about 40 % of Siberia. Light conifers may have an advantage over dark conifers in a predicted dry climate and may cover a larger area in the near future due to their stronger resistance to water stress and wildfire. The SiBCliM also predicted new habitats suited to temperate vegetation (broadleaf forest and forest–steppe) in the south by 2090. More than 40 % of the land will be new habitats suitable for the temperate broadleaf forest. Desertification is expected to affect as much as 10–15 % of extreme southwestern Siberia, expanding from the dry lands of Kazakhstan as a result of a decrease in precipitation while temperatures are increasing dramatically.

Fire and permafrost are considered to be the principal mechanisms affecting the forest's range and structure (Polikarpov et al. 1998). Predicted warm and dry climates enhance the risks of high fire danger and thawing permafrost, both of which challenge contemporary ecosystems. The northern tree line shift is dependent on tree migration rates, permafrost retreat rates, and soil suitability for the future forests. Due to low natural migration rates, forest zones and tree species shifts will require long periods to adjust to the great amount of predicted climate change. While adaptation of the forests and tree species to climate change at the range boundaries should occur by means of migration (Kirilenko and Solomon 1998), within the forest ranges, the genetic means are considered the principal means of adaptation (Rehfeldt et al. 2002). Developing management strategies for seed transfer to locations that are best ecologically suited to these genotypes in future climates could be humanity's contribution toward assisting trees and forests to be harmonized with a changing climate (Rehfeldt et al. 1999, 2002).

The permafrost boundary is expected to retreat to the north and east. An increased thawing of the ALD will have an effect on the forest structure change. Regeneration of dark conifers beneath the larch shelter is expected to result in a northward expansion of their distribution. Evidence of change in the Siberian taiga structure and shifts in the tree line in central Siberia are already occurring, and these changes are not in some distant future projection but already evident. Kharuk et al. (2005) showed that in Evenkia (Central Siberian Tableland), an undergrowth of *P. sibirica, P. obovata,* and *A. sibirica*, which were not previously found on cold permafrost soils, are now emerging in the *Larix gmelinii* taiga possibly due to the increased ALD that allows for the survival of dark-needled seedlings. There is already strong evidence of tree line shifts of 50–120 m during a 50-year span in the mid-20th century, calculated using in situ data taken from the southern mountains in central Siberia.

However, despite predicted increases in temperatures, permafrost will not thaw deep enough across Siberia to support dark coniferous taiga. The expansive East Siberian landscape is expected to continue being populated with larch (*L. dahurica*), which is a taiga that is able to withstand shallow active layer depths. Permafrost thawing also changes hydrology (e.g. greater river discharge, disappearing lakes) and geomorphology (solifluction and thermokarst processes) across broad expanses of the contemporary permafrost zone. In a warmer and dryer climate, larger areas will be affected by solifluction, thermokarst (Abaimov et al. 2002), and windthrow (Vygodskaya et al. 2007) modified by frequent catastrophic fires and deeper active layer thaw. As a whole, retreating permafrost should cause a reduction in the area of forests and their replacement by steppe on well-drained, tilted geomorphology (Lawrence and Slater 2005) or by bogs on poorly drained, flat geomorphology (Velichko and Nechaev 1992).

Additionally, in a future hotter and dryer climate, fire weather would increase and biomass fuels would dry, promoting increased severe fire, which would act to move ecosystems more rapidly toward a new equilibrium with the new climate. One unanswered question is whether seed sources would be available as forests are transitioning. This altered land cover could generate additional regional forcing and feedback to the climate system, potentially resulting in a nonlinear response to changes in climate (Soja et al. 2007).

The southern tree-line shift is controlled by fire. In the last two decades, extreme fire seasons have significantly increased in Siberia (Soja et al. 2007), and catastrophic fire frequency has increased to once in 10 years (Shvidenko et al. 2011). Due to an increased forest fire frequency and shorter fire return intervals, forest regeneration may not be possible in a hotter and drier climate or, if it is possible, may not survive the forest establishment stage. Frequent fires may also change the forest structure. The fire return interval in the light conifer (*Larix spp.* and *P. sylvestris*) middle taiga in central Siberia is currently 20–30 years (Furyaev et al. 2001) compared to 200–300 years in the dark conifer (*P. sibirica* and *A. sibirica*) southern and mountain taiga in southern Siberia (Polikarpov et al. 1998). After fire events, slowly growing dark conifers, not adapted to frequent fires, typically die, both larch and pine, evolutionarily adapted to fire, successfully regenerate after fire events. The impact of natural climate change-caused disturbances (weather, wildfire, infestations) and anthropogenic disturbances (legal/illegal cuttings) on the boreal forest in Siberia has increased for the last three decades (Kukavskaya et al. 2013; Soja et al. 2007; Pleshikov 2002).

Principal forest ecosystem services would also be altered under the impact of climate change. The ecosystem services which Gerasimchuk (2011) studied in southern mountain forests are predicted as follows: both demand and supply of provisioning of timber and firewood would remain the same; both demand and supply of carbon sequestration would increase; demand for prevention of wildfires would increase while the supply of service would worsen; demand for maintaining natural habitats for biodiversity would increase while the supply would worsen; both demand and supply for the provisioning of freshwater would increase; and both demand and supply for the provisioning of land and conditions for farming would improve.

The establishment of agricultural lands may appear in new forest–steppe and steppe habitats because the forests would retreat northward. Currently, food, forage, and biofuel crops primarily reside in the steppe and forest–steppe zones, which are known to have favorable climatic and soil resources. During this century, traditional Siberian crops could be gradually shifted northward and new crops, which are currently nonexistent but potentially important in a warmer climate, could be introduced in the extreme south (Tchebakova et al. 2016).

5 Principal Conclusions

- The SiBCliM simulations of the current Siberian vegetation distribution compare fairly to the current real-world map distributions;
- Simulations using the HadCM3 B1 and A2 projected climates show zonobiomes must shift far northward to reach equilibrium with the change in climate. Under the harsh A2 warmer and dryer climates, ~40 % of Siberia would be suitable for the forest–steppe ecotone and grasslands and another 40 % would be suitable for temperate broad-leaf forest, which is not currently viable in Siberian biomes, and only 17 % of the area could sustain taiga, two-thirds of which would consist of light taiga and one-third of the transitional mixed conifer-broad leaf forests. Dark-conifer taiga of Siberian pine, fir, and spruce would nearly disappear.
- The dryer climate would favor light conifers due to their stronger resistance to water stress and wildfire: Outside the permafrost zone *P. sylvestris, L. sibirica* and *L. sukaczewii* would dominate and within the permafrost zone *L. dahurica.*
- Permafrost would not retreat appreciably, thus Dahurian larch taiga will remain the dominant zonobiome on permafrost in East Siberia.
- Accumulated surface fuel loads due to increased tree mortality from drought, insects, and other factors, especially at the southern forest border and in interior Siberia (Yakutia), together with an increase in severe fire weather are expected to lead to increases in large, high-severity fires, which are expected to facilitate vegetation progression more rapidly toward a new equilibrium with the climate.
- Adaptation of the forest types and tree species to climate change in the south may be based on (1) the genetic means to assist trees and forests by seed transfer from locations that are best ecologically suited for these genotypes in future climates and (2) the introduction of suitable crops that are potentially important in a warmer climate on the areas of failing forests.

References

Abaimov AP, Zyryanova OA, Prokushkin SG (2002) Long-term investigations of larch forests in the cryolithic zone of Siberia: brief history, recent results and possible changes under global warming. Eurasian J For Res 5–2:95–106

Bazhenova OI, Martyanova GN (2003) Evaluation of geocryological conditions in subarid regions of Siberia in current climate warming (Оценка изменений геокриологических условий субаридных районов Сибири при современном потеплении климата). Geogr Nat Resour 4:51–58 (in Russian)

Box EO (1981) Macroclimate and plant forms: an introduction to predictive modeling in phytogeography. Junk, The Hague 258 p

Budyko MI (1974) Climate and life. Academic Press, New York 508 p

Dokuchaev VV (1899) On studies of natural zones: horizontal and vertical soil zones (К учению о зонах природы: горизонтальные и вертикальные почвенные зоны). The City Authorities Printing House, Sankt Petersburg, 28 p (in Russian)

Dostavalov BN, Kudryavtsev VA (1967) Basic permafrost science (Общее мерзлотоведение0. Moscow University Press, Moscow 404 p

Gerasimchuk IV (2011) Adaptation actions: observed and forecasted climate change impacts on population, economy and ecosystem services. Chapter 6. In: Assessment report. Climate change and its impact on ecosystems, population and economy of the Russian portion of the Altai-Sayan ecoregion. WWF Russia ,Moscow, pp 94–131

Gruza GV, Rankova EY (2004) Climate change detection: state, variability and climate extremes (Обнаружение изменений климата: состояние, изменчивость и экстремальность климата). Russ J Met Hydrol 4:50–66

Hogg EH (1997) Temporal scaling of moisture and the forest-grassland boundary in western Canada. Agric Forest Meteorol 84:115–122

Hutchinson MF (2000) ANUSPLIN Version 4.1 user's guide'. (Canberra: Australian National University, Centre for Resource and Environmental Studies), version 4.4 at: http://fennerschool.anu.edu.au/files/anusplin44.pdf. Accessed 20 Jan 2015

IPCC (2007) Climate change 2007: the physical science basis. In: Solomon S, Qin D, Manning M et al (eds) Working group 1 contribution to the IPCC third assessment report, URL: http://www.ipcc.ch. Accessed 20 Jan 2015

IPCC (2013) Climate change 2013: the physical science basis. URL: http://www.ipcc.ch/report/ar5/wg1/. Accessed 20 Jan 2015

Isachenko AG, Shlyapnikov AA, Robozertseva OD, Filipetskaya AZ (eds) (1988) The landscape map of the USSR (Ландшафтная карта СССР). General Ministry of Geodesy and Cartography of the USSR, Moscow, 4 plates. M 1:4000000

Isaev AS (1990) Map "Forests of the USSR" (Карта лесов СССР). M 1:2500000. ed. Isaev A.S

Ivanov NN (1954) On evaporativity calculation (Об определении величин испаряемости). All-Union Geogr Soc Bull 86(2):189–196

Izrael YuA, Sirotenko OD (2003) Modeling on climate change impacts on the agriculture productivity of Russia (Моделирование влияния изменения климата на продуктивность сельского хозяйства России). Russ J Meteorol Hydrol 6:5–17

Kharuk VI, Dvinskaya ML, Ranson KG, Im ST (2005) Invasion of evergreen conifers into the larch dominance zone and climate trends (Проникновение вечнозеленых хвойных деревьев в зону доминирования лиственницы и климатические тренды). Russian J Ecol 3:186–192 (in Russian)

Kirilenko AP, Solomon AM (1998) Modeling dynamic vegetation response to rapid climate change using bioclimatic classification. Climatic Change 38(1):15–49

Kukavskaya EA, Soja AJ, Petkov AP, Ponomarev EI, Ivanova GA, Conard SG (2013) Fire emissions estimates in Siberia: evaluation of uncertainties in area burned, land cover, and fuel consumption. Can J For Res 43(5):493–506

Lawrence DM, Slater AG (2005) A projection of severe near-surface permafrost degradation during the 21st century. Geophys Res Lett 32:L 24401

Malevsky-Malevich SP, Molkentin EK, Nadyozhina ED, Shklyarevich OB (2001) Numerical simulation of permafrost parameters distribution. Cold Reg Sci Tech 32:1–11

Maximov TC (2007) Carbon exchange in the larch forest of the Yakutian sector of the cryolithozone (Круговорот углерода в лиственничных лесах якутского сектора криолитозоны). Diss Thesis, Inst for Biology Problems of the cryolithozone, Yakutsk

Monserud RA, Leemans R (1992) Comparing global vegetation maps with the kappa-statistic. Ecol Model 62:275–293

Nazimova DI, Andreeva NM, Polikarpov NP, Sofronov MA (2006) A concept of the forest zone as a structural unit of biogeocenotic cover (Концепция лесорастительной зоны как структурной части биогеоценотического покрова). Russian J Forest Sci 1:1–19 (in Russian)

Peng C (2000) From static biogeographical model to dynamic global vegetation model: a global perspective on modeling vegetation dynamics. Ecol Model 135:33–54

Pleshikov FI (ed) (2002) Forest ecosystems of the Yenisei meridian (Лесные экосистемы Енисейского меридиана). Publishing House of SB RAS, Novosibirsk, p 356

Polikarpov NP, Tchebakova NM, Nazimova DI (1986) Climate and mountain forests of southern Siberia (Климат и горные леса Южной Сибири). Nauka, Novosibirsk, 225 p (in Russian)

Polikarpov NP, Andreeva NM, Nazimova DI, Sirotinina AV, Sofronov MA (1998) Formation composition of the forest zones in Siberia as a reflection of forest-forming tree species interrelations (Формационный состав лесных зон Сибири как отражение взаимодействия лесообразователей). Russian J For Sci 5:3–11 (in Russian)

Ponomarev VI, Kaplunenko DD, Krokhin VV (2005) Tendencies of climate change in the second half of the XX century in northeastern Asia, Alaska, and the Pacific Тенденции изменений климата во второй половине XX века в Северо-Восточной Азии, на Аляске и северо-западе Тихого океана). Russ J Met Hydrol 2:15–26 (in Russian)

Popov PP (1980) European and Siberian spruce (Ель европейская и сибирская). Nauka, Novosibirsk, 230 p (in Russian)

Pozdnyakov LK (1993) Forest on permafrost (Лес на вечной мерзлоте). Nauka, Novosibirsk, 192 p (in Russian)

Prentice IC, Cramer W, Harrison SP, Leemans R, Monserid RA, Solomon AM (1992) Global biome model: precting global vegetation patters from plant physiology and dominance, soil properties and climate. J Biogeogr 19:117–134

Rehfeldt GE, Tchebakova NM, Barnhardt LK (1999) Efficacy of climate transfer functions: introduction of Eurasian populations of *Larix* into Alberta. Can J For Res 29:1660–1668

Rehfeldt GE, Tchebakova NM, Parfenova EI et al (2002) Intraspecific responses to climate in *Pinus sylvestri*s. Glob Change Biol 8:912–929

Shugart HH, Leemans R, Bonan GB (1991) A systems analysis of the global boreal forest. Cambridge University Press, New York, p 565

Shumilova LV (1962) Botanical geography of Siberia (Ботаническая география Сибири). Tomsk University Publ, Tomsk (in Russian)

Shvidenko AZ, Shchepashchenko DG, Vaganov EA et al (2011) Impact of wildfire in Russia between 1998–2010 on ecosystems and the global carbon budget. Dokl Earth Sci 441(4):544–548

Sirotenko OD, Gruza GV, Rankova EYa, Abashina EV, Pavlova VN (2007) Contemporary climate change of warmth, moisture and productivity of the agrisphere of Russia (Современные климатические изменения теплообеспеченности, увлажненности и продуктивности агросферы России). Russ J Meteorol Hydrol 8:90–103 (in Russian)

Soja AJ, Tchebakova NM, French NF, Flannigan MD, Shugart HH, Stocks BJ, Sukhinin Ai, Parfenova EI, Chapin III FS, Stackhouse Jr PW (2007) Climate-induced boreal forest change: predictions versus current observations. Glob Planet Change Special NEESPI Issue, 56 (3–4): 274–296

Sokolov SY, Svyazeva OA, Kubli BA (1977) Tree and shrub ranges of the USSR (Ареалы деревьев и кустарников СССР). Nauka, Leningrad, 164 p (in Russian)

Stephenson N (1998) Actual evapotranspiration and deficit: biologically meaningful correlates of vegetation distribution across spatial scales. J Biogeogr 25(5):855–870

Tchebakova NM, Parfenova EI (2006) Prediction of forest shifts under climate change at the end of the 20th century in Central Siberia (Прогноз продвижения границ леса при изменении климата к концу 20 века в Средней Сибири(Прогноз продвижения границ леса при изменении климата к концу 20 века в Средней Сибири). Comput Tech 11 part 3: 77–86 (in Russian)

Tchebakova NM, Parfenova E (2012) The 21st century climate change effects on the forests and primary conifers in central Siberia. Bosque. doi:10.4067/S0717-92002012000300004

Tchebakova NM, Parfenova E (2013) Potential land cover change in Siberia predicted in Siberian bioclimatic model (pp 225–231). In: Groisman P, Gutman G (eds) Regional environmental changes in Siberia and their global consequences. Springer, Berlin, http://www.springer.com/environment/global+change+-+climate+change/book/978-94-007-4568-1. Accessed 20 Jan 2015

Tchebakova NM, Rehfeldt GE, Parfenova EI (2003) Distribution of vegetation zones and populations of *Larix sibirica* and *Pinus sylvestris* in central Siberia under climate warming (Перераспределение растительных зон и популяций лиственницы сибирской и сосны обыкновенной в Средней Сибири при потеплении климата). Contemp Probl Ecol 6:677–686 (in Russian)

Tchebakova NM, Parfenova EI, Soja AJ (2009) Effects of climate, permafrost and fire on vegetation change in Siberia in a changing climate. Environ Res Lett 4. doi:10.1088/1748-9326/4/4/045013

Tchebakova NM, Rehfeldt GE, Parfenova EI (2010) From vegetation zones to climatypes: effects of climate warming on Siberian ecosystems. In: Osawa A, Zyryanova OA, Matsuura Y, Kajimoto T, Wein RW (eds) Permafrost ecosystems. Siberian larch forests. Springer, Dordrecht-Heidelberg-London-New York (Chapter 22) pp 427–447

Tchebakova NM, Parfenova EI, Soja AJ (2011) Climate change and climate-induced hot spots in forest shifts in central Siberia at the turn of the 21st century. Regional environmental change. doi:10.1007/s10113-011-0210-4

Tchebakova et al (2016) Evaluating the agroclimatic potential of central siberia. In: L. Mueller et al. (eds) Novel methods for monitoring and managing land and water resources in Siberia. Springer, Cham, pp 287–305 (Chapter 11 of this book)

Thornthwaite CW (1948) An approach toward a rational classification of climate. Geogr Rev 38 (1):55–94

Velichko AA, Nechaev VP (1992) Evaluation of the permafrost zone dynamics in Northern Eurasia under global climate warming. Trans Russian Acad Sci Geogr Ser 324:667–671 (in Russian)

Vygodskaya NN, Groisman PY, Tchebakova NM, Kurbatova JA, Panfyorov O, Parfenova EI, Sogachev AF (2007) Ecosystems and climate interactions in the boreal zone of northern Eurasia. Environ Res Lett 2 (045033)

Walter H (1985) Vegetation on the earth and ecological systems of the geo-biosphere. Third English edition. Springer, Berlin/Heidelberg/New York/Tokyo, 318 p

Chapter 11
Evaluating the Agroclimatic Potential of Central Siberia

Nadezhda M. Tchebakova, Valentina V. Chuprova, Elena I. Parfenova, Amber J. Soja and Galina I. Lysanova

Abstract Human beings have traditionally cultivated the fertile soils of the steppe and forest-steppe for agriculture. Forests are predicted to migrate northward in a warmer climate and are likely to be replaced by forest-steppe and steppe ecosystems. We analysed potential climate change impacts on agriculture in south/central Siberia, hypothesizing that agriculture in traditionally cold Siberia may benefit from warming. Current carbon (C) fluxes in agrosystems have also been analysed, as they are important for the development of land use strategies. Potentials for cropping were evaluated based on simple climate indices such as temperature sums above a base of 5 °C (GDD_5), and an annual moisture index (AMI), which is the ratio of GDD_5 to annual precipitation. Envelope models which determine crop range, and regression models which determine crop yields, were constructed and applied to climate change scenarios for several time frames: 1960–1990, using historic data; and data taken from HadCM3 B1 and A2 scenarios for 2020 and 2090. Analyses of carbon fluxes in agrosystems showed that plant phytomass and soil humus serve as a principal C sink. Mineralization flux forms from phytodetritus

N.M. Tchebakova (✉) · E.I. Parfenova
V.N. Sukachev Institute of Forest, Russian Academy of Sciences, Siberian Branch (SIF SB RAS), Akademgorodok №50, 660049 Krasnoyarsk, Russia
e-mail: ncheby@ksc.krasn.ru

E.I. Parfenova
e-mail: lyeti@ksc.krasn.ru

V.V. Chuprova
Krasnoyarsk State Agricultural University, Mira Str. 90, 660049 Krasnoyarsk, Russia
e-mail: soil-valentina@yandex.ru

A.J. Soja
NASA Langley Research Center, 21 Langley Boulevard, Mail Stop 420, 664033 Irkutsk, Russia
e-mail: Amber.J.Soja@nasa.gov

G.I. Lysanova
V.B. Sochava Institute of Geography, Russian Academy of Sciences, Siberian Branch, Akademgorodok, 664033 Irkutsk, Russia
e-mail: lysanova@irigs.irk.ru

L. Mueller et al. (eds.), *Novel Methods for Monitoring and Managing Land and Water Resources in Siberia*, Springer Water,
DOI 10.1007/978-3-319-24409-9_11

decomposition, and recently formed humus includes portions of "used" mobile humus. Currently, the C balance of agrosystems is slightly in deficit: the C loss is 0.25 t ha^{-1} $year^{-1}$. From 50 to 85 % of central Siberia is predicted to be climatically suitable for agriculture by the end of the century, and only soil potential would limit crop advance and expansion to the north. Crop production could double. Future Siberian climatic resources could provide the potential for a great variety of crops to grow which previously did not exist on these lands. Traditional Siberian crops could gradually shift as far as 500 km northward (about 50–70 km per decade) if soil conditions are suitable, and new crops which are non-existent today may be introduced in the dry south, which would necessitate irrigation. Agriculture in central Siberia would likely benefit from climate warming. Adaptation measures would sustain and promote food security in a warmer Siberia.

Keywords Climate change · Agriculture · Scenario · Carbon balance · Central Siberia

1 Introduction

Climate change impact investigations on Siberian vegetation have shown that forests would shift northward across Siberia by the end of the century (Tchebakova et al. 2009, 2010). At the expense of forests, about 40 % of Siberia was predicted to be covered by forest-steppe and steppe, which could be potentially suitable for agriculture, given suitable soils. In cold Siberia, agriculture as a whole may benefit from climate change (Izrael and Sirotenko 2003). In European Russia, climate change over the last three decades, 1975–2005, favoured agriculture in 70 % of Russia's federal subjects (administrative units), producing 85 % of its trade grain. However, during this same time period West Siberia and East Siberia experienced crop decreases due to the drying climate across arable lands (Sirotenko et al. 2007).

In our study area in central Siberia, within Krasnoyarsky Krai and the adjacent southern Republics of Khakassia and Tyva (Fig. 1) within two windows: (1) 50–56° N and 89–96°E and (2) 56–75°N and 85–105°E, historical-record-based temperatures were found to be 1 to 2 °C warmer in both summer and winter in the first decade of the 21st century, compared to the baseline 1960–1990 (Soja et al. 2007). The change in precipitation was more complicated, increasing regionally by 10 % on average but decreasing by 10–20 % in the south, promoting further local drying in already dry landscapes. Grassland advancement was predicted as far as 56°N latitude from the south into areas where grassland was previously non-existent. Additionally, aridification was predicted over the drylands in Khakassia and Tyva (Tchebakova et al. 2011b). It has also been predicted that extreme weather events (high temperature, droughts, fires, etc.) will occur more frequently (Meshcherskaya et al. 2011; Dronin and Kirilenko 2010; Groisman et al. 2007).

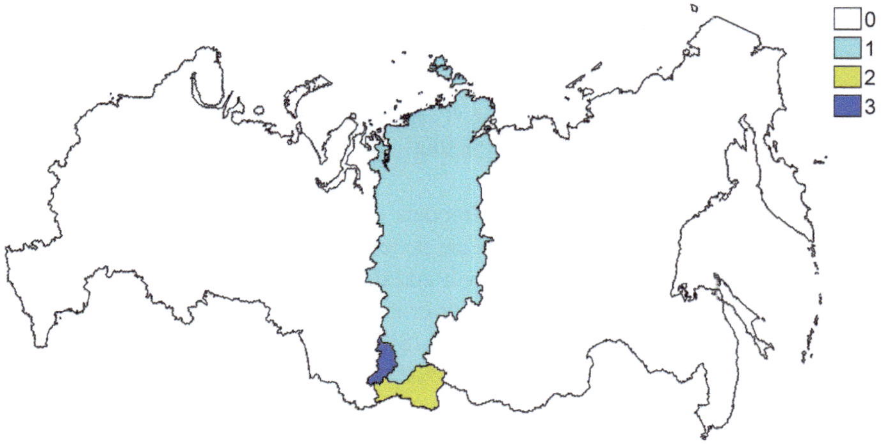

Fig. 1 The study area is the Krasnoyarsk territory (*light blue*) and Republics of Khakassia (*dark blue*) and Tyva (*green*) within 85–105°E and 48–72°N

In view of growing world concerns about food security as the result of observed and projected climate change impacts on agriculture, our goals are: to evaluate the agroclimatic potential of central Siberia using HadCM3 climate change projections; to identify "hot spots" suitable for introducing traditional crops in new habitats and new crops suitable for former farming lands; and to predict crop yields using our crop models, which have been specifically developed for this area (Tchebakova et al. 2011a).

2 Methods of Evaluating Agroclimatic Potential

In Russian applied agriculture, the classification of agricultural regions for warmth and moisture supplies has been extensively studied in agroclimatology, and this has been summarized by Sinitsyna et al. (1973) among others. Actual agricultural regions in the contemporary climate over our study area were mapped by Ivanov (1994), usually based on three climatic indices accepted in the Russian agriculture literature: a sum of temperatures for the season with daily temperatures exceeding 10 °C (called "active temperatures"), a hydrothermic coefficient representing a ratio of precipitation to 1/10 of the sum of active temperatures, and any parameter of wintering (a sum of negative temperatures, a mean of absolute air or soil (or together) temperature minima, etc.). We converted these common Russian climatic indices (Ivanov 1994) to climatic indices used in the western literature: growing degree days above a base of 5 °C (GDD_5), and an annual moisture index (AMI), which is the ratio of GDD_5 to annual precipitation. Agricultural regions in contemporary and future climates were then remapped (Fig. 2). For the future climate, we applied HadCM3 climate change projections from the Hadley Centre, UK (IPCC 2007), which predicts a July temperature increase of 4–6 °C and a January temperature increase of up to 9 °C in the B1 SRES and correspondingly 6–7 °C and >9 °C in the A2 SRES scenarios.

Climatic crop range envelope-type models (Table 1) for a variety of traditional and potential crops in southern Siberia are developed and defined from known crop GDD_5 and AMI requirements for growth and development (Sinitsyna et al. 1973). An AMI <6 is assumed to be critical for crops to thrive, because this defines the border between dry steppe and semidesert (Sinitsyna et al. 1973).

Climatic crop yield models are constructed as multiple linear regressions relating crop yields to GDD_5 and AMI (Tchebakova et al. 2011b):

$$W = a + b\ \mathrm{AMI} + c\,\mathrm{GDD}_5(100\,\mathrm{kg\,ha}^{-1}),$$

where a, b, and c are regression coefficients (Table 2).

Crop data are collated for 30 farming regions across forest-steppe, steppe, and dry steppe in the study region. The area of a farming region varied between 40 and 120 thousand hectares (ha) (Lysanova 2001). Each farming region is characterized by both crop yield and climate means for 1966–2009 (with some breaks in data) that are correlated and representative of the regression yield models. Their statistical characteristics are given in Table 2.

Determination coefficients show that 30–70 % of the crop production in southern Siberia can be attributed to climatic factors ($R^2 = 0.31$–0.68). To assess the "linearity of crop model" assumption, the Fisher criteria are applied at 1 % significance (Table 2). The F-values for all models are larger than theoretical values, demonstrating the viability of the linearity assumption for the crop models. Additionally, our correlation coefficients between the two model regressors, GDD and AMI, of less than 0.6 show that the collinearity is acceptable and provides the basis necessary to permit the use of multiple regression models based on these climatic indices.

Our crop yield models are verified using Sirotenko and Pavlova's (2003) data in European Russia and data collected by Gumenyuk et al. (2010) in the Ukraine, and the models predict crop harvest in the current climates reasonably well (Tchebakova et al. 2011a). These results show that our models are statistically reliable and are applicable for a GDD range up to 2200 °C and an AMI of up to 6.0, which is 30 % greater than the current climatic range in the study area.

We applied our crop range and yield models to climatic indices in 1960–1990 and B1 and A2 2090 climate scenarios to predict potential crop change by the end of the century. Future bioclimatic indices are calculated using anomalies for July and January temperatures, and annual precipitation in 2090, which were derived from two climate change scenarios taken from the HadCM3 moderate B1 (Fig. 2,

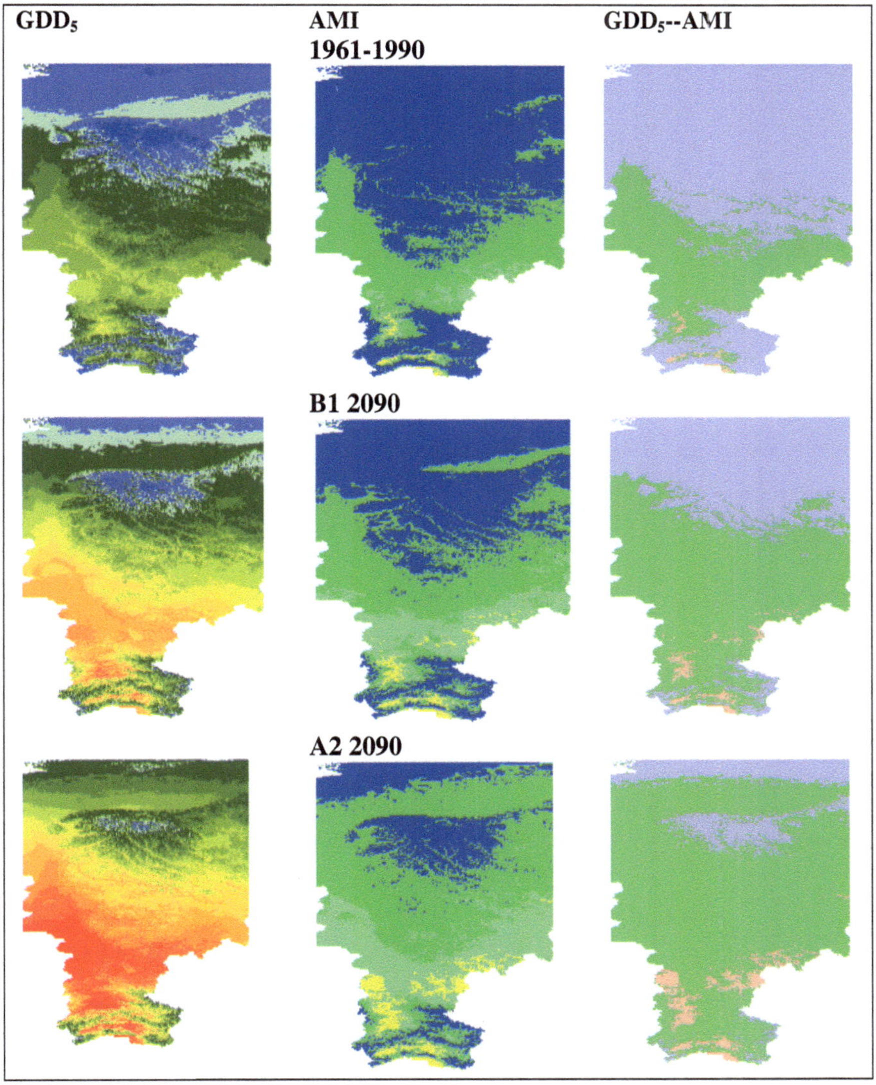

Fig. 2 The baseline 1960–1990 climate (*left column*) and departures between the 2090 climate of HadCM3 B1 (*centre*) and A2 (*right*) and the baseline climate across the study region. July temperature, °C (*upper*): *1*—< 6, *2*—6 to 10, *3*—10 to 14, *4* 14 to 18, *5* >18; January temperature, ° C: (*middle*): *1*—<−35, *2*—30 to 35, *3*—−25 to 30, *4*—−20 to 25, *5* −15 to 20, *6* > −15 °C; annual precipitation, mm (*lower*): *1*—< 200, *2*—200 to 400, *3*—400 to 600, *4*—600 to 800, *5*—800 to 1000, *6*—>1000

middle) and A2 harsh (Fig. 2, right) Special Report on Emission Scenarios, SRES (IPCC 2007). The baseline climate from 1960–1990 is calculated using data we collated from the 90 weather stations across the study region (Fig. 2, left).

Table 1 Envelope models of traditional Siberian agri-crops and some potential crops which may be introduced under warming (*) (Sinitsyna et al. 1973; Gumeniuk et al. 2010)

Crop	GDD_5, °C
Barley (early-ripe), pea (early ripe)	>900
Oats (early-ripe), barley (medium-ripe)	>1050
Spring wheat, winter wheat, maize (silage), oats	>1200
Sunflower (seed), Spring wheat (late-ripe)	>1500
Maize (grain)*, sugar beets*, beans*, millet*	>1650
Rice*, soya bean*, grape*	>1800
Apricot*	>2000

Table 2 Statistical model parameters for various crops

Crops	Inter-cept a	Slope b	Slope c	R^2_{adj}	SE	N	p	F-test	$F_{theoretical}$, 1 %-level
Cereals for grain (C3 type)	7.92	−3.45	0.011	0.43	2.7	29	<0.001	11.718	5.53
Spring wheat	15.05	−7.15	0.017	0.68	3.4	27	<0.001	29.051	5.61
Oats	3.86	−3.94	0.0163	0.52	2.8	29	<0.001	15.712	5.53
Spring barley	14.5	−3.61	0.0073	0.52	2.4	25	<0.001	13.812	5.72
Potato	−29.10	−7.86	0.1062	0.31	17.0	31	<0.002	7.638	5.45
Maize (silage)	−7.45	−96.77	0.356	0.67	30.4	15	<0.001	15.314	6.93
Grasses and alfalfa (hay)	6.9	−4.2	0.0168	0.49	2.0	30	<0.001	15.099	5.49
Berries	−59.0	−11.0	0.077	0.48	5.6	11	<0.031	5.596*	4.46*

Explanations

Statistical data Intercept **a** is the constant term of the regression equation, slope **b** is the regression coefficient of AMI, where AMI is GDD_5/annual precipitation in mm, GDD_5 is the sum of temperatures of all days per year warmer than 5 °C; slope **c** is the regression coefficient of GDD_5; R^2_{adj} is the determination coefficient (square of the correlation coefficient); SE is the standard error of estimate of the crop yield in 100 kg/ha; N is the number of data pairs for computing the regression equation; p is the probability that the regression equation is not significant, F is statistics in an F-test, a measure of the assumption that the model is linear at the 1 % significance level; *—at the 5 %-level; $F_{theoretic}$ is theoretical values of the statistics at the 1 % significance level; *—at the 5 % level

Crops Cereals of C-3 type are small-grain cereals such as barley, wheat, rye and oats. The yields of all cereals are given in 100 kg ha^{-1}, the yield of maize for silage is given in fresh matter. The standard moisture content for cereals and hay was 14 %. Berries include raspberry, blackcurrant and sea buckthorn

The climate-based crop potential is substantially restricted by soil development and fertility. Forest soils are suitable for agriculture in the southern taiga, subtaiga, and forest-steppe (Lysanova 2001). In the north, we assumed the limit for the climatic potential of agriculture using the northern border of the southern taiga, which has suitable soils.

3 Current and Potential Agricultural Regions

In the Krasnoyarsk territory, agrosystems are found south of 60°N on 3233 thousand ha of all Siberian arable lands, with 88 % in forest-steppe and steppe areas. In the taiga forest zone, arable fields are established mainly in areas previously occupied by small-leaved forests. Grain crop is cultivated on 39 and 68 % of the croplands in the taiga zone and in the combined forest-steppe and steppe, respectively. Regarding forage crops, 8–9 % are annuals; 15–20 % of cultivated perennials are alfalfa (lucerne). Potatoes and other vegetable sites are small and make up only 1–4 % of the Krasnoyarsk territory arable land. Agrochernozem (black earth) and agrogray soils prevail in the soil cover.

The agroclimatic potential regions in central Siberia during the 21st century are summarized in Table 3 and Fig. 2. Regular farming becomes possible at GDD_5 >900 °C (Sinitsyna et al. 1973). Currently, 64 % of the study area is colder than 900 °C GDD_5, which is not suitable for open-air farming except in special conditions, such as applying special cultivation techniques or taking advantage of topography (southern slopes or river valleys). By 2090, it is expected cold lands will decrease twofold in the B1 climate and sixfold in the A2 climate (Table 3, Fig. 2, upper).

By the end of the century, the substantially dryer-climate drylands that are not suitable for agriculture will increase, but still only account for up to 6 % of the total land area modelled, even from the harsh A2 scenario (Table 3, Fig. 2, middle). However, the lands with insufficient moisture requiring irrigation would increase up to 15 % (B1 scenario) to 25 % (A2 scenario) of the total area (Tchebakova et al. 2011a). In the absence of irrigation, those lands will likely be unsuitable, hence lost for agriculture.

The lands favourable for farming as determined by GDD_5 >900 °C, and moisture, AMI < 4.0, currently occur on about a third of the study area and would occur

Table 3 Area change (%) of potential agricultural regions in 1960 and the B1 and A2 warmed climates at 2090

Agricultural regions suitable for farming by warmth and moisture	GDD_5 and AMI	1961–1990	B1 2090	A2 2090
Too cold	<900	63.6	33.7	11.0
Optimal	>900 and <4.0	35.6	64.0	83.2
Too dry	>4.0	0.8	2.3	5.7

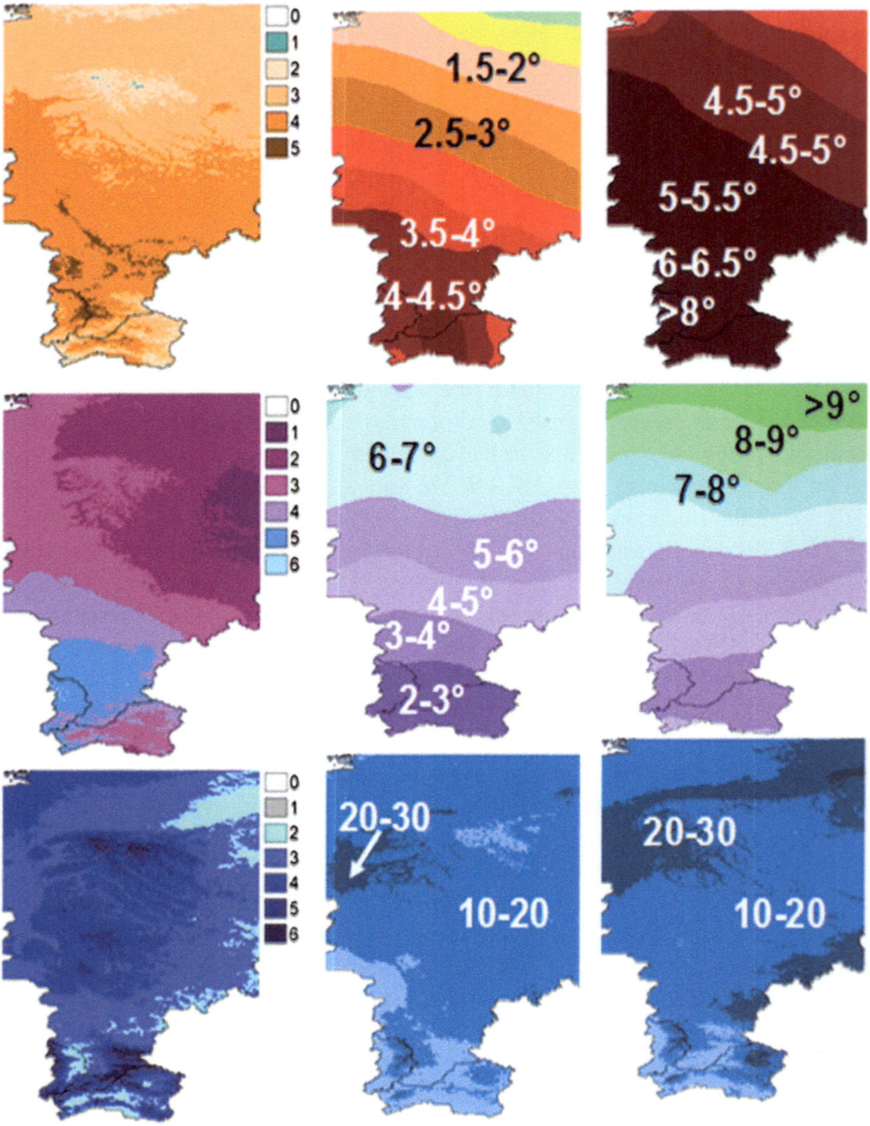

Fig. 3 Potential agricultural regions classified by warmth (*left*) and moisture (*centre*) based on the classification of Ivanov (1994) at the baseline 1960–1990 and HadCM3 B1 and A2 projections for 2090. Warmth-moisture combinations (*right*) suitable/unsuitable for farming

on 64–85 % of the area by 2090 depending on the climate change scenario (Table 3, Fig. 3). Optimal heat and water conditions for crops already exist in the current taiga zone and are predicted to shift northward under climate change scenarios. However, unsuitable forest soils will halt this progression.

Legend of Fig 3:

1. 2. 3. 4. 5. 6. 7. 8. 9. 10. 11. 12. 13. 14.

Left column: contemporary classes: 1. Severe, 2. Very cold, 3. Cold, 4. Moderately cold, 5. Very cool, 6. Cool, 7. Moderately cool, 8. Insufficiently warm, 9. Sufficiently warm; ***Potential classes*: 10. *Warm*, 11. *Very Warm*, 12. *Hot*, 13. *Very hot*, 14. *Extremely hot***

1. 2. 3. 4.

Centre column: 1. excessive moisture, 2. sufficient moisture, 3. insufficient moisture, 4. dry

1. 2. 3.

Right: 1. Cold climates not suitable for farming ; 2. Climates suitable for farming including southern regions under irrigation; 3. Dry climates not suitable for farming

The distributions of some traditional crops (spring wheat, winter wheat, maize (silage), oatmeal, barley, and millet) and some new crops which are non-existent at present, are shown in Fig. 4. By the end of the century, the climatic range of traditional crops would greatly expand, up to the Arctic Circle, about 10° or 1000 km northward. This expansion would allow wheat farming to shift northward, theoretically at a rate of 150 km a decade. However, poor soils would limit the introduction of agriculture to only half of that distance, thus constraining agriculture at about the southern taiga border. The warmer 2090 climate would allow the introduction of sunflower (seed), late wheat, rice, maize (grain), sugar beets, beans, late millet, soybeans, and some exotic crops, such as apricot. Some far eastern varieties of apricots are adapted to the shorter growing season and have been introduced to south-central Siberia. New climate conditions, however, would allow additional varieties due to the longer vegetation season, although their survival will be determined by winter conditions. Due to the dry climate, farming would require irrigation.

Grain crop trends in the contemporary climate show that for the last 45 years warming has resulted in increased crop production of about 5–7 × 100 kg ha^{-1} when there has been sufficient moisture in the forest-steppe and reductions of about 8–10 × 100 kg ha^{-1} as the climate became dryer in steppe regions (Tchebakova et al. 2011a).

In the future B1 and A2 2090 climates, grain crop production may increase from 20 to 35 × 100 kg ha^{-1}; forage silage crop production may increase from 300 to 600 × 100 kg ha^{-1} (Fig. 5); potato crop production may increase from 100 to 200 × 100 kg ha^{-1} by 2090; and berry crop production may increase from 10 to 60 100 kg ha^{-1} (Tchebakova et al. 2011a). Sirotenko et al. (2007) showed an increase in grain by as much as 5 × 100 kg ha^{-1} for the last 30 years, which is comparable to our estimate of a 15 × 100 kg ha^{-1} increase for 1990–2090.

4 Carbon Balance

In order to estimate the carbon balance of agrosystems in the study area, we used two approaches (Titlyanova et al. 1982): we analysed the carbon dynamics in principal components (live phytomass, dead organic matter, and soils) and the

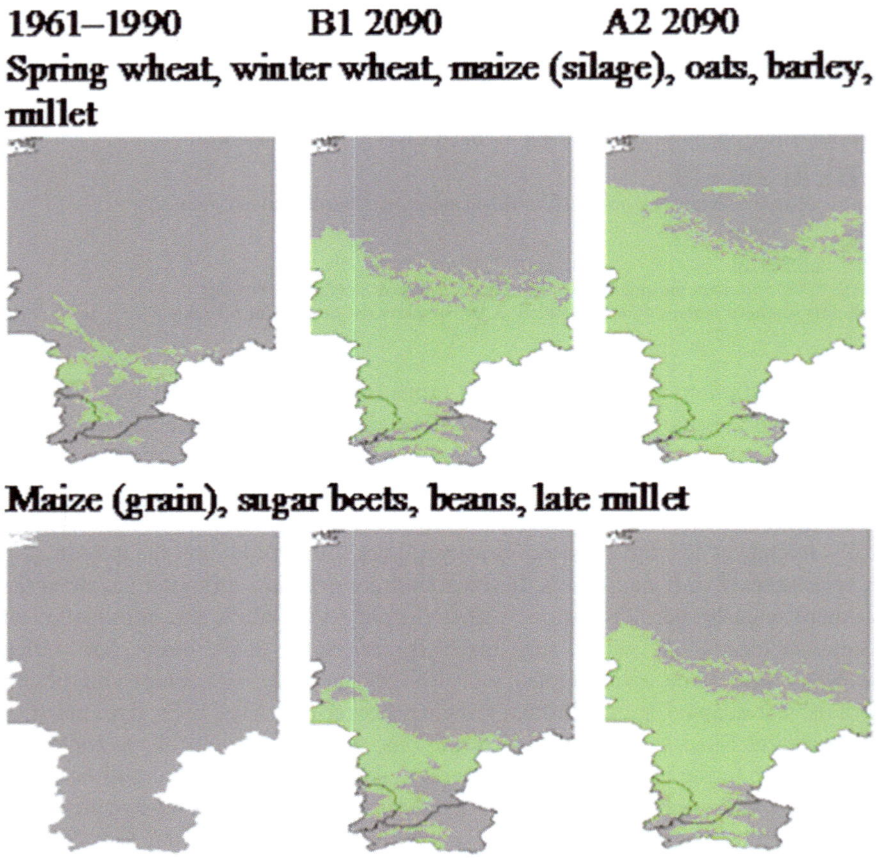

Fig. 4 Potential climatic ranges (*green*) of traditional (*upper*) and new (*lower*) across south-central Siberia in the baseline and HadCM3 B1 and A2 2090 climates

carbon fluxes between these components. These fluxes are the result of biomass production (carbon accumulation) and destruction (carbon release). Combinations of these two processes result in the net carbon balance. The carbon balance is characterized by a net primary production (NPP) which results in ecosystem productivity or biomass and carbon accumulation (incoming), balanced by heterotrophic respiration (Rh), which results in carbon released from an ecosystem to the atmosphere (outgoing). We estimated the above- and below-ground net production (ANP and BNP respectively) of agrosystems based on the results of our multiyear measurements on permanent experimental plots. The NPP for each landscape-climatic zone was obtained by multiplying the NPP (ANP + BNP) by the area of each crop in each zone. Heterotrophic respiration was estimated as production losses due to harvest removal plus losses through plant residue and humus mineralization. The difference between incoming carbon fluxes and their effluxes results in the net ecosystem production (NEP) and the biospheric status of any given

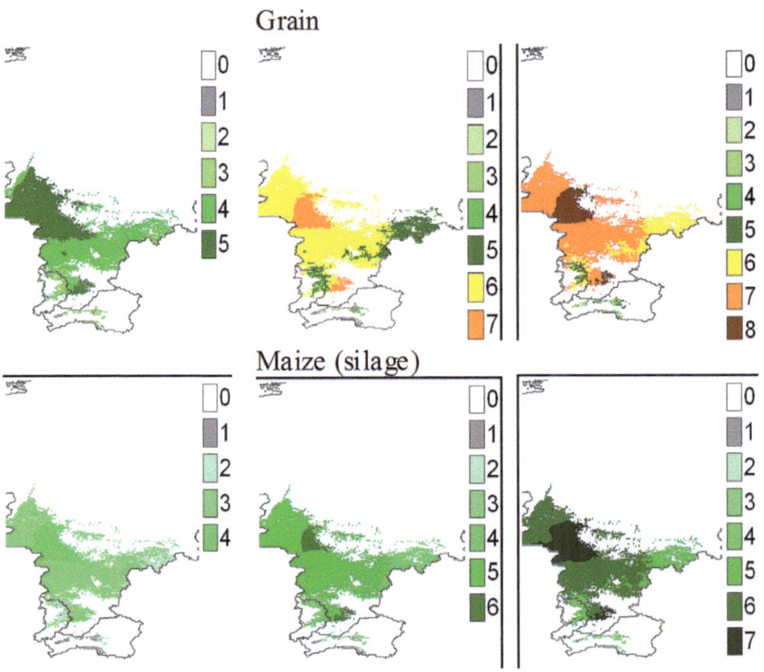

Fig. 5 Crop yields (100 kg ha $^{-1}$) of grain and maize silage on suitable soils across south-central Siberia in the baseline (*left*) and HadCM3 B1 (*centre*) and A2 2090 (*right*) climates. Grain: 0. beyond the area of suitable soils; *1* <5; *2* 5–10; *3* 10–15; *4* 15–20; *5* 20–25; *6* 25–30; *7* 30–35; *8* >35; Silage: 0. beyond the area of suitable soils; *1* <100; *2* 100–200; *3* 200–300; *4* 300–400; *5* 400–500; *6* 500–600; *7* >600

agrosystem, i.e., whether an agrosystem is a CO_2 sink (negative, more stored in the ecosystem) or a source (positive, more flux to the atmosphere).

In agrosystems, carbon was allocated primarily in crop plant and weed biomass and soil humus (Table 4). The aboveground biomass carbon accumulation changed due to crop rotation and had short residence times, i.e., only during growing seasons, but these occurred regularly.

Total biomass increase varied among the crops studied. Alfalfa had the greatest total biomass of all crops across the zone of interest. The aboveground biomass of annuals was higher than the below-ground biomass. In alfalfa, the aboveground biomass was considerably less than the plant root biomass. In root crops, the aboveground biomass was two to five times less than in grain crops. Interestingly, the above-ground-to-below-ground biomass ratio was higher for steppe agrosystems compared to that found in forest-steppe and taiga forest.

However, there were similarities. The maximum aboveground biomass of all crops came immediately prior to harvesting, whereas root biomass reached its maximum during flowering. Root increment was not necessarily accompanied by

Table 4 Carbon storage in agrosystems of the vegetation zones in central Siberia

Carbon characteristics	Middle taiga		Southern taiga		Forest-steppe		Steppe		Total	
	10^3 t C	t ha^{-1}	10^3 t C	t ha^{-1}	10^3 t C	t ha^{-1}	10^3 t C	t ha^{-1}	10^3 t C	t ha^{-1}
Phytomass	50	1.9	850	2.5	7447	3.0	1200	3.2	9547	3.0
Phytodetritus	21	0.8	657	1.9	6156	2.5	1022	2.7	7856	2.4
Humus in 0–50 cm soil layer:	2422	91.7	37,536	108.6	508,102	204.6	52,433	138.5	600,493	85.7
mobile	470	17.8	6273	18.2	80,824	32.5	9825	26.0	97,392	30.1
stable	1952	73.9	31,263	90.5	427,278	172.1	42,608	112.5	503,101	155.6
	2493	94.4	39,043	113.0	521,705	210.2	54,655	144.4	617,896	191.1

an active growth of aboveground biomass. Organic fertilizers enhanced root biomass increment, as opposed to mineral fertilizers.

Annual carbon sequestration in the agrosystem biomass varied from 9547 thousand tonnes in total (Tables 4) or 2.96 t ha^{-1} per unit area. Agrosystems found in the steppe zone appeared to have higher sequestered carbon per unit area: 3.00–3.07 t ha^{-1} versus 1.88–2.46 t ha^{-1} in the middle and southern taiga. Steppes occupy larger areas where nutrients are more readily available to soils; they occur in more comfortable hydrothermal regimes and enjoy a longer frost-free period. Therefore, the carbon allocated becomes higher with warmer and better environmental conditions for plant growth and development.

The maintenance of the amount of carbon allocated to soils depends on the quantity available, chemical composition, and transformation of plant residue, both below the duff and occurring on the soil surface (Pleshikov et al. 2002; Kudeyarov et al. 2007). In agrosystems, plant residue was diverse and included cured aboveground portions of plants, crop roots and crop weeds that died during the growing season, as well as above- and below-ground harvest residue. The amount of plant residues annually sequestered in soils was largely controlled by the production process, which depended, in turn, on the soil and climatic condition of a given area and also on the agrotechnologies applied. Fresh plant residue sequestered in soils adds to the partially decomposed residue that accumulated in previous years, i.e. to mortmass. Overall, mortmass occurrence in soil appears to depend on biomass production and agrosystem destruction processes.

Phytodetritus, the aggregate of the previous and current years' vegetative material, reveals several trends. Our multiyear studies found:

- field crops, aboveground and roots, begin to die back in the second half of June and from mid-July, respectively, across the Krasnoyarsk territory vegetative zones;
- root residues are found mainly in the 0–20-cm soil layer;
- aboveground phytodetritus accumulates on the soil surface during the growing season, and in autumn mixes with fine earth in the plough layer as a result of agronomic treatments;

- perennial herb agrosystems showed the greatest phytodetritus amount, whereas vegetable agrosystems had minimum phytodetritus;
- the aboveground-to-below-ground phytodetritus ratio was much smaller for perennial herbs compared to grain crop agrosystems;
- phytodetritus varied considerably in amount from spring throughout autumn in the agrosystems studied;
- different rates of vegetative residue accumulation led to soil phytodetritus variability in the partially decomposed duff and soils; and
- below-ground phytodetritus consisted largely of coarse fractions of partially to half-decomposed vegetation residue.

In agrosystems, annual phytodetritus carbon was 7856 thousand tonnes or 2.43 t ha^{-1}, and increased 3.3 times progressing from the middle taiga zone to steppe (Table 4). Aboveground crop residue did not exceed 18 % of the total phytodetritus, versus nearly 82 % represented by roots. Coarse fractions of belowground plant residue accounted for 57–62 % of the annual phytodetritus.

The regional soil humus (0–50-cm soil layer) carbon appeared to be up to 600,493 thousand tonnes (185.7 t ha^{-1}) with mobile compounds constituting 16.2 % of the humus. The amount of carbon in the humus layer reflected soil formation processes and was found to increase proceeding from the middle taiga zone (91.7 t C ha^{-1}) to forest-steppe (204.7 t C ha^{-1}) and to decrease in the steppe zone (138.5 t C ha^{-1}).

Carbon accumulated in the agrosystems was estimated to total 617,896 thousand tonnes (Table 4), with 1.5, 1.3, and 97.2 % being allocated in aboveground biomass, phytodetritus, and soil humus, respectively.

The current agrosystem NPP was calculated to be 3.63 t ha^{-1} $year^{-1}$ (Table 5). The production process coincides dynamically with the seasonal rhythms of the field crops of interest. The NPP was controlled by field crop species and type of treatment. In steppe agrosystems, the NPP was twice as much as in the middle taiga. The NPP deviated from the annual mean by as much as 25–30 %, and this was due to weather fluctuations and variations in the amount of the fertilizer applied. The agrosystem NPP pool in the Krasnoyarsk territory was 11,747 thousand t C $year^{-1}$, 55 % of which was aboveground. More than half of the NPP was from grain crops, whereas perennials accounted for 30 %. Annual herbs, corn, and potatoes totalled only 15 % of NPP, and the smallest contribution came from industrial crops and vegetables, which occupy fairly small agricultural fields.

Agrosystem carbon losses consist of removal with crops through their harvesting (44 % of NPP), mineralization of plant residue in soil (53 % of NPP), and humus (9 % of NPP) (Table 5). All grain and 70–80 % of straw are removed with grain-legumes crops. About 90–94 % of aboveground biomass is removed with annual and perennial herbs for green forage and hay. Additionally, a considerable amount of below-ground biomass is removed through harvesting root crops, e.g., potatoes. From cabbage-dominated fields, aboveground biomass is primarily removed. We calculated a total annual carbon loss of 5215.2 thousand tonnes (1.61 t C ha^{-1}) from agricultural lands.

Table 5 Carbon balance of agrosystems in the vegetation zones in central Siberia

C-balance components	Middle taiga		Southern taiga		Forest-steppe		Steppe		Total	
	t C $ha^{-1}y^{-1}$	Total 10^3 t C $year^{-1}$	t C $ha^{-1}y^{-1}$	Total 10^3 t C $year^{-1}$	t C $ha^{-1}y^{-1}$	Total 10^3 t C $year^{-1}$	t C $ha^{-1}y^{-1}$	Total 10^3 t C $year^{-1}$	t C $ha^{-1}y^{-1}$	Total 10^3 t C $year^{-1}$
Input: NPP	2.06	54.5	2.88	993.5	3.67	9118.1	4.18	1580.6	3.63	11746.8
ANP	1.08	28.5	1.52	525.0	2.01	4984.3	2.49	942.2	2.00	6480.1
BNP	0.98	26.0	1.36	468.5	1.66	4133.8	1.69	638.4	1.63	5266.7
Output: total	2.18	57.7	3.00	1036.0	3.92	9752.2	4.48	1697.3	3.88	12543.2
taken with yield	1.02	27.1	1.10	380.0	1.62	4027.4	2.06	780.7	1.61	5215.2
Phytodetritus mineralization	0.78	20.6	1.71	591.1	1.97	4896.6	2.00	757.6	1.94	6265.9
mobile humus mineralization	0.38	10.0	0.19	64.9	0.33	828.2	0.42	159.0	0.33	1062.1
Balance (NEP)	−0.12	−3.2	−0.18	−42.5	−0.26	−634.1	−0.30	−116.2	−0.25	−794.4
% of NPP		5.9		4.3		6.9		7.4		6.8

NPP net primary production; *ANP* aboveground net production; *BNP* below-ground net production; *NEP* net ecosystem production

The rate of phytodetritus decomposition (k) is 0.37–0.44 year^{-1}. Annual soil treatments and agricultural measures enhancing maximal crop production are known to stimulate the decomposition of vegetation residue in agrosystems. Total phytodetritus carbon storage supplies the carbon efflux to the atmosphere and humification flux to maintain soil humus. We found that annual mineralization-caused carbon release ranged from 0.78 t C ha^{-1} in agrosystems of the middle taiga zone to 2.00 t C ha^{-1} in steppe agrosystems, averaging 1.94 t C ha^{-1}. The ratio between the amount of carbon lost due to plant residue mineralization and that lost due to harvest removal varied among agrosystems. Carbon loss due to mineralization dominates in fields of cultivated crops, potatoes, other vegetables, and industrial and perennial crops, whereas carbon loss due to harvest prevails where grain crops dominate.

The carbon flux caused by phytodetritus mineralization in all study areas exceeded that induced by humus development. The less the amount of soil detritus, the less, as a rule, its rate of decomposition (Vedrova 1997; Chuprova 1997). On fallow fields, 7 % of new humus was synthesized due to phytodetritus decomposition, with the values for cultivated, industrial, and green crops being 15 % and those for grain crops, annual and perennials being 25–30 %.

Carbon allocated in humus is released during humus mineralization. Mobile humus compounds are decomposed first. This process varied little in intensity (0.19–0.42 t C ha^{-1} year^{-1}) among the agrosystems studied in the range of vegetation zones. Of the total mineralization flux, mobile humus losses were 9–33 %. Mineralization-caused carbon flux was thus formed by phytodetritus decomposition, rather than by the decomposition of mobile humus forms. Newly formed humus was found to partly compensate mobile humus that "ran out." Additional humus accumulation (0.11–0.19 t C ha^{-1} year) was observed across the natural zone of interest, but only for the fields covered by perennials. Therefore, no mobile humus is lost in such fields. The total mineralization-caused carbon losses appeared to be 1.1–1.7 times those removed by harvesting.

5 Discussion

Russia's agriculture in general benefits from contemporary warming due to increased precipitation in both summer and winter and milder winters resulting in a longer growing season that favours phenological phases (Sirotenko and Gringof 2006). In particular, Central Siberia may benefit from current climate warming: stable positive trends of both summer and winter temperatures have shown increases of 2 and 4 °C, correspondingly, for 50 years, which have prolonged the growing season by 1 month since 1960. Some extreme southern regions experienced 10–20 % less precipitation up to 2010 compared to the baseline precipitation, which in combination with 2–4 °C summer warming causes a stronger aridification. Since 1997, 2–4 t ha^{-1} of grain crops have been collected, and from 2007–2011 some farms in the Krasnoyarsk Krai harvested up to 5–6 t ha^{-1}, particularly in the moist western regions.

As the climate becomes warmer, southern habitats (forest-steppe, steppe, and semidesert) would expand by up to 40 % (Tchebakova et al. 2009) and may become climatically suitable for farming by the end of the century. The northern border of farming may shift northward as far as the Arctic Circle; however, forest soil development will likely impede this progression. Soils, being a product of the interaction between climate and vegetation, need a much longer time to evolve than climate and vegetation change.

Agricultural ecosystems are composed of annual or perennial plants (except multiyear hay grasses) and therefore respond to climate change with no inertia. Human beings may take advantage of this fact and steadily shift farming northward based on agricultural potential change in a changing climate.

As the climate becomes warmer, more warmth-loving varieties of the same Siberian crops may be used for farming. New crops may also be introduced if winter conditions allow. We predicted maize could be planted to yield grain in a warmer Siberian climate. The climate was not obviously taken into account in the 1960s, when the extensive cultivation of virgin lands in West Siberia was undertaken and maize was planted. This experiment failed. The virgin steppe lands degraded under powerful human invasion; they were lost to pastureland for many years and have never fully recovered.

Climatic resources of Siberia provide the potential to grow a great variety of crops, primarily basic crops: spring wheat, winter rye and wheat, cereals (oats, buckwheat, and barley), early-ripe legumes (pea); forage—forage grains (millet, barley, oats, and vetch) and root plants; annuals—maize for silage, sunflower silage, and seed; perennials—forage crops, legumes (lucerne, clover, sainfoin, and melilot); and grasses (couch grass and timothy-grass). In a warmer climate, with a prolonged growing season, productive lands in southern Siberia may become suitable for introducing new warmth-loving crops: melons, gourds, fruits, and berries.

Some benefits and adaptation of Siberian agriculture to climate warming would be farming new crops, which would also decrease their import and transportation costs. Also, southern Siberia may become a land for farming oil crops (rape seed, maize grain, soy beans, etc.), which could be used for biofuel production and thus promote the development of a biofuel industry in Siberia. The future climate is predicted to be dry; thus, farming would require additional water, so another adaptation measure would be the necessity of irrigating farming. This would be the prevailing type of farming in southern Siberia. Wide-ranging irrigation infrastructure development should be started shortly with a view to adapting to climate change. Due to the availability of major Siberian rivers (the Ob, Yenisei, and Lena) with their numerous tributaries, southern Siberia has high irrigation potential; water will not be a limiting factor for irrigating farm land. To adapt to the negative effects of more frequent extreme climate events (floods, droughts, frost, etc.) on food security, it has already been suggested that food and farmland reserves should be moved beyond the risk zone (Sirotenko and Gringof 2006).

Due to the vegetation degradation caused by the intensive influence of agriculture and climate change, some organic carbon will be lost from ecosystems and the greenhouse effect will hence increase. Therefore, studies focused on estimating the role of agrosystems in carbon cycling and the regulation of atmospheric processes are highly important. A combination of herbal and forest ecosystems, which sequester atmospheric carbon, could be used to maintain the carbon balance with agrosystem soil carbon loss. The carbon balance in Siberia is large, and large areas of the region are under agricultural use; these are controlled by crop biomass production and destruction processes.

It has been predicted that forest-steppe and steppe areas will increase in a warmer climate by 2090. On the whole, forest-steppe may increase sixfold in Siberia (see Tchebakova et al., Chap. 10 here). Assuming that agricultural lands and industry would expand in a new climate, then agrosystems would contribute more to the total carbon balance, due to extended lands and to increasing production-destruction processes on those lands. Assuming that phytodetritus destruction rates are the same as current values ($k = 0.37–0.44\ year^{-1}$), the mineralization flux would remain substantially higher than the humification flux; therefore, newly formed humus would replace only a small portion of the removed mobile humus. Thus, the carbon balance of agricultural systems would remain disturbed beyond the equilibrium state.

6 Principal Conclusions

- For half a century, from 1960 to 2010, results from grain crop trends in the contemporary climate demonstrated that warming has favoured crop production in forest-steppe with sufficient moisture and reduced productivity in dryer steppe regions.
- About 65–85 % of central Siberia is predicted to be climatically suitable for agriculture by the end of the century, although potential croplands would be limited by the availability of suitable soils within the steppe, forest-steppe, subtaiga, and southern taiga zones.
- Climatic factors control crop distribution and production in southern Siberia ($R^2 = 0.3–0.7$).
- Crop production may increase twofold as the climate warms during the century.
- Traditional crops (grain, potato, and maize for silage) could gradually shift as far as 500 km northward (about 50–70 km per decade) and new crops (maize for grain, apricot, grape, and gourds) could be introduced in the far south depending on winter conditions; these would necessitate irrigation in the drier 2090 climate.
- Agriculture in central Siberia would likely benefit from climate warming.

- Grain, grain-legume, and perennial herb agrosystems were found to be carbon sinks, whereas cultivated crop, green crop agrosystems, and fallow fields were carbon sources.
- Agrosystems in various natural zones were found to be carbon sources, emitting 794 thousand tonnes of carbon (0.25 t C ha^{-1}) annually.

References

Chuprova VV (1997) Carbon and nitrogen in agrosystems of central Siberia Krasnoyarsk (Углерод и азот в агроэкосистемах Средней Сибири). Krasnoyarsk State University Press. 166 p (in Russian)

Dronin N, Kirilenko A (2010) Climate change, food stress, and security in Russia. Reg Environ Change doi:10.1007/s10113-010-0165-x

Groisman PY, Sherstiykov BG, Razuvaev VN et al (2007) Potential forest fire danger over Northern Eurasia: changes during the 20th century. Glob Planet Change Spec NEESPI Issue 56 (3–4):371–386

Gumeniuk K, Mishchenko N, Fischer G, van Velthuizen H (2010) Agro-ecological assessment for the transition of the agricultural sector in Ukraine. Methodology and results for baseline climate. IIASA Publication, Laxenburg, p 58

IPCC (2007) Summary for policymakers. In: Solomon S, Qin D, Manning M, Chen Z, Marquis M, Averyt KB, Tignor M and Miller HL (eds) Climate change 2007: the physical science basis. Contribution of working group i to the fourth assessment report of the intergovernmental panel on climate change Cambridge, United Kingdom and New York, NY, USA, Cambridge University Press, pp 1–18

Ivanov V (1994) (ed) Atlas of Krasnoyarsky Kray and Republic of Khakassia (Атлас Красноярского края и республики Хакасия). Roskartografia, Novosibirsk, p 84 (in Russian)

Izrael YA, Sirotenko OD (2003) Modeling on climate change impacts on the agriculture productivity of Russia (Моделирование влияния изменения климата на проуктивность сельского хозяйства России). Russ J of Meteorol Hydrol 6:5–17 (in Russian)

Kudeyarov VN, Zavarzin CA, Blagodatsky (2007) Carbon pools and fluxes in Russian terrestrial ecosystems (Пулы и потоки углерода в наземных экосистемах России). Nauka, Moscow, 315 p (in Russian)

Lysanova GI (2001) Landscape analysis of agri-natural potential of geosystems (Ландшафтный анализ природного агропотенциала геосистем). Institute of Geography SB RAN, Irkutsk, 187 p (in Russian)

Meshcherskaya AV, Mirvis VM, Golod MP (2011) The drought of summer 2010 against the background of long-term changes in the aridity over bread-basket region in the European territory of Russia (Засуха лета 2010 г. на фоне долговременных изменений засушливости в хлеборобных районах европейскойтерритории России). Trans Voeikov Main Geophys Obs 563:1–27 (in Russian)

Pleshikov FI, Vaganov EA, Vedrova EF et al (2002) Forest ecosystems of the Yenisei meridian. (Лесные экосистемы Енсейского меридиана). Nauka, Novosibirsk, 356 pp (in Russian)

Sinitsyna II, Goltsberg IA, Strunnikov EA (1973) Agroclimatology (Агроклиматология). Gidrometeoizdat, Leningrad, p 344 (in Russian)

Sirotenko OD, Gringof IG (2006) Estimation of anticipated climate change impacts on agriculture of the Russian Federation (Оценки влияния ожидаемых изменений климата на сельское хозяйство Российской Федерации). Russ J Meteorol Hydrol 8:92–101 (in Russian)

Sirotenko OD, Pavlova VN (2003) Evaluation of climate change impacts on agriculture using the method of spatio-temporary analogs (Оценка влияния изменений климата на сельское хозяйство методом пространственно-временных аналогов). Russ J Meteorol Hydrol 8:89–99 (in Russian)

Sirotenko OD, Gruza GV, Rankova EYa, Abashina EV, Pavlova VN (2007) Contemporary climate change of warmth, moisture and productivity of the agrisphere of Russia (Современные климатические изменения теплообеспеченности, увлажненности и продуктивности агросферы России). Russ J Meteorol Hydrol 8:90–103 (in Russian)

Soja AJ, Tchebakova NM, French NHF, Flannigan MD, Shugart HH, Stocks BJ, Sukhinin AI, Parfenova EI, Chapin III FS, Stackhouse Jr. PW (2007) Climate-induced boreal forest change: Predictions versus current observations. Global and Planetary Change, Special NEESPI Issue, 56: 274–296. 10.1016/j.gloplacha.2006.07.028

Tchebakova NM, Parfenova EI, Soja AJ (2009) Effects of climate, permafrost and fire on vegetation change in Siberia in a changing climate. Environ Res Lett 4: doi:10.1088/1748-9326/4/4/045013

Tchebakova NM, Parfenova EI, Soja AJ (2010) Potential climate–induced vegetation change in Siberia in the twenty-first century. In: Baltzer H (ed) Environmental change in Siberia. Earth observation, field studies and modelling. Springer, Dordrecht, pp 67–82

Tchebakova NM, Parfenova EI, Lysanova GI, Soja AJ (2011a) Agro-climatic potential across central Siberia in an altered twenty-first century. Environ Res Lett 6:45207. doi:10.1088/1748-9326/6/4/045207

Tchebakova NM, Parfenova EI, Soja AJ (2011b) Climate change and climate-induced hot spots in forest shifts in central Siberia from observed data. Reg Environ Change. doi:10.1007/s10113-011-0210-4

Titlyanova AA, Tikhomirova NA, Shtokhina NG (1982) Production process in agrocoenoses (Продукционный процесс в агроценозах). Nauka, Novosibirsk, 184 p (in Russian)

Vedrova EF (1997) Decomposition of organic matter of forest litter (Разложение органического вещества лесных подстилок). Eurasian Soil Sci 2:216–223 (in Russian)

Chapter 12
Probabilistic Assessment of Contemporary Soil Evolution in the South of Western Siberia Based on Analysis of Soil Monitoring Data

Irina V. Mikheeva

Abstract The aim of this article was to introduce a new general and theoretical approach to the probabilistic assessment of contemporary soil evolution (CSE) by analyzing soil monitoring data. The CSE is considered a continuous block process of change in soil conditions over a period extending from 10 to 100 years. The soil condition is considered as its position in the range of *n* soil properties in *k* soil horizons. As previous investigations have shown the essential intrinsic changeability of soil properties in this range, even in homogeneous objects, my proposed idea to assess CSE was to evaluate changes in the probability distribution functions (pdfs) of soil properties at different moments in time. Taking into account soil variability at different scales, I have introduced three categories for the spatial changeability of soil properties. Assessments of the variability of soil properties at the field level are of most importance for the evaluation of CSE. This variance is presented as the sum of variances induced by elementary soil processes, the micro- or meso-heterogeneity of factors of soil formation and elementary landscape processes, distinctions in the anthropogenous factor, and how the soil reacts to them. I developed a method that consists of (1) identifying the pdfs of soil properties, which means a quantitative evaluation of the kind and parameters of pdfs according to data samples resulting from soil investigations; (2) calculating probabilistic indicators such as the statistical entropy of pdfs as probabilistic characteristics of soil status and informational divergence that is a measure of pdf difference. A case study has been done on the large territory in the south of Western Siberia. New findings were the changes in the probability structure of Kastanozem soil properties during CSE under natural processes and anthropogenous influences. Distinctions in pdfs were evaluated from the values of informational divergence and increment in statistical entropy, which were quantitatively different for soils of different granulometric composition, that is, useful to point out the most vulnerable soils in the territory under investigation. It may be concluded that it is necessary to use

I.V. Mikheeva (✉)
Institute of Soil Science and Agrochemistry (ISSA),
Ac. Lavrentieva Av. 8, Novosibirsk 630090, Russia
e-mail: pulya80@ngs.ru

L. Mueller et al. (eds.), *Novel Methods for Monitoring and Managing Land and Water Resources in Siberia*, Springer Water,
DOI 10.1007/978-3-319-24409-9_12

probabilistic indicators to assess CSE from pdf alterations. They characterize a degree of influence of soil-forming factors and anthropogenous influences on the probability structure of the properties of a soil and its stability. Thus, they could be reliable indicators of environmental transformation, that is, important for land resources research, land use policy planning, and basic research.

Keywords Soil · Monitoring · Variability · Pdfs · Probabilistic indicators · Statistical entropy · Informational divergence

1 Introduction

The mathematical analysis of soil monitoring data and modeling of soil development are important for practical decisions in land usage and assessment for bank insurances, cadastres, and so on. Also, they are very important for basic research into the condition, dynamics, and evolution of soil systems and the biosphere as a whole. Assessing contemporary soil evolution (CSE), or soil development, calls for a clear definition of what soil development is. It is reasonable to consider it as a continuous block process with changes in the soil conditions over time (previous, current, and future conditions). The time intervals between conditions may be close to the time of the soil evolution or soil formation, relatively large intervals—hundreds or thousands of years ($\Delta t \approx T$ evolution), but this monitoring investigation is based on relatively small intervals—from 10 to 100 years ($\Delta t \ll T$ formation; Fig. 1).

It is reasonable to consider the soil condition as a set of properties in soil horizons (or in layers) which are basic diagnostic characteristics of soil processes (humus content, texture fractions contents, pH, etc.—Kozlovsky 1991). In other words, the condition of the soil S is its position in the range of n soil properties in soil horizons $h_{i\ (i\ =\ 1\ldots k)}$, where k is the number of horizons in the soil:

$$S = \begin{pmatrix} s_1^{h1} & s_2^{h1} & \ldots & s_n^{h1} \\ s_1^{h2} & s_2^{h2} & \ldots & s_n^{h2} \\ \ldots & \ldots & \ldots & \ldots \\ s_1^{hk} & s_2^{hk} & \ldots & s_n^{hk} \end{pmatrix}.$$

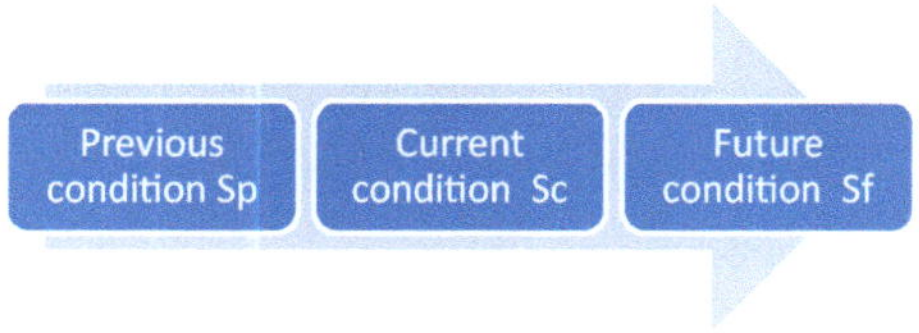

Fig. 1 Soil development as a continuous block process

Accordingly, the change in the soil condition ΔS should be considered as a set of changes in these properties:

$$\Delta S = \begin{pmatrix} \Delta s_1^{h1} & \Delta s_2^{h1} & \ldots & \Delta s_n^{h1} \\ \Delta s_1^{h2} & \Delta s_2^{h2} & \ldots & \Delta s_n^{h2} \\ \ldots & \ldots & \ldots & \ldots \\ \Delta s_1^{hk} & \Delta s_2^{hk} & \ldots & \Delta s_n^{hk} \end{pmatrix}.$$

2 Methodology

Large-scale and detailed scale monitoring investigations have shown that soils are characterized by essential intrinsic changeability in space. Soil properties in soil horizons display stochasticity even in homogeneous soil objects (Dmitriyev 1983; Mikheeva 2001). This means that it is impossible to predict or evaluate their values or changes in any locality *exactly*. For example, there are stochastic fluctuations in the clay content of an extremely homogenous (leveled relief, lithology, vegetation, and usage) transect of loamy sandy Kastanozem, with soil samples taken across 0.25 m (Fig. 2). Stochastic fluctuations in the clay content grow at the bottom of the illuvial horizon. The reasons behind stochastic fluctuations are processes of formation of specific soil structures (pores, cracks, and so on), “fingers” flows, and “tongues,” leaching of substances, etc. These physical processes occur in soils at scales which we do not consider in the conventional field investigation of soil. Different properties reveal the different sizes of these fluctuations, and the rules of their changes in the soil profile also differ (Mikheeva 2013a, b). These fluctuations may be considered as a minimal nugget effect in the total variance, which is usually is by geostatistics.

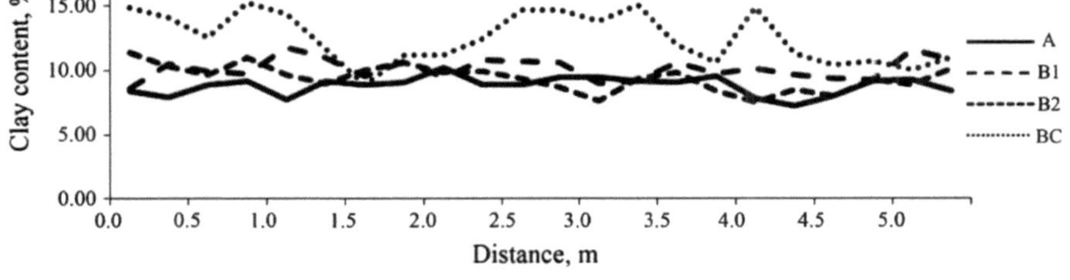

Fig. 2 Clay content (%) in soil horizons in a homogenous transect of loamy sandy Kastanozem

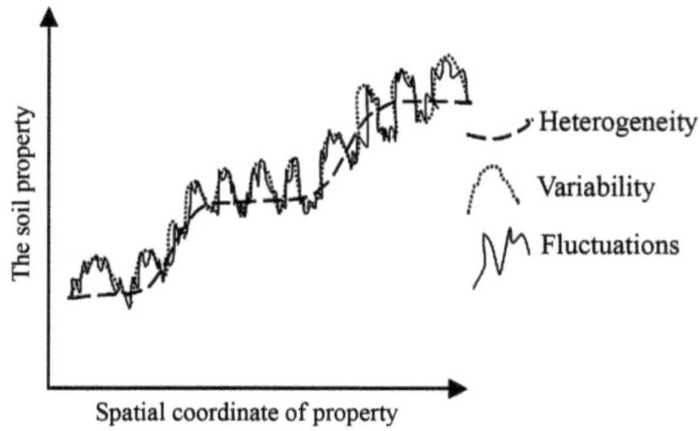

Fig. 3 Categories of spatial changeability of soil properties (schematically)

We have proposed considering three categories for the spatial changeability of soil properties using the concept of nesting (Fig. 3): heterogeneity for significant changes in soil-forming factors, variability for insignificant changes, and fluctuation for leveled soil-forming factors. All three changeability categories are observed at different organization levels of soil and soil cover. In our opinion, relationships between different changeability categories of soil properties also require quantitative evaluation to assess the stability of the soil cover as a hierarchical system (Mikheeva 2005a, b).

In general, the values of soil properties in soil horizons in any point in space are a cumulative effect of various factors and processes at different scales operating mutually, which is the reason for multiscalar soil variation, shown differently at distances from several centimeters to tens and thousands of meters and kilometers. Using the hierarchical approach allows a quantitative consideration of soil cover at various scales. From the point of view of common sense, the variance in the properties of soil at the so-called field level of organization σ^2_{Field} can be presented in the form of the sum of variances (Mikheeva 2001):

$$\sigma^2_{\text{Field}} = \sigma^2_{\text{flu}} + \sigma^2_{\text{fac}} + \sigma^2_{\text{an}} + \sigma^2_{\text{anflu}} + \sigma^2_{\text{anfac}}, \quad (1)$$

where

σ^2_{flu} is variance induced by stochastic fluctuations in the soil properties of soil individuals (pedons);

σ^2_{fac} is variance in soil properties owing to the micro- or meso-heterogeneity of factors of soil formation within an elementary soil area or a combination of elementary soil areas, which form the field as a whole (Fridland 1972). In some cases, this heterogeneity may lead out the soil outside the framework of the taxonomic unit. In other cases, it may not lead out it, but in spite of this, the heterogeneity permits to distinguish some soil contours according to more detailed scales or another classification criterion:

σ^2_{an} is the variance determined by the heterogeneity of soil properties owing to the heterogeneity of an anthropogenous factor (in virgin soil it equals zero).

σ^2_{anflu} is the variance induced by different changes in the stochastic fluctuations of soil properties as a consequence of the effect of anthropogenous factors;

σ^2_{anfac} is defined by differences in the soil owing to the different "lag effect" of separate sites of a soil cover in relation to anthropogenous factors.

Thus, the variability in soil properties at the field level is defined by the total action of elementary soil and elementary landscape processes, the micro- or meso-heterogeneity of factors of soil formation, distinctions between anthropogenous factors, and how the soil reacts to them. Each item in Eq. (1) is of varying qualitative importance and covers a different range of quantitative values. σ^2_{fac} and σ^2_{an} are larger than nother addends. This variability defines a structure of the soil cover of the field. σ^2_{flu}, σ^2_{anflu}, and σ^2_{anfac} are smaller in value, but they are very important for characterizing the quality conditions of soil and its reaction to anthropogenous factors.

3 Probability Distribution Functions of Soil Properties as Characteristics of Soil Conditions and Changes

There is in fact stochasticity at each separate point in the soil, but the soil system as a whole is rigidly determined by the result of the sum of multiple factors that determine the properties of the soil. Integrity is given to this system by the presence of specific and determinative external conditions in which elements of the system exist. It is necessary to consider at least two levels of organization of this system—the macrostate and microstates (Fig. 4).

The macrostate is the condition of the soil cover on the soil object as a whole, and microstates are the conditions of individual soil profiles or pedons. The

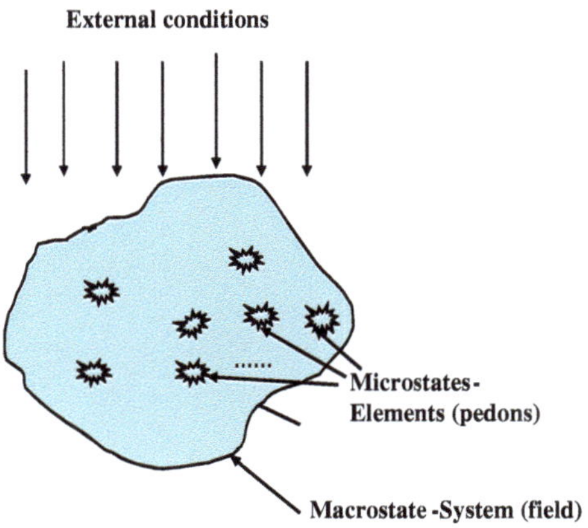

Fig. 4 Macrostate and microstates of soil system (schematically)

macrostate is defined by the probabilities of their microstates. Thus, it is logical to use probability distribution functions (pdfs) of properties that reflect two structural levels—an internal structure of microstates with a different display of soil properties and which determines the macrostate of the system. As a consequence, pdf characterizes the holistic property of the system. The transformation of the macrostate of the soil system arises from the set of individual, chaotic changes in the microstates, and the alteration in pdfs is a quantitative characteristic of this transformation.

Let us consider the pdf of a soil property as a set of two branches—right hand and left hand, which characterize, accordingly, the values above and below the central value. The behavior of these two branches of pdf defines different cases of increasing and decreasing in the property (Fig. 5). The growth of the highest values reflects that a tendency continue to exist over time because as there are large values of the soil property at these specific points and they are growing, it means that conditions remain for more rapid growth in the values (but with some restrictions). The growth of slightly increased values in the case of initially symmetrical pdfs will result in the right-hand asymmetry, an increase in the variance, and, probably, a slight growth in the mean value (Fig. 5a).

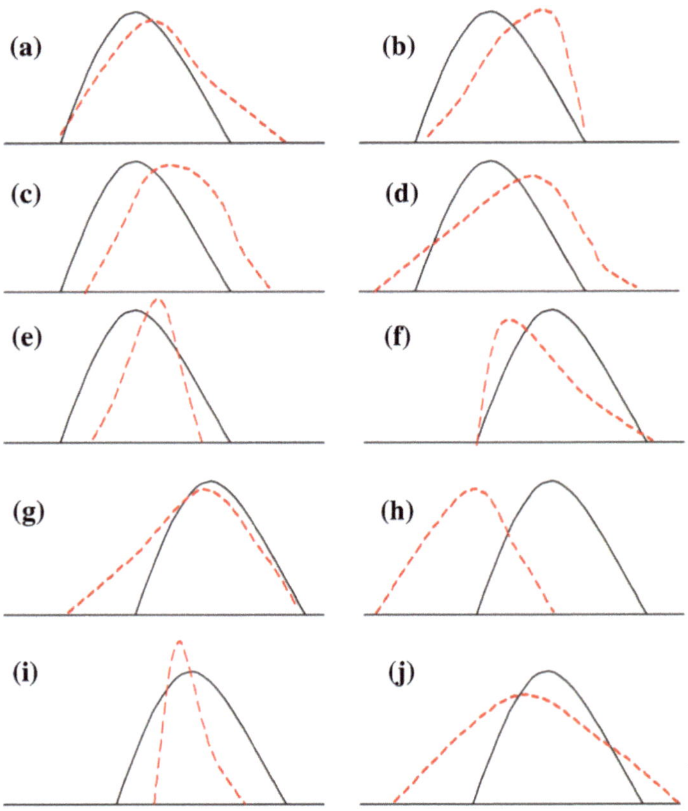

Fig. 5 Typical change in probabilistic distributions (schematically): **a–e** increasing, **f–j** decreasing

The growth of relatively low values produces a reduction in the variance, a small growth in the average, and the appearance of negative asymmetry (Fig. 5b). The growth of all values (Fig. 5c) produces a significant change in the mean value ("a classical case" of growth); other distribution parameters thus cannot change, but in the case of the advancing growth of the increased values, growths in the variance and the right-hand asymmetry are possible. With the growth in the increased values and the simultaneous decrease in lowered values (Fig. 5d), the variance grows. In the case of significant increase in lowered values and some reduction in raised values (Fig. 5e), the variance decreases; the mean value can produce slight growth, and the left-hand asymmetry of distribution is possible.

A decrease in the slightly increased values (Fig. 5f) leads to a slight lowering in the mean value, a slight reduction in the variance, and the appearance of right-hand asymmetry. An essential decrease in the lowered values, i.e., a progressively decreasing process in the lowest values (Fig. 5g), increases the variance, shifts the mean value to the left, and probably produces slight left-hand asymmetry. A simultaneous decrease in all values ("a classical case" of decrease—Fig. 5h) significantly shifts the mean value toward reduction, thus other characteristics of statistical distribution may change or not, but depending on the preferential change in this or that branch of pdfs, both a growth in the variance and the appearance of asymmetry are possible. In the case of an essential decrease in the increased values and a slight growth in lowered ones (Fig. 5i), a slight reduction in the mean value, an essential decrease in the variance, and the appearance of right-hand asymmetry are observed. For example, this case of pdf alteration was observed in the case of changes in the humus content in chestnut arable soils. In the case of a considerable decrease in the lowered values, a slight growth in raised values (Fig. 5j) can take place, and then an increase in the variance, a slight shift in the mean value, and, probably, the appearance of right-hand asymmetry are seen.

The above-mentioned reasonings are the schematic typification of possible different cases of alterations in pdf; however, it is enough general and logical. Thus, analyzing various cases of pdf behavior provides information about the stability of different localities of soil cover with this or that range of soil properties. It is very useful to assess and research mechanisms of interaction between the formation processes of specific soil properties. Further, in the presence of investigated functional or statistical relations of soil properties with factors, it can be used to forecast changes in soil properties in the case of changes in factors and soil-forming conditions.

Analysis of pdf dynamics shows that a certain property can often both weaken and intensify in different localities within one soil series, although, as a rule, a prevailing tendency is observed. Changes in the right-hand or left-hand branches of pdfs are important to predict the possibility of future processes of soil transformation. This means that the data from soil monitoring and CSE modeling should be analyzed by studying the pdfs of soil properties and modeling their alterations in the soil cover of investigated sites.

4 Probabilistic Indicators

The visual analysis of pdf alterations is very useful for assessing changes in the probability structure of soil properties in soil objects subject to anthropogenous or natural processes. However, it is important to have some numerically convoluted values for a quantitative assessment of the difference in these functions at different times or external conditions. Thus, we have introduced the idea of using probabilistic indicators of pdfs and their alterations. Taking the above-mentioned fact into account, we suggested considering probabilistic indicators for the characterization of the soil status and soil alterations (Mikheeva 2005a, b, 2009). For the soil status, these are the pdfs of soil properties and their statistical entropy (Mikheeva 2004). For soil change, they are informational divergence and increment in statistical entropy (Fig. 6), and they are a numerical instrument for assessing and modeling CSE.

Statistical entropy is the numerical indicator of the pdfs of soil properties, being a measure of the randomness and internal uncertainty of micro conditions in the soil system. Moreover, it is reasonable to consider statistical entropy as a reliable characteristic of soil variability because our research has shown the greater stability of statistical entropy than other statistical characteristics of soil variability (Mikheeva 2005a, b). Entropy itself is a characteristic of the system status, but the increment in statistical entropy is one important indicator of a system's stability and behavior over time. In the theory of systems, and also in ecology, it is accepted that the steady state of a system is characterized by a minimum increment in entropy over time $dh/dt \rightarrow \min$. To use this criterion in practice, it is necessary to calculate the entropy value during various moments in time and to estimate its increment, which, if the system is stable, should be small. We have investigated changes in the variability of soil properties affected by agriculture within soil series, where variability cannot be considered as discrete as it is continuous. Therefore, we have used a definition of statistical entropy in continuous variability as a probabilistic

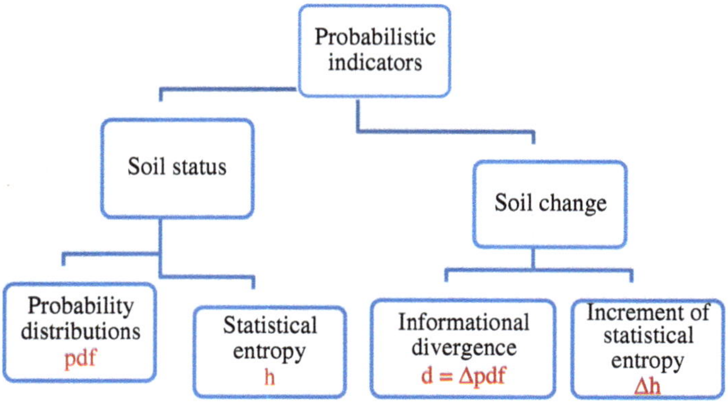

Fig. 6 Probabilistic indicators of soil status and soil alteration

approach, which was proved by Kolmogorov (1965). The value of statistical entropy, using probabilistic distribution, may be calculated by the formula:

$$h = -k \int_{-\infty}^{+\infty} W(x) \ln W(x) \mathrm{d}x + h_0$$

where $W(x)$ = function of probabilistic distributions of random variable x. k and h_0 are constants (Gubarev 1992).

A quantitative comparison of the pdfs of soil properties at different moments provides a holistic estimation of changes in their variability. For this kind of convoluted holistic estimation of changes in soil properties, we proposed using the mathematical value of divergence of pdfs (informational divergence). This value is a quantitative measure of difference in/between pdfs (Mikheeva 2009). The concept of (genetic) divergence is more often interpreted following Darwin in biology. However, in methodological works, the divergence concept is given a more fundamental systemic meaning. All areas of life on Earth can be considered in its whole as one system of divergences; it is believed that an increase in distinctions leads to more and more steady structural parities (Bogdanov 1989). We proposed that it is reasonable to name processes of modern changes in soil properties concerning their start meanings by the term "divergence" and investigate it quantitatively (Mikheeva 2013a, b). As a measure of dissimilarity, divergence should satisfy some conditions. The most important is the scale-invariance property of divergence (Kanamori 2014). To estimate how the pdfs of soil properties differ in different horizons, at different moments in time, or in different soil states (arable, deflated, nondeflated, and irrigated), we used the value of divergence (d) that satisfies this condition (Gubarev 1992)

$$d = \int_{-\infty}^{+\infty} (W1(x) - W2(x)) \ln\left(\frac{W1(x)}{W2(x)}\right) \mathrm{d}x,$$

where $W1(x)$ and $W2(x)$—pdfs of property in compared objects.

5 Statistical Method

Statistical analysis consisted of two stages: (1) the identification of probability distributions of soil properties and (2) the calculation of statistical entropy and information divergence.

The first is a statistical procedure in which the parameters of functions were evaluated in terms of maximal likelihood from factual soil survey data, and then many hypotheses about coincidence were checked with a large set of theoretical functions (37). In each case, the only functions and corresponding parameters chosen were those which are the best approximation coordinated with a set of parametric and nonparametric criteria (for more details see Appendix).

6 Object and Materials of Case Study

We study soil monitoring data over a large territory (16,000 km^2 in area, 200 km from north to south, and 80 km from east to west) of the Kulunda steppe located in the south of Western Siberia. The region is characterized by a droughty continental climate, and its terrain may be defined as gently undulating plains. The soil cover consists of chestnut soils (70 %), meadow–chestnut soils, meadow soils, solonetzes, and solonchaks with different degrees of hydromorphism according to Russian classification. The soils formed on ancient Quaternary lacustrine-alluvial deposits, which is responsible for the primarily light granulometric composition and spatial diversity of the soil fractions. Soils vary significantly in texture, from loose sands to loams. The manifestation of wind erosion, especially, is probable with the dry climate, the flat-plain terrain, the coarse texture of the soils, and absence of forest, which are typical of that territory. This makes it important to use the soils properly. High-input agriculture in the drought zone of Western Siberia requires the development of irrigation, which is known to cause disaggregation and intensify soil weathering.

The initial data were field materials from our own trench investigations of soils, and large-scale (1:25,000) continuous soil surveys, conducted at various times in several takes using standard methods in Russia (All-Union Guidelines 1973). During each survey, one full soil profile with six samples taken was investigated in the 350 × 350 m grid. The technical errors in the information acquisition were removed during the procedure of checking the field and laboratory results. The archive data of the laboratory analyses formed the basis of our data bank. The latter consisted of several databases, which contained information on all the soil profiles surveyed in different time periods. The information records correspond to genetic horizons or soil layers, and the information fields represent the data on the taxonomic category of the soil and its properties: the thickness of the genetic horizons; the contents of humus, particle-size fractions, and salts; pH; etc. We have used archive materials of four soil surveys: a deflationary survey in 1965, a soil survey in 1975, and two ameliorative surveys on irrigated fields in 1982 and 1989. Each database contains information on 1000 soil profiles, and a total of 4000 soil profiles were analyzed. No data were rejected because all the original data have information value.

All the values obtained were grouped within the chestnut soil subtype according to the soil genetic principles of Russian soil classification. As the result, we received data samples, corresponding textural series of this soil. The grouping also involves a consideration of the geomorphologic regions, sampling years (1965, 1975, 1982, and 1989), land use, and soil status (virgin, arable, irrigated, deflated, nondeflated, etc.). Each data sample includes information about one of the soil properties (humus content, content of granulometric fractions, etc.) in a separate soil horizon or layer at a certain time or state in certain soil series. The volumes of these data samples varied between 20 and 25 samples of data and 650, depending on the abundance of certain soil series, which is enough for statistical and probabilistic analysis. We evaluated the change in the probability distributions of properties in chestnut soils with different textures, in different states of ground: (1) undeflated, (2) deflated, and (3) after long-term plowing without deflation. These states followed each other in time during the last half of the twentieth century in the territory investigated and can be considered different conditions of CSE (Fig. 1).

7 Results and Discussion

After the total plowing of virgin soils, deflationary processes began at once. Most part of the chestnut soils in the territory examined was subjected to deflation to some degree. Deflation is a spatially widespread process. Therefore, it is quite difficult to gain an exact quantitative estimation of the general losses of soils. Soil resistance to deflation is determined by texture, so the analysis was executed for each texture variety separately. The greatest interest lies in changes in the humus content of arable soils, which depends on the standard of farming and size of the crops. We show that it also depends on the soil texture. The statistical analysis of soil monitoring data provided us with a data set of soil property probability functions as shown in Table 1 and Fig. 7a–d.

The statistical entropy of the humus content in chestnut soils decreased under the influence of deflation. Long-term, soil-protective agriculture led to growth of it. It is interesting that in fine-loamy sandy chestnut soil, there are practically no changes in statistical entropy. It remains constant under deflation and under soil-protective agriculture. The key point is that this soil is “average” for the territory investigated, representing its central image. And, importantly, the humus content of this soil is stable (Fig. 8).

Table 1 Probability distributions of humus content in Ap horizon of chestnut soils

Soil variety, status	Type of distribution	Parameters of distribution		α*	Figure
Cohesive sand, deflated	Double power	$\theta_0 = 6.71$	$\theta_1 = 146.6$	0.6	7a-1
Cohesive sand, nondeflated	Ln-normal	$\theta_0 = -0.04$	$\theta_1 = 0.28$	0.7	7a-2
Coarse-loamy sand, deflated	Double power	$\theta_0 = 3.94$	$\theta_1 = 84.56$	0.6	7b-1
Coarse-loamy sand, nondeflated	Su-Johnson	$\theta_0 = -3.34$	$\theta_1 = 3.21$	0.7	7b-2
		$\theta_2 = 0.48$	$\theta_3 = 0.59$		
Coarse-loamy sand, protected	Maximum value	$\theta_0 = 0.23$	$\theta_1 = 1.18$	0.8	7b-3
Fine-loamy sand, deflated	Su-Johnson	$\theta_0 = -0.42$	$\theta_1 = 1.64$	0.5	7c-1
		$\theta_2 = 0.4$	$\theta_3 = 1.42$		
Fine-loamy sand, nondeflated	Logistic	$\theta_0 = 1.42$	$\theta_1 = 0.3$	0.4	7c-2
Fine-loamy sand, protected	Nakagami	$\theta_0 = 0.63$	$\theta_1 = 0.89$	0.5	7c-3
		$\theta_2 = 0.98$			
Sandy loam, deflated	Normal	$\theta_0 = 1.66$	$\theta_1 = 0.38$	0.7	7d-1
Sandy loam, nondeflated	Su-Johnson	$\theta_0 = -1.43$	$\theta_1 = 1.76$	0.5	7d-2
		$\theta_2 = 0.48$	$\theta_3 = 1.40$		
Sandy loam, protected	Maximum value	$\theta_0 = 0.38$	$\theta_1 = 1.83$	0.6	7d-3

*Level of significance reached (*p*-value)

Divergence of the humus content in chestnut soils is not very large, but it is different for soils with a different texture. The reverse change takes place under conditions of soil conservation without deflation, but it is small in value. Alterations are observed not only in the A horizon but also in the subsoil. The highest values of divergence are observed in the cohesive sandy and sandy loamy soils. These soils are very vulnerable (Table 2).

None of these alterations change the taxonomic soil category, but they influence the status of the soil and allow us to model long-term trends in soil cover change.

In this article, we show our methodological approach for the analysis of soil monitoring data based on the example of one soil property, mostly in the upper horizon. However, this methodology is also very useful for analyzing deeper soil horizons and providing suitable information to consider not only the CSE but also the historical evolution.

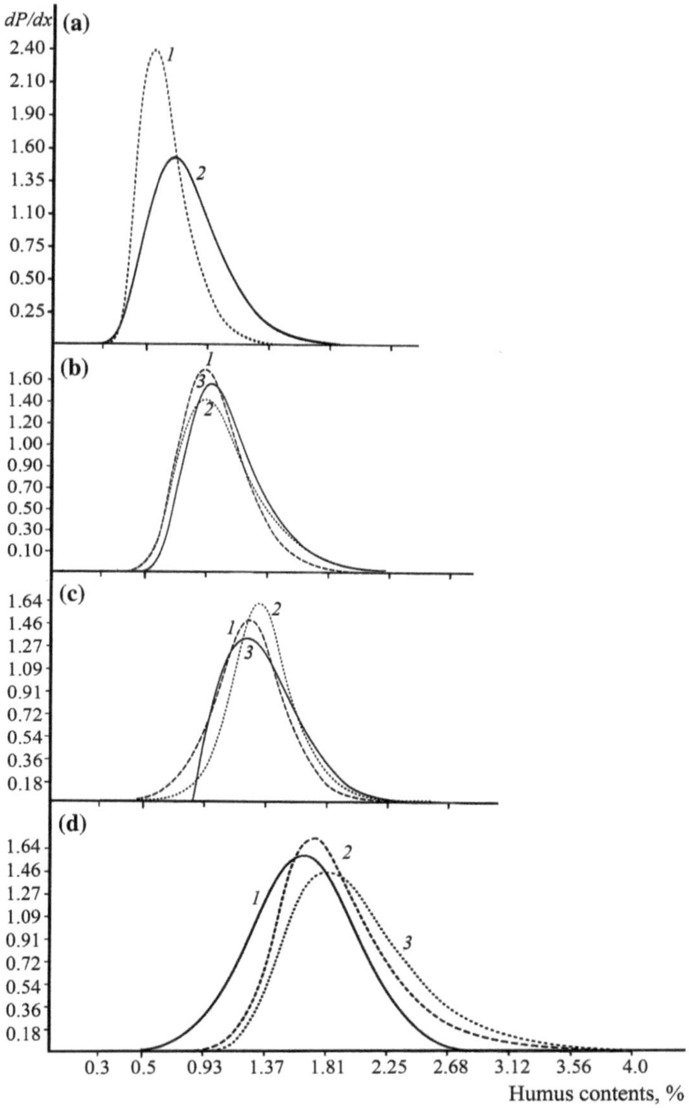

Fig. 7 Probability distribution functions of humus content in texture varieties of chestnut soil: **a** cohesive sandy; **b** coarse loamy sandy; **c** fine loamy sandy **d** sandy loamy. (*1*) deflated, (*2*) nondeflated, (*3*) protected

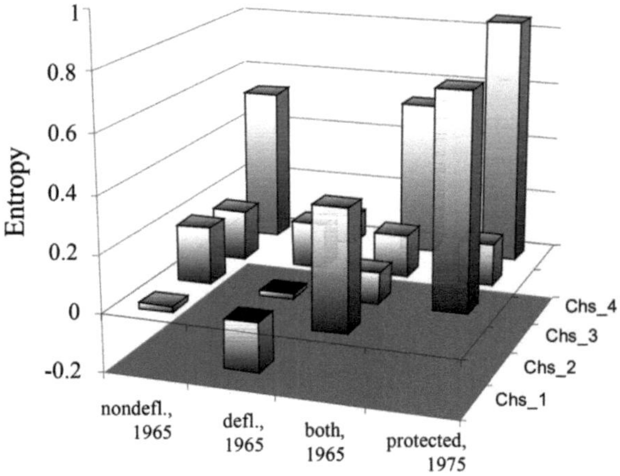

Fig. 8 Entropy of humus content in upper horizon of chestnut soils under the action of deflation and soil-protective agriculture

Table 2 Divergence of humus content in chestnut soils after deflation and prolonged tillage

Variant	(Non-deflated)—(deflated)		Both (1965)—(1975)			
	(Ap)—(Ap)		(Ap)—(Ap)		(B1)—(B1)	
	Divergence	Direction	Divergence	Direction	Divergence	Direction
Chs_1	0.6	− −	N.d.	N.d.	N.d.	N.d.
Chs_2	0.1	0 −	0.1	0 +	0.1	+ +
Chs_3	0.2	− −	0.1	+ −	0.1	0 −
Chs_4	0.6	− −	0.1	+ +	0.7	+ −

Note Designations of soil varieties in this table are the same as those in Fig. 8. The two signs in the column after the value of divergence mean the direction of alterations in the lower and upper limits of variation, respectively

8 Conclusions

The CSE occurs under natural processes and anthropogenous influences; it leads to a change in the probability structure of values of soil properties. Therefore, it is necessary to use probability distributions to estimate their transformations.

Distinctions between probability distributions can be evaluated quantitatively using values of informational divergence and the increment in statistical entropy. They characterize a degree of influence of soil-forming factors and anthropogenous influences on the probability structure of properties of the ground and its stability. Thus, they could be reliable indicators of environmental transformation, that is, important for land resources research, land use policy planning, and basic research.

Appendix: Identification of Probability Distribution Functions

The statistical method of constructing mathematical models of probability distribution involves selecting the type and parameters of distribution functions and fitting them for use as experimental data. Thus, the basic task was to determine pdfs. The standard procedures for verifying hypotheses on how close the statistical distribution studied is to the theoretical one often fail to give a satisfactory, unambiguous answer. This is explained by methodological problems and the ambiguity of hypothesis testing. Thus, we used the approach and software developed by the Department of Applied Mathematics and Informatics at Novosibirsk State Technical University (Lemeshko 2005; Mikheeva and Kuzmina 2000). This has overcome problems in the definition of mathematical functions pdfs using a principle, whereby many hypotheses about coincidence were checked with a large set of theoretical functions. In each case, the only function and corresponding parameters chosen is that which is the best approximation coordinated with the set of parametric and nonparametric criteria. The identification of distribution involved selecting of the best theoretical probability distribution from among the thirty known functions to describe the empirical distribution. An incomplete list of such functions is shown in Table 3.

Table 3 Density functions of statistical probability distributions of soil properties

Function name	Density function
Su-Johnson	$W(x) = \frac{\theta_1}{\sqrt{2\pi}\sqrt{(x-\theta_3)^2+\theta_2^2}} \exp\left\{-\frac{1}{2}\left[\theta_0 + \theta_1 \ln\left\{\frac{x-\theta_3}{\theta_2} + \sqrt{\left(\frac{x-\theta_3}{\theta_2}\right)^2+1}\right\}\right]^2\right\}$
Exponential functions	$W(x) = \frac{\theta_2}{2\theta_1\Gamma(1/\theta_2)} \exp\left\{-\left(\frac{\lvert x-\theta_0\rvert}{\theta_1}\right)^{\theta_2}\right\}$
Double power	$W(x) = \theta_0\theta_1 \exp\{-x\theta_0 - \theta_1 \exp(-x\theta_0)\}$
Maximum value	$W(x) = \frac{1}{\theta_1} \exp\left\{-\frac{x-\theta_0}{\theta_1} - \exp\left(-\frac{x-\theta_0}{\theta_1}\right)\right\}, \quad x > \theta_0$
Minimum value	$W(x) = \frac{1}{\theta_1} \exp\left\{\frac{x-\theta_0}{\theta_1} - \exp\left(\frac{x-\theta_0}{\theta_1}\right)\right\}, \quad x < \theta_0$
Normal	$W(x) = \frac{1}{\theta_1\sqrt{2\pi}} e^{-\frac{(\ln x-\theta_0)^2}{2\theta_1^2}}$
Ln-normal	$W(x) = -\frac{1}{x\theta_1\sqrt{2\pi}} e^{-(\ln x-\theta_0)^2/2\theta_1^2}$
Log-normal	$W(x) = -\frac{1}{x\theta_1 \lg 10\sqrt{2\pi}} e^{-(\lg x-\theta_0)^2/2\theta_1^2}$
Weibull	$W(x) = \frac{\theta_0(x-\theta_2)^{\theta_0-1}}{\theta_1^{\theta_0}} \exp\left\{-\left(\frac{x-\theta_2}{\theta_1}\right)^{\theta_0}\right\}$
Cauchy	$W(x) = \frac{\theta_0}{\pi\left[\theta_0^2+(x-\theta_1)^2\right]}$
Nakagami	$W(x) = \frac{2}{\Gamma(\theta_1)} \left(\frac{\theta_1}{\theta_0^2}\right)^{\theta_1} (x-\theta_2)^{2\theta_1-1} \exp\left\{-\frac{\theta_1(x-\theta_2)^2}{\theta_0^2}\right\}$

For this purpose, the parameters of each of these functions were estimated from the factual data of the samples studied using the method of maximum likelihoods. Then, statistical hypotheses about the agreement between the empirical data and the tested distribution were checked.

Most researchers select the best distribution using a single-fitting criterion for a specified significance level (usually 0.01, 0.05, or 0.1) and accept the hypothesis of agreement when the corresponding statistics do not suggest agreement with other criteria. Our experience shows that several criteria with different measures of agreement should be used, and the decision should be based on their integrity. Our program system used six criteria for testing the hypotheses; the Pearson χ^2 criterion (two modifications), the likelihood ratio test (two modifications), the Kolmogorov criterion, the Smirnov criterion, and two Mises criteria (ω^2 and Ω^2).

The conventional procedure of hypothesis testing involves the nonrejection of the null hypothesis. However, frequently there is no reason to reject several hypotheses; that is, a number of alternatives remain. At the same time, the distribution with the best fit should be selected. Therefore, the probability is calculated for each *i*th criterion and each *j*th theoretical distribution (where i is the index of one of the K criteria used, and j is the index of one of the R tested theoretical distributions):

$$P\{S_i > S_{ij}^*\} = \int_{S_{ij}^*}^{\infty} g_i(s)\mathrm{d}s = \alpha_{ij},$$

where S^* is the corresponding statistics of the criterion used for the *i*th distribution, and $g_i(s)$ the known distribution density function of statistic S_i provided that the H_0 hypothesis is true.

Thus, in testing the hypothesis about the agreement with the *j*th distribution by the *i*th criterion, if $\alpha_{ij} > \alpha$ (where α is the preset significance level), there is no reason to reject the hypothesis of agreement with the *j*th distribution according to the *i*th criterion. There is also no reason to reject the hypothesis about the agreement with many laws noted by indices $R1$ (of R indices). Thus, the distribution law is selected for which $\forall i \alpha_{ij} = \max_{j \subset Rl} \alpha_{ij}$.

An unambiguous conclusion may usually be drawn. However, if any uncertainty remains (e.g., in the case of similar distribution laws), the multicriterion problem of decision making is solved. In this case, a simple compromise criterion is composed in the form $\max_{j \subset Rl} \sum_i \omega_i \alpha_{ij}$, where ω_i is the weighting coefficient of the *i*th criterion, $\sum_i \omega_i = 1$.

We have applied this statistical procedure, which has allowed pdfs to be identified with very high probability values. The usage of large samples (n = 50…600) gives us confidence that the form of probabilistic distribution received by this statistical procedure is close to its real form. We have identified pdfs of soil properties for each chestnut soil series; and as a result, this has given us a database of pdfs of soil properties on the Kulunda steppe at different stages of soil usage (Mikheeva 2001, 2005a, b).

Substituting the specific values of the parameters in a distribution equation in Table 3, we obtain the specific form of the pdfs. For example, the pdf of the clay content in the A horizon of the loamy sandy chestnut soil is the Su-Johnson distribution, with parameters $\theta_0 = -0.43$, $\theta_1 = 1.78$, $\theta_2 = 2.21$ and, $\theta_3 = -7.17$. Substituting the parameter values in the corresponding equation in Table 3 gives us the specific function of pdfs for this property:

$$W(x) = \frac{1.78}{\sqrt{2\pi}\sqrt{(x-7.27)^2 + 2.21^2}} \times \exp\left\{-\frac{1}{2}\left[-0.43 + 1.78\ln\left\{\frac{x-7.27}{2.21} + \sqrt{\left(\frac{x-7.27}{2.21}\right)^2 + 1}\right\}\right]^2\right\}$$

The seeming awkwardness of the equation poses no problem, as all the calculations are performed by a computer.

References

Guidelines All-Union (1973) All-union guidelines for soil survey and compilation of large-scale maps of land tenure, (Общесоюзная инструкция по почвенным обследованиям и составлению крупномасштабных почвенных карт землепользования)1973. Kolos, Moscow, p 48

Bogdanov AA (1989) Tectology: Vseobshyaya organizacionnaya nauka (Тектология : всеобщая организационная наука). V1–2. Moskow. Economika, p 530

Dmitriyev YeA (1983) Soil horizons. Soviet Soil Sci 15(4):62–70

Fridland VM (1972) Structure of soil cover (Структура почвенного покрова). Moscow Publishing, Mysl, p 422

Gubarev VV (1992) Probability models (Вероятностные модели). Part 2. Publishing Novosibirsk State Technical University, Novosibirsk, Russia, p 422

Kanamori T (2014) Scale-invariant divergences for density functions. Entropy 16:2611–2628. doi:10.3390/e16052611

Lemeshko BYu (2005) Statistical analysis of univariant observations of a random variable: a program system. (Статистический анализ одномерных наблюдений случайных величин). Publishing Novosibirsk State Technical University, Novosibirsk, Russia, p 125

Kolmogorov AN (1965) Problems of transfer of information (Проблемы передачи информации). Moscow 1:3–11

Kozlovsky PhI (1991) Contemporary natural and anthropogenous processes of soil evolution (Современные естественные и антропогенные процессы почвенной эволюции). Nauka Publishing, Moscow, p 195

Mikheeva IV (2001) Statistical probability models of soil properties (with Chestnut Soils of the Kulunda Steppe as an Example). (Вероятностно-статистические модели свойств почв на примере каштановых почв Кулундинской степи). Publishing Rossiyskoy Akademii Nauk, Novosibirsk, p 200

Mikheeva IV (2004) Statistical entropy as a criterion for estimation evolution and dynamics of top soil. (Статистическая энтропия как критерий оценки эволюции и динамики почвенного покрова) Sibirskii Ekologicheskii Zhurnal 3: 445–454

Mikheeva IV (2005a) Statistical probability evaluation of sustainability and variability of nature objects under modern processes (with Chestnut Soils of the Kulunda Steppe as an Example). (Вероятностно статистическая оценка устойчивости и изменчивости природных объектов при современных процессах на примере каштановых почв Кулундинской степи). Publishing Rossiyskoy Akademii Nauk, Novosibirsk

Mikheeva IV (2005b) Spatial fluctuations and statistical probability distributions of chestnut soil properties in the Kulunda Steppe. Eurasian Soil Sci 38(3):278–288

Mikheeva IV (2009) Divergence of probability distributions of the soil properties as a quantitative characteristic of the soil cover transformation. Contemp Prob Ecol 2(6):667–670

Mikheeva IV (2013a) Use of soil survey database for the probabilistic evaluation of soil cover transformation in the semiarid zone of Western Siberia. In: SA Shahid et. al (eds) Developments in soil classification, land use planning and policy implications: innovative thinking of soil inventory for land use planning and management of land resources. Springer Science+Business Media Dordrecht. pp 305–314

Mikheeva IV (2013b) Probabilistic models of spatial fluctuations of edaphic properties in native soils in Steppe Zone of Western Siberia. Eurasian. J Soil Sci 2(2):82–89

Mikheeva IV, Kuzmina Ye D (2000) Evaluation of soil properties changes by treatment of their statistical distribution. Technique Recommendation. Institute of Soil Science and Agrochemistry SB RAS, p 32

Part III
Novel Approaches and Technologies of Application Potentials for Siberia

Chapter 13
Study of the Suitability of NIR Spectroscopy for Monitoring the Contamination of Soils with Oil Products

Klara G. Pankratova, Vladimir I. Shchelokov, Galina A. Stupakova and Victor G. Sychev

Abstract The suitability of near-infrared (NIR) spectroscopy for monitoring the contamination of soils with oil products has been studied with a view to expanding the application range of NIR analyzers used in the agrochemical laboratories of Russia. Experiments have been performed on arable soils of various types and varieties differing in texture and the contents of humus and nutrients sampled from different regions of European Russia and artificially contaminated with commercial oil products (gasoline, kerosene, diesel fuel, and motor oil). Laboratory-scale scanning diffusion-reflectance NIR analyzers have been used. It has been shown that the differences in soil types, soil moisture, and humus content, which are reflected in the NIR spectra, affect the results of the NIR analysis of soils. Their effect can be reduced using samples in the entire range of the affecting parameters for the calibration set of NIR analyzers. Using separate calibrations for two soil groups (organomineral and mineral soils) gives better results than the same calibration for all soil types. The effect of particle size distribution can be reduced by unifying the sample preparation procedure used to calibrate the instrument and analyze unknown samples and using spectral derivation. The level at which the soil is supplied with the main nutrients (P, K, Ca, and Mg) has no effect on the results of analysis. The content of a selected oil product in the soil can be determined in the presence of other oil products, if the calibration set of the NIR analyzer includes all expected oil products. The NIR analyzer calibrated for a single oil product will

K.G. Pankratova (✉) · V.I. Shchelokov · G.A. Stupakova · V.G. Sychev
Pryanishnikov All-Russian Research Institute of Agrochemistry, ul. Pryanishnikova 31A, Moscow 127550, Russia
e-mail: kgpan@yandex.ru

V.I. Shchelokov
e-mail: vic371@yandex.ru

G.A. Stupakova
e-mail: vniia@list.ru

V.G. Sychev
e-mail: info@vniia-pr.ru

L. Mueller et al. (eds.), *Novel Methods for Monitoring and Managing Land and Water Resources in Siberia*, Springer Water,
DOI 10.1007/978-3-319-24409-9_13

determine the content of all the oil products in the soil. The relatively high determination limit of about 0.2 % and the calibration on a large number of native soil samples have to be taken into consideration. The obtained results showed that NIR spectroscopy is a promising technique for monitoring the contamination of soils with oil products. When the NIR methods are introduced into further laboratories in Russia, based on our results, standardized procedures for sampling, calibration, and analysis can be developed.

Keywords Soil • Oil • Contamination • NIR spectroscopy • Calibration

1 Introduction

Environmental contamination with compounds that are hazardous for human health is a very serious problem in many countries (Panagos et al. 2013; Science Communication Unit 2013). Oil products are among the numerous hazardous substances of anthropogenic origin contaminating the environment (air, water, soil, vegetation, etc.). Not only the soils of industrial enterprises, cities, and other settlements are subject to contamination but also more and more agricultural and natural lands.

The choice of remediation measures to be undertaken should be based on the assessment of the level and scope of contamination.

The main quantitative chemical analysis methods used at present to determine oil products in soils in laboratories of the agroindustrial complex of the Russian Federation are the gravimetric method (PND F 16.1.41 2004), the IR-spectroscopic method (PND F 16.1:2.2.22 2005), and the fluorimetric method (PND F 16.1:2.21 2012). The most precise method is gas chromatography, which is recognized as an ISO method (ISO 16703 2005); however, it has not yet been introduced into the agrochemical laboratories of Russia. All these methods involve the extraction of all polar and nonpolar hydrocarbons from the sample with an organic solvent, the purification of the extract on a chromatographic column, and the determination of the oil product concentration in the extract using one of the above-mentioned techniques. Their shared disadvantage is their relatively long duration (up to several hours for some of them).

2 Near-Infrared Spectroscopy

2.1 Principle

Near-infrared (NIR) spectroscopy is one of the latest methods suitable for environmental monitoring in a large scale. This fast and cost-effective technique can provide qualitative and quantitative information unavailable from other methods and is finding increasing use in quality control for numerous agricultural and industrial products and materials (Burns and Ciurczak 2008). The main advantage of this method is that there is no need for extraction and purification stages because one of the versions of this method (diffuse reflectance mode) involves measurements of the light diffusely reflected from the surface of a solid sample (rather than the light passed through the extract), which significantly reduces the duration of the analysis.

The NIR spectral region (from 800 to 2500 nm, or from 12,800 to 4000 cm^{-1}) includes bands that are less intense (usually by 10–100 times) and broader than in the other spectral ranges. It is not usually possible to analyze the spectra visually or assign peaks to individual vibrations. Nevertheless, the development of electronic methods and computer technologies has enabled the creation of instruments suitable to analyze various matrices.

Modern NIR analyzers ensure that the analytical procedure is extremely simple. A powdered (or granular) sample is put into the sample holder, and the measurement takes 1–2 min. depending on the instrument model. The NIR analyzers vary in their construction (transmission or reflection modes, wavelength scanning or filter selection, etc.), scope (products and components to analyze), mode of use (in-line, laboratory, or portable), and sample preparation (restrictions on particle size, water content, sample size, etc.). The advantages related to the minimum sample preparation, sample nondisturbance, and fast and precise analytical procedure in combination with the exclusion of toxic chemical reagents and problems of waste disposal make NIR spectroscopy an attractive alternative to conventional chemical analyses.

However, the seeming simplicity of the instruments and their application hides extremely complicated processes of measurements and data processing. The NIR spectroscopy is a secondary analytical technique that requires the calibration of instruments with other (standardized) methods and the application of chemometric methods.

Because of the problems related to the assignment of spectral bands, the statistical processing of spectral data is of special importance. Methods of data processing and calibration are complex, requiring a great deal of time and hard work. Selecting the optimum set of analytical wavelengths and the method of mathematical processing depends on the specific product to analyze and components to be determined; this is a complicated methodological problem.

The calibration equation of a diffusion reflectance NIR analyzer takes the form (Burns and Ciurczak 2008)

$$[X] = \sum_{i=1}^{n} K_i C_i + K_0 \tag{1}$$

where $[X]$ is the concentration of the component to be determined (in our case, oil products); K_0 and K_i are calibration coefficients; C_i is the value of the analytical signal measured at the wavelength λ, and n is the number of analytical wavelengths.

The following main analytical signals are used in NIR analyzers:

(a) $C_i = D_i = \log 1/R_i$ (where R_i is the reflectance) is the absorbance (or optical density) measured at the wavelength λi.
(b) $C_i = dD_i = d(\log 1/R_i)$ is the first derivative of optical density measured at the wavelength λi.
(c) $C_i = d^2D_i = d^2(\log 1/R_i)$ is the second derivative of optical density measured at the wavelength λi.

Optical density $D_i = \log 1/R_i$ is the only signal that is used in NIR analyzers with narrowband interference filters to select analytical wavelengths. Scanning analyzers can also use derivatives of optical density. The first and second derivatives are usually used because the absolute values of higher derivatives become too small, which decreases the signal-to-noise ratio and thus increases the effect of noise.

The quality of calibration and prediction is assessed using the following main statistics (Davies and Fearn 2006).

The standard error of calibration (SEC; also called standard error of estimate, SEE) is the standard error of difference between the reference method values and NIR spectroscopy analysis values in the calibration set of samples. It is calculated as follows:

$$\mathrm{SEC} = \sqrt{\frac{\sum (X_i - Y_i)^2}{(N - p - 1)}}, \tag{2}$$

where X_i is the calculated value for the i-th sample from the calibration set, Y_i is the reference method value for the i-th sample, N is the number of samples in the calibration set, and p is the number of independent variables (calibration coefficients in the calibration equation, excluding the intercept).

To validate the calibration using an independent set of samples, the standard error of prediction, or standard error of performance (SEP) is calculated as follows:

$$\mathrm{SEP} = \sqrt{\frac{\sum (X_i - Y_i)^2}{N}}, \tag{3}$$

where X_i is the calculated value for the i-th sample from the validation set, Y_i is the reference method value for the i-th sample, and N is the number of samples in the validation set.

Every component of the sample affects the entire spectrum in the NIR range; therefore, NIR spectroscopy is characterized by a strong matrix effect on the accuracy of analysis. This effect can be reduced in two ways: either by including samples in the calibration set which have the maximum variation range of each affecting parameter or by carrying out the analysis under conditions excluding this effect. The former approach is used by many producers of NIR analyzers: they calibrate their instruments using large calibration sets to include all possible variations in sample composition, e.g., about 40,000 wheat samples for the calibration of Infratec 1241 (Persson and Sjödin 2004). The latter approach is mainly used to exclude the effect of such external factors as temperature: the working conditions for which the effect of temperature is compensated are indicated in the instrument manual.

2.2 Application of NIR for Soil Analyses

The method of using NIR and other spectrophotometric methods already has a successful tradition in Soil Science and related disciplines. Dalal and Henry (1986) applied Near Infrared Diffuse Reflectance Spectrophotometry for the simultaneous prediction of the moisture, organic C, and total N contents of air-dried soils. The best predictions were achieved within a narrow range of soil color and with moderate amounts of organic matter (0.3–2.5 % C). Later, more and more focus was placed on detecting and analyzing hydrocarbons in soils. For example, Hörig et al. (2001) tested air-borne hyperspectral remote sensing to detect hydrocarbons in the field. When applying NIR measurements in the laboratory, Zwanziger and Förster (1998) showed that it is possible to distinguish among contaminated samples of stone chippings, sand, cultivated soil, humus, and potting soil using the multivariate analysis of NIR data. Chakraborty et al. (2010) used visible NIR (VisNIR, 350–2500 nm) diffuse reflectance spectroscopy (DRS) as a rapid, nondestructive, proximal-sensing technique to predict the total petroleum hydrocarbon content in the soil using partial least squares (PLS) regression and boosted regression tree models. They tested three combinations of moisture content and pretreatment: (i) field moist intact aggregates, (ii) air-dried intact aggregates, (iii) and air-dried ground soil, sieved through a 2-mm sieve (Chakraborty et al. 2010). Chakraborty (2011) developed a method for the rapid identification of oil-contaminated soils in the field using visible NIR DRS. Chakraborty et al. (2012) combined the use of VisNIR prediction and geostatistics to identify the spatial patterns of petroleum contamination in soil. They favor this approach for mapping the spatial variability of petroleum-contaminated soils. Also, Okparanma et al. (2014a, b) found that the NIR methodology may be useful for the rapid assessment of the spatial variability in polycyclic aromatic hydrocarbons in petroleum-contaminated soils. The diffuse reflectance of soil decreased as the oil concentration and clay and moisture contents increased. When studying the combined influence of the oil concentration, moisture content, and clay content on soil reflectance spectra, these interaction effects were significant and thus remained a problem for prediction accuracy in some natural soils (Okparanma and Muazen 2013a, b).

In the assessment of soil contamination with oil products, the main factors that can affect the results of analysis using NIR spectroscopy include particle size distribution, soil moisture, humus and nutrients as well as the procedure of sample preparation and the method of calibration.

3 Studying the Effect of Different Factors on the Assessment of Soil Contamination with Oil Products Using NIR Spectroscopy

3.1 Objects and Apparatus

In the Pryanishnikov All-Russian Research Institute of Agrochemistry, studies were performed on the suitability of diffuse-reflectance NIR spectroscopy to assess the contamination of soil with oil products, with a view to expanding the application ranges of NIR analyzers used in the Russian agrochemical service system.

The objects of study were arable soils of different types and varieties differing in texture and in their humus and nutrient contents. The soil sampling and analysis were performed at the stations of the Agrochemical Service of the Russian Federation within the framework of long-term soil monitoring. Soil samples were taken from the 0- to 20-cm layer of arable soils at more than 200 sites in different regions of European Russia according to the Russian standard (GOST 28168 1989). The soils were analyzed using the standard procedures: pH_{KCl} by means of the potentiometric titration of a salt extract (GOST 26483 1985); total acidity using the Kappen method modified by TsINAO: the potentiometric titration of a sodium acetate suspension (GOST 26212 1991); humus using the Tyurin method modified by TsINAO: the oxidation of organic matter with potassium dichromate and the photocolorimetric determination of trivalent chromium equivalent to the content of organic matter (GOST 26213 1992); available phosphorus and exchangeable potassium using the Kirsanov method modified by TsINAO: the extraction of available phosphorus and exchangeable potassium compounds with hydrochloric acid and the photocolorimetric determination of phosphorus as a blue phosphorus–molybdenum complex and potassium on a flame photometer (GOST 26207 1991); and exchangeable calcium and exchangeable magnesium using the TsINAO method: the extraction of exchangeable calcium and exchangeable magnesium with a potassium chloride solution and the measurements of absorbance at 422.7 and 285.2 nm for Ca and Mg, respectively, on an atomic absorption spectrophotometer (GOST 26487 1985).

The main agrochemical parameters of the studied soils are given in Table 1.

Commercial gasoline (C5–C10), kerosene (C10–C16), diesel fuel (C14–C20), and motor oil (C20–C50) were used as oil products. Experiments were performed on model soil samples artificially contaminated with different concentrations of oil products in the range from 0 to 3–5 %. This range was used because no maximum permissible concentration of oil products in soils has been defined in Russia, but it is generally thought that the level of 3 % roughly corresponds to the lower degree of soil contamination with oil products (Pikovskii 2003).

Measurements were conducted using three spectrometers: a Pacific Scientific Model 6250 (NIRSystems, Silver Spring, MD, USA), an NIRSystems model 6500 (FOSS-NIRSystems, Silver Spring, MD), and an Infrapid-61 (Labor-MIM, Hungary). These are diffuse reflectance NIR analyzers that work in the almost same

Table 1 Agrochemical parameters of the soils studied

No.	Soil[a]	Texture class	pH_{KCl}	Total acidity	Humus (%)	P_2O_5	K_2O	Ca	Mg
						mg/kg		meq/100 g	
1	Luvic Chernozem[c]	Sandy clay	4.7–6.9	0.6–5.9	0.8–8.5	6–541	59–800	13.7–150.0	2.7–5.5
2	Haplic Chernozem	Clay loam	5.1–7.0	0.5–4.8	4.1–8.0	12–252	85–694	18.2–92.0	1.9–5.7
3	Greyzemic Chernozem	Sandy clay	4.8–6.2	1.0–7.1	2.6–8.2	32–1159	63–363	13.7–21.1	2.6–5.1
4	Albic Luvisol (Clayic)	Clay loam	1.4–3.4	168–453	2.2–3.5	213–355	11.2–14.0	2.0–2.9	2.2–3.5
5	Greyzemic Luvic Phaeozem (Clayic)	Clay loam	4.5–6.5	0.7–3.8	2.3–3.1	87–480	120–322	13.3–16.3	2.2–3.5
6	Chernic Luvic Phaeozem (Clayic)	Clay loam	5.1–6.2	1.2–3.4	2.4–5.1	196–422	148–200	12.9–18.0	2.0–4.2
7	Albic Retisol (Humic, Clayic)	Clay loam	4.7–6.8	0.4–2.7	1.1–2.9	34–694	56–414	8.2–13.3	1.3–3.3
8	Albic Retisol (Humic, Siltic)	Loam	4.6–6.4	0.8–4.5	1.3–2.7	45–1173	83–421	6.1–10.6	1.1–2.9
9	Albic Retisol (Humic, Loamic)	Sandy loam	4.7–6.2	0.7–2.7	2.0–2.9	20–239	46–124	5.2–8.8	0.8–1.8
10	Albic Retisol (Humic, Arenic)	loamy sand	5.9–7.0	0.4–1.4	1.0–4.3	105–1251	58–293	7.0–19.7	1.2–4.1
11	Albic Retisol (Humic, Arenic)	Sand	4.9–6.4	0.2–4.6	0–5.9	30–460	42–186	3.2–24.0	1.0–9.0
12	Fluvisol (Humic)	sand to Clay loam[b]	5.1–7.0	0.2–7.8	0–8.2	47–405	62–360	1.7–29.7	0.8–6.5
13	Gleyic Fluvisol (Humic)	Sandy loam to clay loam[b]	5.0–7.3	0.4–8.6	0–12.1	40–485	67–421	4.2–30.0	1.0–6.8
14	Gleyic Histic Fluvisol	–	5.1–6.8	5.2–28.6	–	12–121	58–268	3.6–6.9	0.8–1.8
15	Gleyic Histic Retisol	–	4.3–5.2	6.5–24.2	–	38–86	60–107	1.1–1.3	0.5–0.8
16	Eutric Histosol	–	4.7–5.5	1.6–26.4	–	21–82	40–177	0.6–2.3	0.3–1.0
17	Dystric Histosol	–	6.4–7.1	3.0–16.5	–	9–104	76–160	3.0–3.7	1.0–1.3

[a]Soil names are given in accordance with the World Reference Base for Soil Resources (2014). They are based on field classifications

[b]Several samples of each indicated and intermediate texture class were used

[c]A few samples of transition horizons included

(–) no data

wavelength range (1100–2500 nm, or 1300–2400 nm for Infrapid-61), have analogous software, and give similar results. Infrapid-61 has a slightly inferior performance, but it is most commonly used in the Russian agrochemical service system. The embedded software involved algorithms for processing the data obtained by scanning the entire instrument range as well as those from specified wavelengths (simulating the measurements on filter-based NIR analyzers).

Calibration equations were developed using two methods: stepwise multiple linear regression (SMLR) and PLS regression. In the former case, a linear regression equation is calculated for several analytical wavelengths; the latter algorithm calculates a regression equation based on the spectrum of analyzed material in the entire range of the instrument. The number of calibration coefficients (analytical wavelength) is significantly larger in this case (e.g., 160 against 5).

3.2 *Soil Type and Texture*

Soils may differ in genesis, composition, structure, physical and chemical properties, water and heat conditions, and other parameters. This diversity can be reflected in the spectral parameters of soils, including their NIR spectra.

The effect of differences in the spectral parameters of different soil types and subtypes on the determination of their contamination with oil products by NIR spectrometry was studied using the various soils (listed in Table 1) and oil products. The spectra of two highly different varieties, an organic soil (Dystric Histosol) and an organomineral soil (Albic Retisol), are given in Fig. 1 in comparison to the spectrum of an oil product (diesel fuel).

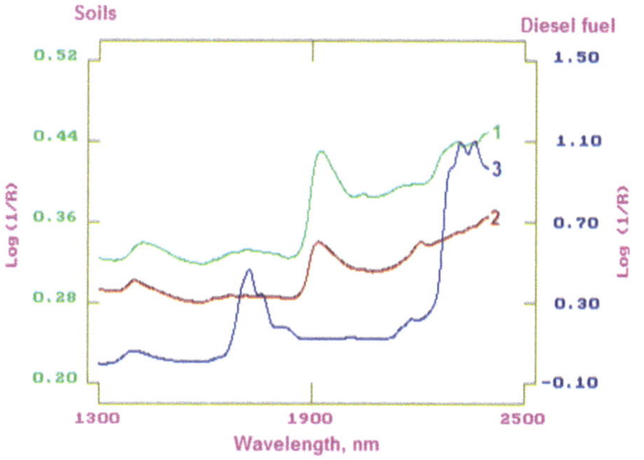

Fig. 1 NIR spectra of *1* Dystric histosol, *2* Albic retisol, and *3* Diesel fuel. The most significant differences between the spectra of soils differing greatly in humus content coincide with the positions of the fuel absorption bands

It can be seen that the most significant differences between the spectra of the soils are in two regions coinciding with the positions of the high absorption bands of the fuel (about 1720 and 2300 nm). This fact suggests that there could be some problems determining oil products in the organic soils because of the superposition of the absorption bands of soil organic matter with those of oil hydrocarbons when present in the soil.

Processing the spectra recorded for the sets of soils contaminated with different amounts of oil products (gasoline, diesel fuel and motor oil) in the concentration range from 0 to 3 % allowed us to calculate the regression equations relating the spectral parameters of the contaminated soils with their oil product contents. The equations differed in statistical terms: SEC varied from 0.12 % for Albic Retisol (Humic, Loamic) to 0.39 % for Gleyic Histic Retisol, and the correlation coefficient R varied from 0.86 for Gleyic Histic Retisol to 0.99 for Albic Retisols ($n = 30$–50, oil product concentration range 0–3 %).

The total set of all samples was also used to calibrate the NIR analyzer. The validation using a set of independent samples resulted in SEP = 0.4 % and $R = 0.92$. The calibration performed for the organic and organomineral soils separately gave better results: SEP was 0.23 % for organomineral soils and 0.28 % for organic soils.

Thus, the use of the same calibration model for all soil types allows the contamination of soils with oil products to be assessed without any preliminary analysis of their taxonomy, but the results are not very exact because the soil matrices have significantly different properties. The use of separate calibration for a specific soil with properties which are very different from those of other soils obviously improves the analysis accuracy of independent soil samples.

The effect of soil texture on the determination of soil contamination with oil products was studied on Albic Retisols of different textures (soils 7–11 in Table 1).

The particle-size distribution is the most significant factor affecting NIR spectra. The amount of light absorbed and diffused by a particle depends on the particle size (Hapke 1993).

In many NIR analyzers, the effect of particle size is corrected using derivative spectra. Our studies showed that when the first derivative spectra were used, the particle size distribution in the soil had no significant effect on the results of analysis: The values of SEP (0.12–0.16 %) differ insignificantly among the soil subtypes as well as from the corresponding value for the united calibration model (SEP = 0.14 %) in the oil product concentration range of 0–3 %.

Thus, the soil contamination with oil products can be assessed on an NIR analyzer using the same calibration model for soils with different textures.

3.3 Humus Content

As was noted earlier (Sect. 3.2), organic and organomineral soils differ in the results of analysis for oil products. Therefore, the effect of humus content on the determination of oil products in soils was studied in more detail. Two types of calibrations in the oil product concentration range of 0–3.5 % were developed: (A) three calibrations for gasoline, diesel fuel, and motor oil in Chernozems, Luvisol, and Phaeozems (1–6 in Table 1) containing less than 5 % humus and (B) analogous calibrations for Albic Retisols (7–11 in Table 1) containing less than 2.5 % humus (n = 70–90). These calibrations were then validated on the sets of independent samples containing more than 5 and 2.5 % humus, respectively.

The obtained results show that the analysis of soil samples containing more than 5 % humus using calibrations (A) resulted in a significant decrease in the accuracy of analysis. The values of SEP increased from 0.35–0.46 to 0.74–0.80 % for gasoline, from 0.15 to 0.16–0.24 % for diesel oil, and from 0.15–0.22 to 0.31–0.33 % for motor oil. A significant underestimation of results was observed: The systematic error was −0.5 to −0.6 % for gasoline, −0.03 to −0.15 % for diesel fuel, and −0.09 to −0.18 % for motor oil.

The obtained results indicate that the errors when determining oil products in soils correlate linearly with the humus content: The negative error values increase with the humus content. The coefficients of correlation vary from 0.37 to 0.50; i.e., the contribution of the soil humus to the error of determining oil products makes up 14–25 % for diesel fuel and motor oil, respectively. Significantly less certain results were obtained for gasoline, which is related to the spread of data because of its high volatility.

Humus was not revealed to have any effect on the results of determining oil products for the samples of Albic Retisols containing more than 2.5 % humus, analyzed using the calibrations based on the samples containing less than 2.5 % humus. Thus, there is a threshold value (about 5 %), above which the content of humus interferes with the determination of oil products in soils.

It is noteworthy that when the calibration set includes samples in the entire range of humus content, the latter has no effect on the determination of oil products.

3.4 Soil Moisture

Soil moisture can significantly interfere with the determination of oil products because water is extremely well absorbed in the NIR region (Büning-Pfaue 2003). However, the drying of soil during sample preparation can result in the loss of volatile oil components.

To study the effect of moisture on the results of analysis, three sets of motor oil contaminated samples of Albic Retisol (soil 7 in Table 1) were prepared in different moisture ranges: (L) low, 1.0–3.9 %; (M) medium, 7.7–10.3 % and (H) high 12.7–15.4 % (n = 50 for each set). The concentration range of oil product was 0–3.5 % in each set of samples.

The standard deviation of the NIR spectra for the wettest set of samples is given in Fig. 2. The curve characterizes the variation in the soil spectra related to the changes in the contents of water and oil products. It can be seen that the regions of maximum variation due to changes in the content of water (zones W) and oil

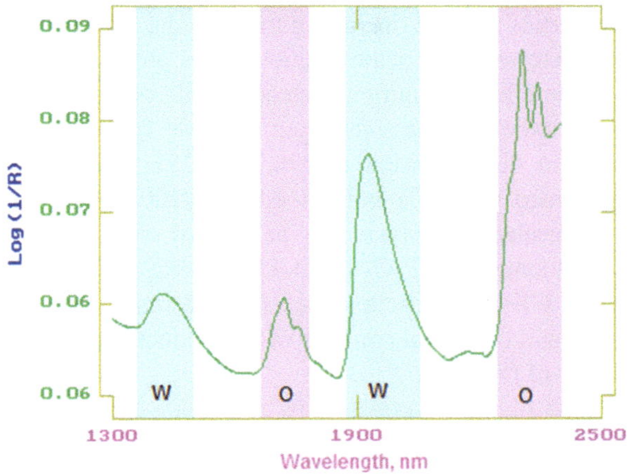

Fig. 2 Standard deviation of NIR spectra for the set of wet samples contaminated with oil products. The regions of maximum variation caused by changes in the content of water (*zones W*) and oil products (*zones O*) are separated in space

products (zones O) are different. This suggests that the effect of water may be lower than expected.

The calibration for oil products developed using sample set L was tested for sample sets M and H. The obtained results show that analyzing wetter soil samples using calibration based on dry soil samples significantly decreases the accuracy of results. The values of SEP increased from 0.12 to 0.44 and 0.82 % for the samples from sets M and H, respectively.

However, this increase in the SEP values was mainly related to the systematic error, which can be decreased by correction based on the results of control soil samples in the corresponding range of water content. Thus, the correction for systematic error decreased the SEP values to 0.17 and 0.16 % for sample sets M and H, respectively.

Further studies showed that the inclusion of wet samples in the calibration set significantly decreases the error of determining oil products (to SEP ~0.14 % for soil samples from all sets).

Thus, the effect of soil moisture on the determination of oil products can be excluded (or significantly decreased) when wet samples are included in the calibration set.

3.5 *Plant Nutrients*

The NIR spectroscopy requires calibration using the set of samples reflecting the natural variability in their properties. Data from the literature indicate that the presence of nutrients such as P, K, Ca, and Mg in the soil, especially when there is a significant variation in their contents because of their uptake by plants and the application of corresponding fertilizers, can significantly affect the spectral properties of soils, including those in the NIR range (Pirie et al. 2005) and, hence, interfere with the determination of oil products in soils using NIR spectroscopy.

The effect of nutrient levels in soils on the results of determining oil products was studied on the soils listed in Table 1; the agrochemical properties of the soils vary in the following ranges: 12–1251 mg/kg P_2O_5, 42–421 mg/kg K_2O, 1.1–30.0 meq/100 g Ca and 0.8–9.0 meq/100 g Mg.

Correlation analysis was performed on data on the agrochemical properties of soils (contents of P_2O_5, K_2O, Ca and Mg) and the error of determining oil products using NIR spectroscopy (i.e., the differences between the results of the NIR analyzer and the actual contents of oil products in soils) in the oil product concentration range of 0–3 %.

The obtained results show that P_2O_5, K_2O, Ca, and Mg in the ranges studied have no significant effect on the results of NIR analysis. The coefficients of determination were statistically insignificant for all nutrients and did not exceed 0.1 in most cases (70–80 % for each parameter). Consequently, the calibration of NIR analyzers and the subsequent analysis of soils contaminated with oil products can be performed without taking into account the soil's supply of P, K, Ca and Mg.

3.6 *Determination of Oil Products Present Simultaneously*

The oil products studied have similar spectra in the NIR range (Fig. 3). However, they also have some visible differences, which can form the basis for the determination of different oil products simultaneously present in the soil.

To study the possibility of determining different oil products simultaneously present in the soil, sets of soil samples were prepared containing the following oil products: (1) kerosene; (2) diesel fuel; (3) motor oil; (4) kerosene + diesel oil; (5) kerosene + motor oil; (6) diesel fuel + motor oil; (7) kerosene + diesel fuel + motor oil (n = 30 for each set, oil product concentration range 0–5 %).

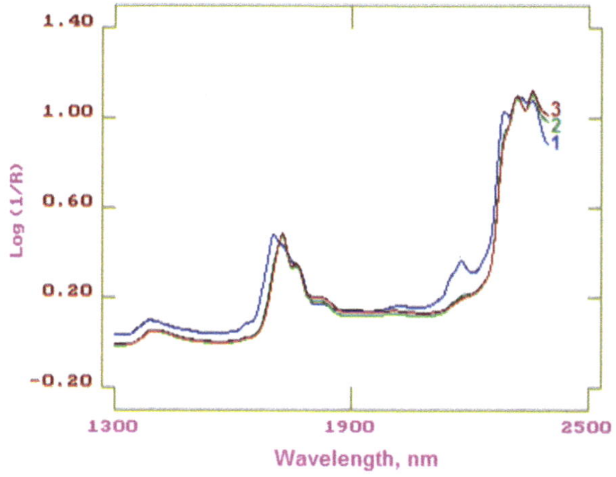

Fig. 3 NIR spectra of *1* kerosene, *2* diesel fuel and *3* motor oil. They are very similar, but show a few visible differences at wavelength ranges of 1700 and 2000 nm

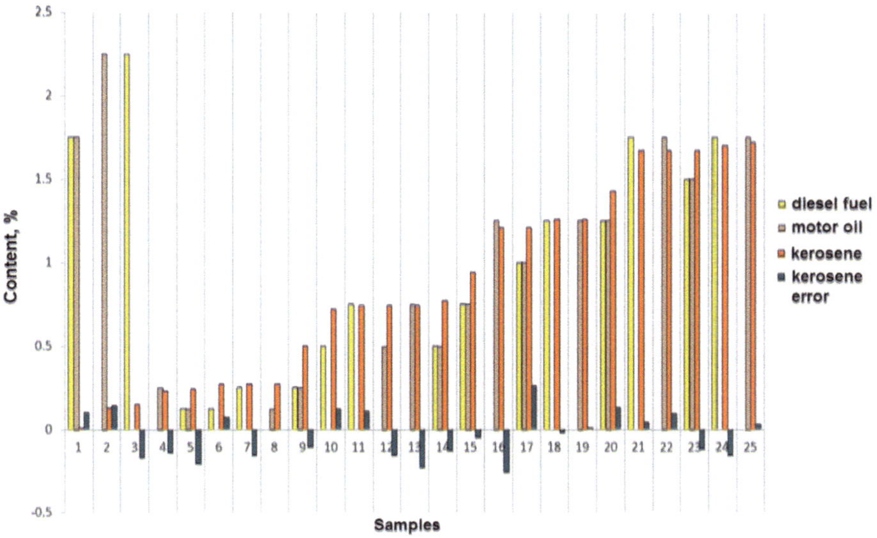

Fig. 4 Error of determining kerosene in the presence of diesel fuel and motor oil. The error of determination of kerosene (*dark columns*) does not depend on the absolute content of any oil product (*light columns*)

Calibration equations were calculated for each oil product and their total content after separating 10 samples from each set for validation.

For each individual oil product, similar determination accuracies were obtained for the samples containing the target oil product alone or mixed with other oil products. For kerosene, the values of SEP were 0.12 % for the samples containing only kerosene and 0.13–0.15 % for its mixtures with diesel fuel, motor oil, or both. For diesel fuel, the corresponding values were 0.19 and 0.16–0.20 % and for motor oil 0.08 and 0.09–0.13 %. For all the oil products, SEP = 0.09 % regardless of the proportions of different oil products.

The comparison of the error of determination for any individual oil product with the contents of other oil products present in the sample showed no correlation and, hence, no effect on the determination of the target product. The error of determining kerosene in the presence of diesel fuel and motor oil is shown as an example in Fig. 4.

However, when the instrument was calibrated using the samples containing only one of the oil products, the presence of other oil products in the analyzed samples resulted in significant errors. An analysis of the results showed that the NIR analyzer calibrated on the single oil product determines the total content of all the oil products present in the sample. This is related to the general similarity of the NIR spectra of all the oil products studied (see Fig. 3). Therefore, in the calibration process for the single oil product, the program selects the wavelengths most closely correlated with the concentration of the target component, i.e., those common for all oil products, while the calibration for one of the oil products in the presence of others selects the wavelengths discriminating the target product.

Thus, the content of a selected oil product in the soil can be determined in the presence of other oil products, if the calibration set of the NIR analyzer includes all possible components. It is not necessary to include all the oil products in each sample. Even when each of the calibration samples contains only one product, the analysis of independent samples containing several oil products gives satisfactory results.

3.7 Sample Preparation and Calibration Strategy

Different methods of sample preparation can be used when determining oil products in soils using NIR spectroscopy. These may be milling in a laboratory mill or triturating in a mortar. However, the main requirement is to use the same, most unified procedure for the calibration and the subsequent analysis of unknown samples. For the samples that may contain volatile oil products, special ball mills suitable for low-temperature milling in sealed working chambers should be used. Some special requirements are imposed upon NIR analyzers used for the determination of oil products in soil:

- the NIR analyzer should work in diffuse reflectance mode;
- a scanning NIR analyzer should be used;
- the wavelength range of the instrument should include the regions of maximum variation related to changes in the content of oil products (1650–1800 and/or 2250–2400 nm).

The following NIR analyzers can be noted as examples: NIRSystems models 4250, 4500, 5000, 6500, SY-3650-II, 3665-II, DS2500 (Foss); DA 7200 (Perten); SpectraStar 2200 and 2400 (Unity Scientific); Infrapid-61 (LaborMIM).

The calibration procedure for determining oil products in soils significantly depends on the nature of the soils and oil products; therefore, no specific method for the mathematical processing of spectral signals can be recommended for developing a calibration model. Several highly promising methods should be tested. The final selection of the calibration model should be based on the results of validation tests.

4 Conclusions

1. Studies performed on a large number of soil types and varieties and different commercial oil products have shown that diffuse-reflectance NIR spectroscopy is a promising technique for monitoring large areas contaminated with oil products.

2. Using the same calibration model for all soil types allows the contamination of soils with oil products to be assessed without a preliminary analysis of the soil taxonomy, but the results are not very exact because the soil matrices have significantly different properties.
3. Using separate calibrations for two soil groups (organomineral and mineral soils) gives better results. To mitigate the effect of humus, the NIR analyzer should be calibrated on samples corresponding to the entire range of humus content in the soils to be analyzed. The use of local calibration for a specific soil also improves the results of analysis.
4. Calibrating the NIR analyzer on a set of air-dried soil samples results in lower accuracy than determining the oil products in wet soils. The error of determination can be reduced by including wet soil samples in the calibration set.
5. The level of soil supply with P, K, Ca, and Mg has no effect on the results of determining oil products using NIR spectroscopy; neither do the differences in particle-size distribution, when the same mill model is used for sample preparation during the calibration and the subsequent analysis of unknown soil samples, and the derivative spectra are processed.
6. NIR spectroscopy can be used to determine separate oil products simultaneously present in the soil. However, the instrument should be calibrated on a set of samples containing all the oil products that can be found in the samples to be analyzed. An NIR analyzer calibrated on a single oil product will determine the content of all the oil products in the soil.
7. The relatively high determination limit of about 0.2 % is a main restriction of the method and has to be taken into consideration.
8. Based on these results, a standardized methodology for the calibration procedure should be developed. This will be required if the NIR method is to be introduced into further Russian laboratories.

References

Büning-Pfaue H (2003) Analysis of water in food by near infrared spectroscopy. Food Chem 82:107–115

Burns DA, Ciurczak EW (eds) (2008) Handbook of near-infrared analysis, third edition (practical spectroscopy), CRC Press. Taylor & Francis Groups, Boca Raton, p 826

Chakraborty S (2011) Rapid identification of oil contaminated soils using visible near infrared diffuse reflectance spectroscopy. Dissertation. Louisiana State University. http://etd.lsu.edu/docs/available/etd-06292011-161018/unrestricted/Chakrabortydiss.pdf. Accessed 12 Mar 2015

Chakraborty S, Weindorf DC, Zhu Y, Li B, Morgan CLS, Ge YF, Galbraith J (2012) Assessing spatial variability of soil petroleum contamination using visible near-infrared diffuse reflectance spectroscopy. J Environ Monit 14:2886–2892. doi:10.1039/C2EM30330B. http://pubs.rsc.org/en/Content/ArticleLanding/2012/EM/c2em30330b#!divAbstract. Accessed 12 Mar 2015

Chakraborty S, Weindorf DC, Morgan CLS, Ge YF, Galbraith JM, Li B, Kahlon CS (2010) Rapid identification of oil-contaminated soils using visible near-infrared diffuse reflectance spectroscopy. J Environ Qual 39:1378–1387. doi:10.2134/jeq2010.0183. https://dl.sciencesocieties.org/publications/jeq/abstracts/39/4/1378. Accessed 12 Mar 2015

Dalal RC, Henry RJ (1986) Simultaneous determination of moisture, organic carbon, and total nitrogen by near infrared reflectance spectrophotometry, SSSAJ 50 (1):120–123. doi:10.2136/sssaj1986.03615995005000010023x. https://dl.sciencesocieties.org/publications/sssaj/abstracts/50/1/SS0500010120. Accessed 12 Mar 2015

Davies AMC, Fearn T (2006) Back to basics: calibration statistics. Spectr Eur 18(2):31–32. http://www2.qta.com/clientimages/43916/pdfs/r2rmsepexplanation.pdf. Accessed 12 Mar 2015

GOST 26207 (1991) Soils. Determination of mobile compounds of phosphors and potassium by Kirsanov method modified by TsINAO (Почвы. Определение подвижных соединений фосфора и калия по методу Кирсанова в модификации ЦИНАО), Moscow, p 7 (in Russian)

GOST 26212 (1991) Soils. Determination of hydrolytic acidity by Kappen method modified by TsINAO (Почвы. Определение гидролитической кислотности по методу Каппена в модификации ЦИНАО), Moscow, p 7 (in Russian)

GOST 26213 (1992) Soils. Methods of determining organic matter (Почвы. Методы определения органического вещества.), Moscow, p 8 (in Russian)

GOST 26483 (1985) Soils. Preparation of salt extract and determination of its pH by the TsINAO method (Почвы. Приготовление солевой вытяжки и определение ее pH по методу ЦИНАО), Moscow, p 6 (in Russian)

GOST 26487 (1985) Soils. Determination of exchangeable calcium and exchangeable (mobile) magnesium by TsINAO methods (Почвы. Определение обменного кальция и обменного (подвижного) магния методами ЦИНАО), Moscow, p 14 (in Russian)

GOST 28168 (1989) Soils. Sampling (Почвы. Отбор проб), Moscow, p 7 (in Russian)

Hapke B (1993) Introduction to the theory of reflectance and emittance spectroscopy. Cambridge University Press, New York

Hörig B, Kühn F, Oschütz F, Lehmann F (2001) HyMap hyperspectral remote sensing to detect hydrocarbons. Int J Remote Sens 22(8):1413–1422

ISO 16703 (2005) Soil quality—Determination of content of hydrocarbon in the range of C10 to C40 by gas chromatography

Okparanma RN, Mouazen AM (2013a) Combined effects of oil concentration, clay and moisture contents on diffuse reflectance spectra of diesel-contaminated soils. Water Air Soil Poll 224:1539. http://link.springer.com/article/10.1007%2Fs11270-013-1539-8. Accessed 12 Mar 2015

Okparanma RN, Mouazen AM (2013b) Combined effects of oil concentration, clay and moisture contents on diffuse reflectance spectra of diesel-contaminated soils. Water Air Soil Poll 224:1539. http://link.springer.com/article/10.1007%2Fs11270-013-1539-8. Accessed 12 Mar 2015

Okparanma RN, Coulon F, Mouazen AM (2014a) Analysis of petroleum-contaminated soils by diffuse reflectance spectroscopy and sequential ultrasonic solvent extraction–gas chromatography. Environ Poll 184:298–305. http://www.sciencedirect.com/science/article/pii/S0269749113004685. Accessed 12 Mar 2015

Okparanma RN, Coulon F, Mouazen AM (2014b) Analysis of petroleum-contaminated soils by diffuse reflectance spectroscopy and sequential ultrasonic solvent extraction–gas chromatography. Environ Poll 184:298–305. http://www.sciencedirect.com/science/article/pii/S0269749113004685. Accessed 12 Mar 2015

Panagos P, van Liedekerke M, Yigini Y, Montanarella L (2013) Contaminated sites in Europe: review of the current situation based on data collected through a european network. J Environ Publ Health Volume 2013, Article ID 158764, p 11. http://dx.doi.org/10.1155/2013/158764. Accessed 12 Mar 2015

Persson JA and Sjödin (2004) Efficiency all the way from grain to flour. In: Focus No. 1: 4–5

Pikovskii YuI, Gennadiev SS, Chernyanskii SS, Sakharov GN (2003) The problem of diagnostics and standardization of the levels of soil pollution by oil and oil products. Eur Soil Sci 36:1010–1017

Pirie A, Singh B, Islam K (2005) Ultra-violet, visible, near-infrared, and mid-infrared diffuse reflectance spectroscopic techniques to predict several soil properties. Austral J Soil Res 43:713–721. http://www.publish.csiro.au/?paper=SR04182. Accessed 12 Mar 2015

PND F 16.1.41 (2004) Procedure for measuring the mass concentration of oil products in soil samples by the gravimetric method. (Методика выполнения измерений массовой концентрации нефтепродуктов в пробах почв гравиметрическим методом). Moscow, p 12 (in Russian)

PND F 16.1:2.2.22 (2005) Procedure for measuring the mass fraction of oil products in soils and bottom sediments by IR spectroscopy. (Методика выполнения измерений массовой доли нефтепродуктов в минеральных, органогенных, органо-минеральных почвах и донных отложениях методом ИК-спектрометрии). Moscow, p 21 (in Russian)

PND F 16.1:2.21 (2012) Procedure for measuring the mass fraction of oil products in soil samples by fluorimetry on a Fluorat-02 liquid analyzer. (Количественный химический анализ почв. Методика измерений массовой доли нефтепродуктов в пробах почв и грунтов флуориметрическим методом на анализаторе жидкости "Флюорат-02" (М 03-03-2012)). Moscow, p 26 (in Russian)

Science Communication Unit, University of the West of England, Bristol (2013) Science for environment policy in-depth report: soil contamination: impacts on human health. Report produced for the European Commission DG Environment. Available at: http://ec.europa.eu/science-environment-policy. Accessed 12 Mar 2015

World Reference Base for Soil Resources (2014) World soil resources report no 106, p 181. ISBN 978-92-5-108369-7. http://www.fao.org/3/a-i3794e.pdf. Accessed 12 Mar 2015

Zwanziger HW, Förster H (1998) Near infrared spectroscopy of fuel contaminated sand and soil. I. preliminary results and calibration study. J. Near Infrared Spectrosc 6(1–4):189–197

Chapter 14
Emerging Measurement Methods for Soil Hydrological Studies

Uwe Schindler, Lothar Mueller, Georg von Unold, Wolfgang Durner and Johann Fank

Abstract Monitoring and protecting the natural resources of soil and water, and their ecosystems, is intended to ensure the long-term conservation of their functions. To understand the reasons for resource degradation or ecosystem alterations and interactions, knowledge is required of processes and parameters on different scales of landscapes. Soil hydrological studies are an essential part of ecosystem and landscape research. The aim of our study was to develop new research methods and technical equipment to understand and monitor soil hydrological processes. The investigations were carried out on different scales, starting with laboratory and lysimeter measurements, followed by investigations in the field. To measure soil hydrological properties, we developed the Extended Evaporation Method (EEM) and the HYPROP device. In this chapter we report on some innovations in this field. Using new cavitation tensiometers and applying the air-entry pressure of the tensiometer's porous ceramic cup as the final tension value allowed us to quantify both hydraulic functions close to the wilting point. Additionally, both soil shrinkage dynamics and soil water hysteresis can now be quantified easily and reliably. The experimental setup followed the HYPROP system, which is a

U. Schindler (✉) · L. Mueller
Leibniz Centre for Agricultural Landscape Research (ZALF) e.V.,
Eberswalder Strasse 84, 15374 Müncheberg, Germany
e-mail: uschindler@zalf.de

L. Mueller
e-mail: mueller@zalf.de

G. von Unold
UMS GmbH, Gmunder Str. 37, 81379 Munich, Germany
e-mail: gvu@ums-muc.de

W. Durner
Braunschweig Technical University, Institute of Geoecology, Langer Kamp 19c,
38106 Brunswick, Germany
e-mail: w.durner@tu-bs.de

J. Fank
Joanneum Research, Elisabethstraße 18/II, 8010 Graz, Austria
e-mail: johann.fank@joanneum.at

L. Mueller et al. (eds.), *Novel Methods for Monitoring and Managing Land and Water Resources in Siberia*, Springer Water,
DOI 10.1007/978-3-319-24409-9_14

commercial device with vertically aligned tensiometers that is optimized to perform evaporation measurements. Depending on the soil and the evaporation rate, the measurement time varied between 2 and at most 10 days. The simultaneous measurement of multiple soil samples was possible with only one balance. Pedotransfer functions (PDFs) were created on the basis of various measured soil water retention and hydraulic conductivity functions. In the next step, a method for quantifying deep seepage and solute leaching under field conditions was developed, tested and applied at more than 40 soil hydrological field plots in Germany. The method is based on tension and soil water content measurements down to a depth of 3 m at arable and grassland sites and a depth of 5 m at forest sites. These data were used to construct a field water retention curve. This pF curve was fitted, the relative hydraulic conductivity function $K(\theta)$ was derived and relative deep seepage rates were calculated based on DARCY's law. To obtain reliable discharge rates, the K function was matched to the water balance. Lysimeter experiments confirmed the validity and reliability of this soil hydrological field method. It works like a virtual lysimeter on sandy to loamy soils which have a deep water table and a zero flux plane above the measurement depth The EEM and the soil hydrological field method have the potential to improve soil hydrological studies, and water and solute transport monitoring systems could be installed in Eurasia.

Keywords Soil hydrology · Soil hydraulic functions · HYPROP · Lysimeter · Soil hydrological field measurements · Deep seepage · Leaching · Soil water balance

1 Introduction

The protection of the soil, water, and other natural resources means the long-term conservation of their functions. These include the soils' filter and buffer properties, and retaining a sustainable function as a habitat for humans, animals and plants. Both the loss and reduction of productive soils through erosion, sealing and other degradation, and the contamination of groundwater and open waters must be avoided. Comprehensive studies of interactive processes between soil, water, plants and the atmosphere to protect the natural resources require knowledge of hydrological processes and their parameters on different scales (Fig. 1).

Laboratory measurements provide basic information about soil data and potential soil water and solute transport processes. These data are required as inputs for simulation models. Lysimeters offer the opportunity to study the interaction between soil, water, plant, management and the atmosphere in large soil columns under defined boundary conditions. They provide a good basis for developing and testing balancing methods and for validating simulation models. Investigations and simulations on the field, catchment and landscape scales are influenced by a large number of separate processes. These studies aim to expand our knowledge and

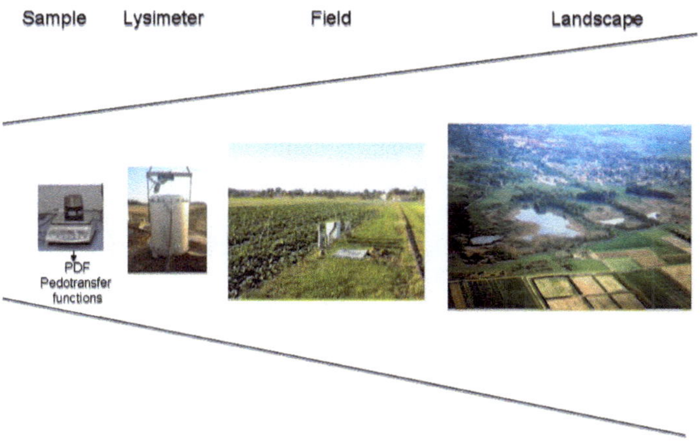

Fig. 1 Soil hydrological studies on different scales

understanding of the process interactions. Hydrological simulation models and models of higher complexity help to understand the response of the resource of land and water, of ecosystems and landscapes to a changing environment in terms of weather, climate and human impacts. The aim of this chapter is to present novel soil hydrological measurement methods which may improve our basic understanding of processes of water storage, water flow and solute transport in soils. Most equipment can also serve as field monitoring tools for water and solute transport.

2 Measurement of Soil Hydraulic Properties in the Laboratory

2.1 Classical Methods and Devices

Knowledge of soil hydraulic properties—the water retention curve or pF curve and unsaturated hydraulic conductivity (K)—is required for soil water modelling and various soil hydrological studies. Traditionally, soil hydraulic properties have been determined using various methods and procedures. Depending on the desired measuring range, different methods and devices are in operation to determine the soil water retention curve. In the low tension range, between 0 and 10 kPa, the sand box (de Rooij et al. 2004; Cresswell et al. 2008) is the common method for quantifying the water retention data points. The sand/kaolin box is mainly used in the tension range between 10 and 50 kPa. For higher tensions (100–1500 kPa), the pressure plate extractor is applied (Dane and Hopmans 2002).

Additionally, various methods are available to estimate the unsaturated soil hydraulic conductivity function of soil samples. The steady state pressure membrane procedure (Henseler and Renger 1969; Boels et al. 1978; Schindler et al. 2012) and the tension disc infiltrometer method (Reynolds and Elrick 1991) only allow hydraulic conductivity values to be measured in the low tension range. Hot-air methods (Arya 2002; Tyner et al. 2006) and centrifugation techniques (Nimmo et al. 2002) provide rapid measurements of soil water diffusivity and unsaturated hydraulic conductivity. The measurement conditions, however, differ markedly from natural conditions for all these methods. The one-step method (Kool et al. 1985), and especially the multistep outflow method (Hopmans et al. 2002; Fujimaki and Mitsuhiro 2003; Durner and Iden 2011), produce reliable hydraulic conductivity data and are widely in use.

Measurements of pF and K with the classical methods are, however, time-consuming, the equipment is costly and measured results are affected by uncertainties. The evaporation method provides an effective alternative (Wind 1968; Schindler 1980; Wendroth et al. 1993; Halbertsma 1996).

2.2 The Evaporation Method

The evaporation method is frequently used for the simultaneous determination of the hydraulic functions of unsaturated soil samples, i.e., the water retention curve and hydraulic conductivity function (Wind 1968; Becher 1970; Schindler 1980; Klute and Dirksen 1986; Plagge 1991: Halbertsma 1996; Wendroth et al. 1993; Bertuzzi and Voltz 1999; Šimůnek et al. 1999; Schindler and Müller 2006). Due to the limited measurement range of common tensiometers, all methodological variations of the evaporation method suffer from the limitation that the hydraulic functions can only be determined up to at most 70 kPa.

Measurement procedure: In a soil sample (250 cm^3, height 5 cm) two tensiometers are installed at depths of 1.25 and 3.75 cm (Fig. 2). The sample is saturated with water from the bottom, sealed at the bottom and placed on a balance. Its surface remains open to free evaporation. Tensions (Ψ) and the sample mass (m) are recorded at consecutive times. Individual points of the water retention curve are calculated on the basis of the water loss per volume of the sample at a certain time t and the geometric mean tension of the sample at that time. The hydraulic conductivity (K) is calculated according to the modified Darcy-Buckingham's law (Eq. 1), where the evaporated water volume per time interval relates to half the sample height versus the hydraulic gradient as determined by the tensiometers (Schindler 1980). The flux (q) is derived from the soil water volume difference ΔV (1 cm^3 of water = 1 g) per surface area (A) and time unit (Δt). The mean hydraulic gradient (i_m) is calculated on the basis of the mean tensions in time intervals.

$$K(\overline{\psi}) = \frac{\Delta m}{2A\rho_{H_2O}\Delta t i_m} \tag{1}$$

where $(\overline{\psi})$ is the mean tension geometric averaged over the upper and the lower tensiometer and the time interval, Δm is the sample mass difference in the time

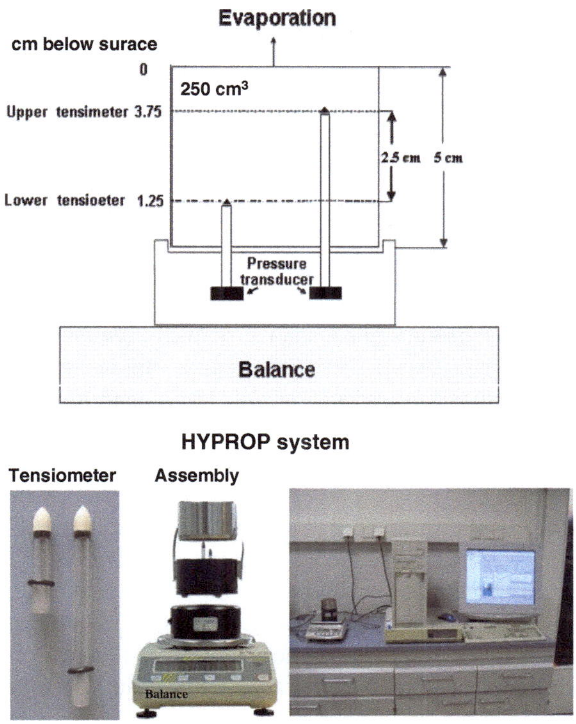

Fig. 2 Schematic illustration of the evaporation experiment and photo of the HYPROP system

interval (assumed to be equal to the total evaporated water volume ΔV_{H_2O} of the whole sample in the interval), ρV_{H_2O} is the density of water and is assumed to be 1 g cm^{-3}, A is the cross sectional area of the sample, Δt is the time interval, and i_m is the mean hydraulic gradient in the interval.

At the end of the measurement, the residual amount of storage water is derived from water loss by drying in the oven (105 °C). The initial water content is determined by total water loss (evaporation part plus residual amount) related to core volume. The dry bulk density is derived from the dry soil mass divided by the core volume. For this reason the volume of the tensiometer holes (1 cm^3) is subtracted from the core.

Assumptions for the validity of Eq. 1 are: (i) that the water flow out of the core can be treated as a "succession of steady states" where the flux and hydraulic gradient are effectively constant within each time interval; and (ii) that the water content difference across the sample height decreases linearly in the measuring interval. Accordingly, the flux through the measuring layer is half of the total flux and can be calculated from the total evaporative soil water volume (mass) difference in the time interval. These assumptions have been found to be valid (Wendroth et al. 1993). Peters and Durner (2008) and Schindler et al. (2010b) reported that linearization in space led to only minor errors, even in the late stage of evaporation where strongly non-linear tension profiles emerge.

2.3 *The Extended Evaporation Method (EEM) and the HYPROP Device*

2.3.1 Basis for Extending the Measurement Range

The extended evaporation method (EEM) overcomes the measurement limitations (Schindler et al. 2010a, b). Using new cavitation tensiometers and applying the air-entry pressure of the tensiometer's porous ceramic cup as the final tension value allow both hydraulic functions to be quantified close to the wilting point.

The dynamics of a tensiometric measurement in a drying soil can be divided into three distinct stages. In the first stage, the measured tension reflects the real matric potential of the surrounding soil. The second stage is the vapour pressure stage. The tensiometers are out of function. The tensiometer readings in this stage are no longer representative of the soil water matric potential. The third and final stage can be called the "air-entry stage". It occurs when the tension in the surrounding soil exceeds the air-entry pressure of the ceramic material. The largest continuous pore of the ceramic drains, and air from the soil enters the tensiometer. At this moment, the measured tension collapses towards zero, which can easily be seen in the tensiometer reading. The basic idea for extending the measurement range is to use the ceramic's air-entry pressure (*Ae*) at the well-defined moment of the tension collapse, i.e., at the initiation of stage three, as an additional measurement of the soil's matric potential. Any smooth function with higher-order continuity, such as polynomial functions or Hermite spline interpolation, can be used for interpolation between tension values of stage 1 and the final tension at stage three, the air-entry value (Schindler et al. 2010b). Using this procedure we are able to quantify the hydraulic functions in the range between saturation and close to the permanent wilting point, as shown as an example in Fig. 3 in a screen shot of the HYPROP fit software (UMS 2012b).

Additionally, both soil shrinkage dynamics and soil water hysteresis can be quantified (Schindler et al. 2013). The experimental setup followed the HYPROP® system (Fig. 2, UMS 2012a), which is a commercial device with vertically aligned tensiometers that is optimized to perform evaporation measurements. The HYPROP software (UMS 2012b) was developed for data recording, calculating, evaluating, fitting and exporting the hydrological data. A methodological comparison of soil hydraulic functions obtained using classical methods and the EEM showed a good agreement between the results. Systematic deviations were not found (Schelle et al. 2010, 2013; Schindler et al. 2012).

2.3.2 Shrinkage Measurements

In general, soils and their pore size systems are assumed to be unchanging (rigid) while they lose water on drying. This is different from reality for many soils, especially for soils with high quantities of clay or organic matter. As the result of shrinking, the bulk density, the porosity and the pore size distribution of these soils

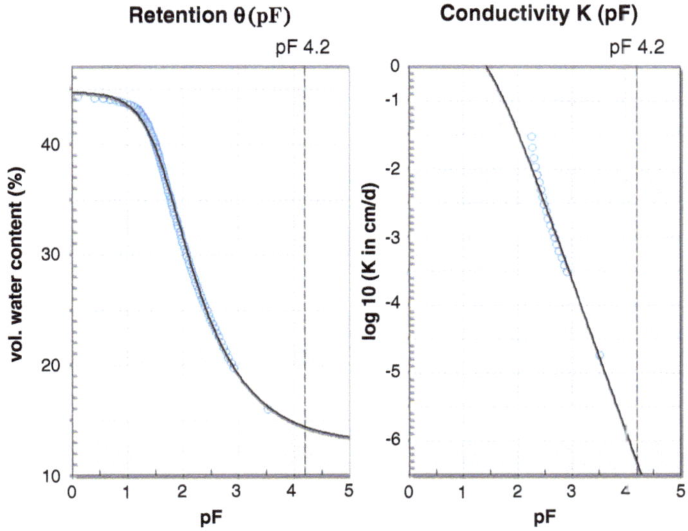

Fig. 3 Data points and fits of the water retention curve, *left*, and hydraulic conductivity function, *middle* and *right*, for a silty loam, Linfen China, screenshot (HYPROP fit software, UMS 2012b)

change. To quantify the hydraulic functions taking shrinkage into account, the HYPROP® evaporative device was combined with a circumference meter (Schindler et al. 2013). In order to make these measurements, the steel cylinder was removed and the sample was coated with a rubber membrane approximately impermeable to water and air. A preliminary investigation confirmed that the sample diameter decreased linearly during evaporation from the bottom to the top. To sum up, recording the perimeter change in the middle position of the sample as it dried out, together with the corresponding tension and water content, was sufficient to determine (i) the increasing dry bulk density and (ii) the hydraulic functions taking into account shrinkage in the range between saturation and close to the permanent wilting point.

Measurements were carried out on 25 samples which were different in texture and origin. The maximum shrinkage (35.4 % by vol. between saturation and 5000 hPa) was measured in the peat samples. The minimum shrinkage was quantified with 2.8 % by vol. for the Chilean silty loam samples (Schindler et al. 2013). Taking soil water content measurements on shrinking soils in the field can lead to an underestimation of soil water content differences if the changing dry bulk density is not considered. However, the degree of misinterpretation depends strongly on the soil and its shrinkage activity and ranges from negligible to high. The unsaturated hydraulic conductivity function was only slightly influenced by shrinkage (Schindler et al. 2013).

The advantages of the method presented are: (i) the water retention curve and the hydraulic conductivity function can be determined simultaneously, taking into consideration shrinkage, with a high resolution over the whole range from wet to dry, (ii) the method and device are simple and robust to use, (iii) little time is

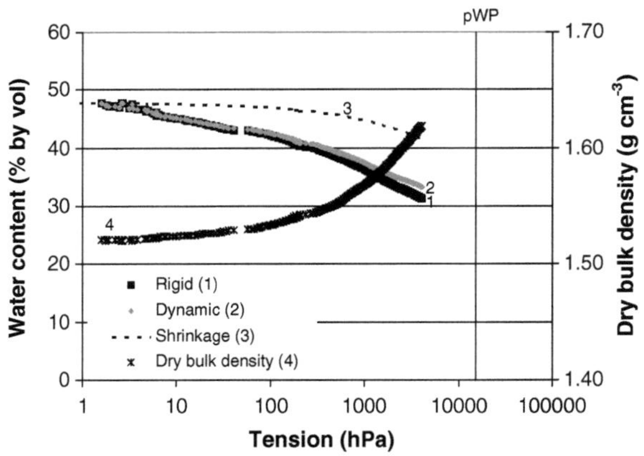

Fig. 4 Water retention function of a loamy clay sample, Seelow site, Oder Valley, Germany, *line 1* rigid soil, *line 2* dynamic soil water retention curve under consideration of increasing dry bulk density, *line 3* shrinkage, *line 4* dry bulk density (*DBD*)

required for measurement, between 3 and 10 days, (iv) the functions are described over the whole tension range, using numerous user-defined data points, (v) the evaluation of the soil water content measurement in shrinking soils is improved and (vi) common data models can be fitted to the hydraulic data as well as to the shrinkage data. Figure 4 shows an example of the change in the water retention curve within the shrinkage dynamics of a clay sample.

2.3.3 Hysteresis of the Hydraulic Functions

Generally, soil hydraulic functions are measured during desorption in the laboratory. Sorption and desorption processes, however, alternate in the field. Tension and water content measurements in the field do not show a clear relationship. Haines (1930) has long recognized that the water retention curve during desorption differs from the curve when the soil is rewetted. The water content is smaller when the soil is rewetted. We call this phenomena hysteresis. Knowledge of the different hydraulic functions of desorption and sorption is important for soil water and solute transport (Luckner et al. 1989; Šimůnek et al. 1999; Abbasi et al. 2012; Likos et al. 2014). There are as yet no practicable methods and devices for the quantification of hysteresis, though they are required. Here, we show the possibility to measure hysteresis with the HYPROP system. No additional technical effort is required.

Procedure: The desorption curve of the hydraulic functions are measured. The measurement is stopped at time *t* and the sample surface is rewetted with free water. A downward water movement starts and the tension of the upper and lower tensiometer decreases. The first step of rewetting is finished when the tensions of the tensiometers are equalized. This rewetting procedure can be repeated as long as the tensiometers are working correctly. At the end of rewetting, desorption by evaporation starts again.

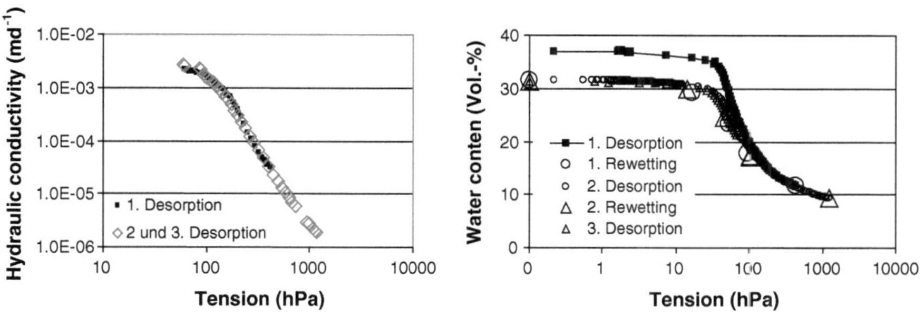

Fig. 5 Water retention curve, *left* and hydraulic conductivity function, *right*, during the desorption and sorption cycle, sand sample, Müncheberg, Germany

Figure 5 shows an example of the cycle of desorption and sorption and the different functions. Hysteresis was observed mainly in the range between saturation and about 100 hPa. That agrees with findings of Ilnicki (1982), Maqsoud et al. (2004) and Malaya and Screedeep (2010).

2.4 Soil Hydraulic Data and Pedotransfer Functions (PDFs)

In the period between 1980 and 2014, the hydrological properties—water retention curve and unsaturated hydraulic conductivity function—of more than 2000 soil samples which were different in texture, organic matter content and origin were measured. From 1980 to 2008, the evaporation method (Schindler 1980) was applied. Afterwards the extended evaporation method (EEM, Schindler et al. 2010a, b) was used. The samples are from different locations in Germany. The database includes soil hydrological properties and additional information on the georeference, the soil type and horizon, the particle size distribution, the dry bulk density and other parameters. The data (raw data) are part of the HYPRES database (Wösten et al. 1999) and the EUHYDI database (Schindler 2013). The HYPRES data are published in Schindler and Müller (2010).

3 Method for Quantifying Deep Seepage and Solute Leaching Based on Soil Hydrological Field Measurements

3.1 Process Analyses

Quantifying deep seepage and solute leaching is necessary to address numerous economic and environmental problems, for example by developing sustainable farm management systems, with a view to providing non-polluted water for different users and determining safe yields of aquifers.

Arable land is the main source of deep drainage and groundwater in humid and sub-humid temperate climates (Müller et al. 1996). Land management practice is a decisive factor for the quantities of seepage flow and solute leaching (Benson et al. 2006; Köhler et al. 2006), which constitute two fundamental aspects of land use characterised by potentially conflictive ecologic implications. Efficient water use and intelligent systems are essential for sustainable land and water management. To meet this requirement, knowledge is required about seepage flow in terms of quantity and quality.

Soil water content and tension are basic hydrologic variables that reflect effects of land surface processes. Soil moisture information, in spite of its importance for land use planning, agriculture and drought monitoring, is not widely available. There is a lack of reliable methods and measured values for predicting the effects of land use change on the soil water and solute status. Evaluating the impact of land use and management practices on soil water dynamics and solute leaching turns out to be difficult. This is because but a few techniques are capable of monitoring the quantity and quality of soil water flow below the root zone without disturbing the soil profile and affecting natural flow processes.

Soil hydrology measurements provide an alternative way of analysing the soil water and matter status in situ under undisturbed soil conditions. Compared with lysimeters, these methods are less expensive and more flexible. However, taking measurements at various sites and with many sensors across the whole profile is also expensive and is not a feasible solution. Utilising knowledge about the process dynamics (Gee and Hillel 1988; Albright et al. 2004; Masarik et al. 2004; Gee et al. 2009; Schindler and Müller 1998; Schindler et al. 2009) of soil water and solute transport in the field requires the development and application of an effective field method for quantifying deep seepage and solute leaching. Soil hydrology measurements may provide the qualified basis for transmitting quantitative results within a particular area (Mölders et al. 2003; Miao et al. 2003; Bryant et al. 2006; Köhler et al. 2006; Perkins et al. 2011).

For decades, lysimeters have been the main devices for monitoring and quantifying soil water and solute balances and transport processes. However, lysimeters are expensive and the results are often influenced by specific lysimeter effects (Kastanek 1995; Schindler et al. 2008). There is a lack of effective and reliable methods for quantifying deep seepage under undisturbed soil conditions in the field. Lysimeters in combination with more flexible and less expensive field methods could be a successful combination for monitoring and managing soil and water resources in future.

3.2 Soil Hydrological Concept

Below the zero flux plane, water flow is driven by gravity only (Fig. 6). Over time, the soil in that zone can be considered as being a pipe with various fillings. The indicator used to assess filling levels is the soil water content. All changes in the

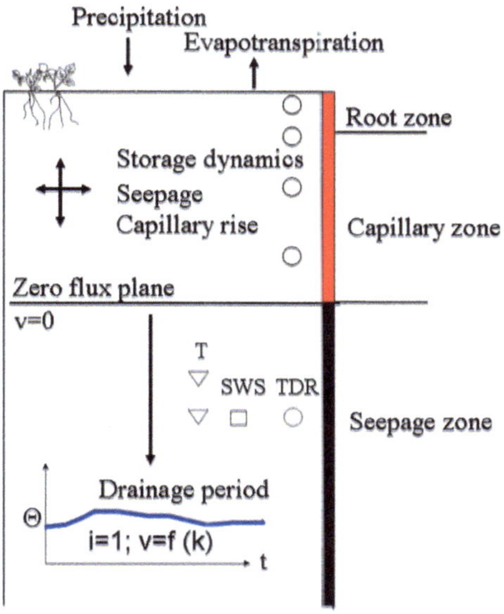

Fig. 6 Soil hydrological concept

water content or tension are induced by changing seepage conditions. The filling level (soil water content) can be transformed into the flux (seepage flow rate v) using the Darcy Buckingham equation (Eq. 2) containing a non-linear scaling factor, where the hydraulic conductivity function $K(\theta)$ is dependent on water content θ and the hydraulic gradient (i) is the driving force.

$$(q = K(\theta) * i) \tag{2}$$

"Steady-state" conditions are assumed at daily intervals and the unit hydraulic gradient is considered to be valid (Schindler and Müller 1998). This method is feasible on soils not influenced by ground water. Daily deep seepage rates (q) are calculated based on water content measurements below the zero flux plane, and an unsaturated hydraulic conductivity function $K(\theta)$ calibrated to the water balance. Further information about soil properties, land management, weather data and other data is not necessary. The prerequisites are high precision, temporally stable tensiometers, TDR probes with no drift or temperature influence and temporally constant flow pathways.

3.3 *Procedure for Quantifying Deep Seepage Rates*

The water content and tensions are recorded with a high temporal resolution below the zero flux plane (Schindler et al. 2009). The recommended measurement depth is 3 m at arable sites in a humid climate, and 5 m at forest sites.

In the second step, the van Genuchten model Eq. 3 (van Genuchten 1980) is fitted to the relationship between tension and water content, providing a field retention function,

$$\theta(\psi) = \theta_r + \frac{\theta_s - \theta_r}{[1 + (\alpha\psi)^n]^m} \tag{3}$$

with

θ_s saturated water content

θ_r residual water content

α, n, m ($m = 1 - 1/n$) scaling parameters

and the relative hydraulic conductivity function K_r (ψ) is predicted Eq. (4) (Mualem 1976).

$$\frac{K(\psi)}{K_s} = K_r(\psi) = \frac{\left\{1 - (\alpha\psi)^{n-1}[1 + (\alpha\psi)^n]^{-m}\right\}^2}{[1 + (\alpha\psi)^n]^{\frac{m}{2}}} \tag{4}$$

with

$K(\psi)$ hydraulic conductivity dependent on tension

K_s saturated hydraulic conductivity

K_r (ψ) relative hydraulic conductivity

If ψ in Eq. 4 is substitutes, as in Eq. 5, this gives the hydraulic conductivity as a function of water content θ. The use of $K(\theta)$ minimises hysteresis effects. Any extrapolation of $K(\theta)$ beyond field recordings is not permitted.

$$\psi = f(\theta) = \left\{\left[\left(\frac{\theta_s - \theta_r}{\theta - \theta_r}\right)^{\frac{1}{m}} - 1\right]^{\frac{1}{n}}\right\}^{-\alpha} \tag{5}$$

In the next step, fictitious daily deep seepage rates (q_f) are calculated (Eq. 2) based on the recorded water content values (θ) and the relative hydraulic conductivity function $K_r(\theta)$.

A unit gradient $i = 1$ is assumed. This assumption has been validated (Schindler and Müller 1998), and prompts only small uncertainties. Inaccuracies in tension measurements, however, can produce larger errors.

To provide reliable seepage rates (q), the relative hydraulic conductivity function has to be converted to a reliable level. First, the total deep seepage (V_c) is determined from the water balance (Eq. 6) for a frost-free autumn/winter calibration period. An autumn/winter period should be chosen because it minimises uncertainties for calculating evapotranspiration. Precipitation (P) should be measured at the site of the experiment. The soil water storage difference ($\Delta\theta$) is derived from water content measurements in the profile from the beginning and the end of the calibration period. The ratio V_c/V_f reveals M as the matching factor (Eq. 7) for transforming the $Kr(\theta)$ function to a reliable level $K(\theta)$. V_f is the sum of the fictive flow rates q_f in the calibration period. Finally, Eq. 8 is used to calculate reliable daily seepage rates. The term $f(\theta)$ in Eq. 8 substitutes ψ according to Eq. 5.

$$V_c = P - ET + \Delta\theta \tag{6}$$

$$M = \frac{V_c}{V_f} \tag{7}$$

$$q = M \cdot K_r(\theta) = \frac{V_c}{V_f} \frac{\left\{1 - (\alpha \cdot f(\theta))^{n-1}[1 + (\alpha \cdot f(\theta))^n]^{-m}\right\}^2}{[1 + (\alpha \cdot f(\theta))^n]^{\frac{m}{2}}} \tag{8}$$

The method was applied at more than 40 soil hydrological plots in North-East Germany and produced reliable values of deep seepage rates and solute leaching (Schindler et al. 2010c)

3.4 Lysimeter Experiments—Validation of the Soil Hydrological Field Method

3.4.1 Lysimeter and Soil Data

Lysimeter data were used from the Dedelow lysimeter station (Schindler et al. 2001) and from the experimental field in Wagna (Steiermark, Austria, Fank and Unold 2005; Klammler and Fank 2014). Validation periods: Dedelow lysimeter 2001–2010, Wagna lysimeter 2004–2012. Some soil physical properties and the stratification are collected in Table 1.

The lysimeters were equipped with TDR instruments and tensiometers comparable to the field measurements. The measured daily discharge rate was compared with the calculations based on the water content and tension measurements as in paragraph 3.3. The Willmott Index of Agreement Eq. 9 (Legates and McCabe 1999) was used to evaluate the comparison of the measured and calculated discharge rates. The index varies between 0 (poor agreement) and 1 (perfect agreement).

$$d = 1 - \frac{\Sigma(O_{\mathrm{i}} - P_{\mathrm{i}})^2}{\Sigma\left(|P_{\mathrm{i}} - O_{\mathrm{m}}| + |O_{\mathrm{i}} - O_{\mathrm{m}}|)^2\right)} \tag{9}$$

O_{i}—observed values, P_{i}—predicted values, O_{m}—mean observed value

Table 1 Stratification and soil physical properties of the lysimeters

Site	Horizon	Depth (cm)	OMC (%)	ρ (g cm^{-3})	Clay (%)	Silt (%)	Sand (%)	Gravel (%)
Dedelow	Ap	0–35	1.1	1.52	4	24	71	1
	Bv	35–115	0.1	1.63	9	28	62	1
	C	115–200	0.1	1.65	1	4	93	2
Wagna KON	Ap	0–30	1.8	1.51	20	33	44	3
	B	30–70	0.9	1.53	18	28	54	0
	C	70–190	0.1		1	2	33	64
Wagna BIO	A	0–30	2.2	1.53	19	34	45	2
	B	30–110	1.3	1.55	14	28	57	1
	C	110–190	0.1		0	1	30	69

OMC: organic matter content, ρ: dry bulk density of the soil, clay: <0.002 mm, silt: 0.2–0.63 mm, sand: 0.63–2 mm; gravel: 2–63 mm

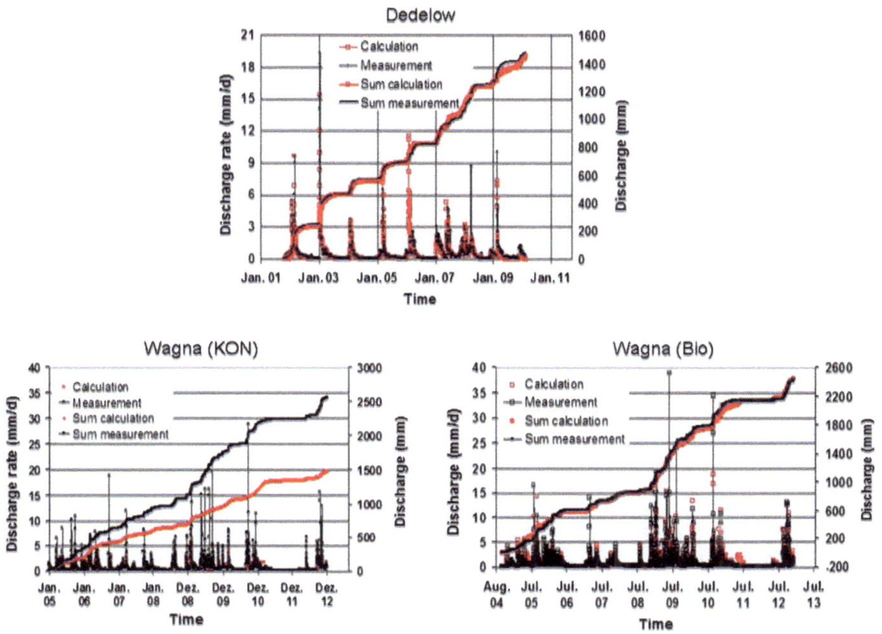

Fig. 7 Comparison of lysimeter discharge measurements with calculations

3.4.2 Results

The comparison (Fig. 7) of the measured and the calculated discharge rates showed a good agreement for the lysimeters in Dedelow (d = 0.97 and Wagna (BIO, d = 0.91). However, differences were recognizable for the Wagna lysimeter (KON, d = 0.64). The agreement in the Wagna KON lysimeter was also good for discharge rates lower than 4 mm d^{-1}. Due to the high amount of gravel in the underground and the thin, fine-textured soil cover (Table 1), discharge rates larger than 4 mm d^{-1} moved downwards as the preferential flow and could not be detected by the TDR sensors (Schindler et al. 2009). In conclusion, the proposed method for quantifying deep seepage rates based on soil hydrological measurements performed well, except for sites with preferential flow. This agrees with simulation findings by Hohenbrink and Lischeid (2014).

4 Conclusions

I. The EEM allows the simultaneous determination of the soil hydraulic functions pF and K in the range between saturation and close to the wilting point. Depending on the soil and the amount of water loss by evaporation, the measurement time ranged between 2 days for clay soils and about 10 days for organic soils. The simultaneous measurement of a large number of samples is possible with only one balance. Additionally, both soil shrinkage dynamics and soil water hysteresis can be quantified.

II. The presented method for estimating deep seepage rates and solute leaching on the basis of soil hydrological measurements in the field was successfully applied and the suitability was tested with lysimeter experiments. The method is simple to handle, is flexible and is less expensive than lysimeters.

III. Both methods (the EEM with the HYPROP device and the soil hydrological field method) have the potential for improving soil hydrological studies in Eurasia.

References

Abbasi F, Javaux M, Vanclooster M, Feyen J (2012) Estimating hysteresis in the soil water retention curve from monolith experiments. Geoderma 189–190:480–490

Albright WH, Benson CH, Gee GW et al (2004) Field water balance of landfill final covers. J Environ Qual 33(6):2317–2332

Arya LM (2002) Wind and hot-air methods. In: Dane JH, Topp GC (eds) Methods of soil analysis. Part 4. Physical methods. SSSA Book Ser. 5. SSSA, Madison, WI, pp 916–926

Becher HH (1970) Ein Verfahren zur Messung der ungesättigten Wasserleitfähigkeit. Z. Pflanzenernaehrung u. Bodenkd. 128:1–12

Benson VS, van Leeuwen JA, Sanchez J, Dohoo IR, Somers GH (2006) Spatial analysis of land use impact on ground water nitrate concentrations. J Environ Qual 35:421–432

Bertuzzi P, Voltz M (1999) The wind method: a standard laboratory method adopted by the French INRA laboratories. Workshop proceedings: the use of pedotransfer in soil hydrology research, pp 51–53. 10–12 Oct. Orleans, France

Boels D, van Gils JBM, Veerman GJ, Wit KE (1978) Theory and system of automatic determination of soil moisture characteristics and unsaturated hydraulic conductivities. Soil Sci 126:191–199

Bryant RB, Gburek WJ, Veith TL, Hively WD (2006) Perspectives on the potential for hydropedology to improve watershed modeling of phosphorus loss. Geoderma 131(3–4): 299–307

Cresswell HP, Green TW, McKenzie N (2008) The adequacy of pressure plate apparatus for determining soil water retention. Soil Sci Soc Am J 55(72):41–49

Dane JH, Hopmans JW (2002) Pressure plate extractor. In. Dane JH, Topp GC (eds) Methods of soil analysis. Part 4. Physical methods. SSSA Book Ser. 5. SSSA, Madison, WI, pp 688–690

de Rooij GH, Kasteel RTA, Papritz A, Flühler H (2004) Joint distributions of the unsaturated soil hydraulic parameters and their effect on other variates. Vadose Zone J 3:947–955

Durner W, Iden SC (2011) Extended multistep outflow method for the accurate determination of soil hydraulic properties near water saturation. Water Resour Res 47:W08526. doi:10.1029/ 2011WR010632

Fank J, Unold G (2005) Wägbare monolithische Lysimeter unter maschineller Feldbewirtschaftung (Wagna-Austria). Bericht 11. Gumpensteiner Lysimetertagung, pp 55–60

Fujimaki H, Mitsuhiro I (2003) Reevaluation of the multistep outflow method for determining unsaturated hydraulic conductivity. Vadose Zone J 2:409–415

Gee GW, Hillel D (1988) Groundwater recharge in arid regions: review and critique of estimation methods. J Hydrol Process 2:255–266

Gee GW, Newman BD, Green SR, Meissner R, Rupp H, Zhang ZF, Keller JM, Waugh WJ, van der Velde M, Salazar J (2009) Passive wick fluxmeters: Design considerations and field applications. Water Resour Res 45:1–18

Haines WB (1930) Studies in the physical properties of soil. V. The hysteresis effect in capillary properties and the modes of moisture distribution associated therewith. J. Agric. Res. 20:97–116

Halbertsma J (1996) Wind's evaporation method, determination of the water retention characteristics and unsaturated hydraulic conductivity of soil samples. Possibilities, advantages and disadvantages. In: Durner W, Halbertsma J, Cislerova M (eds) European workshop on advanced methods to determine hydraulic properties of soils, Thurnau, Germany, June 10–12, 1996, Department of Hydrology, University of Bayreuth, 107 pp

Henseler KL, Renger M (1969) Die Bestimmung der Wasserdurchlässigkeit im wasserungesättigten Boden mit der Doppelmembran-Druckapparatur. Z. Pflanzenernähr. Bodenkd. 122:220–228

Hohenbrink T, Lischeid G (2014) Texture-depending performance of an in situ method assessing deep seepage. J Hydrol 511:61–71

Hopmans JW, Šimunek J, Romano N, Durner W (2002) Simultaneous determination of water transmission retention properties – Inverse methods. In Dane JH, Topp GC (eds) Methods of soil analysis, Part 4: Physical methods, 4th edn. SSSA Book Series No. 5, American Society of Agronomy Soil Science Society of America, pp 963–1008

Ilnicki P (1982) Hysterese der Wasserspannungskurve in organogenen Böden. Z. Pflanzenernährung und Bodenkd. 145:363–374

Kastanek F (1995) Kritische Bemerkungen zur Verwendung von Lysimetern. 5. Gumpensteiner Lysimetertagung, "Stofftransport und Stoffbilanz in der ungesättigten Zone" BAL Gumpenstein 25–26 April 1995, BAL Bericht, pp 93–102

Klammler G, Fank J (2014) Determining water and nitrogen balances for beneficial management practices using lysimeters at Wagna test site (Austria). Sci Total Environ 499(2014):448–462

Klute A, Dirksen C (1986) Hydraulic conductivity and diffusivity: laboratory methods. In: Klute A (ed) Methods of soil analysis. Part 1. 2nd ed. Agron. Monogr. 9. ASA and SSSA, Madison, WI, pp 687–734

Köhler K, Duijnisveld WH, Böttcher J (2006) Nitrogen fertilization and nitrate leaching into groundwater on arable sandy soils. J Plant Nutr Soil Sci 169:185–195

Kool B, Parker JC, van Genuchten MTh (1985) Determining soil hydraulic properties from one-step outflow experiments by parameter estimation: I. Theory and numerical studies. Soil Sci Soc Am J 49:1348–1354

Legates DR, McCabe GJ (1999) Evaluating the use of "goodness-of-fit" measures in hydrologic and hydroclimatic model validation. Water Resour Res 35:233–242

Likos W, Lu N, Godt J (2014) Hysteresis and uncertainty in soil water-retention curve parameters. J Geotech Geoenviron Eng 140(4):04013050

Luckner L, van Genuchten MTh, Nielsen DR (1989) A consistent set of parametric models for the two-phase flow of immiscible fluids in the subsurface. Water Resour Res 25(10):2187–2193

Malaya C, Screedeep S (2010) A stud on wetting-soil water characteristic curve of a sand soil. In: Proceedings of Indian geotechnical conference, 16–18 Dec 2010, IGS Mumbai Chapter and IIT Bombay

Maqsoud A, Bussière B, Mbonimpa M (2004) Hysteresis effects on the water retention curve: a comparison between laboratory results and predictive models. 57th Canadian geotechnical conference, 24–26 Oct 2004, Quebec, Canada

Masarik KC, Norman JM, Brye KR, Baker JM (2004) Improvements to measuring water flux in the vadose zone. J Environ Qual 33(3):1152–1158

Miao Z, Cheplick MJ, Williams MW, Trevisan M, Padovani L, Gennari M, Ferrero A, Vidotto F, Capri E (2003) Simulating pesticide leaching and runoff in rice paddies with the RICEWQ–VADOFT Model. J Environ Qual 32:2189–2199

Mölders N, Haferkorn U, Döring J, Kramm J (2003) Long-term investigations on the water budget quantities predicted by the hydro-thermodynamic soil vegetation scheme (HTSVS)—Part I:

description of the model and impact of long-wave radiation, roots, snow, and soil frost. Meteorol Atmos Phys 84(1/2):115–135
Mualem Y (1976) A new model for predicting the hydraulic conductivity of unsaturated porous media. Water Resour Res 12:513–522
Müller L, Dannowski, Schindler U, Eulenstein F, Meißner R (1996) Gebietsabflüsse aus Agrarlandschaften Nordost- und Mitteldeutschlands. Arch. Acker- Pfl. Boden 40:345–362
Nimmo JR, Perkins KS, Lewis AM (2002) Steady-state centrifuge. In: Dane JH, Topp GC (eds) Methods of soil analysis. Part 4. Physical methods. SSSA Book Ser. 5. SSSA, Madison, WI, pp 903–916
Perkins KS, Nimmo JR, Rose CE, Coupe RH (2011) Field tracer investigation of unsaturated zone flow paths and mechanisms in agricultural soils of northwestern Mississippi, USA. J Hydrol 396(1–2):1–11
Peters A, Durner W (2008) Simplified evaporation method for determining soil hydraulic properties. J Hydrol 356:147–162
Plagge R (1991) Bestimmung der ungesättigten hydraulischen Leitfähigkeit im Boden. Ph.D. thesis, Technical University Berlin, Institute of Ecology, Department Soil Science, 152 pp
Reynolds WD, Elrick DE (1991) Determination of hydraulic conductivity using a tension infiltrometer. Soil Sci Soc Am J 55:633–639
Schelle H, Iden SC, Peters A, Durner W (2010) Analysis of the agreement of soil hydraulic properties obtained from multistep-outflow and evaporation method. Vadose Zone J 9. doi:10.2136/vzj2010.0050
Schelle H, Heise L, Jaenicke K, Durner W (2013) Water retention characteristics of soils over the whole moisture range: a comparison of laboratory methods. Eur J Soil Sci 64:814–821. doi:10.1111/ejss.12108
Schindler U (1980) Ein Schnellverfahren zur Messung der Wasserleitfähigkeit im teilgesättigten Boden an Stechzylinderproben. Arch. Acker- u. Pflanzenbau u. Bodenkd, Berlin 24, 1, 1–7, 1980
Schindler U (2013) Soil data from Germany (North East and Central Germany). JRC technical reports, European hydrological data inventory (EU-HYDI). doi:10.2788/5936. http://eusoils.jrc.ec.europa.eu/ESDB_Archive/eusoils_docs/other/EUR26053EN.pdf. Accessed 10 Jan 2015
Schindler U, Müller L (1998) Calculating deep seepage from water content and tension measurements in the vadose zone at sandy and loamy soils in north-east Germany. Arch Agronomy Soil Sci 43:233–243
Schindler U, Müller L (2006) Simplifying the evaporation method for quantifying soil hydraulic properties. J Plant Nutr Soil Sci 169:623–629
Schindler U, Müller L (2010) Data of hydraulic properties of North East and North Central German soils. Earth Syst Sci Data 3:131–142. doi:10.5194/essdd-3-131-2010
Schindler U, Wolff M, Kühn G (2001) Lysimeterstudie zum Einfluss von Düngung und Bewirtschaftung auf die Ertragsbildung, den Wasserhaushalt und die Nährstoffauswaschung im Trockengebiet der Uckermark. Zeitschrift für Pflanzenernährung und Bodenkunde 164:697–703
Schindler U, Müller L, Eulenstein F, Dannowski R (2008) A long-term hydrological soil study on the effects of soil and land use on deep seepage dynamics in northeast Germany. Arch Agron Soil Sci 54:451–463
Schindler U, Fank J, Müller L (2009) Quantifizierung der Grundwasserneubildung aus bodenhydrologischen Messungen in situ: Methode—Ergebnisse—Trends. In: Hydrologische Systeme im Wandel : Beiträge zum Tag der Hydrologie am 26/27. März 2009 an der Christian-Albrechts-Universität zu Kiel. DWA, Hennef, pp 45–50
Schindler U, Durner W, von Unold G, Müller L (2010a) Evaporation method for measuring unsaturated hydraulic properties of soils: Extending the range. Soil Sci Soc Am J 74 (4):1071–1083
Schindler U, Durner W, von Unold G, Müller L, Wieland R (2010b) The evaporation method—extending the measurement range of soil hydraulic properties using the air-entry pressure of the ceramic cup. J Plant Nutr Soil Sci 173(4):563–572

Schindler U, Müller L, Dannowski R, Barkusky D, Francis G (2010c) Long-term measurements to quantify the impact of arable management practices on deep seepage and nitrate leaching. Long-term ecological research: between theory and application. Springer, Dordrecht, pp 243–252

Schindler U, Müller L, da Veiga M, Zhang Y, Schlindwein SL, Hu C (2012) Comparison of water-retention functions obtained from the extended evaporation method and the standard methods sand/kaolin boxes and pressure plate extractor. J Plant Nutr Soil Sci 174(4):527–534

Schindler U, Müller L, Dörner J (2013) Neue Methode zur Quantifizierung bodenhydraulischer Kennfunktionen unter Berücksichtigung von Schrumpfung und Quellung. In: Böden—Lebensgrundlage und Verantwortung: Jahrestagung der DBG, 7–12 Sept 2013, Rostock. Deutsche Bodenkundliche Gesellschaft, Oldenburg. http://eprints.dbges.de/1054/1/DBG_Rostock_2013__Schindler.pdf. Accessed 10 Jan 2015

Šimůnek J, Šejna M, van Genuchten MTh (1999) The HYDRUS-2D Software package for simulating the two-dimensional movement of water, heat, and multiple solutes in variably-saturated media, version 2.0.—U.S. Salinity Laboratory, Riverside, California

Tyner JS, Arya LM, Wright WC (2006) The dual gravimetric hot-air method for measuring soil water diffusivity. Vadose Zone J 5:1281–1286

UMS (2012a) UMS GmbH Munich, HYPROP©—Laboratory evaporation method for the determination of pF-curves and unsaturated conductivity. http://www.ums-muc.de/en/products/soil_laboratory.html. Accessed 10 Jan 2015

UMS (2012b) UMS GmbH Munich, HYPROP Fit software. http://www.ums-muc.de. Accessed 10 Jan 2015

van Genuchten MTh (1980) A closed-form equation for predicting the hydraulic conductivity of unsaturated soils. Soil Sci Soc Am J 44:892–898

Wendroth O, Ehlers W, Hopmans JW, Kage H, Halbertsma J, Wösten JHM (1993) Reevaluation of the evaporation method for determining hydraulic functions in unsaturated soils. Soil Sci Soc Am J 57:1436–1443

Wind GP (1968) Capillary conductivity data estimated by a simple method. In: Proceedings of UNESCO/IASH symposium. Water in the unsaturated zone, Wageningen, The Netherlands, pp 181–191

Wösten JHM, Lilly A, Nemes A, Le Bas C (1999) Development and use of a database of hydraulic properties of European soils. Geoderma 90:169–185

Chapter 15
Methods for Measuring Water and Solute Balances in Forest Ecosystems

Jürgen Müller

Abstract The environmental monitoring of the state of forests throughout Europe can identify negative developments, allowing targeted countermeasures. In the Forest Monitoring Network, not only are regular representative grid surveys carried out but also ecosystem-relevant energy and solute fluxes are monitored over the whole year. A monitoring plot consists of an open-field plot and a plot in forest stands. An open-field automatic weather station records meteorological parameters. In the forest stands, soil water fluxes are monitored in measuring fields with tensiometers and soil moisture sensors. In addition, suction probes take soil water samples that are tested for quality. Tree increments are checked with dendrometers. The long-term measurement of meteorological, hydrological, and growth-related parameters provides information about the state of forests and also allows an intensive study of causes and effects in forest ecosystems. Some factors relating to the water balance in forests can be measured directly, while others are calculated using models. To validate the model results, it is necessary to investigate water fluxes in forest plots with various structures. Additional innovative instruments and methods are used for special environmental observations. Large-scale lysimeters are used to measure the influence of trees of various ages and species on the groundwater recharge and evapotranspiration (ET). Weighing lysimeters are used to determine the ET of ground vegetation and young trees. The monitoring indicates that the seepage water below forests is clean but affected by periods of summer drought, which also reduce tree increments. Groundwater recharge is influenced by the age and species of forest trees, the vertical structuring and heterogeneity of forests, and the way they are managed. Broad-leaved forests are found to have more groundwater recharge than coniferous forests due to the differences in the interception between the evergreen canopies of coniferous forests and broad-leaved forests which lose their leaves in the winter. Depending on the structure of the investigated stands, a redistribution of precipitation was found with effects on the proportions of the individual components of the ET so that the contribution of

J. Müller (✉)
Johann Heinrich v. Thünen-Institute, Institute of Forest Ecosystems,
A.-Möller-Str. 1, 16225 Eberswalde, Germany
e-mail: juergen.mueller@ti.bund.de

L. Mueller et al. (eds.), *Novel Methods for Monitoring and Managing Land and Water Resources in Siberia*, Springer Water,
DOI 10.1007/978-3-319-24409-9_15

forests to the landscape water balance varies. The findings can be used to assess the future threats to today's forests and to develop strategies for adapting to anticipated climate change.

1 Introduction

1.1 The Importance of Hydrologic Studies in Forests

Water is of key importance for forest ecosystems, and most chemical and biological processes are dependent on its availability. The quantity of water that is available is frequently a controlling factor for the intensity of these processes. The vitality and the growth of forests are influenced by the water cycle and the levels of nutrients, so the measurement of water and solute transport are very important for the investigation of forest ecosystems.

For the investigations, the following aspects are of key importance:

- the components of the water cycle (precipitation, interception, evaporation, transpiration, seepage, and surface runoff);
- deposition rates;
- soil water fluxes to determine solute transport rates and the quality of seepage water;
- soil water reserves and the soil water potential in the evaluation of drought effects;
- water supplies and the nutrients available to vegetation in connection with tree growth; and
- water oversupply and excessive levels of nutrients.

It is necessary to adapt the methods used to measure water and solute balances in order to take into account the dynamic nature of the water cycle components and their spatial variability.

Some of the components of the water cycle and the parameters of the water and solute balances can be measured directly, while others have to be estimated or calculated. This makes it necessary to develop models.

1.2 Water Balance Components

The special structural features of forests, such as the species of trees; the vertical layering of trees, shrubs, and herbs; and the height and the external features of the trees have a considerable influence on the water and solute fluxes in a stand.

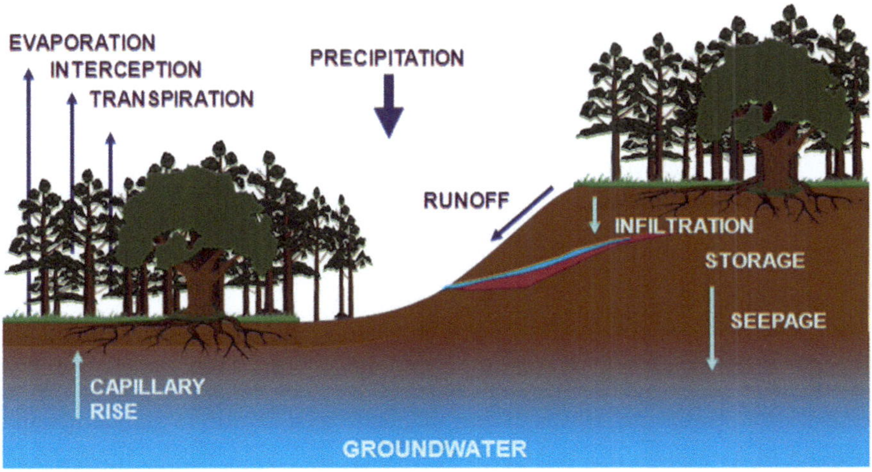

Fig. 1 Components of the water cycle in the forest ecosystem

For example, precipitation is diverted by the structures of the tree, shrub, and herb layers and some of it evaporates. The rest may reach the forest floor as stem flow or through fall, where it runs off, evaporates, or infiltrates into the soil, recharging the soil water store and seeping down beneath the main root zone into the groundwater. On the other hand, there is the interception and transpiration of the tree, shrub, and herb layers and the evaporation of forest soils (Fig. 1).

A proportion of the precipitation is held back on the surfaces of the vegetation layers and evaporates. This is the interception evaporation. Transpiration is the process where water is drawn up by the plant roots in the soil and evaporated through the leaves. Total evaporation is the sum of the interception, transpiration of the tree stand, and evaporation or evapotranspiration (ET) from the forest floor and herb layer vegetation. In order to take the special structural features of the forest ecosystems into account, it is necessary to measure the water and solute fluxes separately for each compartment.

1.3 Forest Monitoring

Forest environment monitoring investigates the state of European forests and registers changes. Forest monitoring in Germany involves a coordinated system of regular grid-based inventories (soil and forest state surveys) and intensive forest surveys. The intensive forest monitoring is part of an international program that was originally intended to investigate damage to forests due to the impact of air pollutants (Ulrich et al. 1981). Currently, some 70 areas in Germany are examining the cause–effect relationships between the state of forest trees (foliage and phenology) and their growth on the one hand and on the other hand the levels of nutrients, the state of the soil, atmospheric deposition, weather conditions, and water balances. The measurements also cover meteorological, hydrological, and phonological parameters.

The program brings together a number of institutions and measures a variety of parameters, with particularly high-quality assurance standards during data collection. In order to establish the uniform evaluation of all observation areas and identification of cause–effect relationships, it is necessary to harmonize methodologies and structured data management.

In addition to the investigation of water and solute fluxes in forests as part of the forest ecosystem monitoring, there are also a number of international and national investigations. The research aims at water balance effects over time for entire catchment areas (Hauhs 1985). In addition, investigations are carried out at the stand level to determine structural and process interrelationships, to study stress reactions of forest trees in phase of water shortage, and to establish parameters for water dynamics models (Rakei 1991).

This article provides an overview of the methods and equipment used to measure water and solute fluxes and to determine components of the water dynamics in forest stands with a variety of vegetation structures. Examples are also provided of the results that are obtained.

2 Methods for Monitoring Forest Ecosystems

2.1 Measurement of Water, Energy, and Solute Fluxes

2.1.1 Precipitation and Deposition

An observation area consists of open-field and stand areas in the forest. The open-field areas are usually close to the forest areas (within 2 km).

In Table 1 and Fig. 2, the relevant parameters are summarized, with the equipment used to measure them in open fields and stands.

On the open field, and if necessary also in the stand, a fully automatic meteorological station is used to register meteorological parameters (air humidity, air temperature, solar radiation, wind speed and direction, precipitation, and soil temperature). It serves to monitor weather conditions, to assess calamities, and to calculate the water balance. The stations are setup in accordance with standards for meteorological measurements (e.g. DWD 2001; VDI 3786-13 2006; WMO 1996) and the guidelines for automatic meteorological stations (Fig. 3a, b).

In addition to the measurements of the meteorological stations, precipitation is also measured on open fields using Hellmann precipitation collectors monitored at various time intervals (day, month). The Hellmann collectors have a defined surface area, as a rule 200 or 100 cm^2.

Table 1 Parameters and equipment to measure the fluxes of water, energy and solute

Parameter	Equipment	Position
Weather	Meteorological station	Open field
Deposition	BULK sampler	Open field and stand
Rain, snow	Precipitation collector, channels	Open field and stand
Stem flow	Stem-flow collector	Tree stand
Leaf, needle loss	Leaf, needle collector	Tree stand
Soil water flux and soil moisture	Tensiometer and soil moisture probe	Tree stand
Seepage water quality	Suction probe	Tree stand
Growth	Dendrometer	Tree stand

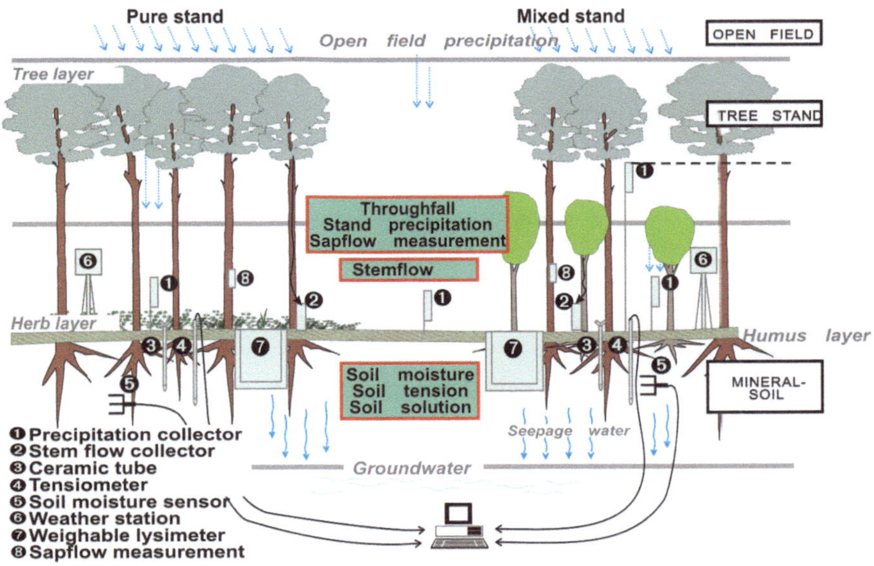

Fig. 2 Measuring water flows in forest stands

At the beginning of the cold weather, a snow baffle is included in the 200 cm^2 precipitation sampler to prevent snow from being swirled out. When precipitation is frozen in the collector, the entire system is exchanged, the frozen precipitation is carefully allowed to thaw in a warm room, and the quantity determined using a measuring cylinder.

Depositions in open fields and stands are determined in accordance with the ICP Forest Manual (2010, 2015). In most cases, funnel bulk collectors are used. The deposition rate is determined from the amount of precipitation collected over a certain period and the solute concentrations. This gives the amount deposited per

Fig. 3 Continuously measuring meteorological stations for logging the weather parameters and a precipitation sampler, (**a**) open field station, (**b**) station in a mixed stand

unit time and unit area. In most cases, the precipitation is measured in combination with the deposition.

The stand precipitation is the sum of through fall and stem flow. The through fall is determined using precipitation collectors—either individual Hellmann collectors or channels, as appropriate (Fig. 4a, b).

In order to investigate how the level of through fall depends on the crown density and to determine the canopy interception channel collectors, individual collectors are used. The number of collectors to be installed depends on the homogeneity of the canopy. Inhomogeneous stands require more collectors than homogenous stands, so the number required in a forest stand may vary between 15 and 30 collectors (Müller 2002, 2006).

In addition to through fall, the stem flow is also measured for individual trees of the stand. Depending on the trunk dimensions, a total of 12–15 trees in three trunk classes are fitted with stem flow measuring devices. The water flowing down the trunk is collected with collars or spirals (Fig. 5a, b). We found the collars most effective because they minimize the damage to the tree bark, and a watertight fit is ensured naturally by the growth of the tree. The stem flow water is collected and measurements taken every day, week, or month. Tipping buckets are used to measure the intensity of the flowing water in comparison with the rain intensity from open-field measurements (Fig. 5c).

Fig. 4 Measuring the throughfall on the forest floor (**a**) Hellman sampler in a mixed stand of Scots pines and beeches, (**b**) channel collector in a willow plantation

Fig. 5 Measuring system for stemflow (**a**) on a beech tree with a spiral, (**b**) on a beech tree with a collar, (**c**) on a beech tree with a collar and a tippic counter (UGT 2014)

The difference between open-field and stand precipitation is the canopy interception (DVWK 1986).

2.1.2 Measuring Soil Water Flux and Extracting Soil Solution

Soil water flux (seepage, soil water extraction) is monitored in the forest stands by measuring water retention and soil moisture levels. Figure 6 shows the principle of the combined installation of suction probes and soil moisture meters in hydrological measuring fields. The measurements are taken in horizontal layers down to a depth of 2 m. Five tensiometers and TDR probes are installed at each level. The readings from the sensors are fed through a central bus system, and the data are recorded by a

Fig. 6 Measuring soil moisture and water retention in a Scots pine forest

data logger or notebook computer. In our investigations, readings are taken at hourly intervals. The systems are powered by batteries. The hydrological measuring fields are positioned to cover representative areas under the tree crowns and also areas close to the trunks (Müller 2002, 2006).

In order to determine the quality of seepage water at varying depths in the root zone and also at a depth below the intensive rooting, in general at a depth of 2–2.5 m, seepage water is extracted by means of suction probes with a ceramic head (Fig. 7a, b). The free seepage water is pumped up to the surface and stored in a bottle. At monthly intervals, the contents are analyzed in a laboratory. The array of suction probes is analogous to the placing of the sensors to measure soil water flux. The solute burden at each level is given by the product of the amount of seepage water and the solute concentration.

(a) (b)

Fig. 7 **a** Measuring field with suction probes to collect soil solution at different soil depths in a pine-beech mixed stand. **b** A suction probe with ceramic cup

2.2 *Other Methods for Assessing Forest Ecosystems*

2.2.1 Phenological Observations

The phenological observation of the development of vegetation has a long tradition. All over Europe, phenological gardens were established in order to collect comparative data (Schirmer et al. 1987). One main interest of the observation program is the investigation of possible consequences of climate changes in forest ecosystems.

For our work, the phenological observations are important for:

- determining the levels and seasonal dynamics of leaf area indices with regard to the modeling of water balances;
- identifying drought stress of trees in combination with declining quantities of soil water and increased needle and leaf fall; and
- determining deposition rates.

2.2.2 Dynamics of Leaf and Needle Loss

Leaves and needles fall from trees into collectors (Fig. 8), with a defined surface area, in our case 0.25 m^2. A total of 13 collectors are installed crosswise on a specified area of the stand, and the amounts collected are then converted to give a quantity of leaves or needles per hectare. In combination with the nutrient content determined by laboratory analysis, this gives the substance load deposited on the forest floor in tonnes per hectare.

The measurements over the year of the leaf or needle loss, taken in combination with the phenological observations, make it possible to draw conclusions about the dynamics of the leaf area index. This is an important parameter for modeling the interception and transpiration as part of a forest water balance model.

Fig. 8 Measuring field with litter fall samplers in a pure pine stand

(a) (b)

Fig. 9 **a** Dendrometer measuring the radial increment on a pine tree (manual reading and continuously measuring). **b** Continuous measurement on a small beech tree

2.2.3 Growth Dynamics of Trees

In the forest area, the growth dynamics of individual trees are monitored. In three different trunk classes, 12–15 trees are fitted with a measuring device.

The aim is to measure the radial increase in the size of the trunk over specified periods of time and depending on various environmental influences. The devices either provide continuous measurements or are read off and recorded manually (Fig. 9a, b).

The choice of measuring intervals and the type of device used depends on the objectives:

- Manual dendrometers are used to determine radial increments in a vegetation period or a month.
- Continuous measurements are taken to determine the growth dynamics over a year, depending on weather influences, usually at hourly intervals.

2.3 Using Lysimeters to Investigate Water Balances

2.3.1 Principle of Lysimeters

Lysimeters are containers filled with soil which can be used to measure the quantity and quality of the seepage water and determine the solute and water balances for vegetation, vegetation cover, or other ground cover. They can contain either disturbed soil or a monolith of undisturbed soil, and a distinction is also made between weighing and nonweighing lysimeters.

The seepage water flowing out of the lysimeter is usually measured directly. Together with the amount of precipitation and the change in soil moisture, the ET over a given period can be determined using the water balance equation:

$$ET_{ges} = P - S \pm \Delta\Theta \quad (1)$$

(P = precipitation; S = seepage; $\Delta\Theta$ = change of water content in soil)

Lysimeters are commonly used to investigate water balances and solute cycles. Over the past decade, there has been an increase in their use as a result of new extraction and installation technologies as well as innovative measuring technologies (Meissner et al. 2014).

Lysimeters are suited to determining the influence of vegetation on the water balance of soils. If the lysimeters are properly constructed and dimensioned, this also applies for forest ecosystems.

There is a long tradition of using lysimeters to study forest hydrology and the influence of forests on evaporation and groundwater recharge in the Eberswalde area 50 km northeast of Berlin. Table 2 shows the various types of lysimeter used in this forest hydrological research.

In 1907, studies of the water balances of small forest trees began in Eberswalde using very small lysimeters. These were replaced in 1929 by a larger weighing setup with three lysimeters, to which four more were added in 1954 (Table 2). To the best of our knowledge, the test station is the oldest lysimeter station for forest hydrology purposes (Müller and Bolte 2009). The weighing lysimeters are used to determine ET and seepage for various sorts of vegetation under defined climatic and soil conditions. A key finding is that the plant species and the degree of cover of the vegetation significantly affect the level and the temporal variation in seepage and ET (Müller and Bolte 2009). For example, less water seeps into deeper layers under 6-year-old Scots pine than under grass cover. These first results were one of the reasons for constructing larger lysimeters.

2.3.2 The Large-Scale Lysimeters

On the basis of this experience with lysimeters, nine large-scale lysimeters were installed in 1972 with a depth of 5 m and a surface area of 100 m^2 (Fig. 10). The "Britz large-scale lysimeters" are unique in Europe because while other lysimeters planted with trees may have an adequate surface area, they only have depths of 2.5, 3, or 3.5 m, which is not sufficient, example, in Castricum in the Netherlands (van der Hoeven 2011), St. Arnold in Lower Saxony (Schroeder 1976), or the Letzlinger Heath in Saxony-Anhalt (Helbig 1988).

Table 2 The different types of lysimeters in forest hydrological research

Year	1907	1929	1966	1972	1994	2005	2009
Location	Eberswalde "Drachenkopf"	Eberswalde "Drachenkopf"	Liepe	Britz	Test areas	Britz "Postluch"	Eberswalde Drylab (outdoors)
Lysimeter type	Very small lysimeter	Small-scale lysimeter	Understorey lysimeter	Large-scale lysimeter	Small-scale lysimeter	Ground-water lysimeter	Small-scale lysimeter
Weighing	Non-weighing	Weighing	Non-weighing	Non-weighing	Weighing	Weighing	Non-weighing
Soil	Disturbed sand	Disturbed sand	Monolith sand	Disturbed sand	Monolith sand	Monolith peat on sand	Disturbed sand
Cover	Diverse	Diverse	Scots pine	Trees	Trees ground veg.	Trees	Trees
Surface area	500 cm^2	1 m^2	500, 1500 cm^2	100 m^2	1 m^2	1 m^2	2 m^2
Depth	1.0 m	1.5 m	5 m	5 m	1.8 m	2 m	1.5 m

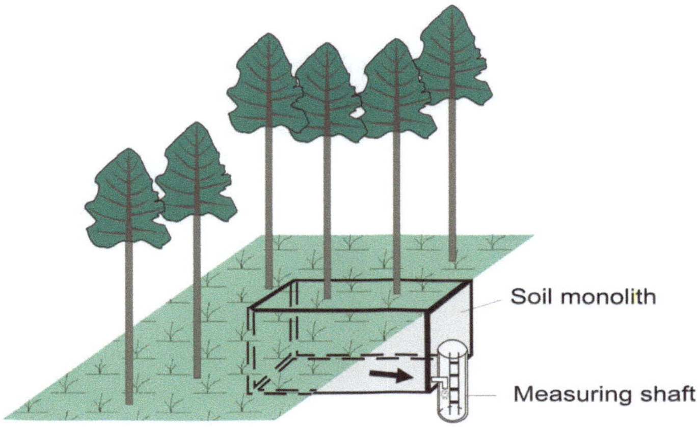

Fig. 10 Schematic diagram of a large-scale lysimeter planted with trees

The station is situated on the broad main terrace of the Eberswalde glacial valley, at 40 m above sea level. The soil is a Cambic Podzol consisting of 75 % medium sand to a depth of 5 m (with 17–19 % fine sand and 4 % clay and silt). The long-term mean annual precipitation is 570 mm and the annual mean temperature is 8.2 °C. The site is representative of large areas of the northeast German lowland.

The large-scale nonweighing lysimeters are filled with disturbed soil (Müller und Bolte 2009). The enclosing walls consist of 0.8 mm polyethylene foil. In 1974, these lysimeters and a 0.3 ha area around them were planted with beech (2 lysimeters), Scots pine (3 lysimeters), larch (2), and Douglas fir (2) in common combinations.

The seepage water from the lysimeter is collected in a measuring shaft and measured mechanically and electronically. The soil moisture is measured by probes at 10 different levels down to a depth of 4.6 m. Precipitation is measured using Hellman's rain gauges. The interception is calculated as the difference between the open field precipitation and that under the stands.

The initial goal of the research was to investigate the influence of the species and age of lysimeter tree stands growing under comparable climatic and soil conditions on groundwater recharge and ET.

In the future, the forests in Germany are to be mixed stands which are as varied in structure as possible. The scientific monitoring of the conversion program and the background research will investigate how hydroecological conditions, which are often a constraint on forest growth in drier regions, are changed by an understorey of beech and oak trees in Scots pine stands and how the conditions can be positively influenced by forest management of the stand structure. Against this background, individual lysimeter stands of pine trees were planted with oak and beech understoreys.

Figure 11 shows a Scots pine stand with an oak understorey on the large-scale lysimeter array.

Fig. 11 A Scots pine stand with oak understorey

With the introduction of new objectives for the lysimeter system, important ecological insights can be gained for successful forest conversion (Müller 2012). At the same time, it becomes possible to assess realistically the long-term consequences of the forest conversion for the landscape water balance.

2.3.3 The Weighing Lysimeter in Forest Stands

Apart from the contribution of forests to the landscape water balance, research is also concentrating on water consumption and the growth of forests in view of the growing shortage of water resources in the vegetation period. Against the background of climate change, the question of dry summers is becoming increasingly important (Müller 2009; Müller and Bolte 2009).

In Scots pine stands, the water consumption of the ground vegetation is an important but inadequately researched factor. However, given low precipitation levels, it is an important controlling factor in the water balance of forests. In order to determine the ET of ground vegetation and its specific water consumption separately from the transpiration of the tree stands, special small-scale weighing lysimeters were developed. Beginning in 1994, these were installed in various Scots pine ecosystems (Fig. 12). The lysimeters had to meet the following constructional criteria:

- Circular lysimeter surface area of 1–2 m^2, which is up to twice the height of the ground vegetation.
- Lysimeter depth of some 2 m so that if water accumulates at the base of the lysimeter, it will not have any effect on the level of transpiration.

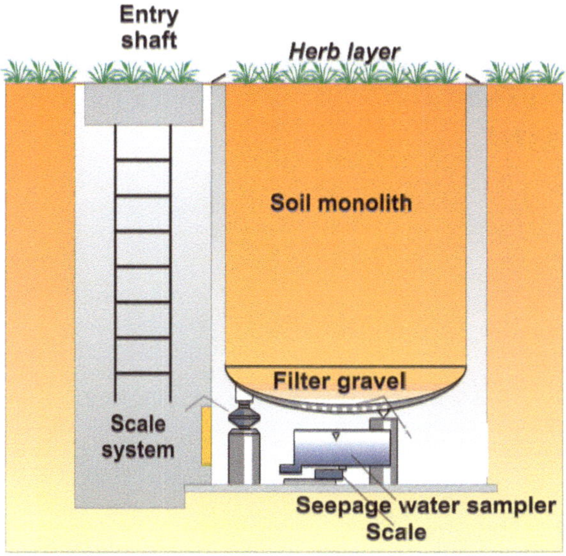

Fig. 12 Schematic diagram of a weighable lysimeter

- Extraction of an undisturbed soil monolith with growing ground vegetation.
- Continuous measurement of changes in soil moisture and seepage by high-precision weighing.
- Mobility of the lysimeter, allowing redeployment in other locations.

At the time, a mobile, weighing monolith unit of this size without a lysimeter cellar was new. It could be used to assess the water balance under various conditions at comparatively low cost. The water consumption of the vegetation was measured using special weighing cells to register the change in soil moisture in the monolith, while the outflow of seepage water was measured with an accuracy of 0.1 mm.

Since 1994, the weighing lysimeters have been used for the following forest hydrology research:

- Measuring the water consumption and possible competition effects of ground vegetation with various species compositions in Scots pine stands (Fig. 13a).
- Measuring the water consumption of understorey trees in mixed stands (Fig. 13b).
- Using the lysimeters to determine the influence of changes in water table level on the water balance and the growth of young black alders.
- Investigating the effects of dryness on the growth and water balance of young oaks. The weighing lysimeters have a sensor-controlled rainout shelter (Fig. 13c). Watering is provided for various scenarios (dry or normal scenario).

Fig. 13 **a** Small-scale lysimeter with ground vegetation in a pure Scots pine stand. **b** Weighable lysimeter in a pine-beech mixed stand, **c** small-scale lysimeter with young oaks to evaluate the drought risk

3 Results

3.1 The Contribution of the Forest to the Landscape Water Balance

The level of groundwater recharge under forests is highly dependent on the tree species (Müller 2005, 2013). The amount of seepage is mainly influenced by the differences in vegetation structure of the tree species.

To understand the species-specific differences in seepage, it is necessary to analyze the interception, transpiration, and ground level ET. As an example, Fig. 14 shows the water balance for the beech and Scots pine. With comparable ET of the stands, the interception is the main reason for the differences in seepage, which represents about a quarter of annual precipitation in the beech stands and a third in the Scots pine stands. The beech foliage varies seasonally (green in summer and bare in winter), the boughs and branches funnel water to the trunk, and the smooth bark indicates low trunk interception and leads to a stem flow of up to 8 % of open-field precipitation and low interception evaporation. The higher interception evaporation of the Scots pine stands in comparison is due to the fact that it keeps its needles throughout the year, it has an open crown (funnel), and rougher bark which inidcates that there is little or no stem flow.

3.2 On the Role of Ground Vegetation in Scots Pine Stands

In order to determine the ET of ground vegetation and thus its specific water consumption and competition effects separately from the transpiration of the tree stands, the weighing lysimeter was installed with various vegetation forms of Scots pine.

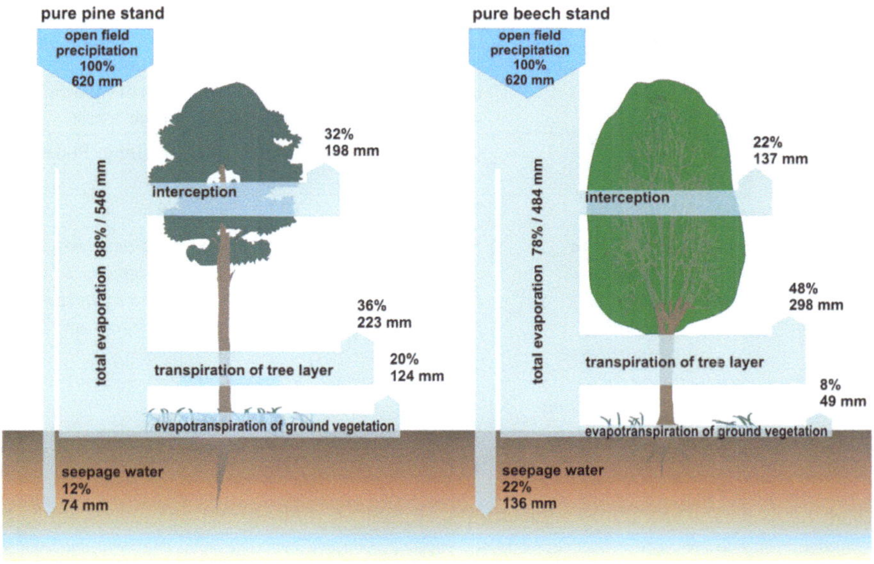

Fig. 14 Water balance of mature pine and beech stands on sandy soil with deep groundwater level

The investigated ground cover (Table 3) represents the characteristic species composition of ground vegetation commonly found in Scots pine ecosystems on sandy locations in the northeast German lowland (Hofmann 1995).

The ET of the ground vegetation depends to a large extent on the species composition. Over the year, closed wood small-reed (*Calamagrostis epigejos*) cover consumed more than a third of the annual precipitation of 620 mm and the wavy hair grass (*Avenella flexuosa*) cover nearly 30 %. Where a low shrub layer was present, the evaporation was sometimes distinctly lower than with the pure grass cover, e.g., with the wavy hair grass/raspberry layer, evaporation was a little more than 25 % of the annual precipitation and with a bilberry/wavy hair grass layer almost 20 % (Fig. 15). Investigations with the weighing lysimeters show that only approximately 10 % of the annual precipitation evaporates from the forest floor free of vegetation under a stand with a closed crown canopy. Under mature trees, the ET of full grass cover is 30 % of the annual precipitation, which is comparable with the interception of the tree stand. In the main growth period, the relative proportion of tree transpiration of the Scots pine stands declines in dry periods with increasing grass cover. This indicates competition with the ground vegetation for groundwater extraction, in turn leading to a reduction in the rate of growth of the tree stand (Müller et al. 1998).

Table 3 Ecological data of the forest ecosystems examined with weighable lysimeters

Vegetation form	Bilberry-Scots pine forest	Raspberry-wavy hair grass-Scots pine forest	Wavy-hair grass-Scots pine forest	Wood small-reed-Scots pine forest
Soil type	Podzol	Cambic Podzol	Cambic Podzol	Cambic Podzol
Stand age (years)	70	77	125	81
Species composition of ground vegetation	*Vaccinium myrtillus*, *Avenella flexuosa*, *Pleurozium schreberi*, *Scleropodium purum*	*Rubus idaeus*, *Avenella flexuosa*, *Sclero-podium purum*	*Avenella flexuosa*, *Pleurozium schreberi*	*Calamagrostis epigejos*, *Brachythecium salebrosum*
Mean evapo-transpiration (mm)	138	154	180	236

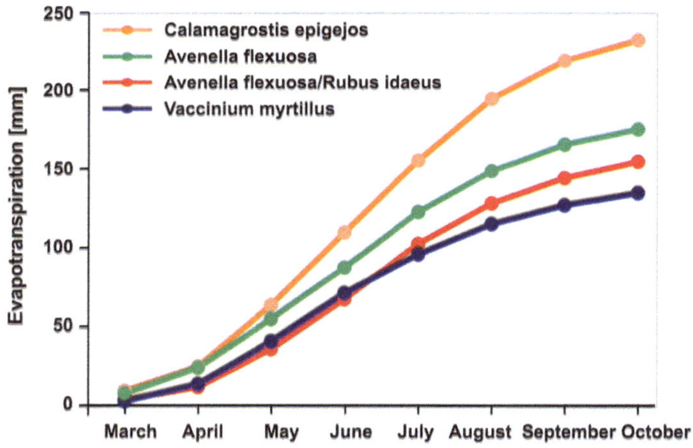

Fig. 15 Cumulative evapotranspiration of different species of ground vegetation in Scots pine stands from March until October

4 Conclusions

A wide range of methods and modern measuring systems are available for forest monitoring and the assessment of the state of forests and forest ecosystems.

Meteorological and hydrological measurements are of particular interest. They can be used to quantify environmental influences and the effects of management measures on the water and solute fluxes in forest ecosystems.

In order to be able to assess the effects of environmental influences on the health of forests, various factors must be taken into account when selecting potential observation areas:

- Long-term observations are essential, and an observation period of at least 10 years should be ensured.
- The plots should be representative of the current distribution of tree species and their age distributions
- Any exposure to emissions that would be harmful to the state of forests and would require additional environmental observations.

One precondition for the establishment of observation areas is agreement on the methods to be applied and the equipment that will be used. An example of the harmonization of European forest monitoring is provided in the ICP Forests Manual (2010).

This manual offers guidelines on establishing and operating areas for environmental observation and presents methods which can be adopted. In order to achieve compliance with the rules documented in the manual, it is advisable to setup demonstration areas for training purposes. It is important that training courses and workshops focusing on specific topics should be carried out at regular intervals.

Against the background of an increase in extreme weather events, environmental observations are facing new challenges regarding the development of sensor systems to warn against the threats that these pose. This involves:

- linking existing national and international measuring stations;
- developing novel, autonomous networks of sensors for the localization, classification, and identification of hazards, e.g., early warning systems for forest fires;
- establishing an information and alarm management system to identify the nature and location of hazards and the level of threat; and
- developing methods and equipment for the investigation of causes and effects in ecosystems, e.g., to identify the effects of drought on tree growth.Questions remain concerning the interactions between plant physiology and hydrology. Lysimeters are a very promising tool for the investigation of this complex issue.

In addition to the development of technical equipment, future work will also have to address quality assurance for data, the use of models, and the extension of data to cover larger areas. Examples are:

- developing principles for the assessment of biodiversity in forests;
- Quality assurance for data: this includes developing routines for checking the plausibility of data and for filling in gaps in the data, and Autonomous data storage is recommended. This ensures flexibility in the national evaluation and offers the possibility of evaluation at a higher level of abstraction;
- recommending models for calculating components of the water and solute balance; and
- approaches to extend the areas covered by data so as to enable the formulation of recommendations for actions in connection with climate scenarios.

References

DVWK (1986) Ermittlung des Interzeptionsverlustes in Waldbeständen bei Regen. –DVWK Merkblätter zur Wasserwirtschaft (11 p), H. 211, Hamburg, Berlin: Parey

DWD (Deutscher Wetterdienst) (2001) Richtlinie für automatische Klimastationen. Offenbach am Main 71 p

Hauhs M (1985) Wasser- und Stoffhaushalt im Einzugsgebiet der Langen Bramke (Harz). Ber des Forschungszentrum Waldökosysteme/Waldsterben A 17:206

Helbig A (1988) Vergleich der Wasserhaushaltskomponenten eines Kiefernbestandes und einer Waldgrasfläche nach Lysimetermessungen. Abhandlungen des Meteorologischen Dienstes der DDR, No.140, pp 123–128

Hofmann G (1995) Wald, Klima, Fremdstoffeintrag - ökologischer Wandel mit Konsequenzen für Waldbau und Naturschutz dargestellt am Gebiet der neuen Bundesländer Deutschlands. Angew. Landschaftsökol. 4:165–189

ICP Forest Manual (2010) Manual on methods and criteria for harmonized sampling, assessment, monitoring and analysis of the effects of air pollution on forests (577 p). UNECE, ICP Forests, Hamburg

ICP Forests Manual (2015) http://icp-forests.net/page/icp-forests-manual. Accessed 22 Feb 2015

Meissner R, Rupp H, Seyfarth M (2014) Advanced technologies in Lysimetry. In: Mueller L, Saparov A, Lischeid G (eds) Novel measurement and assessment tools for monitoring and management of land and water resources in agricultural landscapes of Central Asia (pp. 159–173). Springer, Berlin. http://link.springer.com/chapter/10.1007/978-3-319-01017-5_8. Accessed 22 Feb 2015

Müller J (2002) Wirkungszusammenhänge zwischen Vegetationsstrukturen und hydrologischen Prozessen in Wäldern und Forsten. In: Anders S (ed) Ökologie und Vegetation der Wälder Nordostdeutschlands. Verlag Dr. Kessel, Oberwinter, pp 99–122

Müller J (2005) 30 Jahre forsthydrologische Forschung auf der Großlysimeteranlage in Britz–Zielstellung und Ergebnisse. Tagungsband der 11. Gumpensteiner Lysimetertagung "Lysimetrie im Netzwerk der Dynamik von Ökosystemen" (pp 29–32)

Müller J (2006) Die Ressource Wasser im zweischichtigeb Nadel-Laub-Mischbestand. In: Fritz P (ed) Ökologischer Waldumbau in Deutschland. Fragen, Antworten, Perspektiven. Oekom Verlag München, München, pp 152–183

Müller J (2009) Forestry and water budget of the lowlands in northeast Germany—consequences for the choice of three species and for forest management. J Water Land Dev 13A:133–148

Müller J (2012) Auswirkungen von waldstrukturellen Veränderungen im Zuge des Waldumbaus auf die hydroökologischen Bedingungen in den Beständen. In: Grünewald U, Bens O, Fischer H, Hüttl R.F, Kaiser K, Knierim A (eds) Wasserbezogene Anpassungsmaßnahmen an den Landschafts- und Klimawandel (pp 280–291). Stuttgart

Müller J (2013) Die Bedeutung der Baumarten für den Landschaftswasserhaushalt. In: Bericht/15. Gumpensteiner Lysimetertagung: Lysimeterforschung als Bestandteil der Entscheidungsfindung; 16/17 April 2013. Irdning: Lehr- und Forschungszentrum für Landwirtschaft Raumberg-Gumpenstein, pp 49–56

Müller J, Bolte A (2009) The use of lysimeters in forest hydrology research in north-east Germany. Landbauforschung vTI Agric Forestry Res 59(1):1–10

Müller J, Bolte A, Beck W, Anders S (1998) Bodenvegetation und Wasserhaushalt von Kiefernforstökosystemen (Pinus sylvestris L.). Verh. Ges. Ökol. Berlin 28:407–414

Rakei A (1991) Wasserhaushalt eines Alt- und Jungkiefernbestandes auf Rostbraunerde des Grunewaldes (Berlin). Schrift aus der TU Berlin, Inst. f. Ökologie, Fachgebiet Bodenkunde und Regionale Bodenkunde Heft 4. 134 p

Schirmer H, Buschner W, Cappel A, Mattäus H-G, Schlegel M (1987) Meteorologie—Meyers kleines Lexikon (496 p). Mannheim

Schroeder M (1976) Grundsätzliches zum Einsatz von Lysimetern. Deutsche Gewässerkundliche Mitteilungen 20(1):8–13

UGT (2014) V2A tipping counters. http://www.ugt-online.de/en/produkte/hydrologie/durchflussmessung/kippzaehler-aus-v2a-stahl.html. Accessed 22 Feb 2015

Ulrich B, Mayer R, Khanna PK (1981) Deposition von Luftverunreinigungen und ihre Auswirkungen in Waldökosystemen im Solling. Schriften aus der Forstlichen Fakultät der Universität Göttingen und der Niedersächsischen Versuchsanstalt, Band 58. Sauerländer's Verlag Frankfurt am Main. 291 S

van der Hoeven PCT (2011) Lysimeters Castricum, waterbalans lysimeter-1. Alterra rapport 2053-6, Wageningen, 57 p

VDI (2006) Verein Deutscher Ingenieure. Meteorologische Messungen, Messstation. VDI Richtlinie 3786, Blatt 13. Düsseldorf. 42 p

WMO (1996) World Meteorological Organization. Guide to Meteorological Instruments and Methods of Observation Parts II–III.—WMO, 8. Geneva. 681 p

Chapter 16
Using the Innovative Lysimeter Technology in the German–Russian Research Project "KULUNDA"

Dmitry Balykin, Aleksandr Puzanov, Eckart Stephan and Ralph Meissner

Abstract The interdisciplinary KULUNDA project unites the efforts of German and Russian scientists to tackle the problems of soil degradation and water scarcity in the Kulunda steppe of the Russian Altai Krai. From 1954 to 1963, approximately 42 million ha of the Southern Soviet steppe, of which 6.2 million ha are located in Western Siberia, were converted into a large-scale intensive agriculture area. The affected areas are highly vulnerable to wind erosion, resulting in decreased top soils and humus contents and therefore in decreased concentrations of sequestered carbon. The assessment and management of the soil water and solute balance are of great importance for crop yield potentials and the sustainable development of the territory. In 2013, the first weighable gravitation lysimeter station in Siberia was successfully installed in Altai Krai (Russia) under Kulunda dry steppe conditions. Weighable lysimeters allow the continuous monitoring of changes in soil monolith mass. This is the precondition for calculating actual evapotranspiration (ETa—major component in the terrestrial water cycle) with high precision. Knowledge regarding the development of ETa is essential to evaluate the impact of climate change on the future water balance.

D. Balykin (✉) · A. Puzanov
Institute for Water and Environmental Problems, Siberian Branch of the Russian Academy of Sciences (IWEP SB RAS), 1 Molodyoznaya St., 656038 Barnaul, Russia
e-mail: balykinDN@yandex.ru

A. Puzanov
e-mail: puzanov@iwep.ru

E. Stephan
Martin-Luther-University Halle-Wittenberg, Institute of Geosciences and Geography, Von-Seckendorff-Platz 4, Haus Geographie, 06120 Halle, Germany
e-mail: eckart.stephan@geo.uni-halle.de

R. Meissner
Helmholtz Centre for Environmental Research—UFZ, Department Soil Physics, Lysimeter Station, Dorfstrasse 55, 39615 Falkenberg, Germany
e-mail: ralph.meissner@ufz.de

L. Mueller et al. (eds.), *Novel Methods for Monitoring and Managing Land and Water Resources in Siberia*, Springer Water,
DOI 10.1007/978-3-319-24409-9_16

Keywords KULUNDA project · Lysimeter station · Weighable lysimeter · Soil hydrology · Altai Krai

1 Introduction

The Altai region is one of the top grain-producing regions in Russia. Its total area of cultivated land is 5.4 million ha, of which 3.5 million ha are designated for grain and leguminous crops, including maize. The gross grain yield average for the period 2006–2010 is 4.4 million tonnes per year (Safonov and Safonova 2013). From 1954 to 1963, approximately 42 million ha of the Southern Soviet steppe, of which 6.2 million ha are located in Western Siberia, were converted into a large-scale intensive agriculture area. This area was highly vulnerable to wind erosion, resulting in decreased top soils and humus content and therefore in decreased concentrations of carbon. The adversely influenced soil water and nutrient regime as well as the declining fertility resulted in a decrease in crop yields (BMBF 2011–2016; Morkovkin et al. 2013). Between 1963 and 1965, more than 1 million ha of eroded land were set aside in Altai. Between 1960 and 1970, a set of measures aimed at soil conservation were implemented, including the establishment of plantations and shelter belts, conservative tillage, soil-protecting crop rotation, crop lane placement, mulch with straw, etc. However, the problem of "black storms" (wind erosion), which was resolved in the 1960s and 1970s, is currently reappearing. "Black storms" were recorded in the Altai region in the mid-1990s and in the 2000s (Safonov and Safonova 2013).

The region is part of the Southwest Siberian Kulunda steppe lowland and located between the Central Asian steppe and the North Asian forest–steppe. In the North, the Kulunda steppe borders on the Baraba forest steppe with a vegetation mosaic composed of steppe communities and birch "kolki" (birch islets). The Western part of the Kulunda steppe borders is on the Eastern part of the Irtysh valley in Kazakhstan. The Kulunda depression is located at altitudes of 100–140 m asl. The lowland is covered by a 50–60 m thick layer of pleistocene alluvial and 0.5–10 m of eolian sediments. The soil cover on the dry steppes of Kulunda consists of chestnut soils, meadow–chestnut soils, meadow soils, solonetz, and solonchaks with different degrees of hydromorphism. The chestnut soils significantly vary in texture as a result of the ancient limnic and eolian genesis of the territory. Sandy loams (15–19 % clay, 11–20 % silt, 65–70 % sand) are predominant, and their contents of humus (2–4 %) and carbon (5–8 %) are comparatively high (Bazilevich 1959; Rudaya et al. 2012).

In the continental climate of the Kulunda steppe, there are long, cold winters with little snow and short but hot and dry summers. Due to its open position, the steppe is often affected by cold air masses from the Kara Sea, and warm and dry air masses from Kazakh and Middle Asian steppes and deserts. Thus, the temperature is highly variable: May and September often have night frost, in late snow-free autumn periods the temperature can drop to −20 °C or lower, the spring sometimes has very dry periods, and dry winds are common throughout the year. The mean annual temperature is about 0 °C, the mean temperature of January (the coldest month) is −19 °C, the absolute minimum is −47 °C, the mean temperature of July (the warmest month) is +19 °C, and the absolute maximum is +40 °C. The frostless period lasts 112–120 days per year from May 15–25 to September 10–15. The annual precipitation is about 250–450 mm, and the precipitation in April–October is about 200 mm. The duration of a stable snow cover reaches 140–150 days (from November 10–15 to April 5–10). The mean depth of the snow cover is 15 cm (absolute maximum 35–38 cm). Such a thin snow cover does not protect the soil from frost, so in winter the soil freezes at 2 m deep (and even more). The amount of global radiation is 2–3 times higher than is required for the evaporation of this amount of precipitation (Chernikova 1971).

Kulunda is situated in the Eurasian steppe zone with a prevalence of grass communities (Lavrenko 2000). Numerous salt lakes are surrounded by plant associations that include *Festuca valesiaca*, *Goniolimon speciosum*, *Koeleria gracilis*, *Artemisia spec.*, *Kochia prostrata*, and *Achnatherum splendens*. Pine forests with *Pinus sylvestris*, which spread southward from the Kulunda steppe, are the most xerophytic Siberian forests (Ermakov et al. 2000). The east of Kulunda is covered by forest–steppe with isolated stands of *Betula pendula* and *Populus spec.* This vegetation type, called “kolki”, is botanically closer to European deciduous forests than to subarctic or boreal vegetation (Nimis et al. 1994).

“KULUNDA” is 1 of 12 regional projects financially supported by the funding measure “Sustainable Land Management” (Module A) provided by the Project Management Agency (PT-DLR) on behalf of the German Federal Ministry of Education and Research (BMBF; www.kulunda.eu). It unifies the knowledge of 16 partner institutions (universities and research facilities), which are organized in 11 subprojects (SPs). These are SP0—Scientific coordination and project management, SP1—Effect of land use, land cover and climate change on soil degradation, SP2—Soil water and solute balance, SP3—Land use impact on carbon sequestration, SP4—Interaction of vegetation with changing land use and climate, SP5—Geo-database and satellite image classification, SP6—Process-based modeling of carbon cycle and impact of land use change, SP7—Agronomic and technical solutions for innovative tillage and cropping system, SP8—Assessment of farm level costs and socioeconomic requirement of providing ecosystem services, SP9—Social and institutional drivers of land use change, and SP10—Stakeholder involvement and implementation. The German partner institutions of the project are Martin-Luther-University Halle-Wittenberg (MLU), Leibniz University Hannover, Georg-August-University Goettingen, Helmholtz Centre for Environmental Research—UFZ, Leibniz Institute of Agricultural Development in Transition Economies (IAMO), Leibniz Institute for Regional Geography (IfL), Potsdam Institute for Climate Change Impact Research (PIK), the Senckenberg Community for Natural Research; and the medium-sized enterprise Amazonen-Werke H. Dreyer GmbH & Co. KG. The Russian partner institutions are Altai State University, Altai State Agrarian University and Institute for Water and Environmental Problems,

Siberian Branch of the Russian Academy of Sciences. The head of the project is Prof. Dr. Dr. h.c. Manfred Fruehauf and the scientific coordinator is Dr. M. Kasarjyan (Martin-Luther-University Halle-Wittenberg, Faculty of Natural Sciences III, Institute for Geosciences and Geography; BMBF 2011–2016).

The main goal of the interdisciplinary KULUNDA project is to mitigate degradation and desertification processes in the Southern Siberian steppe sites used for agriculture, to stimulate, and, in the long run, to enhance carbon sequestration in soils of the Kulunda steppe (Haerdle 2013). The project also aims to increase crop yields and to implement sustainable land management practices for agricultural areas and thereby to contribute to rural and regional development. The knowledge gained and the results of the KULUNDA project will largely contribute to the research on climate change, sustainable land management practices, and rural and regional development (BMBF 2011–2016).

The assessment and management of water and solute balance of soils are of great importance for crop yield potentials and sustainable development of the territory. The overall objective of SP2 is to evaluate land management practices according to their impact on the soil water and solute balance in the Kulunda steppe of Altai Krai (BMBF 2011–2016). SP2 focuses on the scientifically based improvement in the monitoring techniques and the evaluation modules for decision-making processes toward sustainable land management. The results will help to establish and use sophisticated ecosystem models, as such models are excellent tools for forecasting the impacts of climate and land use changes on water and solute balances in agricultural landscapes (Nendel 2014). They require reliable, high-resolution input data on soil hydrological processes. Precision lysimeters may provide valuable data for the calibration of such models.

The objectives of this paper are (i) to inform readers about the design and construction of a weighable lysimeter station in the Kulunda steppe and (ii) to demonstrate the accuracy of the first measurements.

2 Materials and Methods

2.1 Hypotheses of Lysimeter Research in the Kulunda Steppe

It is well known that a weighable gravitation lysimeter is a sufficient tool to measure the relevant parameters of the water balance equation with high precision. No such data exist for the dry steppe conditions of Siberia, but they are essential for the evaluation of the former development of this region. Climate change, especially, is a main driver regarding the direction of future land use. The lysimeter measuring results will help to investigate and establish sustainable land management strategies for this region. The aim is to deepen the understanding of relevant processes concerning the interaction of land degradation and climate change to quantify fluxes of water, carbon, and other relevant matters. The results serve as the basis for simulating the effects of land management on the soil and matter balance using adequate models (for example the hydrological model HYDRUS). Beyond these central questions, it is possible to gain further information depending on corresponding scientific questions: In particular, information can be gained about the functionality of lysimeter technology under extreme climatica conditions, and comparative analyses can be carried out between different crops regarding their responses to water availability between frost and dry periods or during heavy rainfall events or fast snow melting in the spring, etc.

2.2 *Purpose of Lysimeters*

Lysimeters are useful devices to investigate water and solute transport, as well as transformation processes, using undisturbed representative soil monoliths (Meissner et al. 2007, 2014). They represent the link between laboratory- and field-scale studies and offer the opportunity to reproduce natural conditions in model systems. The lysimeter station that was installed in the Kulunda steppe consists of two vessels, which allows comparative analyses to be carried out between arable land which was converted from grassland and unconverted (nearly pristine) grassland. The vessels are weighable, which ensures that mass changes at the study site can be monitored continuously for any time period. Thus, it is possible to calculate water fluxes between the pedosphere and the atmosphere, in particular precipitation and actual evapotranspiration (ETa). Based on the measured parameters, ETa (in mm) can be derived using the following equation:

$$\mathrm{ETa} = P - S \pm \Delta W \tag{1}$$

where

P precipitation (mm),

S amount of seepage water (mm),

ΔW change in the quantity of stored water [mm] as determined from the mass change in the lysimeter over time (1 kg ≈ 1 $\mathrm{L/m^{-2}}$ = 1 mm)

If the water balance is calculated correctly, the solute load (L in mg $\mathrm{m^{-2}}$) can be determined with high accuracy from the following relationship:

$$L = C_s \times S \tag{2}$$

where

C_s solute concentration in the seepage water (mg $\mathrm{L^{-1}}$),

S amount of seepage water (L $\mathrm{m^{-2}}$ = mm)

Furthermore, the lysimeter delivers reliable information regarding seepage water quantity and quality. Only lysimeters permit a direct determination of the amount of water percolating through a soil profile and of the type and amount of solutes contained in it. Hence, they allow more reliable calculation of solute loads carried toward the groundwater than any other method (Gee et al. 2009). Although those processes may be not relevant to typical steppe soils under current climates, they could emerge under an altered climate and hydrological regime. Laterally inserted probes facilitate the measurement of the soil moisture content, soil moisture tension, and soil temperature. The provision of high-resolution data allows the detailed tracking of soil moisture development at different depths. This information is also important for other project partners who are trying to develop sustainable soil tillage techniques and procedures for this region. Furthermore, it is possible to extract soil water samples at different depths in order to get precise information about the water and solute transports in representative undisturbed soil monoliths. During the first measuring period, the lysimeter investigations focus on comparing water balance parameters for soils that are or are not agriculturally used. Based on these findings, specifications can be derived regarding crop rotation, management, fertilization, irrigation, etc.

2.3 Soil Properties

In order to operate comparative analyses between arable land and soils which are not used for agriculture, the monoliths were extracted at two sites with different land uses (Table 1). Following the FAO guidelines, the soils are identified as Calcic Chernozems (FAO 2006). The first monolith consists of a 25-cm-thick humic horizon including a plowed layer at the bottom. It is followed by a crossing AC horizon that is 50 cm deep at the bottom. Beneath, there is a subsoil C-horizon composed of parent material including calcareous deposits. The grain size distribution in the upper part of the profile (0–50 cm) indicates sandy loamy silt, beneath which there is silt loam and, below 70 cm, loamy sand. The site was under intensive agricultural usage for 60 years. The second monolith consists of a 30-cm-thick humic Ah horizon, followed by a 15-cm-thick crossing AC horizon and by subsoil composed of parent material which is interspersed with calcareous deposits. The upper 30 cm consists of sandy loamy silt, below which there is silty loam underlain by loamy sand at a depth greater than 70 cm.

2.4 Lysimeter Extraction and Installation

In 2013, the two-fold containerized (Polyethylene PE-HD) lysimeter station (UGT 2014) with two weighable soil monoliths provided by the company UGT—Environmental Measurement Devices Ltd. and the Helmholtz Centre for

Table 1 Soil properties of the lysimeter extraction sites

Name/No.	Extraction site lysimeter 1			
Date	28.06.2013			
Position	N52 04.012–E79 54.526			
Altitude	138 m			
Slope/exposition	<1			
Usage	Field/wheat			
Note	Amazone test field; Poluyamki/conventional tillage			
Soil type	Calcic Chernozem			
Horizon	Ah	AC	Ckc	C
Lower boundary (cm)	25	50	70	120+
Grain size fraction	SLSi (Uls)	SLSi (Uls)	SiL (Lu)	LS (Sl4)
Name/No.	Extraction site lysimeter 2			
Date	28.6.2013			
Position	N52 03.778–E79 55.868			
Altitude	142 m			
Slope/exposition	<1			
Usage	Natural steppe vegetation			
Note	No tillage for decades			
Soil type	Calcic Chernozem			
Horizon	Ah	AC	Ckc	C
Lower boundary (cm)	30	45	70	110+
Grain size fraction	SLSi (Uls)	SiL (Lu)	LS (Sl4)	LS (Sl4)

Environmental Research—UFZ was installed at a KULUNDA project test farm in Poluyamki, in the Mikhaylovskij region of Altai Krai, under Kulunda dry steppe conditions (Fig. 1). The location was chosen because the South–Western Kulunda steppe region is most affected by soil degradation processes compared to other parts of the administrative unit Altai Krai (Paramonov et al. 2003). In addition, severe effects of global climate change should be detected to a great extent in this vulnerable region. Other lysimeter stations in different climatic subzones are envisaged to get a reliable network regarding the exact measurement of water balance parameters and the calculation of ETa.

In cooperation with German and Russian project partners, the soil monoliths (surface area of 1 m^2, 2 m depth) were extracted from both an arable land site (lysimeter 1) and a fallow site which was plowed once in the 1950s but had since then been covered with natural, nearly pristine steppe vegetation (lysimeter 2) for the lysimeter-based investigations (Fig. 2). After the extraction of the soil column, a 20-cm-thick filter layer (geotextile over fine–coarse sand over gravel) was placed at the bottom of the lysimeter vessels to minimize disturbances to the natural flux. The depths of the soil columns are greater than the natural flux plane, which under the environmental and management conditions at the extraction sites lies between 0.9 and 1.3 m below the surface. Hence, it can be assumed that seepage flow in the

Fig. 1 Installation of the PE-HD lysimeter station at the Poluyamki test farm (*photo* E. Stephan)

lysimeters is not affected before it reaches the filter layer. The composition of the filter layer was chosen to allow free drainage of seepage water. The vessels were transported to Poluyamki and positioned in the lysimeter station on load cells using a three-legged steel frame.

The lysimeter station measures mass changes with sufficiently high precision (±20 g) to enable the calculation of water fractions input (dew, rime, and the water equivalent of snow) and the rates of ETa (Meissner et al. 2014). The total mass of each lysimeter vessel is approximately 4 t (the mass changes depending on the water content of the vessel). Both lysimeter vessels are equipped at different depths (0.30, 0.50, and 1.20 m) with time domain reflectrometry stick probes to measure the soil moisture content, with combined tensiometer and temperature probes to measure the matrix potential as well as the temperature changes in the soil and with suction cups to extract the soil solution (Fig. 3). The lysimeter is specifically adapted to the cold winter conditions in Siberia by adding a ring to cut the snow on the lysimeter surface from the adjacent snow.

The measured values are consolidated and stored in a data logger with a recording interval chosen by the user. This permits a very high temporal resolution (<1 min). The amount of seepage water is measured with a tipping bucket (values are stored by the data logger) and collected in a storage container, from which water

Fig. 2 Extraction sites: the arable land with young spring wheat (*left*) and the natural steppe vegetation (*right*) (*photo* E. Stephan)

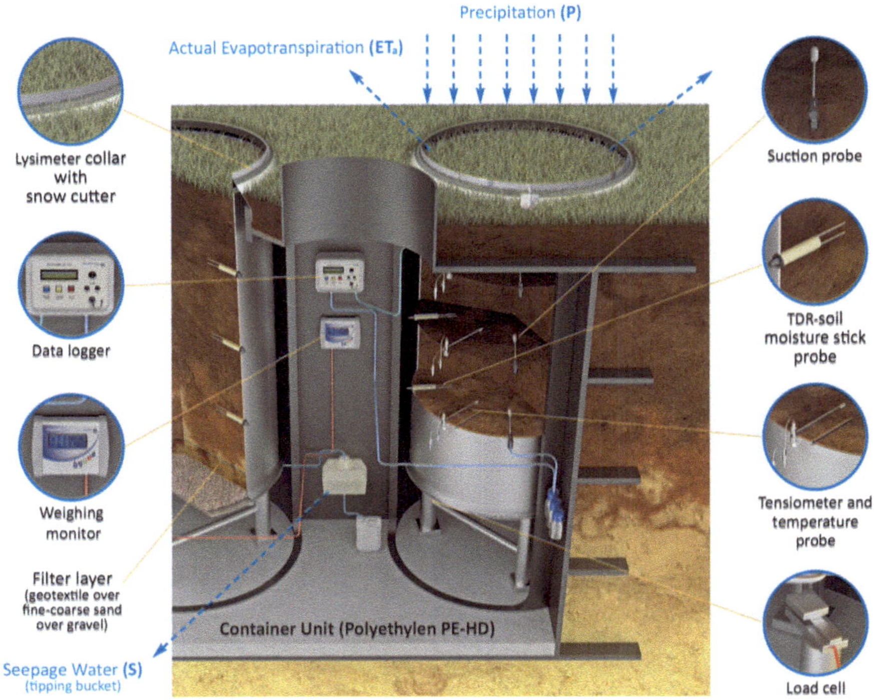

Fig. 3 Scheme of a weighable gravitation lysimeter installed at the Poluyamki site, Siberia

samples can be taken for chemical analysis. The installed computer software makes it possible to present all the parameters measured in different ways (e.g., average, minimum, maximum, or all values).

3 Results

Lysimeters are typically used to quantify rainfall, drainage, and ETa. According to the water balance equation, the changing masses (weights) represent the water storage within the soil monoliths. These changes can be displayed as positive or negative values. A mass gain in the soil monolith means precipitation; a mass reduction must be interpreted as seepage or as a release of gaseous water from plants and soil in the form of evaporation or transpiration, respectively. The rising and falling masses of the soil monoliths are thus the basis for calculating precipitation and ETa, which can be displayed as quantities in mm. The weighing precision of the presented lysimeter station allows small mass changes to be detected, even those generated by dew, fog, rime, or snow. To demonstrate the operational reliability of the lysimeter station, Fig. 4 shows the mass of the installed lysimeter recorded over a 3-day period in August 2013 based on a 1-hour measurement. There was no drainage from the lysimeters in the period mentioned. While the mass

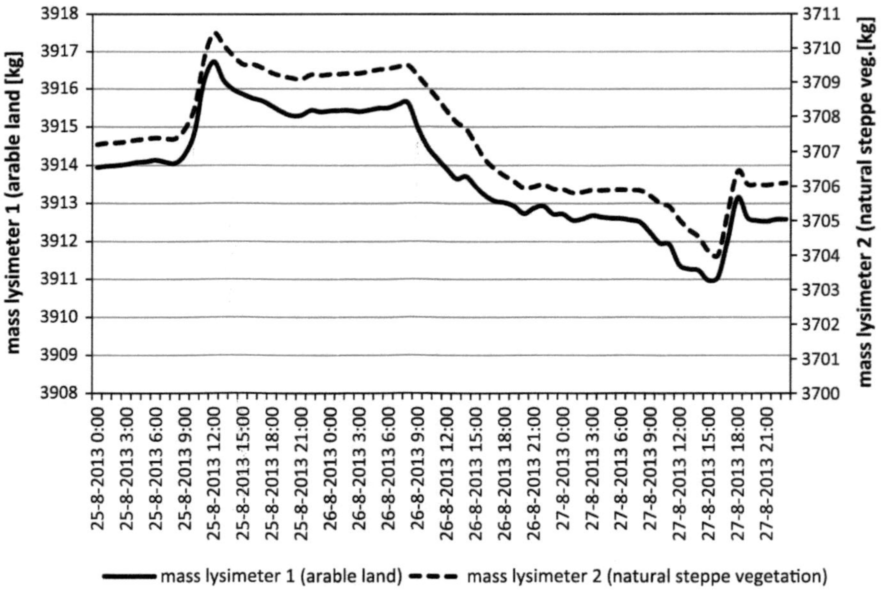

Fig. 4 Example of the diurnal mass changes of two gravitation lysimeters planted with spring wheat and a pristine plant cover

of lysimeter 1 (arable land, spring wheat) on August 25, 2013, 08:00 was 3914.1 kg, the lysimeter 2 mass (natural steppe vegetation) was 3707.4 kg.

Both soil monoliths registered a precipitation event on August 25, 2013, 08:00–12:00 resulting in mass increases about 2.7 kg (=2.7 mm; lysimeter 1) and 3.1 kg (=3.1 mm; lysimeter 2), respectively. The differences between the two monoliths are mainly based on the different vegetation covers and associated interception. Between 12:00 and 21:00, both monoliths registered a mass reduction of about 1.4 kg (=1.4 mm), which has to be interpreted as ETa. On the following day, the monolith mass decreased with the rising sun; between 08:00 and 12:00, ETas of 2.9 mm (lysimeter 1) to 3.6 mm, respectively were measured. Also on August 27, 2013, between 08:00 and 15:00 ETa values of 1.5 mm (lysimeter 1) and 1.7 mm (lysimeter 2) were determined. The precipitation event after 16:00 (lysimeter 1: 2.2 mm, lysimeter 2: 2.4 mm) stopped the ongoing ETa. From 18.00 until the end of the measuring period ETa occurred, causing small mass losses of 0.5 kg (=0.5 mm, lysimeter 1) and 0.4 kg (=0.4 mm, lysimeter 2). Considering the entire study period (August 25, 2013, 00:00–August 27, 2013, 23:00), it was seen that on lysimeter 1 ΔW was reduced by about −1.4 kg during a precipitation amount of 4.9 mm, leading to an ETa of 6.3 mm. In comparison, for lysimeter 2 a mass change in −1.6 kg and a precipitation amount of 5.5 mm were recorded, based on this data the calculated ETa was 7.1 mm (Fig. 4).

As already mentioned and demonstrated in the example earlier, the weighing precision of the lysimeter is very high. One item of interest is a detailed view for the time period between August 25, 2013, 20:00 and August 26, 2013, 08:00 for lysimeter 1 (see Fig. 4). During this time slot the mass increased by 340 g. Because no precipitation was measured and ETa stopped due to a lack of daylight (Xiao 2013), the formation of dew (amounting to 0.34 mm) must be the reason for the mass increase. This assumption is supported by measuring the relative humidity in a neighboring weather station, which increased in the late afternoon from 65 to 90 % at 23:00, remained at that level until 08:00 the next morning, and then decreased to 70 %. Regarding the amount and time-dependent occurrence of dew, this result fits well with our own measurements in Northern Germany (Meissner et al. 2007). A more detailed investigation of these phenomena will be the focus of future work. Information regarding the amount of precipitation caused by dew, fog, rime, or snow is especially important in dry areas that are prone to wind erosion because the establishment of a vegetation cover is a prerequisite for protecting the soil against further degradation.

4 Conclusions

The use of the unique weighable gravitation lysimeter station on the German–Russian KULUNDA project will enable us to carry out a thorough study of the water balance under dry steppe conditions, thus forming the basis for a highly accurate solute balance calculation and for modeling hydrological processes in the

Kulunda steppe of Altai Krai (Russia). In combination with meteorological data and soil hydrological field measuring stations, the lysimeter delivers reliable information regarding water balance input factors (precipitation, snow, etc.) and output factors (seepage water quantity and quality, changes in soil moisture at different depths, Eta, etc.). Sustainable land management practices for the Kulunda steppe will be developed to tackle the problems of soil degradation, carbon sequestration, and water scarcity.

Acknowledgments This article is based on the results of the research work carried out in the scope of the German–Russian cooperation project KULUNDA. The project is financed by the Federal Ministry of Research and Education of Germany (FK 01LL0905D). We would like to thank everyone working at the German and Russian partner institutions for their cooperation and support during the investigations. We would particularly like to thank the Koshanov farming family in Poluyamki for their support during the installation of the lysimeter station.

References

Bazilevich NI (1959) Soils of the Kastanozem Zone in the dry steppe. In: Soils of the Altai Krai. (Почвы каштановой зоны сухой степи in: Почвы Алтайского края). Academy of Sciences of the USSR, Moscow, pp 46–65 (in Russian)

BMBF (2011–2016). www.kulunda.eu. Accessed 2014/06/05

Chernikova MI (1971) Agroclimatic resources of the Altai Krai (Агроклиматические ресурсы Алтайского края) Leningrad, Gidrometeoizdat, 156 p (in Russian)

Ermakov N, Dring J, Rodwell J (2000) Classification of continental hemiboreal forests of North Asia. Braun-Blanquetia 28, Camerino 131 p

FAO (2006) World reference base for soil resources. Food And Agriculture Organization of the United Nations, Rome, pp 76–84

Gee GW, Newman BD, Green SR, Meissner R, Rupp H, Zhang ZF, Keller JM, Waugh WJ, van der Velde M, Salazar J (2009) Passive wick fluxmeters: design considerations and field applications. Water Resour Res 45:1–18

Haerdle B (2013) New life for the steppe. Science Portrait 8 (Kulunda): http://modul-a.nachhaltiges-landmanagement.de/fileadmin/user_upload/DOCUMENTS/RPs/KULUNDA/KULUNDA_SciencePortrait_2013-11-19.pdf. Accessed 2015/03/05

Lavrenko EM (2000) Steppes of the USSR. In: Selected papers (Степи СССР. In: Избранные труды) St. Petersburg, pp 11–223 (in Russian)

Meissner R, Seeger J, Rupp H, Seyfarth M, Borg H (2007) Measurement of dew, fog, and rime with a high-precision gravitation lysimeter. J Plant Nutr Soil Sci 170:335–344

Meissner R, Rupp H, Seyfarth M (2014) Advanced technologies in lysimetry. In: Mueller L, Saparov A, Lischeid G (eds) Novel measurement and assessment tools for monitoring and management of land and water resources in agricultural landscapes of Central Asia. Springer international, environmental science and engineering 2014, pp 159–173. http://link.springer.com/chapter/10.1007/978-3-319-01017-5_8. Accessed 2015/03/05

Morkovkin G, Litvinenko Y, Maksimova N (2013) Dynamics of soil cover state and degradation processes intensity in natural soil zones of the Altai Region. EURASIAN SOIL WORKSHOP 2013, "The biophysical attributes of soil quality". 29–31 May 2013, Ondokuz Mayis University, Samsun, Turkey, pp 25–29

Nendel C (2014) MONICA: a simulation model for nitrogen and carbon dynamics in agro-ecosystems. In: Mueller L, Saparov A, Lischeid G (eds) Novel measurement and assessment tools for monitoring and management of land and water resources in agricultural landscapes of Central Asia. Springer international, environmental science and engineering 2014, pp 389–405. http://link.springer.com/chapter/10.1007/978-3-319-01017-5_23. Accessed 2015/03/05

Nimis PL, Malyshev LI, Bolognini G (1994) A phytogeographic analysis of birch woodlands in the southern part of West Siberia. Plant Ecol 113(1):25–39

Paramonov E G, Ischutin JN, Simonenko A P (2003) The Kulunda Steppe. Problems of desertification (Кулундинская Степ': Проблемы опустынивания. Алтайский государственный университет). Altai State University, Barnaul, 137 p

Rudaya N, Nazarova L, Nourgaliev D, Palagushkina O, Papin D, Frolova L (2012) Mid-late Holocene environmental history of Kulunda, southern West Siberia: vegetation, climate and humans. Quatern Sci Rev 48:32–42

Safonov G, Safonova Y (2013) Economic Analysis of the impact of climate change in agriculture in Russia. http://www.oxfam.org/sites/www.oxfam.org/files/rr-economic-impacts-climate-change-agriculture-russia-010413-en.pdf. Accessed 2015/03/05

UGT (2014) Novel Lysimeter-Techniques. http://www.ugt-online.de/en/produkte/lysimetertechnik/lysimeter.html. Accessed 2015/03/05

Xiao H, Meissner R, Seeger J, Rupp H, Borg H, Zhang Y (2013) Analysis of the effect of meteorological factors on dewfall. Sci Total Environ 452–453:384–393

Chapter 17
Measuring Major Components of the Terrestrial Carbon Balance

Marek Urbaniak, Bogdan H. Chojnicki, Radosław Juszczak, Jürgen Augustin, Jacek Leśny, Klaudia Ziemblińska, Natalia Kowalska, Karolina Sakowska, Paweł Siedlecki, Alina Danielewska and Janusz Olejnik

Abstract The carbon balance issue is directly related to global warming, since the presence of carbon dioxide and methane in the atmosphere is the key factor behind the greenhouse effect. As observing the exchange of carbon compounds between

---

M. Urbaniak (✉) · B.H. Chojnicki · R. Juszczak · J. Leśny · K. Ziemblińska · N. Kowalska · K. Sakowska · P. Siedlecki · A. Danielewska · J. Olejnik
Department of Meteorology, Poznan University of Life Sciences, Piątkowska 94, 60-649 Poznań, Poland
e-mail: marek.urbaniak@up.poznan.pl

B.H. Chojnicki
e-mail: bogdan.chojnicki@gmail.com

R. Juszczak
e-mail: radjusz@up.poznan.pl

J. Leśny
e-mail: jlesny@up.poznan.pl

K. Ziemblińska
e-mail: klaudiaziem@wp.pl

N. Kowalska
e-mail: natkonieczna@gmail.com

K. Sakowska
e-mail: sakowska.karolina@wp.pl

P. Siedlecki
e-mail: siedlecki.pawel@gmail.com

A. Danielewska
e-mail: alinkadanie@poczta.onet.pl

J. Olejnik
e-mail: janusz.olejnik@up.poznan.pl

J. Augustin
Leibniz Centre for Agricultural Landscape Research (ZALF) e.V., Institute for Landscape Biogeochemistry, Eberswalder Strasse 84, 15374 Müncheberg, Germany
e-mail: jaug@zalf.de

L. Mueller et al. (eds.), *Novel Methods for Monitoring and Managing Land and Water Resources in Siberia*, Springer Water,
DOI 10.1007/978-3-319-24409-9_17

the ecosystem and the atmosphere has become important in ecological studies, several methods have been developed over recent decades in order to estimate carbon fluxes at different scales. The following text contains a description of the most commonly applied techniques, such as chambers, eddy covariance, and relaxed eddy accumulation. The authors of this chapter have gained experience in designing, building, and using the techniques described below over the past several decades. These systems have been successfully applied in wetlands, forests, and crop ecosystems and have been reliable sources of ecological data until now. The presented overview of the measurement methods is focused on providing an insight into the theoretical basis, as well as the advantages and limitations of all these techniques. This chapter is intended to help a potential user to decide what approach could be applied in their own investigations.

Keywords Carbon balance · Eddy covariance · Relaxed eddy accumulation · Chamber method

1 Introduction

In recent years, the issue of the carbon cycle has become one of the main research topics in the world (Grace 2001). The carbon balance, in the context of climate changes, plays a significant role. Carbon dioxide absorption by terrestrial ecosystems and the oceans could be one of the solutions that would save our planet from uncontrolled global warming. For this reason, increasing the rate of carbon sequestration is a great hope in the processes of climate change mitigation.

Carbon, in the form of carbon dioxide (CO_2), methane (CH_4), and volatile organic carbon (VOC), circulates between the atmosphere and the Earth's surface. A lot of researchers are attempting to fully understand the role of individual components in the global carbon cycle (Falkowski et al. 2000). The carbon cycle is driven by a number of natural processes such as photosynthesis, respiration, chemical processes, or dissolution. "An understanding of these processes is essential to our understanding of life because all biomass is carbon based" (Grace 2001).

Based on the global CO_2 budget reported by Le Quéré (2009), it is easy to see that, according to up-to-date assessments, the land sink efficiency is about

K. Sakowska
Department of Sustainable Agro-Ecosystems and Bioresources, Research and Innovation Centre—Fondazione Edmund Mach, Via E. Mach 1, 38010 San Michele all'Adige, TN, Italy

J. Olejnik
Department of Matter and Energy Fluxes, Global Change Research Centre, Czech Academy of Sciences, Vvi, 60300 Brno, Czech Republic

Table 1 Global carbon dioxide budget comparison for 1990–2000 and 2000–2008

Global carbon dioxide budget (gigatonnes of carbon per year)					
Period	Fossil fuel and cement	Atmospheric growth	Land use change	Land sink	Ocean sink
1990–2000	⇧ 6.4 ± 0.4	3.1 ± 0.1	⇧ 1.6 ± 0.7	⇩ 2.6 ± 0.9	⇩ 2.2 ± 0.4
2000–2008	7.7 ± 0.5	4.1 ± 0.1	1.4 ± 0.7	2.7 ± 1.0	2.3 ± 0.5

The *arrows* pointing up stand for CO_2 release while those pointing down indicate carbon dioxide absorption from the atmosphere (Le Quéré 2009)

0.5 Gt year^{-1} higher than that of the oceans, accounting for 2.7 ± 1.0 Gt of C annually (Table 1). One of the most important messages implied by this study is that the amount of carbon released to the atmosphere due to fossil fuel burning and cement production has increased more than the increase in the carbon sequestration capacities of the terrestrial and marine ecosystems. These two reasons lead researchers to focus on terrestrial ecosystems' carbon balance because of their important role in climate change mitigation. Nonetheless, comparing two periods, 1990–2000 and 2000–2008, it is worth pointing out that the absorption of carbon dioxide has slightly increased, by 1 Gt on average, in both types of ecosystems (Le Quéré 2009).

In the atmosphere, most carbon-based components behave like greenhouse gases (GHGs), i.e., gases that trap the heat in the atmosphere. After their emission from the Earth's surface (natural or anthropogenic) each of these gases can remain in the atmosphere for a different period of time, varying from a few to thousands of years. They also have different impacts on global temperature increase. The so-called Global Warming Potential (GWP) is calculated for each GHG. This reflects the total energy that an individual gas absorbs in comparison to carbon dioxide absorption over a particular period of time (usually 100 years), e.g., CO_2 has a GWP of 1, while methane's GWP equals 28 (IPCC AR5 2013).

The current atmospheric concentration of GHGs depends not only on their lifetime but also on ecosystems' absorption potential. For example, different ecosystems can absorb different amounts of CO_2, which depend, among other things, on plant age, plant species composition, the length of the growing season, soil type, climate conditions, etc. According to these factors, some ecosystems absorb a large amount of carbon, which can significantly slow down the increase in the CO_2 concentration in the atmosphere (Ardö and Olsson 2004). It has been estimated that in terrestrial ecosystems 2100 Gt of carbon has already been accumulated (GRID-Arendal/UNEP 2013). Therefore the potential of ecosystems to absorb or emit any GHGs is crucial in debates on future climate change.

To estimate the carbon balance of terrestrial ecosystems and to describe CO_2 fluxes in any ecosystems, the following common terms are used: gross primary production (GPP), or gross ecosystem production (GEP), net primary production (NPP), net ecosystem exchange (NEE), or net ecosystem production (NEP), and ecosystem respiration (R_{eco}). In practice GPP and GEP refer to the total amount of carbon fixed by plants in the process of photosynthesis. According to Kirschbaum et al. (2001), R_{eco} is the sum of the autotrophic respiration (R_a) of below- and above-ground parts of plants and the heterotrophic respiration of soil microorganisms (R_h) Eq. (1):

$$R_{eco} = R_a + R_h \tag{1}$$

NPP refers to the net production of organic carbon by plants in an ecosystem, usually measured over a period of a year or more (Eq. 2). It is GPP reduced by the amount of carbon respired by plants in autotrophic respiration, R_a:

$$\mathrm{NEP} = \mathrm{GPP} - R_a \tag{2}$$

NEE and NEP refer to NPP reduced by the heterotrophic respiration of soil microorganisms (R_h) (Eq. 3):

$$-\mathrm{NEE} = \mathrm{NEP} = \mathrm{NPP} - R_h \tag{3}$$

or simply:

$$-\mathrm{NEE} = \mathrm{NEP} = \mathrm{GPP} - R_{eco} \tag{4}$$

There is a kind of agreement among scientists that the signs standing in front of the above symbols may vary depending on the perspective taken. From the ecosystem perspective, R_{eco} will have a negative value (there is a loss of CO_2 in the ecosystem), while GPP has a positive value (the gain of CO_2 in the process of photosynthesis). On the other hand, from the perspective of the atmosphere, R_{eco} will be marked with a plus (CO_2 loss from the ecosystem but CO_2 gain in the atmosphere) while GPP is marked with a minus (gain of CO_2 in a process of photosynthesis for the ecosystem, but loss of CO_2 from the atmosphere). Therefore, NPP and NEP absolute values are the same but may have reversed signs depending on the approach considered.

Different ecosystems vary in terms of carbon absorption and release. Moreover, absorption/release properties can differ within the same vegetation type (e.g., they are not equal for boreal, temperate, or rain forests). There are several studies focusing on various ecosystem comparisons in terms of their assimilation potential (e.g., Schulze et al. 2010). Nonetheless, many scientists focus on carbon balance assessments of a particular type of vegetation. On the basis of scientific study it is agreed that the world forests are a significant sink of CO_2. In particular, Pan et al. (2011) reported that boreal forests accounted for 21 % (0.50 ± 0.08 Pg C year^{-1}) of the global carbon sink of all established forests (2.41 ± 0.42 Pg C year^{-1}). Grassland ecosystems in different parts of the world can be the sink or the source of CO_2. In grasslands NEE varies from −800 to 521 g C m^{-2} year^{-1} and, in most of the grasslands studied, CO_2 fluxes and balances are within the range of ±200 g C m^{-2} year^{-1} (Novick et al. 2004).

Apart from forests and grasslands, croplands also play a significant role in the carbon cycle, because in Europe they occupy about 45 % of the area (Wattenbach et al. 2010). In croplands, CO_2 is assimilated during the process of photosynthesis and non-harvested plant biomass is accumulated in the soil, which may contribute to the increase in soil carbon stock. Whether croplands are more a sink or a source of CO_2 depends on several factors mentioned above, but cultivation techniques and weather conditions also have to be taken into account. Therefore, it is very difficult to assess the global CO_2 balance for croplands.

Besides carbon dioxide, the second important carbon-based GHG is methane. It is assumed that the total annual methane emission from anthropogenic and natural sources accounts for about 600 Tg (Whalen 2005). Recent studies by Kirschke et al. (2013) have shown that in the years 2000–2009 the CH_4 emission from natural sources amounted to 347 Tg and wetlands were dominant natural ecosystems emitting CH_4 into the atmosphere (51–82 % of the total emission). Peatlands, which cover about 3 % of the global land area (Gorham 1991), accumulate about one-third (Pg 329–528) of the total volume of the organic carbon present in the soil. Based on a review of the literature, Blodau (2002) concludes that the methane uptake in this type of ecosystem is very small and that emission could be even greater than 91 t CO_2 and (CO_2 equivalent) ha^{-1} $year^{-1}$. The variability of methane fluxes strongly depends on the wetland type, the season of the year, and the ecosystem location (latitude), e.g., typical average emissions from northern peatlands account for 0.5–7.3 t CO_2 and ha^{-1} $year^{-1}$ (5.5–80 mg CH_4 m^{-2} d^{-1}) (Morison 2012).

The carbon balance is one of the main issues discussed in the context of climate change. Estimating the amount of carbon absorbed/emitted by the ecosystem has been a real challenge for scientists for years. The development of methods for measuring GHG fluxes continuously follows the technological development, which facilitates more precise measurements and consequently adequate assessment of the global carbon balance. The data obtained allow for the parameterization of appropriate models used to predict future climate changes. Therefore, continuous measurements of GHG fluxes over different ecosystems and climatic zones are of the highest priority to environmental researchers. The following chapter presents a brief overview of the most commonly used carbon flux measuring methods.

2 Chamber Method

Historically, chamber- or enclosure-based techniques are the oldest approaches used to estimate the gas exchange between terrestrial ecosystems and the atmosphere. In 1922 Swedish researcher Lundegårdh developed the first chamber technique which was used to estimate CO_2 fluxes. Since then, enclosure techniques have been commonly applied to estimate a broad range of trace gas fluxes such as those of CO_2, methane (CH_4), nitrous oxide (N_2O), and ammonia (NH_3). These techniques are ideal for small-scale studies and can be applied in nearly all terrains. They offer low cost and portability and allow for measurements at different site types or land use treatments. In short, it is recommended to use this technique under conditions where other micrometeorological techniques cannot be applied.

Originally, to measure soil respiration, the released CO_2 was absorbed by alkali chemicals inside an enclosure and then the fluxes were calculated from the amount of trapped CO_2. Even with a great deal of work, only a very low temporal resolution of the CO_2 flux was possible and a lot of effort was required to get reliable flux rates (Bekku et al. 1997). Currently, the enclosure techniques are mostly combined with a direct determination of the gas concentration of the chamber headspace air, which enables an estimation of the flux rates with a much higher temporal resolution. In the case of the so-called *static* chamber approach [known as *non-steady-state non-flow-through chambers* (Livingston and Hutchinson 1995)], trace gas fluxes are determined on the basis of gas concentration changes in the chamber headspace over a certain period of time when the chamber is closed. The gas samples are taken from the chamber every 5–20 min, usually from the top of the chamber, and then they are analyzed with a gas chromatographer to determine the gas concentration in each sample. This chamber method is most commonly used to measure soil CO_2 efflux or CH_4 and N_2O fluxes. Especially for CO_2 exchange measurements (NEE, R_{eco}) *static* systems were further developed into *dynamic* systems [known as *non-steady-state flow-through chambers* (Livingston and Hutchinson 1995)], where the chamber is connected to a precise infrared gas analyzer (IRGA, e.g., Li-Cor or 820 or 840, Li-Cor, USA) and the air circulates in a closed system from the chamber to the analyzer and back to the chamber (Fig. 1). Therefore, in this system, the gas concentration changes inside the chamber headspace can be monitored continuously with high frequency (e.g., every 1–10 s, depending on the analyzer capabilities). Since then, sensitive analyzers for measuring fluxes of other trace gases (e.g., CH_4 and N_2O) have also been developed (Christiansen et al. 2011;

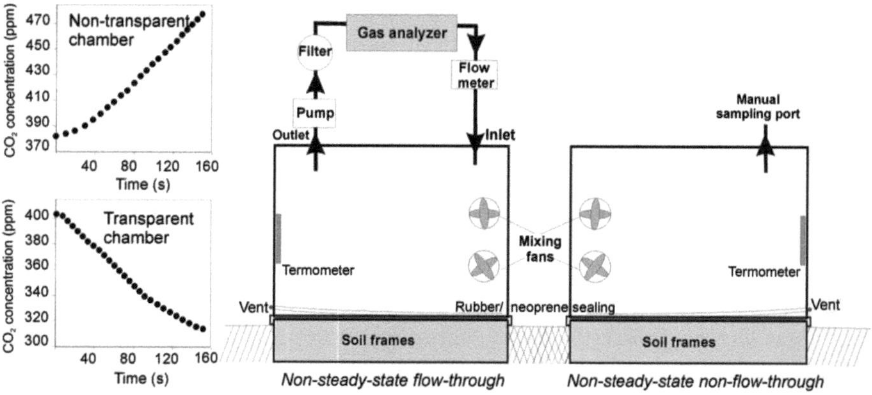

Fig. 1 Scheme of the most commonly applied closed dynamic (*non-steady-state flow-through*) and closed static (*non-steady-state non-flow-through*) chambers. On the *left-hand side* there is an example of the CO_2 concentration development in the dynamic non-transparent chamber applied for soil or ecosystem respiration measurements and in the dynamic transparent chamber used for NEE measurements

Juszczak 2013), but their price and dimensions still limit their applicability at field conditions on a larger scale. Another type of enclosure system is *open* chambers [known as *steady-state flow-through chambers* (Livingston and Hutchinson 1995)], where the gas fluxes are determined from the difference between the gas concentrations at the inlet and outlet of the chamber. In this chamber system accurate estimations of the flux rates can be obtained only with extremely sensitive gas analyzers.

The gas flux (F) from the *non-steady-state* chambers can be calculated from the gas concentration change in the chamber headspace $\left(\frac{\Delta C}{\Delta t}\right)$, the chamber volume ($V$), and the enclosed soil area (A) from Eq. (5):

$$F = \frac{\Delta C}{\Delta t} \cdot \frac{V}{A \cdot M_v} \tag{5}$$

where M_v ($m^3\ mol^{-1}$) is the molar volume of air at chamber air temperature and pressure. In *steady-state flow-through* chamber systems, where air with a known gas concentration continuously passes through the chamber at a known, constant flow rate (f; $m^3\ s^{-1}$), the gas flux (F) inside the chamber headspace is determined from the difference between the gas concentration at the chamber inlet and outlet (ΔC), the through-flow rate (f), and the covered soil area (A) from Eq. (6):

$$F = \frac{\Delta C}{M_v} \cdot \frac{f}{A} \tag{6}$$

The choice of the chamber material is mainly determined by the type of gas flux which is going to be examined. Enclosures used for soil CO_2 efflux, R_{eco}, CH_4, and N_2O flux measurements are usually made of non-transparent materials (e.g., white PVC) which insulate the chamber headspace well from the outside air and solar radiation (Juszczak et al. 2013a). Despite the fact that, in the case of CH_4 flux measurements, transparent chambers were recommended for a long time (Livingston and Hutchinson 1995), comparison measurements with transparent and non-transparent chambers did not show different flux rates (Minke et al. 2014). However, the chambers applied for NEE measurements must be made of transparent material. The most common *Plexiglas®* (Evonik Industries, Darmstadt, Germany) is recommended to be applied for this purpose, as this material is robust and has a high solar radiation transmittance of approximately 92 % (Drösler 2005; Chojnicki et al. 2010b).

Using manual chamber systems, a limited number of measurements are possible during the year, so the data gaps must by filled with modeling approaches in order to get annual flux rates (Drösler 2005). This problem may be solved by newly emerging automated chamber systems allowing continuous measurements of gas fluxes. Today there are only a few automated chamber systems for R_s, R_{eco}, and even NEE measurements commercially available from different manufacturers (e.g., from Li-Cor, USA, PP Systems, USA, ADC BioScientific Ltd., UK). The main constraint of such systems is that the size and dimensions of the chambers are relatively small and therefore they cannot be applied for most ecosystem types with tall vegetation at the moment. Hence, lots of researchers have developed their own automated chamber systems, which vary greatly in shape, dimension, design, equipment, and measuring approach. For more than 20 years, there has been debate on how the chamber systems should be designed and how the measurements and flux calculations should be conducted to minimize biases and uncertain results (Nay et al. 1994; Livingston and Hutchinson 1995; Davidson et al. 2002; Pumpanen et al. 2004; Pihlatie et al. 2013). Nonetheless, there are still no clear standards for the chamber design or flux calculation method. However, based on the recommendations coming from several chamber inter-comparison studies and detailed knowledge of the literature, we can state that:

- The application of "push-in" chambers inserted into the soil at the time of measurement should be avoided due to soil disturbances associated with the insertion of the chamber and a high risk of leakage (Pumpanen et al. 2009; Juszczak et al. 2013b). The soil frames should be applied instead and they should be inserted into the soil at least a few days before the measurements. The depth to which chamber collars should be inserted to minimize the lateral flow of the gas beneath the soil frame depends on the soil porosity. For chamber deployments lasting a few minutes, Hutchinson and Livingston (2001) proved that 2.5 cm insertion depth into soils of low-to-moderate porosity and 9 cm depth into highly porous soils should be enough. For the longer deployment times typical in static chamber measurements, the collar insertion depth should increase.
- Chambers should be well sealed once they are installed on soil frames to avoid leakage. The chamber may have soft rubber or neoprene gaskets or the soil frame may have a furrow containing water as a seal.
- In order to minimize pressure changes inside the closed chambers, which may happen, e.g., while the chamber is being placed on the collar (Davidson et al. 2002; Christiansen et al. 2011), and to transmit changes in external atmospheric pressure to the chamber headspace, chambers should be equipped with venting tubes. However, improper installation of the vents may depressurize the chamber, especially in windy conditions, as a result of the Venturi effect, which pulls the air out of the chamber. Therefore, the venting tubes should be installed as near the soil surface as possible, because that is where the wind speed is at the lowest (Rochette and Hutchinson 2005).
- On wetland sites with a high groundwater level, the frames should be surrounded by wooden boardwalks. Otherwise the measurements may induce the release of gas bubbles due to mechanical disturbance of the site, which can heavily distort the measured fluxes.
- For non-steady-state enclosures, the gas concentration inside the chamber headspace increases, the diffusion gradient between soil surface and the air in the chamber decreases, and the flux begins to decline with time, which may eventually lead to significantly underestimated gas fluxes (Pumpanen et al. 2004; Pihlatie et al. 2013). The risk is particularly great in the case of small chambers a few liters in volume, a low height (<0.2 m), and a small surface area

(<0.1 m^2) (Pihlatie et al. 2013). Moreover, the presence of the chamber on the soil may also bias the measured flux through soil and air temperature and humidity changes inside the chamber, which is the case especially in transparent chambers. Hence, the deployment time should be as long as necessary to detect the gas fluxes, but on the other hand as short as possible so as not to change the gas gradient inside the chamber significantly. The length of the deployment time should be adjusted in situ conditions and depends on the measuring approach and expected flux rates of the measured gas (Pumpanen et al. 2009). In practice, in the case of static chambers, at least three gas samples should be taken (the greater the number of samples the better) from the chamber headspace during the deployment time. The first sample should be taken immediately after closing the chamber. The closure time should not be longer than 30–60 min. In dynamic chambers used for CO_2 flux measurements, by contrast, the closure time should not be longer than 2–5 min. In general, the use of static chambers should be limited as much as possible, while the dynamic approach in combination with fast-responding, precise analyzers should be preferred because of its higher accuracy.

- When calculating fluxes, different calculation approaches may be used. If the gas concentration development inside the chamber headspace is curvilinear, the application of linear methods will lead to an underestimation of the measured flux. In such cases non-linear methods should be used instead (Pihlatie et al. 2013). It is recommended to use the freely available R-based script that has been proved successful for calculating flux rates accurately and precisely (Jurasinski et al. 2012).
- The determination of gas fluxes in closed chambers is based on the assumption that the gas concentration is homogeneous within the chamber's headspace. Recent studies by Christiansen et al. (2011) have proved that chambers without any air mixing underestimated the measured fluxes by as much as 36 %. Therefore, the air inside the chamber has to be well mixed. This can be ensured by the air circulating between the chamber and analyzer (which is the case in some small chambers holding only a few liters), or by the application of fans. The application of fans inside the chamber is highly recommended but the fact must be taken into account that excessive mixing may over- or underpressurize the chamber headspace, which may lead to biased gas flux rates (Pumpanen et al. 2009).
- In order to avoid excessive heating of the transparent chamber headspace, an effective cooling system should be applied in the chambers. Drösler (2005) showed that the changes of higher than 1.5 °C in the air temperature inside the chamber headspace have a significant impact on the measured flux.
- Owing to the extremely high temporal variability of the gases, accurate and precise data can only be collected if the measurements are conducted with a high temporal resolution throughout the year. The minimum requirements for the different gases are as follows: soil CO_2 efflux, CH_4 and N_2O exchange every 14 days (at least one measurement per campaign) completed with event-based measurements (e.g., after fertilization); R_{eco} and NEE every four weeks (full-day measuring campaigns) completed with event-based measurements (e.g., tillage or harvest). Moreover, for the correct interpretation of the flux rates there is an

urgent need to record all information about factors such as the site and weather conditions, plant development, and land use activities which may influence the gas exchange. The gas measurement sites should be equipped with different types of monitoring systems such as weather stations or groundwater wells.

It is generally recommended to select the type of chamber approach and the measuring design after a thorough analysis of the research question in the context of the site conditions (spatial heterogeneity, ecosystem, and land use type). If all these factors are taken into account, you may get results of similarly high quality with enclosure-based techniques as with micrometeorological approaches (Myklebust et al. 2008)

3 Eddy Covariance (EC) Method

In recent years the eddy covariance (EC) technique has been developed the most when compared to other measuring methods for the assessment of mass and energy fluxes. "Eddy covariance is typically measured in the surface boundary layer, which is approximately 20–50 m high in the case of unstable stratification and a few tens of meters in stable stratification" (Stull 1988; Foken 2008). Fluxes are approximately constant with height in the surface layer, hence the measurements taken in this layer are representative of fluxes from underlying surfaces. The atmospheric turbulence is the dominant transport mechanism, justifying the use of eddy covariance approach to measure the fluxes (Aubinet et al. 2012).

To put it simply, the theory of the eddy covariance method assumes that in the case of a notionally separated volume (e.g., a portion of air above a canopy) the molecules of gas such as CO_2 move through constantly in both directions (upward and downward), and at the same time the concentration of gas in that volume remains unchanged. In these conditions one is dealing with a stationary flow, which means that it does not change in time and is equal at different heights inside the volume investigated. The above idea is presented in Fig. 2 and could be interpreted as follows: if at one moment we know that three molecules of CO_2 carried by eddy

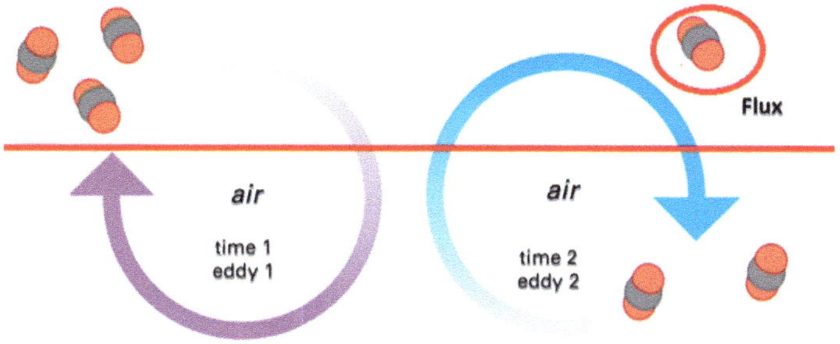

Fig. 2 The concept of turbulent transport of gas in the surface boundary. The *red line* refers to the height of eddy flux measurements

1 have gone up, and a moment later only two molecules of CO_2 (with eddy 2) have gone down, then the net flux over this time was directed upwards and was equal to one molecule of CO_2 (Burba 2013).

In a real-life situation a turbulent flow in the atmosphere is much more complicated and the eddies creating turbulence are different in size and speed.

The principles of the EC method include an assumption that, in an average time interval, the product of the vertical wind velocity and the quantity of substance, heat or momentum carried by the air should be equal to the substance, heat or momentum flux, respectively. This statement can be expressed as in the following equation (Swibank 1951):

$$F = \overline{w \cdot \rho} \tag{7}$$

where F is the flux value of quantity of interest [energy (W m^{-2}) or mass (μmol m^{-2} s^{-1})], w is the vertical wind speed component (m s^{-1}), ρ is the quantity of interest [energy density (J m^{-3}) for energy or gas concentration (μmol m^{-3})], and $\overline{}$ is a symbol of averaging over time.

As mentioned earlier, the values listed above change dynamically in a constantly mixing atmosphere. In such conditions, in order to observe their variability, measurements must be applied at an adequately high frequency. In the specified averaging time, the values of the measured parameters, such as temperature, heat or gas density, will be subject to change and their fluctuations may be considered as a deviation from some arbitrary values which are equal to the mean values of these parameters over a certain time. This issue was noticed and used by Swibank in heat and water vapor flux calculation in the early 1950s (Swibank 1951). Theoretically, the specified averaging time should be long enough to set the mean vertical wind velocity value to zero. However, because of the changes in air density in the vertical profile, this assumption cannot be met. Nowadays, a 30-min integration time is mostly applied in practice for eddy covariance measurements.

In that context, one of the most important steps in EC development and popularization was the application of the so-called Reynolds decomposition approach (Reynolds 1895; Baldocchi 2003). Introducing Reynolds decomposition has resulted in an ability to present the value of measured fluxes as a covariance of the vertical air flow velocity and a quantity of mass, heat, or momentum "carried in" that portion of air (Eq. 10). It is based on an assumption that instantaneous physical quantity values can be "decomposed" into the sum of their mean values and fluctuation within the time of measurement. The equation for the vertical component of wind speed is presented as follows:

$$w = \overline{w} + w' \tag{8}$$

where w is the vertical wind speed (m s^{-2}), $\overline{w}$ is the mean value of the vertical wind speed (m s^{-2}), and w' is the fluctuation of the vertical wind speed (m s^{-2}).

The same operations are applied to scalar values (e.g., CO_2, H_2O concentrations); if they are included together into Eq. 7, this results in

$$F = (\bar{w}+w')(\bar{\rho}+\rho') = (\bar{w}\bar{\rho}+\bar{w}\rho'+w'\bar{\rho}+w'\rho') \\ = \underbrace{\bar{w}\bar{\rho}}_{I} + \underbrace{\bar{w}\bar{\rho}'}_{II} + \underbrace{\bar{w}'\bar{\rho}}_{III} + \underbrace{\bar{w}'\bar{\rho}'}_{IV} \tag{9}$$

As can be seen, as a result of this calculation, the mean value of deviation from the mean for all samples in the statistical population of data equals zero ($\Sigma\bar{w}' = 0,\ \Sigma\rho'$). Parts II and III in Eq. 9 also equal zero. Then the formula is simplified to the following:

$$F = \underbrace{\bar{w}\bar{\rho}}_{I} + \underbrace{\bar{w}'\bar{\rho}'}_{II} \tag{10}$$

As Eq. (10) implies, the total vertical flux of any scalar value is a sum of its mean vertical mass/energy flow (part I) and turbulent flux (part II) (Moncrieff et al. 1997; Webb et al. 1980). The second part of the above equation is a covariance of the vertical wind speed and scalar value. This flux is of interest in the EC method. This part (II) represents the mean value of the product of instantaneous vertical wind speed and scalar value fluctuations (deviations) from their mean values and thus describes the turbulent mass and energy exchange between an active surface and the atmosphere. The first part of Eq. 10 is the mass/energy flow induced by a vertical non-zero wind speed.

Reynolds decomposition is the first of several corrections applied in EC measuring practice. The effective and widespread application of the EC method became possible due to a recent development in the field of IT and the miniaturization of spectroscopic methods. In the early 1990s, the EC method became widely used to measure the exchange of CO_2 and H_2O between the ecosystem and the atmosphere (Wofsy et al. 1993; Aubinet et al. 2012). It is worth mentioning that the development of new gas analyzers allowed the range of trace gases investigated to be extended. Nowadays, it is possible to measure the fluxes of CH_4 (e.g., Kowalska 2013), N_2O (e.g., Kroon et al. 2010), volatile organic compounds (VOC) (e.g., Spririg et al. 2005), and ozone (O_3) (e.g., Gerosa et al. 2003) using the EC technique.

The subsequent development of measurement networks started after 1990: CarboEurope (Aubinet et al. 2000; Valentini et al. 2000) and AmeriFlux (Running et al. 1999). Nowadays, long-term investigations using EC as a general methodology are carried out at almost 600 (ORNL DAAC 2013) measuring stations worldwide as part of the FLUXNET project (Baldocchi et al. 2001, Danielewska et al. 2013a, b).

3.1 Hardware Challenges

The EC measurement set consists of two instruments: a sonic anemometer and a spectrometric gas analyzer. Using the anemometer it is possible to measure fluctuations in wind speed components in three dimensions, two of which are horizontal (*u*, *v*) and one vertical (*w*). They are also used to measure the speed of sound, from which the air temperature is derived (it allows fluctuations in the energy density ($J \cdot m^{-3}$) to be calculated). The other instrument measures the changes in concentration of the gas of interest. As the sensors used in that method have a very short time constant, measurement with a frequency of a few readings per second became available, and nowadays the most common frequencies of data acquisition are within the range of 10–20 Hz.

Almost all high-frequency 3D anemometers can be used in EC measurements, but there are some crucial differences in gas analyzers related to the rules of their operation. In general, concentration fluctuations are measured with either open-path (OP) or closed-path (CP) gas analyzers. The values of computed fluxes measured with OP or CP analyzers are the same, but the choice of using a particular type of analyzer is usually determined by several factors (user requirements or abilities to maintain the whole EC system):

- An EC system consisting of an OP gas analyzer is less power-demanding, which is why it can be used in locations with power limitations.
- As opposed to CP, an OP gas analyzer is more vulnerable to harmful disturbances (dust, bird droppings, rainfall, snow, or white frost on the measuring path), which can affect the data set quality, as OP analyzers generate larger data series gaps than CP.
- Measurements obtained by means of OP analyzers are carried out in unstable air temperature and pressure conditions which change continuously in the open air. To compensate for that, more corrections than for CP type should be applied (e.g., Webb-Pearman-Leuning correction (WPL), Burba corrections, etc.).
- A CP analyzer requires a power-consuming pump, drawing the air into the measuring path, thus ensuring a steady flow. Otherwise there will be a risk of creating artificial fluctuations in the investigated gas concentration.
- In a CP analyzer the air flows through a tube, which can result in signal attenuation (fluctuations declining). Some materials, e.g., polypropylene (PP), may absorb H_2O, balancing out air humidity. This phenomenon can cause an underestimation of measured fluxes, thus more attention should be paid to spectral correction.
- For calibration, the OP analyzer should be demounted; unfortunately, the OP automatic calibration cannot be realized, which is not the case with a CP analyzer.

3.2 *Spatial Representativeness of EC Measurements*

When using the EC technique it is necessary to take special care that the characteristics of the measurement site are appropriate and the results obtained are properly corrected (Foken and Wichura 1996; Massman and Lee 2002). However, if the measurements of turbulent fluxes are conducted in the specified atmospheric layer above non-homogeneous vegetation (contrary to the assumption made at the beginning), it is necessary to define the spatial representativeness of the measured values. In this case the signal obtained depends on which part of the area has a stronger influence on the sensor. As mentioned above, the measurements at a certain height above the active surface should be representative of the investigated area. For this reason, the probability of the origin of certain air portion is assessed. Modeling studies showed that the surface of the area around the sensor, where the consecutive investigated portions of air came from, is called a "footprint" (Finn et al. 1996; Gash 1986; Leclerc and Thurtell 1990). The magnitude and shape of the footprint depend on the measuring sensor's height above the active surface, its aerodynamic properties (canopy roughness), and the dynamic properties of the atmosphere (wind speed and atmospheric stability). In the literature there are several basic ways to designate the size of the area of influence; these can be divided into two main groups: the analytic solution of a two-dimensional advection–diffusion equation (Gash 1986; Schmid 1994; Korman and Meixner 2001; Horst 2001) and the three-dimensional stochastic Lagrangian model of dispersion (backward trajectories) (Leclerc and Thurtell 1990; Flesch 1996; Kljun et al. 2002; Kljun et al. 2003).

3.3 Averaging Period and Measurement Frequency

When using the EC approach, the time in which the obtained results are averaged is a key issue. Too short a period of integration leads to mass or energy fluxes being underestimated. This is caused by the fact that the measured parameter changes carried by larger eddies with diameters of dozens to hundreds of meters are simply not measured (are omitted), because their "transition" time through the sensors is longer than the averaging period. On the other hand, too long a period results in disturbances in the assumption that the air flow is stationary (e.g., too large temperature changes) and can level out fluctuations in measured fluxes overtime. Taking the above into consideration as well as scientists' experience using the EC method worldwide, the suggested integration time varies from 30 to 60 min (Aubinet et al. 2000). Only in particular cases can it be extended to several hours. One special case is, for instance, night-time measurements, when turbulence is much weaker and large eddies are dominant in the surface boundary layer.

3.4 Time Series Displacement Due to Spatial Sensor Separation

Even though dimensions and shapes of the two main instruments used in the EC method are fitted well in order to minimize air turbulence disturbances, nowadays most EC systems do not sample exactly the same air portion at the same time. The distance between these two sensors is usually about 0.4 m. It can lead to serious disturbances in air turbulence when this distance is shortened (the sensors influence one another). The spatial separation of two main instruments in the EC technique implies another problem in flux calculations. Depending on the wind speed, the same portion of air is sampled by the two sensors with a certain time displacement (the higher wind speed the shorter the time). Moore (1986) noticed that the calculated covariance between fluctuations in vertical wind speed and the quantity of interest is substantially lowered as the investigated portion of air moves first through one of the sensors and then, after some time, through the next one. The presented situation lowers the covariance between the signals from the two sensors, calculated directly from the spectra of obtained data, especially when low wind speed values occur. This problem can be solved by shifting the measurement series relative to each other according to the wind direction (the wind direction determines which sensor will get the investigated air portion first) to obtain a maximum covariance value.

3.5 Coordinate System Rotation

Theoretical assumptions for the EC method provide special rules for the location of the sonic anemometer. It should be placed above a flat, horizontal terrain, perpendicular to geo-potential lines, but these requirements are very difficult to meet in practice. A slight deviation from the vertical placement caused by, e.g., local terrain conditions, meteorological factors, or mechanical inclination overtime results in a non-zero value of $\bar{w}$ being recorded by the anemometer. This disadvantage of the EC method has been solved by rotating the anemometer's coordinate system, or by using appropriate software to set the projection of the wind speed resultants to zero on two axes (vertical and horizontal) of the rotated coordinate system. In most cases double rotation around the vertical axis "z" is applied, resulting in $\bar{v} = 0$, and around the y axis, obtaining $\bar{w} = 0$. The rotation is carried out for each 30-min averaging period. Some approaches use so-called triple rotation (one additional rotation around a new axis $z \rightarrow z'$), resulting in: $\overline{u'v'} = 0$ (covariance between the horizontal wind speed components u and v is "0") (Aurela 2005; Baldocchi and Meyers 1998; Massman and Lee 2002; Tanner and Thurtell 1969). This procedure, performed by software and used for mass and energy calculation, solves the issue of the anemometer frequently recording non-zero values for the vertical wind speed.

3.6 Advantages and Limitations of the EC Technique

The most important advantage of this method is the fact that it is not based on any empirical approximations of the flux–gradient relation (as opposed to the heat balance method, which uses the Bowen ratio). Secondly, the EC technique allows an assessment of the mass and energy flux exchange for relatively large areas (depending on the footprint size). Recently developed measurement techniques enable the EC method to be used nearly all year round (without disturbing the vegetation) (Wofsy et al. 1993; Baldocchi et al. 2001). Moreover, EC enables turbulent flux measurements above very rough surfaces (forest), where the vertical gradients of measured parameters (air temperature, water vapor concentration) are relatively low and it thus becomes much more complicated to apply the profile method. Additionally, the method discussed is relatively precise, with errors estimated to occur in fewer than 5 % (Goulden et al. 1996; Moncrieff et al. 1996).

The EC approach has also some limitations. In order to conduct measurements properly with this method, some basic assumptions have to be made, which are often very difficult to perform. The ideal conditions for carrying out micrometeorological measurements of fluxes using the EC method are as follows:

- Flat and homogeneous surface, at a sufficiently large distance from the upwind direction (within the footprint);
- No convergence or divergence of measured fluxes near the surface air layer (below the instrument placement level),
- Slightly changing mean value of investigated parameters in time (the time stationarity of scalars, e.g., CO_2, H_2O)

As a consequence of the requirements described above, the eddy covariance method is not easy to apply in highland areas or in places with a sharp transition from one type of ecosystem to another, e.g., in close vicinity to lakes (Baldocchi et al. 2001). The breach of the above rules can be the source of systematic errors in measurement interpretation obtained when using the EC technique (Foken and Wichura 1996; Massman and Lee 2002). Nonetheless, the advantages of the method largely outnumber its weaknesses and limitations. However, two types of errors can be still distinguished which may occur while using the eddy covariance technique. The heat balance of the investigated surface is often not closed (Aubinet et al. 2000; Twine 2000; Wilson 2002). The other error is connected with an underestimation of night-time CO_2 effluxes in low turbulence conditions (Black et al. 1996; Moncrieff et al. 1996; Aubinet et al. 2000). Nowadays, a commonly used technique to overcome this problem is night-time measurement filtering. This procedure is based on the analysis of friction velocity (u*) (Gu et al. 2005).

To conclude, the EC method offers a great opportunity to assess almost any vertical mass and energy fluxes exchanged between ecosystems and the atmosphere with high temporal and spatial resolution. Contrary to other micrometeorological techniques (e.g., profile methods), the eddy covariance system directly measures fluxes with no empirical coefficients. This method has become a standard in ecological studies of the ecosystem (Figs. 4 and 5).

4 Relaxed Eddy Accumulation (REA) Method

When a fast gas analyzer (at least 10 Hz sampling rate) is not available, the relaxed eddy accumulation (REA) method can be used as an alternative to EC. REA is a modified eddy accumulation method (Desjardin 1972) and it is based on the application of the w dead-band, whose range is defined by the threshold w value (w_0) (Ammann 1998). The air samples are collected in the so-called "updraft" and "downdraft" reservoirs, when $w > w_0$ and $w < -w_0$, respectively. The air is usually collected at 30-min intervals and the concentration of studied scalars in the air located in each reservoir can be measured using slow gas analyzers or gas chromatographs. The net scalar exchange can be estimated using REA on the basis of the following formula (Businger and Oncley 1990):

$$F = \beta \sigma_w \left(\rho^{\mathrm{up}} - \rho^{\mathrm{dn}} \right) \tag{10}$$

where β is the empirical coefficient (dimensionless); σ_w is the w standard deviation ($\mathrm{m} \cdot \mathrm{s}^{-1}$); $\rho^{\mathrm{up}}/\rho^{\mathrm{dn}}$ are the updraft/downdraft scalar average concentrations ($\mu\mathrm{mol} \cdot \mathrm{m}^{-3}$ or $\mathrm{g} \cdot \mathrm{m}^{-3}$).

The absolute w_0 value ($|w_0|$) is estimated dynamically on the basis of th σ_w value and this strategy makes the β value independent of the air stability.

Different REA systems are being developed today, but the general idea of the REA system structure is presented in Fig. 3. The valves (V) are controlled by an electronic module that analyzes real-time w values measured by a sonic anemometer. The air is stored in updraft (UR) and downdraft (DR) reservoirs, which are usually made of non-reactive materials, e.g., Tedlar. The tubing system that comes into contact with the sampled air is usually made of Teflon or stainless steel. The pump (P) and pressure tank (PT) in the presented system create the under pressure necessary to fill the sample bags.

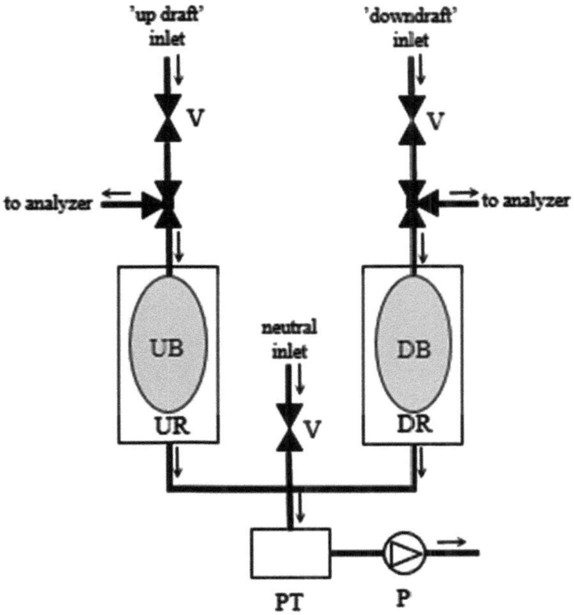

Fig. 3 REA system scheme. The *arrows* indicate the air flow direction; *V* valve; *UB* updraft bag; *DB* downdraft bag; *UR* updraft reservoir; *DR* downdraft reservoir; *PT* pressure tank; *P* pump

REA systems have proven their applicability when measuring the net fluxes of such carbon compounds as isoprenoids (Graus et al. 2006) and methane (Chojnicki et al. 2010a) infield conditions.

The limitations and representativeness of REA are the same as those of EC, but the application of β during the flux calculations makes it less direct. The use of empirical coefficients increases the uncertainties of flux estimation in this system. REA can be competitive with chamber techniques, since it can be used above the tall vegetation, e.g., in forests.

Appendix: Eddy Flux Measurement Stations In Situ

Figures 4 and 5.

Fig. 4 Eddy flux measurement station in a grassland ecosystem

Fig. 5 Constructing an eddy flux measurement station in a forest ecosystem above the forest stand

References

Ammann C (1998) On the applicability of relaxed eddy accumulation and common methods for measuring trace gas fluxes. Diss. ETH No. 12795, Swiss Federal Institute of Technology Zurich, Zurich, 239 pp

Ardö J, Olsson L (2004) Soil carbon sequestration in traditional farming in sudanese drylands. Environ Manage 33(1):318–329

Aubinet M, Grelle A, Ibrom A, Rannik Ü, Moncrieff J, Foken T, Kowalski AS, Martin PH, Berbigier P, Bernhofer Ch, Clement R, Elbers J, Granier A, Grünwald T, Morgenstern K, Pilegaard K, Rebmann C, Snijders W, Valentini R, Vesala T (2000) Estimates of the annual net carbon and water exchange of forests: the EUROFLUX methodology. Adv Ecol Res 30:113–175

Aubinet M, Vesala T, Papale D (eds) (2012) Eddy covariance: a practical guide to measurement and data analysis. Springer, Dordrecht 438 pp

Aurela M (2005) Carbon dioxide exchange in subarctic ecosystems measured by a micrometeorological technique. Finnish Meteorological Institute Contributions, No. 51, 39 pp

Baldocchi DD (2003) Assessing the eddy covariance technique for evaluating carbon dioxide exchange rates of ecosystems: past, present and future. Global Change Biol 9:479–492

Baldocchi DD, Meyers TP (1998) On using eco-physiological, micrometeorological and biogeochemical theory to evaluate carbon dioxide, water vapor and gaseous deposition fluxes over vegetation. Agric For Meteorol 90:1–26

Baldocchi DD, Falge E, Lianhong G, Olson R, Hollinger D, Running S, Anthoni P, Bernhofer Ch, Davis K, Evans R, Fuentes J, Goldstein A, Katul G, Law B, Lee X, Malhi Y, Meyers T, Munger W, Oechel W, Paw UKT, Pilegaard K, Schmidt HP, Valentini R, Verma S, Veselaq T, Wilson K, Wofsy S (2001) FLUXNET: a new tool to study the temporal and spatial variability of ecosystem-scale carbon dioxide, water vapor, and energy flux densities. Bull Am Meteorol Soc 82(11):2415–2434

Bekku Y, Koizumi H, Oikawa T, Iwaki H (1997) Examination of four methods for measuring soil respiration. Appl Soil Ecol 5:247–254

Black TA, den Hartog G, Neumann HH, Blanken PD, Yang PC, Russell C, Nesic Z, Lee X, Chen SG, Staebler R, Novak MD (1996) Annual cycles of water vapour and carbon dioxide fluxes in and above a boreal aspen forest. Glob Change Biol 2:101–111

Blodau C (2002) Carbon cycling in peatlands—a review of processes and controls. Environ Rev 10:111–134. doi:10.1139/A02-004. Published on the NRC Research Press Web site at: http://er.nrc.ca/ on 21 June 2002

Burba G (2013) Eddy covariance method for scientific, industrial agricultural and regulatory applications. A field book on measuring ecosystem gas exchange and areal emission rates. LI-COR Biosciences, Lincoln, Nebraska 331 pp

Businger JA, Oncley SP (1990) Flux measurements with conditional sampling. J Atmos Oceanic Technol 7(04):349–352

Chojnicki BH, Siedlecki P, Rinne J, Urbaniak M, Juszczak R, Olejnik J (2010a) Methane emission measurements using relaxed eddy accumulation technique—preliminary results from Rzecin wetland. In: Chojnicki BH (ed) Acta agrophysica, vol 179, pp 102–112

Chojnicki BH, Michalak M, Acosta M, Juszczak R, Augustin J, Droesler M, Olejnik J (2010b) Measurements of carbon dioxide fluxes by chamber method at Rzecin wetland ecosystem in Poland. Polish J Environ Stud 19(2):283–291

Christiansen JR, Korhonen J, Juszczak R, Giebels M, Pihlatie M (2011) Assessing the effects of chamber placement, manual sampling and headspace mixing on CH_4 fluxes in a laboratory experiment. Plant Soil 343:171–185

Danielewska A, Paoletti E, Clarke N, Olejnik J, Urbaniak M, Baran M, Siedlecki P, Fischer R (2013a) Towards integration of research and monitoring at forest ecosystems in Europe. Forest Syst 22(3):535–545

Danielewska A, Clarke N, Olejnik J, Hansen K, de Vries W, Lundin L, Tuovinen J-P, Paoletti E (2013b) A meta-database comparison from various European research and monitoring networks dedicated to forest sites. IForest 6:1–9

Davidson EA, Savage K, Verchot LV, Navarro R (2002) Minimizing artefacts and biases in chamber-based measurements of soil respiration. Agric For Met 113:21–37

Desjardins RL (1972) A study of carbon dioxide and sensible heat fluxes using eddy correlation technique. Ph.D. dissertation, Cornell University, 189 pp

Drösler M (2005) Trace gas exchange and climatic relevance of bog ecosystem, Southern Germany. Ph.D. dissertation, Lehrstuhl für Vegetationsokologie, Department für Ökologie, Technischen Universität München, 182 pp

Falkowski P, Scholes RJ, Boyle E, Canadell J, Canfield D, Elser J, Gruber N, Hibbard K, Högberg P, Linder S, Mackenzie FT, Moore B, Pedersen T, Rosenthal Y, Seitzinger S, Smetacek V, Steffen W (2000) The global carbon cycle: a test of our knowledge of earth as a system. Science 290(5490):291–296

Finn D, Lamb B, Leclerc MY, Horst TW (1996) Experimental evaluation of analytical and Lagrangian surface-layer flux footprint models. Bound-Layer Meteorol 80:283–308

Flesch TK (1996) The footprint for flux measurements, from backward Lagrangian stochastic models. Bound-Layer Meteorol 78:399–404

Foken T (2008) Micrometeorology. Springer, Berlin 308 pp

Foken Th, Wichura B (1996) Tools for quality assessment of surface-based flux measurements. Agric For Meteorol 78:83–105

Gash JHC (1986) A note on estimating the effect of a limited fetch on micrometeorological evaporation measurements. Bound-Layer Meteorol 35:409–414

Gerosa G, Cieslik S, Ballarin-Denti A (2003) Ozone dose to a wheat field determined by the micrometeorological approach. Atmos Eviron 37:777–788

Gorham E (1991) Northern peatlands: role in the carbon cycle and probable responses to climatic warming. Ecol Appl 1(2):182–195

Goulden ML, Munger JW, Fan SM, Daube BC, Wofsy SC (1996) Measurement of carbon storage by long-term eddy correlation: methods and a critical assessment of accuracy. Global Change Biol 2:169–182

Grace J (2001) Carbon cycle. Encyclopedia of biodiversity 1:609–629

Graus M, Hansel A, Wisthaler A, Lindinger Ch, Forkel R, Hauff K, Klauer M, Pfichner A, Rappengluck B, Steigner D, Steinbrecher R (2006) A relaxed-eddy-accumulation method for the measurements of isoprenoids canopy-fluxes using an online gas-chromatographic technique and PTR-MS simultaneously. Atmos Environ 40:43–54

GRID-Arendal/UNEP (2013) http://www.grida.no/publications/rr/natural-fix/page/3724.aspx. Accessed on 6 Mar 2015

Gu L, Falge EM, Boden T, Baldocchi DD, Black TA, Saleska SR, Suni T, Verma SB, Vesala T, Wofsy SC, Xu L (2005) Objective threshold determination for nighttime eddy flux filtering. Agric For Meteorol 128:179–197

Horst TW (2001) Comment on "footprint analysis: a closed analytical solution based on height-dependent profiles of wind speed and eddy viscosity". Bound-Layer Meteorol 101:435–447

Hutchinson GL, Livingston GP (2001) Vents and seals in non-steady-state chambers used for measuring gas exchange between soil and the atmosphere. Eur J Soil Sci 52:675–682

IPCC (2013) Special report of the intergovernmental panel on climate change. http://www.climatechange2013.org/. Accessed on 6 Mar 2015

Jurasinski G, Koebsch F, Hagemann U (2012) Flux: flux rate calculation from dynamic closed chamber measurements. R package version 0.2-1, Rostock. Available online at http://cran.r-project.org/web/packages/flux. Accessed on 6 Mar 2015

Juszczak R (2013) Biases in methane chamber measurements in peatlands. Intern Agrophys 27:159–168

Juszczak R, Humphreys E, Acosta M, Michalak-Galczewska M, Kayzer D, Olejnik J (2013a) Ecosystem respiration in a heterogeneous temperate peatlands and its sensitivity to peat temperature and water table depth. Plant Soil 366(1–2):505–520

Juszczak R, Sakowska K, Leśny J, Uzdzicka B, Chojnicki B, Urbaniak M, Olejnik J (2013b) Is it possible to measure CO_2 effluxes in an arable ecosystem using chambers without soil frames? Global change and resilience. In: Stojanow R, Žaluda Z, Cudlin P, Farda A, Urban O, Trnka M (eds) From impacts to responses, Brno 2013. ISBN: 978-80-904351-9-3, 168–172 pp

Kirschbaum MUF, Eamus D, Gifford RM, Roxburgh SH, Sands PJ (2001) Definitions of some ecological terms commonly used in carbon accounting. Workshop proceedings. Cooperative research centre for greenhouse accounting, Canberra, Australia, pp 2–5

Kirschke S, Bousquet P, Ciais P, Saunois M, Canadell JG, Dlugokencky JE, Bergamaschi P, Bergmann D, Blake DR, Bruhwiler L, Cameron-Smith P, Castaldi S, Chevallier F, Feng L, Fraser A, Heimann M, Hodson EL, Houweling S, Josse B, Fraser PJ, Krummel PB, Lamarque JF, Langenfelds RL, Le Quéré C, Naik V, O'Doherty S, Palmer PI, Pison I, Plummer D, Poulter B, Prinn RG, Rigby M, Ringeval B, Santini M, Schmidt M, Shindell DT, Simpson IJ, Spahni R, Steele LP, Strode SA, Sudo K, Szopal S, van der Werf GR, Voulgarakis A, van Weele M, Weiss RF, Williams JE, Zeng G (2013) Three decades of global methane sources and sinks. Nat Geosci 6:813–823. doi:10.1038/ngeo1955

Kljun N, Rotach MW, Schmid HP (2002) A three-dimensional backward Lagrangian footprint model for a wide range of boundary-layer stratifications. Bound-Layer Meteorol 103:205–226

Kljun N, Kormann R, Rotach MW, Meixer FX (2003) Comparison of the Lagrangian footprints. Bound-Layer Meteorol 106(2):349–355

Kormann R, Meixner FX (2001) An analytic footprint model for neutral stratification. Bound-Layer Meteorol 99:207–224

Kowalska N, Chojnicki BH, Rinne J, Haapanala S, Siedlecki P, Urbaniak M, Juszczak R, Olejnik J (2013) Measurements of methane emission from a temperate wetland by the eddy covariance method. Int Agrophysics 27(3):283–290

Kroon PS, Hensen A, Jonker HJJ, Ouwersloot HG, Vermeulen AT, Bosvelt FC (2010) Uncertainties in eddy covariance flux measurements assessed from CH4 and N_2O observations. Agric For Meteorol 150:806–816

Le Quéré C (2009) Closing the global budget for CO_2. Global Change 74:28–31

Leclerc MY, Thurtell GW (1990) Footprint predictions of scalar fluxes using a Markovian analysis. Bound-Layer Meteorol 52:247–258

Livingston GP, Hutchinson GL (1995) Enclosure-based measurement of trace gas exchange: application and sources of error. In: Matson PA, Harriss RC (eds) Biogenic trace gases: measuring emissions from soil and water. Blackwell Science, Oxford, pp 14–51

Massman WJ, Lee X (2002) Eddy covariance flux corrections and uncertainties in long-term studies of carbon and energy exchanges. Agric For Meteorol 113:121–144

Minke M, Augustin J, Hagemann U, Joosten H (2014) Similar methane fluxes measured by transparent and opaque chambers point at belowground connectivity of *Phragmites australis* beyond the chamber footprint. Aquat Bot 113:63–71

Moncrieff JB, Mahli Y, Leuning R (1996) The propagation of errors in long term measurements of land atmosphere fluxes of carbon and water. Global Change Biol 2:231–240

Moncrieff JB, Massheder JM, de Bruin H, Elbers J, Friborg T, Huesunkveld B, Kabat P, Scott S, Soegaard H, Verhoef A (1997) A system to measure surface fluxes of energy, momentum and carbon dioxide. J Hydrol 188(189):589–611

Moore C (1986) Frequency response corrections for eddy correlation systems. Bound-Layer Meteorol 37:17–35

Morison (2012) Sea ice death spiral driving atlantic water into arctic causing wild weather. Get energy smart! Now. Blogging for a sustainable energy future. http://getenergysmartnow.com/2012/01/20/sea-ice-death-spiral-driving-atlantic-water-into-arctic-causing-wild-weather/. Accessed on 6 Mar 2015

Myklebust MC, Hipps LE, Ryel RJ (2008) Comparison of eddy covariance, chamber, and gradient methods of measuring soil CO_2 efflux in an annual semi-arid grass, *Bromus tectorum*. Agric For Meteorol 148:1894–1907

Nay SM, Mattson KG, Bormann BT (1994) Biases of chamber methods for measuring soil CO_2 efflux demonstrated with a laboratory apparatus. Ecology 75:2460–2463

Novick KA, Stoy PC, Katul GG, Ellsworth DS, Siqueira MBS, Juang J, Oren R (2004) Carbon dioxide and water vapor exchange in a warm temperate grassland. Oecologia 138:259–274

Oak Ridge National Laboratory Distributed Active Archive Center (ORNL DAAC) (2013) FLUXNET Maps and Graphics Web Page. Available online http://fluxnet.ornl.gov/maps-graphics. From ORNL DAAC, Oak Ridge, Tennessee. Accessed 5 Nov 2013

Pan Y, Birdsey RA, Fang J, Houghton R, Kauppi PE, Kurz WA, Phillips OL, Shvidenko A, Lewis L, Canadell JG, Ciais P, Jackson RB, Pacala SW, McGuire AD, Piao S, Rautiainen A, Sitch S, Hayes D (2011) A large and persistent carbon sink in the world's forests. Science 333 (6045):988–993

Pihlatie M, Christiansen JR, Aaltonen H, Korhonen J, Nordbo A, Rasilo T, Benanti G, Giebels M, Helmy M, Hirvensalo J, Jones S, Juszczak R, Klefoth R, Vale Lobo do R, Rosa AP, Schreiber P, Serça D, Vicca S, Wolf B, Pumpanen J (2013) Comparison of static chambers to measure CH4 emissions from soils. Agric For Met 171–172:124–136

Pumpanen J, Kolari P, Ilvesniemi H, Minkkinen K, Vesala T, Niinisto S, Lohila A, Larmola T, Morero M, Pihlatie M, Janssens I, Yuste JC, Grunzweig JM, Reth S, Subke JA, Savage K, Kutsch W, Ostreng G, Ziegler W, Anthoni P, Lindroth A, Hari P (2004) Comparison of different chamber techniques for measuring soil CO_2 efflux. Agric For Met 123:159–176

Pumpanen J, Longdoz B, Kutsch WL (2009) Field measurements of soil respiration: principles and constraints, potentials and limitations of different methods. In: Kutsch WL, Bahn M, Heinemeyer A (eds) Soil carbon dynamics: an integrated methodology. Cambridge University Press, Cambridge, pp 16–33

Reynolds O (1895) On the dynamical theory of incompressible viscous fluids and the determination of criterion. Philos Trans R Soc Lond A174:935–982

Rochette P, Hutchinson GL (2005) Measurement of soil respiration in situ: chamber techniques. In: Micrometeorology in agricultural systems. Agronomy monograph, vol 47, pp 247–280
Running SW, Baldocchi DD, Turner D, Gower ST, Bakwin P, Hibbard K (1999) A global terrestrial monitoring network, scaling tower fluxes with ecosystem modeling and EOS satellite data. Remote Sens Environ 70:108–127
Schmid HP (1994) Source areas for scalars and scalar fluxes. Bound-Layer Meteorol 67:293–318
Schulze ED, Ciais P, Luyssaert S, Schrumpf M, Janssens IA, Thiruchittampalam B, Theloke J, Saurat M, Bringezu S, Lelieveld J, Lohila A, Rebmann C, Jung M, Bastviken D, Abril G, Grassi G, Leip A, Freibauer A, Kutsch W, Don A, Nieschulze J, Bőrner A, Gash JH, Dolman AJ (2010) The European carbon balance. Part 4: integration of carbon and other trace-gas fluxes. Glob Change Biol 16:1451–1469
Spirig C, Neftle A, Ammann C, Dommen J, Gramber W, Thielmann A, Schaub A, Beauchamp J, Withaler A, Hansel A (2005) Eddy covariance flux measurements of biogenic VOCs during ECHO 2003 using proton transfer reaction mass spectroscopy. Atmos Chem Phys 5:465–481
Stull RB (1988) An introduction to boundary layer meteorology. Kluwer Academic Publishers, Dordrecht, p 666
Swinbank WC (1951) Measurements of vertical transfer of heat and water vapor by Eddies in the lower atmosphere. J Meteorol 8:135–145
Tanner CB, Thurtell GW (1969) Anemoclinometer measurements of Reynolds tress and heat transport in the atmospheric surface layer. Department of Soil Science, University of Wisconsin, Madison, WI, Research and Development Technical Report ECOM 66-G22-F to the US Army Electronics Command, 200 pp
Twine TE, Kustas WP, Norman JM (2000) Correcting eddy-covariance flux underestimates over a grassland. Agric For Meteorol 103:279–300
Valentini R, Matteucci G, Dolman AJ (2000) Respiration as the main determinant of carbon balance in European forests. Nature 404:861–864
Wattenbach M, Sus O, Viuchard N, Lehuger S, Gottschalk P, Longhui L, Leip A, Williams M, Tomelleri E, Kutsch WL, Buchmann N, Eugster W, Dietiker D, Aubinet M, Ceschia E, Béziat P, Grünwald T, Hastings A, Osborne B, Ciais P, Cellier P, Smith P (2010) The carbon balance of European croplands: a cross-site comparison of simulation models. Agric Ecosyst Environ 139:419–453
Webb EK, Pearman G, Leuing R (1980) Correction of flux measurement for density effects due to due to heat and water vapor transfer. Q J R Meteorol Soc 106:85–100
Whalen SC (2005) Biogeochemistry of methane exchange between natural wetlands and the atmosphere. Environ Eng Sci 22(1):73–94. doi:10.1089/ees.2005.22.73
Wilson KB, Goldstein AH, Falge E (2002) Energy balance closure at FLUXNET sites. Agric For Meteorol 113:223–243
Wofsy SC, Goulden ML, Munger JW (1993) Net exchange of CO_2 in a mid-latitude forest. Science 260:1314–1317

Chapter 18
Assessment and Measurement of Wind Erosion

Roger Funk

Abstract Wind erosion has become an important soil degradation process on arable land, caused by land use techniques that leave a disturbed, temporarily bare surface or insufficient soil cover material at the surface. Soils that have been formed by aeolian processes over centuries are endangered to be destroyed by the same processes within a very short time. Wind erosion is not only a soil-removing process but also a very effective sorting process. Coarse particles remain in the field, whereas the finest and most valuable parts of the soil are lost, such as particles of the silt and clay fractions and the soil organic matter. This chapter introduces advanced methods to assess wind erosion and to quantify the corresponding soil losses. Evaluation schemes generally consider two categories to determine the extent of wind erosion: the erosivity of the climate and the erodibility of the soil, divided into few classes and linked in simple matrices to derive the wind erosion risk in a comparative way. The German standard DIN 19706 "Soil quality—Determination of the soil exposure risk from wind erosion" was a basis for a Geographic Information System (GIS) risk map of all agricultural fields in Germany. The quantification of wind erosion is based on the measurements of the horizontal fluxes that can be used further to derive soil losses/dust emissions or the deposition of the transported particles. Sufficient depositions can be measured in their thickness and extent to calculate the relocated volume or mass. The comparison of the grain size distribution and the organic carbon content of the original soil, the redistributed material, and the depositions enable the losses of fine particles and organic matter to be calculated. Models of wind erosion have been calibrated to specific conditions of the soil surface and plant cover and refined by wind tunnel experiments. The Fallout-Radionuclide method is suited to identify wind erosion and dust deposition pattern at larger spatial and temporal scales. Finally, remote-sensing and GIS procedures are used to present areas for wind erosion and

R. Funk (✉)
Leibniz Centre for Agricultural Landscape Research (ZALF) e.V.,
Eberswalder Strasse 84, 15374 Müncheberg, Germany
e-mail: rfunk@zalf.de

L. Mueller et al. (eds.), *Novel Methods for Monitoring and Managing Land and Water Resources in Siberia*, Springer Water,
DOI 10.1007/978-3-319-24409-9_18

dust deposition for large landscape units. The methods presented here have been shown to be proven in important agricultural regions around the globe. Conclusions include recommendations for the installation of wind erosion monitoring plots.

Keywords Wind erosion · Methods · Measurement · Evaluation · Modelling · Monitoring

1 Introduction

A significant global process: Wind erosion is an important process of both progressive and regressive pedogenesis in arid and semiarid environments around the world in particular. Associated dust emissions influence physical and chemical processes in the atmosphere, affecting other ecosystems far away from the source areas (Bristow et al. 2010). Dust and its chemical and biological components have negative effects on air quality and are risks to human health (Perez et al. 2008; Karanasiou et al. 2012).

Wind erosion on arable land is usually caused by land use techniques that leave a disturbed, temporarily bare surface or insufficient soil cover material at the surface (Fig. 1). On agricultural land, wind erosion is a soil-degrading process, but quantitative data are rather approximate estimates. Agricultural soils and their tillage may also significantly contribute to PM10 dust concentrations of the air (Sharratt et al. 2007; Singh et al. 2012) which are strongly debated issues related to

Fig. 1 Intensive conventional soil tillage in steppe regions leads to significant soil deflation including dust emission. Topsoils of formerly well-structured Chernozems are converted into skeletised and cloddy surfaces, which are unfavourable for germination and the emergence of crops. *Photos* Lothar Mueller

environmental quality and human health. Emissions of black carbon from agricultural burning (Romanenkov et al. 2014) contribute to and enhance fine dust problems (Wagenbrenner et al. 2013). Consequently, there is a close relationship between the cycles of dust (D-cycle), carbon (C-cycle), and energy (E-cycle) in the global context (Shao et al. 2011). In longer time scales, dust also has modifying influences on the climate (Martínez-Garzia et al. 2011).

The spatial extent of wind erosion has increased in the recent decades, mainly due to the changes in agricultural practices. The spectrum of growing crops has changed to greater proportions of arable land crops, to the disadvantage of grassland. Further factors include frequent disturbances of the soil surface by plowing and a multitude of tillage operations. The time of highest mechanical stress by tillage operations coincides with the time of highest climatic erosivity in spring. Other factors influencing wind erosion are:

- The higher level of mechanisation requires larger fields and in consequence the removal of hedges and other landscape structures.
- The drainage of arable land causes a faster and more frequent drying of the soil surface, resulting in decomposition of organic matter and decreasing soil aggregate stability.
- Overgrazing is a significant causative factor in the semiarid and arid regions, where no other type of land use is possible (van Lynden 1995, Riksen et al. 2003).

Besides soil degradation, the effects of wind erosion are crop damage and pollution of adjacent areas. Soil deterioration includes the loss of fine material and organic matter, the degradation of soil structure, and the loss of fertilizers and herbicides. The loss of topsoil is the predominant impact of wind erosion in Europe (van Lynden 1995). In the worst case, the productivity has declined so substantially that arable land has been removed from production as reported in Sweden and Poland (Jönsson 1992; Veen et al. 1997). The most severe damage is reported in the former Soviet Union, where wind erosion removes an estimated 1.5 million ha of cropland from cultivation each year and a much larger area is damaged to some extent (Schroeder and Kort 1989).

An insidious process The main difficulty is noticing wind erosion in the field site because both erosion and deposition processes affect thin layers and large areas. Chepil (1960) already pointed out that annual average soil losses up to 40 tonnes/ha are possible without any visible indications of soil movement (Table 1).

This descriptive schema of the terms by which the "quantity of erosion" is assessed is intended as an example only. Besides the difficulty to perceive even larger soil losses, it demonstrates the extremely high tolerance against soil losses from agricultural lands in the USA 55 years ago, very likely influenced by the impressions of the disastrous "dust bowl" events in the 1930s. From a contemporary view, this scheme trivializes the soil degradation process. Annual soil losses of 2.5 cm or 125–375 tonnes a year can no longer be considered "moderate." A completely redefined and much more refined evaluation scheme must be developed

Table 1 Relationships between quantity of wind erosion, effects of erosion and annual soil loss (from Chepil 1960)

Quantity of erosion	Description of erosion	Annual soil loss (t/ha)
None to insignificant	No distinct visible effects of soil movement	<40
Slight	Soil movement not sufficient to kill winter wheat in boot stage	40–125
Moderate	Removal and associated accumulations to about 2.5 cm depth, sufficient to kill wheat in boot stage	125–375
High	About 2.5–5 cm removal and associated accumulations	375–750
Very high	5–7.5 cm removal with small dune formations	750–1125
Exceedingly high	More than 7.5 cm removal with appreciable piling into drifts or dunes	More than 1125

with a focus below the "invisible" limit of 40 tonnes/ha. This requires measurements and sharing of methods and results. More awareness of the severity of the problem and of sustainability in land use practices needs to be built up. A study by Nordstrom and Hotta (2004) underlined the fact that because of these apparently undetected losses of soil, many farmers still do not consider wind erosion as a significant problem at all.

Focus on agricultural lands Agricultural lands are hotspots of wind erosion in terms of their significance and consequences for the human civilization. This has been identified and addressed as a process of soil degradation and land desertification and thus a risk for the existence of the human civilization (Montgomery 2007).

Wind erosion removes predominantly the finest and most valuable particles of a soil such as silt and clay and the soil organic matter. In addition to the gradual soil degradation, individual wind erosion events may result in soil losses of more than 100 t ha^{-1} and cause considerable on-site and off-site damage (Funk et al. 2004a; Goossens and Riksen 2004; Hoffmann et al. 2011). In the recent decades, the spatial extent of wind erosion has increased, mainly caused by changes in agricultural land use and inappropriate farming practices. Increasing demands on food production have expanded arable land use to marginal sites such as natural grassland or forest (FAO 2009). Wind erosion has been mainly regarded as a soil-removing process, but soil particle sorting, long distance nutrient transport, fertilizing aquatic and terrestrial ecosystems, and the increase in soil heterogeneity are also elements of the problem with local to global consequences on soil properties and the carbon cycle as well (Shao et al. 2011). These direct and indirect effects are difficult to evaluate in their entirety and so the debate is currently ongoing on whether soils that are affected by erosion act as a source or sink for carbon.

On agricultural soils, organic carbon is concentrated in the top layer (0–30 cm) and therefore vulnerable to wind erosion in its particulate form because of its lower density compared to the mineral particles (Lal 2003; Buschiazzo and Funk 2015).

Thus, the most susceptible soils are cultivated organic soils which are also subject to massive decomposition processes. Mineral soils cover the wide range from highly erodible sandy soils to nonerodible loam and clay soils.

The erodibility of a soil is mainly attributed to the texture and organic matter content, which influence a soil's resistance to wind erosion, such as its water holding capacity and ability to produce aggregates or crusts (Chepil and Woodruff 1961; Zobeck et al. 2003b). In general, sandy soils are highly erodible because they dry quickly; form only few, weak aggregates and contain a large number of particles in the most erodible fraction of 80–200 μm in diameter. Loamy soils are more resistant to wind erosion but have a greater potential for dust production (particles < 50 μm) if they erode.

Temporal factors are soil moisture and roughness of the surface. To the same extent that slight increases in both factors can prevent or reduce wind erosion, small decreases can cause or raise the processes considerably (Chen et al. 1996; Cornelis et al. 2004; Ravi and D'Odorico 2005). Due to the rapid changes, neither factor is regarded in simple assessment schemes, but both are an essential part of most wind erosion models.

The above-mentioned facts show that basic processes and consequences of wind erosion have been ascertained in numerous studies over the past decades. Hundreds of scientific articles a year are increasing the pool of knowledge. It has become clear that the process of wind erosion has local, regional, and global dimensions that nearly every eco- and land use system is affected and that the human impact on ecosystem use and management may accelerate it significantly.

The need for monitoring and risk assessment How much deflation occurs with a certain land use system in a specific region in a certain time? How will climate change and intensified land use practices affect erosion intensity? How much deflation is tolerable? How efficient are specific conservation measures? What windbreak structure and design in the landscape is viable and effective? These and other questions are currently being posed in rural regions. Answering them requires knowledge and reliable data. There is a constant need to monitor and assess wind erosion on agricultural land. Process models are powerful tools for this purpose, but they need to be calibrated (Youssef et al. 2012) and amended (Zou et al. 2015) because possible results may be largely uncertain or biased (Li et al. 2014). This requires basic data measured using internationally proven procedures and standards. Borrelli et al. (2014) point to the need for a new phase of field measurements and local monitoring operations in order to make further modeling of wind soil erosion more reliable. This is important for fulfilling the goals of the EU Thematic Strategy for Soil Protection (European Commission 2006).

The aim of this chapter The following chapter introduces common methods to assess wind erosion and to quantify the corresponding soil losses. From its basic structure and content, it follows another book chapter (Funk et al. 2014) but considers both our own updates regarding work about methods conducted at the Leibniz Centre for Agricultural Landscape Research in Müncheberg, Germany, and also the latest findings given in the scientific literature.

The methods and tools developed for this purpose are specific standards, models, wind tunnels, and field measurement devices. Assessments are often based on available models, parts of them, or key parameters to gain a comparative analysis over large areas. Wind tunnel experiments enable the quantification of flux parameters and calibration of models. Field measurements of both deflation and deposition at key sites are required for model tests, outreach, and demonstration, e.g., convincing specialists and stakeholders about the severity of the problem and reliability of results achieved.

The quantification of wind erosion is generally based on the measurements of the horizontal fluxes, which can be used further to derive soil losses or dust emissions from the affected areas. Sufficient depositions can be measured in their thickness and extent to calculate the relocated volume or mass. A comparison of the grain size distribution and the organic matter content of both the original soil and the depositions can be used to derive the losses of fine particles and organic matter. The Fallout-Radionuclide method is a further possibility to identify wind erosion and dust deposition patterns at larger spatial and temporal scales (Funk et al. 2012).

2 Methods for Assessing Wind Erosion

Erosivity and erodibility Wind erosion occurs when the erosive forces of the wind exceeds the resistant forces of the soil stabilizing factors. Thus, two categories have to be regarded which determine the extent of wind erosion by their coincidence: the erosivity of the climate and the erodibility of the soil. Evaluation schemes regularly use one or both categories divided into classes and linked in simple matrices to derive the wind erosion risk in a comparative way.

The erosivity describes tendencies of the climate to produce favorable conditions for wind erosion. In particular, erosivity depends on the wind velocity and the amount and distribution of precipitation and evaporation (Skidmore 1986). Their interactions determine the intensity, frequency, and duration of wind erosion events on susceptible surfaces. The erosivity is mainly determined by the wind velocity and modified by the moisture conditions (Chepil et al. 1962). This can be expressed generally by an erosivity index, calculated from the cube of the wind or friction velocity above a certain threshold. The ratio of potential evaporation and precipitation determines the soil water content within defined periods and can be integrated by a reductive factor or by increasing the threshold velocity (Bagnold 1943). A compilation of various equations that describe the transport intensity of sand particles related to the wind or friction velocity can also be found in Greeley and Iversen (1985).

Erosivity indices are generally grouped into classes and useful to compare differences over large regions. An example for deriving an erosivity index for wind erosion in Europe is shown in Fig. 2, based on meteorological data from 1958 to 2001 of the ERA40 reanalysis (ECMWF 2005). The parameters considered are u- and v-components of the surface wind, precipitation, and evaporation in a

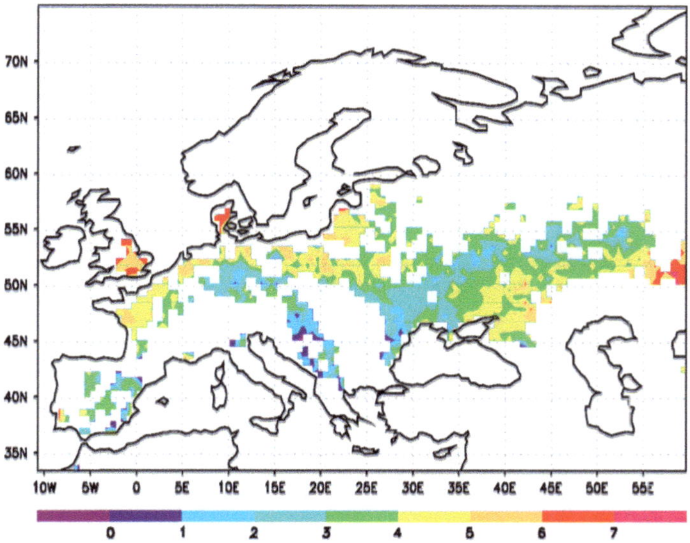

Fig. 2 Climatic erosivity indices for wind erosion in Europe based on WFIs of the period 1958–2001

temporal resolution of 6 h and a grid size of 1.12°. The comparison is based on the derivation of wind force integrals (WFIs), which are calculated if the following preconditions are met: precipitation within the last 6 h < 0.3 mm; precipitation < evaporation within the last 6 h; and wind velocity (u) above the texture depending threshold velocity (u_{thr}). The dominating soil texture determines the threshold wind velocity, which was set to: 5 m s^{-1} for "coarse", 7 m s^{-1} for "medium", 9 m s^{-1} for "fine" and 11 m s^{-1} for "very fine" as classified in the European Soil Data Base (ESDB 2013).

$$\mathrm{WFI} = \sum_{1}^{n} (u - u_{\mathrm{thr}}) * u^2 \tag{1}$$

After the calculation, the WFIs were grouped equidistantly into 8 classes from very low to very high climatic erosivity.

The erodibility describes the potential of a soil to erode or, the reverse, the ability to resist the acting wind forces. This is mainly attributed to the texture and organic matter content, which influence the water-holding capacity and the ability of the soil to produce aggregates or crusts (Chepil 1955a, b). The stability of the surface roughness is also related to the soil texture.

There is a distinct relationship between the sand content of a soil and its potential to be eroded by the wind. The influence of the soil texture on threshold wind velocity (u_{thr}, in 10 m height) of a dry, noncrusted soil can be calculated in a simple approximation from the sand content (S) in % with Eq. 2, estimated from experiments with soils ranging between 70 and 95 % in the sand content:

$$u_{thr} = 27.6 - 0.22S \quad r^2 = 0.51 *** \quad n = 23 \tag{2}$$

The soil loss (SL, t ha^{-1} h^{-1}) is correlated to the sand content by:

$$\mathrm{SL} = 4.2 * 10^{-7} e^{0.2S} \quad r^2 = 0.61 \quad n = 23 \tag{3}$$

Equation 3 is valid for a wind erosion event with a duration of 1 h and an average wind velocity of 8 m s^{-1}. Each increase in the wind velocity at 2 m s^{-1} will halve the time to erode the same amount of soil, i.e., 10 m s^{-1} −30 min, 12 m s^{-1} −15 min, 14 m s^{-1} −7.5 min, and so on (Funk and Frielinghaus 1997).

Wind erosion risk More complex evaluation schemes combine several factors to assess the wind erosion risk. The German standard DIN 19706 "Soil quality—Determination of the soil exposure risk from wind erosion" represents a simple method to combine soil information, crop rotation, wind velocity, and landscape structure to derive the field-specific wind erosion risk. In a step-by-step analysis, matrices and maps are used to derive the wind erosion risk of a single field in five classes (very high to none) by combining the above-mentioned parameters (Fig. 3). The scheme represents the minimum number of parameters to make a state-wide, consistent, and fair evaluation. The importance of each parameter is shown by the decrease in the affected area in the Federal State of Brandenburg in Germany (Table 2). The region has sandy soils and a subhumid climate. Landscape elements,

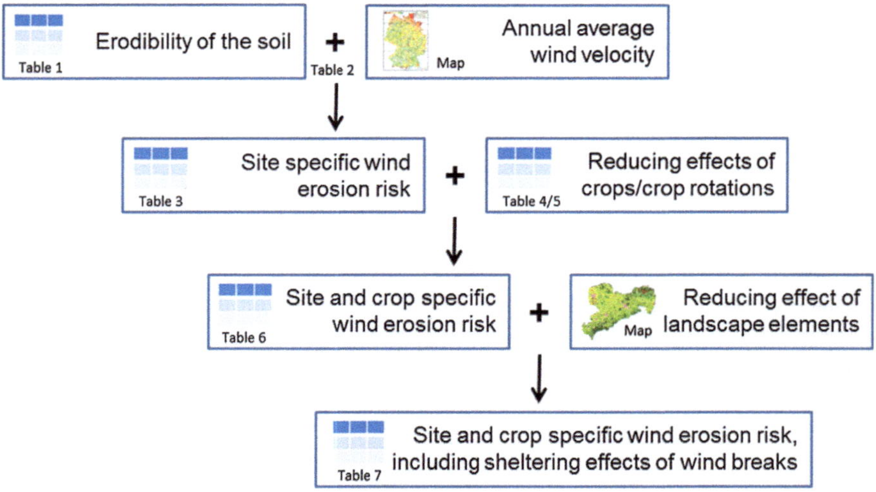

Fig. 3 Evaluation scheme to derive the wind erosion risk of a field with the German standard DIN 19706 "soil quality—determination of the soil exposure risk from wind erosion"

Table 2 Changes in the wind erosion risk in the Federal State of Brandenburg, Germany, based on the included parameters

Wind erosion risk based on	Area in ha	Area in %
Erodibility of the soil (E)	743,035	100
(E) + average wind velocity (AV)	624,371	84
E + AV + Sheltering effect of landscape elements (LE)	150,012	20
E + AV + LE + land use	122,979	17

especially, reduce the wind erosion risk considerably (Funk et al. 2004b). The table also indicates that the result depends on the input parameter.

Based on the algorithm of Table 2, all agricultural fields in Germany were assessed regarding their wind erosion risks. Data are available on a Geographic Information System (GIS) platform for all authorities responsible for soil protection in the federal states of Germany. They are a basis for controlling the status of land management according to rules of a "good agricultural practice," fulfilling the standards of soil protection. Data and risk classification are of great importance for associated tillage restrictions or compensatory payments.

Models of wind erosion Methods to estimate soil losses by wind erosion include models such as the Wind Erosion Equation (WEQ, Woodruff and Siddoway 1965), the Revised Wind Erosion Equation (RWEQ), the Single-event Wind Erosion Evaluation Program (USDA-ARS 2008), and the Wind Erosion Prediction System (WEPS). They were developed for wind erosion prediction and the conservation planning of wind erosion control practices on the field scale (Hagen 1991). All models and extensive documentation are available on the Internet (USDA 1995, 2002, 2008). The order of the models corresponds with their increasing complexity, which is already reflected by the input parameters. The WEQ has some limitations due to the strong correlation with climatic conditions in the Middle West of the USA of only a few years and the estimate of an annual average. The RWEQ has some substantial improvements, such as the calculation of the sediment transport over different field lengths, estimates of the maximum and average soil loss, the consideration of changing surface conditions by crust formation or management practices, and a variable temporal resolution (Webb and McGowan 2009). Appropriately, a comparison of soil losses computed with WEQ and RWEQ resulted in much better estimates by RWEQ (Fryrear et al. 1999; Van Pelt and Zobeck 2004; Van Pelt et al. 2004). The WEPS was developed on process-based algorithms and simulates weather, field conditions, management, and erosion with continuous daily time steps. The WEPS has been permanently updated and well documented (Wagner 2013). The applicability of the WEPS model to regions outside the USA was proven by Funk et al. (2004a). In Argentina, WEQ, RWEQ, and WEPS were tested for different arable land management systems (Buschiazzo and Zobeck 2008). In China, the mentioned models were also implemented (Chen et al. 2013; Zou et al. 2015). Chen et al. (2013) combined WEPS with an air quality model for assessing environmental effects of wind erosion in China.

Zhang et al. (2012) created an approach based on computational fluid dynamics (CFD), modeling the spatial variability of wind-velocity–dependent erosion or deposition patterns at the landscape scale. The CFD-WEM model by Zhang et al. (2012) provided the assessment of regional wind erosion risks on different grasslands in northern China.

In Australia, some examples of recent developments are the Integrated Wind Erosion Management System (Lu and Shao 2001), the Australian Land Erodibility Model (Webb et al. 2009), which predicts land susceptibility to wind erosion in western Queensland on a daily time step, and the wind erosion monitoring network "DustWatch" (McTainsh et al. 2009).

The approach for assessing land susceptibility to wind erosion on a Pan-European scale (Borrelli et al. 2014) is based on the combination of digital soil and terrain information with temporally variable factors of land use and climate, leading to an Index of Land Susceptibility to Wind Erosion. Its status of information comes close to that of map 6 within Fig. 3 whereby the resolution is coarser.

Assessment of wind erosion effects on land quality Evaluating the consequences of wind erosion for the farmers and human society will be an ongoing issue, as economical and societal framework conditions and attitudes of stakeholders groups underlie a permanent dynamics. It is thus important to develop and update tools such as risk assessment procedures (Funk et al. 2004b), decision support systems (DSS; Mirschel et al. 2016, Chap. 23 of this book), and frameworks of environmental impact procedures (Helming 2014). Those tools have to be based on clear natural scientific data, thresholds, and response curves. Wind erosion diminishes the fertility of soils and their productivity. A crucial relationship for decisions about land use strategy, land versatility, and possible conservation measures is that between the degree of erosion and crop yield. A number of field studies about the degree of erosion related to crop yields have been conducted in the USA (Lyles 1975), in Russia (Kuznetsov and Glazunov 1996; Bezuglov et al. 2008), and in other regions (Stocking 2003). They show that erosion has the potential to diminish crop yields significantly or even liquidate the land suitability for cropping completely (Jönsson 1992). As crop yields are influenced by many other factors, the crop yield potential is important for assessments. Soil or land quality figures may serve as scaling factors for assessing crop yield limiting parameters over different spatial scales. The Muencheberg Soil Quality Rating (Mueller et al. 2014) is one such framework enabling soil quality and crop yield potential evaluations. It has the potential for quantifying several soil quality restrictions on crop yield potentials, the effects of wind erosion included. It may thus serve as a compartment of DSS (see Mirschel et al. 2016) or as a basis for different other trade-offs that deal with wind erosion control or further measures of soil conservation and land use planning (Mueller et al. 2016).

3 Methods of Measuring Wind Erosion

3.1 Process-Based Particle Measurements in the Field

Generally, wind erosion initiates three major modes of particle motion: creeping, saltation and suspension. Because each mode is related to specific grain sizes, wind erosion is also a very effective sorting process. Generally, there are two quantities to measure within one erosion event: the horizontal saltation flux and the vertical dust flux. Reliable measurements of both fluxes are the most problematic procedures in aeolian research (Goossens et al. 2000).

The basic principle is to measure the horizontal transport intensity in different heights to enable the calculation of a complete vertical profile for a specific height range and to derive vertical fluxes. This can be achieved by detecting the intensity of the particle movement by optical or acoustic sensors as the SANTRI, the SENSIT or the Saltiphon (Etyemezian 2015; Spaan and van den Abeele 1991; Zobeck et al. 2003a). Sampling the moving particles has the advantage of further analyses of the material regarding particle sizes and composition. Two widely used sampler types are the Big Spring Number Eight (BSNE, Fryrear 1986) and the Modified Wilson and Cooke (MWAC) sampler (Kuntze et al. 1990). The latter are also highly efficient at trapping dust particles (Goossens and Offer 2000). One newly developed trap is the Basaran and Erpul Sediment Trap (Basaran et al. 2011), which samples sand and dust based on the cyclone principle.

Most of the measurements are focused on quantifying soil losses by wind erosion, which is dominated by the horizontal saltation flux. One of the most common methods to measure wind erosion in field studies are sets of passive BSNE or MWAC traps to measure the horizontal soil transport at different heights (Fig. 4). Generally, more than 3 traps are arranged vertically at a pole between heights of about 0.05–2.0 m. The trapped amounts of transported soil can be used to calculate the entire transport profile and to integrate over a certain range of height. Usually, 99 % of the total amount is transported below 1 m in height by creep and saltation, but in fine-textured soils suspension transport can predominate especially and result in complete different height distributions (Zobeck et al. 2003a; Hoffmann et al. 2008). The MWAC samplers have a good trapping efficiency both for sand and for dust particles independent of the wind speed and have been used in many projects with a well-described trapping efficiency for a wide range of soil textures (Goossens and Offer 2000; Goossens et al. 2000; Funk 1995; Sterk and Stein 1997; Mendez et al. 2011). Furthermore, the trap is inexpensive, easy constructed, measures reliably, and can be reconstructed with locally available material (Figs. 4 and 5).

Measurements of wind erosion or dust deposition across an area or along a transect require the installation of at least two traps, one at the incoming and one at the outgoing side of the area under examination. To balance the spatial distribution of soil losses or gains on a measuring field, a certain number of sediment traps are needed. Sterk and Stein (1997) used 21 traps on a field of 0.24 ha, Funk et al. (2004a) 15 traps on 2.25 ha, and Visser et al. (2004) 17 traps on a 1.6 ha field in

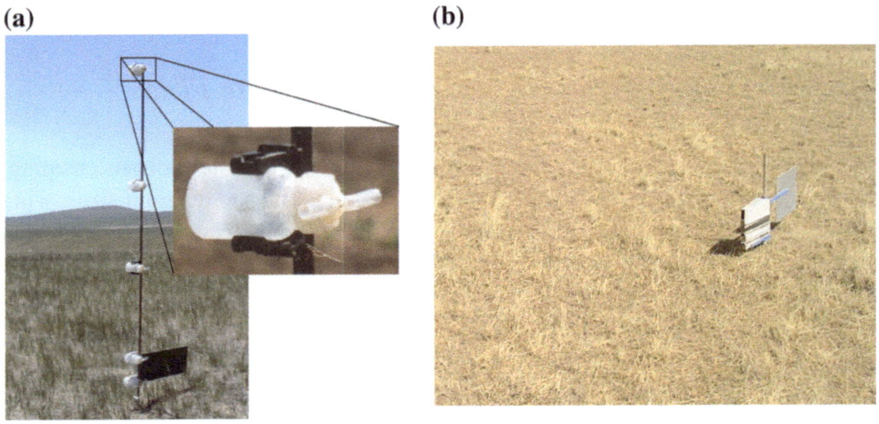

Fig. 4 **a** Modified Wilson and Cooke sediment trap (MWAC) as a set of 5 traps on a pole; **b** Big Spring Number Eight (BSNE) as a set of 3 traps; both traps are always aligned to the wind direction by a wind vane

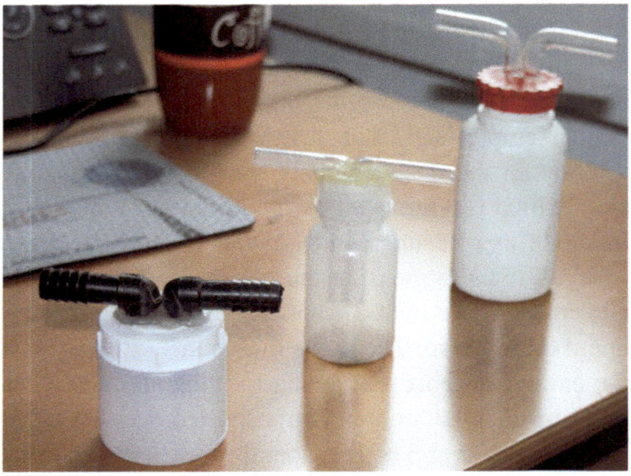

Fig. 5 Different kinds of Modified Wilson and Cook sediment traps (MWAC)

regular and unregular grids. Sterk et al. (2012) measured the sediment transport on a 100-m long transect with 8 MWAC traps. This trap density is necessary on arable land to derive important transport parameters (such as the vertical flux density, the particle composition, the organic matter content), their spatial distribution and dependency on the transported distance, and finally to calculate the spatial variability of the soil loss (Funk et al. 2004a). In steppe regions, the distance between the traps can be wider. Hoffmann et al. (2008) used 20 MWAC traps on a grazing site of 65 ha to calculate these parameters.

Figure 6 gives an example of the spatial variability of wind erosion and soil losses on a sandy soil based on measurements with 15 MWAC traps arranged in a regular grid on the field. The left side shows the total transported soil, which was

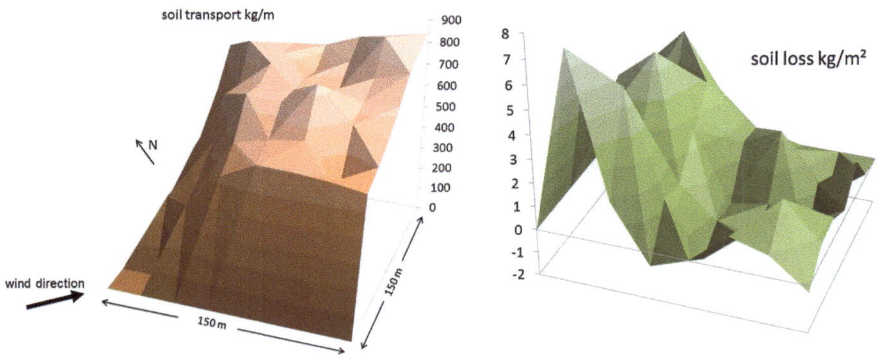

Fig. 6 Spatial distribution of wind erosion on a measuring field of 150 × 150 m; *left side* total transported soil per m width; *right side* soil loss per m^2 (negative values = deposition)

calculated from measured profiles and spatially interpolated. The characteristics of wind erosion on a sandy soil are the rapid increase of the transported amount and the saturation/limitation after a relatively short distance. The wind already reaches the maximum transport capacity at 50 m and stays constant for the rest of the field. The consequences for the spatial variability of the soil loss are: a distinct soil loss appears in only the first half of the measuring field, then erosion and deposition are in equilibrium or deposition even prevails locally. Thus, wind erosion not only causes soil losses but it also increases the soil heterogeneity of a field by temporarily depositing the coarser fractions in the field and blowing out the finer ones.

Further techniques to measure the horizontal fluxes are electronic impact sensors such as the Saltiphon or the Sensit (Spaan and van den Abeele 1991, www.sensit.com) or recording traps such as the SUSTRA (Janssen and Tetzlaff 1991; UGT 2014). Together with a meteorological station they all enable to estimate the threshold wind or friction velocity, the correlation between wind speed and transport intensity to be estimated, and allow information to be gained on the duration of an event.

Based on measurements of the horizontal fluxes as shown in Fig. 7, vertical dust fluxes can be derived. Using the dust concentrations at least two different heights and the logarithmic wind profile, the vertical flux F (up- or downward) can be calculated using Eq. 4:

$$F = \frac{-ku_*(C_2 - C_1)}{\ln(z_2/z_1)} \tag{4}$$

With
F vertical dust flux (g m^{-2} s^{-1})
k von Karman constant 0.4, dimensionless
u_* friction velocity (ms^{-1})
C_{12} concentrations at height z_1 and z_2 (g m^{-3})

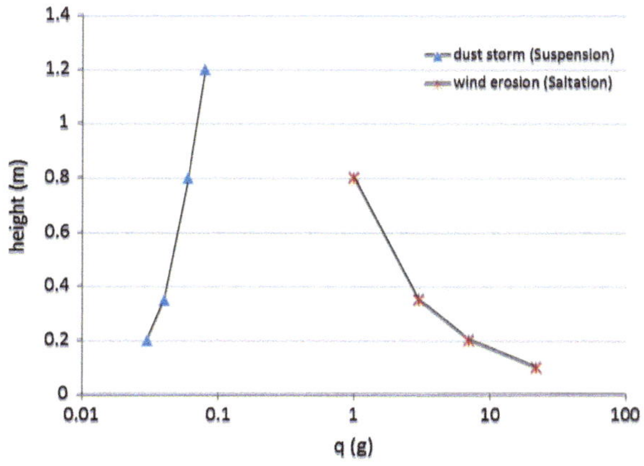

Fig. 7 Measured transport profiles of a dust storm crossing grassland (*blue triangles*) and of a wind erosion event on arable land (*red crosses*)

The direction of the vertical fluxes is generally from the higher to the lower concentration. In the example of Fig. 7, both cases are shown, deposition (vertical downward flux) as a dust storm crosses a grassland and dust emission (vertical upward flux) caused by local wind erosion on arable land.

One of the determining factors for wind erosion, dust emission, or deposition at an area is the roughness. A rough surface increases the turbulence and therefore decreases the wind velocity close to the ground. On arable land, roughness is influenced by the tillage operations and can be used to reduce the wind erosion susceptibility. Methods to measure roughness include pin meters, chains, or laser scanners (Jester and Klik 2005). In wind erosion studies, the wind profile can be used to easily gain additional information about the surface roughness (Helming and Funk 2001). Figure 8 shows the effect of four tillage tools on the wind profiles below 1 m height. It can be seen that a rough surface, as produced by plowing, reduces the wind velocity considerably compared to a smooth surface after rolling. Thus, a ploughed surface can resist a much higher wind velocity to initiate wind erosion than on a surface which is prepared for a seedbed.

The processes of wind erosion and dust emissions are not limited to arable land. The effects of both horizontal sediment flux and PM10 vertical dust emission in a

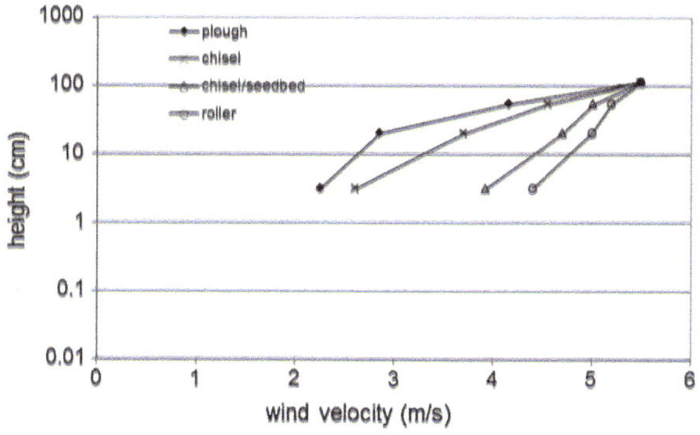

Fig. 8 Measured wind profiles over different rough surfaces produced by tillage operations

post-wildfire landscape were studied by Wagenbrenner et al. (2013) using a field measurement station, where different sediment traps and PM10-sensors were arranged over a height of some meters.

3.2 Mass Balances

Another method to quantify wind erosion in the field, especially severe erosion events or over longer time span, is the quantification of eroded and deposited material by mass balances (Fig. 9). The basic principle is to measure the height differences of a surface before and after erosion or calculate depositions by their area and thickness. The tools needed are simple and the soil relocation can be measured with erosion sticks (erosion and deposition) or with a ruler (only deposition). The reduction in the A-horizon of eroded arable land in comparison to an adjacent site without wind erosion indicates soil losses over longer times (Larionov 1993). Depositions can also be measured using RTK-GPS (real time kinematic), which have a vertical resolution of few millimeters.

The sorting process of material transported by wind makes the identification of the depositions quite easy. These are generally composed of a well-sorted, lighter layer above the original soil. Using the bulk density of the soil, the thickness and the dimensions of the deposition the volume and mass of the eroded soil can be calculated. A comparison of the grain size distributions and the soil organic carbon contents of the soil and the depositions also allow a qualitative analysis of the dust emissions and carbon losses (Hoffmann et al. 2011). This method is, however, limited to wind erosion with a clearly defined source area and a manageable area of depositions.

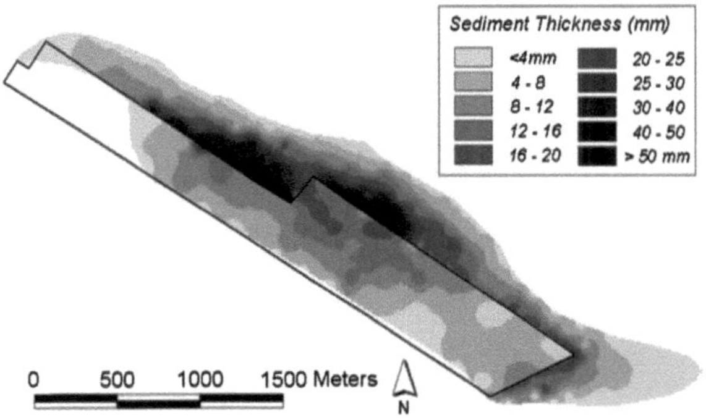

Fig. 9 Thickness of the sedimentation from an arable land to the adjacent grassland caused by wind from the west, corresponding to a soil relocation of 45,900 m^3

3.3 Indirect Measures to Quantify Wind Erosion Processes

Fallout environmental radionuclides, in particular ^{137}Cs, are a useful tracer and chronometer in soil erosion and sedimentation as shown in many studies in different regions of the world (Zapata et al. 2002; Ritchie and Ritchie 2007; Van Pelt 2013). Compared to conventional monitoring studies, ^{137}Cs technique enables soil redistribution processes to be estimated with relatively little effort (Zapata 2003; Mabit et al. 2008). The ^{137}Cs originates from nuclear weapon tests from the 1950s to the early 1970s and is distributed across the land surface in uniform patterns depending on latitude and regional rainfall (Davis 1963). It is strongly adsorbed by the fine soil particles and nearly nonexchangeable (Tamura 1964). In steppe landscapes where leaching processes do not occur, those natural tracer methods are particularly reliable. Therefore, any later movement of ^{137}Cs across the landscape can be related to the movement of these fine soil particles. The ^{137}Cs technique can be used to document erosion rates over the last 50 years, beginning with the nuclear bomb tests in the atmosphere in the 1950s. The comparison of the measured ^{137}Cs inventory of a sampling point with a reference inventory indicates either erosion, if it is smaller, or deposition, if it is higher than the reference value.

Due to the half-life of 30.2 years of ^{137}Cs, more than two-thirds of the original inventory is gone by natural decay processes. Current research offers promising surrogate tracers for wind erosion studies such as $^{239+240}Pu$, which have the same binding affinity to soil particles but a much longer half-life (Van Pelt and Ketterer 2013).

3.4 Wind Tunnel Experiments

An important link between laboratory and field experiments comes from the use of wind tunnels. There are mobile and stationary wind tunnels, each with specific advantages and disadvantages regarding their application possibilities (Fig. 10). Scaling laws, the matching of nondimensional parameters, and the proper

Fig. 10 Mobile and stationary wind tunnel used by the author

development of the boundary layer are the main challenges if wind tunnels are used (Hagen 2001). Though the expense seems to be high, wind tunnels are very effective for the local calibration of wind erosion models and for studies of land surface and vegetation on deflation processes. In experiments at the field site, wind tunnels enable more flexibility with a high rate of replication. Separate or combined effects of influencing parameters can be investigated under controlled conditions. Hagen (1996) and more recently Singh et al. (2012) studied crop residue effects on deflation in wind tunnel experiments. The effect of soil crusts on deflation processes and parameters was investigated by Ries et al. (2014). For this purpose, they used a portable wind tunnel combined with a rainfall simulator (Fister et al. 2012). Hong et al. (2014) used tunnel experiments for evolving and testing WEQ and CFD modeling for simulations of short-term wind erosion processes.

Funk and Engel (2015) tested the wind erosion risk of the row crops of sugar beet (*Beta vulgaris*) and maize (*Zea mays*) on sandy soils. The plants' influence on wind erosion was calculated as the ratio between the measured soil flux of a plot covered by plants and the soil flux of the same plot without plants (soil flux ratio, RQ). This ratio was then tested for correlation with vegetative soil cover, silhouette area, dry mass, or row orientation. The results demonstrate the high susceptibility of row crops to wind erosion in conventional tillage systems. As a new approach, an empirical vegetation parameter was introduced into a sand transport equation of the form

$$q = \frac{c\sqrt{\mathrm{MWD}/D_0}}{g}\rho(u_* - u_{*t})u_*^2 \qquad (5)$$

with

q	sediment flux (kg m^{-1} s^{-1})
ρ	air density (kg m^{-3})
g	gravitational acceleration (m s^{-2})
MWD	mean weighted diameter of texture (mm)
D_0	reference diameter 0.25 mm
C	empirical factor

The factor C has a good correlation to most of the parameters describing the plant development, so it can easily be replaced by plant cover, silhouette area, dry mass or others.

4 Remote-Sensing Methods

Remote-sensing methods (RSMs) are powerful tools for understanding, monitoring, and modeling wind erosion. Monitoring includes not only the conditions of soil, vegetation, landscape structure, and climate factors responsible for wind erosion

(Clark et al. 2010) but also the process in terms of aeolian features such as dunes (del Valle et al. 2009) and dust emissions. The RSMs provide spatiotemporal information over large regions. Huge amounts of data can be effectively handled in combination with GIS technologies (Saha 2004). Depending on the above-mentioned specific targets of object identification and desired spatial resolution, different sensors can be used. Yang and Leys (2014) used data from a Moderate Resolution Imaging Spectroradiometer to map wind erosion hazards in Australia. Reiche et al. (2012) used data from the Advanced Spaceborne Thermal Emission and Reflection Radiometer in combination with field measurements of vegetation features, surface roughness. and sediment transport. These few examples may indicate further potentials of RSMs within regular monitoring programs. This potential can only be utilized if reliable field monitoring data provide ground truth and validation of the results.

5 Outlook: Conclusions for Research and Monitoring

The review revealed some focal points for better understanding, monitoring and controlling wind erosion. They are:

Assessment of wind erosion

- There are various methods available to estimate wind erosion and dust emissions or depositions, from single-point measurements to large-scale mapping of the processes. Evaluation schemes of erosion parameters are inconsistent and biased. Some of them do not meet the demands of sustainable land use. There is a need to refine, harmonize, and standardize methods of wind erosion assessment, measurement methods, and technologies.
- Available spatiotemporal information based on different sources and measurement technology has to be integrated into land information systems. Their basic data and algorithms need to be well documented.
- For more objective economical trade-offs about the efficiency of protection and mitigation measures, wind erosion risk assessment should be integrated into decision support systems and their underlying assessment frameworks.

Monitoring

- Following the suggestion of the European Commission, Joint Research Centre, Institute for Environment and Sustainability (Borrelli et al. 2014), representative field monitoring plots at different locations across Europe should be installed.
- Measurement technologies have to meet advanced international rules and standards, as do data storage and processing systems.

- Measuring the horizontal fluxes or depositions at different heights is the most recommended method because of its easy realization and the derivation of further important parameters. Most of the samplers used have been tested and calibrated by many researchers under various conditions. Besides the quantification of wind erosion, the reliability and comparability of the measurements are also important criteria.
- Minimum data requirements for field studies need to be specified. They should include:

(1) field data: location and dimensions, direction of tillage, occurrence, and description of any upwind obstructions;
(2) sediment sampler data: number, type, placement, efficiency, sampling frequency, fetch distance, and time of sampling;
(3) soil surface data: soil type and classification, texture, soil moisture, organic matter and calcium carbonate content, random and oriented roughness, dry aggregate size distribution, presence of crust and estimate of stability and loose erodible material, soil cover type, and amount;
(4) meteorological data: wind speed and direction with averaging times during storms, aerodynamic roughness, friction velocity, duration of storms, antecedent rainfall, relative humidity, solar radiation, air temperature, and wind direction variability during storm sampling (Zobeck et al. 2003a).

Prevention and rehabilitation measures

- Actions need to be intensified to minimize wind erosion by conservation tillage, windbreak networks, plowless grassland renovation, limitation of grazing stock intensity, and afforestation (see Nordstrom and Hotta 2004; Guo et al. 2014).
- Effective technologies for soil conservation are available (Meinel et al. 2014) but need to be fostered by joint research projects and technology transfer in some regions (BMBF 2011–2016).

References

Bagnold RA (1943) The physics of blown sand and desert dunes. William Morrow & Co., New York 265 p

Basaran M, Erpul G, Uzun O, Gabriels D (2011) Comparative efficiency testing for a newly designed cyclone type sediment trap for wind erosion measurements. Geomorphology 130:343–351. doi:10.1016/j.geomorph.2011.04.016

BMBF (2011–2016) KULUNDA. How to prevent the next “Global Dust Bowl”? Ecological and economic strategies for sustainable land management in the russian steppes: a potential solution to climate change. www.kulunda.eu. Accessed 25 Jan 2015

Bezuglov VG, Gomachadze GD, Sinigovets ME (2008) Status of soil erosion in Russia. (Состояние с эрозией почв в России) Ecologiya 2008 No. 1. http://agroecoinfo.narod.ru/journal/STATYI/2008/1/st_01.doc. Accessed 25 Jan 2015

Borrelli P, Panagos P, Ballabio C, Lugato E, Weynants M, Montanarella L (2014) Towards a Pan-European assessment of land susceptibility to wind erosion. Land degrad develop (2014). Published online in Wiley Online Library (www.wileyonlinelibrary.com). doi:10.1002/ldr.2318

Bristow CS, Hudson-Edwards KA, Chappell A (2010) Fertilizing the Amazon and equatorial Atlantic with West African dust. Geophys Res Lett 37(14):L14807. doi:10.1029/2010GL043486

Buschiazzo DE, Zobeck TM (2008) Validation of WEQ, RWEQ and WEPS wind erosion for different arable land management systems in the Argentinean Pampas. Earth Surf Proc Land 33 (12):1839–1850

Buschiazzo DE, Funk R (2015) Wind erosion of agricultural soils and the carbon cycle. In: Banwart SA, Noellemeyer E, Milne E (eds) Soil carbon: science, management and policy for multiple benefits. CABI, Wallingford, pp 161–168

Chen W, Dong Z, Li Z, Yang Z (1996) Wind tunnel test of the influence of moisture on the erodibility of loessial sandy loam soils by wind. J Arid Environ 34(4):391–402

Chen L, Zhao H, Han B, Bai Z (2013) Combined use of WEPS and models-3/CMAQ for simulating wind erosion source emission and its environmental impact. Sci Total Environ 466–467(2013):762–769

Chepil WS (1955a) Factors that influence clod structure and erodibility of soil by wind: V. Organic matter at various stages of decomposition. Soil Sci 80(5):413–421

Chepil WS (1955b) Factors that influence clod structure and erodibility of soil by wind: IV. Sand, silt, and clay. Soil Sci 80(5):155–162

Chepil WS (1960) Conversion of relative field erodibility to annual soil losses by wind. Soil Sci Soc Am Proc 24:143–145

Chepil WS, Woodruff NP (1961) The physics of wind erosion and its control. Adv Agron 15:211–302

Chepil WS, Siddoway FH, Armbrust DV (1962) Climatic factor for estimating wind erodibility of farm fields. J Soil Water Conserv 17:162–165

Clark R, MacEwan R, Robinson N, Jonathon Hopley J (2010) Remote sensing of land cover and land management practices affecting wind erosion risk in NW Victoria, Australia. http://www.iuss.org/19th%20WCSS/Symposium/pdf/0575.pdf. Accessed 25 Jan 2015

Cornelis W, Gabriels D, Hartmann R (2004) A parametrisation for the threshold shear velocity to initiate deflation of dry and wet sediment. Geomorphology 59:43–51

Davis JJ (1963) Caesium and its relationship to potassium in ecology. In: Schultz V, Klement AW Jr (eds) Radioecology: proceedings of the first national symposium on radioecology. Reinhold Pub. Corp, New York, pp 539–556

del Valle HF, Blanco PD, Metternicht GI, Zinck JA (2009) Radar remote sensing of wind-driven land degradation processes in northeastern Patagonia. J Environ Qual 39(1):62–75. doi:10.2134/jeq2009.0071

DIN 19706 (2004) Soil quality—determination of the soil exposure risk from wind erosion. Deutsches Institut Für Normung e.V. (German National Standard), 05/01/2004, Beuth, Verlag

ECMWF (2005) European Centre for Medium-Term Weather Forecasts, ERA 40 Re-analysis data set. The climate data guide: ERA40. Last modified 20 Nov 2014. https://climatedataguide.ucar.edu/climate-data/era40. Accessed 25 Jan 2015

Etyemezian VR (2015) Vertical profiles, size distributions, and intermittency in aeolian sediment transport: field measurements with a novel instrument (Sand Sensor I). Research project. Desert Research Institute (DRI). http://www.dri.edu/das-research/3811-vertical-profiles-size-distributions-and-intermittency-in-aeolian-sediment-transport–field-measurements-with-a-novel-instrument-sand-sensor-i. Accessed 25 Jan 2015

European Commission (2006) Thematic strategy for soil protection. COM (2006), 231. http://eur-lex.europa.eu/LexUriServ/LexUriServ.do?uri=COM:2006:0231:FIN:EN:PDF. Accessed 25 Jan 2015

ESDB (2013) European soil database. European Commission—Joint Research Centre. http://eusoils.jrc.ec.europa.eu/ESDB_Archive/ESDB/Index.htm

FAO (2009) Global agriculture towards 2050. High level expert forum "How to feed the world in 2050, Office of the Director, Agricultural Development Economics Division Economic and Social Development Department, FAO, Rome. http://www.fao.org/wsfs/forum2050/wsfs-background-documents/hlef-issues-briefs/en/. Accessed 25 Jan 2015

Fister W, Iserloh T, Ries JB, Schmidt R-G (2012) A portable wind and rainfall simulator for in-situ soil erosion measurements. Catena 91(2012):72–84

Fryrear DW (1986) A field dust sampler. J Soil Water Conserv 41(2):117–120

Fryrear DW, Sutherland PL, Davis G, Hardee G, Dollar M (1999) Wind erosion estimates with RWEQ and WEQ. In: Stott DE, Mohtar RH, Steinhardt GC (eds) Sustaining the global farm. Selected papers from the 10th international soil conservation organization meeting held 24–29 May 1999 at Purdue University and the USDA-ARS National Soil Erosion Research Laboratory, pp 760–765

Funk R (1995). Quantifizierung der Winderosion auf einem Sandstandort Brandenburgs unter besonderer Berücksichtigung der Vegetationswirkung. ZALF-Bericht Nr. 16, Müncheberg

Funk R, Engel W (2015) Investigations with a field wind tunnel to estimate the wind erosion risk of row crops. Soil Tillage Res 145:224–232. doi:10.1016/j.still.2014.09.005

Funk R, Frielinghaus M (1997) Berechnung potentieller Bodenabträge durch Wind für ausgewählte Böden Mecklenburg-Vorpommerns. Mitt Dt Bodenkundl Ges 83:107–110

Funk R, Skidmore EL, Hagen LJ (2004a) Comparison of wind erosion measurements in Germany with simulated soil losses by WEPS. Environ Model Softw 19(2):177–183

Funk R, Deumlich D, Völker L, Steidl J (2004b) GIS application to estimate the wind erosion risk in the Federal State of Brandenburg. In: Goossens D, Riksen M (eds) Wind erosion and dust dynamics: observations, simulations, modelling. ESW Publications, Wageningen, pp 139–149

Funk R, Li Y, Hoffmann C, Reiche M, Zhang Z, Li J, Sommer M (2012) Using 137Cs to estimate wind erosion and dust deposition on grassland in Inner Mongolia—selection of a reference site and description of the temporal variability. Plant Soil 351:293–307. doi:10.1007/s11104-011-0964-y

Funk R, Hoffmann C, Reiche M (2014) Methods for quantifying wind erosion in steppe regions. In: Novel measurement and assessment tools for monitoring and management of land and water resources in agricultural landscapes of central Asia. Environmental science and engineering 2014, pp 315–327. http://link.springer.com/chapter/10.1007/978-3-319-01017-5_18. Accessed 25 Jan 2015

Goossens D, Offer ZY (2000) Wind tunnel and field calibration of six aeolian dust samplers. Atm Environ 34:1043–1057

Goossens D, Offer Z, London G (2000) Wind tunnel and field calibration of five aeolian sand traps. Geomorphology 35:233–252

Goossens D, Riksen M (2004) Wind erosion and dust dynamics at the commencement of the 21st century. In: Goossens D, Riksen M (eds) Wind erosion and dust dynamics, simulations, modelling. ESW Publications, Wageningen, pp 7–13

Greeley R, Iversen JD (1985) Wind as a geological process on Earth, Mars, Venus and Titan. Cambridge Planetary Science Series 4. Cambridge University Press, Oxford, p 333

Guo Z, Huang N, Dong Z, Van Pelt RS, Zobeck TM (2014) Wind erosion induced soil degradation in northern China: status. Measures and perspective. Sustainability 6(12):8951–8966. doi:10.3390/su6128951

Hagen LJ (1991) A wind erosion prediction system to meet user needs. J Soil Water Cons 46 (2):106–111

Hagen LJ (2001) Assessment of wind erosion parameters using wind tunnels. In: Scott DE, Mohtar RH, Steinhardt GC (eds) Sustaining the global farm. Selected papers from the 10th international soil conservation organization meeting held 24–29 May 1999 at Purdue University and the USDA-ARS National Soil Erosion Research Laboratory

Hagen LJ (1996) Crop residue effects on aerodynamic processes and wind erosion. Theor Appl Climatol 54(1996):39–46

Helming K, Funk R (2001) Eine Methode zur kontinuierlichen Bestimmung der Rauhigkeit von Ackeroberflächen. Mitt Dt Bodenkd Ges 96(2):611–612

Helming K (2014) Impact assessment for multifunctional land use. In: Novel measurement and assessment tools for monitoring and management of land and water resources in agricultural landscapes of central Asia. Springer, Environmental Science and Engineering 2014, pp 223–234. http://link.springer.com/chapter/10.1007/978-3-319-01017-5_12. Accessed 25 Jan 2015

Hoffmann C, Funk R, Wieland R, Li Y, Sommer M (2008) Effects of grazing and topography on dust flux and deposition in the Xilingele grassland, Inner Mongolia. J Arid Environ 72:792–807

Hoffmann C, Funk R, Reiche M, Li Y (2011) Assessment of extreme wind erosion in Inner Mongolia. Aeolian Res, China, pp 343–351

Hong S-W, Lee I-B, Seo I-H, Kwon K-S, Kim T-W, Son Y-H, Kim M (2014) Measurement and prediction of soil erosion in dry field using portable wind erosion tunnel. Biosyst Eng 118:68–82. doi:10.1016/j.biosystemseng.2013.11.003

Janssen W, Tetzlaff G (1991) Entwicklung und Eichung einer registrierenden Suspensionsfalle. Z. f. Kulturtechn. Landentwickl. 32:167–180

Jester W, Klik A (2005) Soil surface roughness measurements—methods, applicability, and surface representation. Catena 64:174–192

Jönsson P (1992) Wind erosion on sugar beet fields in Scania, southern Sweden. Agric Forest Meteorol 62:141–157

Karanasiou A, Moreno N, Moreno T, Viana M, de Leeuw F, Querol X (2012) Health effects from Sahara dust episodes in Europe: literature review and research gaps. Environ Int 15(47):107–114. doi:10.1016/j.envint.2012.06.012

Kuntze H, Beinhauer R, Tetzlaff G (1990) Quantification of soil erosion by wind, I. Final Report of the BMFT project. Project No. 0339058 A, B, C. Institute of Meteorology and Climatology, University of Hannover, Germany

Kuznetsov MS, Glazunov GP (1996) Erosion and soil conservation (Эрозия и охрана почв). Publisher Moscow State University (MGU) 1969, 335 p. ISBN: 5211033817 (in Russian)

Lal R (2003) Soil erosion and the global carbon budget. Environ Int 29:437–450

Larionov GA (1993) Erosion and deflation of soils (Эрозия и дефляция почв: основные закономерности и количественные оценки). Moscow University Press, 200 p. ISBN 5-211-02467-2 (in Russian)

Li J, Okin GS, Tatarko J, Webb NP, Herrick JE (2014) Consistency of wind erosion assessments across land use and land cover types: A critical analysis. Aeolian Research 15:253–260

Lu H, Shao Y (2001) Toward quantitative prediction of dust storms: an integrated wind erosion modelling system and its applications. Environ Model Softw 16:233–249

Lyles L (1975) Possible effects of wind erosion on soil productivity. J Soil Water Conserv 30 (6):279–283

Mabit L, Benmansour M, Walling DE (2008) Comparative advantages and limitations of the fallout radionuclides ^{137}Cs, ^{210}Pbex and ^{7}Be for assessing soil erosion and sedimentation. J. Env. Rad. 99:1799–1807

Martínez-Garzia A, Rosell-Mele A, Jaccard SL, Geibert W, Sigman DM, Haug GH (2011) Southern ocean dust-climate coupling over the past four million years. Nature 476:312–316

McTainsh G, Leys J, Bastin G, Tews K, Strong C, McGowan H (2009) Wind erosion risk management for more environmentally sustainable primary production, Research Report 24, Desert Knowledge Cooperative Research Centre, Alice Springs. http://www.nintione.com.au/resource/DKCRC-Report-24-Wind-erosion-risk-management.pdf. Accessed on 25 Jan 2015

Meinel T, Grunwald L-C, Akshalov K (2014) Modern technologies for soil management and conservation in Northern Kazakhstan. In: Novel measurement and assessment tools for monitoring and management of land and water resources in agricultural landscapes of Central Asia. Springer, Environmental Science and Engineering 2014, pp 455–464, http://link.springer.com/chapter/10.1007/978-3-319-01017-5_27. Accessed on 25 Jan 2015

Mendez M, Funk R, Buschiazzo DE (2011) Field wind erosion measurements with Big Spring Number Eight (BSNE) and Modified Wilson and Cook (MWAC) samplers. Geomorphology 129:43–48

Mirschel W, Wenkel K-O, Berg M, Wieland R, Nendel C, Köstner B, Topazh AG, Terleev VV, Badenko VL (2016) A spatial model-based decision support system for evaluating agricultural

landscapes under the aspect of climate change. In: L. Mueller et al. (eds) Novel methods for monitoring and managing land and water resources in Siberia. Springer, Cham, pp 519–540 (Chapter 23 of this book)
Montgomery DR (2007) Dirt: the erosion of civilizations. University of California Press, California 285 p
Mueller L, Schindler U, Shepherd TG, Ball BC, Smolentseva E, Pachikin K, Hu C, Hennings V, Sheudshen AK, Behrendt A, Eulenstein F, Dannowski R (2014) The Muencheberg Soil Quality Rating for assessing the quality of global farmland. In: Novel measurement and assessment tools for monitoring and management of land and water resources in agricultural landscapes of Central Asia. Springer, Environmental Science and Engineering 2014, pp 235–248. http://link.springer.com/chapter/10.1007/978-3-319-01017-5_13. Accessed on 25 Jan 2015
Mueller L, Schindler U, Hennings V, Smolentseva EN, Rukhovich OV, Romanenkov VA, Sychev VG, Lukin S, Sheudshen AK, Onishenko L, Saparov A, Pachikin K, Behrendt A, Mirschel W, Eulenstein F (2016) An emerging method of rating global soil quality and productivity potentials. In: L. Mueller et al. (eds) Novel methods for monitoring and managing land and water resources in Siberia. Springer, Cham, pp 573–595 (Chapter 26 of this book)
Nordstrom KF, Hotta S (2004) Wind erosion from cropland in the USA: a review of problems, solutions and prospects. Geoderma 121(2004):157–167
Perez L, Tobias A, Querol X, Künzli N, Pey J, Alastuey A, Viana M, Valero N, González-Cabré M, Sunyer J (2008) Coarse particles from Saharan dust and daily mortality. Epi-demiology 19 (6):800–807. doi:10.1097/EDE.0b013e31818131cf
Ravi S, D'Odorico P (2005) A field-scale analysis of the dependence of wind erosion threshold velocity on air humidity. Geophys Res Lett 32:L21404
Reiche M, Funk R, Zhang Z, Hoffmann C, Reiche J, Wehrhan M, Li Y, Sommer M (2012) Application of satellite remote sensing for mapping wind erosion risk and dust emission-deposition in Inner Mongolia grassland, China. Grassland Science 58:8–19
Riksen M, Brouwer F, Spaan W, Arrue JL, Lope MV (2003) What to do about wind erosion. In: Warren A (ed) Wind erosion on agricultural land in Europe. European Commission, EUR 20370, pp 39–54
Ries JB, Marzen M, Iserloh T, Fister W (2014) Soil erosion in Mediterranean landscapes—experimental investigation on crusted surfaces by means of the portable wind and rainfall simulator. J Arid Environ 100–101:42–51
Ritchie JC, Ritchie CA (2007) Bibliography of publications of ^{137}Cs studies related to erosion and sediment deposition. http://www.ars.usda.gov/Main/docs.htm?docid=15237. Accessed on 25 Jan 2015
Romanenkov V, Rukhovich D, Koroleva P, McCarty JL (2014) Estimating black carbon emissions from agricultural burning. In: Novel measurement and assessment tools for monitoring and management of land and water resources in agricultural landscapes of Central Asia. Springer, Environmental Science and Engineering, pp 347–364
Saha SK (2004) Water and wind induced soil erosion assessment and monitoring using remote sensing and GIS. Satellite Remote Sensing and GIS Applications in Agricultural Meteorology, pp 315–330
Schroeder WR, Kort J (1989) Shelterbelts in the Soviet Union. J Soil Water Cons 44(2):130–134
Shao Y, Wyrwoll KH, Chappell A, Huang J, Lin Z, McTainsh GH, Masao M, Tanaka TY, Wang X, Yoon S (2011) Dust cycle: an emerging core in earth system science. Aeolian Res 2:181–204
Sharratt B, Feng G, Wendling L (2007) Loss of soil and PM10 from agricultural fields associated with high winds on the Columbia Plateau. Earth Surf Proc Land 32:621–630
Singh P, Sharratt B, Schillinger WF (2012) Wind erosion and PM10 emission affected by tillage systems in the world's driest rainfed wheat region. Soil Tillage Res 124:219–225. http://dx.doi.org/10.1016/j.still.2012.06.009. Accessed on 25 Jan 2015
Spaan WP, van den Abeele GD (1991) Wind borne particle measurements with acoustic sensors. Soil Technol 4:51–63
Skidmore EL (1986) Wind erosion climatic erosivity. Clim Change 9:195–208

Sterk G, Stein A (1997) Mapping wind-blown mass transport by modelling variability in space an time. Soil Sci Soc Am J 61:232–239
Sterk G, Parigiani J, Cittadini E, Peters P, Scholberg J, Peri P (2012) Aeolian sediment mass fluxes on a sandy soil in Central Patagonia. Catena 95:112–123
Stocking M (2003) Erosion and crop yield. Encyclopedia of soil science. Marcel Dekker Inc., New York, pp 1–4. doi:10.1081/E-ESS120006664
Tamura T (1964) Consequences of activity release: selective sorption reactions of caesium with soil minerals. Nucl Saf 5:262–268
UGT (2014) Umwelt-Geräte-Technik GmbH. Wind erosion sediment trap. http://www.ugt-online.de/en/produkte/bodenkunde/erosions-messtechnik/winderosions-sedimentfalle.html. Accessed on 25 Jan 2015
USDA (1995) The wind erosion prediction system. Homepage http://www.weru.ksu.edu/weps/wepshome.html. Accessed on 25 Jan 2015
USDA (2002) Revised wind erosion equation. http://www.csrl.ars.usda.gov/wewc/rweq/readme.htm. Accessed on 25 Jan 2015
USDA (2008) NRCS wind erosion equation (WEQ) site. http://www.weru.ksu.edu/nrcs/weq.html. Accessed on 25 Jan 2015
USDA-ARS (2008) Single-event wind erosion evaluation program SWEEP. User manual draft February 2008, USDA-ARS Wind Erosion Research Unit Manhattan, Kansas, USA. http://www.weru.ksu.edu/new_weru/ftp_site/wagner/IECA_10/SWEEPUserGuide.pdf. Accessed on 25 Jan 2015
Van Lynden GWJ (1995) European soil resources. Current status of soil degradation, causes impacts and need for action. Nat Environ, No. 71, Council of Europe
Van Pelt RS (2013) Use of anthropogenic radioisotopes to estimate rates of soil redistribution by wind I: historic use of 137Cs. Aeolian Res 9:89–102
Van Pelt RS, Ketterer ME (2013) Use of anthropogenic radioisotopes to estimate rates of soil redistribution by wind II: the potential for future use of $^{239+240}$Pu. Aeolian Res 9:103–110
Van Pelt RS, Zobeck TM (2004) Validation of the wind erosion equation (WEQ) for discrete periods. Env Model Softw 19:199–203
Van Pelt RS, Zobeck TM, Potter KN, Stout JE, Popham TW (2004) Validation of the wind erosion stochastic simulator (WESS) and the revised wind erosion equation (RWEQ) for single events. Env Model Softw 19:191–198
Veen PH, Kampf H, Liro A (1997) Nature development on former state farms in Poland. Report within the framework of the memorandum of understanding for nature conservation between the Polish Ministry of Environmental Protection, Natural Resources and Forestry and the Dutch Ministry of Agriculture, Nature Management and Fisheries, 48 pp
Visser SM, Sterk G, Snepvangers J (2004) Spatial variation in wind-blown sediment transport in geomorphic units in northern Burkina Faso using geostatistical mapping. Geoderma 120:95–107
Wagenbrenner NS, Germino MJ, Lamb BK, Robichaud PR, Foltz RB (2013) Wind erosion from a sagebrush steppe burned by wildfire: measurements of PM10 and total horizontal sediment flux. Aeolian Res 10:25–36
Wagner LE (2013) A history of wind erosion prediction models in the United States Department of Agriculture: the wind erosion prediction system (WEPS). Aeolian Res 10:9–24
Webb NP, McGowan HA (2009) Approaches to modelling land erodibility by wind. Prog Phys Geogr 2009:1–27
Webb NP, McGowan HA, Phinn SR, Leys JF, Grant H, McTainsh GH (2009) A model to predict land susceptibility to wind erosion in western Queensland, Australia. Environ Model Softw 24 (2):214–227
Woodruff NP, Siddoway FH (1965) A wind erosion equation. Soil Sci Soc Am Proc 29:602–608
Yang X, Leys J (2014) Mapping wind erosion hazard in Australia using MODIS derived ground cover, soil moisture and climate data. 35th international symposium on remote sensing of environment (ISRSE35). IOP Publishing, IOP conference series: earth and environmental science, vol 17, p 012275. doi:10.1088/1755-1315/17/1/012275. http://iopscience.iop.org/1755-1315/17/1/012275/pdf/1755-1315_17_1_012275.pdf. Accessed on 25 Jan 2015

Youssef F, Visser S, Karssenberg D, Bruggeman A, Erpul G (2012) Calibration of RWEQ in a patchy landscape; a first step towards a regional scale wind erosion model. Aeolian Res 3 (4):467–476

Zapata F (2003) The use of environmental radionuclides as tracers in soil erosion and sedimentation investigations: recent advances and future developments. Soil Till Res 69:3–13

Zapata F, Garcia-Agundo E, Ritchie JC, Appleby PG (2002) Introduction. In: Zapata F (ed) Handbook for the assessment of soil erosion and sedimentation using environmental radionuclides, IAEA, pp 1–13

Zhang Z, Wieland R, Reiche M, Funk R, Hoffmann C, Li Y, Sommer M (2012) Identifying sensitive areas to wind erosion in the Xilingele grassland by computational fluid dynamics modelling. Ecol Inform 8(2012):37–47

Zobeck TM, Sterk G, Funk R, Rajot JL, Stout JE, Van Pelt RS (2003a) Measurement and data analysis methods for field-scale wind erosion studies and model validation. Earth Surf Process Land 28:1163–1188

Zobeck TM, Popham TW, Skidmore EL, Lamb JA, Merrill SD, Lindstrom MJ, Mokma DL, Yoder RE (2003b) Aggregate-mean diameter and wind-erodible soil predictions using dry aggregate-size distributions. Soil Sci Soc Am J 67:425–436

Zou XY, Zhang CL, Cheng H, Kang LQ, Wu YQ (2015) Cogitation on developing a dynamic model of soil wind erosion, Science China Earth Sciences. http://link.springer.com/article/10.1007%2Fs11430-014-5002-5. Accessed on 25 Jan 2015

Chapter 19
Multi-Scale Vegetation and Water Body Mapping of the Northern Latitudes in Siberia with Optical Remote Sensing

Marcel Urban, Michael Voltersen, Stefan Poecking, Soeren Hese, Martin Herold and Christiane Schmullius

Abstract The Arctic ecosystem is highly vulnerable to modification within the climate system. Increasing greenhouse gas emissions, e.g. from melting permafrost soils, are assumed to result in positive feedback mechanisms within the global climate system. Water bodies and the seasonal dependent freeze/thaw dynamics of the uppermost permafrost layer (active layer) are the major carbon and methane source in the Arctic regions. Impacts induced by climate change are resulting in the transformation of the existing Arctic landforms, such as the destruction of settlements or changes in the terrain, hydrology and vegetation cover. This paper presents the potential of using Earth observation data from various sources and time steps to monitor land cover characteristics and changes in the Arctic regions. Information on vegetation structure types and physiognomy is commonly incorporated into spatial models predicting the permafrost distribution. The MODIS land cover, the GlobCover land cover map, SYNMAP and MODIS VCF (vegetation continuous field) have been combined in a product describing the fractional vegetation cover. The dataset, with a spatial resolution of 1 km, consists of four layers providing percentage cover information for trees, shrubs, herbaceous areas and barren areas. Additional information, such as the CAVM (Circumpolar Arctic

M. Urban (✉) · M. Voltersen · S. Poecking · S. Hese · C. Schmullius
Department for Earth Observation, Institute of Geography, Friedrich-Schiller-University, Löbdergraben 32, 07737 Jena, Germany
e-mail: Marcel.Urban@uni-jena.de

M. Voltersen
e-mail: michael.voltersen@uni-jena.de

S. Poecking
e-mail: stefan.poecking@gmail.com

S. Hese
e-mail: soeren.hese@uni-jena.de

C. Schmullius
e-mail: c.schmullius@uni-jena.de

M. Herold
Centre for Geo-Information, Wageningen University, 6700AA Wageningen, The Netherlands
e-mail: martin.herold@wur.nl

L. Mueller et al. (eds.), *Novel Methods for Monitoring and Managing Land and Water Resources in Siberia*, Springer Water,
DOI 10.1007/978-3-319-24409-9_19

Vegetation Map), has been integrated into the harmonization approach. Local land cover and water body changes have been analyzed using high spatial resolution earth observation information from Landsat, RapidEye and Corona Keyhole. This analysis was carried out for a test site in central Yakutia and the Lena river delta system in Siberia, Russia. High-resolution land cover information was mapped using an object-oriented classification approach. Object characteristics, such as the shape, spectral properties and information within different hierarchical object levels, are utilized to identify individual vegetation class properties for assignment to a thematic class. Water body changes are identified using historical earth observation data from the 1960s and recent RapidEye data.

Keywords MODIS · Globcover · RapidEye · Corona keyhole · Land cover · Thermokarst · Permafrost · Vegetation

1 Introduction

The cold regions of the northern hemisphere were subject to substantial changes during the last century (Grace et al. 2002; Moritz et al. 2002; Nelson 2003). The Arctic ecosystem is highly vulnerable to modification within the climate system due to increasing greenhouse gas emissions (Chapin et al. 2005; Crowley 2000; Overpeck et al. 1997). This is also assumed to result in positive feedback mechanisms within the global climate system (Cuevaz-Gonzalez et al. 2009; Schuur et al. 2008). Arctic temperatures have never been as high as during the last decades (Moritz et al. 2002). The modification and degradation of permafrost soils have been identified for many regions on a pan-Arctic scale. The impacts induced by climate change are resulting in the transformation of existing landforms in Arctic regions, such as the destruction of settlements, and changes in terrain, hydrology and vegetation (French 2008; Haeberli et al. 2002; Nelson et al. 2002). The rise in temperatures, in particular, is causing an intensification of vegetation activity and changes in land surface structures and phenological dynamics, resulting in a decrease in the surface albedo, triggering the increasing absorption of solar radiation (Hinzman et al. 2005; Langer et al. 2013; Myneni et al. 1997).

The land surface structure in permafrost regions is formed and characterized by the uppermost soil layer, known as the active layer. The active layer dynamics are strongly connected to seasonal freeze/thaw dynamics, which have been subject to substantial changes during recent years (Anisimov and Reneva 2006). Several studies have emphasized the monitoring of vegetation in the Arctic regions (Laidler and Treitz 2003; Peddle et al. 1993; Stow et al. 2004). As thawing is the primary characteristic in a changing permafrost environment, thermokarst effects also need to be taken into account. Moreover, water bodies and seasonally dependent freeze/thaw dynamics are the major carbon and methane source in the Arctic regions (Christensen 2004; Frohn et al. 2005; Kozlenko and Jeffries 2000).

This paper presents the outcome of land cover and water body monitoring as part of the Data User Element (DUE) Permafrost project, funded by the European Space Agency (ESA). The main goal is to describe the methodologies of utilizing Earth observation data and products from various sources with different spatial and temporal resolutions. The DUE Permafrost project aims to define, demonstrate and validate permafrost information services to address climate modeling and climate change studies. Moreover, the project focuses on the connection between the permafrost user and the remote sensing community to support the integration of both instances for future climate research (Bartsch et al. 2012). The ESA User Requirement Documentation (URD) has defined the parameters of land cover and water bodies as the key observation variable for developing a modeling framework of the permafrost state. Additional, variables such as soil moisture, land surface temperature, snow cover, elevation and subsidence have also been addressed within the DUE Permafrost project (Bartsch et al. 2012).

Information about the state of the vegetation structure, water bodies and non-vegetated areas is required in order to characterize the permafrost distribution and dynamics in modeling approaches. Commonly, coarse-scale information data are used to model the permafrost distribution. The URD emphasizes the need for information on the percentage of vegetation cover and non-vegetated areas within a resolution of 1 km for integration into permafrost models. The coarse-resolution land cover approach described in this paper harmonizes different state-of-the-art land cover classification products (Sect. 3.1). The aim of this harmonization was to extract vegetation cover information based on the individual class description of each legend.

Additional outcomes of the URD have shown the need for local-scale information about the vegetation types and water body changes. Furthermore, this paper presents a land cover classification of the Lena Delta (Sect. 3.2) and water body change detection using historical and current high-resolution optical remote sensing data for the area around Yakutsk (Sect. 3.3). Local-scale approaches allow fine-scale changes to be detected in the characteristics of thermokarst lakes as well as in vegetation structure and distribution.

2 Materials and Methods

2.1 Coarse Resolution Land Cover Assessment—Harmonization

Land cover is one of the essential climate variables. Remote sensing information and techniques from various sources have shown high potential in monitoring the state of the land surface in Arctic regions (Bartsch et al. 2012; Urban et al. 2010).

Requirements from the modelling and data user community, defined within the DUE Permafrost project, have shown the need for percentage cover information for different vegetation types, such as trees, shrubs and herbaceous areas, as well as non-vegetated areas. This information is mandatory for the different modelling

Table 1 Overview of global land cover products

Product	Input data	Temporal resolution	Spatial resolution	Time series	Reference
MODIS Land Cover (MCD12Q1)	MODIS (Aqua and Terra)	Yearly	500 m	2001–2010 (Version 5)	(Friedl et al. 2002, 2010)
GlobCover	MERIS (ENVISAT)	–	300 m	2005/2006	(Arino et al. 2007)
Synmap	AVHRR MODIS SPOT	–	1 km	1990/2000	(Jung et al. 2006)
MODIS VCF (MOD44B) Version 3/4	MODIS (Aqua and Terra)	Yearly	500 m	2001	(Defries et al. 2000; Hansen et al. 2002)

approaches, as land cover is an important indicator and parameter for predicting permafrost distributions (Gruber and Hoelzle 2001). The aim of this paper is to use existing global land cover products, which provide the required information, to harmonize and regionalize patterns and processes by integrating earth observation information from multiple sources. The major goal is to extract cover percentage information from the legends of different land cover products (Table 1) such as the LCCS Code (Land Cover Classification System) (Gregorio and Jansen 2005).

The harmonization was carried out utilizing MODIS Land Cover (Moderate Resolution Imaging Spectroradiometer), GlobCover, SYNMAP and MODIS VCF (vegetation continuous fields collection).

The MODIS Land Cover product (MCD12Q1) has a spatial resolution of 500 m. The 17 land cover classes, which are defined and described by the IGBP legend (International Geosphere-Biosphere Programme), are derived from multi-temporal satellite data using the MLCCA (MODIS Land Cover Classification Algorithm). This algorithm is based on a supervised classification approach utilizing existing land cover information from a training database. The MLCCA also integrated different parameters, such as BRDF, texture, land surface temperature, etc., into the classification methodology (Friedl et al. 2002, 2010).

The GlobCover product is derived from multi-temporal MERIS (Medium Resolution Imaging Spectrometer) satellite imagery. MERIS has a spatial resolution of 300 m and is onboard ENVISAT (Environmental Satellite). Satellite data with an acquisition time window between 2005 and 2006 were used for the GlobCover classification. The algorithm is based on supervised and unsupervised classification approaches, developed separately for different "equal reasoning areas". The GlobCover legend is based on the LCCS, which was developed by (Gregorio and Jansen 2005) and covers 22 land cover classes (Arino et al. 2007).

Compared to the other land cover products used in this paper, SYNMAP has the highest number of land cover classes (48 classes). This product presents a best-estimate land cover approach, by a synergetic combination of existing global products, such as GLCC (Global Land Cover Characterization), GLC2000 (Global Land Cover 2000) and MODIS (Jung et al. 2006).

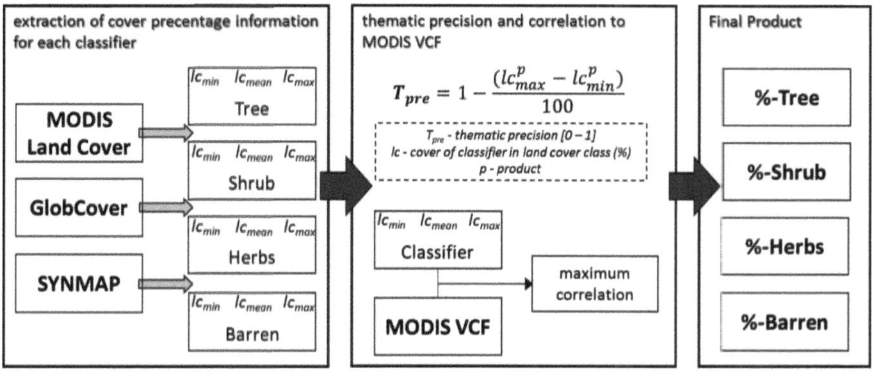

Fig. 1 Extraction and harmonization of percentage land cover information

Information on the percentage cover of trees and herbaceous and barren areas are provided by MODIS VCF. Product Version 5, which was used for this study, has a spatial resolution of 500 m. The product development is based on a regression tree approach using MODIS reflectance data and integrating training datasets and NDVI (Normalized Differenced Vegetation Index) information. A detailed description of the methodology and VCF processing chain can be found in (Defries et al. 2000; Hansen et al. 2002).

The combination of land cover products from different sources is a challenging issue. In general, harmonization is commonly used to emphasize similarities between existing datasets so that inconsistencies are reduced (Herold et al. 2006). In a pre-processing step, all land cover products were rescaled to a common spatial resolution of 1 km. Based on the legend description of each land cover product, information was extracted on the percentage cover for each classifier (trees, shrubs, herbaceous areas and barren areas) (Fig. 1).

The resulting minimum (lc_{min}) and maximum (lc_{max}) information is used to calculate the mean cover percentage (lc_{mean}) in a further step of the methodology. The thematic precision (T_{pre}), which is a measure of the range between lc_{min} and lc_{max}, was calculated for each classifier of each land cover class. For example, MODIS (IGBP—International Geosphere-Biosphere Programme) defines a tree cover threshold from 60 to 100 %. This results in a factor of thematic precision (T_{pre}) of 0.6, which is higher than the definition for forest classes in LCCS with a tree canopy >15 % (T_{pre} = 0.15). The thematic precision is used as weighting factor within the harmonization of different estimates from different products. We assumed that the narrower the class description, the higher the thematic precision of the land cover class which is integrated by the value of T_{pre} during the product combination.

After the conversion of land cover products to percentage cover information, the datasets are linked to MODIS VCF. The classifier (lc_{min}, lc_{max} or lc_{mean}) which shows the highest correlation with MODIS VCF is used for the harmonization process. Additionally, we extract information on shrub cover using the percentage relationship between the classifiers. The final synergy product has a spatial

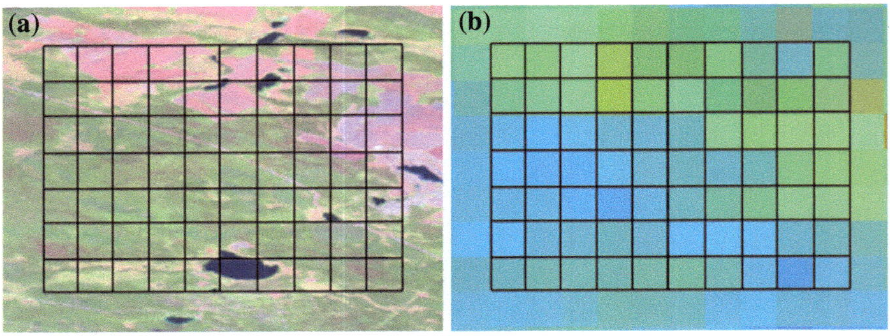

Fig. 2 Zonal statistics analysis for the validation approach using a 1 km regular grid based on the fractional vegetation cover remote sensing product. **a** Landsat 7—ETM + (12/08/2002) 30 m spatial resolution RGB = channel 5/4/3. **b**Fractional vegetation cover remote sensing product—1 km spatial resolution RGB = % herb /% shrub /% trees

resolution of 1 km and consists of four components providing information on the percentage cover of each classifier. By summarizing all four layers, each pixel ends up with a value of 100 %.

The first version of the fractional vegetation cover remote sensing product was improved using the Circumpolar Arctic Vegetation Map created by Walker et al. (2005). Due to the misclassification of the forest classes in the land cover products, we excluded cover percentage misclassification from trees in the Arctic tundra regions to improve the representation of cover information within the taiga tundra transition zone. This is necessary as the modeling group requests an exact definition of land cover types distinguishing between forest and non-forest information in the high-latitude regions.

The accuracy assessment was carried out for three different zones covering the tundra, the taiga-tundra transition zone and the taiga areas using high-resolution Landsat data. The high-resolution satellite imagery was classified into the four classifiers of the fractional vegetation cover remote sensing product. The accuracy was assessed by calculating the cover percentage of each classifier for a regular grid of 1 km cell size for both the Landsat classification and the fractional vegetation cover remote sensing product (Fig. 2). The 1 km regular grid was used for the extraction of sampling points. A minimum number of 200 samples were selected for each Landsat scene.

2.2 High-Resolution Land Cover Mapping

The user requirements within the DUE Permafrost project have been indicating a need for spatial high-resolution land cover information for different regions, e.g. Central Yakutsk, the Lena river delta (Siberia) or the North Slope (Alaska) region. Specific land cover classes and legends are needed to address the characteristics at each site, as the regions are covered by different vegetation structures and types, such as mosses, grass and shrub-dominated areas, water bodies and boreal forest.

This paper presents an object-oriented land cover image classification approach using high-resolution RapidEye data for the Lena river delta system. Object features extracted from individual band values are used to create image objects based on information from spectral properties, e. g. the mean, standard deviation and maximum difference. Other supporting features, such as textures, variables, thematic attributes and the object hierarchy are applicable to improve the class description and structure the hierarchical classification process (Definiens 2007a, b). In contrast to the fractional vegetation cover remote sensing product, the high-resolution legend is based on various land cover types and differentiations between different types of vegetation.

The methodology for the development of a high-resolution land cover classification can be summarized into the following processing steps (Fig. 3). Different pre-processing steps, such as data importing, conversion from raw data format and atmospheric correction, are mandatory before the creation of layer stacks and geo-coding. The classification was carried out using an unsupervised approach integrating ground data and auxiliary information, such as maps. A validation was carried out to assess the accuracy of the land cover classification.

The process chain for the extraction of land cover information from RapidEye data is based on (1) pre-processing, (2) segmentation, (3) object-oriented classification and (4) validation. In this study, the methodology was developed for the land cover classification of the Lena Delta. However, the individual processing steps are transferable to other regions, with the use of site-specific ground measurements and auxiliary data as well as the adaption of some parameters within the segmentation procedure.

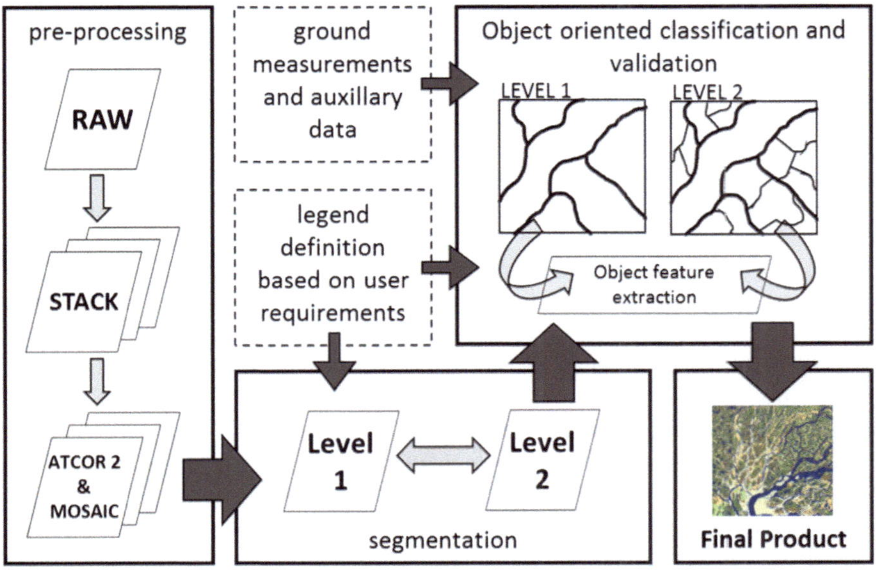

Fig. 3 Object-based land cover image classification methodology

1. An essential processing step for the analysis of Earth observation data is the conversion of raw satellite data to a common file format with geographic information. RapidEye data were provided by DLR/RESA (German Aerospace Center /RapidEye Science Archive) in NITF format (National Imagery Transmission Format) including RPCs (Rational Polynomial Coefficients). The conversion and georeferencing is carried out using the information from RPCs. The individual RapidEye bands, with 16-bit radiometric resolution, have a spatial resolution of 6.5 m and are co-registered to the coordinate system UTM (Universal Transverse Mercator) WGS84 (World Geodetic System 1984). After layer stacking, the data is corrected for atmospheric effects using ATCOR2 (Richter 1996). This processing step is followed by mosaicking, where the histogram of each RapidEye scene is matched using one master image.
2. The multispectral segmentation of the RapidEye data was carried out for 2 levels to integrate object characteristics from different object features and object hierarchies into the classification process. The coarse-scale segmentation level (Level 1) was created using a scaling factor of 100. Additional parameters, such as object shape (weighting between spectral information and shape) and compactness (object characteristics: smooth or compact), are integrated with the default settings (shape = 0.1; compactness = 0.5) (Definiens 2007b). The Level 1 product was used to identify large objects, such as water bodies and non-vegetated areas. The fine-scale segmentation level (Level 2) was created using the unclassified objects from Level 1. Fine-scale land cover classes, such as vegetation patches and small water bodies, were identified in the hierarchy. The segmentation calibration was strongly dependent on the land cover legend definition, which was based on the user requirements and defined prior to the processing steps.
3. The utilization of additional information sources, such as ground measurements, classifications from other Earth observation data, brightness layers (the sum of normalized reflectance from all spectral channels), NDVI layers (used to determine vegetation conditions) as well as the standard deviation of the brightness (to distinguish between spectral homogeneous and heterogeneous areas) were used within the class description of the land cover classification approach. Ground-based measurements of different vegetation classes were limited in spatial coverage. This information source as was not used as a training data set as the sample size is not adequate. However, additional expert knowledge was used as auxiliary information within the unsupervised classification approach. The derived characteristics, such as shape, spectral properties and information on hierarchical object levels were used to analyze vegetation class properties and assign each image object to a thematic class (Hájek 2005; Lewinski and Bochenek 2008; Lucas et al. 2007).
4. The accuracy the classification results was assessed using ground measurements and auxiliary data, which was provided by the user core group of the DUE Permafrost project. A stratified random sampling strategy was used to compare the classification results with the reference information (50 randomly distributed samples per class compared with an expert vegetation classification provided by AWI Potsdam).

2.3 High-Resolution Water Body Change Mapping

The natural ecosystem of Arctic permafrost regions underlies changes caused by climate-driven variations in freeze/thaw processes. In this context, changes in thermokarst water bodies are quantified for central Siberia, near the town of Yakutsk, using multi-temporal high-resolution satellite data. The water body changes were analyzed using an object-based classification approach and historical panchromatic Corona data as well as recent RapidEye imagery. Grosse et al. (2005) used a threshold approach to detect water bodies at the northeast Siberian Coast with Corona data. Cuevaz-Gonzalez et al. (2009) mapped thaw lakes and drained thaw lake basins at the North Slope Region of Alaska using an object-oriented concept.

The main goal in generating a water body change product is to develop fine-scale thermokarst lake change information, to provide detailed information for up-scaling analysis for regional and pan-boreal level and to improve the understanding of permafrost processes, such as permafrost degradation or possible related water body density changes. The methodology is based on (1) pre-processing, (2) co-registration and (3) segmentation and object-oriented classification (Fig. 4).

The historical KeyHole Corona data were acquired using analogue panchromatic Kodak Eastman film. The spectral information content is limited and not important as a primary information source. Using such data, relative grey level changes, feature shape and context information are relevant for analyzing and interpreting panchromatic airborne remote sensing data. The robustness and transferability of data analysis concepts should also be important aspects for the classification of water object structures in Corona data (Hese 2008). The raw Corona data are delivered through the USGS and available in strips of 13.8 km × 188 km ground coverage. The ground resolution is dependent on the KeyHole mission. The KH-4 Corona data which were used for this study have a spatial resolution of 1.8 m. After the data import, the Corona stripes were mosaicked (Richard and Jia 2006). The pre-processing of the RapidEye data is identical to the methodology described for

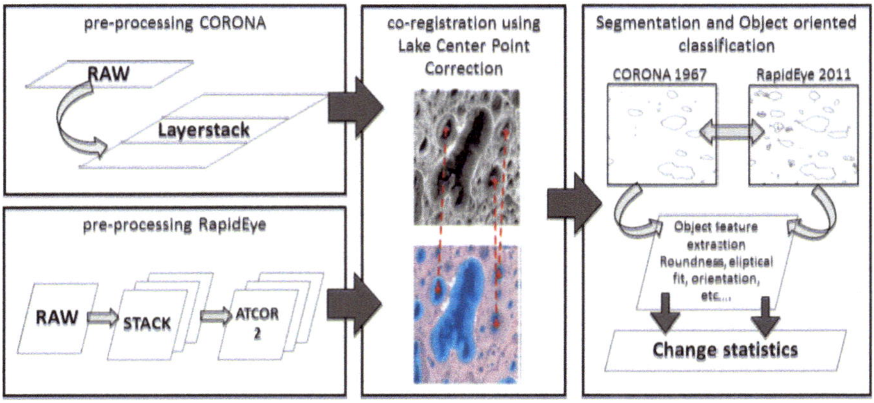

Fig. 4 Object-based lake change analysis using high-resolution historical and recent satellite data

the local land cover product (Sect. 2.2). The RapidEye data were used as a spatial reference dataset for co-registering and georeferencing the historical Corona data. In order to obtain spatially accurate image matching for the lake change analysis, the co-registration is carried out using the lake center point correction approach (LCPC) (Hese 2008; Sheng et al. 2008). However, trends due to shifts in lake objects cannot be measured and detected with this preprocessing method, as lake center points are used as features for relative co-registration. However, important, high-precision water area change statistics can be derived with this method.

The detection of water bodies and the analysis of lake changes were carried out using an object-oriented data classification approach. The resolution of the data is resampled to 5 m. Lake area changes are extracted, as are various structural lake object features (border index, elliptical fit, length to width ratio of objects, lake direction and roundness).

Transferring the presented method to other regions has shown the occurrence of problems in detecting the exact border of lakes from the panchromatic Corona data. The integration of Landsat MSS (79 m spatial resolution) from the 1970s might be feasible by using the multi-spectral information to increase ability to detect lakes. However, the different resolutions of Corona and Landsat MSS as well as the different acquisition times limit this kind of data fusion approach.

3 Results and Discussion

3.1 *Coarse-Resolution Land Cover Assessment—Harmonization*

The fractional vegetation cover remote sensing product presents four LCCS classifiers providing information on the percentage cover for trees, shrubs (including low to tall-shrubs), herbaceous tundra (including the erect-dwarf and prostrate-shrub tundra) and barren and graminoid tundra (including surface water) in each pixel (Fig. 5). Each pixel equals 100 % when all components are added together. The product is feasible for use as a boundary condition when modelling permafrost distributions integrating different parameters such as the land surface temperature, soil moisture and disturbances (e.g. fire). The dataset can be downloaded via the PANGAEA platform (Data Publisher for Earth & Environmental Science) (Urban et al. 2012).

One of the major advantages of the fractional vegetation cover remote sensing product in comparison to other land cover data is the possibility to rescale the product to the required spatial resolution without losing any information. A rescaling of thematic land cover information from global land cover datasets is a challenging issue due to the regionalisation of different land cover classes. Moreover, the product can be converted to thematic land cover classes by using the LCCS to generate an adapted land cover legend. This is of great importance for regional and local land cover monitoring and analysis as well as for scaling

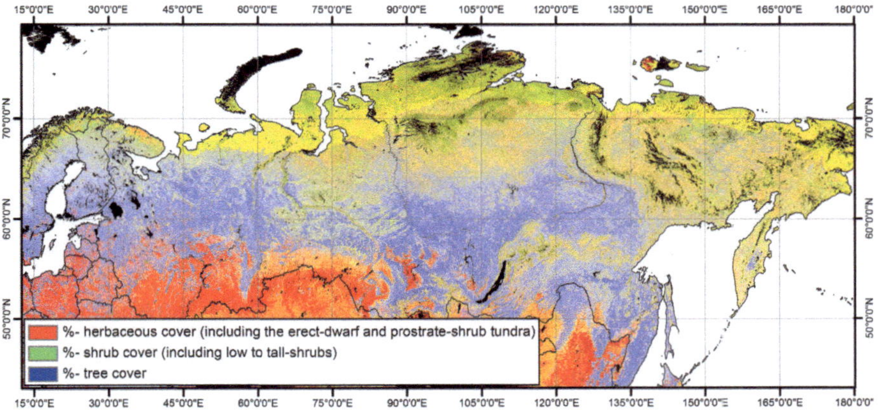

Fig. 5 ESA DUE fractional vegetation cover remote sensing product. Version 2

Table 2 Results of the accuracy assessment for the fractional vegetation cover remote sensing product (overall accuracy)

Validation zone	Description	Trees	Shrubs	Herbs	Barren
Zone 1	Arctic coastline (Tundra)		86.5–91.5 %	65.5–78 %	74–85.4 %
Zone 2	Taiga-Tundra transition zone	45 %	55 %	67.5 %	88 %
Zone 3	Boreal Region (Taiga)	83–97 %			

approaches. Moreover, the use of the LCCS improves the comparability between land cover products and legends from various sources.

The fractional vegetation cover remote sensing product was validated for (1) the tundra region, (2) the taiga-tundra boundary and (3) the taiga region. A summary of the resulting agreement with the Landsat reference is given in Table 2.

1. As the tundra shows no signs of forest cover, the validation was carried out for the parameters of shrubs, herbaceous areas and barren areas. Different case regions were defined in the pan-Arctic region. The accuracy was assessed for the North Slope region in Alaska as well as for Taymir and Chokurdakh in Russia. Considerable agreement with the Landsat reference was found for the North Slope area. The overall accuracy for this site was 91 % for shrubs, 78 % for herbaceous areas and 85 % for barren areas. The test site Chokurdakh results in accuracies of 87 % for shrubs, 66 % for herbaceous areas and 74 % for barren areas. The validation of the test area at the Taymir peninsula has shown an overall accuracy of 90 % for shrubs, 72 % for herbaceous areas and 80 % for bare soil.
2. The validation of the taiga-tundra transition zone is carried out for three selected test sites in Canada, the northernmost part of Yakutia in Russia, and Alaska. The agreement between the fractional vegetation cover remote sensing product and

the Landsat reference ranges between 45 and 89 % for the tree line region. The classifier which represents tree cover shows the lowest agreement, whereas barren areas result in the highest correlation. Moderate agreement can be found for the herbaceous layer. The cover information for shrubs has a high correlation for all test regions, ranging between 87 and 92 %. As this validation site represents heterogenic landscapes, low accuracies are expected for the individual classifiers. These transition zones represent a mixture of different vegetation classes and are frequently classified as mosaic classes in global land cover products. Bicheron et al. (2008) pointed out that "the Globcover land cover map contains a significant amount of mosaic classes, which may limit the thematic sharpness of the Globcover product and its relevancy to derive very specific products" (Bicheron et al. 2008).
These challenging issues cause low accuracies in these regions. Moreover, a validation of a global land cover product is always a challenging issue, which is influenced by mixed pixels and the spatial resolution and precise geo-location of the reference data (Herold et al. 2006).
3. The accuracy assessment of the boreal forest region was carried out for two test sites in Canada and two test sites in Russia. The comparison between the fractional vegetation cover remote sensing product and the Landsat reference was carried out only for the tree cover classifier. The highest correlations were found for the Russian test site (87–97 %). The test sites in British Columbia and Quebec showed an agreement between the reference and the fractional vegetation cover remote sensing product of 83–86 %.

The agreement between the Landsat reference, the fractional vegetation product and GlobCover 2009 is shown in Fig. 6. The classifier trees and shrubs display similar structures in both the Landsat classification and fractional vegetation product. The remaining differences are caused by the different spatial resolution, as the high-resolution Landsat data highlights the heterogeneity of the land surface. A comparison of the GlobCover 2009 with Landsat and the fractional vegetation product showed a very weak agreement. The GlobCover 2009 classification shows large errors in the assignment of different land cover types seen in the other product and satellite data.

3.2 *High-Resolution Land Cover Mapping*

The land cover for the Lena river delta was classified using an object-based image analysis approach utilizing high-resolution RapidEye data (Fig. 7). The product covers the entire delta system of approx. 27,600 km^2 and differentiates between various types of tundra vegetation classes as well as non-vegetated areas such as water bodies and bare soil with 5 m spatial resolution.

Nearly 30 % of the entire Lena river delta is covered by surface water bodies such as rivers or lakes. The lakes which have the largest dimensions are situated in

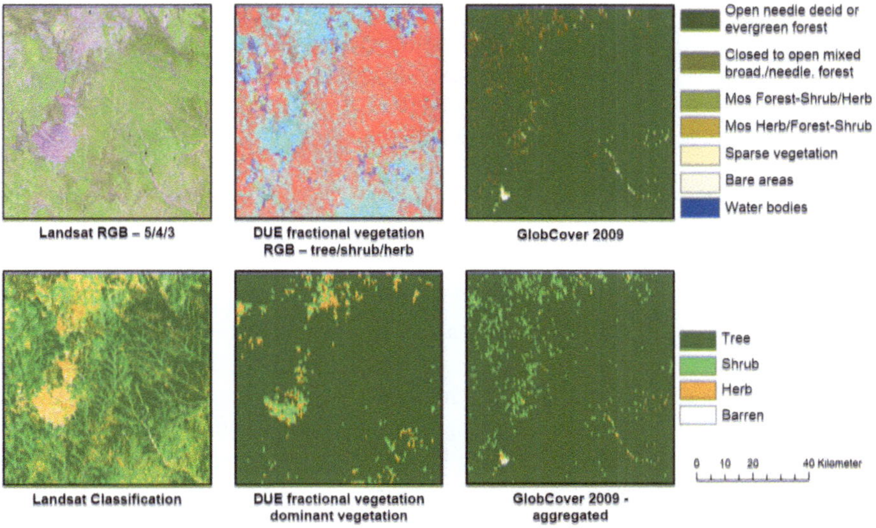

Fig. 6 Agreement between the Landsat reference, the fractional vegetation product and GlobCover 2009

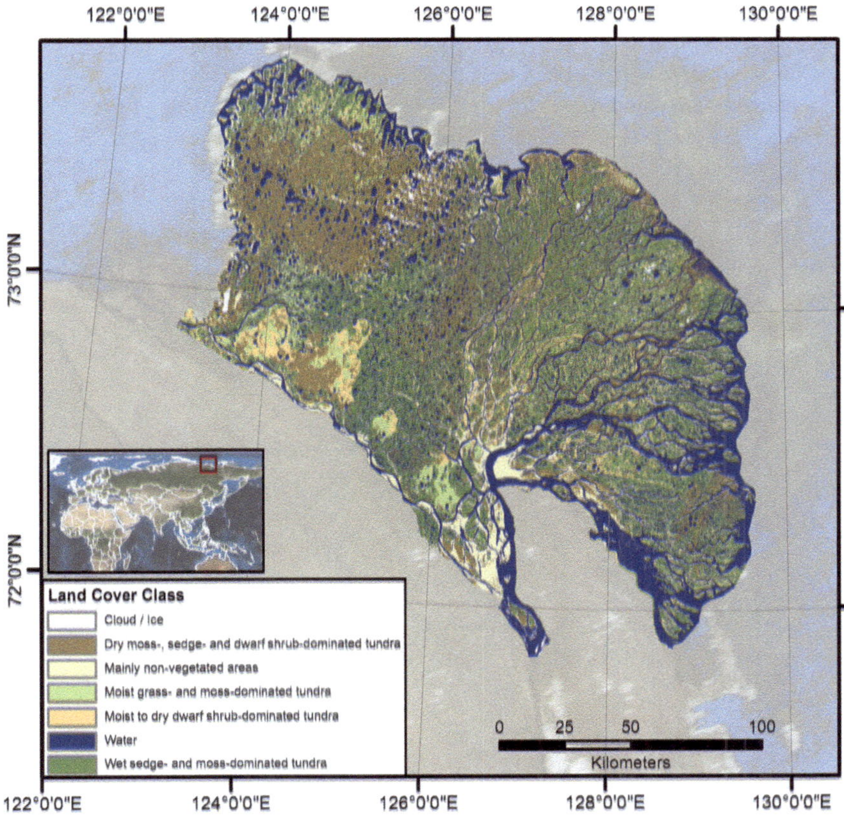

Fig. 7 High-Resolution land cover product of the Lena river delta system

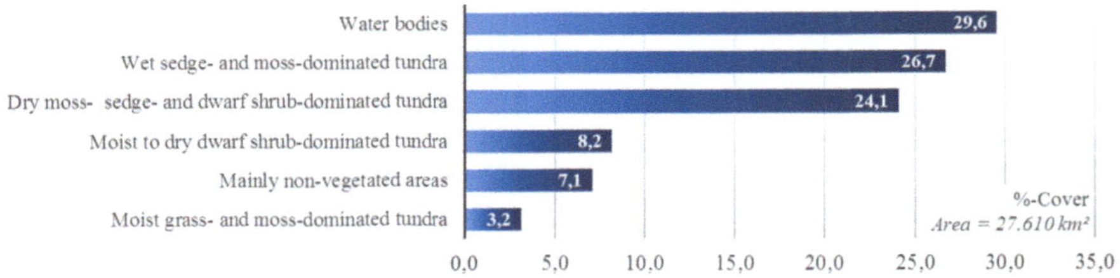

Fig. 8 Area statistics derived from the land cover classification for the Lena Delta

the north-western part of the delta. Dry and wet mosses as well as sedges appear to be the main vegetation types, representing nearly 50 % of the entire land cover classification. The moist to dry dwarf shrub-dominated tundra makes up 8.2 % of the study area. The majority of non-vegetated areas occur around the riverbanks. Additionally, grasslands are the minor land cover type. A summary of the individual cover percentage can be found in Fig. 8.

The accuracy assessment of the land cover classification was carried out using a stratified sampling design with 50 sampling points per class defined randomly. A vegetation classification provided by the AWI (Alfred Wegner Institute) user core group in Potsdam, Germany, and based on expert knowledge, was used as reference information (Schneider et al. 2009). The overall accuracy of the high-resolution land cover classification is 76.3 %. Figure 9 shows the errors of commission and omission for each individual land cover class. Water bodies and wet land cover types, such as sedges and mosses, show the highest accuracy. The user accuracies of the different vegetation types range between 0.62 for moist to dry dwarf shrubs and 0.88 for wet sedges and mosses. The highest differences between omission and commission errors are found for non-vegetated areas. The sedimentary deposition of the Lena during the interval between the reference and classified data, changes in vegetation composition and cloud and ice coverage are the main sources of misclassification.

The validation is an initial estimate of the accuracy of the high-resolution land cover classification, as the majority of reference information was extracted from a

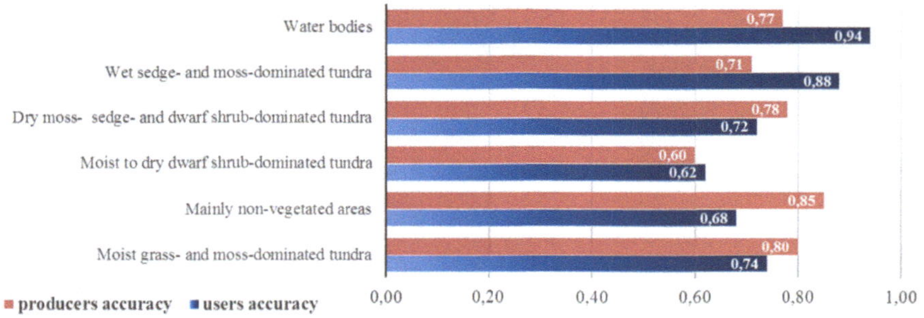

Fig. 9 Overview of the class-specific producers and user accuracies of the Lena Delta land cover classification

supervised classification of the Lena Delta, carried out by the user core group at the AWI. Ground measurements are limited to only very few regions, covering only few land cover classes.

3.3 High-Resolution Water Body Change Mapping

The methodology developed for analyzing water body changes between historical Corona and recent RapidEye data was applied to a region near the town of Yakutsk, Russia (Fig. 10). The change analysis was carried out for a 45-year time interval between the two acquisition times. The water bodies were classified using an object-oriented classification approach, which used hierarchical connections of objects from different scales. Changes in thermokarst lakes were identified by applying a spatial intersect of the water objects classified in 1967 and 2011 within a GIS (GeoInformation System). Lake area statistics, as well as object shape features

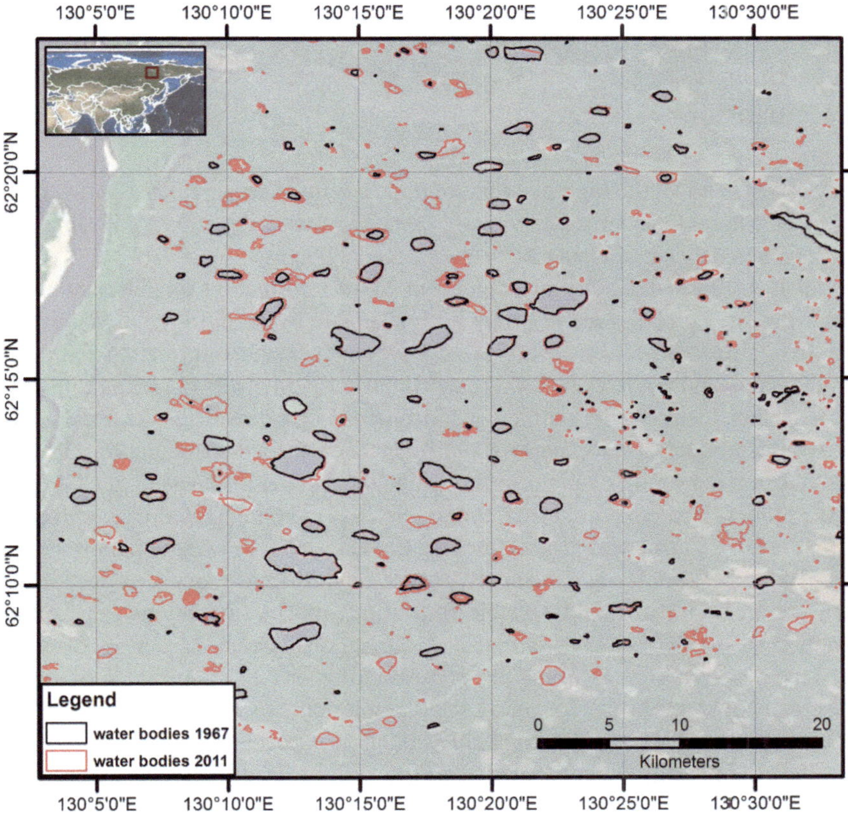

Fig. 10 Lake change analysis using Corona (1967) and RapidEye (2011) data for the area near Yakutsk, Russia

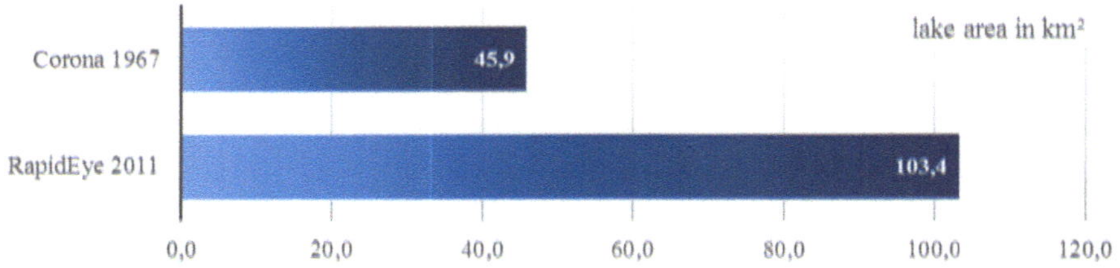

Fig. 11 Comparison of the lake area derived from high-resolution remote sensing data between 1967 and 2011

such as the lake border index, elliptical fit, lake density, direction and the roundness of water objects, were calculated and compared.

Substantial lake changes were identified for Yakutsk between 1967 and 2011. The water body area nearly doubled (Fig. 11). (Iijima et al. 2009) found this region to be highly affected by significant changes in the permafrost conditions as well as changes in temperature and precipitation. Lakes which existed in 1967 were found to have increased significantly in size during the subsequent 45 years.

4 Conclusions

This work has shown the high potential of optical remote sensing data from various sources at different spatial scales when it comes to extracting and monitoring land cover characteristics and changes for the Arctic regions.

Until now, the availability of percentage cover information for different vegetation types on a global scale has only been available by MODIS VCF. Within the DUE Permafrost project, an estimate of cover percentage information extracted from thematic land cover classes was carried out for permafrost modeling purposes with a special focus on the northern hemisphere. Future investigations and Earth observation product developments should concentrate on the integration of additional information sources and upcoming global land cover products (e.g. CCI Land Cover Product). Moreover, the user community has to be integrated in the process. Within the DUE Permafrost project, the requirements have shown the importance of detailed land cover information for the tundra regions (e.g. mosses, sedges, etc.), which are an important input variable for various model simulations (e.g. Arctic vegetation models). The standardized classification system LCCS is of great importance in terms of assuring the comparability and harmonization of future global land cover products. Upcoming satellite missions, such as Sentinel-3, will be a source for the continued sustainable monitoring of the land surface on a coarse-resolution scale (land cover, vegetation activity and fire-affected areas).

Climate-induced modifications to the permafrost environment were identified by analyzing water body changes using very high-resolution historical and recent remote sensing information in some regions. The high potential of high-resolution

Earth observation data from various sources and time steps is clearly of value for permafrost monitoring. The identification of different land cover types to analyze the heterogeneity of the tundra vegetation characterizing the Lena delta was demonstrated utilizing multi-temporal RapidEye data.

The cloud cover and short growing season are the major limiting factors using multispectral optical remote sensing data for mapping the very high latitudes. To reduce the influence of these issues, future observation strategies should focus on high repetition rates of the observation systems for successful data acquisitions for land surface mapping in boreal and Arctic latitudes. Hence, monitoring concepts using multiple platforms (constellations) and very high spatial resolution data coverage with daily revisits, such as the RapidEye constellation, are the preferred systems in the future. The availability of suitable reference information from ground measurements is an additional issue. Within the DUE Permafrost project, vegetation information from ground surveys was only available for a small area. More representatives spatially distributed and thematic detailed local-scale vegetation information is needed to analyze local land cover characteristics with remote sensing data in the future.

Acknowledgements The authors would like to thank the entire project team and user groups of the ESA Data User Element Permafrost project (ESA ITT "AO/1-5688/08/I-EC STSE DUE Permafrost"). Special thanks go to the ESA project coordinator Frank-Martin Seifert. Additional acknowledgements are given to the student assistants from FSU Jena, who were involved in the coarse-resolution land cover validation. Finally, the authors would like to thank the team of the RapidEye Science Archive (RESA) at the German Aerospace Center for providing RapidEye images (RESA projects 275 and 344).

References

Anisimov O, Reneva S (2006) Permafrost and changing climate: the Russian perspective. Ambio 35:169–175

Arino O, Leroy M, Ranera F, Gross D, Bicheron P, Nino F, Brockmann C, Defourny P, Vancutsem C, Achard F, Durieux L, Bourg L, Latham J, Gregorio A Di, Witt R, Herold M, Sambale J, Plummer S, Weber J-L, Goryl P, Houghton N, Brockman C (2007) Globcover—a global land cover service with MERIS. Envisat Symp. 2007. http://www.geogr.uni-jena.de/~c5hema/telecon/Arino_Globcover_ENVISATSYMPOSIUM.pdf. Accessed on 8 Mar 2015

Bartsch A, Sabel D, Naeimi V, Paulik C, Ressl C, Notegger C, Melzer T, Mistelbauer T, Schlaffer S, Reschke J, Zwieback S, Park SE, Glira P, Wagner W, Strozzi T, Wiesmann A, Duguay C, Soliman A, Hachem S, Urban M, Pöcking S, Voltersen M, Hese; S, Heim B, Elger K, Schwamborn G (2012) ESA DUE Permafrost—Final Report v2. http://geo.tuwien.ac.at/permafrost/images/publications/ESA_DUE_Permafrost_Final_report_v2.1_standardsize.pdf. Accessed on 8 Mar 2015

Bicheron P, Defourny P, Brockmann C, Schouten L, Vancutsem C, Huc M, Bontemps S, Leroy M, Achard F, Herold M, Ranera F, Arino O (2008) GLOBCOVER—products description and validation report. Toulouse. http://dup.esrin.esa.int/files/p68/GLOBCOVER_Products_Description_Validation_Report_I2.1.pdf. Accessed on 8 Mar 2015

Chapin FS, Sturm M, Serreze MC, McFadden JP, Key JR, Lloyd AH, McGuire AD, Rupp TS, Lynch AH, Schimel JP, Beringer J, Chapman WL, Epstein HE, Euskirchen ES, Hinzman LD,

Jia G, Ping CL, Tape KD, Thompson CDC, Walker DA, Welker JM (2005) Role of land-surface changes in Arctic summer warming. Science 310:657–660
Christensen T (2004) Thawing sub-Arctic permafrost: effects on vegetation and methane emissions. Geophys Res Lett 31:1–4
Crowley TJ (2000) Causes of climate change over the past 1000 years. Science 289:270–277
Cuevaz-Gonzalez M, Gerard F, Baltzer H, Rianos D (2009) Analysing forest recovery after wildfire disturbance in boreal Siberia using remotely sensed vegetation indices. Glob Change Biol 15:561–577
Definiens (2007a) Definiens developer 7. Reference book. Definines AG, München. http://ecognition.cc/download/userguide.pdf. Accessed on 8 Mar 2015
Definiens (2007b) Definiens developer 7. User guide. Document version 7.0.5.968. Definines AG, München. http://ecognition.cc/download/userguide.pdf. Accessed on 8 Mar 2015
Defries RS, Hansen MC, Townshend JRG, Janetos A, Loveland TR (2000) A new global 1-km dataset of percentage tree cover derived from remote sensing. Glob Change Biol 6:247–254
French H (2008) Recent contributions to the study of past permafrost. Permafrost Periglac Process 19:179–194
Friedl MA, Sulla-Menashe D, Tan B, Schneider A, Ramankutty N, Sibley A, Huang X (2010) MODIS Collection 5 global land cover: Algorithm refinements and characterization of new datasets. Remote Sens Environ 114:168–182
Friedl M, McIver D, Hodges J, Zhang X, Muchoney D, Strahler A, Woodcock C, Gopal S, Schneider A, Cooper A, Baccini A, Gao F, Schaaf C (2002) Global land cover mapping from MODIS: algorithms and early results. Remote Sens Environ 83:287–302
Frohn RC, Hinkel KM, Eisner WR (2005) Satellite remote sensing classification of thaw lakes and drained thaw lake basins on the North Slope of Alaska. Remote Sens Environ 97:116–126
Grace J, Beringer F, Laszlo N (2002) Impacts of climate change on the tree line. Ann Bot 90:537–544
Gregorio A Di, Jansen LJM (2005) Land cover classification system (LCCS). Classification concepts and user manual. FAO, Rome. http://www.fao.org/docrep/008/y7220e/y7220e00.HTM. Accessed on 8 Mar 2015
Grosse G, Schirrmeister L, Kunitsky VV, Hubberten H-W (2005) The use of CORONA images in remote sensing of periglacial geomorphology: an illustration from the NE Siberian coast. Permafrost Periglac Process 16:163–172
Gruber S, Hoelzle M (2001) Statistical modelling of mountain permafrost distribution: local calibration and incorporation of remotely sensed data. Permafrost Periglac Process 12:69–77
Haeberli W, Burn C, Sidle R (2002) Natural hazards in forests: glacier and permafrost effects as related to climate change. In: Sidle RC (ed) Environmental changes and geomorphic hazards in forests. IUFRO Research Series, vol. 9. Report No. 4, pp 167–202
Hájek F (2005) Object-oriented classification of remote sensing data for the identification of tree species composition. In: Proceedings of ForestSat 2005. May 31–June 3, 2005, Boras, Sweden, pp 16-20. http://shop.skogsstyrelsen.se/shop/9098/art92/4646092-23ff72-1740-1.pdf. accessed on 8 Mar 2015
Hansen M, Defries RS, Townshend JRG, Sohlberg R, Dimiceli C, Carroll M (2002) Towards an operational MODIS continuous field of percent tree cover algorithm: examples using AVHRR and MODIS data. Remote Sens Environ 83:303–319
Herold M, Member S, Woodcock CE, Gregorio A, Mayaux P, Belward AS, Latham J, Schmullius CC (2006) A joint initiative for harmonization and validation of land cover datasets. Geosci Remote Sens IEEE Trans 44:1719–1727
Hese S (2008) Strategies for object-oriented Thermokarst Lake change analysis in Siberia using very high resolution satellite data. Digital earth summit geoinformatics: tools for global change 2008. Potsdam 12.-14.11.2008, conference proceedings
Hinzman L, Bettez N, Bolton W, Chapin F, Dyurgerov M, Fastie C, Griffith B, Hollister R, Hope A, Huntington H, Jensen A, Jia G, Jorgenson T, Kane D, Klein D, Kofinas G, Lynch A, Lloyd A, McGuire A, Nelson F, Oechel W, Osterkamp T, Racine C, Romanovsky V, Stone R, Stow D, Sturm M, Tweedie C, Vourlitis G, Walker M, Walker D, Webber P, Welker J,

Winker K, Yoshikawa K (2005) Evidence and implications of recent climate change in Northern Alaska and other Arctic Regions. Climatic Change 72:251–298
Iijima Y, Fedorov AN, Park H, Suzuki K, Yabuki H, Ohata T (2009) Abrupt increases in soil temperatures following increased precipitation in a Permafrost Region, Central Lena River Basin, Russia. Permafrost Periglac Process 21:30–41
Jung M, Henkel K, Herold M, Churkina G (2006) Exploiting synergies of global land cover products for carbon cycle modeling. Remote Sens Environ 101:534–553
Kozlenko N, Jeffries M (2000) Bathymetric mapping of shallow water in thaw lakes on the North Slope of Alaska with spaceborne imaging radar. Arctic 53:306–316
Laidler GJ, Treitz P (2003) Biophysical remote sensing of Arctic environments. Prog Phys Geogr 27:44–68
Langer M, Westermann S, Heikenfeld M, Dorn W, Boike J (2013) Satellite-based modeling of permafrost temperatures in a tundra lowland landscape. Remote Sens Environ 135:12–24
Lewinski S, Bochenek Z (2008) Rule-based classification of SPOT imagery using object-oriented approach for detailed land cover mapping. 28th EARSEL Symp. 28th EARSEL Symposium: „Remote Sensing for a Changing Europe". 2–5 June 2008, Istanbul, Turkey. http://www.ecognition.com/sites/default/files/1098_lewinski_bochenek_rule_based_classification_of_spot_imagery_for_landcover_mapping.pdf. Accessed on 8 Mar 2015
Lucas R, Rowlands A, Brown A, Keyworth S, Bunting P (2007) Rule-based classification of multi-temporal satellite imagery for habitat and agricultural land cover mapping. ISPRS J Photogramm Remote Sens 62:165–185
Moritz RE, Bitz CM, Steig EJ (2002) Dynamics of recent climate change in the Arctic. Science 297:1497–1502
Myneni RB, Keeling CD, Tucker CJ, Asrar G, Nemani RR (1997) Increased plant growth in the northern high latitudes from 1981 to 1991. Nature 386:698–702
Nelson F, Anisimov O, Shiklomanov N (2002) Climate change and hazard zonation in the circum-Arctic permafrost regions. Nat Hazards 26:203–225
Nelson FEF (2003) (Un)frozen in time. Science 69:8–10
Overpeck J, Hughen K, Hardy D, Bradley R, Case R, Douglas M, Finney B, Gajewski K, Jacoby G, Jennings A, Lamoureux S, Lasca A, MacDonald G, Moore J, Retelle M, Smith S, Wolfe A, Zielinski G (1997) Arctic Environmental Change of the Last Four Centuries. Science 278:1251–1256
Peddle DR, Franklin S, Franklin E (1993) Classification of permafrost active layer depth from remotely sensed and topographic evidence. Remote Sens Environ 44:67–80
Richard J, Jia X (2006) Remote sensing digital image analysis: an introduction. 4th ed, vol 25. Springer, New York, 439 p
Richter R (1996) A spatially adaptive fast atmospheric correction algorithm. Int J Remote Sens 17:1201–1214
Schneider J, Grosse G, Wagner D (2009) Land cover classification of tundra environments in the Arctic Lena delta based on Landsat 7 ETM + data and its application for upscaling of methane emissions. Remote Sens Environ 113:380–391
Schuur EAG, Bockheim J, Canadell JG, Euskirchen E, Field CB, Goryachkin SV, Hagemann S, Kuhry P, Lafleur PM, Lee H, Mazhitova G, Nelson FE, Rinke A, Romanovsky VE, Shiklomanov N, Tarnocai C, Venevsky S, Vogel JG, Zimov SA (2008) Vulnerability of Permafrost carbon to climate change: implications for the global carbon cycle. Bioscience 58:701–714
Sheng Y, Shah C, Smith L (2008) Automated image registration for hydrologic change detection in the lake-rich Arctic. Geosci Remote Sens Lett 5:414–418
Stow DA, Hope A, McGuire D, Verbyla D, Gamon J, Huemmrich F, Houston S, Racine C, Sturm M, Tape K, Hinzman L, Yoshikawa K, Tweedie C, Noyle B, Silapaswan C, Douglas D, Griffith B, Jia G, Epstein H, Walker D, Daeschner S, Petersen A, Zhou L, Myneni R (2004) Remote sensing of vegetation and land-cover change in Arctic tundra ecosystems. Remote Sens Environ 89:281–308

Urban M, Hese S, Herold M, Pöcking S, Schmullius C (2012) A fractional vegetation cover remote sensing product on pan-Arctic scale. Version 2. Download–Address. http://doi.pangaea.de/10.1594/PANGAEA.779575. Accessed on 8 Mar 2015

Urban M, Hese S, Herold M, Pöcking S, Schmullius C (2010) Pan-Arctic Land Cover mapping and fire assessment for the ESA data user element Permafrost. Photogramm Fernerkundung Geoinf 4:283–293

Walker AD, Raynolds MK, Daniels FJA, Einarsson E, Elvebakk A, Gould WA, Katenin AE, Kholod SS, Markon CJ, Melnikov ES, Moskalenko NG, Talbot SS, Yurtsev BA (2005) The circumpolar Arctic vegetation map. J Veg Sci 16:267–282

Chapter 20
Multi-Source Data Integration and Analysis for Land Monitoring in Siberia

Jonas Eberle, Marcel Urban, Anna Homolka, Christian Hüttich and Christiane Schmullius

Abstract Land monitoring is a key issue in Earth system sciences analyzing environmental changes. To generate knowledge about changes, e.g., by decreasing uncertainties in the products and to build confidence in land change monitoring, multiple information sources are needed. Earth observation (EO) satellites and in situ measurements are available for operational monitoring of the land surface. As the availability of well-prepared geospatial time-series data for environmental research is limited, user-dependent processing steps with respect to the data source and formats pose additional challenges. Further steps are necessary for the analysis of time-series data. In most cases, it is possible to support science with spatial data infrastructures (SDI) and web services to provide data in a processed format and to provide time-series plots for further interpretation. Data processing middleware is proposed as a technical solution to improve interdisciplinary research using multi-source time-series data and standardized data acquisition, pre-processing, updating and analyses. This solution is being implemented within the Siberian Earth System Science Cluster (SIB-ESS-C), which combines various sources of EO data and climate data with a focus on vegetation and temperature data. Products from the Moderate Resolution Imaging Spectroradiometer (MODIS), in situ data from meteorological stations and high spatial resolution Landsat data are available in the processing middleware that is connected to different data providers. Analytical tools have been integrated and can be used for time-series plotting,

J. Eberle (✉) · M. Urban · A. Homolka · C. Hüttich · C. Schmullius
Department for Earth Observation, Institute for Geography, Friedrich-Schiller-University Jena, Loebdergraben 32, 07743 Jena, Germany
e-mail: jonas.eberle@uni-jena.de

M. Urban
e-mail: marcel.urban@uni-jena.de

A. Homolka
e-mail: anna.homolka@uni-jena.de

C. Hüttich
e-mail: christian.huettich@uni-jena.de

C. Schmullius
e-mail: c.schmullius@uni-jena.de

L. Mueller et al. (eds.), *Novel Methods for Monitoring and Managing Land and Water Resources in Siberia*, Springer Water,
DOI 10.1007/978-3-319-24409-9_20

phenological dates, trend calculations, break point detection, and data comparison using existing open-source software packages. The development of this SDI is based on the definition of automated and on-demand tools for data searching, ordering and processing, implemented along with standard-compliant web services. Therefore, open-source software is used to build up this system. The tools developed, consisting of a user-friendly data access, download, analysis and interpretation infrastructure, are available within SIB-ESS-C for operational use.

Keywords Middleware · Land monitoring · Web portal · MODIS · Climate data · Data integration · Time-series data · Standard-compliant data provision · Time-series analysis

1 Introduction

The availability of well-prepared geospatial time-series information for environmental research is limited. Individual processing, depending on the data source, as well as changing data formats, needs to be carried out by the user, posing additional challenges. Working with time-series data is especially time-consuming, due to the large amount of data. For environmental studies, time-series analyses are an important information source to identify impacts and changes. Several initiatives, such as the Global Observation of Forest and Land Cover Dynamics (GOFC-GOLD) and the Northern Eurasia Earth Science Partnership Initiative (NEESPI), focus on monitoring the impact of global climate changes on the Earth's surface (Townshend et al. 2006; Groisman and Bartalev 2007). This is important for land-cover and land-use change detection, as well as for the management of disasters, including fires, droughts and floods (Lentile et al. 2006; Franklin 2001; Asner and Alencar 2010; Rogan and Chen 2004; Justice et al. 1998; Giri 2012). To do this, spatial time-series data from multiple data sources are needed. A lot of complementary data are available for environmental research. Remote sensing satellites can provide this time-series data, as they are able to provide spatial and temporal views of environmental parameters, especially for large areas. However, in situ data from meteorological stations are also useful, supporting the analysis of remote sensing data and giving an overview of the climate and the environment being studied.

As many data distributors provide data through web-based systems and programming interfaces, research needs to find ways to automate the steps of finding, downloading and processing data. One example is the Giovanni tool for interactive time-series data exploration and analysis (Acker and Leptoukh 2007). In addition to the processing needs of users, it is important to establish a system that provides access to multiple data sources. The most important aim of this work is to reduce the entry barriers for multidisciplinary applications, as well as providing functionalities that give data added value via steps implementing advanced data discovery, data pre-processing, data transformation, and data analysis.

The Siberian Earth System Science Cluster (SIB-ESS-C, http://www.sibessc.uni-jena.de) was developed with the aim of providing operational tools for multi-source data access, analysis and time-series monitoring for Siberia. The system comprises a metadata catalog allowing for data searching, as well as interoperable interfaces for data visualization, downloading and processing. Within the SIB-ESS-C, data from remote sensing satellites, climate data from meteorological stations and outcomes of research projects are stored. The aim is to provide a wide variety of operational information products free of charge. The advantage of representing different products within a single system is that users' needs can be integrated into web-based processing services. Concerning climate change and land monitoring, the SIB-ESS-C focuses on land-based information products.

There are several other web-based systems that provide tools to search for, order, and download data [NASA Reverb Client (Cechini et al. 2011), USGS Earth Explorer (Aundeen et al. 2002), NCDC Climate WebGIS (National Climatic Data Center 2013)] or to analyze it [NCDC Climate WebGIS (National Climatic Data Center 2013), NASA Giovanni (Acker and Leptoukh 2007), virtual laboratory of remote sensing time-series (Freitas et al. 2011)]. However, if the data are not provided in a format that can be handled by the user, they need to be transformed into another format. Further steps are necessary when time-series data are analyzed, e.g., to calculate trends or detect break points. In such cases, there is a need for data processing steps that can be automated using programming languages and provided as standard-compliant web services.

These activities support the core principle of the Global Earth Observation System of Systems (GEOSS) to establish a "global and flexible network of content providers allowing decision makers to access an extraordinary range of information" (Group on Earth Observations 2013). According to the GEOSS architecture principles (Christian 2008), component systems can be scaled from national to global networks and from in situ to remote sensing data, which is also implemented in the SIB-ESS-C. The challenge of generating standardized and operational multi-source data handling structures has not yet been completely addressed in geo-information science. Therefore, the provision of standard-compliant data is a key component of data distribution; however, the combination of data distributed from different sources is also important. Another need is to overcome the continued lack of up-to-date time-series data that are acquired operationally, preprocessed and provided in common data formats. The interoperability between data providers, data application engineers and policy makers has to be strengthened to make the large amount of valuable information accessible to experts in diverse fields. Data availability, in general, is not an issue, but what kind of data are available for specific dates and areas is a frequently asked question.

The objective of the middleware within the SIB-ESS-C is to build up an operational web-based system where data from different sources are provided and updated automatically. The middleware collects data from integrated resources to provide standard-compliant web services for data access and visualization. Datasets are then available for further analysis. This article describes the development of processing middleware to build up a multi-source database to support land-monitoring research by:

- establishing a multi-source data processing middleware for land observations,
- implementing additional and individual processing steps for integrated data,
- providing standard-compliant visualization, access and download services for time-series data,
- and fostering near-real-time monitoring of land processes.

The description of the integrated data sources and their datasets is given in Sect. 2. Section 3 shows the framework that was developed for the middleware, including data integration and provision. Section 4 describes cases showing different applications of the middleware for multi-source data visualization, as well as single- and multi-source time-series analysis. Section 5 offers a conclusion based on the developments and experiences discussed in the previous sections.

2 Data Sources

Finding suitable spatial data in the area of interest is an essential task in many environmental studies. Different types of data from EO satellites are available, but there are some drawbacks for very high-resolution satellite data; these data are cost-intensive and not available to the general public. However, some high- and medium-resolution satellite data are freely available, such as data from the Moderate Resolution Imaging Spectroradiometer (MODIS) sensor from the National Aeronautics and Space Administration (NASA) as well as data from Landsat satellites. Another important data source is climate data from meteorological stations. The National Oceanic and Atmospheric Administration (NOAA) provides global climate data which can be used for non-commercial purposes, according to Resolution 40 (Cg-XII) of the World Meteorological Organization (WMO 1995).

The data sources used for operational land monitoring in the SIB-ESS-C are listed in Table 1. MODIS data are received from two satellites, *Aqua* and *Terra*, both of which were launched through NASA's Earth Observation System program. Continuous observations are available for Terra since 2000 and for Aqua since 2002. Data from MODIS are provided as land, atmospheric and ocean products on a systematic basis by different science teams (Justice et al. 2002) and are available on NASA and USGS data servers. The products are pre-defined datasets with different spatial (250, 500 m, 1 km, 0.05°) and temporal (daily, every 8 or 16 days, monthly) resolutions.

Data from Landsat satellites provide a higher spatial resolution (30 m) but a lower temporal resolution (only every 16 days). Due to the higher resolution (in comparison to MODIS data) Landsat data can be used better for detecting fine-scale land changes. The automated access is implemented via Google Earth Engine (Google 2012), which provides visualization, processing and download services for the complete Landsat archive.

Table 1 Data sources used for operational land monitoring

Data	Provider	Available time ranges
Moderate resolution imaging spectroradiometer	NASA/USGS	2000–present
Landsat	USGS/Google earth engine	1972–present
Global surface summary of the day	NOAA	1929–present (depending on station)
Integrated surface database	NOAA	1929–present (depending on station)
Global historical climatology network—daily	NOAA	1832–present (depending on station)

Climate data from meteorological stations are available from NOAA National Climatic Data Center (NCDC), which provides different datasets, e.g., hourly synoptic measurements (Integrated Surface Database, ISD; Lott et al. 2008), daily summaries from synoptic measurements (Global Surface Summary of the Day, GSOD; Lott 2006) and daily measurements from different climate data networks (Global Historical Climatology Network, GHCN; Menne et al. 2012).

3 Framework for the Multi-source Data Processing Middleware

3.1 Concept

The middleware service developed within the SIB-ESS-C (Fig. 1 gives a system overview) integrates external data sources, e.g., Landsat raster data from Google Earth Engine, time-series data from the Land Processes Distributed Active Archive Center (LPDAAC), the National Snow and Ice Data Center (NSIDC) and the NCDC. These datasets are downloaded, processed, and finally managed using a spatial database. Standard-compliant metadata are used to describe the integrated datasets. Datasets are published via web services that are compliant with Open Geospatial Consortium (OGC) specifications. In addition to these web services, the SIB-ESS-C web portal acts as a client, accessing the data that are processed by the middleware service. The web portal contains a search engine, a dataset viewer, a time-series plotter and functions to initiate new, on-demand data integration requests for the middleware database.

Open-source tools were used to develop the middleware services. PostgreSQL, with the PostGIS extension, provides the database with the ability to store raster and vector data. Data integration is carried out with Python scripting (e.g., to execute command-line tools for data downloading and for raster time-series data processing), and R script is used to plot the integrated time-series data. On the service level, MapServer (data visualization and downloading), istSOS (climate data provision),

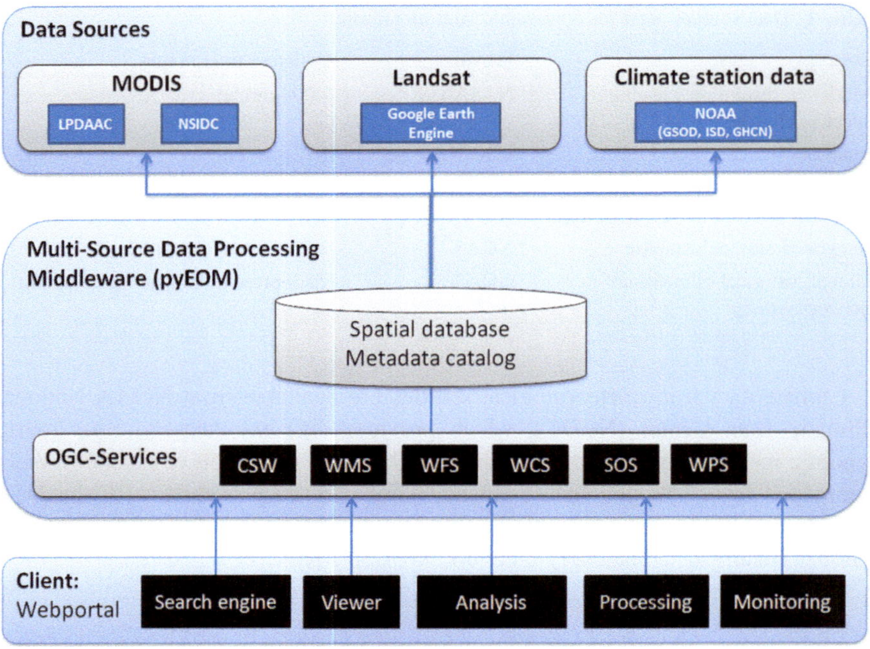

Fig. 1 System overview for the framework developed within the Siberian Earth System Science Cluster (SIB-ESS-C). *OGC* Open Geospatial Consortium; *CSW* Catalogue Service for Web; *WMS* Web Map Service; *WFS* Web Feature Service; *WCS* Web Coverage Service; *SOS* Sensor Observation Service; *WPS* Web Processing Service

pycsw (metadata provision) and PyWPS (time-series plotting) are used to publish OGC-compliant services.

The middleware component aims to enable the integration of datasets from external data sources. Therefore, downloading and processing steps have to be automatic. All necessary methods for the complete processing chain—download preparation, downloading, processing and publishing—were implemented with the programming language Python and published as the Python package *pyEOM*. The functions and processing chains for different data sources provided in this package are shown in Fig. 2 and described in the following sections.

3.2 Data Integration

Depending on each individual data source, the datasets have to be processed in different ways. Specific processing chains were developed for each individual data source, e.g., for different MODIS products, climate station data, and a connection to the Google Earth Engine. The integration part can be split up into download preparation, download and product-specific data processing (Fig. 2).

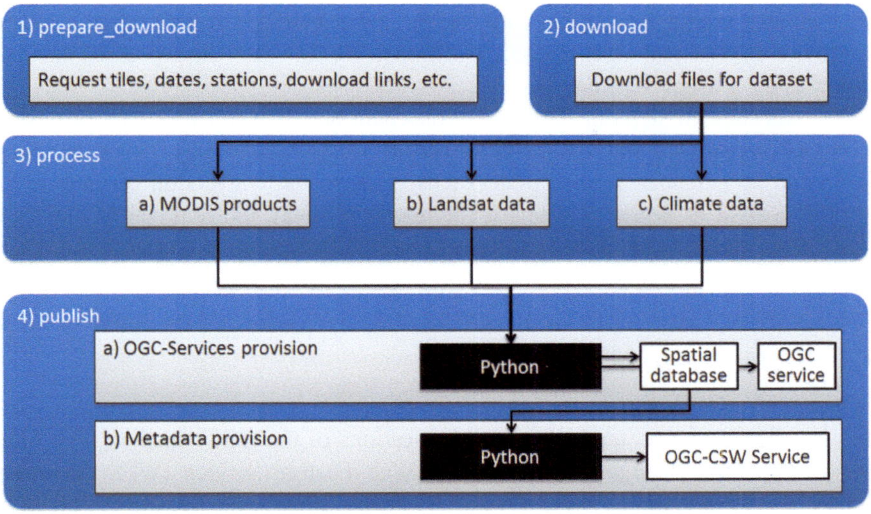

Fig. 2 Processing chain for the integration of external products

MODIS data can be downloaded from NASA file servers; their products are only available in the HDF-EOS format. In that case data download links have to be prepared and the data have to be converted to GeoTIFF after downloading. To integrate a specific MODIS product, further information is needed:

- Data server and directory where files are stored (e.g., FTP or HTTP server)
- Raster type of product (swath, tile, CMG)
- Whether 5 min swaths or tiles are needed (if raster type is equal to swath/tile)
- Dataset names to be extracted
- No-data and scale information (if necessary for processing)

Landsat data are processed and downloaded from Google Earth Engine (Google 2012). The middleware automatically searches for suitable datasets in the Landsat archive. The processed data are then available for downloading as a GeoTIFF file.

After the data download, these raster data are processed further with commands from the Geospatial Data Abstraction Library (GDAL; Warmerdam 2008). GDAL is a library for reading and writing various spatial raster data formats. It provides command-line tools for data translation and processing. GDAL commands used in data integration are *gdal_translate* (data translation), *gdal_merge.py* (mosaic-building) and *gdalwarp* (transforming an image into a new coordinate system).

The automatic integration and processing of the climate datasets from NOAA NCDC is implemented by a specially developed Python script that relies on the abstract scheme in Fig. 2. Each climate dataset (e.g., GSOD, ISD, and GHCN) has its own format, and some do not provide standardized parameter units. Thus, the data need to be converted into a standardized comma-separated-value (CSV) format, and unit conversions are needed after the data download. Furthermore, if dates are not available within the original data file, these dates are added with no-data values to provide a consistent time-series. In addition to the CSV file, the processed data are stored in the PostgreSQL database.

The subsequent steps in the integration chain are data and metadata provision with OGC-compliant services (as described in Sects. 3.2 and 3.3). This processing chain can be extended by additional steps, such as the conversion of units (e.g., degrees Kelvin to degrees Celsius for temperature data), the removal of the scale factor from data values or resampling to a coarser temporal or spatial resolution.

3.3 Data Provision

Services based on the standards and specifications of the OGC, such as Web Map Services (WMS; Beaujardiere 2006), the Web Feature Service (WFS; Vretanos 2005), Web Coverage Services (WCS; Baumann 2010), and Sensor Observation Services (SOS; Bröring et al. 2012) are used to provide access to the middleware database. In accordance with OGC specifications, these services for visualization and downloading can be implemented and published.

OGC-based services for data visualization and raster data downloading can be accomplished using different software packages, such as the open-source software MapServer (http://mapserver.org). MapServer provides OGC-compliant services for data visualization (WMS) and download (WFS, WCS). For time-series data, these services have to be published using the TIME parameter. Using this parameter, users can visualize and download data from a specific date within the time-series.

In situ data can be made available as a web service following the OGC's Sensor Observation Service (SOS) specification. A web service of this kind was set up using the Python-based open-source software istSOS (http://istgeo.ist.supsi.ch/software/istsos/) (Cannata and Antonovic 2010; Cannata et al. 2013). The istSOS tutorial (Cannata and Antonovic 2013) proposes a structure whereby any meteorological station is defined as an SOS procedure linked with observed properties (sensor parameters). Observations can be added either by using transactional SOS with XML code or by including a Python script that parses text files that contains new observations. The observation data are formatted in the proposed CSV format and then inserted into the database. A detailed description of the setup of these services is published in Eberle et al. (2013).

3.4 Metadata

The available time-series data, meteorological stations and additional data layers are described using metadata records. These records list information about the data, visualization and download services, as well as the available time positions and time interval. Based on this, a client can build requests for accessing the data. The middleware service generates a metadata record for each time-series dataset following ISO 19115 (International Organization for Standardization 2003) specification. The Python-based open-source software pycsw (http://pycsw.org) is used to provide metadata services. pycsw provides the OGC-compliant Catalogue Service for Web (CSW; Nebert et al. 2007), as well as transactional CSW for metadata insertion and updating.

The metadata records are the main entry point for accessing the multi-source data middleware database. In addition to general information such as the title, abstract, keywords and lineage, a wide range of metadata information can be used to describe the time-series data (Table 2). To link to OGC-compliant services for visualization and download, *DigitalTransferOptions* are provided within the metadata record. A client can retrieve metadata records based on a specific identifier or a search result, then can parse the information and visualize or download the data through the services provided. With the metadata information, the client knows which time positions are available and which services can be used (e.g., WMS, WCS or any other http link), in accordance with user needs. Parsing the metadata record, the client can further distinguish between time-series raster data as physical measurements or as classifications. This distinction is important for aspects such as providing the correct analysis processes, which differ for classification results (e.g., burned area), as opposed to continuous data, such as land surface temperature, vegetation indices and snow cover.

3.5 Web Portal

The middleware services are contained within the SIB-ESS-C web portal. This web portal (Fig. 3) includes functions allowing users to administer and manage the middleware services; it also allows easy access to the integrated datasets. Users are expected to interact closely with the data to receive the best information. Visiting the web portal, the user can go through the data catalog that contains the available data from the middleware database. The metadata catalog can be searched, and the resulting records can be investigated. The data can then be visualized and downloaded.

Open-source software was used to develop the web portal. In the back end, Drupal CMS (http://drupal.org) provides a proxy to external web services, converts XML code to JSON code for better processing within the web-front-end programming language, JavaScript, and provides RESTful services for user registration and authentication. The frontend was developed using the jQuery library (http://jquery.org) and extensions of jQuery. The map viewer for visualizing the data was created using the OpenLayers library (http://openlayers.org).

Table 2 Excerpt of information stored in the metadata records (example of MODIS Terra land surface temperature time-series data)

General metadata	
File identifier	MODIS_MOD11_C3_LST_Day_Series
Title	Monthly daytime land surface temperature from MODIS Terra
Abstract	Time-series of monthly Terra MODIS daytime land surface temperature in Kelvin at 0.05° spatial resolution. To retrieve actual values in Kelvin, a scale factor of 0.02 has to be applied. The unscaled no-data value is encoded as 0. Original MODIS data are retrieved from the Land Processes Distributed Active Archive Center (http://e4ftl01.cr.usgs.gov/MOLT/)
Keywords	MODIS, Terra, siberia, temperature, global, monthly, series, daytime
Lineage	MODIS HDF Level 2 product was converted to GeoTIFF with *gdal_translate* (Version 1.9)
Data information	
Description	Land surface temperature
Data type	RASTER
Coverage content type	Physical measurement
SRS	EPSG:4326
BBOX	57.1301270 81.2734985 179.8292847 42.2901001
Columns	2454
Rows	780
Resolution	0.05
Scale factor	0.02
No data value	0
Time begin	2000-03-01
Time end	2012-09-01
Time interval	P1M
Dates	2000-03-01, 2000-04-01, 2000-05-01,..., 2012-08-01, 2012-09-01
Services	
WMS URL	http://artemis.geogr.uni-jena.de/sibessc/modis
WMS protocol	WebMapService:1.3.0:HTTP
WMS description	MODIS Terra LST Day Monthly
WMS name	mod11c3_lst_day
WCS URL	http://artemis.geogr.uni-jena.de/sibessc/modis
WCS protocol	WebCoverageService:1.1.0:HTTP
WCS description	MODIS Terra LST Day Monthly
WCS name	mod11c3_lst_day

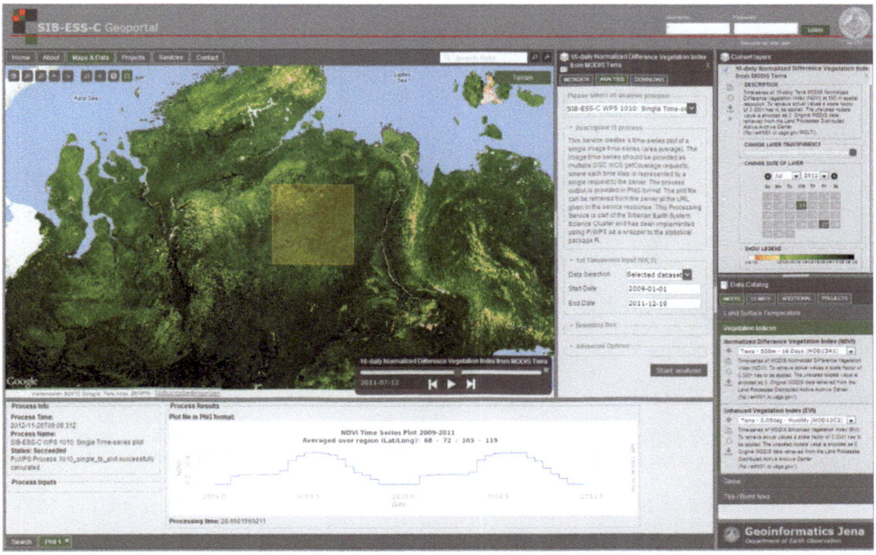

Fig. 3 Screenshot of the SIB-ESS-C web portal (http://www.sibessc.uni-jena.de) (Eberle et al. 2013)

4 Applications for the Multi-source Data Processing Middleware

4.1 Visualization of Global Earth Observation Data for Siberia

The middleware services developed can be used for different applications supporting research initiatives, such as GOFC-GOLD and NEESPI, or for research projects, providing access to a wide range of datasets. For Siberia, a few coarse resolution products from MODIS, as well as daily summaries of the climate data (GSOD), are integrated and updated automatically. These data can be visualized, downloaded and plotted within the web portal developed. Users have the possibility to identify specific areas for further research, e.g., by combining burned area data with vegetation indices. Figure 4 shows the effect of burned areas on vegetation as a pre- and post-fire visualization.

Beside the visualization of MODIS vegetation products with a spatial resolution of 250 m, users have the possibility to search the Landsat archive and visualize selected Landsat scenes with a 30 m spatial resolution before and after the event (e.g., fire disturbance).

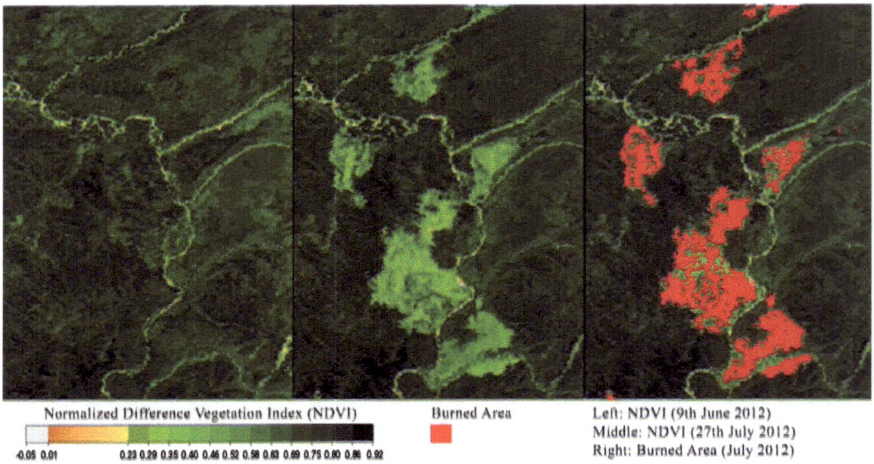

Fig. 4 Combination of MODIS NDVI and MODIS burned-area datasets (*Left* pre-fire NDVI; *Middle* post-fire NDVI, *Right* burned areas) near Yakutsk, Russia (Eberle et al. 2013)

4.2 Time-Series Analysis for Land Monitoring and Change Detection

The assessment and monitoring of forest resources in Siberia, as one of the aims of the EU FP7 ZAPÁS project (Hüttich et al. 2014), is being implemented in the SIB-ESS-C. Multi-source earth observation data are being analyzed in order to detect and characterize forest change dynamics in the Siberian Taiga. An example of detecting clear-cuts is given in Fig. 5. The map shows deforestation areas at the Angara River in the Krasnoyarsk region, mapped using inter-annual high-resolution Phased Array type L-band Synthetic Aperture Radar (PALSAR) data onboard the Japanese ALOS satellite platform. The red areas are patches of large-scale clear-cutting activities between 2007 and 2010. Further confidence and knowledge about the temporal dynamics, such as the detection of deforestation dates, can be

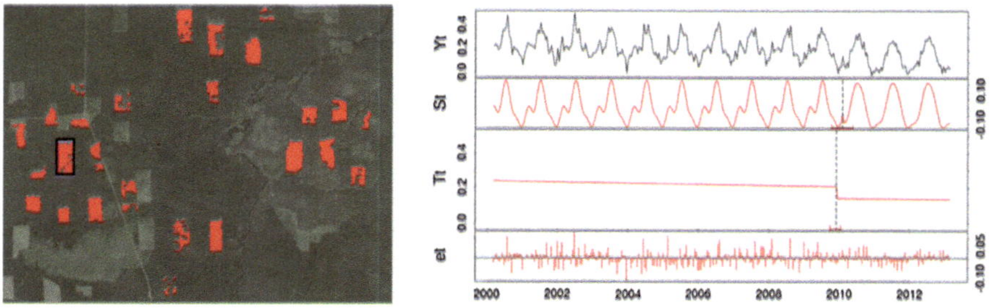

Fig. 5 Deforestation monitoring using the SIB-ESS-C multi-source database (*Left* Classified clear-cuts with very high-resolution data; *Right* BFAST plot of a selected clear-cut)

achieved using the SIB-ESS-C time-series analysis tools using the 16-day MODIS 250 m resolution EVI time-series data. By applying the BFAST algorithm (Verbesselt et al. 2010) at the detected clear-cut, the data of January 2010 can be documented by analyzing the break point function (Fig. 5, right). The break point marks an abrupt decrease in the EVI values. The seasonality series plotted above also detected a change in the phenology from one dominated by evergreen needle leaves to a deciduous phenology, indicated by the pronounced amplitude of the vegetation index after the logging event.

4.3 Multi-source Data Comparison

As time-series data from different sensors are available, a further step in the multi-source data analysis is to compare these products. A first example was produced using daily mean land-surface temperature data from the MODIS sensor on the Aqua and Terra satellites in comparison with daily mean air temperature measurements from meteorological stations, which were based on the GSOD dataset (Urban et al. 2013). Each plot shows the correlation between the two temperature time series (Fig. 6). A kernel density plot is calculated for the comparison and plotted with regression lines and Pearson's correlation coefficient (R) divided into temperatures above and below 0 degrees Celsius (°C).

This analysis was implemented into the SIB-ESS-C and provided as an OGC Web Processing Service accessing the previously integrated daily MODIS LST data and daily in situ measurements. Users have the possibility to compare data from

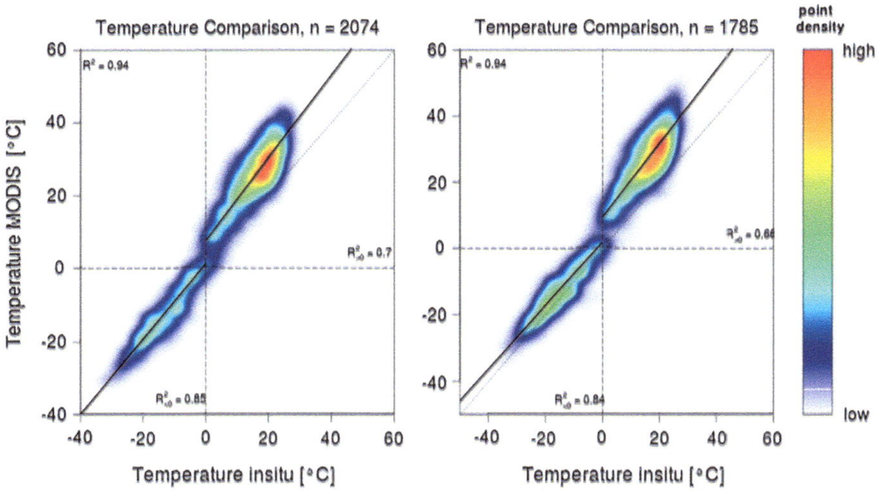

Fig. 6 Comparison of MODIS Terra (*left*) and Aqua (*right*) LST estimates and air temperature records from an individual meteorological station (Aleiskaya). Each plot shows the complete available time period (2000/2002–2013)

different sensors without the need to process any data. The web portal provides a simple user interface as well as metadata information about data processing and data conversion.

5 Conclusions

In accordance with the aims of the study, an operational multi-source data processing middleware for land monitoring was established with standard-compliant services for data visualization and distribution. Data from MODIS Land Team and from NSIDC, Landsat data, as well as climate datasets from NCDC, were connected with the middleware. Additional processing steps were integrated to generate common data formats, such as GeoTIFF and standardized SI-based units for climate data. Since it provides standard-compliant services, the data can be used with other GIS clients or within the web portal developed. The present article showed applications for how the middleware services within the SIB-ESS-C can be used to foster the monitoring of land processes using remote sensing and in situ climate data. Without any processing by the user, datasets can be investigated according to land changes; additional information, such as climate data, can easily be added, making data middleware services powerful in combining datasets from multiple sources. Different datasets can then be compared visually within the SIB-ESS-C web portal.

Standard-compliant data provision is no longer a critical topic, as software packages exist that can handle most common data formats. In combination with a spatial database, sub-selections of data based on parameters, such as time-series data with timestamps, are no longer prohibitively complex to implement. The open-source tools used facilitate the development of such SDI, especially for handling spatial data and making sure that they comply with standards. A more complex task is the provision of styling information for the visualization of data, as this information is not stored in the original data file and has to be generated manually for each dataset.

The advantages of the methods implemented in the system are the flexibility of how data are provided to the user using OGC-based services and data in GIS-common formats, and the possibility to integrate further time-series analysis tools. For the visualization of integrated data, it is necessary for data also to be stored in the system. However, this leads to a main problem: data have to be downloaded from external systems and need disk storage in the owner's system. The speed of downloading depends also on the external server, and for big datasets, this needs the most time of data integration and processing. As an example, the data download for the ZAPÁS project took around two days; the processing, just around four hours. To overcome this main issue, the only solution could be that these processing steps (dataset extraction, format conversion, clipping, OGC-compliant data provision, etc.) are implemented on systems with direct access to the data. Ideally, the user or the data analysis system should just have to download a time-series file and could, then, for example, provide further analysis tools.

State-of-the-art web technologies make it possible to develop web systems to support science, consulting and policy making, especially in the area of additional data processing. Such tools also provide experts in various fields with easy data access and visualization tools. Easy-to-use web systems can provide data processing and visualization procedures; in this case, users do not need to download any datasets if they are only interested in getting an overview of land observations. This is made possible as data are provided through interfaces that allow for automated processing. With the automated data integration, time-series databases can be kept up to date. Data that are provided free of cost and are allowed to be distributed through other systems are a major driver of this movement. Further datasets, such as NPP VIIRS, Spot Vegetation or the ESA Sentinel data, need service-based access to integrate them into the middleware database. This can be realized by web-based search, order and download services or just providing the data via FTP. Additional metadata and processing information—depending on the dataset—have to be added to the system to extract the requested datasets and to provide OGC-compliant services.

Acknowledgments The authors would like to thank the Friedrich Schiller University, Jena, for funding the development of the Siberian Earth System Science Cluster. We also thank the NASA, USGS, and NOAA for the provision of satellite data and data from climate stations.

References

Acker JG, Leptoukh G (2007) Online analysis enhances use of NASA earth science data. EOS, Trans Am Geophys Union 88:14

Asner GP, Alencar A (2010) Drought impacts on the amazon forest: the remote sensing perspective. New Phytol 187:569–578

Aundeen JLF, Anengieter RLK, Uswell MDB (2002) U.S. geological survey spatial data access. J Geospatial Eng 4:145–152

Baumann P (2010) OGC WCS 2.0 interface standard—core. Open Geospatial Consortium Inc., USA. http://www.opengeospatial.org/legal/. Accessed 8 Mar 2015

Beaujardiere JD (2006) OpenGIS Web Map Server implementation specification. Document OGC 06-042. Open. Geospatial Consortium, Inc., USA. http://www.opengeospatial.org/legal/. Accessed 8 Mar 2015

Bröring A, Stasch C, Echterhoff J (2012) OGC Sensor observation service interface standard. Version 2.0, candidate standard (12-006). Wayland, MA, USA, Open Geospatial Consortium Inc. http://www.opengeospatial.org/legal/. Accessed 8 Mar 2015

Cannata M, Antonovic M (2010) istSOS: investigation of the sensor observation service. In: Proceedings of 1st international workshop on pervasive web mapping, geo-processing and services. Como, Italy. http://www.isprs.org/proceedings/XXXVIII/4-W13/ID_04.pdf. Accessed 8 Mar 2015

Cannata M, Antonovic M (2013) Tutorial: using istSOS. https://geoservice.ist.supsi.ch/projects/istsos/index.php/Main_Page. Accessed 8 Mar 2015

Cannata M, Antonovic M, Molinari M, Pozzoni M (2013) istSOS sensor observation management system: a real case application of hydro-meteorological data for flood protection. In: Proceedings of international workshop of the role of geomatics in hydrogeological risk. Padua, Italy 27–28 Feb 2013, pp 111–117. http://www.int-arch-photogramm-remote-sens-spatial-inf-sci.net/XL-5-W3/111/2013/isprsarchives-XL-5-W3-111-2013.pdf. Accessed 8 Mar 2015

Cechini MF, Mitchell A, Pilone D (2011) NASA reverb: standards-driven earth science data and service discovery. AGU Fall Meet Abs 1:1406

Christian EJ (2008) GEOSS architecture principles and the GEOSS clearinghouse. IEEE Syst J 2:333–337

Eberle J, Clausnitzer S, Hüttich C, Schmullius C (2013) Multi-source data processing middleware for land monitoring within a web-based spatial data infrastructure for Siberia. ISPRS Int J Geo-Inf 2:553–576

Franklin SE (2001) Remote sensing for sustainable forest management. Lewis Publishers, Boca Raton, 407 p

Freitas RM, Arai E, Adami M, Ferreira AS, Sato FY, Shimabukuro YE, Rosa RR, Anderson LO, Friedrich B, Rudorff T (2011) Virtual laboratory of remote sensing time series: visualization of MODIS EVI2 data set over South America. J Comput Interdisc Sci 2:57–68

Giri CP (2012) Remote sensing of land use and land cover: principles and applications. CRC Press, Boca Raton, 477 p

Google (2012) Google earth engine. http://earthengine.google.org. Accessed 8 Mar 2015

Groisman PY, Bartalev SA (2007) Northern Eurasia Earth Science Partnership Initiative (NEESPI), science plan overview. Glob Planet Change 56:215–234

Group on Earth Observations (2013) GEO—Group on Earth Observations. What is GEOSS? http://www.earthobservations.org/geoss.shtml. Accessed 8 Mar 2015

Hüttich C, Eberle J, Schmullius C, Bartalev S, Emelyanov K, Korets M, Shvidenko A, Schepaschenko D (2014) Supporting a forest observation system for Siberia: EO for monitoring, assessing and providing forest resource information. http://earthzine.org/2014/07/16/supporting-a-forest-observation-system-for-siberia-earth-observation-for-monitoring-assessing-and-providing-forest-resource-information/. Accessed 8 Mar 2015

International Organization for Standardization (2003) ISO 19115:2003 geographic information—metadata. http://www.iso.org/iso/catalogue_detail.htm?csnumber=26020. Accessed 8 Mar 2015

Justice CO, Vermote E, Townshend JRG, Defries R, Roy DP, Hall DK, Salomonson VV, Privette JL, Riggs G, Strahler A, Lucht W, Myneni RB, Knyazikhin Y, Running SW, Nemani RR, Zhengming W, Heute AR, Van Leeuwen W, Wolfe RE, Giglio L, Muller JP, Lewis P, Barnsley MJ (1998) The moderate resolution imaging spectroradiometer (MODIS): land remote sensing for global change research. IEEE Trans Geosci Remote Sens 36:1228–1249

Justice CO, Townshend JRG, Vermote EF, Masuoka E, Wolfe RE, Saleous N, Roy DP, Morisette JT (2002) An overview of MODIS land data processing and product status. Remote Sens Environ 83:3–15

Lentile LB, Holden ZA, Smith AMS, Falkowski MJ, Hudak AT, Morgan P, Lewis SA, Gessler PE, Benson NC (2006) Remote sensing techniques to assess active fire characteristics and post-fire effects. Int J Wildland Fire 15:319

Lott N (2006) Global surface summary of day. http://www.ncdc.noaa.gov/cgi-bin/res40.pl?page=gsod.html. Accessed 8 Mar 2015

Lott N, Vose R, Del Greco SA, Ross TF, Worley S, Comeaux J (2008) The integrated surface database: partnerships and progress. In: Proceedings of the 24th Conference in IIPS, 3B.5. https://ams.confex.com/ams/88Annual/techprogram/paper_131387.htm. Accessed 8 Mar 2015

Menne MJ, Durre I, Vose RS, Gleason BE, Houston TG (2012) An overview of the global historical climatology network-daily database. J Atmos Oceanic Technol 29:897–910

National Climatic Data Center (2013) Climate.gov: data and services. http://www.climate.gov/#dataServices. Accessed 8 Mar 2015

Nebert D, Whiteside A, Vretanos PP (2007) OpenGIS catalogue services specification. Open Geospatial Consortium Inc. http://portal.opengeospatial.org/files/?artifact_id=20555. Accessed 8 Mar 2015

Rogan J, Chen D (2004) Remote sensing technology for mapping and monitoring land-cover and land-use change. Prog Planning 61:301–325

Townshend JR, Brady MA, Csiszar IA, Goldammer JG, Justice CO, Skole DL (2006) GOFC-GOLD Report No. 24: a revised strategy for GOFC-GOLD. Download-access. http://www.fao.org/gtos/doc/pub42.pdf. Accessed 8 Mar 2015

Urban M, Eberle J, Hüttich C, Schmullius C, Herold M (2013) Comparison of satellite-derived land surface temperature and air temperature from meteorological stations on the pan-arctic scale. Remote Sens 5:2348–2367

Verbesselt J, Hyndman R, Newnham G, Culvenor D (2010) Detecting trend and seasonal changes in satellite image time series. Remote Sens Environ 114:106–115

Vretanos PA (2005) Web feature service implementation specification. Technical report, Open Geospatial Consortium Inc. OGC 04-094, 117 p. http://portal.opengeospatial.org/files/?artifact_id=8339, Accessed 8 Mar 2015

Warmerdam F (2008) The geospatial data abstraction library. Open Source Approaches in Spatial Data Handling 2:87–104

World Meteorological Organization (1995) Resolution 40 (Cg-XII). http://www.wmo.int/pages/about/Resolution40.html. Accessed 8 Mar 2015

Chapter 21
Analytical and Cartographic Predictive Modeling of Arable Land Productivity

Peter A. Shary, Olga V. Rukhovich and Larisa S. Sharaya

Abstract An inventory-based method is suggested to form spatial models of agricultural crop yields for extended regions (size: about $4° \times 5°$). The method is based on analyzing links between the long-term characteristics of crop yields and environmental variables, such as climate, topography, and soils. The response variable in multiple regressions was the maximal crop yield surplus by fertilization. This is the difference between the maximal crop yield obtained for an optimal dose of fertilizers and the control without fertilizers. This addition to yield appeared closely related to environmental variables, and on the other hand, it was relatively independent of previous crops. The environmental variables were climatic data on long-term annual means of temperature and precipitation for each month and certain periods, a digital elevation model and 18 basic topographic attributes from it, and soil type data. The topographic attributes were transformed nonlinearly to obtain normally distributed residuals. Multiple regression models included validation using an empirically founded criterion, tests for multicollinearity, and determining the statistical significance of each environmental variable. The method was tested based on long-term data on winter wheat yields for the western part of the Oka River basin. Analysis showed that the topography-generated microclimate is a major factor determining the maximum addition to the yield. The microclimate variable is the relative slope insolation, which is one of the topographic attributes that takes into account the slope steepness and exposure, and 2 angles that describe

P.A. Shary (✉)
Institute of Physicochemical and Biological Problems in Soil Science, RAS, Institutskaya Street 2, Pushchino, 142290 Moscow Region, Russia
e-mail: p_shary@mail.ru

O.V. Rukhovich
Pryanishnikov All-Russian Research Institute of Agrochemistry (VNIIA), Pryanishnikova St. 31a, 127550 Moscow, Russia
e-mail: o_ruhovich@mail.ru

L.S. Sharaya
Institute of Ecology of the Volga Basin, RAS, Komzina Street 10, Togliatti 445003 Samara Region, Russia
e-mail: l_sharaya@mail.ru

L. Mueller et al. (eds.), *Novel Methods for Monitoring and Managing Land and Water Resources in Siberia*, Springer Water,
DOI 10.1007/978-3-319-24409-9_21

an effective position for the Sun. Relative insolation characterizes the relative power flux of solar radiation on the land surface and allows for statistical comparisons at different azimuths and angles of declination of the Sun above the horizon. On the most heated southwestern slopes of the basin, the largest increase in the maximum addition is observed for winter wheat. The addition also increases during the relatively high precipitation in February, which is the windiest cold winter month in this region. An increase in annual precipitation in the humid climate of the Oka basin leads to a decrease in the addition. Introducing a nonquantitative or indicator variable, soil types, into the multiple regression model results in increasing the strength of the relationships and shows that the maximum addition to the sod-podzolic soils in the northern region depends more strongly on climate than it does for Gray Forest and Chernozem soils in the southern part of the region. The model validated using the criterion introduced explained 74 % of spatial variability in the addition to winter wheat by means of the environmental variables used. Based on the model, a gridded map was constructed for the entire region (4° × 5°). The results of the analysis indicate that the topography, along with environmental factors, may have the largest influence and should also be taken into account when making crop yield prognoses in the conditions of a changing climate.

1 Introduction

To study the influence of environmental factors (variables) on the productivity of agricultural crops, researchers use spatially widely distributed long-term data about crops and the weather. Such studies indicate that climatic factors have a strong influence on the average yield and that economic effects are overestimated (Lobell and Asner 2003) or are smaller compared to the influence from the climate (Adams et al. 2006; Cabas et al. 2009). In these papers, soil types and their fertility were taken into account considering corresponding data groups separately, not within a single model.

It has been shown in several studies that topography has a significant effect on the yield of winter wheat, which is explained by idea that "energy flows that determine the wheat development are modified by elevation, slope steepness and exposure" (Reuter et al. 2005; Ferrara et al. 2010). In evaluations of the influence of topography by means of a micrometeorology model combined with a typical crop growth model in the project STAMINA, it has been shown that slopes with southern exposure may increase the yield of winter wheat in wet years (Acutis et al. 2007). For Northwest Europe, in unfavorable years, the topography may define up to 12 % of the yield (up to 1 t ha^{-1}; Acutis et al. 2007).

In the present study, relationships were analyzed between multiyear averages of winter wheat (*Triticum aestivum* L.) and environmental factors. Information on crop yields was obtained from the database GeoNetwork "Agrogeos." This database contains data on experimental crops grown over the last 40 years by a single method with a variation of fertilizers in different climatic conditions of the former USSR and covering tens of thousands of experimental plots (Sychev et al. 2008). Three yield features were used in the experimental fields: crop yield without fertilizers (the control), an optimized dose of fertilizers (the experiment), and the maximum addition to the yield calculated as the difference between the experiment and the control. The addition is chosen because both the experiment and the control depend on the history of the field: This history may be compensated for in the addition as the difference between them so that the addition is expected to be less dependent on the field's history (e.g., on previous cultures). The 30 climate variables, 18 topographic attributes, and 3 soil types were used as environmental factors. The soil types are nonquantitative features, so they were transformed into two independent indicator variables (Montgomery and Peck 1982). The aim of this paper is to expose the approaches and results of the statistical analysis of the spatial variability of different crop yield parameters, and it takes into account land surface, climate, and soil features.

2 Materials and Methods

If a yield parameter correlates well with environmental variables such as topography, climate, and soils, the spatial distribution of this parameter can be directly predicted from these variables. This approach has acquired currency in ecology and soil science as ***predictive modeling***. The popularity of this approach is confirmed by numerous reviews in ecology (e.g. Guisan and Zimmermann 2000; Austin 2007), soil science (McBratney et al. 2000; Scull et al. 2003), and some papers on agriculture.

Statistical analysis

Methods of multiple linear regression (Montgomery and Peck 1982) are the best-elaborated statistical analysis tools for predictive modeling. A more detailed consideration of the statistical analysis methodology has been reported in the past (Shary et al. 2011; Shary and Pinskii 2013); therefore, we confine ourselves to a brief description of these methods.

The dependent variable (yield parameter) is usually referred to as the *response*, while the independent (environmental) variables are called *predictors*. The significance of the entire regression model is estimated using the Fisher F-test and can be described by the probability P of making mistakes when establishing a statistical relationship (this probability should usually be lower than 0.01), and the significance of separate predictors in the model is estimated using Student's t-statistics. The equation of multiple linear regression for N predictors can be written as follows:

$$R = a_1 P1_{t1} + a_2 P2_{t2} + \cdots + a_N PN_{tN}, \tag{1}$$

where R is the response, $P1$, …, PN are the predictors, a_1, …, a_N are the regression coefficients, and t_1, …,t_N are the values of t-statistics. Generally, the nonlinear functions of predictors can be considered, rather than the predictors themselves. Then, we speak of transformed predictors and mark them with superscript $T\left(P_i^T\right)$. A transformation table for topographic attributes has been reported previously (Shary and Pinskii 2013). If the absolute value of t-statistics for a given predictor is lower than its critical value, this predictor is statistically nonsignificant in the regression model.

It is important that the predictor can describe not only quantitative environmental variables (such as those related to the topography or climate) but also qualitative variables, such as the soil or land use types. In this case, the predictor is called an *indicator* and has only two values (usually 0 and 1). If there are K soil types in the given area, $K - 1$ indicators are needed for their description. For example, two indicators are needed to describe three soil types:

Indicator value	Soil types	
I_1	I_2	
0	0	Chernozems
1	0	Grey forest soils
0	1	Soddy-podzolic soils

Since indicators are binary, their use gives several solution branches to Eq. (1).

Regression analysis is based on the following five assumptions (the error or residual is the difference between the observed and model-predicted response values):

(A) No systematic error (random errors in the predictor should be lower than its variation range; correction by removing too noisy predictors).
(B) Constant error variance (determined from the residual plot; correction by nonlinear transformation of predictors).
(C) No significant autocorrelation of errors (statistical tests are known; correction by adding a missing significant predictor to the model or using special autoregressive models).
(D) Normal distribution of errors (determined from the plot of normal probability of residuals; correction by nonlinear transformation of predictors).
(E) No significant multicollinearity, i.e., approximately linear relationship between predictors (statistical tests are known; correction by removing or substituting dependent predictors or applying compressed regression models).

The regression models meeting these assumptions are called the models *adequate* to the data. For such models, the coefficient of determination R^2 is interpreted as the proportion of response variability *explained* by the variability in predictors. For example, if $R^2 = 0.57$ and the model is adequate to the data, then 57 % of the spatial variability in yield parameter is explained by the spatial variability in the environmental variables. In many works, these basic assumptions are not verified (and even not explicitly listed); in these cases, the derived model is not erroneous, but R^2 can no longer be interpreted as mentioned earlier. When the relationship between the response and the predictors is nonlinear, but the nonlinear terms are not included in the model, this situation is considered a mistake (Montgomery and Peck 1982). Such situations can be seen in plots of residuals (errors, i.e., differences between the observed and predicted response values).

This interpretation of R^2 is of importance, and all basic assumptions of regression analysis are verifiable and can be corrected; therefore, we use nonlinear transformations of environmental variables, especially topographic attributes, the statistical distribution of which can significantly deviate from the normal law and cause a failure of assumption (D). Methods for the verification and correction of regression models have been described before (Shary et al. 2011).

Assumption (C) was not verified in this work because the mean distance between the observation points was 45 km, while autocorrelation in errors (residuals) is usually negligible for such long distances (Guisan and Zimmermann 2000).

Along with these assumptions, the following restrictions are known (Guisan and Zimmermann 2000):

(a) ***quasi-equilibrium*** between the response and the environmental variables. This assumption can fail, e.g., at the invasion of the given plant species from a specific geographical position;
(b) limited applicability of results to ***other physicogeographical regions***; and
(c) ***continuity*** of the response type: nonlinear (logistic) multiple regression is used for binary response (e.g., for the description of the presence/absence of specific plant species or soil horizon at given sites).

In this work, we use only continuous responses, and the obtained results are applied to the same area.

Models with four predictors were used in this work. To select them, all possible combinations of four environmental variables without significant multicollinearity (i.e. with variance inflation factors below 5) were worked out. Four variables with the highest determination coefficient R^2 were selected.

The models obtained were verified using Allen's cross-validation procedure (Montgomery and Peck 1982; Shary and Pinskii 2013). In this procedure, the so-called "prediction" determination coefficient R^2_{Pred} characterizes the predictive power of the model when applied to new observation points within the same area. The value of R^2_{Pred} is usually lower than that of R^2, and a decrease in R^2_{Pred} by 1.5 times and more compared to R^2 corresponds to a significant degradation of model prediction performance.

Land surface description

The applied approach uses methods of geomorphometry, the science of quantitative land surface description (Shary 2006; MacMillan and Shary 2009; Pike et al. 2009). Gedymin (1992), Lastochkin (1991), Simonov (1998), and other authors reported quantitative approaches to land surface description with no digital land surface attributes in each map point of the kind necessary for statistical comparisons. To study relationships in agro landscapes, it is not only the frequently used variables of elevation, slope steepness, and exposure which are important; in fact, they are not as important as a sufficiently complete set of morphometric variables (MVs). Therefore, we used the extended system of 18 basic MVs (i.e., those not composed of others) developed in Shary (1995) and Shary et al. (2002), and no more than 6 basic MVs were previously used in ecology and soil science. "Secondary" MVs are obtained by combining basic ones (Wilson and Gallant 2000).

Note another important statement in the context of this work. Slope exposure is frequently used to describe the thermal regime of slopes. However, Sibirtsev (1900, p. 263) noted that the thermal regime of slopes is determined by the perpendicularity of incident solar rays to the land surface; i.e., the heat depends on both slope steepness and exposure. Correspondingly, MV is referred to as (relative) slope insolation (Shary et al. 2002). It is equal to 100 % for the perpendicular incidence of solar radiation and to zero on the shady slopes. Insolation $F(a, b)$ depends on two angles describing the position of the sun: height above the horizon a and azimuth (from the north clockwise) b. The correlation of any landscape feature with insolation usually depends little on a and more significantly on the azimuth b. Therefore, the value of a was not changed in this work, and it was taken as equal to 35°.

Digital elevation data were taken from the NASA SRTM30 grids with a resolution of 30 arc seconds (900 m on the equator) and transformed to the Kavraisky projection for the studied area (western Oka River basin, European Russia) to obtain a grid with spacing of 600 m. All MVs were calculated from this grid according to procedures described in Shary et al. (2002) using the "Analytical GIS Eco" software developed by P.A. Shary (Wood 2009).

There is no "correct" grid spacing, and it is selected depending on the research objective. Some authors (Currie 1991; Davies et al. 2007) successfully used grid spacings of 100–400 km, with elevation as the main predictor for the 100-km grid (Davies et al. 2007); in general, grid spacings from $\sim$1 m to 400 km are used.

Study area and observation points

In general, the correlation between crop yield parameters and environmental variables may depend on the selection of the area to be studied. For example, when extensive lowlands and uplands are simultaneously present in the selected area, the hydrochemical differences between them should be taken into consideration. Therefore, the orographic map of Russia created with a resolution of 2.7 km using elevations (Shary 2011) was used. Observation points were selected for the western part of the Oka River basin, which partly includes the Central Russian and Moscow–Smolensk uplands and the western region of the Klin–Dmitrov Ridge, without the Meshchera lowlands and Volga uplands. In this area (about 250 km × 350 km), 41 observation points were collected. Each experimental plot (considered as an observation point) was usually 50 m × 50 m in size, which was sufficiently small to fall into a grid cell of 600 m × 600 m. Among the 41 observation points, 27, 8, and 6 points refer to gray forest soils, Soddy-podzolic soils, and Chernozems, respectively. The shortest distance between two observation points was 2.35 km, and the average distance was about 45 km. The locations of observation points and corresponding soil types are shown in Fig. 1. The frost-resistant winter wheat cultivar “Mironovskaya-808” dominated in each soil type (28 out of 41 observation points).

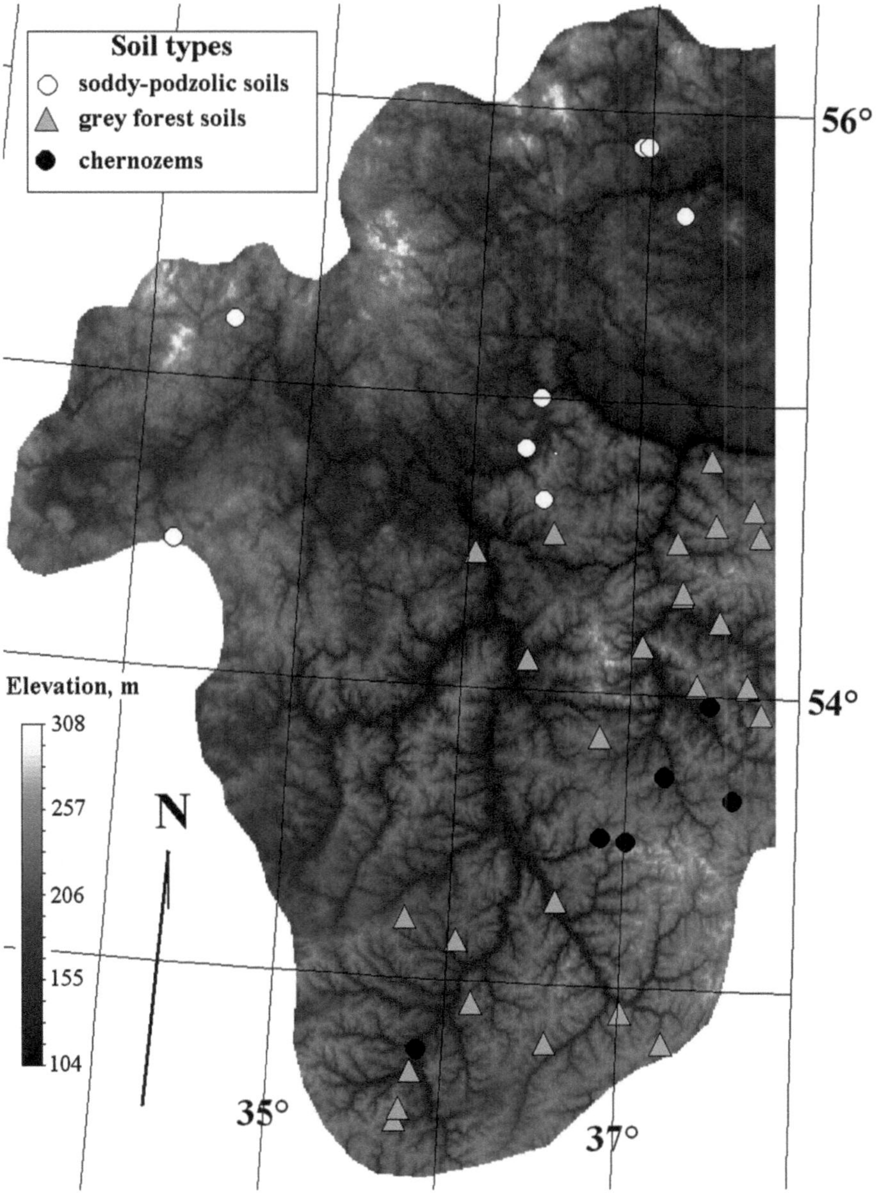

Fig. 1 Location of observation points and corresponding soil types in the study area, that is, in the western part of the Oka River basin

Climate description
Data on the long-term mean climate parameters averaged over 50 years (1950–2000) were described by Hijmans et al. (2005) and are available with a resolution of 30 arc seconds (900 m on the equator) on the Internet. From these data, grids were composed in the same projection and with the same resolution (600 m), which describe average temperature and precipitation for each month as well as their long-term annual means. The average annual sums of active temperatures (above 10 °C) were also calculated.

Soil description
In this work, soil description was based on two indicators, I_1 and I_2, which discriminated, as noted earlier, between three main soil types: Chernozems, Gray forest soils, and Soddy-podzolic soils. There were no observation points in swamps and large river valleys so that these areas were excluded from consideration. Predictive maps used a digitized soil map of Russia at a scale of 1:2,500,000.

Yield parameters
The Agrogeos database (Sychev et al. 2008) includes data on the rates of different fertilizers applied in the *experiment* treatment (no fertilizers were applied to the *control* treatment) in different years. So the difference between the experiment and the control yields, which we term the *addition* (to the yield), characterizes the effect of fertilizers. In each observation point, the maximum addition for a given year was taken and then averaged over the years. No significant correlation was found between additions and corresponding controls. The correlation between additions and environmental variables was primarily studied. Averages and standard deviations for the controls, experiment, and additions are shown in Fig. 2.

The maximum addition averaged for 41 observation points was equal to 60 % of the control.

3 Results

The equation of multiple linear regression describing the relationship between the maximum addition A to winter wheat yields and environmental variables in the study area is as follows:

$$\begin{aligned} A = {} & 4.292 \times F(235)_{+7.16} + 3.411 \times P_{\mathrm{FEB}+5.73} + 0.202 \times I_2 F(235)_{+5.64} \\ & -0.299 \times P_{\mathrm{YEAR}-4.10} - 151.61 \left(R^2 = 0.740; P < 10^{-6}; R^2_{\mathrm{Pred}} = 0.693\right), \end{aligned} \tag{2}$$

where $F(235)$ is the insolation of slopes from the southwest (at an angle from the horizon of 35° and an azimuth of 235°), P_{FEB} is the average precipitation in February, I_2 is the indicator discriminating Soddy-podzolic soils ($I_2 = 1$) from other soil types ($I_2 = 0$), and P_{YEAR} is the average annual precipitation. In the model, environmental variables are arranged in the order of decreasing absolute values of *t*-

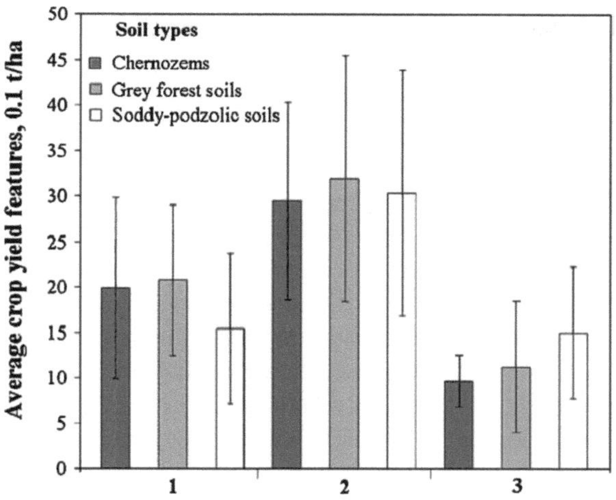

Fig. 2 Averages and standard deviations for parameters of winter wheat yield. *1* control, *2* experiment, *3* addition. "Control" means the crop yield with no fertilizers, while "experiment" means that with optimal fertilizers; addition is the difference between experiment and control

statistics, which are denoted as subscripts. The critical level of *t*-statistics for $P = 0.05$, $n = 41$, and 4 predictors is equal to 2.03 (as found from statistical tables). Absolute values of all *t*-statistics exceed this level so that all predictors in the models are significant.

Model (2) is adequate to the data as was shown by Shary et al. (2011); therefore, 74 % of spatial variability in the maximum addition for winter wheat is explained by the spatial variability of the environmental variables included in the model. A similar determination coefficient (0.740) was obtained when P_{YEAR} was replaced with $P_{YEAR} - P_{FEB}$. On average, *A* is 60 % of the control, and it is regulated by the application of fertilizers; therefore, the close link between *A* and environmental variables as independent of weather conditions in a specific year deserves attention. Now consider the leading environmental variables included in model (2).

The insolation of slopes from the southwest *F*(235), i.e., a land surface feature, is the main environmental variable in the model (2). The positive correlation with this parameter indicates that the maximum addition *A* increases on the well-illuminated southwestern slopes. The southwestern azimuth is related to the fact that the root-inhabited soil layer is heated with some delay; hence, the southwestern slopes, rather than the southern ones, are the most heated (e.g., see Hwang et al. 2011). The selected grid spacing (600 m) implies that slope insolation in model (2) is described by mesotopography rather than by microlandforms. The comparatively high value of *t*-statistic for *F*(235) indicates that the optimal rate of fertilizer is reasonable to apply to well-heated southwestern slopes to obtain sustainable additions to the yield.

Average February precipitation P_{FEB} is the second leading environmental variable. The positive correlation between the maximum addition *A* and P_{FEB} indicates that *A* increases as P_{FEB} grows, i.e., with increasing depth of protecting snow cover in February. Thus, the snow retention measures undertaken, such as snow-retention forested belts (in steppe and forest–steppe regions) and coulisse planting of short-season crops with strong stems (in open and elevated fields), cannot be considered as efficiently organized. Although many works are devoted to the effect of frost and snow cover on winter wheat (Kuperman 1969a), the role of snow retention is difficult to assess from the results obtained for one or several fields, without consideration of a set of geographically different sites.

Soil type is the third leading environmental variable. More precisely, the product of indicator I_2 by insolation $F(235)$ is included in model (2). In other words, the role of soil types is mediated by topography that redistributes heat, light, and moisture. Substituting $I_2 = 1$ to Eq. (2), we obtain the solution branch for Soddy-podzolic soils, and substituting $I_2 = 0$, we obtain the solution branch for other soil types. Indicator I_1 was not included in the list of the leading predictors, probably because the number of observation points in Chernozems is too small (six). The positive correlation between the addition A and $I_2 \cdot F(235)$ implies that, for the Soddy-podzolic soils located in the northern part of the area, the positive role of insolation is more significant than for the other soil types.

Average annual precipitation is the fourth leading environmental variable. The negative correlation with this variable indicates that A decreases as P_{YEAR} grows. This may be related to the known phenomenon of wheat damping (Kuperman 1969b).

The total effect of these four environmental variables may be expressed as a predictive map calculated using Eq. (2). Values of all environmental variables are known for all grid points; therefore, the superposition of grids with their weights found from Eq. (2) can be used.

No observation points were installed on peat-bog and large river valley soils; therefore, their locations should be excluded from the model.

Now let us consider 27 observation points located in gray forest soils. Indicators I_1 and I_2 are not used because the single soil type is considered, but indicator I_3, discriminating the frost-resistant cultivar Mironovskaya-808 from the others (Akhtyrchanka, Kuntsevskaya-45 and -46, Eritrospermus-917), is significant. The regression model is as follows:

$$A = 5.392 \times F(230)_{+7.05} - 0.989 \times P_{\mathrm{JUN+JUL}-3.82} - 3.365 \times k\underset{-3.58}{\overset{T}{\max}} + 2.433 \times I_3 k \underset{2.45}{\overset{T}{\max}} - 138.869, R^2 = 0.784; P < 10^{-5}; R^2_{\mathrm{Pred}} = 0.675, \quad (3)$$

where $F(230)$ is slope insolation from the southwest, $P_{\mathrm{JUN+JUL}}$ is the average precipitation in June and July, k max is the maximal curvature that describes ridge landforms, and indicator I_3 describes wheat cultivars: Mironovskaya-808 ($I_3 = 1$) and others ($I_3 = 0$). In this case, the critical level of t-statistics is equal to 2.074 so that all predictors are significant.

R^2 is somewhat higher in this case than in model (2), but the significance level is slightly lower ($P < 10^{-5}$) because the sample size in model (3) is smaller than that in model (2). Nonetheless, the links are relatively close, and the variability of the environmental factors explains 78 % of variability in the maximum addition A to yields.

Now let us consider the environmental variables included in Eq. (3). Slope insolation from the southwest $F(230°)$ is the main environmental variable, similar to model (2). This confirms the conclusion from model (2) that slope insolation from the southwest calculated for mesoscale landforms is the leading environmental variable for A.

Average precipitation in June and July $P_{\mathrm{JUN+JUL}}$ is the second leading environmental variable. A negative correlation is observed between this variable and A, i.e., A decreases as precipitation grows during this period, when wheat damping is quite possible. The significant role of summer precipitation is also confirmed by the results of Cabas et al. (2009) who showed a negative correlation between the winter wheat yield and precipitation in August in Canada; however, these authors did not report the critical level of t-statistics, and the statistical significance of their results is therefore unclear.

The maximal curvature of land surface k max is the third leading environmental variable. Positive values of this MV describe ridge landforms such as pronounced near-water divide areas, convex terraces, and some other mesoscale landforms (see Shary et al. 2002). The negative correlation between A and k max means that the maximum addition to yields A decreases on ridge landforms and increases outside them.

Indicator I_3 discriminates between the frost-resistant cultivar Mironovskaya-808 and other cultivars, or more precisely, the product $I_3 \cdot k$ max is the fourth leading variable. Since the link between A and $I_3 \cdot k$ max is positive, the relation of the cultivar Mironovskaya-808 to ridge landforms is weaker than for other cultivars. Indeed, let us consider the solution branch of Eq. (3′) for Mironovskaya-808; i.e., let $I_3 = 1$. This results in the equation for this branch:

$$A = 5.39 \times F(230) - 0.989 \times P_{\mathrm{JUN+JUL}} - 0.932 \times k\mathrm{max}^{T} - 138.9. \qquad (3')$$

where the regression coefficient at k max^T is also negative but lower in the absolute value than that in Eq. (3′). For other cultivars $I_3 = 0$, and model (3″) takes the form (for another branch)

$$A = 5.39 \times F(230) - 0.989 \times P_{\mathrm{JUN+JUL}} - 3.37 \times k\mathrm{max}^{T} - 138.9. \qquad (3'')$$

It can be seen that the regression coefficient at k max^T in the case of other cultivars is also negative, but its absolute value is several times greater; i.e., ridge landforms play an essentially more significant role for A of these cultivars.

Note that the average February precipitation for gray forest soils was not included in the list of four leading predictors. This may be primarily related to the fact that the variation ranges of precipitation and temperature for these soils are narrower than those for all three soil types in model (2) because there are no gray forest soils in the north of the study area (Fig. 1). Second, the list of the leading predictors in model (3) includes the indicator of wheat cultivar, which is related to frost resistance, and the frost-resistant cultivar Mironovskaya-808 was used in most of the observation points (18 of 27). Third, gray forest soils may have their own specific features.

4 Discussion

In most agricultural studies, the role of topography was studied, if at all, on a very limited level based on one to three morphometric variables such as the elevation, slope steepness, and exposure. It was shown in this study that these MVs may play no significant role at all, in contrast to the relative insolation of southwestern slopes, which describe the microclimate and play the most important role in the study area. Similar results were obtained by many authors for forest ecosystems (see Hwang et al. 2011; Shary and Smirnov 2013) and by several authors for soils (Shary and Pinskii 2013).

We restricted the environmental variables by morphometric variables, average climate features, and soil types (e.g., not considering weather or quantitative soil properties); hence, the fundamental relations of the maximum addition to winter wheat yield to environmental variables in all our models only slightly depend on time at ~1-year time scales. So in models (2) and (3), we did not study an immediate response to weather changes but rather some basic relations of these additions to environmental variables that change relatively slowly with time. It has been found that topography plays the main role, while climate plays the second role. When soil types are considered in model (2), they occupy the third position in this sequence. If wheat cultivars are considered in model (3), they occupy the fourth position.

The consideration of other soil properties (such as particle-size distribution, acidity, and humus) could enhance our evaluation of the relative roles of environmental variables, but this is a much more labor-intensive and costly way to produce predictive maps.

5 Conclusions

1. Methods of predictive modeling offer great opportunities for the quantification of impacts of environmental variables on crop yields. These methods are data driven and require sophisticated and reliable monitoring systems and databases.
2. In our concrete study in the Oka basin, the use of an extended set of topographic attributes, climate, and soil indicator variables permits us to explain 74 % of variance in the maximum addition to winter wheat yield by environmental variables in the regression model.

3. The microclimate, as described by relative slope insolation from the southwest (a topographic attribute), was of great importance for the addition to winter wheat yield. Generally, the use of the extended system of topographic attributes, together with climate and soils, allows close links to be found in regression equations that describe winter wheat yield features.

References

Acutis M, Rana G, Trevisiol P, Bechini L, Laudato M, Ferrara R, Richter GM (2007) Integrating a spatial micrometeorological model into the risk assessment for arable crops in hilly terrain. In: Kersebaum K et al (eds) Modelling water and nutrient dynamics in soil–crop systems. Springer, Berlin, pp 51–57

Adams RM, Wu JJ, Houston LL (2006) Changes in crop yields and irrigation demand. In: Smith JB, Mendelsohn R (eds) The impact of climate change on regional systems. A comprehensive analysis of California. Bodmin, Cornwall, pp 147–164

Austin M (2007) Species distribution models and ecological theory: a critical assessment and some possible new approaches. Ecol Model 200:1–19

Cabas J, Weersink A, Olale E (2009) Crop yield response to economic, site and climatic variables. Clim Change. doi:10.1007/s10584-009-9754-4

Currie DJ (1991) Energy and large-scale patterns of animal- and plant-species richness. Am Nat 137:27–49

Davies RG, Orme CDL, Storch D, Olson VA, Thomas GH, Ross SG, Ding TS, Rasmussen PC, Bennett PM, Owens IPF, Blackburn TM, Gaston KJ (2007) Topography, energy and the global distribution of bird species richness. Proc Royal Soc Ser B 274(1614):1189–1197

Ferrara RM, Trevisiol P, Acutis M, Rana G, Richter GM, Baggaley N (2010) Topographic impacts on wheat yields under climate change: two contrasted case studies in Europe. Theor Appl Climatol 99:53–65

Gedymin AV (1992) The use of contour method for large-scale soil maps (Использование изображения рельефа горизонталями при создании почвенных карт крупного масштаба). Pochvovedenie 5:5–14

Guisan A, Zimmermann NE (2000) Predictive habitat distribution models in ecology. Ecol Model 135:147–186

Hijmans RJ, Cameron SE, Parra JL, Jones PG, Jarvis A (2005) Very high resolution interpolated climate surfaces for global land areas. Int J Climatol 25(15):1965–1978

Hwang T, Song C, Vose JM, Band LE (2011) Topography-mediated controls on local vegetation phenology estimated from MODIS vegetation index. Landscape Ecol 26:541–556

Kuperman FM (1969a) Physiology of wheat persistency (Физиология устойчивости пшеницы). In: Genkel PA (ed) Physiology of agricultural plants. Physiology of wheat (Физиология сельскохозяйственных растений. Том IV. Физиология пшеницы), vol 4. Moscow State University, Moscow, pp 401–497

Kuperman FM (1969b) Physiology of wheat water exchange (Физиология водообмена пшеницы). In: Genkel PA (ed) Physiology of agricultural plants. Physiology of wheat (Физиология сельскохозяйственных растений. Том IV. Физиология пшеницы), vol 4. Moscow State University, Moscow, pp 204–239

Lastochkin AN (1991) Land surface properties, principles and methods of the statistical geomorphology (Рельеф земной поверхности-принципы и методы статической геоморфологии). Nedra, Leningrad, 340 p

Lobell DB, Asner GP (2003) Climate and management contributions to recent trends in U.S. agricultural yields. Science 299:1032

MacMillan RA, Shary PA (2009) Landforms and landform elements in geomorphometry. In: Hengl T, Reuter HI (eds) Geomorphometry: concepts, software, applications. Developments in soil science, vol 33, Chapter 9. Elsevier, Amsterdam, pp 227–254
McBratney AB, Odeh IOA, Bishop TFA, Dunbar MS, Shatar TM (2000) An overview of pedometric techniques for use in soil survey. Geoderma 97:293–327
Montgomery DC, Peck EA (1982) Introduction to linear regression analysis, 1st edn. Wiley, New York 504 p
Pike RJ, Evans IS, Hengl T (2009) Geomorphometry: a brief guide. In: Hengl T, Reuter HI (eds) Geomorphometry: concepts, software, applications. Developments in soil science, vol 33, Chapter 1. Elsevier, Amsterdam, pp 3–30
Reuter HI, Kersebaum KC, Wendroth O (2005) Modelling of solar radiation influenced by topographic shading—evaluation and application for precision farming. Phys Chem Earth 30:143–149
Scull P, Franklin J, Chadwick OA, McArthur D (2003) Predictive soil mapping: a review. Progr Phys Geogr 27:171–197
Shary PA (1995) Land surface in gravity points classification by a complete system of curvatures. Math Geol 27:373–390
Shary PA (2006) Geomorphometry in earth sciences and ecology, an overview of methods and applications (Геоморфометрия в науках о Земле и экологии, обзор методов и приложений). Izv Sam Nauchn Tsentra RAN 8(2):458–473
Shary PA (2011) Orographic zoning (Орографическое районирование). In: Shoba SA (ed) The national atlas of russian soils (Национальный атлас почв Российской Федерации). Part 2. Factors of soil formation. Astrel, Moscow, p 62
Shary PA, Pinskii DL (2013) Statistical evaluation of the relationships between spatial variability in the organic carbon content in gray forest soils, soil density, concentrations of heavy metals, and topography. Eurasian Soil Sci 46:1076–1087
Shary PA, Smirnov NS (2013) Mechanisms of the effects of solar radiation and terrain anisotropy on the vegetation of dark conifer forests in the Pechora-Ilych state biosphere reserve. Russ J Ecol 44:9–17
Shary PA, Sharaya LS, Mitusov AV (2002) Fundamental quantitative methods of land surface analysis. Geoderma 107:1–32
Shary PA, Rukhovich OV, Sharaya LS (2011) Methodology of analysis of the spatial variation in wheat yield depending on agricultural landscape conditions (Методология анализа пространственной изменчивости характеристик урожайности пшеницы в зависимости от условий агроландшафта). Agrokhimiya 2:57–81
Sibirtsev NM (1900) Soil science: lectures (Почвоведение: курс лекций, читанных студентам Института сельского хозяйства и лесоводства в Новой Александрии). In: Sibirtsev NM (ed) Selected works (Избранные сочинения), vol 1. Soil science, Selkhozgiz, 1951, 472 p
Simonov YuG (1998) Morphometric analysis of land surface (Морфометрический анализ рельефа). Smolensk Humanitarian University, Moscow-Smolensk, 271 p
Sychev VG, Rukhovich OV, Romanenkov VA, Belichenko MV, Listova MP (2008) Creation of the unified systematized database of field experiments of the agrochemical service and the Agrogeos geonetwork (Опыт создания единой систематизированной базы данных полевых опытов Агрохимслужбы и Геосети « Агрогеос »). Probl Agrokh Ekol 3:35–38
Wilson JP, Gallant JC (eds) (2000) Terrain analysis: principles and applications, 1st edn. Wiley, New York 479 p
Wood J (2009) Overview of software packages used in geomorphometry. In: Hengl T, Reuter HI (eds) Geomorphometry: concepts, software, applications. Developments in soil science, vol 33. Elsevier, Amsterdam, pp 257–267 Chapter 10

Chapter 22
Simulating Temperature Impacts on Crop Production Using MONICA

Claas Nendel

Abstract Process-based simulation models that predict crop growth, evapotranspiration, nitrate leaching or other environmental variables are commonly applied to assess their impact on agricultural crop production or the environment. Model inter-comparisons across a wide range of environments suggest that temperature relations are the most crucial for the success of individual models in capturing crop growth and yield formation at a specific site. For Siberia, where the annual temperature amplitude can easily exceed 80 K at some locations, temperature extremes are the most important challenge to the application of agro-ecosystem models. In this chapter, temperature-related algorithms of the dynamic simulation model MONICA are presented, including temperature dependencies of soil organic matter turn-over, plant photosynthesis and respiration, ontogenesis and the impacts of extremely high or low temperatures on crop growth. MONICA was developed to demonstrate the impact of the climate and management on crop yields and environmental variables on the plot scale and in smaller regions in Central Europe. Based on known biophysical processes, MONICA has the potential to assess the impacts of climate change and land management on crop yields, carbon balance and nitrogen efficiency in Siberia.

Keywords Crop modelling · Sensitivity · Climate change impact

1 Objectives

Siberia is a challenging environment for crop production. Harsh winters alternate with hot summers and the periods between the two extremes in which favourable conditions for crop growth prevail are short. Winter cropping is not possible as minimum temperatures frequently drop below the lethal temperatures of the crops

C. Nendel (✉)
Leibniz Centre for Agricultural Landscape Research, Institute of Landscape Systems Analysis, Eberswalder Straße 84, 15374 Müncheberg, Germany
e-mail: nendel@zalf.de

L. Mueller et al. (eds.), *Novel Methods for Monitoring and Managing Land and Water Resources in Siberia*, Springer Water,
DOI 10.1007/978-3-319-24409-9_22

grown in Siberia, and even though snow may cover the crops and protect them against extremely cold temperatures, the risk of black frosts when snow is absent is high. During spring, when the crops are sown, the transition from frozen to completely dried soil due to quickly rising temperatures takes only a few weeks. Strip tillage with narrow strip width is used to absorb as much solar radiation as possible in the cleared, bare soil strips to increase soil temperature and accelerate seed emergence and crop development. In the counter-strips, where the remaining straw of last year's crop is heaped in swaths, the soil is protected against evaporation, which keeps the moisture in the ground. Later on in the season, drought and high temperatures, including occasional extreme heat, frequently hamper crop growth.

The projected climate change is expected to have a considerable impact on agricultural crop production (Lobell et al. 2008; Godfray et al. 2010) and makes future food security for the steadily growing population uncertain. The nature and dimension of the impact, however, are expected to differ greatly in many regions across the world (IPCC 2014). Process-based simulation models are used to assess the impacts of climate and management on crop growth, yield formation or other environmental variables, such as nitrate leaching or soil organic matter development (Reidsma et al. 2010; Challinor et al. 2014). They use the input of global circulation models (Lucarini et al. 2007; Ramirez-Villegas et al. 2013) whose results have previously been scaled down to the spatial level of the impact models (Enke et al. 1997; Christensen et al. 2007; Déqué et al. 2007). Process-based models simultaneously explain the dynamics of a large set of variables and their interactions, and account for feedback relations. Many of them are available, with different strengths and weaknesses (Kersebaum et al. 2007; Rötter et al. 2012). To reduce the uncertainty of the simulations, process-based simulations models can be run in ensembles (Asseng et al. 2013; Martre et al. 2015). Since temperature in general, and extreme temperature at both ends in particular, has been identified as the most crucially changing driver for crop production in the future, much emphasis has been put on temperature relations in agro-ecosystem simulation models (Kumudini et al. 2014; Wang et al. 2014).

In this chapter, we present the algorithms of the dynamic, process-based simulation model MONICA (Nendel et al. 2011) which describe process responses to temperature in the crop and soil. The MONICA model was designed to simulate crop growth, water and nitrogen uptake, and the matter dynamics in the soil for applied purposes. Its simple and robust philosophy enables MONICA to operate under restricted data availability. MONICA is a 1D point model which works on a pedon with variable depth (default: 2 m). Results are interpreted for an area of 1 m^2. The temporal resolution for MONICA is 1 day, which is automatically split into smaller time steps when high water and nutrient fluxes occur, to ensure that the numerics are stable. The overall module concept is presented in Fig. 1. Table 1 shows the input variables. Many of the processes considered in the model are linked in a regulating manner, and affected by temperature. An offspring of the SUCROS model, MONICA follows a generic approach that describes features of different crops using only different sets of parameters. The described temperature functions are therefore valid for all crops that MONICA simulates.

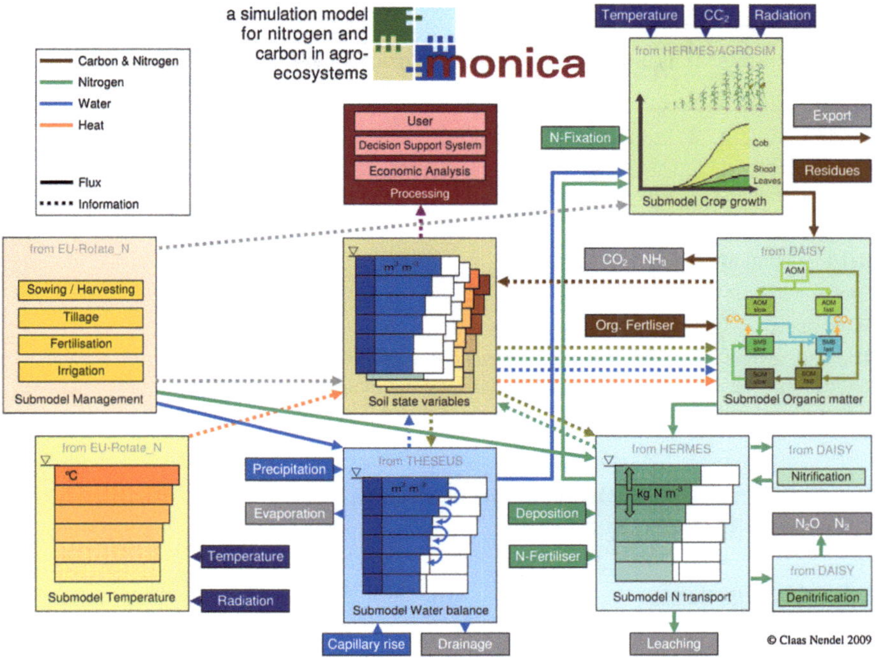

Fig. 1 The MONICA model concept

Table 1 MONICA input variables

Minimum requirement		Optional input	
Variable	Unit	Variable	Unit
Site			
Latitude	°	*N* deposition	kg N ha^{-1} a^{-1}
Altitude	m	Soil stone content	kg kg^{-1}
Horizon boundaries	m	Soil pH	
Soil sand content	kg kg^{-1}	Max. groundwater table	m
Soil clay content	kg kg^{-1}	Min. groundwater table	m
Soil bulk density	kg m^{-3}	Slope	m m^{-1}
Soil organic matter content	kg kg^{-1}	Texture class[a]	
Crop			
Sowing date	ddmmyyyy	Irrigation dates	ddmmyyyy
Crop type		Irrigation amount	mm d^{-1}
Tillage dates	ddmmyyyy		
Tillage depth	m		
N fertilising dates	ddmmyyyy		
N fertiliser type			
Harvest date	ddmmyyyy		
Fraction residue take-off	%		

(continued)

Table 1 (continued)

Minimum requirement		Optional input	
Variable	Unit	Variable	Unit
Weather			
Precipitation	mm	CO_2 concentration	ppm
Min. air temperature 2 m	°C	Sunshine hours[b]	h
Max. air temperature 2 m	°C		
Relative humidity	Pa Pa^{-1}		
Global radiation[b]	MJ m^{-2}		
Wind speed	m s^{-1}		

[a]The soil texture class following the German classification system (Ad hoc-AG Boden 2005 2649/id) may be used if particle size fractions are not available
[b]Either global radiation or sunshine hours are required

2 Temperature Functions in the Crop Part of the MONICA Model

2.1 Ontogenesis

Plant development is simulated using the principle of heat summation. The effective temperature is limited by a minimum temperature, which is referred to as the base temperature. A suitable level of soil moisture (at least 30 % of available water, but less water than would cause ponding at the soil surface) is required for seed emergence. If the soil moisture is ideal, the temperature of the top soil layer can be used for heat summation as

$$DD_{0,t} = DD_{0,t-1} + (T_{S10} - T_{B0}) \cdot \Delta t \tag{1}$$

where $DD_{0,t}$ is the actual temperature sum at the developmental stage 0; $DD_{0,t-1}$ is yesterday's temperature sum at the developmental stage 0; T_{S10} is the soil temperature at 0–10 cm depth; T_{B0} is the base temperature at the developmental stage 0; and Δt is the time step. As soon as the crop-specific temperature sum for seed emergence is reached, the subsequent developmental stage is initiated. From this moment the daily mean air temperature will be summed up. Stress factors considering drought and N deficiency accelerate the summation, while vernalisation and day length factors decelerate it. Degree days are summed as

$$DD_{n,t} = DD_{n,t-1} + (T_{av} - T_{Bn}) \cdot b_S \cdot b_V \cdot b_D \cdot \Delta t \tag{2}$$

where

$$b_S = \max(1 + (1 - \zeta_W)^2, 1 + (1 - \zeta_N)^2) \tag{3}$$

and $DD_{n,t}$ is the actual temperature sum at the developmental stage n; $DD_{n,t-1}$ is yesterday's temperature sum at the developmental stage n; T_{av} is the daily mean air temperature at 2 m above the ground; T_{Bn} is the base temperature at developmental stage n; b_S is the environmental stress acceleration factor; b_V is the vernalisation factor; b_D is the day length factor; ζ_W is the drought stress factor; and ζ_N the stress factor for N deficiency.

2.2 Crop Growth

The modelling of crop growth follows a generic approach used by SUCROS model (van Keulen et al. 1982). Daily net dry matter production by photosynthesis and respiration is driven by radiation and temperature. Gross CO_2 assimilation is calculated by estimating the sky cover duration. [CO_2] has an impact on the crop's photosynthesis rate and stomata resistance, which in turn influences transpiration (Nendel et al. 2009). Mitchell et al. (1995) presented a set of algorithms for calculating the maximum rate of photosynthesis, based on the ideas of Farquhar and von Caemmerer (1982) and Long (1991)

$$A = \frac{(C_i - \Gamma*) \cdot V_{c\max}}{C_i + K_c \cdot \left(1 + \frac{O_i}{K_o}\right)} \quad (4)$$

where A is the CO_2 assimilation rate; C_i is the inter-cellular CO_2 concentration; $\Gamma*$ is the compensation point of photosynthesis, related to C_i in the absence of dark respiration; O_i is the inter-cellular O_2 concentration; $V_{c\max}$ is the maximum saturated Rubisco carboxylation rate; and K_C and K_O are the Michaelis-Menten constants for CO_2 and O_2, respectively. Temperature dependencies of C_i, O_i, K_c, K_O, and $V_{c\max}$ and its parameters were described by Long (1991). Accordingly, C_i is calculated from the atmospheric CO_2 concentration C_a as

$$C_i = C_a \cdot 0.7 \cdot \frac{\left(1.674 - 6.1294 \cdot 10^{-2} \cdot T_{\mathrm{av}} + 1.1688 \cdot 10^{-3} \cdot T_{\mathrm{av}} - 8.8741 \cdot 10^{-7} \cdot T_{\mathrm{av}}^3\right)}{0.73547} \quad (5)$$

and O_i is calculated as

$$O_i = 210 + \frac{\left(0.047 - 1.3087 \cdot 10^{-4} \cdot T_{\mathrm{av}} + 2.5603 \cdot 10^{-6} \cdot T_{\mathrm{av}}^2 - 2.1441 \cdot 10^{-8} \cdot T_{\mathrm{av}}^3\right)}{2.6934 \cdot 10^{-2}} \quad (6)$$

The algorithm used for light intensities below saturation, presented by Mitchell et al. (1995), is not applied in the model. Instead, $A_{\max}$ is adapted to light interception according to Goudriaan and van Laar (1978). Mitchell et al. (1995)

proposed the following algorithm for the transition between photosynthetic quantum use efficiency and light-saturated photosynthesis:

$$\varepsilon_L = \frac{0.37 \cdot (C_i - \Gamma *)}{4.5 \cdot C_i + 10.5 \cdot \Gamma *} \quad (7)$$

where ε_L is the radiation use efficiency of CO_2 assimilation. The compensation point of photosynthesis is obtained from

$$\Gamma * = \frac{0.5 \cdot 0.21 \cdot V_{c\max} \cdot O_i}{V_{c\max} \cdot K_o} \quad (8)$$

where

$$V_{c\max} = 98 \cdot \frac{A_{\max}}{34.668} \cdot k(T)_{vc\max} \quad (9)$$

and $A_{\max}$ denotes the plant-specific maximum CO_2 assimilation rate and $k(T)_{vc\max}$ is the temperature function for $V_{c\max}$. The different temperature dependencies finally result in a temperature response of the CO_2 assimilation rate as portrayed in Fig. 2.

For crops with C4 metabolism, we assume that the atmospheric CO_2 concentration has no direct impact on photosynthesis. The crop-specific maximum CO_2 assimilation rate is merely modified by a simple temperature function (Fig. 3).

Maintenance respiration is calculated for the daylight period (photoperiod) and darkness using AGROSIM algorithms (Mirschel and Wenkel 2007):

$$M_{\text{photo}} = \sum (W_i \cdot m_i) \cdot 2^{a \cdot (T_{\text{photo}} - b)} \cdot (2 - L_N) \quad (10)$$

$$M_{\text{dark}} = \sum (W_i \cdot m_i) \cdot 2^{a \cdot (T_{\text{dark}} - b)} \cdot L_N \quad (11)$$

where

$$L_N = 2 - \left(\frac{L_P}{12}\right) \quad (12)$$

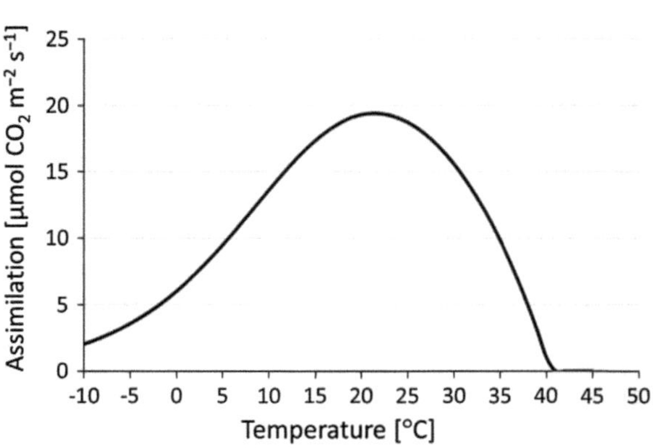

Fig. 2 Temperature response of the CO_2 assimilation rate for winter wheat (Long 1991)

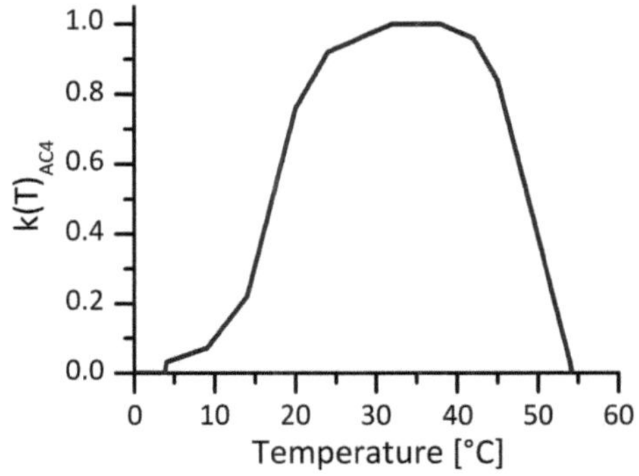

Fig. 3 Temperature function for the CO_2 assimilation rate of C4 crops (Sage and Kubien 2007)

and

$$T_{\text{photo}} = T_{\max} - \left(\frac{T_{\max} - T_{\min}}{4}\right) \tag{13}$$

$$T_{\text{dark}} = T_{\max} - \left(\frac{T_{\max} - T_{\min}}{4}\right) \tag{14}$$

where M_{photo} and M_{dark} are the maintenance respiration during the photoperiod and darkness, respectively. W_i is the dry mass of organ I; m_i is the specific maintenance respiration of organ I; L_N is the normalised day length; L_P is the photoactive day length; and T_{photo} and T_{dark} are the mean temperatures during the photoperiod and dark period, respectively. $T_{\max}$ is the daily maximum and $T_{\min}$ the daily minimum air temperatures. Growth respiration is calculated accordingly.

Assimilates produced in the photosynthesis module are distributed to individual plant organs. The partitioning coefficients are taken from a matrix which includes the development stages of the crop. Daily partitioning is calculated using linear regression between the elements of the matrix and the relative crop development.

2.3 *Heat Stress*

Heat stress affects the production of ovules during bloom, and is defined in this model as heat that exceeds a plant-specific threshold during the photoperiod. According to Challinor et al. (2005), during a defined plant-specific phase of increased sensitivity, extreme temperatures affect the crop following

$$F_H = 1 - \left(\frac{T_{\text{photo}} - T_{\text{critH}}}{T_{\text{limH}} - T_{\text{critH}}}\right) \cdot r_{\text{F}} \tag{15}$$

where F_H is the heat impact on ovules; T_{critH} is the critical temperature for the heat stress effect; $\mathrm{T_{limH}}$ is the maximum temperature for heat stress effect; and r_{F} is the daily flower emergence rate during bloom, which is calculated, following an idea by Moriondo et al. (2011), as

$$r_{\mathrm{F}} = p_{F,d} - p_{F,d-1} \tag{16}$$

where $p_{F,d}$ denotes the fraction of flowers that opened today and $p_{F,d-1}$ the fraction of flowers that opened the day before. The fraction of open flowers p_F is calculated as

$$p_F = \frac{1}{1 + \left(\frac{1}{0.015} - 1\right) \cdot \mathrm{e}^{-1.4 \cdot D_{\mathrm{BF}}}} \tag{17}$$

where D_{BF} denotes the number of days after the start of bloom. The reduction factor ζ_H is then derived from the smallest value of F_H during the sensitive phase

$$\zeta_H = \min(F_{H1,\ldots,}F_{Hn}) \tag{18}$$

where F_{H1} and F_{Hn} are the levels of heat impact on the first day and last days of the sensitive phase, respectively. Heat stress reduces the assimilate flow to the storage organ

$$W_s = A_g \cdot a_s \cdot \zeta_H \tag{19}$$

where W_i is the biomass of the storage organ; A_{g} is the gross CO_2 assimilation; and a_{s} is the assimilate partitioning coefficient for the storage organ.

2.4 Frost Kill

In cold winter environments, frost kill is a frequent mechanism that determines crop rotation decisions. The simulation of frost kill is based on a simulation of the threshold temperature below which 50 % of the crop dies from frost injuries ($\mathrm{LT_{50}}$). The algorithms used in MONICA have been developed based on the work of Fowler et al. (1999), who recently presented a revised version of their winter wheat model (Fowler et al. 2014). However, the principle used in this model is valid for almost all crops. Here, only the most important equations of this approach are presented; more details are found in Fowler et al. (2014).

$\mathrm{LT_{50}}$ is calculated from sowing onwards as:

$$\mathrm{LT}_{50t} = \mathrm{LT}_{50t-1} - H + D + \zeta_S + \zeta_R \tag{20}$$

where H denotes hardening and D denotes dehardening, ζ_S describes stress caused by exposure to very low temperatures and ζ_R stress in conditions where the ground is covered with snow, resulting in a soil temperature around 0 °C. LT_{50t} is restricted to values ≤0. Hardening is calculated as

$$H = H_0 \cdot (T_i - T_C) \cdot (LT_{50t-1} - LT_{50c}) \tag{21}$$

for $T_C < T_i$. LT_{50c} is the maximum frost tolerance of the cultivar, and H_0 is a cultivar-independent constant, set at 0.014 (Fowler et al. 2014). T_i is the cultivar's acclimation threshold induction temperature. Its value depends on LT_{50c} (Fowler et al. 2014) and it is calculated as

$$T_i = 3.72135 - 0.401124 \cdot LT_{50c}. \tag{22}$$

T_C is the crown temperature which is calculated as

$$T_C = \frac{(3 \cdot T_S + 2 \cdot T_{10})}{5} \tag{23}$$

for developmental stages before heading, with T_S being the temperature at the soil surface and T_{10} the temperature in the 0–10 cm layer, assuming that the plant's growing point is at 2 cm soil depth. Hardening ends as soon as the crop's vernalisation requirement is satisfied. Dehardening occurs as

$$D = \frac{5.05}{1 + e^{4.35 - 0.28 \cdot T_C}} \tag{24}$$

for $T_C \geq T_i$ during vernalisation or $T_C \geq -4$ °C after the crop's vernalisation requirement is satisfied (Fowler et al. 2014). After heading, T_C is considered equal to the air temperature overnight T_{dark}, where a damping factor of 0.8 considers the protecting character of the canopy.

Loss of cold tolerance due to cold stress at near-lethal temperatures is considered as

$$\zeta_S = \frac{(LT_{50t-1} - T_C)}{e^{\zeta_{S0}(LT_{50t-1} - T_C) - 3.47}} \tag{25}$$

where ζ_{S0} is a constant independent of the cultivar, set at 0.654 (Fowler et al. 2014), $T_C < -3$ °C and $|LT_{50} - Tc| < 12$ K. The effect of snow cover on crop respiration is considered as

$$\zeta_R = \zeta_{R0} \cdot \frac{(e^{0.84 + 0.051 \cdot T_C} - 2)}{1.85} \cdot f(s) \tag{26}$$

where ζ_{R0} is a constant independent of the cultivar, set at 0.54 (Bergjord et al. 2008) and $f(s)$ is a function of snow depth which takes a value between 0 and 1, increasing linearly with the snow depth up to 125 mm and remaining constant thereafter.

The death of 50 % of the population is simulated by reducing assimilate production by 0.5 each day when the condition $T_C < LT_{50}$ is met.

3 Temperature Functions in the Soil Part of the MONICA Model

The calculation of organic matter turn-over is based on algorithms used in the DAISY model (Hansen et al. 1991). The soil carbon dynamics is described using three pairs (rapid and slow turn-over) of conceptional pools (soil organic matter, microbial biomass and freshly added organic matter; Fig. 4). The rate coefficients of decomposition depend on the soil temperature (Fig. 5) and moisture and describe the environmental conditions of the simulated site. The same coefficients are used for the processes of nitrification and denitrification, which govern the production of NO_3 and, consequently, NO_2 and N_2O in the soil.

Additional temperature effects are related to the calculation of evapotranspiration, where the difference between saturation vapour pressure and actual vapour pressure drives the vaporisation of surface-near soil water and the water loss

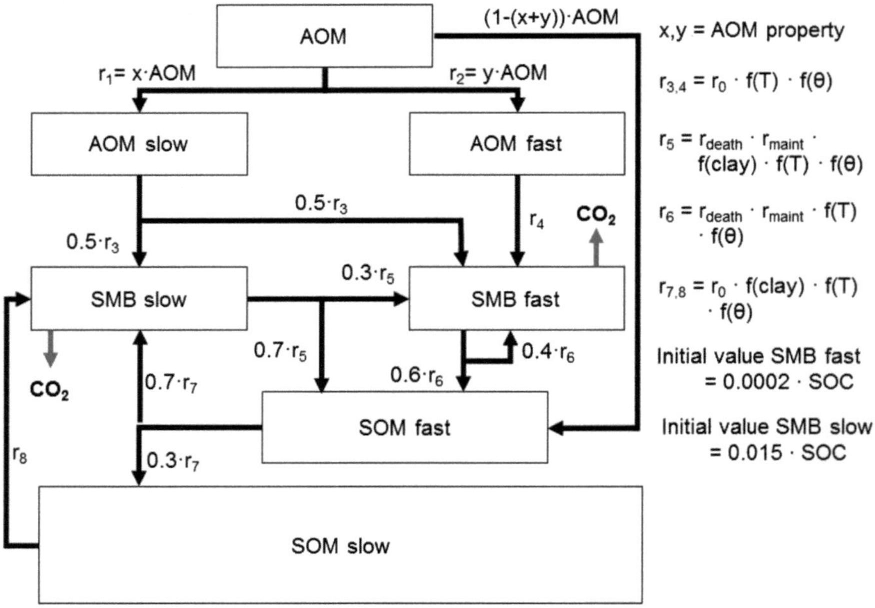

Fig. 4 Scheme of the organic matter turn-over module. *AOM* added organic matter pool; *SMB* soil microbial biomass pool; *SOM* soil organic matter pool, each of which is characterised by a rapid and a slow decomposition rate. r_0 turn-over rate at standard conditions; r_{death} death rate of microbial biomass; r_{maint} maintenance respiration of microbial biomass; f(clay), $f(T)$, $f(\theta)$ = decomposition rates for soil organic matter depending on soil clay content, temperature and moisture (Abrahamsen and Hansen 2000)

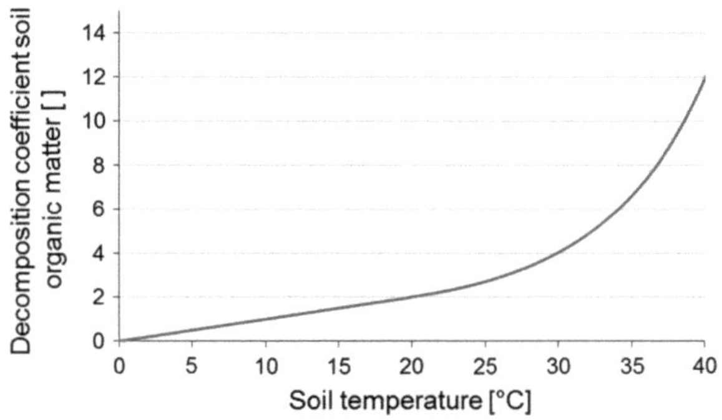

Fig. 5 *f*(*T*)—Decomposition coefficient for soil organic matter depending on soil temperature (Abrahamsen and Hansen 2000). The same coefficient is used for the processes of nitrification and denitrification in the soil

through the plant's stomata. The atmosphere's capacity to store water vapour depends on the air temperature. The temperature also governs the freeze-thaw dynamics in the soil and the accumulation or depletion of snow at the soil's surface, and thus significantly influences soil water movement on some occasions.

4 Temperature Sensitivity

MONICA's sensitivity to temperature and the model's ability to predict crop growth in response to temperature has been subject to a number of analyses. In general, MONICA showed an above-average performance in blind simulations of wheat yield—that is, without prior calibration to the environmental conditions of the respective dataset—across largely different locations, starting from a hot and dry environment in Western Australia with yield expectations below 2 t ha^{-1} and ending at a high-yielding ($\sim$10 t ha^{-1}) temperate and moist environment in the Netherlands (Asseng et al. 2013). Around-average performance was demonstrated in a similar model performance test for maize, with locations in France, the USA, Brazil and Tanzania (Bassu et al. 2014). Both studies also included a sensitivity analysis to temperature, which demonstrates how temperatures both lower and higher than current temperatures affect the yield of locally adapted varieties negatively across most simulation models employed in any ensemble, including MONICA (Fig. 6; Bassu et al. 2014).

Going one step further, Asseng et al. (2015) challenged an ensemble of crop models with a free-air temperature enhancement experiment in Maricopa, AZ, USA, and a data set of two wheat varieties monitored across different hot climate locations across the world, in which the models were tested for their ability to simulate yield penalties and complete crop failure due to extreme heat during

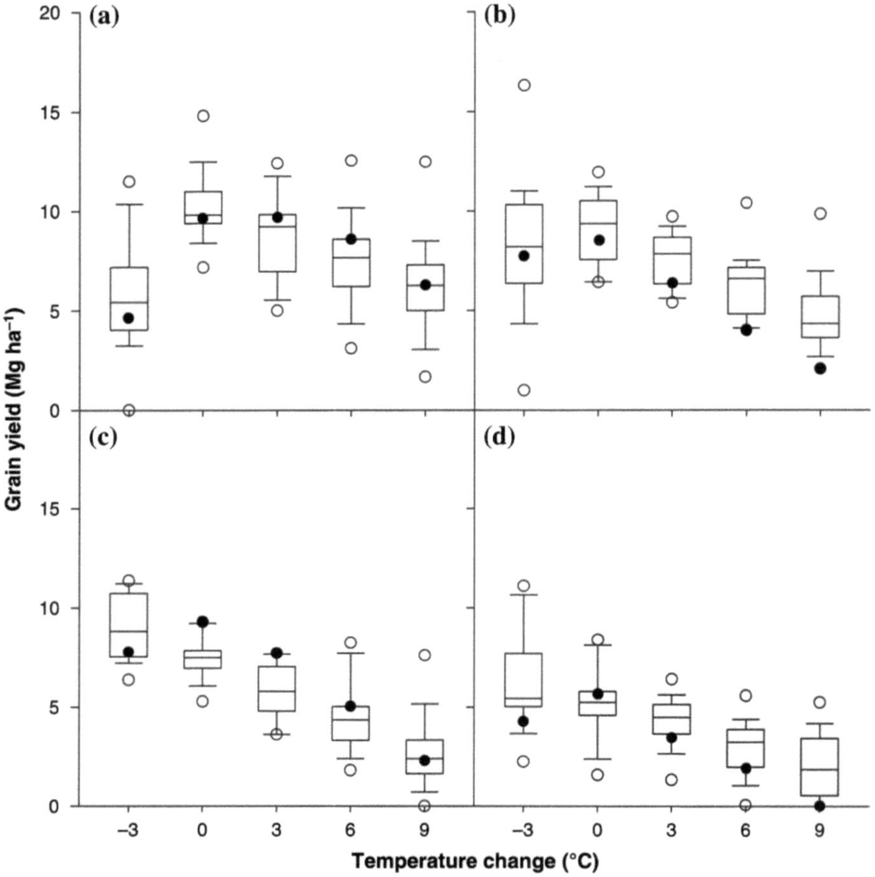

Fig. 6 Grain maize yield sensitivity to temperature (means of 30-year scenarios) as simulated by an ensemble of crop models, including MONICA (•), at four locations across the globe (Bassu et al. 2014)

different developmental phases (Fig. 7). Here, too, MONICA was found to be amongst the above-average performers.

The general pattern of how MONICA responds to combinations of temperature and rainfall was investigated in a study by Pirttioja et al. (2015), in which the response surfaces of a number of simulation models were compared. Along a temperature and rainfall gradient across three locations in Europe, MONICA showed a reasonable response to both climate variables, with the temperature significantly limiting the yield in Finland and rainfall having the same effect in Spain (Fig. 8). Subsequently, MONICA showed a mixture of both patterns for the German location. All in all, across the most recent model inter-comparisons, MONICA demonstrated its fitness for purpose to simulate crop growth in cold, temperate and warm environments.

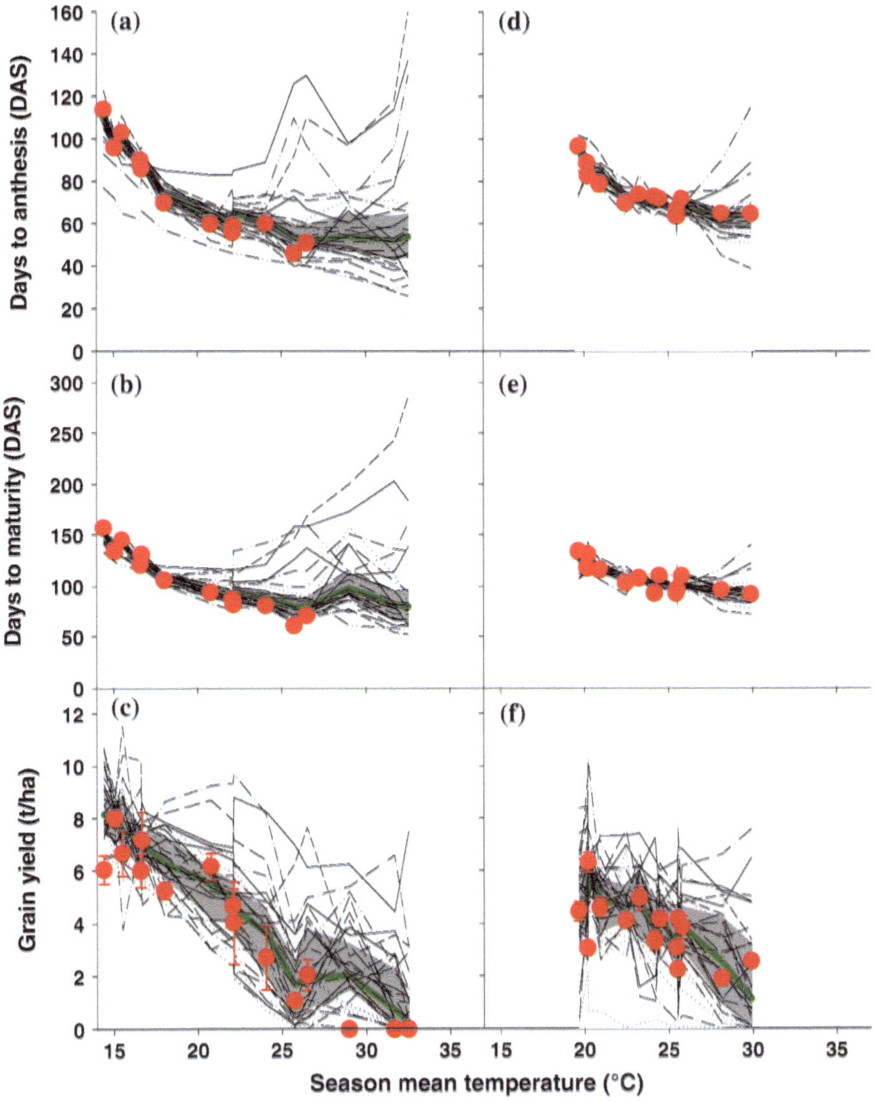

Fig. 7 Ensemble simulation of two heat stress data series of wheat (Asseng et al. 2015). Observed values ±1 standard deviation (s.d.) are shown by *red symbols* with 30 simulated values shown by *black lines*. **a–c** Hot–Serial–Cereal experiment on *Triticum aestivum* L. cultivar Yecora Rojo with days after sowing (*DAS*), time of sowing and infrared heat treatments. **d–f** CIMMYT multi-environment temperature experiments on *T. aestivum* L. cultivar Bacanora with time-of-sowing treatments. Multi-model ensemble medians are shown by *green lines*. Intervals between the 25th and 75th percentiles are shaded *gray*

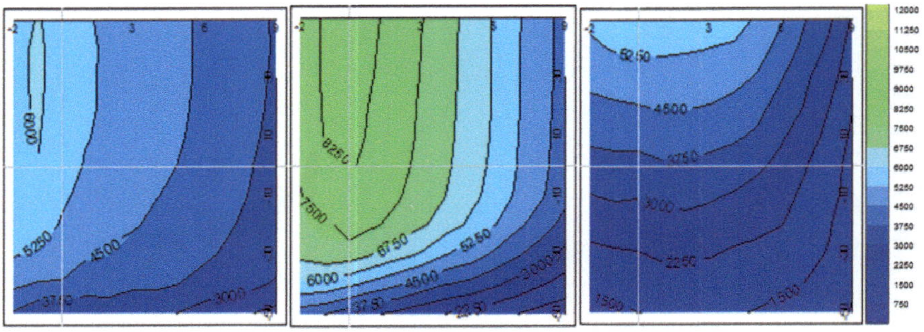

Fig. 8 MONICA temperature (*x*-axis) and precipitation (*y*-axis) response surfaces for winter wheat grain yield at three locations across Europe. *Left* Finland, *centre* Germany, *right* Spain (Pirttioja et al. 2015). Temperature and precipitation changes are relative to current conditions, marked with a *grey line*

References

Abrahamsen P, Hansen S (2000) Daisy: an open soil-crop-atmosphere system model. Environ Mod Software 15:313–330

Ad-hoc-AG Boden (2005) Bodenkundliche kartieranleitung. verbesserte und erweiterte Auflage, Hannover 5:438

Asseng S, Ewert F, Rosenzweig C, Jones JW, Hatfield JL, Ruane A, Boote KJ, Thorburn P, Rötter RP, Cammarano D, Brisson N, Basso B, Martre P, Aggarwal PK, Angulo C, Bertuzzi P, Biernath C, Challinor A, Doltra J, Gayler S, Goldberg R, Grant R, Heng L, Hooker J, Hunt T, Ingwersen J, Izaurralde C, Kersebaum KC, Müller C, Naresh Kumar S, Nendel C, O'Leary G, Olesen JE, Osborne TM, Palosuo T, Priesack E, Ripoche D, Semenov M, Shcherbak I, Steduto P, Stöckle C, Stratonovitch P, Streck T, Supit I, Tao F, Travasso M, Waha K, Wallach D, White J, Williams JR, Wolf J (2013) Quantifying uncertainties in simulating wheat yields under climate change. Nat Clim Change 3:827–832

Asseng S, Ewert F, Martre P, Rötter RP, Lobell DB, Cammarano D, Kimball BA, Ottman MJ, Wall GW, White JW, Reynolds MP, Alderman PD, Prasad PVV, Aggarwal PK, Anothai J, Basso B, Biernath C, Challinor A, De Sanctis G, Doltra J, Fereres E, Garcia-Vila M, Gayler S, Hoogenboom G, Hunt T, Izaurralde C, Jabloun M, Jones CD, Kersebaum KC, Koehler A-K, Müller C, Naresh Kumar S, Nendel C, O'Leary G, Olesen JE, Palosuo T, Priesack E, Eyshi Rezaei E, Ruane AC, Semenov M, Shcherbak I, Steduto P, Stöckle C, Stratonovitch P, Streck T, Supit I, Tao F, Thorburn P, Waha K, Wang E, Wallach D, Wolf J, Zhao Z, Zhu Y (2015) Rising heat reduces global wheat production. Nat Clim Change 5(2):143–147

Bassu S, Brisson N, Durand JL, Boote KJ, Lizaso J, Jones JW, Rosenzweig C, Ruane AC, Adam M, Baron C, Basso B, Biernath C, Boogard H, Conijn S, Corbeels M, Deryng D, de Sanctis G, Gayler S, Grassini P, Hatfield JL, Hoek SB, Izaurralde C, Jongschaap R, Kemanian AR, Kersebaum KC, Naresh Kumar S, Makowski D, Müller C, Nendel C, Priesack E, Pravia MV, Kim SH, Sau F, Shcherbak I, Tao FL, Teixeira E, Timlin D, Waha K (2014) How do various maize crop models vary in their responses to climate change factors? Glob Change Biol 20(7):2301–2320

Bergjord AK, Bonesmo H, Skjelvåg AO (2008) Modelling the course of frost tolerance in winter wheat I. Model development. Eur J Agron 28:321–330

Challinor AJ, Wheeler TR, Slingo JM (2005) Simulation of the impact of high temperature stress on the yield of an annual crop. Agric. Forest Meteorol 135: 180–189

Challinor AJ, Watson J, Lobell BD, Howden SM, Smith DR, Chhetri N (2014) A meta-analysis of crop yield under climate change and adaptation. Nat Clim Change 4:287–291

Christensen JH, Carter TR, Rummukainen M, Amanatidis G (2007) Evaluating the performance and utility of regional climate models: the PRUDENCE project. Clim Change 81:1–6

Déqué M, Rowell DP, Luthi D, Giorgi F, Christensen JH, Rockel B, Jacob D, Kjellstrom E, de Castro M, van den Hurk B (2007) An intercomparison of regional climate simulations for Europe: assessing uncertainties in model projections. Clim Change 81:53–70

Enke W, Spekat A (1997) Downscaling climate model outputs into local and regional weather elements by classification and regression. Clim Res 8(3):195–207

Farquhar GD, von Caemmerer S (1982) Modelling of photosynthetic response to environmental conditions. In: Lange OL, Nobel PS, Osmond CB, Ziegler H (eds) Encyclopedia of plant physiology. New series. Vol 12B. Physiological plant ecology. II. Water relations and carbon assimilation. Springer, Berlin, pp 549–587

Fowler DB, Limin AE, Ritchie JT (1999) Low-temperature tolerance in cereals: Model and genetic interpretation. Crop Sci 39:626–633

Fowler DB, Byrns BM, Greer KJ (2014) Overwinter low-temperature responses of cereals: analyses and simulation. Crop Sci 54:2395–2405

Godfray HCJ, Beddington JR, Crute IR, Haddad L, Lawrence D, Muir JF, Pretty J, Robinson S, Thomas SM, Toulmin C (2010) Food security: the challenge of feeding 9 billion people. Science 327:812–818

Goudriaan J, van Laar HH (1978) Relations between leaf resistance, CO_2 concentration and CO_2 assimilation in maize, beans, lalang grass and sunflower. Photosynthetica 12(3):241–249

Hansen S, Jensen HE, Nielsen NE, Svendsen H (1991) Simulation of nitrogen dynamics and biomass production in winter-wheat using the Danish simulation-model DAISY. Fert Res 27 (2–3):245–259

IPCC (2014) Climate change 2014: impacts, adaptation, and vulnerability. part A: global and sectoral aspects. In: Field CB, Barros VR, Dokken DJ, Mach KJ, Mastrandrea MD, Bilir TE, Chatterjee M, Ebi KL, Estrada YO, Genova RC, Girma B, Kissel ES, Levy AN, MacCracken S, Mastrandrea PR, White LL (eds) Contribution of working group II to the fifth assessment report of the intergovernmental panel on climate change. Cambridge University Press, Cambridge, p 1132

Kersebaum KC, Hecker J-M, Mirschel W, Wegehenkel M (2007) Modelling water and nutrient dynamics in soil-crop systems: a comparison of simulation models applied on common data sets. In: Kersebaum KC, Hecker J-M, Mirschel W, Wegehenkel M (eds) Modelling water and nutrient dynamics in soil crop systems. Springer, Stuttgart, pp 1–17

Kumudini S, Andrade F, Boote KJ, Brown GA, Dzotsi KA, Edmeades GO, Gocken T, Goodwin M, Halter AL, Hatfield JL, Jones JW, Kemanian AR, Kim SH, Kiniri J, Lizaso JI, Nendel C, Nielsen AR, Parent B, Stöckle C, Tardieu F, Timlin DJ, Wallach D, Yang HS, Tollenaar M (2014) Predicting maize phenology: Intercomparison of functions for developmental response to temperature. Agron J 106(6):2087–2097

Lobell DB, Burke MB, Tebaldi C, Mastrandrea MD, Falcon WP, Naylor RL (2008) Prioritizing climate change adaptation needs for food security in 2030. Science 319:607–610

Long SP (1991) Modification of the response of photosynthetic productivity to rising temperature by atmospheric CO_2 concentrations—Has its importance been underestimated. Plant Cell Environ 14(8):729–739

Lucarini V, Calmanti S, Dell'Aquila A, Ruti PM, Speranza A (2007) Intercomparison of the northern hemisphere winter mid-latitude atmospheric variability of the IPCC models. Clim Dynam 28(7–8):829–848

Martre P, Wallach D, Asseng S, Ewert F, Rosenzweig C, Jones JW, Hatfield JL, Ruane AC, Boote KJ, Thorburn PJ, Rötter RP, Cammarano D, Aggarwal PK, Angulo C, Basso B, Bertuzzi P, Biernath C, Brisson N, Challinor AJ, Doltra J, Gayler S, Goldberg R, Grant R, Heng L, Hooker J, Hunt LA, Ingwersen J, Izaurralde RC, Kersebaum KC, Müller C, Naresh Kumar S, Nendel C, O'Leary G, Olesen JE, Osborne TM, Palosuo T, Priesack E, Ripoche D, Semenov MA, Shcherbak I, Steduto P, Stöckle CO, Stratonovitch P, Streck T, Supit I, Tao F,

Travasso M, Waha K, White JW, Wolf J (2015) Multimodel ensembles of wheat growth: Many models are better than one. Glob Change Biol 21(2):911–925

Mirschel W, Wenkel KO (2007) Modelling soil-crop interactions with AGROSIM model family. In: Kersebaum K-C, Hecker J-M, Mirschel W, Wegehenkel M (eds) Modelling water and nutrient dynamics in soil crop systems: proceedings of the workshop on "Modelling water and nutrient dynamics in soil-crop systems" held on 14–16 June 2004 in Müncheberg, Germany. Springer, Dordrecht, pp 59–73

Mitchell RAC, Lawlor DW, Mitchell VJ, Gibbard CL, White EM, Porter JR (1995) Effects of elevated CO_2 concentration and increased temperature on winter-wheat—Test of ARCWHEAT1 simulation model. Plant Cell Environ 18(7):736–748

Moriondo M, Giannakopoulos C, Bindi M (2011) Climate change impact assessment: the role of climate extremes in crop yield simulation. Clim Change 104:679–701

Nendel C, Berg M, Kersebaum KC, Mirschel W, Specka X, Wegehenkel M, Wenkel K-O, Wieland R (2011) The MONICA model: testing predictability for crop growth, soil moisture and nitrogen dynamics. Ecol Mod 222:1614–1625

Nendel C, Kersebaum KC, Mirschel W, Manderscheid R, Weigel HJ, Wenkel K-O (2009) Testing different CO_2 response algorithms against a FACE crop rotation experiment. NJAS—Wageningen J Life Sci 57(1):17–25

Pirttioja N, Fronzek S, Bindi M, Carter TR, Hoffmann H, Palosuo T, Ruiz Ramos M, Trnka M, Acutis M, Asseng S, Baranowski P, Basso B, Bodin P, Buis S, Cammarano D, Deligios P, Doro L, Dumont B, Ewert F, Ferrise R, Destain MF, Gaiser T, Hlavinka P, Kersebaum KC, Kollas C, Krzyszczak J, Lorite Torres I, Minet J, Mínguez Tudela I, Montesino M, Moriondo M, Müller C, Nendel C, Öztürk I, Perego A, Rodríguez A, Ruget F, Sanna M, Semenov M, Slawinski C, Stratonovitch P, Supit I, Tao F, Wang E, Wu LH, Rötter RP (2015) Examining wheat yield sensitivity to temperature and precipitation changes in Europe for a large crop model ensemble using impact response surfaces. Clim Res. http://dx.doi.org/10.3354/cr01322

Ramirez-Villegas J, Challinor AJ, Thornton PK, Jarvis A (2013) Implications of regional improvement in global climate models for agricultural impact research. Env Res Lett 8 (2):024018

Reidsma P, Ewert F, Lansink AO, Leemans R (2010) Adaptation to climate change and climate variability in European agriculture: the importance of farm level responses. Eur J Agron 32 (1):91–102

Rötter RP, Palosuo T, Kersebaum KC, Angulo C, Bindi M, Ewert F, Ferrise R, Hravlinka P, Moriondo M, Nendel C, Olesen JE, Patil R, Ruget F, Tacáè J, Trnka M (2012) Simulation of spring barley yield in different climatic zones of Northern and Central Europe: a comparison of nine crop models. Field Crop Res 133:23–36

Sage RF, Kubien DS (2007) The temperature response of C3 and C4 photosynthesis. Plant Cell Environ 30(9):1086–1106

van Keulen H, Penning de Vries FWT, Drees EM (1982) A summary model for crop growth. In: Penning de Vries FWT, van Laar HH (eds) Simulation of plant growth and crop production. PUDOC, Wageningen, pp 87–97

Wang E, Martre P, Asseng S, Ewert F, Rötter RP, Alderman PD, Zhao Z, Cammarano D, Kimball BA, Ottman MJ, Wall GW, White JW, Reynolds MP, Prasad PVV, Aggarwal PK, Basso B, Biernath C, Challinor AJ, De Sanctis G, Doltra J, Fereres E, Gayler S, Goldberg R, Hoogenboom G, Hunt LA, Ingwersen J, Izaurralde RC, Jabloun M, Kersebaum KC, Koehler A-K, Lobell D, Müller C, Naresh Kumar S, Nendel C, O'Leary G, Olesen JE, Palosuo T, Priesack E, Eyshi Rezaei E, Ruane A, Semenov MA, Shcherbak I, Steduto P, Stöckle C, Stratonovitch P, Streck T, Supit I, Tao F, Thorburn P, Vignjevic M, Waha K, Wallach D, Wolf J, Zhu Y (2014) Causes for uncertainty in simulating wheat response to temperature. In: Rötter RP, Ewert F (eds) FACCE MACSUR CropM international symposium and workshop "Modelling climate change impacts on crop production for food security", Oslo, Norway, 10–12 February 2014, Abstract Book 32–33

Chapter 23
A Spatial Model-Based Decision Support System for Evaluating Agricultural Landscapes Under the Aspect of Climate Change

Wilfried Mirschel, Karl-Otto Wenkel, Michael Berg, Ralf Wieland, Claas Nendel, Barbara Köstner, Alexandre G. Topazh, Vitaly V. Terleev and Vladimir L. Badenko

Abstract Decision support for developing practicable, resilient climate change adaptation strategies for the sustainable use of agro-landscapes encompasses a wide range of options and issues. So far, only a few suitable tools and methods have been available to farmers, regional planners and other stakeholders to support decision-making processes in this direction. The model-based interactive spatial

---

W. Mirschel (✉) · K.-O. Wenkel · M. Berg · R. Wieland · C. Nendel
Leibniz-Centre for Agricultural Landscape Research (ZALF) e.V.,
Eberswalder Str. 84, 15374 Müncheberg, Germany
e-mail: wmirschel@zalf.de

K.-O. Wenkel
e-mail: wenkel@zalf.de

M. Berg
e-mail: Michael.Berg@zalf.de

R. Wieland
e-mail: rwieland@zalf.de

C. Nendel
e-mail: nendel@zalf.de

B. Köstner
Department of Meteorology, Technical University Dresden,
Pienner Straße 23 and 21, 01737 Tharandt, Germany
e-mail: Barbara.koestner@tu-dresden.de

A.G. Topazh
Agrophysical Research Institute, Grazhdansky Prospect 14,
195220 St. Petersburg, Russia
e-mail: topaj@hotmail.ru

V.V. Terleev · V.L. Badenko
State Polytechnical University St. Petersburg, Politekhnicheskaja ul. 29,
195251 St. Petersburg, Russia
e-mail: vitaly_terleev@mail.ru

L. Mueller et al. (eds.), *Novel Methods for Monitoring and Managing Land and Water Resources in Siberia*, Springer Water,
DOI 10.1007/978-3-319-24409-9_23

information and decision support system, LandCaRe-DSS, closes this methodical gap. This system does not only support interactive scenario simulations and multi-ensemble and multi-model simulations at the regional level by providing information about the complex long-term impacts of climate change. It also helps different stakeholders to find suitable, sustainable agricultural adaptation strategies to climate change (crop rotation, soil tillage, fertilisation, irrigation, price and cost changes etc.) at the local or farm level. LandCaRe-DSS uses different ecological impact models, including for crop yield, erosion risk, regional evapotranspiration, total water flow-out and irrigation water demand. At the local level, a farm economy model is directly coupled with both the biophysical-based agro-ecosystem model MONICA and the statistical-based crop yield model YIELDSTAT to simulate the economic consequences of regional climate change and of proposed agricultural adaptation strategies. Due to the modular architecture and innovative design of LandCaRe-DSS, alternative or new impact models can easily be incorporated into the system. Scenario simulation runs can be realised in a reasonable amount of time. The interactive LandCaRe-DSS prototype offers a variety of data analysis and visualisation tools and an information system for climate adaptation in agriculture. This article describes the conceptual framework, the structure, the methodology and basic principles of operating LandCaRe-DSS. A number of selected examples demonstrate the versatility of LandCaRe-DSS applications. Using different scales and regions as examples, the impact of climate change is shown on: the ontogenesis of winter wheat for Müncheberg, Germany; the start, end and duration of the vegetation period in two German regions Uckermark (dry lowlands, 2600 km^2) and Weisseritz (humid mountain area, 400 km^2); irrigation water demand in Thuringia, Germany and the winter wheat yield in the Prenzlau region, Germany. Using LandCaRe-DSS up to 2075 for the Uckermark und Weisseritz regions, the effects and impacts of different agricultural adaption strategies were analysed taking into account irrigation, the absence of soil tillage and two different cropping ratios (actual cropping ratio vs. cropping ratio enriched with energy maize). Thanks to the modular structure of LandCaRe-DSS, little effort is required to adapt the whole system to geo-data valid for other regions or countries; incorporate other static or dynamic impact models; switch to other climate scenarios and implement other interface communication languages. The LandCaRe-DSS is constantly being developed, updated and adapted in different research projects such as the REGKLAM project for agricultural regions of Saxony, Germany, and the CARBIOCIAL project for regions within the Mato Grosso and Pará states of Brazil. It has already been used in a number of climate scenario studies for the Federal States of Thuringia, Brandenburg and Saxony. In the years ahead, international cooperative activities will be initiated with institutions from St. Petersburg, Russia, and Puławy, Poland, in order to use, adapt and advance this system.

V.L. Badenko
e-mail: vbadenko@gmail.com

Keywords Decision support system • Productivity of arable land • Climate change • Regional impact assessment • Simulation models • Agricultural adaptation strategies

1 Introduction

One of today's and tomorrow's fundamental task for mankind is to evaluate the productivity of agricultural landscapes and develop sustainable regional land management strategies for the production of feed, fodder and energy. It is also important that we preserve landscape and biodiversity, conserving soil and water resources. Agricultural land use is a key aspect of this task. However, it is not easy to identify sustainable concepts for agriculture. Land management for food production is a fundamental human activity, supporting the existence of mankind. Approximately 10 % of the earth's 14 billion or so hectares of ice-free land is used to grow crops, and 25 % is used for pasture. Over 2 billion tons of grain are produced annually for food and feed, providing roughly two-thirds of our total direct and indirect protein intake, and a mere 10 % of this total, or 200 million tons, is traded internationally (Tubiello et al. 2007).

Across the globe, agricultural production is primarily dependent on the weather, the general patterns of which are determined by annual fluctuation. Agricultural production is also dependent on future climate change and its variability. Current research confirms that, although many crops would respond positively to elevated atmospheric CO_2 concentration, the associated impacts of high temperatures, altered patterns of precipitation and possibly an increased frequency of extreme events such as drought and floods would probably combine to depress yields and increase production risks in many regions of the world, widening the gap between rich and poor countries (IPCC 2007).

To date, most discussions about climate change have focussed on mitigation measures. Only little attention has been given to climate change adaptation, which will be critical for many countries on all continents, including several regions of Europe, such as the eastern part of Russia up to the Urals. Owing to the expected climate change, conditions for agriculture in northern parts of Europe will improve, unlike in southern, Mediterranean parts of Europe, where conditions for agriculture will be critical due to higher temperatures in summer and less precipitation. Each country or region must find out how they can reduce their vulnerability to climate change and increase desirable outcomes at the lowest possible cost, taking into account the respective regional and local economic and soil–climate conditions.

Adaptation to climate change requires sound knowledge of the potential regional and local impacts of climate and weather extremes. The effects of climate change on agriculture may be positive or negative, depending on the variability of weather conditions, site quality, land use and management. Adaptation must take account of sustainability with regard to high plant production without losing ecosystem services such as soil protection, the purification and recycling of water and the maintenance of biodiversity. In addition, it must be ensured that adaptation measures reduce rather than increase greenhouse gas emissions. This implies that decision making must consider not only the socio-economic consequences of adapted management but also the ecological impact. Various options are available for the site-specific adaptation of agricultural management to climate change. These options include using alternative crop rotations and new drought-resistant varieties; adapting sowing dates to the new climate; reducing or avoiding tillage methods and improving fertilisation, irrigation and plant protection regimes. In practice, however, the number of options is limited to just a few alternatives.

Recent developments in geographical information systems, robust climate impact models and data acquisition technologies have led to progress in modelling. Today's models are capable of identifying the potential of agricultural landscapes and the environmental constraints surrounding crop production at the regional and national level depending on expected climate change. In addition, models can be integrated in interactive usable decision support systems (DSS), helping farmers and other stakeholders to determine the best regional management practices for adjusting agriculture to climate change, taking account of the potential of agricultural landscapes.

However, it is very difficult to develop a model-based interactive and user-friendly DSS for this purpose. An ideal DSS needs to address agricultural regions as a whole, providing up-to-date scientific knowledge and regionalised soil and climate information, the latter being derived from the most recent global climate scenarios. At the same time, it must meet users' demand for transparency, interactivity and user-friendliness without any loss of information (Wenkel et al. 2013). Until now, only a few suitable spatial DSSs and analogous tools have been available. These include LADSS (LADSS 2005), ADSS (Prased et al. 2008) and GPFARM (Ascough et al. 2001), which were developed for specific regions only and are not available to the general public.

To close this gap, LandCaRe-DSS was developed by the Institute of Landscape Systems Analysis of the Leibniz-Centre for Agricultural Landscape Research (ZALF) Müncheberg, Germany, as a real-time response system using mainly regional models of intermediate complexity (REMICs; Wenkel et al. 2008). Other dynamic process-based agro-ecosystem models such as MONICA (Nendel et al. 2011) and AGROSIM (Mirschel and Wenkel 2007), both developed at ZALF, or AGROTOOL (Poluektov et al. 2002; Poluektov and Terleev 2007), developed at the Agrophysical Research Institute (ARI), St. Petersburg, Russia, can be employed at the field plot or farm level only.

This paper presents the conceptual framework, the basic structure and functionality of LandCaRe-DSS, a number of results of LandCaRe-DSS scenario studies and ways in which the system can be adapted for other regions or countries.

2 The Information and Decision Support System LandCaRe-DSS

2.1 Philosophy and Performance

The information and DSS LandCaRe-DSS for climate change impact assessment and climate adaptation of agriculture was developed within the LandCaRe 2020 (Land, Climate and Resources) project (Köstner et al. 2008) as part of the German research programme klima*zwei* (Mahammadzadeh et al. 2009). The main philosophy of LandCaRe-DSS is based on the assumption that there is not just one single solution to climate change-related problems. The system aims to provide all of the expertise required to resolve various issues. One of the advantages of computerised decision support is the ability to use a sheer unlimited amount of stored data, geo-simulation and rule-based models as innovative approaches towards data analysis and visualisation.

LandCaRe-DSS is a user-friendly, interactive, model-based and spatially oriented information and DSS for strategic planning. It also helps farmers and regional stakeholders to find cost-effective solutions for adapting regional agricultural production to climate change at different spatial scales. LandCaRe-DSS provides:

- greater knowledge about past and future climate change in the respective region;
- an evaluation of the productivity of agricultural landscapes;
- an estimation of the potential ecological and economic consequences of climate change;
- the ability to prepare and visualise all available knowledge on potential agricultural climate adaptation strategies; and
- suitable tools for interactive simulations in order to compare alternative land use systems and options for action.

2.2 Conceptual Framework and Methodology

LandCaRe-DSS consists of a general information and advisory system related to climate and climate change, a number of databases, various statistical and process-based simulation models for different spatial scales and a zooming user interface (ZUI) that connects all system levels. The climate change impact for agriculture can be assessed at three different spatial levels: the national level, the regional level and the farm level.

A central database administered by climate specialists contains historical climate data and regionalised future climate projections. Additional local databases contain parameters for different agricultural crops, detailed management and economic data for 188 different cropping procedures according to Münch et al. (2014) as well as geo-referenced data concerning land use, topography, hydrology and soil characteristics. Upon request, climate data are visualised in a condensed manner in order to highlight regional climate trends and weather statistics from the recent past (since 1950) as required by the user. Likewise, future climate scenarios based on simulations of the ECHAM5/MPI-OM Global Circulation Model (GCM) by the Max Planck Institute for Meteorology Hamburg (Roeckner et al. 2004) are shown. Each scenario has been scaled down to the regional level using statistical regionalisation methods WETTREG (Enke et al. 2005) and STAR2 (Gerstengarbe et al. 2003), which produce weather station-based data sets. It is also possible to use climate data created using the process-based regionalisation method CLM (Böhm et al. 2006), which produces weather data in grids of 5–18 km in length.

The system's modular structure also enables regionalised climate data from different emission scenarios of other global climate models to be linked with little effort. One example of such a global climate model is HadCM3, developed by the Met Office Hadley Centre, one of the UK's foremost climate change research centres. Regionalised climate data are also available for the European part of Russia up to the Urals and for Siberia, for example (Ivanov and Kirjushin 2009).

In order to perform a quick climate data analysis, user-friendly algorithms can be used in LandCaRe-DSS to undertake a trend analysis, an inner yearly analysis (daily and monthly) and a frequency analysis. Trend analysis can also be performed for the temperature (minimum, mean and maximum), annual precipitation, the climatic water balance, heating days, cooling days, the start and end of vegetation, the thermal sum, the chill sum and the climatic index for fruit trees according to Schwärzel (2000).

For the regional scale, the latest LandCaRe-DSS version currently includes a range of simulation and impact models and algorithms for calculating different climate and landscape indicators:

BAGLUVA is a regionalised water balance model for calculating the long-term averages of actual evapotranspiration and total flow-out (precipitation minus actual evapotranspiration, thus ground water recharge plus surface runoff). It is used to assess the hydrological impacts of changes in climate, vegetation and land use structures (ATV-DVWK-Regelwerk 2002).
EROSION is an empirical water erosion model that describes the impacts of farm management and climate on the potential soil erosion risk at the regional scale. It is based on the widely used Universal Soil Loss Equation, developed by Wischmeier and Smith (1978). The model is modified in accordance with DIN 19708 (2005). Since rainfall intensity in a 30-min resolution is not available for the climate scenarios, rainfall event lengths of 2 h for daily precipitation exceeding 10 mm were assumed when calculating rainfall intensity.
GL-PROD is a statistical model for grassland ecosystems (Käding et al. 2005) that calculate the impacts of changes in climate and grassland management on grassland yields and forage quality.
ONTO is a temperature sum model for calculating different stages and phases of plant development (ontogenesis) between sowing/planting and ripening for major

agricultural crops. The model is parameterised for 18 different climate regions of eastern Germany (Mirschel 2010).

PHAENO is a model for calculating the starting dates of typical phenological phases of different wild plants (indicator types) such as snowdrop (*Galanthus nivalis*), elder (*Sambucus nigra*), gooseberry (*Ribes grossularia*), apple (*Malus silvestris*), lime tree (*Tilia europaea*) and forest oak (*Quercus pedunculata*). The temperature sum models used, which calculate "chilling" and "forcing" units, were realised according to Henniges (2008) and Chmielewski and Henniges (2007).

VEGPER calculates the length of the vegetation period between the start of vegetation in spring and the end of vegetation in autumn. The model was realised according to Chmielewski (2003).

YIELDSTAT is a statistical-based hybrid model to estimate biomass, carbon fixation and yield of more than 15 agricultural crops, considering site characteristics, weather/climate, atmospheric CO_2 concentration, progress in agro-technology and plant breeding as well as management options, such as soil tillage and irrigation (Mirschel et al. 2014).

BERBEDUE is a model for identifying irrigation poverty. It can be used to identify general site-specific irrigation poverty depending on site conditions, agricultural land use and the weather/climate.

ZUWABE is an empirical model for calculating site-specific irrigation water demand depending on soil characteristics, crop types and rotations, crop-specific irrigation periods and site-specific climate conditions according to Roth (Roth 1993; TGL 1990).

A land use distribution model for computing the stochastic distribution of agricultural crops on arable land, taking into account scenario-related cropping ratios, the compatibility between crops and soil types, and the economic excellence of crops in a given region (model LANUDIS), can also be used at the regional scale.

At the larger regional or national scale, LandCaRe-DSS enables results to be viewed for crop yield, irrigation worthiness and revenues pre-calculated by the model RAUMIS (Gömann et al. 2005; Offermann et al. 2010). Due to its complexity, the RAUMIS model has not been integrated into LandCaRe-DSS. Only pre-calculated scenario results can be displayed in LandCaRe-DSS using its visualisation tools.

At the local or farm scale, both the statistical-based yield estimation model YIELDSTAT (Mirschel et al. 2014) and the process-based dynamic agro-ecosystem model MONICA (Nendel et al. 2011) can be used to calculate crop yields. The Farm Economy Model (FEM, Münch et al. 2014) uses output data from YIELDSTAT and MONICA to calculate farm-level cost accountancy items. It also helps simulate the economic consequences of regional climate change and proposed adaptation strategies. In addition, MONICA—a soil–plant–atmosphere model with a daily resolution—calculates field-specific crop and soil values, such as biomass, crop yield, nitrogen uptake, soil nitrogen, soil water and soil organic matter. The dynamics of all these values can be visualised within the vegetation year or over a

30-year period. An overview and detailed descriptions of all impact models contained in LandCaRe-DSS are given in Wenkel et al. (2013).

Thanks to the modular structure of LandCaRe-DSS, different impact models for describing landscape indicators can easily be integrated into the system. This is crucial if a system needs to be adapted to other regions or countries linked to the inclusion of static or dynamic impact models that have been parameterised and validated for these regions or countries.

The conceptual framework of LandCaRe-DSS is presented in Fig. 1.

LandCaRe-DSS supports long-term and ensemble simulations on a high spatial resolution (grid size: 100 m × 100 m), using coupled climate and agro-economic scenarios. The temporal resolution of models is no lower than 1 day, while the results of climate scenarios are typically evaluated for time periods of 30 years, providing robust information on the variability of the selected variable.

Unlike other DSS, LandCaRe-DSS offers the following special features (Wenkel et al. 2013):

Interactivity: The user defines the problem and the scenario, selects the models and runs them himself in real time. Data exchange between all system components occurs automatically.
Dynamic: A large variety of parameter constellations and simulations can be designed, analysed and compared with each other. New simulation results are created from the chosen preconditions.
Spatial dimension: The desired level of detail is achieved by zooming between the national, regional and farm scale. Results are returned as maps for the selected area.

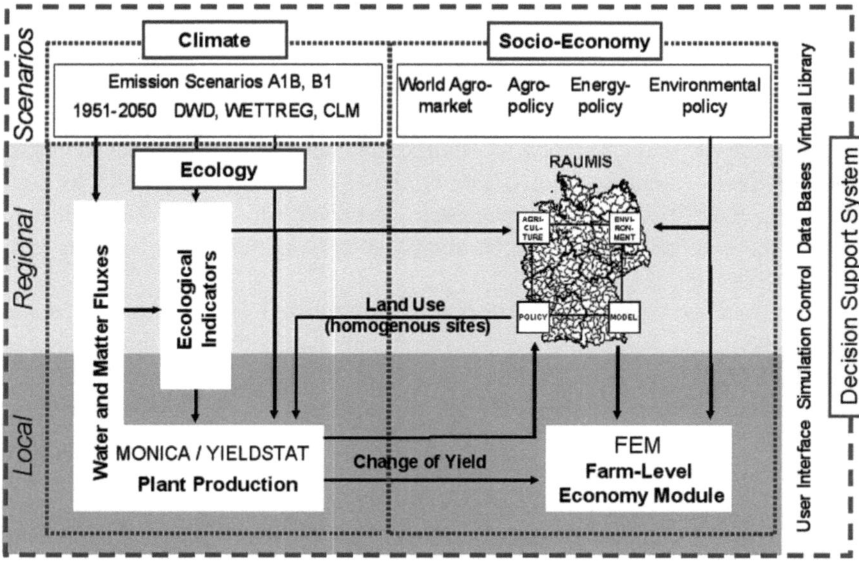

Fig. 1 Conceptual framework and levels of integration of different modules in LandCaRe-DSS (modified from Wenkel et al. 2013)

Scenario simulation: The temporal dimension is extended to the future by using climate projections based on emission scenarios and socio-economic scenarios (policies and markets) for creating response scenarios on ecological and economic impacts.
Web-usability: Central support, control and update of the LandCaRe-DSS software and the database. Positively tested model components are transferred step by step into a web application.
Extendibility: The system is open to further add-ons with a frequent update of knowledge, models, model parameters and data.
Adaptability: The whole system is easy to adapt to other regions or countries taking into account country-specific geo-data, other static and/or dynamic impact models incorporated, special parameterized and validated, other climate and/or emission scenarios and country-specific interface communication languages.

2.3 *Basic Principle of Operation*

Figure 2 shows the basic principle of operating LandCaRe-DSS. Operation can be characterised as an iterative process from defining the scenario and evaluating different agricultural farm management adaptation strategies to deciding, which is the best adaptation strategy for the particular farm.

The first step of the process involves the user defining of a scenario. The impacts of climate change are then simulated and evaluated. Next, the ecological and economic impacts of different agricultural adaptation strategies at the regional and farm scale are investigated using scenario techniques. Finally, the simulation results are analysed, graphically visualised and stored. Specialised tools are available to help users analyse and interpret the data.

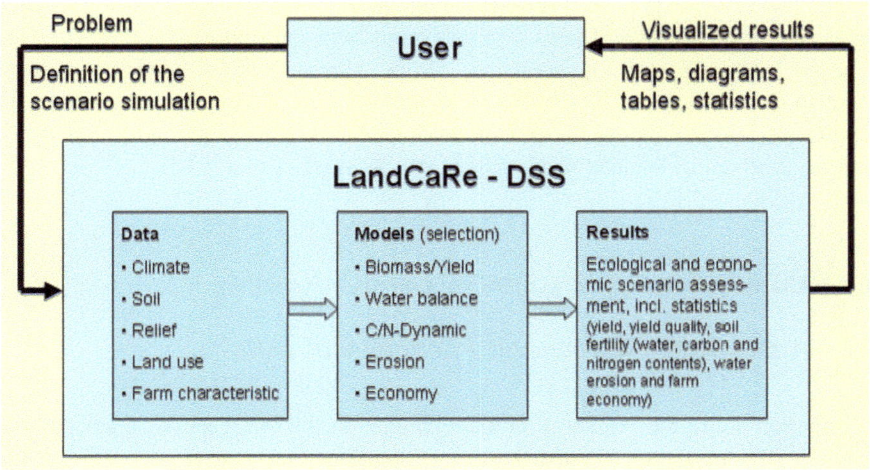

Fig. 2 Basic principle of operating LandCaRe-DSS

2.4 System Validation, Further Development and Availability

The prototype of LandCaRe-DSS was parameterised and validated for two different regions of Eastern Germany, the dry lowlands of the Federal State of Brandenburg (Uckermark region, 2600 km^2) and the humid mountainous area of the Free State of Saxony (Weisseritz region, 400 km^2).

Thanks to the modular structure of LandCaRe-DSS, little effort is required to adapt the system to geo-data bases valid for other regions or countries; incorporate other static or dynamic impact models; switch to other climate and emission scenario data and implement other interface communication languages.

LandCaRe-DSS is constantly being developed, updated and adapted in a number of research projects such as:

REGKLAM: Regional Adaptation Program to Climate Change for the Dresden Region, Saxony, Germany (REGKLAM 2013; REGKLAM-KONSORTIUM 2013),
CARBIOCIAL: Carbon sequestration, biodiversity and social structures in Southern Amazonia: models and implementation of carbon-optimised land management strategies for regions within the Mato Grosso and Pará states of Brazil (Carbiocial 2014),
KIT LandCaRe-DSS: Model-based tools for strategic and operational irrigation measures under climate change for the model region Uelzen, Lower Saxony, Germany (KIT LandCaRe-DSS 2014).

LandCaRe-DSS has been used in practical climate scenario studies for assessing the impact of climate change on agricultural productivity and irrigation water demand up to 2050 for three federal states of Germany: Saxony (Mirschel et al. 2009, 2010), Brandenburg (Mirschel et al. 2013a) and Thuringia (Mirschel et al. 2012, 2013b). Institutions in St. Petersburg, Russia, and an institution in Puławy, Poland, have also expressed a serious interest in the wish to use, adapt and advance this system on the basis of joint research projects in the future.

LandCaRe-DSS was converted into an operative web-based version as a basis for wider distribution (http://www.landcare-dss.de). A detailed description of *LandCaRe-DSS,* including the models and databases it contains and initial applications, is given in Wenkel et al. (2013) and Köstner et al. (2012).

3 Examples of Applications of LandCaRe-DSS

3.1 Analysis of Climate and Phenological Data

Information about the impact of climate change on the ontogenesis of agricultural crops is very important when it comes to planning agro-management. Using the example of winter wheat, Fig. 3 shows the lengths of different ontogenesis stages

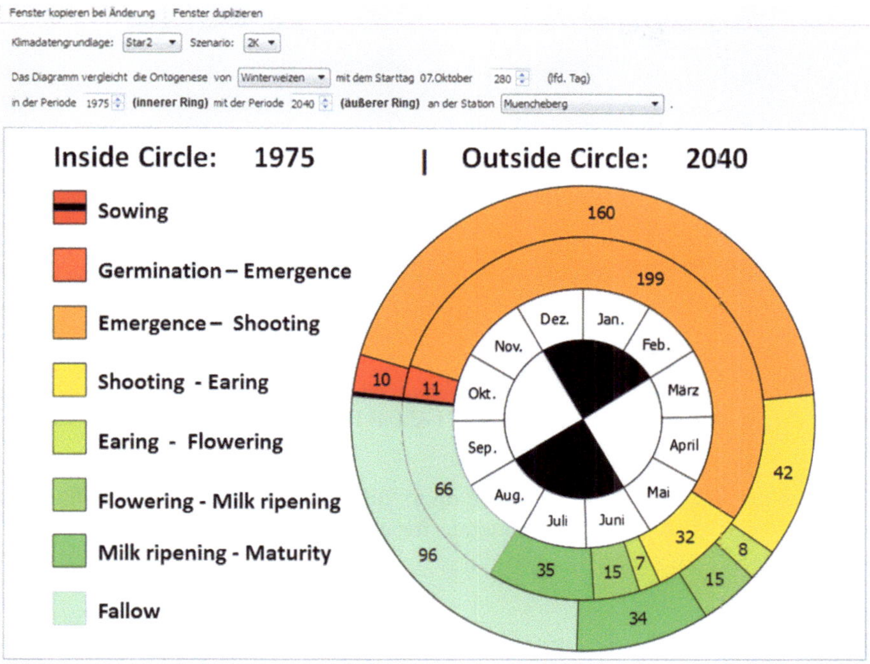

Fig. 3 Length of ontogenesis stages for winter wheat (in days) between sowing and harvest for 1975 (*inside circle*) compared to 2040 (*outside circle*), calculated using the ONTO model (site: Müncheberg; climate regionalisation: STAR2; emission scenario: 2 K; sowing date: 7 October)

between sowing and harvest for two climate time periods (1975 vs. 2040). The DSS user can choose different regionalisation methods, emission scenarios, time periods and sowing dates for nine important agricultural crops and numerous weather stations in order to run the ONTO model. The simulation results show the crop reaction, enabling the user to draw conclusions about the consequences for agro-management, e.g. all measures should start earlier in spring.

The start and length of the vegetation period are also vital to farmers when planning agro-management, commencing with the sowing process. Figure 4 shows for two regions (*Uckermark* and *Weisseritz,* located in north-east and south-east Germany, respectively) that vegetation will continue to start earlier up to 2075 and that the vegetation period will end later. As a result, the length of the vegetation period will increase up to 2075, which could cause an increase in biomass accumulation, especially for spring crops. More effective cropping systems are possible in the case of longer vegetation periods.

The results given in Fig. 4 show that the vegetation period in the Uckermark region will be 60 days longer by 2075 according to the WETTREG A1B scenario and 32 days longer according to the WETTREG B1 scenario. The vegetation period in the Weisseritz region will be 73 and 51 days longer, respectively, by 2075.

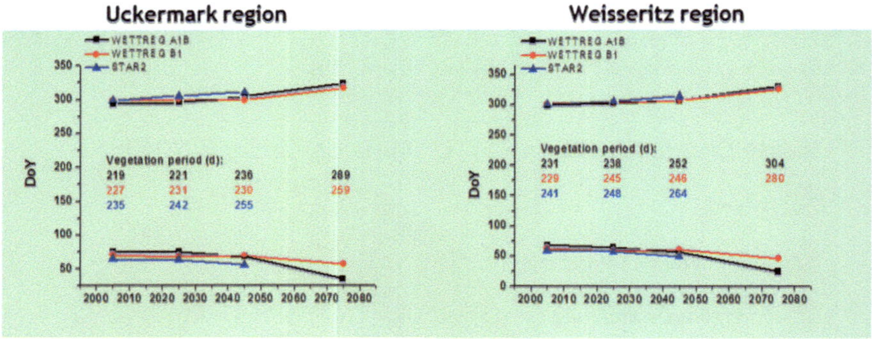

Fig. 4 The start, end and length of the vegetation period for two German regions *Uckermark* and *Weisseritz* according to climate scenarios WETTREG/A1B, WETTREG/B1 and STAR2/2K

Owing to a small temperature increase up to 2050, the increase in the length of the vegetation period will be quite moderate for all climate scenarios.

3.2 Climate Change Impact Assessment on a Regional Scale

At the regional scale, the ecological impact assessment of climate and land use changes is realised on a high spatial resolution, i.e. a minimum pixel size of 1 ha (100 × 100 m). Although most of the models mentioned earlier can be activated, they cannot be linked to the economic model at this scale. Different models can be used to calculate the expected impacts of climate change on aspects such as arable and grassland yield, the potential risk of erosion, regional evapotranspiration and regional total water flow-out and irrigation water demand. At this regional scale, statistical analysis (average, median, histogram, etc.) is performed automatically.

Figure 5 shows the example of irrigation water demand for winter wheat in the Free State of Thuringia, Germany, in 1981–2010 (average for Thuringia: 53.4 mm) compared to the situation in 2021–2050 (average for Thuringia: 95.1 mm). This comparison is based on 30-year averages. Figure 5 shows that the potential irrigation water demand will be significantly higher in 2021–2050 and that the area requiring irrigation will be larger than in 1981–2010.

The average irrigation water demand required for silage maize in Thuringia was 73.8 mm in 1981–2010 compared to 94.2 mm, which will be required in 2021–2050. Irrigation water demand for spring barley was 44.2 mm and 81.4 mm, respectively.

The average yield increase for winter wheat caused by irrigation in Thuringia was 0.81 t ha^{-1} in 1981–2010 compared to 1.43 t ha^{-1} in 2021–2050. For silage maize, the yield increase caused by irrigation was 8.81 and 11.30 t ha^{-1}, respectively. For spring barley, the irrigation-induced yield increase was 0.66 and 1.22 t ha^{-1}, respectively.

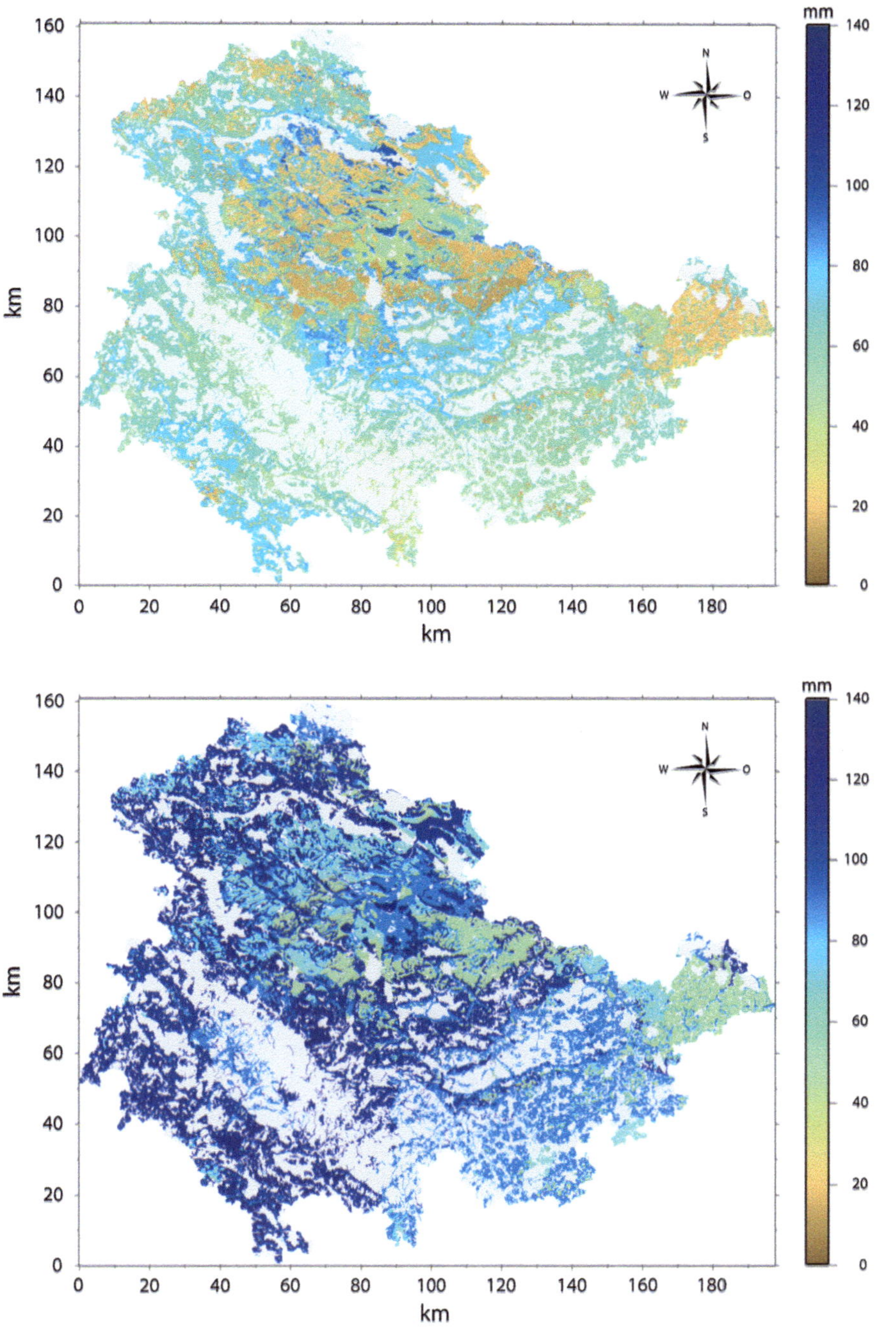

Fig. 5 Irrigation water demand for winter wheat in 1981–2010 (*top*) versus 2021–2050 (*below*) for the Free State of Thuringia, Germany, calculated using the model ZUWABE (climate scenario WETTREG A1B, *grey areas*—non-arable land)

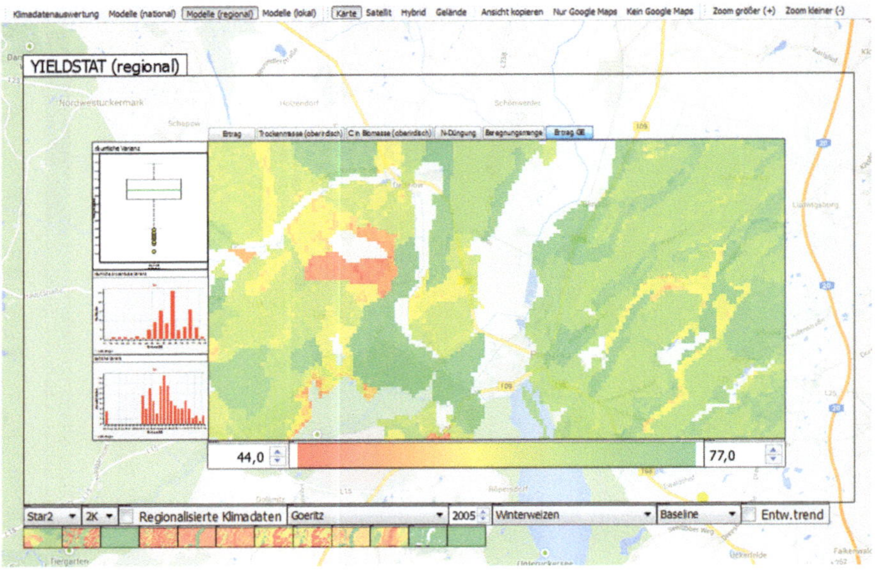

Fig. 6 Regional winter wheat yield calculated for the Prenzlau region, north-east Germany, using the model YIELDSTAT (climate regionalisation method: Star2, climate scenario: 2 K, climate period: 1991–2020; colour scale between *red* and *green*—winter wheat crop yield, *white*—grassland or urban area, *grey*—forests, *blue*—lakes)

It is very important for landscape planners, farmers and authorities connected with agro-landscapes to know the productivity potential of agricultural land, expressed by its crop yield. This aspect is also important in connection with the impact assessment of climate change on crop yields. The models YIELDSTAT and GLPROD were developed for arable and grass lands, respectively, and incorporated into the LandCaRe-DSS system in order to assess regional crop yields for different agricultural crops and types of grassland.

Figure 6 gives the example of average winter wheat yield for the Prenzlau region (mainly arable and grassland) located in north-east Germany, calculated using the YIELDSTAT model within LandCaRe-DSS for the time period 1991–2020. Depending on soil heterogeneity, yield ranges from 4.4 to 7.7 t ha^{-1}. The regional crop yield distribution map corresponds with the soil distribution map. The snapshot taken from the LandCaRe-DSS interface also provides statistical information, i.e. the yield boxplot for the chosen region and time- and space-related yield histograms. In the upper part, the data of the actual scenario run can be chosen and all of the input information can be activated and presented.

3.3 Climate Change Impact Assessment at the Local and Farm Scale

LandCaRe-DSS offers the interactive simulation and integrated impact assessment of agricultural adaptation strategies to climate change (crop rotation, soil tillage, fertilisation, irrigation, price and cost changes etc.) at the local and farm scale. Users are informed about changes in crop productivity (yield and yield quality in the case of grassland), soil fertility (water, carbon and nitrogen contents), water erosion and farm economy. At the farm level, the dynamic agro-ecosystem model MONICA and the statistical-based hybrid model YIELDSTAT for yield estimation are coupled with the farm economy model (Münch et al. 2014). LandCaRe-DSS users gain information about different economic parameters, fertiliser quantities and costs, irrigation water demands and costs and, finally, crop yields and sales profits. The variances in results based on up to 90 simulation runs within a time period of 30 years are given for all output information. The results are visualised using normalised bar graphs, making comparison between different scenario runs easier. Figure 7 gives an example of the visualised results simulated by the MONICA model for a small part of a farm, based on the Google-map background. The bar graphs are arranged around the fixed part of the farm, which is subdivided into 1 hectare (100 × 100 m) pixels. The data of the actual scenario run can be chosen in the upper part. Then all input information can be activated and presented. The left

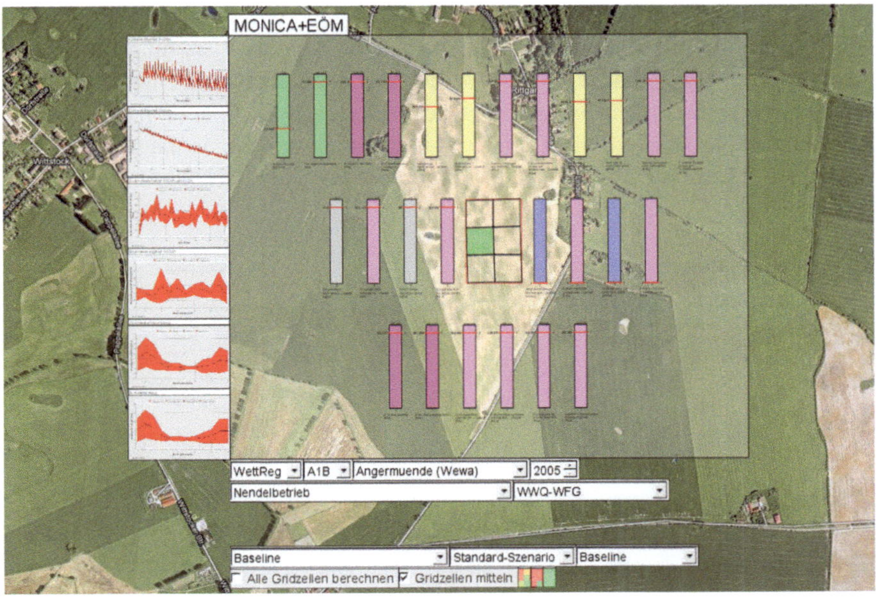

Fig. 7 Visualisation of simulation results for combined MONICA and FEM models at the farm level for a small part of a farm within the Uckermark district, Germany (*green* = ecology, *pink* = economy and costs, *yellow* = yield, *grey* = fertiliser demand, *blue* = irrigation water demand)

part of Fig. 7 shows parts of the dynamic results generated by the MONICA model as 30-year averages for the cropping period and as time courses (example.g. the soil carbon dynamic) over the chosen 30-year period.

3.4 Adaptation of Agriculture to Climate Change

Agriculture has multiple options for adapting the land use management to climate change. These include adjusting cropping ratios and crop rotations, using new agricultural crops and new selected varieties, applying irrigation and drainage, changing soil tillage methods and fertilisation or plant protection regimes. The ecological and economic effects and impacts of these adaption strategies can be analysed using LandCaRe-DSS.

Using the example of the Uckermark and Weisseritz regions of Germany, an investigation was carried out into how the regional productivity of crops would change taking into account future climate change (the climate scenario WETTREG A1B in this case) if farmers were urged to grow energy plants, causing the cropping ratio to change in favour of silage maize in these regions. In this case, it is assumed that the silage maize cropping ratio would increase 2.5 times for the Uckermark region and would double in the Weisseritz region, mainly at the expense of winter and spring cereals in both cases. An analysis was also made of how the irrigation of economically important crops or a change from conventional to non-tillage methods for winter rape and winter cereals would impact regional productivity. The simulation results in Fig. 8 show that the increased integration of more productive energy crops (energy maize in this case) in crop rotations would enhance regional productivity, assuming that silage maize does not exceed 25 % in crop rotations and that the possible negative yield effects caused by pests such as the Western corn root-worm (*Ostrinia nubilalis*) or the European corn borer (*Diabrotica virgifera*

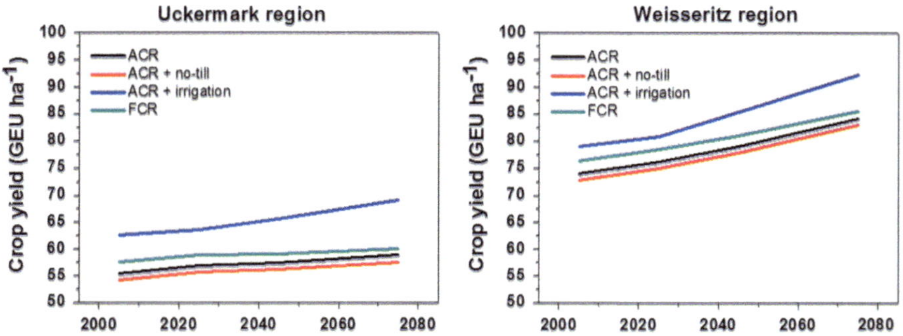

Fig. 8 Crop yield in grain equivalent units (*GEU*) for the Uckermark and Weisseritz regions for the actual cropping ratio (*ACR*) and under possible adaptation measures concerning irrigation and non-tillage, as well as a possible future cropping ratio (*FCR*) enriched with silo maize as an energy crop generated by the climate scenario WETTREG A1B

virgifera) can be neglected. When soil tillage is changed in favour of non-tillage for winter crops, the models used produce a small decrease in yield, as attested by field and on-farm experiments on German farms (Kahle et al. 2012; Gruber et al. 2012). Since process knowledge regarding soil tillage effects is incomplete, however, these results must be considered with caution.

The impacts of climate change on agriculture and agro-landscapes may be positive or negative, depending on the variability of weather conditions, site quality, land use and management. Assuming that recent agricultural cropping and management conditions remain unchanged in the future up to 2050, simulations generated using the YIELDSTAT model of LandCaRe-DSS demonstrate that the impacts of regional climate change on yields in Eastern Germany in the near (2025) and medium (2045) future are relatively small. There are winners (winter crops) and losers (spring crops).

The best way to adapt to climate change is to select a good mix of different cropping systems with various management options, including a wide range of agricultural crops (crop diversity), irrigation technologies, conservation of a high soil fertility level, highly productive and stress-tolerant crop varieties and new agricultural technologies such as strip cropping, precision agriculture, energy plantations and others.

4 Conclusions and Outlook

- The expected climate change will influence agriculture and the entire agro-landscape in many ways. So far, only a few suitable tools have been available to farmers and regional stakeholders to support decision-making processes in this area. The newly developed information and decision support system called LandCaRe-DSS—a new generation of interactive usable DSS—closes this gap.
- The LandCaRe-DSS supports interactive spatial scenario simulations, multi-ensemble and multi-model simulations using scenario techniques, uncertainty analysis and easy-to-follow visualisation methods connected with the advantages of a high simulation speed, a broad functionality, an open system design and architecture and that users do not necessarily need to be skilled at GIS in order to use the system.
- The LandCaRe DSS was developed to facilitate the impact assessment of regional climate change in agriculture and agricultural ecosystems and to help farmers and regional stakeholders to develop and quantify effective climate adaptation strategies (crop rotation, soil tillage, fertilisation, irrigation, price and cost changes and others) by virtually testing potential adaptation options and different climate projections.
- Adaptation of agricultural land use systems to climate change takes place at different spatial scales and has to deal with uncertainties and different kinds of risk. The LandCaRe-DSS operates at both the regional and the local scale.

- Thanks to the modular structure of LandCaRe-DSS, little effort is required to extend the system to other regions or countries. The system can be adapted by incorporating region-specific geo-data and static and/or dynamic models that are valid for the regions of adaptation; switch to different climate and emission scenario data and implementing other interface communication languages.
- The LandCaRe-DSS shows that the development and practical use of such a system is a process and not merely a product. Improvements to LandCaRe-DSS and users' learning progress must therefore be seen as an interactive and iterative process. The more applications users deploy, the greater demands they place on the system.
- Despite all of the progress made in climate and climate impact modelling in recent decades, many uncertainties remain. These uncertainties stem from climatic scenarios; incomplete agronomic and ecological knowledge and the agronomic and ecologic impact models used, which influence the accuracy of statements and results generated by LandCaRe-DSS.
- The system will constantly be advanced, updated and used in a number of research projects such as REGKLAM (project for agricultural regions of Saxony, Germany) and CARBIOCIAL (project for regions within the Mato Grosso and Pará states of Brazil) and different climate scenario studies for the Federal States of Thuringia, Brandenburg and Saxony.
- As the next step, Müncheberg Soil Quality Rating (M-SQR, Mueller et al. 2014) for assessing the quality of global farmland will be incorporated into LandCaRe-DSS.
- In addition, there are plans to apply the system to real-world case studies in order to test its performance, practicability and limitations at the farm and regional scale. There are also plans to use the system as a "learning and knowledge sharing tool" to help students, advisors, authorities and other interested persons to gain a better understanding of the processes and feedbacks within complex agro-landscape systems.
- In collaboration with two Russian institutions from St. Petersburg—the ARI and the State Polytechnical University—LandCaRe-DSS will be adapted to Russian conditions in the future years. In this case, the geo-databases with their special soil classification system (Shishov et al. 2004) and cadastral land register (Badenko et al. 2003) will have to be adapted to incorporate additional impact models for agro-landscapes developed in and for Russia. These include the agro-ecosystem model AGROTOOL (Poluektov and Terleev 2007; Poluektov et al. 2012) and models for soil nitrogen (Poluektov and Terleev 2010) and different soil properties (Terleev et al. 2010, 2014). It will also be necessary to switch to climate and emission scenario data for Russia using a special weather generator (Topaj et al. 2000) and to implement a system interface in Russian.
- In addition, future cooperative activities will involve the use, adaptation and further development of the system on the basis of joint research projects with institutions from Germany (Hohenheim) and Poland (Puławy).

Acknowledgements This article was funded by the German Federal Ministry of Education and Research (BMBF) within the *klimazwei* research programme (grant: 01 LS 05109). The authors would like to acknowledge the support they received from the German Federal Ministry of Food and Agriculture, the Brandenburg Ministry of Sciences, Research and Cultural Affairs (Germany) and the Russian Academy of Agricultural Sciences.

References

Ascough JC, Shaffer MJ, Hoag DL, Mc Master GS, Dunn GH, Ahuja LR, Weltz MA (2001) GPFARM: an integrated decision support system for sustainable great plain agriculture. In: Scott DE, Mohtar RH, Steinhardt GC (eds) sustaining the global farm selected papers from the 10th international soil conservation organization meeting held from 24 to 29 May 1999 at Purdue University and the USDA-ARS National Soil Erosion Research Laboratory, pp 951–960

ATV-DVWK-Regelwerk (2002) Verdunstung in Bezug zu Landnutzung, Bewuchs und Boden. ATV-DVWK-Regelwerk: Merkblatt; M 504: 144 pp

Badenko VL, Garmanov, VV, Osipov GK (2003) National cadastral land register. 320 p (Государственный земельный кадастр. СПб:Питер, 320 с)

Böhm U, Kücken M, Ahrens W, Block A, Hauffe D, Keuler K, Rockel B, Will A (2006) CLM—the climate version of LM: brief description and long-term applications. COSMO Newsl 6:225–235

Carbiocial (2014) Carbon sequestration, biodiversity and social structures in Southern Amazonia: models and implementation of carbon-optimized land management strategies. http://www.uni-goettingen.de/de/211024.html. Accessed 9 Mar 2015

Chmielewski F-M (2003) Phenology and Agriculture (chapter 7). In: Schwartz MD (ed) Phenology: an integrative environmental science. Kluwer Academic Publishers, Boston, Dordrecht, London, pp 505–522

Chmielewski F-M, Henniges Y (2007) Phänologische Modelle als Grundlage zur Abschätzung des Klimaimpakts. Berichte Meteorologisches Institut Freiburg 16:229–235

DIN 19708 (2005) Bodenbeschaffenheit—Ermittlung der Erosionsgefährdung von Böden durch Wasser mit Hilfe der ABAG (Soil quality—Predicting soil erosion by water by means of ABAG). DIN 19708:2005-02, Normenausschuss Wasserwesen (NAW) im DIN, 25 p

Enke W, Schneider F, Deutschländer T (2005) A novel scheme to derive optimized circulation pattern classifications for downscaling and forecast purposes. Theoret Appl Climatol 82:51–63

Gerstengarbe F-W, Badeck F, Hattermann F, Krysanova V, Lahmer W, Lasch P, Stock M, Suckow F, Wechsung F, Werner PC (2003) Studie zur klimatischen Entwicklung im Land Brandenburg bis 2055 und deren Auswirkungen auf den Wasserhaushalt, die Forst- und Landwirtschaft sowie die Ableitung erster Perspektiven. PIK-Report No. 83, Potsdam, 79 p

Gömann H, Kreins P, Herrmann S, Wechsung F (2005) Impacts of global changes on agricultural land-use in the German Elbe region: results of an operational modelling tool for planning, monitoring and agri-environmental policy counselling [CD-ROM]. In: 21st European Regional Conference—ERC 2005—integrated land and water resources management: towards sustainable rural development, 15 19 May 2005, Frankfurt (Oder), Germany and Slubice, Poland, p 11

Gruber S, Pekrun C, Möhring J, Claupein W (2012) Long-term yield and weed response to conservation and stubble tillage in SW Germany. Soil and Tillage Res 121:49–56

Henniges Y (2008) Entwicklung phänologischer Modelle für ausgewählte Pflanzen für die Naturräume 74 (Quillow-Einzugsgebiet) und 42 (Weißeritzkreis), Humboldt Universität Berlin, 9 pp

IPCC (2007) Impacts, adaptation and vulnerability. Contribution of working group II in the fourth assessment report of the intergovernmental panel on climate change, summary for policemakers. Cambridge University Press, Cambridge, pp 7–22

Ivanov AL, Kirjushin VI (2009) Global climate change and risk prediction in Russian agriculture. (Глобальные изменения климата и прогноз рисков в сельском хозяйстве России). Russian Academy of Agricultural Science, Moscow, 517 p

Käding H, Keiser T, Werner A (2005) Model for calculating grassland yields and forage quality in North-East Germany on the basis of site and management characteristics. Arch Agron Soil Sci 51(4):417–431

Kahle P, Korn G, Pach F, Lehmann E (2012) Auch ohne Pflug gute Erträge. Landwirtschaft ohne Pflug 6/2012, pp 35-41

Köstner B, Berg M, Bernhofer Ch, Franke J, Gömann H, Kersebaum KC, Kuhnert M, Lindau R, Manderscheid T, Mengelkamp H-T, Mirschel W, Nendel C, Nozinski E, Pätzold A, Simmer S, Stonner R, Weigel H-J, Wenkel K-O, Wieland R (2008) Land, climate and resources (LandCaRe) 2020—foresight and potentials in rural areas under regional climate change. Ital J Agron/Riv Agron 3(3):743–744

Köstner B, Wenkel K-O, Bernhofer Ch (2012) New model-based information systems for climate impact scenarios in agriculture. Promet 38(1/2):42–52

KIT LandCaRe-DSS (2014) KMU-innovativ—joint project—LandCaRe-DSS: Model-based tools for strategic and operational irrigation measures under climate change. http://www.zalf.de/de/forschung/projekte/Seiten/detail.aspx?fpid=1543&tpid=832. Accessed 9 Mar 2015

LADSS (2005) Land Allocation Decision Support System. http://www.macaulay.ac.uk/LADSS. Accessed 9 Mar 2015

Mahammadzadeh M, Biebler H, Bardt H (2009) Klimaschutz und Anpassung an die Klimafolgen —Strategien, Maßnahmen und Anwendungsbeispiele. Institut der deutschen Wirtschaft Köln Medien GmbH, Köln, 310 p

Mirschel W (2010) ONTO—Modell zur Beschreibung der Pflanzenentwicklung (Ontogenese) von landwirtschaftlichen Fruchtarten. In: Wenkel K-O, Berg M, Wieland R, Mirschel W, Vorsorge und Gestaltungspotenziale in ländlichen Räumen unter regionalen Wetter- und Klimaänderungen (LandCaRe 2020): Modelle und Entscheidungsunterstützungssystem zur Klimafolgenabschätzung und Ableitung von Adaptationsstrategien der Landwirtschaft an veränderte Klimabedingungen (AGROKLIM-ADAPT); Müncheberg (Leibniz-Zentrum für Agrarlandschaftsforschung), pp A3/1–A3/9

Mirschel W, Wenkel KO (2007) Modelling soil-crop interactions with AGROSIM model family. In: Kersebaum KC, Hecker JM, Mirschel W, Wegehenkel M (eds) Modelling water and nutrient dynamics in soil-crop-systems: proceedings of the workshop on "Modelling water and nutrient dynamics in soil-crop systems" held from 14 to 16 June 2004 in Müncheberg, Germany. Springer, Dordrecht, pp 59–73

Mirschel, W. Wenkel K.-O, Wieland R, Albert E, Köstner B (2009) Klimawandel und Ertragsleistung: Auswirkungen des Klimawandels auf die Ertragsleistung ausgewählter landwirtschaftlicher Fruchtarten im Freistaat Sachsen - eine landesweite regionaldifferenzierte Abschätzung -. In: SÄCHSISCHES LANDESAMT FÜR LANDWIRTSCHAFT, UMWELT UND GEOLOGIE (eds) Schriftenreihe des Landesamtes für Umwelt, Landwirtschaft und Geologie (LfULG Sachsen), Issue 28/2009, 61 p, Dresden

Mirschel W, Wenkel K-O, Wieland R, Albert E, Köstner B (2010) Impact assessment of climate change on agricultural crop yields using the hybrid model YIELDSTAT—a case study for the Free State of Saxony, Germany. In: Materialy vserossiskoi konferencii "Matematiceskie modeli i informachionnye tekhnologii v sel'skokhozaistvennoi biologii: itogi i perspektivy" (14.-15.10.2010, ARI St. Petersburg, Russia), St. Petersburg, pp 200–203. http://www.agrophys.ru/upload/poluektov/sbornik.pdf. Accessed 9 Mar 2015

Mirschel W, Wieland R, Guddat C, Michel H, Wenkel K-O (2012) Crop yield estimation on arable land under climate change for the Free State of Thuringia, Germany, using the model YIELDSTAT. In: Tendencii razvitija agrofiziki v uslovijakh izmenjajushhegosja klimata (k 80-letiju Agrofiziceskogo NII): Materialy Mezhdunarodnoj konferencii, Sankt-Peterburg, 20-21 sentjabrja 2012.- SPb: Ljubovich, pp 169–180. http://www.agrophys.ru/Thesys-collection. Accessed 9 Mar 2015

Mirschel W, Wieland R, Gutzler C (2013a) Ertragsleitung. In: Gutzler C, Helming K (eds) Folgenabschätzung von Szenarien der landwirtschaftlichen Produktion in Brandenburg 2025: Feldberegnung und Energiemaisanbau. Leibniz-Zentrum für Agrarlandschaftsforschung, Müncheberg, pp 32–38

Mirschel W, Wieland R, Wenkel K-O, Guddat CH, Michel H (2013b) Modellgestützte Abschätzung der Auswirkungen des Klimawandels auf Ertrag und Zusatzwasserbedarf im Freistaat Thüringen bis 2050. In: Nguyen Xuan Thin (ed) Modellierung und Simulation von Ökosystemen: Workshop Kölpinsee 2012. (Berichte aus der Umweltinformatik), Shaker Verlag Aachen, pp 1–19

Mirschel W, Wieland R, Wenkel K-O, Nendel C, Guddat C (2014) YIELDSTAT—a spatial yield model for agricultural crops. Eur J Agron 52:33–46

Mueller L, Schindler U, Shepherd TG, Ball BC, Smolentseva E, Pachikin K, Hu C, Hennings V, Sheudshen AK, Behrendt A, Eulenstein F, Dannowski R (2014) The Muencheberg Soil Quality Rating for assessing the quality of global farmland. In: Mueller L, Saparov A, Lischeid G (eds) Novel measurement and assessment tools for monitoring and management of land and water resources in agricultural landscapes of Central Asia, Springer Cham Heidelberg New York Dordrecht London, pp 235–248

Münch T, Berg M, Mirschel W, Wieland R, Nendel C (2014) Considering cost accountancy items in crop production simulations under climate change. Eur J Agron 52:57–68

Nendel C, Berg M, Kersebaum KC, Mirschel W, Specka X, Wegehenkel M, Wenkel KO, Wieland R (2011) The MONICA model: testing predictability for crop growth, soil moisture and nitrogen dynamics. Ecol Model 222(9):1614–1625

Offermann F, Gömann H, Kleinhanß W, Kreins P, von Ledebur O, Osterburg B, Pelikan J, Salamon P, Sanders J (2010) vTI-Baseline 2009—2019: Agrarökonomische Projektionen für Deutschland. Landbauforschung vTI Agriculture and Forestry Research, Sonderheft 333, 88 pp. http://literatur.ti.bund.de/digbib_extern/dk043068.pdf. Accessed 9 Mar 2015

Poluektov RA, Terleev VV (2007) Crop simulation of the second and the third productivity levels. In: Kersebaum KC, Hecker JM, Mirschel W, Wegehenkel M (eds) Modelling water and nutrient dynamics in soil-crop-systems: proceedings of the workshop on "Modelling water and nutrient dynamics in soil-crop systems" held from 14 to 16 June 2004 in Müncheberg, Germany. Springer, Dordrecht, pp 75–89

Poluektov RA, Terleev VV (2010) Computer model of nitrogen dynamics in the rooting zone. Agrokhimiya 10:68–74

Poluektov RA, Fintushal SM, Oparina IV, Shatskikh DV, Terleev VV, Zakharova ET (2002) Agrotool—a system for crop simulation. Arch Agron Soil Sci 48(6):609–635

Poluektov RA, Topazh AG, Jakushev VP, Medvedev SA (2012): Usage of a dynamic agro-ecosystem model for impact assessment of climate change on crop yield productivity (theory and realization). (Использование динамической модели агроекосистемы для оценки влияния климатических изменений на продуктивность посевов (теория и реализация). Vestnik Russian Acad Agric Sci 2:7–12)

Prased JR, Prasad RS, Kulkami UV (2008) A decision support system for agriculture using natural language processing (ADSS). In: Proceedings of the international multiconference of engineers and computer scientists (IMECS 2008, 19–21 March 2008, Hong Kong), vol. I, pp 1–5

REGKLAM (2013) REGKLAM – Regionales Klimaanpassungsprogramm Modellregion Dresden, http://www.regklam.de

REGKLAM-KONSORTIUM (2013) Integriertes Regionales Klimaanpassungsprogramm für die Region Dresden: Grundlagen, Ziele und Maßnahmen, REGKLAM-Publikationsreihe 7, Rhombus-Verlag Berlin, 344 p

Roeckner E, Brokopf R, Esch M, Giorgetta M, Hagemann S, Kornblueh L, Manzini E, Schlese U, Schulzweida U (2004) The atmosphere general circulation model ECHAM5. Part 2: Sensitivity of simulated climate to horizontal and vertical resolution. MPI-Report 354, Hamburg: Max Planck Institute for Meteorology, 64 p. http://www.mpimet.mpg.de/fileadmin/publikationen/Reports/max_scirep_354.pdf. Accessed 9 Mar 2015

Roth D (1993) Richtwerte für den Zusatzwasserbedarf in der Feldberegnung. In: Schriftenreihe LUFA Thüringen, Issue 6 (Richtwerte Pflanzenproduktion), pp 53–86

Schwärzel H (2000) Untersuchungen zu Kriterien der Standort- und Gehölzauswahl bei extensiven Anpflanzungen von Obstbäumen. Dissertation, Humboldt University Berlin, 86 p

Shishov LL, Tonkonogov VD, Lebedeva II, Gerasimova MI (2004) Classification and diagnostic of Russian soils. (Классификация и диагностика почв России. Смоленск: Ойкумена)

Terleev VV, Mirschel W, Schindler U, Wenkel KO (2010) Estimation of soil water retention curve using some agrophysical characteristics and Voronin's empirical dependence. Int Agrophysics 24(4):381–387

Terleev VV, Narbut MA, Topazh AG, Mirschel W. (2014): Modelling hydrophysical properties of soils…) (Моделирование гидрофизических свойств почвы как капиллярно-пористого тела и усовершенствование метода Муалема-Ван Генухтена: теория). Agrofisica (Агрофизика) St. Petersburg 2(14):35–44) (in Russian)

TGL (1990) Verfahren der Pflanzenproduktion, Beregnung – Ermittlung des Zusatzwasserbedarfs. TGL 46200/03, 30 pp

Topaj AG, Mirschel W, Künkel KJ, Hoffmann J (2000) Methodik zur Identifikation von Klimatrends aus langjährigen meteorologischen Datenreihen. Meteorol Z 9(6):339–350

Tubiello FN, Soussana JF, Howden MS (2007) Crop and pasture response to climate change. Proc Nath Acad Sci USA 104:19686–19690

Wenkel K-O, Wieland R, Mirschel W, Schultz A, Kampichler C, Kirilenko A, Voinov A (2008) Regional models of intermediate complexity (REMICs): a new direction in integrated landscape modelling. In: Environmental modelling, software and decision support: state of the art and new perspective, Amsterdam (Elsevier), pp 285–295

Wenkel K-O, Berg M, Mirschel W, Wieland R, Nendel C, Köstner B (2013) LandCaRe DSS—an interactive decision support system for climate change impact assessment and the analysis of potential agricultural land use adaptation strategies. J Environ Manage 127:168–183

Wischmeier WH, Smith DD (1978) Predicting rainfall erosion losses—a guide to conservation planning. Agriculture Handbook, vol. 537, Washington D.C., U.S. Department of Agriculture, 60 pp

Chapter 24
Monitoring of Soil Fertility (Agroecological Monitoring)

Victor G. Sychev, Evgeny N. Yefremov and Vladimir A. Romanenkov

Abstract Monitoring the ecological status of agricultural land is a fundamental precondition for controlling its sustainable functions for human society and for maintaining the ecosystem's capacity. We analyze fundamentals, developments, and trends and present results of agroecological monitoring in Russia. This system has been developed and operated by the Pryanishnikov Institute of Agrochemistry in Moscow. Agroecological monitoring in Russia was installed in the 1970s and is based on a regular 5-year agrochemical survey of agricultural lands all over the country, more than 300 field experiments in all bioclimatic zones of the country, and more than 1000 reference monitoring plots. In trials with different inputs of fertilizers, the focus is on analyzing soil fertility indicators and their impact on productivity. Some of these experiments are long-term experiments and part of international networks. Their results are of fundamental importance for monitoring, modeling, and controlling the status of soils in future despite climate change. In a regular survey, we found tendencies toward decreasing soil fertility in some regions, for example with decreased contents of humus and plant-available minerals, and topsoil acidification. Nutrient withdrawals must be compensated for by regular fertilization regimes, nutrient mining must be avoided. We detected some gaps in knowledge on the topic of balancing elements and modeling the agroecosystem's response to climate and land use changes. We conclude that there is a need to implement modern measurement and modeling systems in some key long-term trials. The Pryanishnikov Institute has taken responsibility for coordinating running programs in different regions and administrative units of the Russian Federation, and for elaborating methodical guidelines and highly advanced monitoring technologies. National and international cooperation, research programs and

V.G. Sychev · E.N. Yefremov · V.A. Romanenkov (✉)
Pryanishnikov All-Russian Research Institute of Agrochemistry,
Pryanishnikova st. 31a, Moscow 127550, Russia
e-mail: geoset@yandex.ru

V.G. Sychev
e-mail: info@vniia-pr.ru

E.N. Yefremov
e-mail: info@vniia-pr.ru

L. Mueller et al. (eds.), *Novel Methods for Monitoring and Managing Land and Water Resources in Siberia*, Springer Water,
DOI 10.1007/978-3-319-24409-9_24

networks are key for agroecological monitoring systems of the twenty-first century in addressing challenges for a highly productive, stable, sustainable, and environmentally safe food production.

Keywords Agriculture · Soil · Fertilization · Soil fertility · Experiments · Russia · Monitoring

1 Introduction

Contemporary agricultural production systems are crucial to ensure that food security is achieved for the global population (Borlaug 2007; Lal 2009). However, they have the potential to degrade land, water, and air resources, thus limiting their functions for future generations (Gordeev and Romanenko 2008; Lichtfouse et al. 2009; Mueller et al. 2014). Understanding and controlling human impacts on resources such as soil, water, and air, and on the ecosystem's functioning, requires the monitoring of agricultural land.

In Russia, the sustainable development of the country and the agrarian sector of its economy, as well as the country's food safety, can be ensured only by preserving natural systems and maintaining an adequate quality of the natural environment.

The State Program of Agricultural Development and Regulation of Agricultural Products (2013–2020) (State Program 2013–2020) and Ecological Doctrine of the Russian Federation (Ecological Doctrine 2002) defined pathways and means for implementing state policy in the fields of agriculture, ecology, and environmental management:

- development of the state management system for nature conservation and nature use;
- monitoring of all environmental components, including agricultural lands; and
- scientific support for nature management.

The main concept of environmental monitoring was defined as early as the 1970s as a system for the observation, assessment, and prediction of anthropogenic changes in the state of abiotic components of the biosphere, ecosystems' responses to these changes, and anthropogenic changes in ecosystems related to the impact of economic activity (Gerasimov 1978). In accordance with this definition, monitoring includes three main lines of activity:

- assessing the actual environmental state;
- observing the affecting factors and the environmental state; and
- predicting the natural environment state and assessing the predicted environment state.

In the Russian Federation, several systems are in operation at different levels and locations for controlling and observing the environment. These are:

- the emergency monitoring system of the Russian Federation Ministry of Civil Defense and Emergency Response;
- the forest, water, and mineral resources monitoring networks of the Russian Ministry of Natural Resources and Environmental Protection;
- the soil fertility monitoring network (agrochemical and agroecological state of soils of agricultural lands) of the Russian Ministry of Agriculture;
- the environmental pollution monitoring network of the Russian Federal Service for Hydrometeorology and Environmental Monitoring; and
- the sanitary and hygienic monitoring system of the Russian Agency for Health and Consumer Rights.

2 Soil Agroecological Monitoring

Soil monitoring systems are in operation in different countries. The analysis of their experience has allowed a number of statements to be formulated are taken into consideration when improving the Russian system of soil agroecological monitoring. This is primarily the optimization of the cost-to-information value ratio. The concept of information value is insufficiently clear, but, in our opinion, it means the sufficiency and reliability level of information acquired in the course of monitoring observations for solving management problems.

Monitoring observation data should match the following criteria:

(a) completeness of information (objects of observation, their attributes and relationships);
(b) logical agreement (degree of conformity of data structure, attributes and relationships to logical rules);
(c) position accuracy (accuracy of object positions);
(d) time accuracy (accuracy of the time attributes and time relationships of objects); and
(e) topic accuracy (accuracy of quantitative attributes and correctness of non-quantitative attributes and classifications of objects and their relationships).

Because of the spatial and temporal variability of all environmental parameters, including soils of agricultural lands, monitoring data are always characterized by an uncertainty which does not allow the observation data to be used without probabilistic interpretation. The monitoring results are a time series, in which the trend, cyclic, and random components can occur in any ratios and at any levels; the isolation or exclusion of each component is determined by requirements for data interpretation (System of agroecological monitoring 2006).

Depending on the prioritized tasks, requirements for soil monitoring can be different, but the organizational and methodological principles of monitoring

Fig. 1 Aim of soil fertility monitoring: the basis for effective and environmentally-friendly application of fertilizers for producing healthy food on healthy soils

remain the same: the fulfillment of scheduled observations and the standardization of methods for data processing and analysis (Sychev and Romanenkov 2009; Sychev et al. 2013a).

Soil fertility monitoring, which is performed by organizations within the agrochemical service of the Russian Ministry of Agriculture, with the participation of research institutes, should allow numerous observation results to be transformed into specific parameters suitable for assessing the fertility status of soils of agricultural lands, and predicting its changes under different farming practices (Fig. 1).

This system is based on long-term experience and innovations (Fig. 2).

There is no uniform monitoring system of agricultural lands in Russia. The existing system and alternatives proposed for the development of soil fertility monitoring in combination with the monitoring of agricultural land use by the Russian Ministry of Agriculture are not perfectly adequate for the current requirements stated in federal and regional legislations. Soil fertility monitoring is one of the numerous tools for assessing the fertility of soil and its agroecological potential (Sychev 2003). Local monitoring on the reference plots, test areas, and plots of long-term field experiments run by the geographical network, as well as the periodical agrochemical soil survey of agricultural lands and models of soil fertility, are complementary methods within the integral approach to assess the effect of economic activities and natural factors on the agricultural soils.

The organizational principles of agricultural soil fertility monitoring are as follows:

- objectivity of information;
- comparability of data derived from different system modules;

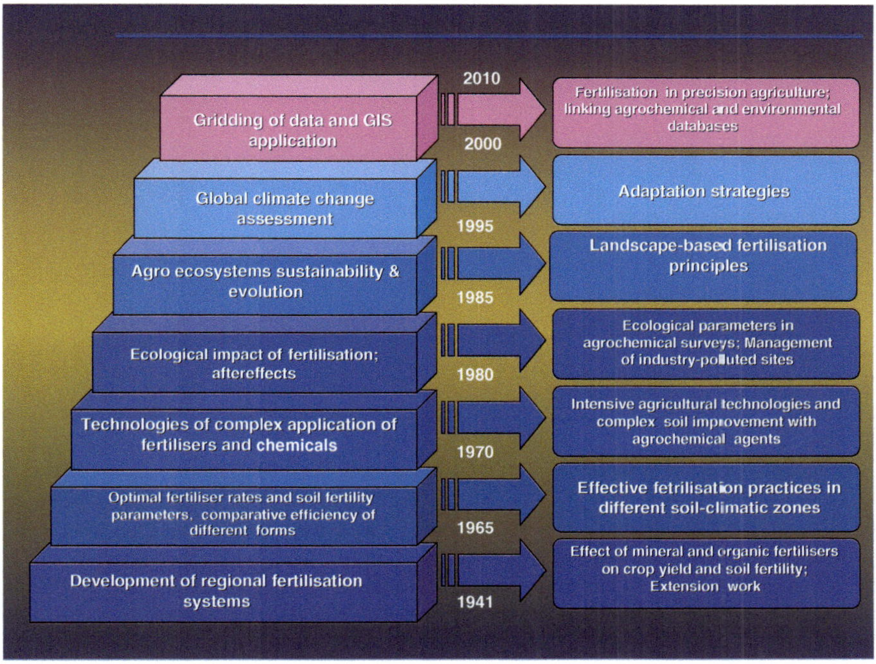

Fig. 2 Changes in the main tasks and practical results of the geographical network in Russia since its foundation

- adequacy of information for specified goals;
- efficiency of information collection and processing; and
- permanence of observations.

3 Soil Fertility Database

An important element determining the performance of soil fertility monitoring is its database. This concept involves not only the list of parameters to be monitored, but also the principles of data acquisition, transfer, and use. The latter is particularly important as the network covers numerous experiments and analyzes over different climate and time zones (Fig. 3).

In our opinion, the soil monitoring database can be divided into three blocks. The first block of the database includes data on the periodical agrochemical and agroecological survey of agricultural soils. The second block includes information acquired during observations on local reference plots for soil and vegetation monitoring. The third block includes information from long-term stationary field experiments (LTFEs) and agroecological polygons of the geographical network (Table 1).

Fig. 3 Location map of local soil fertility monitoring. The current network (*red dots*) includes 115 institutions and 356 trials, of which 320 are more than 25-years old. The network will be completed in cooperation with further agricultural research institutions and stations in different administrative units of the Russian Federation. Green dots indicate main stations of Agrochemical Service (overall: 102) performing local monitoring on 1380 reference sites since the 1970s

Table 1 Main differences between survey and monitoring in the National Monitoring network of agricultural soils in Russia (after EEA 2001, with amendments)

Issue	Local monitoring: reference plots	Local monitoring: LTFEs	Soil survey
Site selection	Based on a representative distribution due to characteristic land use, landscape units, and pollution exposure	Based on a representative distribution due to characteristic soil types and flat ploughing areas	Based on a regular field revisiting
Number of sites	1380	356	15–18 million ha per year
Monitoring activities	Soil nutrients (P, K), general soil properties (C, pH), pesticides, organo-chemicals, radionuclides, deposition of nutrients, heavy metals, trace elements	Soil nutrients (P, K), general soil properties (C, pH), soil chemistry (acid-base properties, soil carbon), microbiology, heavy metals, radionuclides	Soil nutrients (P,K) and general soil properties (C, pH)
Interval of measurements	Mainly yearly measurements	Yearly or end-of-rotation measurements	5-yearly

LTFEs run by the geographical network are an interesting and informative investigative tool. The field experiment is the most representative method of studying the theoretical and practical principles for reproducing soil fertility, increasing crop yields, and improving crop quality. The network of similar experiments simultaneously performed in several geographical and soil-climatic zones according to the standardized procedure allows us to trace the changes in any soil parameter or the effect of the elements of agricultural technologies (rate and type of fertilizer, time of fertilization, type of cultivation, chemical crop protection) under different soil-climatic conditions and reveals general geographical regularities. Long-term stationary experiments form the main scientific basis for adequately assessing the changes in slow processes, including those in the system of parameters composing the agrochemical Pryanishnikov triangle: plant–soil–fertilizer (Fig. 1). Measurements within the geographical network of long-term field experiments supplement the observation design with a new information block: an assessment of variations in the agrochemical soil parameters under the impact of multiple factors.

The dynamics of such stable parameters as the mineralogical composition, organic matter content, nitrogen pool, and particle-size distribution, which determine soil fertility, can thus be assessed. Results of the integrated observations of agrochemical, agrophysical, and meteorological parameters on the same plot have already been accumulated in long-term field experiments run by the geographical network. Long-term field experiments allow the environmentally optimal fertilizing systems to be determined experimentally and the role of separate practices in positive and negative environmental impacts to be assessed. In the 1990s, a methodology for launching new field experiments was developed, which includes the extension of experimental studies in agroecosystems to include information about adjacent environments by organizing a new type of experimental network: agroecological polygons on sloped lands. The organization of polygons allows the effect of intensification factors in agrolandscapes to be studied by taking into account the transfer of water, nutrients, and xenobiotics. The main objective of the polygons was their integration into the network of land observations on the state of arable soils as a component of agrochemical and agroecological monitoring. The experimental treatments were located along the main slopes and allowed the treatments within catenae to be compared by taking into account the geochemical conjugation and the control of soil material, nutrient, and water transfer both within and beyond the experimental plots.

The practical implementation of this task revealed a number of contradictions because of the differences in the requirements for the organization of long-term experiments and monitoring principles. The implementation of monitoring implies as wide as possible a range of all soil-environmental conditions typical for the landscape studied. The main requirement for the establishment of long-term experiments is, by contrast, the maximum reduction of the initial heterogeneity of the properties studied, and the studies were performed on small plots of 50–100 m^2 in area. A significant spatial heterogeneity of the standard agrochemical parameters was revealed on the established polygons.

The solution to this problem required the development of research techniques by allowing the data obtained in long-term experiments to be interpolated to large agrolandscape units. This problem was solved, in particular, by the involvement of special techniques for the statistical treatment of experimental results (geostatistics methods and the analysis of local trends), which allowed the spatial heterogeneity of the parameters studied to be characterized. The analysis of local trends in the design of experiments based on the test sowing data allowed the areas to be distinguished which had relatively low variation in crop yield, as well as those with similar trends little dependent on the crop species, and the weather conditions of the year. In an agroecological field experiment by the Central Experimental Station of the Pryanishnikov Research Institute of Agrochemistry (Moscow oblast), the use of geostatistics methods allowed, in particular, natural acidification processes to be revealed, which reproduce the spatial heterogeneity on the prelimed soil and the main mechanism of potassium loss with the surface runoff along the slope beyond the field, without the formation of local minimums or maximums (Romanenkov and Kuzyakova 2000).

The principles of the agroecological polygon study affected the level of modern research in the following predominant fields:

- expansion of the set of parameters in stationary experiments and the analysis of interconnections between different agrolandscape elements;
- assessment of the manageability of soil fertility in the long term;
- analysis and classification of the data obtained to reveal the most informative parameters; the use of simulation as a method to design economically efficient and environmentally safe technologies for managing soil productivity and fertility, including the estimation of optimal parameters for zonal soil types.

The mandatory use of agroecological approaches became a general rule in the analysis of modern agricultural technologies.

4 Soil State Parameters

The justified selection of parameters set to be regularly observed in time and space is the basis for rational monitoring strategies. Technologies for the acquisition and integrated processing of multi-aspect information in combination with scientifically substantiated models are necessary tools.

The minimum set of parameters to be traced in the agroecological soil monitoring system should include the following items:

- basic parameters: characteristics of soil fertility of agricultural lands;
- background parameters reflecting the peculiar features of soil composition in the region of observations;
- parameters characterizing the effect of agricultural activity on the status of agricultural soils; and
- parameters characterizing the effect of technogenic impact on the soil and agricultural crops.

European soil monitoring programs usually monitor general soil properties, the content of macronutrients and heavy metals, with less emphasis on organic compounds and biological properties (EEA 2001). The existing network of local monitoring and regular soil agrochemical surveys allows a different set of parameters to be applied in the national monitoring network of agricultural soils in Russia (Table 1).

A set of soil agrochemical parameters which is presently used in agroecological monitoring was significantly extended compared to the early period of the overall agrochemical survey of soils or the period of development of the local agroecological monitoring network. Analytical methods for agroecological monitoring are based on national standards, so analytical records are exactly the same at different laboratories, and this activity is under the permanent control of the institute (Methodological guidelines 1992, 2003). Table 2 includes a short description of the standard methods. All the national standards mentioned were prepared by the institute based on accepted analytical procedures with the necessary modifications to improve the performance of automated laboratories.

On reference plots of the local monitoring of soils and crops, the contents of heavy metals (total and mobile forms), residual pesticides, and radionuclides ^{90}Sr and ^{137}Cs are also determined (Methods of agroecological monitoring on reference plots 2002). There are plans to extend the set of ecological parameters and to start determining the content of oil products in the soils in the near feature.

5 Data Flow

A satisfactory data flow from the data producers to the main clients requires adequate data transfer. The design, development, and implementation of the monitoring network were based on the development of suitable data storage and exchange, as well as on an aggregation of the data collected (Fig. 4). In the soil survey system this was conceived as a shared format for data exchange in the form of annual reports, and based on clustering each land use class according to soil acidity, humus content, and available nutrient content. Regular 5-year updates of land clustering data allowed the current state of agricultural soils to be recorded, processed, and analyzed on a regular basis. The national reports provide a summary based on regional (oblast) and federal district level. The latest printed edition is the "Register of Soil Fertility" by Sychev et al. (2013b)—an example of aggregated data reporting on the state and reproducing soil fertility.

The assessment of changes in soils over time in connection with changing agricultural inputs is a subject of special reviews, including the analysis of temporal trends in clusters of lands based on soil acidity, available nutrients, and SOC content (Sychev 2007; Sychev et al. 2010a).

The modern system of data storage and processing is based on an electronic format of regional databases with georeferencing of sampling points and field boundaries. The national database is supported by the portal of the Central Computer

Table 2 Standard analytical methods for soil state parameters

Parameter	Standard	Short method description
Soil organic matter	GOST 26213-91	Wet combustion in acid-dichromate mixture and titration of dichromate excess with ferrous ammonium sulfate
pH $_{KCl}$	GOST 26483-85	1n KCl extract, pH meter
Available P and K	GOST 26207-91	0.02 N HCl (Kirsanov method) in taiga zone
	GOST 26209-91	0.5 N CH_3COOH (Chirikov method) in forest steppe zone
	GOST 26205-91	1 % $(NH_4)_2CO_3$ solution in steppe and semi-desert zone with subsequent flame-photometric determination
Soil salinity	GOST 26424-85, GOST 26425-85, GOST 26426-85, GOST 26427-85, GOST 26428-85	HCO_3^-, CO_3^{2-}, Cl^-, SO_4^{2-}, Na^+, K^+, Ca^{2+}, Mg^{2+} in water extract,
Cation exchange capacity	GOST 17.4.4.01-84	Barium chloride-acetate buffered at pH 6.5, triethanolamine method, and magnesium acetate buffered at pH 7.0 for calcareous, gypseous, and saline soils
Available B	GOST R 50688-94	Hot water extraction and photometry with quinalizarin or azomethines
Available Zn	GOST R 50686-94	Ammonium acetate buffer (pH 4.8) extraction and photometry with dithizone or AAS
Available Cu	GOST R 50684-94, GOST R 50683-94	1 M HCl (in taiga and forest steppe zone) or ammonium acetate buffered at pH 4.8 (for calcareous soils in steppe and semi-desert zone) extraction and photometry with Pb diethyl-dithiocarbamate or AAS
Available Mo	GOST R 50689-94	Ammonium oxalate extraction (pH 3.3) (Grigg method) and photometry with Zn dithiol
Available Mn	GOST R 50685-94, GOST R 50682-94	0.1 N H_2SO_4 extraction (in taiga and forest steppe zone) or ammonium acetate buffered at pH 4.8 extraction for calcareous soils in steppe and semi-desert zone) and photometry or AAS
Available Co	GOST R 50687-94, GOST R 50683-94	1 N HNO_3 (in taiga and forest steppe zone) or ammonium acetate buffered at pH 4.8 (for calcareous soils in steppe and semi-desert zone) extraction and photometry with nitroso-R salt or pyridylazo naphthol in soil extract

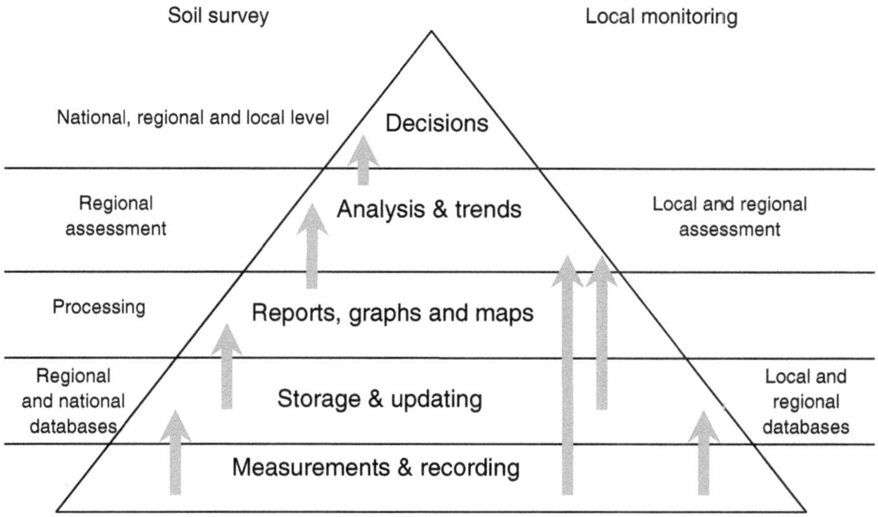

Fig. 4 Data flow in the soil monitoring system

Center in the Ministry of Agriculture with data flow line access for data providers and automatic integration and processing of the scheduled updates. The system was launched in 1998–2003. Georeferenced information can be aggregated to larger dimensions according to land use, soil type, or a group of indicators and can be used for policy-relevant decisions at local, regional, or national level, as, for example, with the calculation of the soil fertility index for subsidy assistance in territorial subjects of the Russian Federation, introduced since 2013.

Local monitoring data provide an opportunity to integrate an expanded set of indicators with data sets obtained during the soil survey. Reference plots provide additional information about acid–base indicators, nitrate dynamics, and the atmospheric input of acidity, heavy, and trace metals aggregated according to soil type (Sychev and Kuznetsov 2009; Sychev et al. 2010b).

The storage of data demands a unified format of databases for practical data management—at a local level (for example, the European format for Russian LTFEs (Smith et al. 2001) or at a regional level (Sychev et al. 2008). A database of long-term experiments can be an indispensable tool for integration with Geographic Information Systems, modeling tools, and the development of analytical methods (Barre et al. 2010; Smith et al. 2007).

However, a satisfactory data flow from local monitoring for regular reporting and decision-making still has to be developed. At the moment, all analysis and trends are the subject of scientific research and not updated on a regular basis, and we currently have much better opportunities for manipulating data using excellent techniques, such as modeling, for forecasting, prediction, and scenario analysis (Sychev et al. 2011).

In order to make the soil information collected relevant for different categories of end users, the following activities may be helpful:

- Giving users access to archived information through regular reports and websites.
- Regularly comparing the soil survey and local monitoring data.
- Performing statistical analysis to obtain confidence limits for interpretation, ensuring that the data collected is validated, and verified through automatic processing.
- Integrating data sets obtained from other environmental monitoring networks, including remote sensing domains, into our monitoring system.
- Widening the utilization of georeferencing information and spatial analytical tools to identify "hot spot" areas.

6 Main Results of Fertility Monitoring on Agricultural Soils

Soil organic matter (SOM): Soil organic matter and humus (its component) are important parameters, determining the genesis and potential fertility of soils. The reserves of soil organic matter play a decisive role in the reproduction of soil fertility. Monitoring data of agricultural soils based on regular surveys indicates that a decrease in the area of ploughland with a low humus content was observed during the years of intensive chemicalization (1975–1990). Since the 1990s, an inverse process has been observed (an increase in the area of ploughland with a low humus content, especially in the southern natural–agricultural zones). The results of the survey and data from local agroecological monitoring show that the weighted average content of organic matter in arable soils is 4.1 % for the entire Russian Federation. Arable soils with an average organic matter content of 2–4 and 4–6 % are predominant (39.1 and 28.9 % of the surveyed area, respectively). Arable lands containing less than 2.0 % prevail in the Northwestern (21.4 %) and Central (22.6 %) federal districts, where most ploughland is located in the zone of podzolic and soddy-podzolic soils (Table 3).

A significant increase in the area of ploughland with a low humus content has been noted during the past 25 years. Especially, significant changes occurred in the dry-steppe zone, where their increase exceeded 20 %.

A comparison of the total humus content with different SOC fractions reveals the role of organic matter as a factor connected with crop productivity and soil functions. Sharkov (2009) generalized the results of this kind of study in arable and native Leached Chernozems from the forest-steppe zone of the Novosibirsk Ob' region, and showed that the decrease in the soil organic carbon pool under long-term use as ploughland is due to the predominant loss of detritus (nonhumified fraction of organic matter) and easily mineralizable organic matter (Table 4). The contents of

Table 3 Proportions of ploughlands with low humus content (% of the sample set)

Natural–agricultural zone	Years				
	1981–1985	1986–1990	1991–1995	1996–1998	1999–2009
Forest–tundra–northern taiga	76.4	74.6	71.6	69.5	69.3
Middle taiga	24.3	23.9	19.6	18.8	18.5
Southern taiga–forest	50.4	47.8	58.4	55.9	55.7
Forest-steppe	42.6	38.0	35.7	41.2	41.1
Steppe	35.6	34.0	37.8	38.1	38.0
Dry steppe	69.0	52.7	58.4	73.9	75.0

Table 4 Contents of total and labile organic matter in the 0–25 cm layer of Leached Chernozem, wt%, mean of 7 determinations (Sharkov 2009)

Plot	C_{org}		Detritus*		Labile C**		Portion in C_{org}, %	
	Mean	Range	Mean	Range	Mean	Range	Detritus C	Labile C
Ploughland under cereals for not less than 60 years	3.00	2.65–3.35	0.147	0.125–0.170	0.514	0.366–0.758	4.9	17.1
Native or long-abandoned soil	3.37	2.65–3.85	0.563	0.319–0.845	0.634	0.471–0.967	16.7	18.8

*1.8 g cm^{-3} density NaJ extraction after Ganzhara (1997)
**0.1 n NaOH extraction after Ponomaryova and Plotnikova (1975)

detritus, labile C, and total organic C in native and abandoned soils were 3.8, 1.2, and 1.1 times higher than those in the old arable soil, respectively.

Soil acidity: In most of European Russia, soil acidity (low pH) is the main factor limiting the obtainment of high crop yields. In the 1960s, natural soils with an acid reaction comprised up to 80 % of the area in northern European Russia and up to 45 % in the central part. During the period of intensive chemicalization, the area of soils with a pH < 5.0 decreased to 20 % due to periodical liming. However, the volume of liming has now decreased by more than 20 times. Therefore, soil acidification has become an urgent problem in some regions. The highest portion of acid soils (more than 70.0 %) is observed in the Komi Republic; Orel, Kirov, Tambov, and Penza Oblasts; and Perm Krai. More than half of arable soils have an acid reaction in some regions of the Russian Federation: 58.4 % in Adygeya, 60.6 % in Karelia, 68.8 % in Mordovia, 54.7 % in Vologda Oblast, 55.5 % in Yaroslavl Oblast, 63.2 % in Smolensk Oblast, 65.7 % in Lipetsk Oblast, 65.4 % in Kostroma Oblast, 69.6 % in Tula Oblast, 69.2 % in Ryazan Oblast, and 68.3 % in Nizhni Novgorod Oblast.

In the past 25 years, the area of acid arable soils in the Tambov Oblast (zone of chernozemic soils) has increased by 27.8 %, with the annual increase exceeding

1.1 %. The weighted average pH value decreased from 5.7 in 1971 to 5.4 in 2009. Same processes are taking place in other regions of the Central Chernozemic zone (Penza, Lipetsk and Belgorod Oblasts).

At the same time, the generalized data for the Non-Chernozemic zone of Russia obtained during 30 years of monitoring observations indicate that agriculture is significantly efficient when acidic conditions are optimized in both light- and heavy-textured soils.

Plant-available nutrients: Under natural conditions, most ploughland has an insufficient content of available phosphorus. Until 1990, the areas of soils with a low phosphorus content decreased by 2–3 times due to the application of significant amounts of phosphorus-containing fertilizers. A tendency for these areas to decrease is also observed presently. However, this is due to the removal of low-productivity lands from agricultural use rather than to the long-term after-effect of phosphorus fertilizers. Presently, the arable soils of Russia are mainly characterized by a moderate phosphorus availability. The percentage of arable soils with medium and increased phosphorus availability is 57.0 % of the area. Data on the dynamics of soils with a low phosphorus availability are given in Table 5.

In some regions of the Russian Federation, the reduction of the ploughland area with a high phosphorus availability is accompanied by an increase in the areas of soils with a very low or low phosphorus content. In total, the area of arable soils with a low phosphorus availability has increased by 478.7 thousand ha. In the Republic of Kalmykia and the Lipetsk and Tambov Oblasts, the area of these soils has increased significantly (from 10 to 20 thousand ha); in three regions, the areas of arable soils with a low phosphorus availability have increased by more than 40 thousand ha: by 47.2, 43.0, and 49.7 thousand ha in Stavropol Krai, Orel Oblast, and Volgograd Oblast, respectively. In Rostov Oblast, the area of ploughland with very low phosphorus availability has increased by 56.3 thousand ha.

The results of the agrochemical survey of agricultural soils, local agroecological monitoring, and long-term field experiments of the geographical network revealed two main tendencies of changes in soil fertility:

- a slow but permanent decrease in the contents of humus and available phosphorus and potassium (the rates of these processes vary among the regions, but negative tendencies are observed everywhere);

Table 5 Percentage of soils with low phosphorus availability during different time intervals

Natural–agricultural zone	1965–1970	1976–1980	1986–1990	1996–1998	1999–2005
Northern taiga	56.6	56.9	46.1	47.5	48.0
Middle taiga	50.3	35.9	16.0	12.9	13.0
Southern taiga–forest	66.8	42.2	17.6	13.1	13.3
Forest-steppe	36.1	25.2	11.5	8.6	8.9
Steppe	59.1	52.0	32.8	29.0	29.0
Dry steppe	41.8	37.0	45.2	27.5	27.6

- movement of the boundary of acid soils to the south, to the zone of Chernozems, the most valuable soils.

To preserve and reproduce soil fertility, the contents should not only of major mineral nutrients, but also of microelements, should be considered. Data on soil fertility monitoring indicate that the soils of all agricultural lands are insufficiently supplied with mobile forms of essential microelements (B, Mo, Zn, Cu, Mn, Co). The area of soils where microfertilizer application is necessary is increasing. Presently, the microelement deficit needs to be made up for in almost all natural–agricultural zones of Russia: by 26.8 % for boron, by 84.5 % for molybdenum, by 96.9 % for zinc, by 71.6 % for copper, by 74.5 % for manganese, and by 92.6 % for cobalt.

7 Tendencies of Soil Fertility in Siberian Soils

Clearer temporal changes in the agrochemical parameters of agricultural soils are traced in the analysis of soil fertility monitoring data in different regions. Data on the agrochemical parameters of soils in the *Omsk Oblast* are given in Table 6. Monitoring data show that the most significant changes have occurred in the content of humus and available potassium.

Chelyabinsk Oblast is a complex site for monitoring because of its specific geographical location along the Ural Range, significant natural zonality, and developed industry. During the 15-year period between the survey rounds, the acidity of the main subtypes of zonal soils at the monitoring sites did not change significantly, except in the zone of dark Gray Forest and Mountain-Forest Soils, where the pH value decreased from 5.0 to 4.9 and total acidity increased from 7.38 to 9.97, which was because acid soils were no longer limed after 1993 (Scientific principles 2013) (Table 7).

Studies performed by the Chelyabinsk Research Institute of Agriculture revealed that the content of exchangeable bases in soils increases from north to south, from the mountain-forest zone to forest-steppe and open steppe. Ordinary and Southern Chernozems are most saturated with bases; while Dark Gray Forest and Mountain-Forest Soils are least saturated.

From the monitoring data, the reserves of available phosphorus in Leached Chernozems and Meadow-Chernozemic Soils exceed those in the other subtypes of

Table 6 Dynamics of changes in the agrochemical parameters of soils in the Omsk Oblast

Year	Humus, %	Available forms, mg/kg		pH
		Phosphorus	Potassium	
1972	–	83	175	5.8
1982	–	88	173	5.8
1990	5.34	110	173	5.7
2000	5.10	103	170	5.4
2010	5.00	95	170	5.4

Table 7 Values of pH and total acidity (TA, meq) in the main subtypes of zonal arable soils (0- to 20-cm layer) in the Chelyabinsk Oblast (Scientific principles 2013)

Survey round (years)	Dark gray forest and mountain-forest soils		Chernozems					
			Leached		Ordinary		Southern	
	pH	TA	pH	TA	pH	TA	pH	TA
I (1993–1997)	5.01	7.38	5.59	4.32	6.40	1.61	7.30	0.45
II (1998–2002)	5.07	5.68	5.69	3.82	6.46	1.39	7.08	0.41
III (2003–2007)	4.92	7.07	5.64	3.50	6.74	1.30	7.10	0.36
IV (2008–2011)	4.90	8.97	5.51	4.01	6.80	1.60	7.22	0.43

zonal soils (Table 8). An increase in the content of available phosphorus was also observed in the fourth round compared to the first one. At the same time, the content of available phosphorus varies among the different land types. This is related to soil cultivation. For example, Leached Chernozems of ploughlands and fallow areas contain significantly higher contents of available phosphorus than those of native soils. The phosphorus content in the profile of arable Leached Chernozems exceeds that in the native soil by 1.5 times (4.17 and 2.71 mg/kg in the first round and 6.34 and 2.79 mg/kg in the fourth round, respectively). A similar tendency is observed for Southern Chernozems and Meadow-Chernozemic and Solonetzic Soils; only in Ordinary Chernozems, the reserves of mobile phosphorus are almost similar in both types of land, which can be explained by the relatively small period of ploughing compared to the other soil subtypes. Thus, the content of available phosphorus in the soil strongly depends on its degree of cultivation and is related to climatic factors (alternation of droughty and perhumid years).

The high and elevated content of available potassium of regional chernozemic soils contributes to the maintenance of the soil potassium status for a long time without changes even when significant amounts of this element are removed with the crop harvest. As for ploughland, a decrease in the content of potassium by 7–30 % depending on soil type until the fifth round was universally observed.

The pool of soil humus depends on tillage practice, erosion processes, and the content of organic substances returned to the soil with manure and plant residues. The average content of humus for all subtypes of zonal soils in the Chelyabinsk Oblast is significantly lower in native soils than in arable ones. The most significant differences were noted for Solonetzic Soils and Ordinary and Calcareous Southern Chernozems. In particular, the content of humus in native soils was higher than in arable soils by 2.91 % for Meadow-Chernozemic Solonetzs, by 2.41 % for Chernozemic Solonetzs, by 1.44 % for Ordinary Chernozems, and by 1.31 % for Calcareous Southern Chernozems. The mean annual losses of humus after conversion to ploughing land were 0.99–1.11 t/ha for Solonetzic Soils and 0.52–0.45 t/ha for Ordinary and Calcareous Southern Chernozems. The trends in soil humus content between the first and fourth survey rounds are related to the input of organic substances. Humus loss in arable soils was revealed in the last survey round. This is due to the lack of sufficient input of fresh organic matter to the soil because of a drop

Table 8 Humus content in soils (0- to 20-cm layer) of the reference monitoring plots in the Chelyabinsk Oblast (Scientific principles 2013)

Soils	Sample set	Humus				Decrease in Ploughland against native soil		Mean annual humus loss after ploughing native soil, t/ha
		Native soil		Ploughland				
		%	t/ha	%	t/ha	%	t/ha	
Dark-Gray Mountain-Forest	99	6.86 ± 2.30	156.41	6.20 ± 1.16	130.2	−0.66	−26.21	0.19
Meadow-chernozemic	8	6.99 ± 0.81	162.17	6.74 ± 0.30	164.46	−0.25	2.29	–
Meadow-chernozemic Solonetzes	33	6.99 ± 0.63	149.59	4.08 ± 0.66	93.02	−2.91	−55.57	1.11
Chernozemic Solonetzs	22	6.67 ± 1.19	142.74	4.26 ± 0.69	103.09	−2.41	−39.65	0.99
Leached Chernozems	132	7.93 ± 1.04	169.70	7.09 ± 0.76	167.32	−0.84	−2.38	0.01
Ordinary Chernozems	110	7.36 ± 0.77	160.45	5.92 ± 0.62	129.06	−1.44	−31.39	0.52
Calcareous Southern Chernozems	11	5.10 + 0.33	120.36	3.79 + 0.28	102.33	−1.31	−18.03	0.45

in organic fertilization, the low percentage of perennial grasses in crop rotation and erosion processes. At the same time, some humus loss was also observed for fallow and virgin lands, with a steady input of fresh organic matter.

Generalization of the results of soil fertility monitoring in the Chelyabinsk Oblast revealed the following:

- soil carbon stock in arable lands is significantly less than those in virgin and fallow lands for all soil types and natural–agricultural zones of the Oblast;
- there are differences in the content of available phosphorus among the types of lands; in particular, the phosphorus content in the arable of Leached Chernozems and Southern Chernozems, Meadow-Chernozemic and saline soils is greater than those in their native analogs, while the available potassium content, on the contrary, is greater in native soils than in arable soils, which is related to both natural factors and the tillage system;
- a gradual decrease in the content of available potassium in arable soils and its insignificant increase in fallow and virgin lands are observed;
- the content of calcium and magnesium in the soil exchange complex of Dark Gray Forest and Mountain-Forest Soils decreases because of the removal of these elements with crop harvest and leaching, which results in an increase in soil acidity;
- the native soils have lower exchangeable and total acidities than the arable soils.

8 Ongoing Work

The perfection of the methodological supply of agroecological monitoring involves the development of new organization charts, methodological guidelines, and reference–information instruments (analytical information databases and data pools). Modernization is performed with consideration for the implementation of international programs such as HELCOM, the European Monitoring and Evaluation Program (EMEP), and the Acid Deposition Monitoring Network in East Asia (EANET) (Tørseth et al. 2012), as well as data on monitoring observations of soil state in the EU countries and the USA.

9 Conclusions

- Agroecological monitoring is an efficient tool for solving problems in the maintenance of soil fertility, as well as for preventing and controlling the negative anthropogenic impact on agroecosystems.
- Local monitoring on the reference plots, test areas, and plots of long-term field experiments of the Geographical Network, as well as the periodical agrochemical soil survey of agricultural lands and models of soil fertility, are

complementary methods within the integral approach to assess the effect of economic activities and natural factors on the agricultural soils.

- Analytical methods for agroecological monitoring are based on national standards, thus analytical protocols are exactly the same at different laboratories under the permanent control of the institute.
- Data from the current monitoring program are a fundamental basis for ecosystem modeling and scenario analysis.
- Long-term field experiments can be used to determine environmentally optimal fertilizing systems and assess the role of separate practices in the positive and negative environmental impacts; they are an indispensable tool for integration with GIS and developing analytical records.
- Local monitoring data are still underrepresented in regular reports and decision making, including those of Siberia and the Far East, and a satisfactory data flow from local monitoring still has to be developed.
- There is some ongoing work to integrate new modern monitoring methods and information systems into the already successful running monitoring program.

References

Barre P, Eglin T, Christensen BT, Ciais P, Houot S, Katterer T, van Oort F, Peylin P, Poulton PR, Romanenkov V, Chenu C (2010) Quantifying and isolating stable soil organic carbon using long-term bare fallow experiments. Biogeosciences 7(11):3839–3850. doi:10.5194/bg-7-3839-2010

Borlaug N (2007) Feeding a hungry world. Science 318:359

Ecological Doctrine (2002) Ecological Doctrine of the Russian Federation (2002) http://www.asser.nl/upload/eel-webroot/www/documents/Ecological%20Doctrine%20SUMMARY_en.pdf. Accessed on 9 Mar 2015

EEA (2001) European Environmental Agency. Proposal for a European soil monitoring and assessment framework. Technical report No 61, Copenhagen, 58 pp

Ganzhara NF, Borisov BA (1997) Humus formation and agronomic rating of soil organic matter (Гусообразование и агрономическая оценка органического вещества почв). Agroconsult, Moscow, 82 pp (in Russian)

Gerasimov IP (1978) Scientific principles of current environmental monitoring (Научные основы современного мониторинга окружающей среды). Izv AN SSSR Ser Geogr 3:13–25 (in Russian)

Gordeev AV, Romanenko GA (2008) (Eds) Problems of agricultural land degradation and rehabilitation of its productivity in Russia (Проблемы деградации и восстановления продуктивности земель сельскохозяйственного назначения в России) Ministry of Agriculture of the Russian Federation and Russian Academy of Agricultural Sciences 2008. Rosinformagrotex, Moscow, 67 p. Online at: http://agro.geonet.ru/publications/degradation.pdf . Accessed on 9 Mar 2015

Lal R (2009) Soils and food sufficiency, a review. Agron Sustain Dev 29:113–133

Lichtfouse E, Navarrete M, Debaeke P, Souchère V, Alberola C (2009) Sustainable agriculture, 1st ed. Springer, New York, 645 p. ISBN: 978-90-481-2665-1

Methodological guidelines (1992) Methodological guidelines for the determination of heavy metals in soils of agricultural lands and crops (1992) (Методические указания по определению тяжелых металлов в почвах сельхозугодий и продукции растениеводства). TsINAO, Moscow 62 pp (in Russian)

Methodological guidelines (2003) Methodological guidelines for the integrated monitoring of soil fertility on agricultural lands (2003) (Методические указания по проведению комплексного мониторинга плодородия почв земель сельскохозяйственного назначения). Rosinformagrotekh, Moscow 240 pp (in Russian)

Methods of agroecological monitoring on reference plots (2002) (Методы агроэкологического мониторинга на реперных участках). Rosselkhozakademia, Moscow, 58 pp (in Russian)

Mueller L, Schindler U, Ball BC, Smolentseva E, Sychev VG, Shepherd TG, Qadir M, Helming K, Behrendt A, Eulenstein F (2014) Productivity potentials of the global land resource for cropping and grazing. In: Novel measurement and assessment tools for monitoring and management of land and water resources in agricultural landscapes of central Asia. Environmental Science and Engineering 2014, pp 115–142. http://link.springer.com/chapter/10.1007%2F978-3-319-01017-5_6. Accessed on 9 Mar 2015

Ponomaryova VV, Plotnikova TA (1975) Methodology instructions for determination of content and composition of soil humus (Методические указания по определению содержания и состава гумуса в почвах). Leningrad 212 pp (in Russian)

Romanenkov VA, Kuzyakova IF (2000) New methodological approaches to agroecological field experiments in landscape agriculture (Новые методологические подходы к проведению агроэкологических полевых опытов в ландшафтном земледелии). In: International scientific and practical conference "current problems of experimental studies", vol 1, St. Petersburg, pp 140–146 (in Russian)

Scientific principles (2013) Scientific principles for the prevention of degradation of agricultural soils (lands) in Russia and the development of systems for the reproduction of their fertility in landscape-adaptive agriculture (2013) (Научные основы предотвращения деградации почв (земель) сельскохозяйственных угодий России и формирования систем воспроизводства их плодородия в адаптивно-ландшафтном земледелии) vol 2, Moscow, pp 388–422 (in Russian)

Sharkov IN (2009) Fertility in terms of the current concepts of soil organic matter. In: Gamzikov GP (ed) Agrochemical properties of soils and methods of their regulation. (Плодородие в свете современных представлений об органическом веществе почвы В: Агрохимические свойства почв и приемы их регулирования) IV Pryanishnikov Siberian lectures: International scientific and practical conference, Siberian branch of the Russian Academy of Agricultural Sciences, Novosibirsk, pp 60–72 (in Russian)

Smith P, Falloon PD, Körschens M, Shevtsova LK, Franko U, Romanenkov V, Coleman K, Rodionova V, Smith JU, Schramm G (2001) EuroSOMNET—a European database of long-term experiments on soil organic matter: the WWW metadatabase. J Agric Sci 33(3, 1):233–239

Smith JU, Smith P, Wattenbach M, Gottschalk P, Romanenkov VA, Shevtsova LK, Sirotenko OD, Rukhovich DI, Koroleva PV, Romanenko IA, Lisovoi NV (2007) Projected changes in cropland soil organic carbon stocks in European Russia and the Ukraine, 1990–2070. Glob Change Biol 13:342–356

State Programme (2013–2020) The State Programme of agricultural development and regulation of agricultural products, commodities and food markets, 2013–2020 http://government.ru/en/docs/3360/. Accessed on 9 Mar 2015

Sychev VG (2003) The main resources of crop productivity and their interrelation (Основные ресурсы урожайности сельскохозяйственных культур и их взаимосвязь). Pryanishnikov Research Institute of Agrochemisty, Moscow, 228 p (in Russian)

Sychev VG (2007) Trends of agrochemical parameters of soil fertility in the European Russia (Тенденции изменения агрохимических показателей почвенного плодородия Европейской территории России). Bulletin of the geographical network of fertilizer experiments no. 3, Pryanishnikov Research Institute of Agrochemisty, Moscow 28 pp (in Russian)

Sychev VG, Lunev MI et al. (2010a) Agrochemical review of plowing soils in natural-economic zones and provinces of Russia (based on agrochemical soil survey) (Агрохимическая характеристика пахотных почв природно-сельскохозяйственных зон и провинций России (по данным сплошного агрохимического обследования). Bulletin of the geographical network of fertilizer experiments no. 9. Pryanishnikov Research Institute of Agrochemisty, Moscow 34 pp (in Russian)

Sychev VG, Kuznetsov AV (2009) Effect of snow atmospheric deposition on contamination of agricultural lands ((local monitoring data) (Влияние атмосферных осадков виде снега на загрязнение сельскохозяйственных угодий (по данным локального мониторинга)). Bulletin of the geographical network of fertilizer experiments no. 8. Pryanishnikov Research Institute of Agrochemisty, Moscow 28 pp (in Russian)

Sychev VG, Romanenkov VA (2009) Mineral fertilizers in Russian agriculture: demands and realness//Fertilizers and Fertilization: stewardship for food security, food quality, environment and nature conservation. In: Proceedings: Chiang Mai, Thailand, 22–27.01.2006. Iasi: Terra Nostra, Romania. ISBN 978-973-1888-224

Sychev VG, Rukhovich OV, Romanenkov VA, Belichenko MV, Listova MP (2008) The problem of formation unified and structured database "Agrogeos" of agrochemical service and geographic network of field experiments (Опыт создания единой систематизированной базы данных полевых опытов агрохимслужбы и Геосети « Агрогеос »). Probl Agrochemistry Ecol 3:35–38 (In Russian)

Sychev VG, Lunev MI, Kusnetsov AV, Pavlichina AV, Lobas NV, Demina NA, Vasileva NM et al. (2010b) Fertility of plowing soils in Russian Federation (local monitoring data) (Плодородие пахотных почв Российской Федерации (по данным локального мониторинга). Bulletin of the geographical network of fertilizer experiments no. 8. Pryanishnikov Research Institute of Agrochemisty, Moscow 52 pp (in Russian)

Sychev VG, Yefremov EN, Romanenkov VA (2011) Monitoring of agricultural lands and rational use of soil fertility potential: current status and future prospects (Состояние и перспективы мониторинга земель сельскохозяйственного назначения и рационального использования потенциала почвенного плодородия). Probl Agrochemistry Ecol 4:42–46 (in Russian)

Sychev VG, Yefremov EN et al (2013a) Register of Soil Fertility. Agrochemical characteristics of agricultural soils in Russian Federation (Реестр плодородия почв. Агрохимическа яхароактеристика почв сельскохозяйственных угодий Российской Федерации). Pryanishnikov Research Institute of Agrochemisty, Moscow 208 pp (in Russian)

Sychev VG, Yefremov EN, Romanenkov VA (2013b) Agrochemical science: results and outlook (Итоги и перспективы развития агрохимии). Probl Agrochemistry Ecol 4:11–16 (in Russian)

System of agroecological monitoring (2006) System of agroecological monitoring of agricultural lands (Система агроэкологического мониторинга земель сельскохозяйственного назначения). Russian Academy of Agricultural Sciences, Moscow 79 pp (in Russian)

Tørseth K, Aas W, Breivik K, Fjæraa AM, Fiebig M, Hjellbrekke AG, Lund Myhre C, Solberg S, Yttri KE (2012) Introduction to the European monitoring and evaluation programme (EMEP) and observed atmospheric composition change during 1972–2009. Atmos Chem Phys 12:5447–5481. doi:10.5194/acp-12-5447-2012

Chapter 25
The International Soil Classification System WRB, Third Edition, 2014

Peter Schad

Abstract The international soil classification system World Reference Base for Soil Resources (WRB) is edited by a Working Group of the International Union of Soil Sciences and published by the Food and Agriculture Organization of the United Nations. The third edition was released in 2014. The WRB has two hierarchical levels. The first level allocates the soils of the world to 32 Reference Soil Groups (RSGs), which are identified using a key. Many RSGs represent specific soil-forming processes, are representative of major soil regions, or reflect special parent materials. In the second level, the soil names are constructed by adding a set of qualifiers to the name of the RSG. In total, 186 qualifiers are defined. Some can be combined with many RSGs, others with only a few or even with just one. For every RSG, a list of the possible qualifiers is provided. These possible qualifiers are subdivided into principal and supplementary qualifiers. Principal qualifiers are regarded as being the most significant for a further characterization of soils of the particular RSG. They are ranked and given in an order of importance. Supplementary qualifiers give some further details about the soil. They are not ranked but are used in alphabetical order. The WRB uses diagnostic horizons, diagnostic properties, and diagnostic materials. Diagnostic horizons and properties reflect widespread, common results of the processes of soil formation or indicate specific conditions of soil formation. In addition, diagnostic horizons require a certain thickness, thus forming a recognizable layer in the soil. Diagnostic materials are materials that significantly influence pedogenetic processes or are indicative of them. The definitions of many RSGs and qualifiers refer to the presence or absence of certain diagnostics at a certain depth. In addition, many definitions refer to individual features such as the base saturation or clay content. To name a soil, the RSG has to be provided with all applying principal and supplementary qualifiers. For map legends, the number of qualifiers depends on scale. Detailed rules are established to achieve comprehensive names for map units at different scale levels.

P. Schad (✉)
Lehrstuhl für Bodenkunde, Technische Universität München, 85350 Freising-Weihenstephan, Germany
e-mail: schad@wzw.tum.de

L. Mueller et al. (eds.), *Novel Methods for Monitoring and Managing Land and Water Resources in Siberia*, Springer Water,
DOI 10.1007/978-3-319-24409-9_25

Keywords Soil classification · Map legends · WRB

1 Introduction

The third edition of the international soil classification system World Reference Base for Soil Resources (WRB) was published in June 2014. It is based on the second (2006) and the first (1998) editions. The WRB follows the Legends (1974, 1988) of the Soil Map of the World (Food and Agriculture Organization of the United Nations [FAO]—UNESCO 1971–1981). The body responsible for the WRB is a Working Group of the International Union of Soil Sciences (IUSS). In 1998, the IUSS Council endorsed the WRB as its officially recommended terminology to name and classify soils. In 2015, an Update of the third edition of the WRB was published with some minor corrections and amendments.

The WRB comprises two levels. The first level has 32 Reference Soil Groups (RSGs), which are identified using a key. In the second level, the soil names are constructed by adding a set of qualifiers to the name of the RSG.

2 The Architecture of the System

2.1 The First Level

Many Reference Soil Groups are the result of specific soil-forming processes or are representative of major soil regions or reflect special parent materials. Table 1 presents a list of the 32 RSGs in the sequence of the key, together with a brief and simplified description and the suggested codes.

2.2 The Second Level

For the lower level, 186 qualifiers have been defined. Some can be combined with many RSGs, others with only a few or even with just one. For every RSG, a list with the possible qualifiers is provided. These possible qualifiers are subdivided into principal and supplementary qualifiers. Principal qualifiers are regarded as being the most significant for a further characterization of soils of the particular RSG. They are ranked and given in an order of importance. Supplementary qualifiers give some further details about the soil. They are not ranked but are used in alphabetical order. The Cambisols have the longest list with 68 possible qualifiers and the Nitisols the shortest list with 33 possible qualifiers. Many qualifiers are mutually exclusive, and most soils have far fewer than 10 qualifiers.

Table 1 Simplified guide to the Reference Soil Groups (RSGs) with suggested codes (IUSS Working Group WRB 2014)

	RSG	Code
1. Soils with thick organic layers	Histosols	HS
2. Soils with strong human influence		
Soils with long and intensive agricultural use	Anthrosols	AT
Soils containing significant amounts of artefacts	Technosols	TC
3. Soils with limitations to root growth		
Permafrost-affected soils	Cryosols	CR
Thin soils or soils with many coarse fragments	Leptosols	LP
Soils with a high content of exchangeable Na	Solonetz	SN
Alternating wet-dry conditions, shrink-swell clays	Vertisols	VR
High concentration of soluble salts	Solonchaks	SC
4. Soils distinguished by Fe/Al chemistry		
Groundwater-affected soils, underwater soils and soils in tidal areas	Gleysols	GL
Allophanes or Al-humus complexes	Andosols	AN
Subsoil accumulation of humus and/or oxides	Podzols	PZ
Accumulation and redistribution of Fe	Plinthosols	PT
Low-activity clay, P fixation, many Fe oxides, strongly structured	Nitisols	NT
Dominance of kaolinite and oxides	Ferralsols	FR
Stagnating water, abrupt textural difference	Planosols	PL
Stagnating water, structural difference and/or moderate textural difference	Stagnosols	ST
5. Pronounced accumulation of organic matter in the mineral topsoil		
Blackish topsoil, secondary carbonates	Chernozems	CH
Dark topsoil, secondary carbonates	Kastanozems	KS
Dark topsoil, no secondary carbonates (unless very deep), high base status	Phaeozems	PH
Dark topsoil, low base status	Umbrisols	UM
6. Accumulation of moderately soluble salts or non-saline substances		
Accumulation of, and cementation by, secondary silica	Durisols	DU
Accumulation of secondary gypsum	Gypsisols	GY
Accumulation of secondary carbonates	Calcisols	CL
7. Soils with a clay-enriched subsoil		
Retic properties	Retisols	RT
Low-activity clays, low base status	Acrisols	AC
Low-activity clays, high base status	Lixisols	LX
High-activity clays, low base status	Alisols	AL
High-activity clays, high base status	Luvisols	LV

(continued)

Table 1 (continued)

	RSG	Code
8. Soils with little or no profile differentiation		
Moderately developed soils	Cambisols	CM
Sandy soils	Arenosols	AR
Soils with stratified fluviatile, marine and lacustrine sediments	Fluvisols	FL
Soils with no significant profile development	Regosols	RG

2.3 The Diagnostics

The WRB uses diagnostic horizons, diagnostic properties, and diagnostic materials. Diagnostic horizons and properties are characterized by a combination of attributes that reflect widespread, common results of the processes of soil formation or indicate specific conditions of soil formation. Their features can be observed or measured, either in the field or in the laboratory, and require a minimum or maximum expression to qualify as diagnostic. In addition, diagnostic horizons require a certain thickness, thus forming a recognizable layer in the soil. Diagnostic materials are materials that significantly influence pedogenetic processes or are indicative of them. Their features may stem from the parent material or be the result of pedogenetic processes.

The definitions of many RSGs and qualifiers refer to the presence or absence of certain diagnostics at a certain depth. In addition, many definitions refer to individual features such as the base saturation or clay content.

3 Soil Designation

3.1 Naming a Soil (a Profile)

The first step is detecting diagnostic horizons, properties, and materials. In the second step, the RSG is identified using the key. In the third step, all applying principal and supplementary qualifiers are allocated. The principal qualifiers are added before the name of the RSG without brackets and without commas. The sequence is from right to left, i.e., the uppermost qualifier in the list is placed closest to the name of the RSG. The supplementary qualifiers are added in brackets after the name of the RSG and are separated from each other by commas. The sequence is from left to right, i.e., the first qualifier according to the alphabet is placed closest to the name of the RSG.

Example: A Chromic Stagnic Leptic Luvisol (Cutanic, Differentic, Humic, Ruptic) is a soil with clay migration, high cation exchange capacity, and high base saturation (Luvisol), with continuous rock starting within 100 cm of the soil surface (Leptic), effects of water stagnation (Stagnic), intense reddish color (Chromic), clay coatings (Cutanic), significantly higher clay contents in the subsoil (Differentic), high concentrations of organic matter in the mineral topsoil (Humic), and a lithic discontinuity (Ruptic).

Constructing the second level by adding qualifiers to the RSG has several advantages compared with a dichotomic key:

- For every soil, the RSG has the appropriate number of associated qualifiers.
- The WRB is capable of indicating most of the soil's properties, which are incorporated into an informative soil name.
- The system is robust. Missing data do not necessarily lead to a dramatic error in the soil name. If one qualifier is erroneously added or erroneously omitted based on incomplete data, the rest of the soil name remains correct.

To achieve higher precision, qualifiers may be combined with specifiers to form subqualifiers. The most common ones are those referring to depth criteria, such as Epi- (the feature is only present from 0 to 50 cm) or Endo- (the feature is only present from 50 to 100 cm), e.g., Epistagnic or Endochromic. Subqualifiers are optional.

Two examples of classification are given in Fig. 1 and the explanations subsequently.

Fig. 1 Classification examples. Two soils from Sakha (Yakutia), Siberia. Soil **a** Cambic Turbic Cryosol (Protocalcic, Humic, Endoloamic, Amphisiltic), Soil **b** Fluvic Sodic Endogleyic Solonchak (Alcalic, Carbonatic, Ochric, Pantosiltic)

3.2 *Examples for Naming Soils*

Soil (a) Cambic Turbic Cryosol (Protocalcic, Humic, Endoloamic, Amphisiltic)

The soil developed on river terraces west of Yakutsk, 220 m asl, and is covered by larch forest. Average annual precipitation is 200 mm. Permafrost starts at a depth of 170 cm. In the WRB, permafrost within 200 cm in combination with cryoturbation within 100 cm qualifies for a Cryosol. Cryoturbation, here best visible from 80 to 120 cm, implies the Turbic qualifier. In situ weathering leads to a cambic horizon from around 10 to 120 cm and thus to the Cambic qualifier. Secondary carbonates start at 25 cm. They are sufficient not for a calcic horizon but for protocalcic properties and, hence, for the Protocalcic qualifier. The Humic qualifier indicates the intense accumulation of organic matter in the mineral topsoil. The texture is silty from 0 to 65 cm and further down loamy, which is expressed by the Amphisiltic and Endoloamic qualifiers.

Soil (b) Fluvic Sodic Endogleyic Solonchak (Alcalic, Carbonatic, Ochric, Pantosiltic)

The soil was formed in lacustrine sediments of Lake Abalakh, east of Yakutsk at 160 m asl. It is covered by discontinuous halophytic vegetation. Annual precipitation is around 200 mm. The lake is very salty and lies in a big alas that is characterized by a deep talik. In the first meter, the pH (in aqua dest.) is above 10, and sodium saturation is between 25 and 45 %. At a depth of 0–60 cm, the electric conductivity (saturation extract) is above 8, the requirements of the salic horizon are fulfilled, and the soil is a Solonchak according to WRB. The high sodium saturation implies the Sodic qualifier, the high pH value the Alcalic qualifier, and the dominance of the carbonate anion the Carbonatic qualifier. The stratification of the lacustrine sediments is visible below 70 cm and in the first 20 cm (very recent sedimentation), which is expressed by the Fluvic qualifier. In between, the stratification was lost due to in situ soil formation. Ascending groundwater from the lake causes redox processes in the subsoil, and, therefore, the Endogleyic qualifier applies. The texture is silty throughout, explaining why the Pantosiltic qualifier is used. The Ochric qualifier indicates a topsoil carbon concentration between 0.2 and 0.6 %.

4 Creating Map Legends

While for naming a soil, all applying qualifiers must form part of the soil name, for maps, the number of qualifiers depends on scale. The WRB recognizes four scale levels:

- for very small map scales, only the RSG used;
- for next larger map scales, the RSG plus the first applicable principal qualifier is used;

- for next larger map scales, the RSG plus the first two applicable principal qualifiers are used; and
- for next larger map scales, the RSG plus the first three applicable principal qualifiers are used.

(If there are fewer qualifiers applicable than described above, the lesser number is used.) These principal qualifiers are placed before the name of the RSG according to the rules for naming a soil. Depending on the purpose of the map or according to national traditions, at any scale level, further qualifiers may be added optionally. These may be additional principal qualifiers from further down the list and not already used in the soil name, or they may be supplementary qualifiers. They are placed using the above-mentioned rules for supplementary qualifiers.

We use the example from above: Chromic Stagnic Leptic Luvisol (Cutanic, Differentic, Humic, Ruptic). The map units according to the scale are:

Luvisol
Leptic Luvisol
Stagnic Leptic Luvisol
Chromic Stagnic Leptic Luvisol

If we want to add additional qualifiers we may have, e.g., the following names in the map units:

Luvisol (Leptic)
Leptic Luvisol (Stagnic, Humic)
Stagnic Leptic Luvisol (Humic)
Chromic Stagnic Leptic Luvisol (Humic, Ruptic)

For most maps, it is desirable that a map unit not only shows one soil but an association of soils. The WRB distinguishes:

- dominant soils represent ≥50 % of the soil cover;
- codominant soils represent ≥25 and <50 % of the soil cover; and
- associated soils represent ≥5 and <25 % of the soil cover.

For codominant or associated soils, fewer numbers of qualifiers than indicated above (or even no qualifier) may be appropriate.

5 Major Changes in WRB 2014 Compared to WRB 2006

The WRB 2006 is suitable for naming soils (profiles). For creating map legends, an addendum was published in 2010. Both documents use the same definitions but have different rules for using the qualifiers. The WRB 2014 serves both purposes. This implies new rules for qualifier usage.

The only change at the RSG level is to replace Albeluvisols by Retisols. Retisols have a broader definition and include the former Albeluvisols. Retisols show clay migration and an eluvial horizon that penetrates into the illuvial horizon forming a net (retic properties).

Fluvisols have moved down in the key to be the second last RSG, which changes some former Fluvisols into Solonchaks, Gleysols, and others. This gives soil properties (salinity, redox conditions, etc.) priority over the mere position in the landscape. The Umbrisols are now placed directly after Phaeozems. The following RSGs switched their positions: Solonetz and Vertisols, Durisols and Gypsisols, Cambisols and Arenosols. The soils characterized by an argic horizon are now in the following order: Acrisols—Lixisols—Alisols—Luvisols.

Base saturation—used to separate Acrisols from Lixisols, Alisols from Luvisols and the Dystric qualifier from the Eutric qualifier—is now calculated by dividing the sum of exchangeable bases by the sum of exchangeable bases plus exchangeable Al (exchangeable bases by 1 M NH_4OAc, pH 7; Al by 1 M KCl, unbuffered). Base saturation—used to separate mollic from umbric horizons and defining Chernozems, Kastanozems and Phaeozems—is still calculated from the CEC at pH 7.

Three new diagnostic horizons have been defined. The chernic horizon is now required for Chernozems. It replaces the voronic horizon but the requirements for the chernic horizon are in between the requirements for the mollic and the former voronic horizon. The pretic horizon allows a better accommodation of 'Terra preta de Indio' within the Anthrosols. The protovertic horizon (the former vertic properties) describes layers with weakly expressed shrink–swell features.

The anthric, takyric, and yermic horizons have been changed to diagnostic properties. The albic horizon was replaced by albic material.

The diagnostic material 'soil organic carbon' has been introduced to separate pedogenetic organic carbon from organic carbon that satisfies the diagnostic criteria of artefacts (e.g., in mine spoil).

Several new qualifiers have been added.

The WRB should be able to express characteristics that are regarded to be important in national systems. Some amendments have been made to allow for the better representation of soil units from national systems that have not been well represented before, for example from the Australian and the Brazilian systems.

Some parts of the world were not well represented in the WRB system before, e.g., ultracontinental permafrost regions. The system has been enlarged to allow a better classification of soils from these regions.

Acknowledgments The editors of the third edition of the WRB are: Peter Schad (Germany), Cornie van Huyssteen (South Africa) and Erika Micheli (Hungary). Members of the WRB Board are: Lúcia Anjos (Brazil), Carlos Cruz Gaistardo (Mexico), Seppe Deckers (Belgium), Stefaan Dondeyne (Belgium), Einar Eberhardt (Germany), Maria Gerasimova (Russia), Ben Harms (Australia), Arwyn Jones (European Commission), Pavel Krasilnikov (Russia), Thomas Reinsch (USA), Ronald Vargas (FAO) and Ganlin Zhang (China). Numerous other soil scientists contributed to the third edition of the WRB.

References

IUSS Working Group WRB (2014) World reference base for soil resources 2014. Edited by Schad P, van Huyssteen C, Micheli E. World Soil Resources Reports No. 106. FAO, Rome. 189 p. ISBN 978-92-5-108369-7

Chapter 26
An Emerging Method of Rating Global Soil Quality and Productivity Potentials

Lothar Mueller, Uwe Schindler, Volker Hennings, Elena N. Smolentseva, Olga V. Rukhovich, Vladimir A. Romanenkov, Victor G. Sychev, Sergey Lukin, Askhad K. Sheudshen, Ludmila Onishenko, Abdulla Saparov, Konstantin Pachikin, Axel Behrendt, Wilfried Mirschel and Frank Eulenstein

Abstract This chapter provides information about an emerging approach for rating agricultural soil quality (SQ) and crop yield potentials consistently over a range of spatial scales. We developed and tested the Muencheberg Soil Quality Rating (M-SQR), a straightforward, indicator-based overall method for agricultural SQ assessment. The aim of this chapter is to improve the precision and consistency of

L. Mueller (✉) · U. Schindler · A. Behrendt · W. Mirschel · F. Eulenstein
Leibniz Centre for Agricultural Landscape Research (ZALF), Eberswalder Str. 84, 15374 Müncheberg, Germany
e-mail: mueller@zalf.de

U. Schindler
e-mail: schind@zalf.de

A. Behrendt
e-mail: abehrendt@zalf.de

W. Mirschel
e-mail: mirschel@zalf.de

F. Eulenstein
e-mail: feulenstein@zalf.de

V. Hennings
Strousbergstraße 9, 30449 Hannover, Germany
e-mail: v.hennings@kabelmail.de

E.N. Smolentseva
Institute for Soil Science and Agrochemistry, 8/2, Ac. Lavrentieva ave, 630090 Novosibirsk, Russia
e-mail: parabraunerde@rambler.ru

O.V. Rukhovich · V.A. Romanenkov · V.G. Sychev
Pryanishnikov All-Russian Research Institute of Agrochemistry, Pryanishnikova St. 31a, 127550 Moscow, Russia
e-mail: o_ruhovich@mail.ru

L. Mueller et al. (eds.), *Novel Methods for Monitoring and Managing Land and Water Resources in Siberia*, Springer Water,
DOI 10.1007/978-3-319-24409-9_26

final ratings by updating the rating frames of most crop-yield-relevant indicators. M-SQR is a framework covering aspects of soil texture, structure, topography and climate which is based on 8 Basic Indicators and more than 12 Hazard Indicators. Ratings are performed by visual methods of soil evaluation and supported by monthly climate data. A field manual is then used to provide ratings from tables based on indicator thresholds. Finally, overall rating scores are given, ranging from 0 (worst) to 100 (best) to characterise crop yield potentials. The current approach is valid for grassland and cropland. Field tests in the main global agricultural regions have confirmed the practicability and reliability of the method. Many experimental sites have been assessed in Russia (Siberia included) and Central Asia. We found that at the field scale, soil texture and structure are most important criteria of agricultural SQ. At the global scale, climate-controlled hazard indicators of drought risk and the soil thermal regime are crucial for soil functioning and crop yield potentials. We present new rating tables for indicators that are most relevant to crop yields globally: a too-cold soil thermal regime (Hazard indicator 12) and agricultural drought (Hazard indicator 7). Final rating scores are well correlated with crop yields of cereals and grass. Regression equations express the relationships between overall M-SQR rating numbers and crop yield potentials at defined levels of farming inputs. We conclude that the combination of the Muencheberg Soil Quality Rating (M-SQR) with the World Reference Base of soil resources (WRB 2014) provides key information about main soil functions and processes. This system

V.A. Romanenkov
e-mail: geoset@yandex.ru

V.G. Sychev
e-mail: info@vniia-pr.ru

S. Lukin
Institute for Organic Fertilization and Peat (VNIIOU), 601390 Vyatkino 492.2, Sudogotskiy Rayon, Vladimir Oblast, Russia
e-mail: vnion@vtsnet.ru

A.K. Sheudshen
Russian Scientific Research Institute of Rice, Bjelosernoye Settlem, 350921 Krasnodar, Russia
e-mail: kubagrohim@mail.ru

A. Saparov · K. Pachikin
Kazakh Research Institute of Soil Science and Agrochemistry named after U.U. Uspanov, Al Faraby Ave 75b, 050060 Almaty, Kazakhstan
e-mail: ab.saparov@yahoo.com

K. Pachikin
e-mail: kpachikin@yahoo.com

L. Mueller · A.K. Sheudshen · L. Onishenko
Kuban State Agrarian University, 13 Kalinin Str, 350044 Krasnodar, Russia
e-mail: kubagrohim@mail.ru

could be evolved for ranking and controlling agricultural SQ on a global scale. It should become a basis for more objective monitoring of global land quality, promoting sustainable land use and management, serving as one of the decision tools (decision support systems, impact assessment procedures) for economic trade-offs and land use planning. As a first step, the current concepts and data have led to a new crop yield potential map of Germany. The method and data given in this chapter could provide the basis for creating a similar map of Russia using the same methodology.

Keywords Soil quality · Indicators · Crop yield · Productivity · Muencheberg Soil Quality Rating · Russia

1 Introduction

Global key issues of the 21st century such as feeding the world population as it grows without control, the scarcity of water and energy, desertification, environmental pollution and loss of biodiversity are raising the pressure on existing and potential agricultural land. Sustainable land use strategies and systems are in demand. This requires a reliable characterisation and assessment of soils, their properties and functions on a trans-national scale. One of their key functions is "food and other biomass production" (Blum 1993). This productivity function is related to agricultural soil quality (SQ). A multitude of approaches are available for the quantification of aspects of agricultural SQ. Specific soil and land evaluation schemes already exist for use on a local or national basis. However, their soil data inputs differ; evaluation ratings are not transferable and are not universally applicable to international studies. Babylonian confusion in soil classification terms is preventing international communication and hindering conventions for sustainable soil use and management. There have been some successful efforts to create a framework for the international classification, correlation and communication of soils (WRB 2006, 2014) over the past 15 years. This international classification system, like most national soil classification systems, provides soil names. These names include information about typical processes, features and properties of most soils but do not provide enough information about the overall soil quality and crop yield potentials (Mueller et al. 2010). Suitable classification systems should be developed, tested and established. Dokuchaev long ago pointed out the need to classify soils in a way that comprised information about crop yield potentials (In: Dokuchaev 1951).

There are as yet no overall soil quality rating systems that are practicable over different scales, ranging from the field scale to the global scale. The Muencheberg Soil Quality Rating (M-SQR, Mueller et al. 2007) is intended to close this gap. It shall provide a reliable and simple evaluation of soil quality and crop yield potentials in terms of good, moderate, poor and very poor. Potential applications of the method are soil resource planning, agro-environmental land monitoring, guiding land purchase, and assessing the sustainability and environmental impacts of land use. It is a tool which can be implemented into the next generations of decision support systems (Mirschel et al. 2016) and impact assessment procedures (Helming 2014). The aim of this paper is to give a short description of the Muencheberg Soil Quality Rating, to show the reliability and potential of this approach, and to present updated rating tables for the globally most crop-yield-relevant indicators.

2 The Principle of the Muencheberg Soil Quality Rating

The concept of the M-SQR is that most terrestrial crops require appropriate seedbed conditions and optimum soil quality for a deep and well-established rooting zone. Productivity-relevant indicators (Fig. 1) characterize the quality of this soil zone, and their scoring provides a functional coding of soils. The approach includes

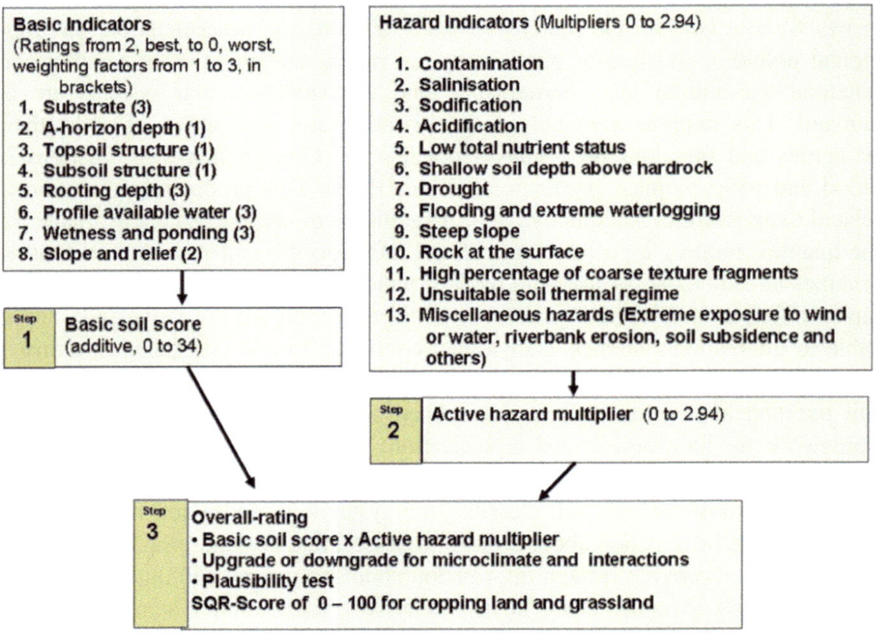

Fig. 1 Rating scheme of M-SQR (Mueller et al. 2007). First, each of the 8 Basic Indicators is rated on a scale from 2 (best) to 0 (worst), multiplied by a weighting factor from 1 to 3 and then summed. Then the occurrence of Hazard Factors is checked and summed as necessary to give a similar rating. The most crop-yield-limiting Hazard Indicator is used to estimate a multiplier which may range from 0 to 2.94. The Basic Score times the active multiplier yields an overall M-SQR rating between 0 and 100. More than 100 agricultural research sites worldwide have been rated and classified

indicators of the inherent (soil substrate) and dynamic (soil structure) agricultural soil quality, of the topography in terms of slope and of the climate in terms of the soil thermal and moisture regimes.

Two types of indicator have been identified and defined in scoring tables. The first is Basic Indicators, which relate mainly to the soil's textural and structural properties relevant to plant growth. They are the soil substrate, depth and characteristics of the A horizon, size and shape of topsoil aggregates, features of subsoil structure and compaction, depth of rooting, water supply, wetness and ponding, slope and relief. These are weighted, with extra weight given to rooting and water factors, then the indicators are summed. The Basic Score (ranging from 0–34) or the Upscaled Basic Score (UBS = Basic Score*2.94, maximum score of 100) of M-SQR mainly reflects properties of the texture and structure of soils. Very high Upscaled Basic Scores (>80) are typical for Loess soils or loess-like soil material and medium or low scores (<60) for sandy, stony or waterlogged soils.

The second type of indicators is Hazard Indicators, relating to the most severe restrictions of soil function identified at the site. The most common Hazard Indicators are a lack of water in the main vegetation period (agricultural drought) or drought in combination with an unsuitable temperature regime (soils too cold, too-short vegetation period). The sum of weighted basic indicator ratings and multipliers derived from ratings of the most severe (active) Hazard Indicator yield an overall SQ rating index, i.e. the M-SQR score. If no Hazard Indicators occur, the UBS and M-SQR score are identical. The M-SQR provides a rating of the overall soil quality on a 100-point scale. Loess soils in a temperate climate or under irrigation have the highest overall soil quality (M-SQR scores >80).

Indicator ratings are based on a field manual (Mueller et al. 2007) and utilise soil survey classifications (FAO 2006a; WRB 2006), soil structure diagnosis tools and local or regional soil and climate data. Matching tables provide a fast orientation to commonly used current assessments of individual indicators. These are mainly documented in the FAO Guidelines for soil description (FAO 2006a) the German "Bodenkundliche Kartieranleitung" (AG Boden 2005) and the U.S. National Soil Survey Handbook (USDA/NRCS 2005a).

The philosophy of the rating procedure is to provide a result based on a minimum of data, but to utilise more detailed information if available. Data need to be allocated to scoring tables, suggested values and sample photographs in the field manual (Mueller et al. 2007). If, for example, analyses of soil density or plant-available water are available and plausible, they should be used instead of the suggested values given in the manual.

Soil quality ratings are restricted to the soil's suitability for cropping and grazing. The focus is on rainfed cropping in temperate zones and rotations with a dominance of cereals, mainly wheat. A growing number of sample ratings have provided a data basis for the adjustment of individual ratings and the constant improvement of the framework and indicator thresholds.

3 The Field Rating Procedure

The field procedure requires a minimum of equipment. This consists of spade + borer + foot rule + knife + M-SQR field guide (Mueller et al. 2007).

Additionally, some equipment can be useful to detect soil properties of particular interest and to document the work. These are:

- A probe for testing the pH (or pH test strips) and electrical conductivity if acidification, sodification or salinisation are expected
- A GPS and camera for geo-referencing and documentation of the visual soil data
- Common soil survey equipment such as the WRB 2006 brochure (WRB 2006), Munsell Colour Charts and 0.1 n HCl if soil rating is being done in combination with a soil taxonomic classification
- A stable plastic box, a larger plastic bag and the field guide "Visual Soil Assessment" (Shepherd 2009) to perform VSA analysis if soil structure restrictions are expected, these being crucial soil quality limiting factors

The field procedure for M-SQR consists of digging a small pit of 0.4–1 m depth and augering a hole down to 1.6 m to detect any layering, a shallow water table or other root-impeding soil properties. A regular soil pit of the kind which is common in soil surveys (Fig. 2) does the job better. It is recommended to perform the M-SQR jointly with a soil taxonomic classification if this is not yet available. The latter gives the soil being studied a name, whilst M-SQR provides quality scores.

The exact sampling point should have been determined using available information from soil maps, airborne data and current or former vegetation patterns. The method requires some experience in soil surveys, such as estimating the soil texture, organic matter content, and soil water balancing and vegetation ecology.

Next, the soil profile is scanned to assess the set of indicators shown in Fig. 1 using visual tactile examination, expert knowledge and minimum equipment. A basic rating score ranging from 2 (best) to 0 (poorest) is given for every indicator with the help of scoring tables related to soil attributes (Table 1).

Hazard Indicators in Table 2 are rated in the same way. Not all growth-limiting factors are visible in the soil profile. The globally dominating hazard factors of agricultural drought (H7) and unsuitable thermal regime (H12) are controlled by the climate. Monthly data on the temperature, precipitation and potential evapotranspiration are required to assess the soil temperature regime and the drought risk at a site. Some potential evapotranspiration data are inconsistent and unreliable, depending on the method of calculation. Thus, the FAO-Penman-Monteith reference evapotranspiration (Allen et al. 1998) should be used. Climate data from the local climate estimator New Loc_Clim 1.10 (FAO 2006b) are reliable in flat to undulating areas.

The time required for the field rating procedure depends largely on the experience and skills of the expert, but also on the study site and availability of support data. It may range from about 5 to 40 min.

Fig. 2 Examples of soil pits showing different Loess soils that have been classified and rated. **a** Krasnoobsk location (Novosibirsk region, Russia), WRB 2006: Haplic Chernozem (Siltic). Very high potential fertility but limitations due to drought in combination with a sub-optimal thermal regime, M-SQR: 42 Rating points, grain yield 3.5 t/ha spring wheat with 70 kg/ha N fertiliser. **b** Haus Duesse location (North Rhine-Westphalia, Germany), WRB 2006: Stagnic Luvisol (Siltic). Temporal wetness in spring, No extreme limitations, M-SQR: 81 Rating points, grain yield 7.5 t/ha winter wheat with 220 kg/ha N fertiliser. **c** Grushevka location (Novosibirsk region, Russia), WRB 2006: Calcic Chernozem (Arenic). Topsoil degraded by wind erosion, extreme limitations by thermal regime and drought, M-SQR: 20 Rating points, grain yield 1.0 t/ha spring wheat without N fertilisation. **d** Krasnodar location (Krasnodar Krai, Russia), WRB 2006: Haplic Chernozem (Pachic, Clayic), minor limitations by drought, M-SQR: 88 Rating points, grain yield 5.2 t/ha winter wheat with moderate N fertilisation, 7.3 t/ha with high fertilisation

Table 1 Main soil attributes used for the basic rating (Mueller et al. 2014b)

Indicator	Main attributes of scoring	Additional attributes for modifying the score	Relevant depth cm
B1. Soil substrate (WF[a]3)	Soil texture class, parent material	Strong gradients of texture (layering), content of coarse material, low organic matter (SOM), proportion of artefacts	0–80 (crop land), 0–50 (grassland)
B2. Depth of A- horizon and depth of humic soil (WF1)	Depth of A horizon	Abrupt boundary between topsoil and subsoil, SOM content <4 % (grassland)	0–25
B3. Topsoil structure (WF1)	Type and size of aggregates and pores	Redoximorphic feature	0–25
B4. Subsoil structure (WF1)	Type and size of aggregates and pores, increased soil strength or density	Redoximorphic feature	25–50

(continued)

Table 1 (continued)

Indicator	Main attributes of scoring	Additional attributes for modifying the score	Relevant depth cm
B5. Rooting depth and depth of biological activity (WF3)	Occurrence of roots, effective rooting depth	Barriers to rooting and their intensity	150 (cropland), 80 (grassland)
B6. Profile-available water (WF3)	Field capacity minus wilting point, rooting depth, capillary supply	Soil texture, stoniness	Rooting zone (<=150)
B7. Wetness and ponding (WF3)	Ponding, depth of ground or perched water table, redoximorphic features, vegetation	Soil position in a depression, wetness due to a perched water table	
B8. Slope and relief (WF2)	Slope at the pedon position	Microrelief and slope aspect at the profile position	

[a]WF = weighting factor of indicator, relevant to crop yield of small grain cereals

Table 2 Checklist of Hazard Indicators and criteria for identification (Mueller et al. 2007, 2014b)

Indicator[a]	Thresholds for orientation		
	Direct soil parameters	Indirect parameters of vegetation, climate or others	Reference soil groups (RSG) or *qualifiers* of WRB 2006, examples
1. Contamination	Specific for each pollutant according to international thresholds	High-risk areas: cities, waste-affected soils, vicinity of industrial plants, floodplains	*Toxic, (Garbic, Spolic)*
2. Salinisation	EC >4 mS/cm in topsoil	White crusts on soil aggregates or surface, occurrence of halophytes, S-number acc. to Ellenberg >3	*Salic, Hypersalic, Puffic, Chloridic*
3. Sodification	ESP >15 % (SAR >13), pH >8.2 in topsoil	High pH indicating plants, R-number acc. to Ellenberg of 9	*Sodic, Alcalic, Natric*
4. Acidification	pH <5.2 (cropping) or <4.5 (grassland) in topsoil	Low pH indicating plants, R-number acc. to Ellenberg of 3 or lower	*Hyperdystric, Hyperthionic*
5. Low total nutrient status	Clear deficit of nutrients, cannot be compensated by fertilisation within one year		*Hypergypsic, Hypercalcic*
6. Soil depth above hard rock	Hard rock or permafrost <120 cm (arable land) or <70 cm (grassland)		*Leptic, Lithic, Petric*

(continued)

Table 2 (continued)

Indicator[a]	Thresholds for orientation		
	Direct soil parameters	Indirect parameters of vegetation, climate or others	Reference soil groups (RSG) or *qualifiers* of WRB 2006, examples
7. Drought	Water budget in the main vegetation period of 4 months <500 mm, ustic, xeric or aridic soil water regime, total soil water balance in the main vegetation period <50 mm	Climatic water balance in the main vegetation period of 4 months < −100 mm, probability of the occurrence of a dry month >10 %, aridity index acc. to De Martonne <30, benefit of irrigation for cereals	*Aridic*
8. Flooding and extreme waterlogging	Flooding probability >5 %, peraquic soil water regime	Delay of beginning of farming on Cropland >20 d, Grassland mF of Ellenberg >8, clear benefit of land drainage	*Floatic, Gelistagnic, Subaquatic, Tidalic*
9. Steep slope	Arable land gradient >12 %, grassland gradient >30 %		
10. Rock at the surface	Rock outcrop on arable land >0.01 %, on grassland >0.05 %		Leptosols; *Ekranic, Hyperskeletic*
11. High percentage of coarse soil texture fragments	Coarse fragments (>2 mm) on arable land >15 % by mass of fragments in topsoil, grassland >30 %		Leptosols; *Hyperskeletic, Skeletic*
12. Unsuitable soil thermal regime[2)]	Cryic or pergelic soil thermal regime, Frigid regime with mean annual temperatures <5 °C	Tundra and Taiga regions	Cryosols; *Cryic, Glacic*

[a]An important characteristic of all indicators is that they have rising response curves as the crop yields rise, i.e. a higher rating correlates with higher crop yields. We avoided indicators where response curves have an inner optimum or minimum. For example, if Hazard Indicators 3 (Sodification) and 4 (Acidification) were combined into an indicator "Soil reaction, pH", the overall rating procedure would work as well as before. However, the approach would lose its potential to define capability classes (for example: Acid soils of moderate productivity potential), as soils of both low and high pH would get low ratings. The functional coding would be not clear without ambiguity

[b]New orientation values are given in this chapter. They replace the former preliminary values from the field manual (Mueller et al. 2007)

4 Rating of Particular Crop-Yield-Limiting Factors (Hazard Indicators)

4.1 *Agricultural Drought (H7)*

Agricultural drought is a complex phenomenon depending mainly on climate and can be influenced by agro-management (drought-resistant varieties, irrigation, soil tillage). We focus on site-specific medium drought intensity, which depends on the climate and soil. In doing so, we assume that there is a correlation between common drought intensity and drought risk. Sites of permanent high drought intensity also face a higher risk of crop failure by drought, e.g. are more unreliable agricultural sites.

Agricultural sites differ in their ability to supply growing plants with water both temporally and spatially. Rating agricultural drought risk aims at characterising spatial differences. The SQR field manual (Mueller et al. 2007) proposes some criteria and indexes for their characterisation, such as the soil water budget in the main vegetation period or climatic drought indexes.

Here we propose some updated rating scales of Hazard Indicator H7 (drought risk, Table 3). Two approaches are best suited to characterise the common drought risk of a site. These are (a) the Budget approach, and (b) the Balance approach. They comprise soil and climate properties.

Budget approach: this is based on the assumption that 500 mm of water in the main vegetation period of 4 months provide a very high crop yield of small-grain cereals and other crops. Lower values indicate sub-optimum conditions as shown in Table 3.

The water budget (W_{Bud}) can be calculated as the sum of plant-available water stored in the soil at the beginning of the vegetation period and added to the water supply during the main vegetation period of 4 months, mainly by precipitation, irrigation, and groundwater recharges. The months of May to August can be considered as the main vegetation period in most regions of the Northern hemisphere; in Siberia it is the period of June to September.

$$W_{\mathrm{Bud}} = \mathrm{PAW} + P + \mathrm{Irri} + \mathrm{GWR} \tag{1}$$

where PAW is the plant-available water at the rooting depth, P is the precipitation during the vegetation period, Irri is the irrigation water amount, and GWR is the recharged plant-available groundwater.

PAW is revealed as a product of rooting depth times average plant-available field capacity in this rooting depth. In regions of low winter precipitation where the high potential soil water store (field capacity minus wilting point) is not filled up over the winter period, for example on Loess soils in a sub-humid or drier climate, the

Table 3 Rating the drought intensity of a pedon or site based on climate and soil parameters over the main vegetation period of 4 months

Score	Drought intensity[a]	Water budget (W_{Bud}), mm[b]	Water balance (W_{Bal}), mm[c]	Precipitation mm	Climatic water balance deficit[d]	De Martonne index[e]	Examples of regions
2	None	>500	>50	>280	<50	>38	Atlantic regions of Denmark, Germany, Netherlands, France etc., irrigated lands in Central Europe and of the North China plain
1.75	Very low	450–500	25–50	250–280	50–110	35–38	Sandy soils in humid regions of Europe, Southern Taiga of Russia
1.5	Low	390–450	−25 to 25	220–250	110–170	32–35	Northeast Germany, Chernozem regions in Europe, Forest steppe of Siberia, Irrigated lands of Central Asia
1.25	Medium	320–390	−100 to −25	180–220	170–230	28–32	Sandy soils in subhumid regions of Europe, Steppes of Siberia
1	High	250–320	−200 to −100	140–180	230–300	24–28	Dry steppes of Russia, Sand-Loess and Loess soils of the Northern plains of the USA and in Inner Mongolia of China
0.75	Very high	180–250	−400 to −200	100–140	300–400	18–24	Dry semiarid regions of the Columbia plateau of the USA, Semi-Deserts of Central Asia
0.5		110–180	−600 to −400	60–100	400–600	13–18	
0.25	Extreme	<110	<−600	<60	>600	<13	Deserts of Central Asia

[a]Scales have been updated and are no longer in accordance with former scales of the M-SQR field guide of 2007 (Mueller et al. 2007)
[b]See formula (1), over the main 4 months of the vegetation period
[c]See formula (2), over the main 4 months of the vegetation period, note that this scale is based on ETp data from the local climate estimator New Loc_Clim 1.10 (FAO 2006b)
[d]ETp minus precipitation over the main 4 months of the vegetation period
[e]$AI_{DM} = [P/(T + 10) + 12\ p/(t + 10)]/2$, P annual precipitation sum, T annual mean temperature, p precipitation of the driest month, t temperature of the driest month, data are given by the local climate estimator New Loc_Clim 1.10 (FAO 2006b)

amount of winter precipitation can be considered as the maximum of PAW. GWR can also be an important or dominant source of plant water supply in many soils with a shallow water table (Mueller et al. 2005).

The balance approach: the water balance (W_{Bal}) considers the water budget (W_{Bud}) and the potential evapotranspiration ($\mathrm{ET_p}$) in the main vegetation period of 4 months.

$$W_{\mathrm{Bal}} = W_{\mathrm{Bud}} - \mathrm{ET_p} \tag{2}$$

It starts out from the assumption that the water budget has to cover the potential evapotranspiration in the vegetation period to avoid growth limitations caused by drought. No drought risk exists if W_{Bal} is greater than the $\mathrm{ET_p}$ during these 4 months. The empirical addition of 50 mm ensures this at a probability of higher than 50 % (Table 3).

Because the establishment of annual crops is based on processes that occur close to the soil surface, and most plant roots are located in this topsoil, solely climate parameters may also serve as indicators of agricultural drought. Examples of updated rating scales for those climate parameters, such as the precipitation and the climatic water balance over the main vegetation period, and the De-Martonne index, are given in Table 3.

All approaches have their advantages and disadvantages. The water budget approach is robust over all regions and should be preferred. The balance approach is more sensitive for Central Europe but may fail in other regions because of unreliable or not specifically adapted ("effective") ETp data. Climate data alone provide only a rough orientation; the climatic water balance deficit may also fail because of unreliable ETp data. We recommend calculating the specific drought intensity values given by all indexes in Table 3 with particular weighting of the water budget approach.

Orientation values for calculating all elements of the water budget and balance are given in soil survey handbooks (AG Boden 2005; FAO 2006a; USDA/NRCS 2005a). Estimating the rooting depth (Basic Indicator 5) is a crucial M-SQR indicator and the most sensitive one. It determines both plant-available water and drought risk assessment. If a field survey does not provide clear information about rooting depth, and soil profiles have a relative homogeneous texture over the depth, data from Table 7 (Appendix) can be applied. Monthly P and ETp data can be taken from the local climate estimator New Loc_Clim 1.10 (FAO 2006b). Orientation data of all components should be dealt with as "effective values", e.g. up- or downgraded due to specific local or regional conditions.

Table 3 also gives some examples of drought for regions where we carried out detailed studies. Most examples refer to "Zonal soils", where soil hydrological processes are dominated by the climate, and not to soils that benefit from shallow water tables or irrigation. The latter soils are characterized by a zero to low drought risk.

Table 4 Rating of the soil thermal conditions (Hazard Indicator 12) common cropping scale, rotations dominated by small-grain cereals

Rating of H12, cropland	Annual mean temperature °C	January temperature °C	Days with T_{mean} > 0 °C	Days with T_{mean} > 5 °C	Days with T_{mean} > 10 °C
2 (Suitable thermal regime)	>8	>−4	>290	>220	>170
1.75	6–8	−4 to −8	230–290	190–220	150–170
1.5 (Slightly too cold)	4–6	−8 to −12	210–230	170–190	135–150
1.25	2–4	−12 to −15	200–210	160–170	125–135
1 (Cold)	0–2	−15 to −20	180–200	150–160	110–125
0.75	<0	<−20	160–180	130–150	100–110
0.5 (Very cold)	<0	<−20	<160	<130	<100
0.25	<0	<−20	<140	<100	<60

4.2 Soil Thermal Regime (H12)

Low temperatures above or in the soil are most important plant-growth-limiting factors in northern latitudes and higher mountain regions. On many Siberian sites, an unsuitable soil thermal regime is the main crop-yield-restricting factor.

Hazard Indicator 12 "Soil thermal regime" is determined by the climate and can be estimated from climate data. Temperatures, mean number of days with positive temperatures (T_{mean} > 0 °C), number of growing days (T_{mean} > 5 °C), and number of days with T_{mean} > 10 °C can help assess and rate the soil thermal regime. The newly introduced Maize scale seems to be useful for Maize and other thermophile grasses requiring distinctly higher temperatures in the vegetation period than for common small grain cereals or other grasses of the temperate zone. All data are available from the local climate estimator New Loc_Clim 1.10 (FAO 2006b). Table 4 and Appendix Tables give orientation values of 4–5 relevant criteria. It is unreliable to use only one or two of them for H12 rating. All criteria should be considered with particular weighting of the number of growing days (>5 °C) to get a reliable result.

5 Rules and Updated Orientation Values of the Overall Rating

Having identified the most serious hazard indicator, a multiplier is derived, which may range from 0 to 2.94. The M-SQR field guide (Mueller et al. 2007) provides conceptual orientation values for scores of all indicators in Fig. 1. Hazard Rating multipliers given there have broad ranges and may lead to overall scores that still include large subjective variability. Meanwhile, it is possible to confine these intervals if drought (H7) or too-cold climate and soils (H12) are the critical Hazard Indicators.

Practical tests in different regions have shown that it is not only the rating value of the most serious (active) Hazard Indicators that provides the most plausible results for the multiplier and the overall score. The number and sometimes the rating score of sub-critical hazard indicators are also significant. This holds particularly true for regions outside of Central Europe.

Table 5 gives updated orientation values of multipliers for drought (H7) and unsuitable thermal regime (H12), which are the factors most limiting soil productivity potentials worldwide. Recommendation values of multipliers consider the number of Hazard Indicators with sub-optimum ratings. If, for example, drought is the dominating Hazard Indicator at a typical location and has been rated at 1.25 on the basis of Table 4, and additionally Hazard Indicators 3 and 12 are less than 2, the multiplier has to be downgraded to 1.7 using Table 5.

If other Hazard Indicators than drought (H7) or thermal regime (H12) are critical, this is not yet underpinned by enough data on the soil, crop yield and other factors to give more detailed recommendations for multipliers. Besides conceptual orientation values given in the field manual (Mueller et al. 2007), this Table 5 could

Table 5 Orientation values for ratings and multipliers of hazard indicator H7 "drought" and H12 "unsuitable thermal regime" (Universal scale, valid for crops, grassland and Maize if H7 or H12 are the most serious crop-yield-limiting factors)

Rating of indicator H7 or H12	Orientation value of multiplier for number of H factors, viz. Hazard Indicators with values <2[a]				
	0	1	2	3	4
2	2.94				
1.75		2.8	2.5	2.3	2.1
1.5		2.6	2.3	2.1	1.9
1.25		2.3	1.9	1.7	1.4
1		1.9	1.5	1.2	0.9
0.75		1.5	1.2	0.9	0.7
0.5		1.0	0.8	0.5	0.3
0.25		0.5	0.3	0.2	0.1
0		<0.2	<0.1	<0.1	<0.1

[a]Number of Hazard Indicators having ratings <2; note that this scale may provide acceptable results of relative overall scores in the field. However, the application of separate tables given in the Appendix may sometimes correlate better with effective crop and grassland yields and could also be taken into consideration
On the other hand, this Table 5 can be considered as a universal table for estimating multipliers. As long as such tables do not yet exist for other critical Hazard Indicators, this table could serve as a preliminary decision basis for other active Hazard Indicators. It should be noted that these values are better proven than those preliminary ones given in the SQR Manual of 2007 (Mueller et al. 2007)

be used as a preliminary work basis to select a multiplier for calculating the M-SQR score.

Data from Table 5 show that active Hazard Indicator ratings of less than 1 lead to multipliers of also less than 1 in most cases. It is highly probable that those soils will fall into the classes of very low overall soil quality (M-SQR scores <20) or low overall soil quality (M-SQR scores of 20–40).

The final rating procedure proposes to check the plausibility of the results and to upgrade or downgrade the result by about 3–15 points, but within the limit of 100 points. The reasons for up- or downgrades are interactions between Hazard Indicators, meso- and microclimate and the temporal uniformity of the soil moisture regime within the upper 10 cm.

6 Rating Scores and Crop Yields

Following the basic rules of site assessment given by Dokuchaev (1951), the relationships between soil quality and crop yields had to be tested. We dug soil profiles on experimental sites in the main regions of cereal cropping: Europe, Western Siberia and Kazakhstan, Northern China and North America. Profiles were classified according to national keys and the World Reference Base for Soil Resources (WRB 2006). As sites were located on agricultural research stations or experimental fields, crop yield data were provided by research reports. In the case of experimental fields at a practical farm level, we accepted the estimates given by the farm owners or local managers, knowing well that there is a difference between the research station yield level and the practical farm yield level. Data on cereal yield were stratified according to the level of fertiliser input, 0) non-fertilised, (1) low- to medium-fertilised (<100 kg/ha N) and (2) highly fertilised (integrated farming with >100 kg/ha N).

The overall M-SQR score is well correlated with the cereal crop yield over a range of scales. Some data had been given by Smolentseva et al. (2014) and Mueller et al. (2014a, d) . Table 6 shows overall regression lines between M-SQR scores on the basis of a dataset enlarged by recent fieldwork from the two last years and other available data from agricultural research stations. This information can be used to check the plausibility of ratings or to estimate yield potentials.

Figure 3 demonstrates that M-SQR ratings reflect the crop yield potentials of grassland ecosystems well. This figure shows also that the yield gap between yield potentials and current yields is significant due to poor grassland management. Grassland degradation is a great threat to the ecosystems of Central Asia and to other dry regions. It seriously affects global biological cycles and desertification. Tendencies to desertification have become significant for the Steppe regions of Siberia (Schreiner and Meyer 2014). Recovering ecosystem functions by proper management and others based on a comprehensive grassland basic inventory and monitoring including M-SQR would be a decisive step towards sustainability in handling land resources (Mueller et al. 2014c).

Table 6 Regression equations of M-SQR rating scores with crop yields

Management category	Regression equation[a]	n	B	SE t/ha
Cropland, unfertilised	$y = 0.048x$	36	0.59*	0.74
Cropland, moderate input (<100 kg/ha of N fertiliser)	$y = 0.072x$	167	0.69*	1.08
Cropland, high input (>100 kg/ha of N fertiliser)	$y = 0.092x$	267	0.65*	2.08
Grassland, unfertilised	$y = 0.07x$	251	0.67*	1.09
Grassland, moderate input (<100 kg/ha of N fertiliser)	$y = 0.085x$	64	0.78*	1.58
Grassland, high input (>100 kg/ha of N fertiliser)	$y = 0.099x$	35	0.80*	2.36

[a]Best linear fit without constant term, Cropland y = grain yield of small-grain cereals in t/ha at 14 % moisture content, x = M-SQR score of cropland scale, Grassland y = Dry matter yield in t/ha, x = M-SQR score of grassland scale

*All regression equations are highly significant at 0.001 %; *n* number of plots, B (r^2) = degree of estimate, *SE* standard error of estimate

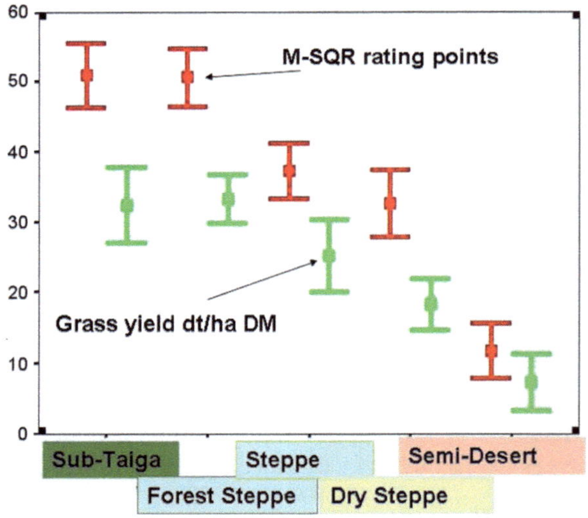

Fig. 3 Grassland rating scores and effective grassland yield (EGY) of different soil ecological zones of Russia and Kazakhstan. EGY is the annual above-ground biomass minus non-palatable plants in decitonnes of dry matter per hectare

7 The Potential of M-SQR for Creating Soil Quality Maps

One of the aims of the M-SQR framework is to provide soil quality and productivity potential assessments consistently over different spatial scales. When this approach was developed, the prime focus was on making the field procedure operable as a basis for monitoring and soil and water management at the field or regional scale. More and more focus could be put on the cross-regional (Helbig, In: Mueller et al. 2013), national (Richter et al. 2009; BGR 2014, Fig. 4) and

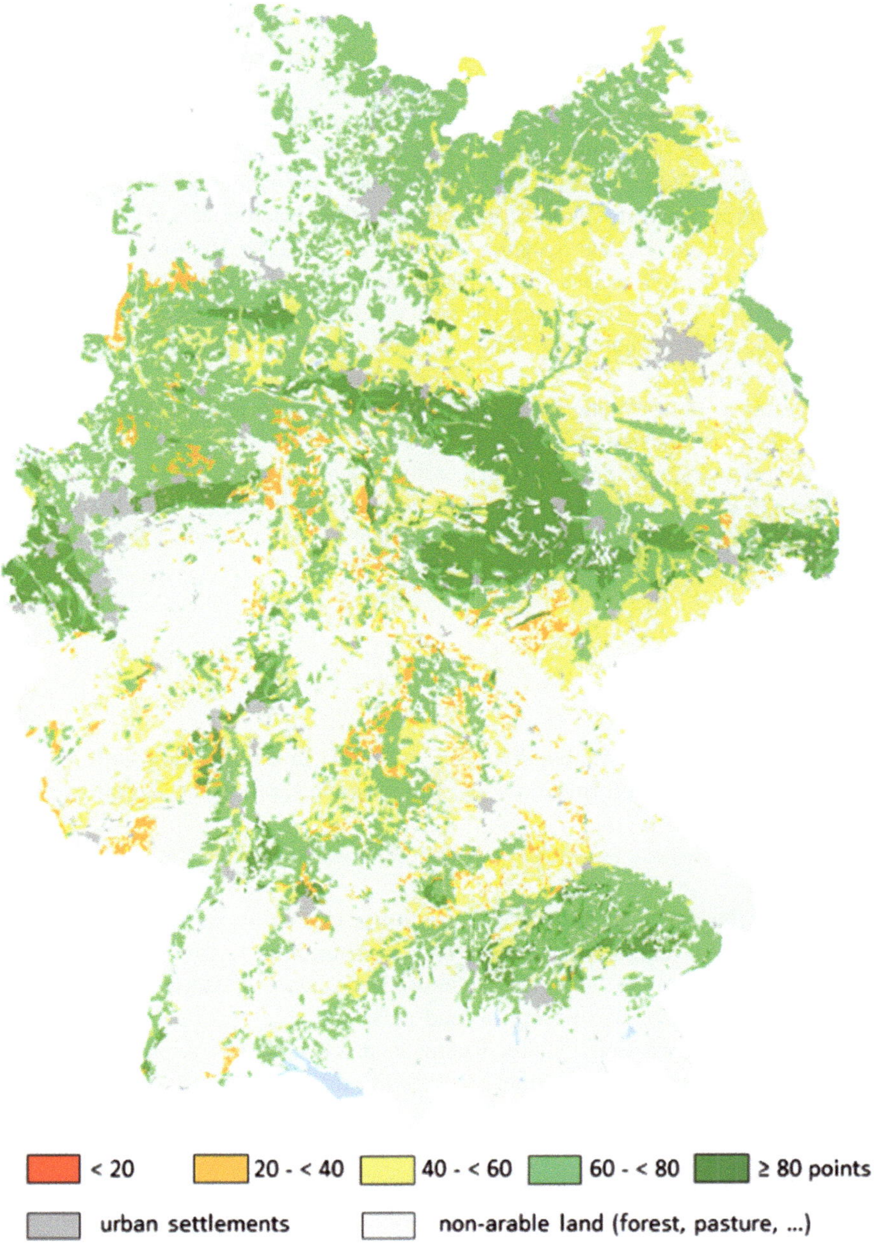

Fig. 4 Draft of the German soil quality map of cropland based on M-SQR (Richter et al. 2009). It shows five soil quality classes, providing a mapping of agricultural soil quality over larger regions based on available soil and climate data. Meanwhile, the German Federal Institute for Geosciences and Natural Resources (BGR) in Hannover has developed slightly modified maps (Hennings et al. 2016). Users interested in the underlying database may get more information from the homepage of the BGR (BGR 2014)

trans-national scale when it was shown that it is not necessary to start new high-resolution soil surveys to map M-SQR data. Available soil categories and databases or existing national soil rating maps can be utilised based on correlations (Mueller et al. 2011b) or parameterizations of mapping categories. This is particularly easy to install for countries which already have sophisticated soil information systems, such as Germany or the USA. In Germany, for example, scores of M-SQR are significantly correlated with scores of the official traditional soil rating system ("Ackerzahl" of the "Bodenschätzung", Mueller et al. 2011b; Hennings et al. 2016). Those data, or soil series data from the US Soil Conservation Service (USDA/NRCS 2005a, b), provide the best preconditions for creating soil quality maps at field and regional scales, which will have high conformity with small-scale cross-regional data.

The same could be done for the agricultural lands of Russia, beginning with small-scale mapping. Fieldwork has shown that soil types and texture classes of the Russian soil taxonomy are associated with typical ranges of Basic and Hazard Indicators of the M-SQR system. All data are available in databases of Russian soil information systems (Stolbovoi and Fischer 1997; Afonin et al. 2008; Mikheeva 2013) and climate databases (FA0 2006b).

8 Conclusions

- A classification of soils by WRB (2006, 2014) in combination with M-SQR provides sufficient information about soil properties, processes and productivity potentials.
- Soil quality scores characterise productivity potentials of sites at a defined level of inputs.
- Drought (lack of plant-available water in the vegetation period) and an insufficient thermal regime are the most crop-yield-limiting factors worldwide.
- We developed rating tables to evaluate these and other factors related to the productivity potentials of cropland and grassland.
- The Muencheberg Soil Quality Rating is practicable and reliable. It has the potential to be applied as a global soil quality reference system both in the field and for mapping purposes.
- This rating system has the potential to be included in the monitoring, decision support, impact assessment and management systems of Eurasian grassland and cropland ecosystems. These are important for halting land degradation and initiating sustainable land use.
- As a first concrete step, our available data and knowledge would allow us to create a crop yield potential map of Russia and neighbouring countries using the M-SQR methodology.

Acknowledgements The authors are grateful for the support of many colleagues and their research teams: Aleksandr Syso, Boris Smolentsev, Yuri Ermolov and teams (ISSA Novosibirsk, Russia); Aleksandr Puzanov and team (IWEP Barnaul, Russia); Sergey Kisinek and team (State farm Krasnoarmejskij, Krasnodar Krai, Russia), Azimbay Otarov and team (Kazakh Research Institute of Soil Science and Agrochemistry Almaty, Kazakhstan); Chunsheng Hu and team (Center for Agricultural Resources Research, Shijiazhuang, Hebei, China); Yong Li and team (Institute of Agro-Environment and Sustainable Development, CAAS, Beijing, China); Robert Evans and team (Agricultural Systems Research Unit (ASRU) und Northern Plains Agricultural Research Laboratory, Sidney, MT, USA); Steve Petrie, Daniel Long and teams (Columbia Basin Agricultural Research Center (CBARC) and USDA- Columbia Plateau Conservation Research Center Pendleton, OR, USA), William Schillinger and team (Washington State University, Lind Dryland Res. Station, Lind, WA, USA); Brenton Sharratt and team (Washington State University und USDA Land Management and Water Conservation Research Unit, Pullman, WA, USA), Lars Munkholm and team (Aarhus University, Denmark), Bev Kay (†), Bill Deen and teams (University of Guelph, Canada); Graham Shepherd (Bio-AgriNomics Ltd, Palmerston North, New Zealand), Bruce C. Ball (SRUC Edinburgh, UK), Ralf Dannowski, Sabine Baur (ZALF Müncheberg, Germany).

The Federal Office for Agriculture and Food (BLE, Bonn, Germany) provided travel funding for research stays with foreign partners. Special thanks go to Mr. Christian Walther. The International Bureau of the Federal Ministry of Education and Research (BMBF, Bonn, Germany) funded contacts and workshops in Novosibirsk and Almaty. The working group "Soil assessment and evaluation" of the German Soil Science Society (DBG) enabled the classification and evaluation on numerous agricultural sites in the framework of scientific excursions. Authors would like to thank all colleagues, teams and funding bodies for their support and engagement.

Appendix: Further Updated Rating Tables

See Tables 7, 8, 9, 10, 11 and 12

Table 7 Potential rooting depths[a] of homogeneous mineral soils of deep ground water table

Coarse sand and coarse material (>0.63 mm, g/100 g)	Clay and silt content (<0.063 mm, g/100 g)				
	0	15	30	60	>60
0	0.70 (0.75)	0.90 (1)	1.05 (1)	1.30 (1.5)	1.50 (2)
20	0.45 (0.25)	0.65 (0.5)	0.85 (0.75)	1.05 (1.25)	1.25 (1.5)
30	0.35 (0)	0.55 (0.25)	0.75 (0.75)	0.95 (1)	1.10 (1.25)
40	0.30 (0)	0.45 (0.25)	0.60 (0.5)	0.85 (0.75)	0.95 (1)
60	0.30 (0)	0.30 (0)	0.40 (0)	0.60 (0.5)	0.75 (0.75)
>60	0.30 (0)	0.30 (0)	0.30 (0)	0.30 (0)	0.40 (0)

[a]The first number is the rooting depth in meters, the second number in parentheses is the rating score for Basic Indicator 5 (Rooting depth, 0 = Minimum, 2 = Maximum), clay and silt content are related to the soil texture fraction <2 mm, coarse sand and coarse material are related to the overall mass of the soil

Table 8 Orientation values for ratings and multipliers of Hazard Indicator 7 (drought), cropland

Rating of cropland drought risk	Orientation value of multiplier for number of H factors, viz. Hazard Indicators with values <2[a]			
	0	1	2	3 or more
2 (None)	2.94			
1.75 (Very low)		2.8	2.4	2.1
1.5 (Low)		2.6	2.3	2.0
1.25 (Medium)		2.1	1.9	1.7
1 (High)		1.8	1.6	1.5
0.75 (Very high)		1.5	1.3	1.1
0.5		1	0.8	0.6

[a]Number of Hazard Indicators having ratings <2

Table 9 Rating of the hazard indicator 12 (soil thermal conditions), scale for grassland

Rating of H12 for grassland	Annual mean temperature (°C)	Mean temperature in January (°C)	Days with $T_{mean} > 0$ °C	Days with $T_{mean} > 5$ °C	Days with $T_{mean} > 10$ °C
2 (suitable thermal regime)	>8	>−4	>280	>220	>170
1.75	6–8	−4 to −8	230–280	190–220	150–170
1.5 (slightly too cold)	4–6	−8 to −12	210–230	170–190	135–150
1.25	2–4	−12 to −15	195–210	160–170	125–135
1 (cold)	0–2	−15 to −20	170–195	150–160	105–125
0.75	0 to −5	−20 to −25	140–170	120–150	80–105
0.5 (very cold)	−5 to −10	<−25	120–140	60–120	35–80
0.25	<−10	<−25	<120	<60	<35

Table 10 Rating of the hazard indicator 12 (soil thermal conditions), scale for maize cropping

Rating of H12 for Maize	Annual mean temperature (°C)	Mean temperature in July (°C)	Days with $T_{mean} > 5$ °C	Days with $T_{mean} > 10$ °C
2 (Suitable thermal regime)	>10	>20	>230	>200
1.5 (Slightly too cold)	8–10	18–20	220–230	160–200
1 (Cold)	6–8	16–18	200–220	120–160
0.5 (Very cold)	0–6	14–16	140–200	80–120
0	<0	<14	<140	<80

Note that this scale is designed to assess restrictions for maize biomass; it does not distinguish between the potentials of maize for corn from maize for silage. A rule of thumb is that maize for corn requires about 150 days with temperatures >10 °C

Table 11 Orientation values for multipliers of hazard indicator 12 (unsuitable thermal regime), scale for cropland

Rating of indicator H12 for cropland	Orientation value of multiplier for number of H factors, viz. Hazard indicators with values <2[a)]				
	0	1	2	3	4 or more
2	2.94				
1.75		2.8	2.3	2.1	2
1.5		2.7	2.2	2.0	1.7
1.25		2.3	1.9	1.7	1.3
1		2	1.4	1	0.8
0.75		1.5	1	0.7	0.5
0.5		1	0.7	0.5	0.4

[a]Number of Hazard Indicators having ratings <2

Table 12 Orientation values for multipliers of hazard indicator 12 (unsuitable thermal regime), scale for grassland

Rating of indicator H12 for grassland	Orientation value of multiplier for number of H factors, viz. Hazard Indicators with values <2[a]				
	0	1	2	3	4 or more
2	2.94				
1.75		2.8	2.6	2.3	2
1.5		2.7	2.4	2.1	1.7
1.25		2.4	2.1	1.8	1.4
1		2.1	1.7	1.5	1.1
0.75		1.7	1.4	1.1	0.7
0.5		1.4	1	0.7	0.4

[a]Number of hazard indicators having ratings <2

References

Afonin AN, Greene SL, Dzyubenko NI, Frolov AN (eds) (2008) Interactive agricultural ecological Atlas of Russia and Neighboring Countries. Economic plants and their diseases, pests and weeds [Online]. Available at http://www.agroatlas.ru. Accessed on 9 Mar 2015

AG Boden (2005) Bodenkundliche Kartieranleitung, 5th edn. (KA5), Hannover, 432 p

Allen RG, Pereira LS, Raes D, Smith M (1998) Crop evapotranspiration—guidelines for computing crop water requirements. FAO Irrigation and Drainage Paper, 56, Rome. http://www.fao.org/docrep/x0490e/x0490e00.htm. Accessed on 9 Mar 2015

BGR (2014) German Federal Institute for Geosciences and Natural Resources (BGR) Hannover. Ackerbauliches Ertragspotential der Böden in Deutschland. Bewertet nach dem Müncheberger Soil Quality Rating—final rating (1:1.000.000) auf Basis der BÜK1000 N http://www.bgr.bund.de/DE/Themen/Boden/Ressourcenbewertung-management/Ertragspotential/Ertragspotential_node.html. Accessed on 9 Mar 2015

Blum WEH (1993) Soil protection concept of the council of europe and integrated soil research. In: Eijsackers HJP, Hamers T (eds) Integrated soil and sediment research: a basis for proper protection. Soil and environment, vol 1. Kluwer Academic Publishers, Dordrecht, pp 37–47

Dokuchaev VV (1951) Isbrannye sotchynenija, vol VI, M-L 1951, 515 p
FAO (2006a) Guidelines for soil description, 4th edn. FAO, Rome, 95 p
FAO (2006b) New_LocClim 1.10. Local Climate Estimator. Download–Address: FTP-Verzeichnis/SD/Reserved/Agromet/New_LocClim/auf ext-ftp.fao.org. Accessed on 9 Mar 2015
Helbig H (2013) Entwurf eine Bodengütekarte für Sachsen-Anhalt, Landesamt für Geologie und Bergwesen Sachsen-Anhalt, Halle an der Saale, In: Mueller L, Schindler U, Hennings V, Smolentseva E, Pachikin K, Rukhovich O, Lukin S, Dong W, Sheudshen AK, Jabro J, Sauer S, Helbig H, Eulenstein F (2013) Bewertung landwirtschaftlicher Produktivitätspotentiale der globalen Landressource. Berichte der DBG 2013, 5p. http://eprints.dbges.de/935/1/DBGRostock2013_5SeitenLMueller.pdf. Accessed on 9 Mar 2015
Helming, K (2014) Impact assessment for multifunctional land use. http://link.springer.com/chapter/10.1007/978-3-319-01017-5_12. Accessed on 28 Oct 2014
Hennings V, Höper H, Mueller L (2016) Small-scale soil functional mapping of crop yield potentials in Germany. In: L. Mueller et al. (eds) Novel methods for monitoring and managing land and water resources in Siberia. Springer, Cham, pp 597–617 (Chapter 27 of this book)
Mikheeva I (2013) Use of soil survey database for the probabilistic evaluation of soil cover transformation in the semiarid zone of western Siberia developments in soil classification, land use planning and policy implications 2013, pp 305–313. http://link.springer.com/chapter/10.1007/978-94-007-5332-7_15. Accessed on 9 Mar 2015
Mirschel W, Wenkel K-O, Berg M, Wieland R, Nendel C, Köstner B, Topazh AG, Terleev VV, Badenko VL (2016) A spatial model-based decision support system for evaluating agricultural landscapes under the aspect of climate change. In: L. Mueller et al. (eds) Novel methods for monitoring and managing land and water resources in Siberia. Springer, Cham, pp 519–540 (Chapter 23 of this book)
Mueller L, Behrendt A, Schalitz G, Schindler U (2005) Above ground biomass and water use efficiency of crops at shallow water tables in a temperate climate. Agric Water Manage 75 (2):117–136
Mueller L, Schindler U, Behrendt A, Eulenstein F, Dannowski R (2007) The Muencheberg Soil Quality Rating (SQR). Field guide for detecting and assessing properties and limitations of soils for cropping and grazing. Online http://www.zalf.de/de/forschung/institute/lwh/mitarbeiter/lmueller/Documents/field_mueller.pdf. Accessed on 9 Mar 2015
Mueller L, Schindler U, Mirschel W, Shepherd TG, Ball B, Helming K, Rogasik J, Eulenstein F, Wiggering H (2010) Assessing the productivity function of soils: a review. Agron Sustain Dev 30(3):601–614
Mueller L, Smolentseva E, Rukhovich O, Lukin S, Hu C, Li Y, Schindler U, Behrendt A, Hennings V, Sauer S, Vorderbrügge T (2011) Zur Bewertung von Bodengüte und Ertragspotentialen in Agrarlandschaften. In: Böden verstehen—Böden nutzen—Böden fit machen: DBG-Jahrestagung, 3–9 Sept 2011, Berlin; Kommission VI: Bodenerosion durch Wind und Wasser. Deutsche Bodenkundliche Gesellschaft, Oldenburg, pp. 1–4. http://eprints.dbges.de/680/1/LMueller_DBG11.pdf. Accessed on 9 Mar 2015
Mueller L, Schindler U, Ball BC, Smolentseva E, Sychev VG, Shepherd TG, Qadir M, Helming K, Behrendt A, Eulenstein F (2014a) Productivity potentials of the global land resource for cropping and grazing. In: Mueller L, Saparov A, Lischeid G (eds) Novel measurement and assessment tools for monitoring and management of land and water resources in agricultural landscapes of Central Asia. Environmental science and engineering 2014, pp 115–142 http://link.springer.com/chapter/10.1007%2F978-3-319-01017-5_6. Accessed on 9 Mar 2015
Mueller L, Schindler U, Shepherd TG, Ball BC, Smolentseva E, Pachikin K, Hu C, Hennings V, Sheudshen AK, Behrendt A, Eulenstein F, Dannowski R (2014b) The Muencheberg Soil Quality Rating for assessing the quality of global farmland. In: Mueller L, Saparov A, Lischeid G (eds) Novel measurement and assessment tools for monitoring and management of land and water resources in agricultural landscapes of Central Asia. Springer International Publishing, Environmental science and engineering 2014, pp. 235–248. http://link.springer.com/chapter/10.1007/978-3-319-01017-5_13. Accessed on 9 Mar 2015

Mueller L, Saparov A, Lischeid G (2014c) Executive summary and conclusions. In: Mueller L, Saparov A, Lischeid G (eds) Novel measurement and assessment tools for monitoring and management of land and water resources in agricultural landscapes of Central Asia. Springer International Publishing, Environmental science and engineering 2014, pp 683–714. http://link.springer.com/chapter/10.1007/978-3-319-01017-5_44. Accessed on 9 Mar 2015

Mueller L, Schindler U, Smolentseva EN, Rukhovich O, Lukin S, Sheudshen AK, Pachikin K, Saparov A, Dannowski R, Eulenstein F (2014d) Soil classification and assessment of their productivity potentials in important regions of cereal cropping. Pochvovedenie i Agrokhimija 2 (2014):5–16

Richter A, Hennings V, Mueller L (2009) Anwendung des Müncheberger Soil Quality Ratings (SQR) auf bodenkundliche Grundlagenkarten. In: Jahrestagung der DBG: Böden—eine endliche Ressource; Kommission VIII, September 2009, Bonn: 1–4. http://eprints.dbges.de/243/1/Beitrag_RICHTER_HENNINGS_MUELLER.pdf. Accessed on 9 Mar 2015

Schreiner V Meyer BC (2014) Indicators of land degradation in steppe regions: soil and morphodynamics in the Northern Kulunda. In: Mueller L, Saparov A, Lischeid G (eds) Novel measurement and assessment tools for monitoring and management of land and water resources in agricultural landscapes of Central Asia. Environmental science and engineering 2014, pp 535–548. http://link.springer.com/chapter/10.1007/978-3-319-01017-5_33. Accessed on 9 Mar 2015

Shepherd TG (2009) Visual soil assessment. Volume 1. Field guide for pastoral grazing and cropping on flat to rolling country, 2nd edn. Horizons Regional Council, Palmerston North, New Zealand, 118 p

Smolentseva E, Smolentsev B, Pachkin K, Mueller L (2014) Assessing the soil quality and crop yield potentials of some soils of Eurasia. In: Mueller L, Saparov A, Lischeid G (eds) Novel measurement and assessment tools for monitoring and management of land and water resources in agricultural landscapes of Central Asia. Environmental science and engineering 2014, pp 505–517. http://link.springer.com/chapter/10.1007%2F978-3-319-01017-5_31. Accessed on 9 Mar 2015

Stolbovoi V, Fischer G (1997) A new digital georeferenced database of soil degradation in Russia. IIASA Laxenburg, INTERIM REPORT IR-97-084/November. http://citeseerx.ist.psu.edu/viewdoc/download?doi=10.1.1.56.689&rep=rep1&type=pdf Accessed on 9 Mar 2015

USDA/NRCS (2005a) United States Department of Agriculture, Natural Resources Conservation Service, Soil Survey Manual 1993, updated 2005. Online source: http://www.nrcs.usda.gov/wps/portal/nrcs/detail/soils/ref/?cid=nrcs142p2_054262. Accessed on 9 Mar 2015

USDA/NRCS (2005b) Soil series description. Online: https://soilseries.sc.egov.usda.gov/osdname.asp. Accessed on 9 Mar 2015

WRB (2006) World Reference Base for Soil Resources 2006, a framework for international classification, correlation and communication. FAO, Rome, World Soil Resources Reports 103, 145 p

WRB (2014) World Reference Base for Soil Resources, World Soil Resources Report No 106, 181p. ISBN 978-92-5-108369-7. http://www.fao.org/3/a-i3794e.pdf. Accessed on 9 Mar 2015

Chapter 27
Small-Scale Soil Functional Mapping of Crop Yield Potentials in Germany

Volker Hennings, Heinrich Höper and Lothar Mueller

Abstract The Muencheberg Soil Quality Rating (M-SQR) is a new approach for assessing soil suitability for arable and grassland farming, and estimating crop yield potential on a global scale. We utilized this approach to construct a small-scale map of crop yield potentials covering arable lands in Germany. M-SQR rules and algorithms were adapted to the terminology and classification of soil parameters as defined by German soil mapping guidelines and were applied to soil profile-related data sets of the land-use stratified soil map of Germany at a scale of 1:1,000,000. According to the resulting thematic map, soils in Germany show a high yield potential for grain; the nationwide mean score accounts for 64 out of 100 possible points. Moderate drought risk is the main crop yield limiting factor in Germany, shallow soil depth, and other crop yield limiting factors may also be locally important. The approach has been validated and tested in comparison with the traditional German Bodenschätzung (German soil assessment). In comparison to the Bodenschätzung classification scheme, which is conventionally applied to assess the yield potential of agriculturally used soils, the M-SQR method incorporates additional climatic variables and considers site-specific drought risks. The two methods, the M-SQR approach and the German Bodenschätzung system, were compared and evaluated on the basis of grain yield data from 79 sites from different Federal States of Germany. In general, the M-SQR method provides reliable estimates of the yield potential for cereals. As the Basic Rating of M-SQR correlates significantly with scores (Ackerzahl) of the Bodenschätzung, both systems are

V. Hennings (✉)
Strousbergstraße 9, 30449 Hannover, Germany
e-mail: v.hennings@kabelmail.de

H. Höper
State Authority for Mining, Energy and Geology (LBEG), Stilleweg 2, 30665 Hannover, Germany
e-mail: heinrich.hoeper@lbeg.niedersachsen.de

L. Mueller
Leibniz Centre for Agricultural Landscape Research (ZALF), Eberswalder Str. 84, 15374 Müncheberg, Germany
e-mail: mueller@zalf.de

L. Mueller et al. (eds.), *Novel Methods for Monitoring and Managing Land and Water Resources in Siberia*, Springer Water,
DOI 10.1007/978-3-319-24409-9_27

potentially convertible. On average, M-SQR point scores even correlate slightly better to measured yields than Bodenschätzung scores. We conclude that this may enable the creation of reliable medium- and large-scale crop yield potential maps within Germany using the M-SQR methodology, available digital soil maps, and climate databases. The M-SQR methodology also enables the consistent incorporation of the newly created small-scale German crop yield potential map into a potential small-scale global soil quality map.

Keywords Soil quality · Crop yield potential · Mapping · Algorithm · Validation · Muencheberg Soil Quality Rating

1 Introduction

Soils can only provide natural habitats, living space for human beings, raw materials, and food to a growing world population if humans are successful in establishing sustainable kinds of land use and sustainable soil resource management on the global scale. Achieving these goals and implementing practical soil protection measures requires area-wide information about soil properties, pedotransfer rules for soil function mapping, and tools for land evaluation. There is a special need for, and emphasis should be laid on, methods to evaluate how suitable soils are for agriculture and to assess their yield potential for the most important agricultural crops.

When planning the agricultural use of soil resources on the basis of any soil productivity rating, decision-making support is required throughout the world. Within the European Union there is an additional need for such tools when identifying “intermediate less favoured areas for agriculture” (IEEP 2006) to receive special subsidies. In Germany, soil functions are protected by a federal law, the Soil Protection Act (BBodSchG 1998); according to this act the ability of the soil to serve as living space is one of these functions and soil fertility is one of several criteria to evaluate this specific function [see also Table 67 of the 5th edition of the German Soil Mapping Guidelines (Ad hoc-AG Boden 2005)]. The Soil Information System of the Federal Institute for Geosciences and Natural Resources (FISBo BGR; Adler et al. 1998; Hennings and Eckelmann 2004) offers nationwide data and maps about soil properties and land qualities in Germany; information on the yield potential of agriculturally used soils is in particular demand among many customers in the fields of science, economy, and politics. However, when a national working group of experts from the soil survey institutions of the federal states were commissioned to collect and standardize pedotransfer rules for land evaluation, their final report did not list any method to derive soil fertility from basic soil characteristics that was applicable on a national scale (Ad hoc-AG Boden 2003).

In general, there is no lack of methods to assess the agricultural yield potential at the national scale. In Germany, the “Bodenschätzung” (soil assessment)

classification system was established and has been approved over many decades. The system offers several advantages: it is applied on the basis of standardized rules throughout the country, leads to reproducible results, and provides soil-related information of full coverage for agriculturally used areas (Roesch and Kurandt 1950; Oelkers and Selge 1993; Pfeiffer et al. 2003). In the international context, Germany and Austria are the only two countries that can make use of soil survey results at such high spatial resolution for all agricultural land. However, when an amendment was to be made to the national Soil Assessment Act, Germany failed to integrate climatic aspects into the existing soil assessment scheme. Because of the well-known weak points of the system, some of the federal states' soil survey institutions prefer alternative approaches to assess the agricultural yield potential of the soil. For example, when the pedotransfer rules database of the soil information system of Lower Saxony (Müller 2004) is used for soil protection purposes, a locally developed algorithm serves as the standard (Richter and Eckelmann 1993). In the Federal State of Baden-Württemberg, another method is applied to assess the soil's suitability for agricultural crops. This local method places emphasis on the soil's moisture characteristics and incorporates climatic site properties (Umweltministerium Baden-Württemberg 1995). In contrast to these methods, the nationwide thematic map of the soil quality of agricultural land within the framework of the National Atlas of Germany (Institut für Länderkunde 2003) is based on German Bodenschätzung scores. For this map, the full range of scores was subdivided into three classes.

In the international context, there is already a broad range of methods to evaluate the productivity function of the soil. Many of these approaches take into account completely different input parameters. A universal framework for land evaluation that is accepted worldwide can be taken from guidelines as published by the FAO (1976, 1980). A very early approach to classify the agricultural yield potential which is in some aspects similar to the Bodenschätzung scheme in Germany and Austria is Storey's index (1933). The method developed by Shepherd (2000) is mainly based on macroscopically visible diagnostic criteria and places emphasis on structural properties of the soil. The paper by Mueller et al. (2010) offers a broad overview of existing approaches.

The M-SQR (Mueller et al. 2007a, b) is a new approach for assessing the soil's suitability for arable and grassland farming and estimating the crop yield potential. Four demands are fulfilled: the method should be applicable to soil data between the field scale and the world soil map, should provide reproducible results, should work as a simple and reliable field method, and should also provide recommendations for good agricultural practice. Under similar conditions, i.e., within climatic subzones, and in the case of similar agricultural management, the final rating score should correlate with the crop yield potential and should therefore act as a reliable yield estimator. Detailed algorithms of the method have been published as part of a manual (Mueller et al. 2007a) which can be downloaded from the homepage of the Leibniz Center for Agricultural Landscape Research (ZALF). Practical applications on test sites in many countries such as Russia, Kazakhstan, China, and the United States have shown good correspondence between estimated and observed yields of cereals (Mueller et al. 2009, 2012, 2013).

2 Methodology

2.1 *Principal Operating Mode of the Muencheberg Soil Quality Rating*

Functionalities of the Soil Quality Rating method have been documented in several publications (Mueller et al. 2007a, b). M-SQR consists of a series of pedotransfer rules, and a set of indicators is scored in terms of good, medium, or poor. First, basic soil indicators such as soil substrate and rooting depth are scored, while individual scores are ranked from best conditions (2) to worst (0). Basic indicators are weighted and their scores are summarized, resulting in a cumulative value ranging from 0 to 34. In a second step, the rating system also considers soil hazard indicators such as shallow soil depth or drought risk. Again, hazard indicators are scored on a scale between 0 (minimum yield potential or maximum limiting effect) and 2 (maximum yield potential or minimum limiting effect). Hazard indicators are considered as multipliers for the basic soil score. The rating scores of all hazard indicators are compared, and the lowest score is considered the valid one. At the end, the basic soil score is multiplied by the multiplier which has been derived from the score of the most limiting hazard indicator (Fig. 1). Further adjustments to the relevant multiplier have not yet been made in the case of several hazard indicators with sub-optimum ratings as proposed by Mueller et al. (2016). The rating system results in a final M-SQR score ranging from 0 to 100 points.

In its current version, the M-SQR method integrates up to 13 hazard indicators; not all of them were taken into account for this investigation. Two processes (salinization, alkalinization; hazard indicators 2, 3) are not of any relevance under the humid climatic conditions of Germany. Information about the present contents of nutrients or contaminants (hazard indicators 1, 5) is not yet part of the profile descriptions used to describe soil-mapping units and is not available from BGR's soil information system; against this background, potential limiting effects caused by these influencing factors could not be evaluated at the national scale. The effects of rock at the surface (hazard indicator 10) can be neglected when the contents of coarse texture fragments (hazard indicator 11) and soil depth above hard rock (hazard indicator 6) have already been interpreted. Under the moderate climatic conditions of Germany, the agricultural yield potential is not expected to be limited by an unsuitable soil thermal regime (hazard indicator 12). Hazard indicator 9 only leads to reductions in the crop yield when the slope gradient exceeds 12 %; those areas are generally not used as arable land in Germany due to business management conditions. The only relevant hazard indicator which is still not included in Fig. 1

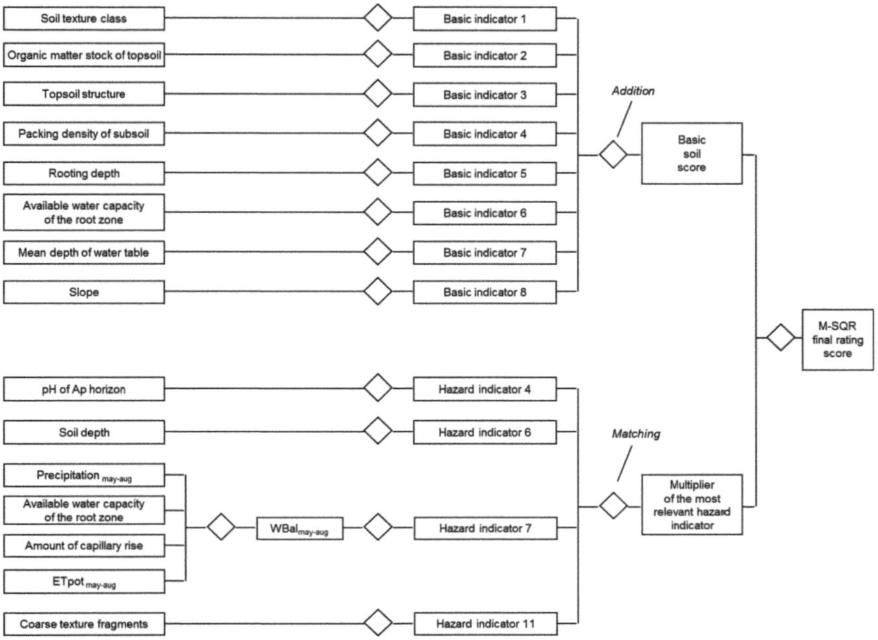

Fig. 1 Flow chart illustrating the operating mode of the Soil Quality Rating. For the purpose of the small-scale mapping procedure, and from the available data basis, not all globally important hazard indicators were significant for Germany

should be incorporated into the mode of operation which is hazard indicator 8 ("flooding and extreme waterlogging"). Within the scope of this study, all eight basic indicators and four selected hazard indicators (acidification, soil depth to hard rock, drought risk, content of coarse texture fragments; hazard indicators 4, 6, 7, 11) were taken into account.

2.2 *Preliminary Work and Modifications to the M-SQR Algorithms with Regard to Their Application on German Soil Maps*

Before the M-SQR method could be applied on soil profile-related data sets of German soil maps, internal rules, and algorithms had to be adapted to the terminology and classification of soil parameters as defined by German soil mapping guidelines (Ad hoc-AG Boden 2005). In 2010, in this modified version, the method was incorporated into the common pedotransfer rules collection of the soil survey institutions of the German federal states (Ad hoc-AG Boden 2000). In two cases, algorithms had to be changed considerably. The first revision was to basic indicator 2; in the original version of the M-SQR method, the depth of the A horizon acts as the determining factor. Usually, however, this kind of information cannot be deduced from the depth of the Ap horizon when soil profiles have been eroded and shortened under agricultural use. Against this background, the organic matter stock of the topsoil was selected as an alternative criterion for assessing basic indicator 2. The second modification affects hazard indicator 11. In the original version of the M-SQR manual, a moderate content of coarse texture fractions already leads to a low multiplier. This evaluation was changed to a new classification; now, moderate contents of coarse fractions are no longer regarded as a strongly yield-limiting factor.

The algorithms of the M-SQR were parameterized for application on soil profile data from the land-use stratified soil map of Germany at a scale of 1:1,000,000 (BÜK 1000N 2014). Furthermore, they were used to compile a nationwide land quality map showing the yield potential of cropland in Germany (Richter et al. 2009).

On the national scale, drought risk acts as the most determining factor limiting the yield potential for cereals. The M-SQR manual from 2007 (Mueller et al. 2007a) offers two criteria for assessing hazard indicator 7: the "water budget," resulting from precipitation, available water capacity of the root zone and capillary rise, and the "climatic water balance," calculated from precipitation minus reference evapotranspiration. When the climatic water balance is used as the only criterion and soil water retention properties are neglected, the resulting agricultural yield potential is underestimated in some areas of the country, particularly in the Eastern Federal State of Brandenburg. When the water budget serves as the influencing factor but reference evapotranspiration is not taken into account, the performance of the M-SQR application is improved and the resulting thematic map shows a more realistic spatial pattern, but this time the yield potential is underestimated in some morainic landscapes from the Weichsel glacial stage and in the Munsterland Basin. Against this background, a new criterion for assessing drought risk was introduced: the "water balance" of the main period of growth between May and August, calculated as water budget minus reference evapotranspiration, combines soil-related and climate-related site characteristics (Richter et al. 2009; Mueller et al. 2016). Although the biomass production of cereals does not take place in August under German climatic conditions, the month of August has been incorporated into the water balance because the M-SQR method is to be applicable on a broad range of crop sequences that include cereals as one of several crops. Because Germany is characterized by a humid to sub-humid climate and the drought risk is comparatively low on the global scale, factor scores of hazard indicator 7 in Germany do not fall below 2.0 (see Table 2).

2.3 Methodology of M-SQR Validation on Yield Data

A comparison of M-SQR point scores with real yield data of cereals was carried out on the basis of data from two different institutions. The State Authority for Mining, Energy, and Geology (LBEG) of Lower Saxony provided data from 47 soil-monitoring sites in Lower Saxony; time series are available from 1992 to 2008. The dominant crops are winter wheat and winter barley. In most cases, local crop rotation patterns are characterized by an annual change of crop. The Leibniz Center for Agricultural Landscape Research (ZALF) owns a collection of published yield data from different institutions. The inventory covers 32 sites from different federal states; most sites are located in Brandenburg. Yield data are available for individual years between 1994 and 2009. Many of these sites are characterized by a continuous cultivation of a specific crop, the dominant crop being winter rye. Some general characteristics of these two samples are presented in Table 1. As can be seen from Fig. 2, most sites from these two samples are located in Northern and Eastern Germany, while sites from the Federal States of Bavaria and Baden-Württemberg are underrepresented.

The final M-SQR rating score has already been calculated for all 32 sites from the ZALF database. For all 47 soil monitoring sites from Lower Saxony, rating scores were determined with little difficulty from available information about soil, relief, and climate. Bodenschätzung point scores were part of the two databases from ZALF and LBEG. This means that it was possible to compare the results of site assessments using Bodenschätzung and Soil Quality Rating methods for all 79 sites as shown in Table 1 and Fig. 2. A validation of the M-SQR approach requires measured yields of cereals in kg/ha or dt/ha and should be based on multi-year means. Ideally, these means should be calculated from at least three years of varying meteorological conditions. Because both data sources predominantly offer information on yields of different crops, and the LBEG database mainly includes sites with annually changing crops, no common subset from the LBEG and ZALF database fulfills all demands, and could be used for statistical analysis. When the same crop in ≥3 reference years is required, the maximum subset that could be generated from the LBEG database consists of only 4 data sets (Table 1). This limited database does not allow the M-SQR approach to be validated on yield data from Lower Saxony. If the same demands are applied to the ZALF database, subsets can be generated that contain 9 data sets (Table 1); for these sites a simple correlation analysis can be carried out between rating scores and yields (Figs. 12 and 13, Table 3). All other considerations, e.g., as illustrated in Fig. 11, are restricted to individual years. A correlation analysis that is based on a sufficient number of sites is also possible for point scores of the M-SQR method and Bodenschätzung.

Table 1 General characteristics of available yield data

Data source: LBEG	Data source: ZALF
47 sites	32 sites
Dominant crops: winter wheat, winter barley	Dominant crop: winter rye
Predominantly crop sequences with an annual change of crop	Predominantly crop sequences with continuous cultivation of a specific crop
When the same crop is required in ≥3 reference years: maximum database n = 4	When the same crop is required in ≥3 reference years: maximum database n = 9
Measures of fertilization are documented, local differences in management	Measures of fertilization are documented, mostly optimal N supply

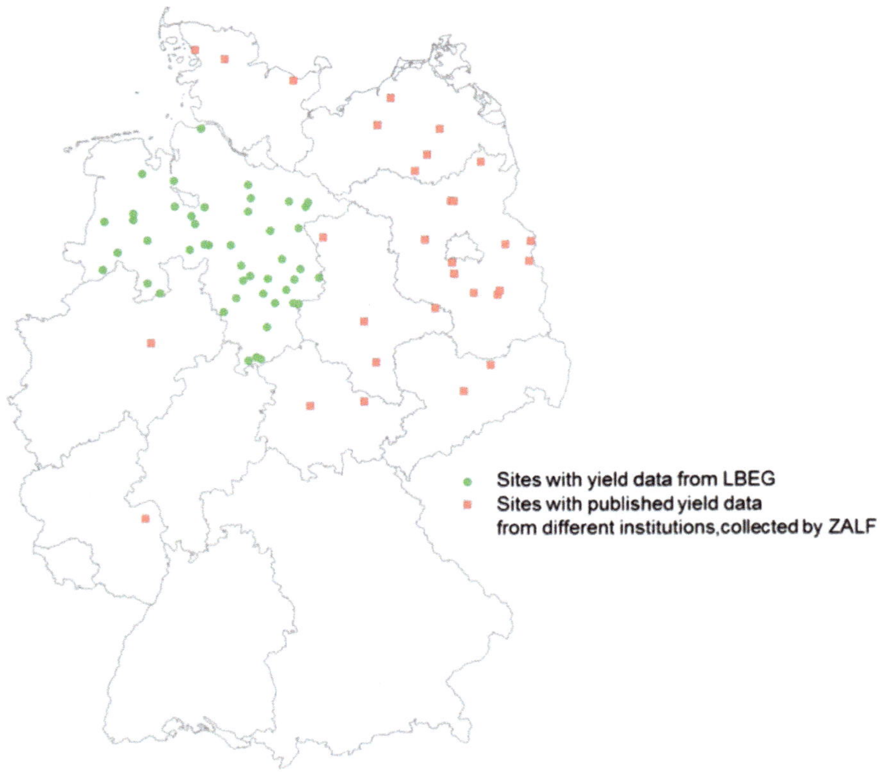

Fig. 2 Spatial distribution of sites with available yield data

3 Results

3.1 Yield Potential of Agriculturally Used Soils in Germany

In the first step, hazard indicators can be disregarded and soils on arable land in Germany can be evaluated on the basis of those soil properties that act as basic indicators within the M-SQR scheme (Fig. 1). Basic soil scores in Germany range from 9.5 to 33 points. The spatial pattern in Fig. 3 meet expectations and correlate to major landscapes. Sites on arable land that are assessed as scoring more than 24 points can be mainly found in loess regions and owe their positive evaluation to their favorable soil texture and soil structure properties.

In order to assess drought risk in an optimal way, the water balance of the main period of growth between May and August was introduced as a new criterion for hazard indicator 7. There had been no verification at that time of which classification fits best to the climatic site conditions of Central Europe. Available yield data for cereals was then used to identify the most suitable multipliers. This was done using an iterative procedure: the associated final M-SQR score was determined for

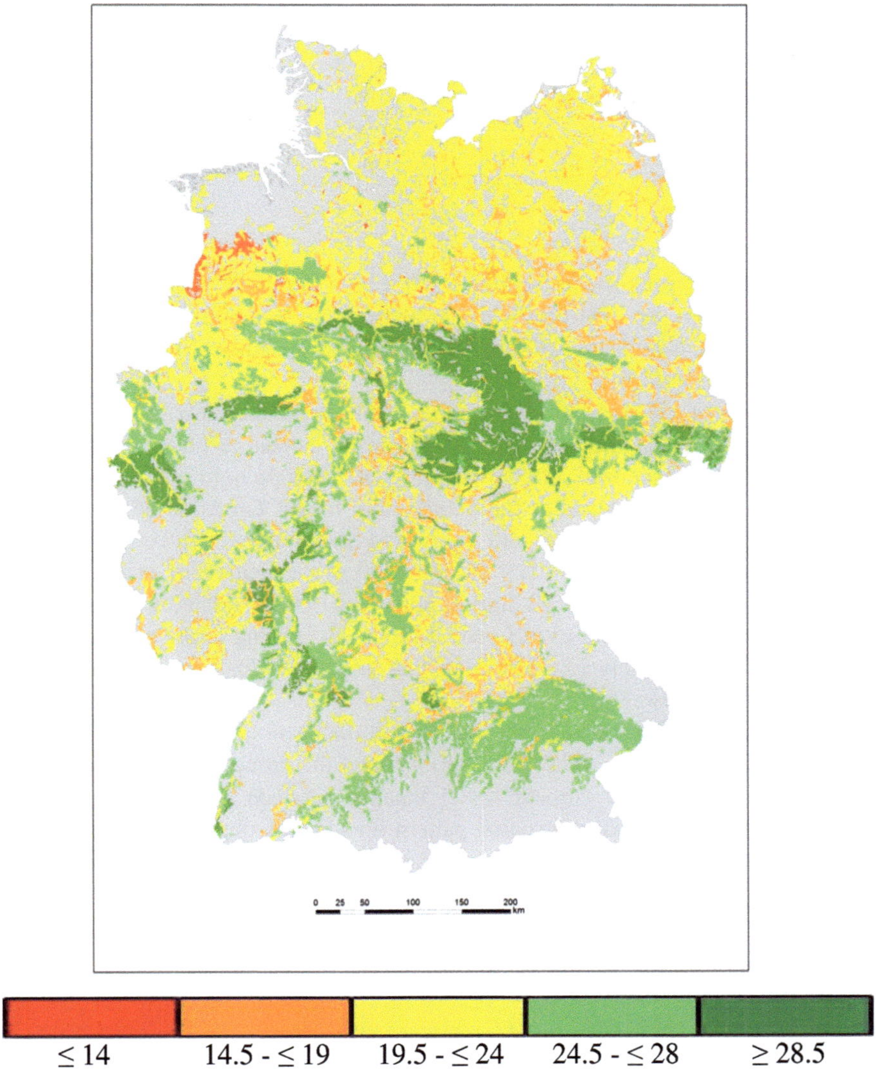

Fig. 3 Basic M-SQR soil scores on arable land in Germany

every possible classification of the water balance, and the correlation coefficient between the M-SQR rating and yield was calculated for every variant. The best possible fit was reached using the classification method presented in Table 2. This classification is addressed to users in Germany, while the one given by Mueller et al. (2016) was designed as a global approach. The classification from Table 2 was incorporated into the M-SQR scheme and from then on used for routine applications. When this algorithm is applied on soil mapping units of the land-use stratified soil map of Germany at a scale of 1:1,000,000 (BÜK 1000N), factor scores of

Table 2 Factor scores of hazard indicator 7 (drought risk) on the basis of the water balance between May and August ($WBal_{May-Aug}$)

$WBal_{May-Aug}$	Multiplier
>50 mm	3
25–50 mm	2.7
0–25 mm	2.5
−25–0 mm	2.3
<−25 mm	2

hazard indicator 7 reach the maximum of 3.0 (Fig. 4) on 74 % of all the areas under consideration; here, no long-time yield limitation due to drought risk is expected. As seen in Fig. 4, factor scores <3 can be observed in the Rhine-Main region and in parts of Eastern Germany.

On 61.5 % of all areas of arable land, there is no limitation by any hazard indicator (Fig. 5), i.e., factor scores of all hazard indicators under consideration reach the maximum score of 3.0. On 19.6% of all areas drought risk and on 15.3% of all areas depth of soil acts as the determining factor. On 2.2% of all areas the suitability of soils to agriculture and their yield potential are limited by high contents of coarse fragments. On 0.8 % of all areas the degree of acidification is regarded as the locally relevant hazard indicator. This evaluation refers to the distribution area of raised bogs in Lower Saxony. According to the present state of knowledge, comparatively low pH values are postulated for these Histosols even under agricultural use.

When the M-SQR method is applied on soil data sets of the land-use stratified soil map of Germany at a scale of 1:1,000,000 and the drought risk is evaluated on the basis of the effective climatic water balance between May and August, the resulting map shows a realistic spatial pattern for the agricultural yield potential of soils on arable land in Germany (Fig. 6). Figure 6 corresponds well to the empirical state of knowledge and existing thematic maps, e.g., maps as part of the National Atlas of Germany (Institut für Länderkunde 2003). The nationwide mean score, weighted according to the spatial proportions of soil mapping units, accounts for 64 points. The associated histogram is presented in Fig. 7.

3.2 Comparison of M-SQR and Bodenschätzung Point Scores with Yield Data

Before the results of an M-SQR application are validated based on yield data for cereals, M-SQR point scores shall be compared with Bodenschätzung point scores. As illustrated by Figs. 8 and 9, on average the point scores using the M-SQR method are higher than the Bodenschätzung ones. This statement is valid for sites from both samples: for sites from the LBEG database as well as for sites from the ZALF database. Particularly, on sites that have been evaluated at 30 to 50 points according to Bodenschätzung, the M-SQR rating score is significantly higher. The Bodenschätzung system reflects the scientific state of knowledge from the 1920s: at

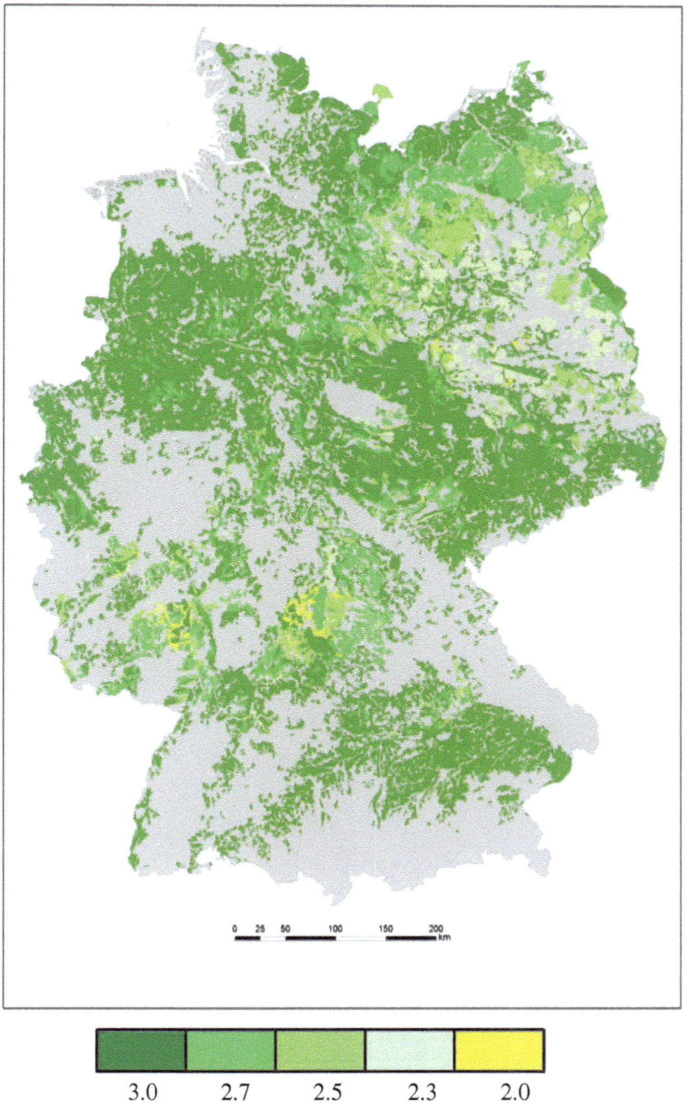

Fig. 4 Factor scores of hazard indicator 7 (drought risk) in Germany

those times the productivity of clay soils was evaluated as much higher than the productivity of sandy soils. The M-SQR tends to equalize these differences.

The convertibility of the traditional German Bodenschätzung into the international M-SQR approach can also be provided on the basis of correlations between Ackerzahl (AZ) and the M-SQR Basic Score (BS). Based on comparable actual field studies at 159 soil profiles in Germany, Mueller et al. (2011) found strong and highly significant ($B = 0.75^{***}$) regressions

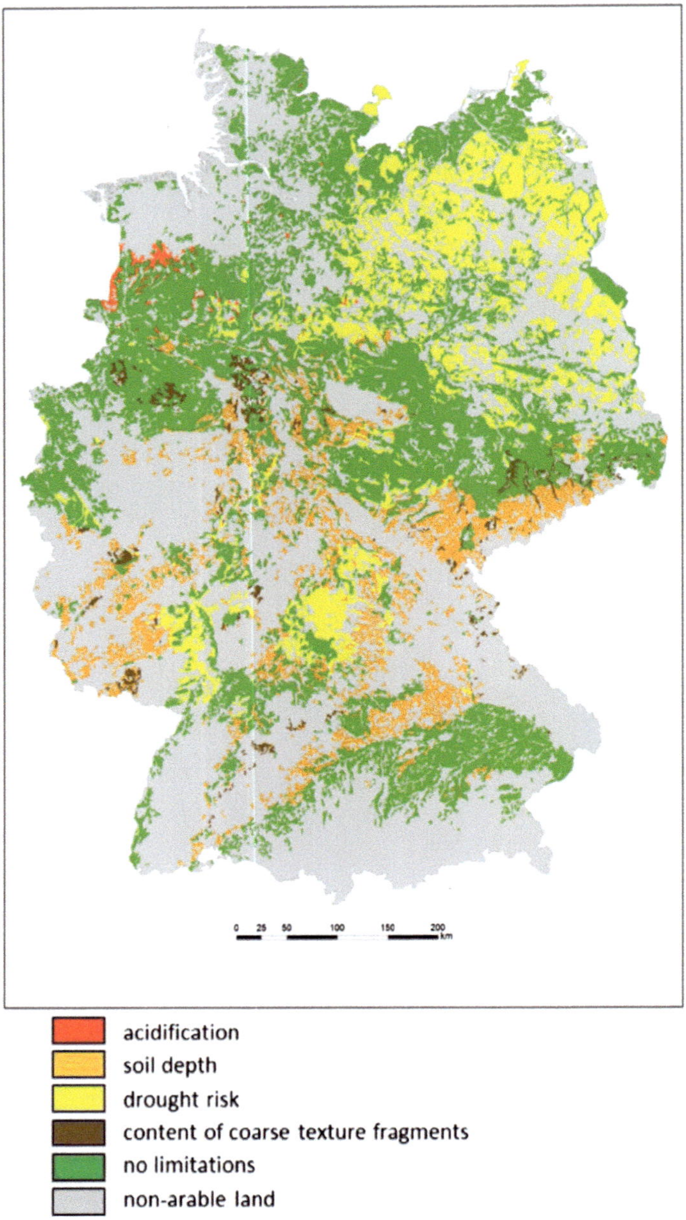

Fig. 5 Locally relevant M-SQR hazard indicators in Germany

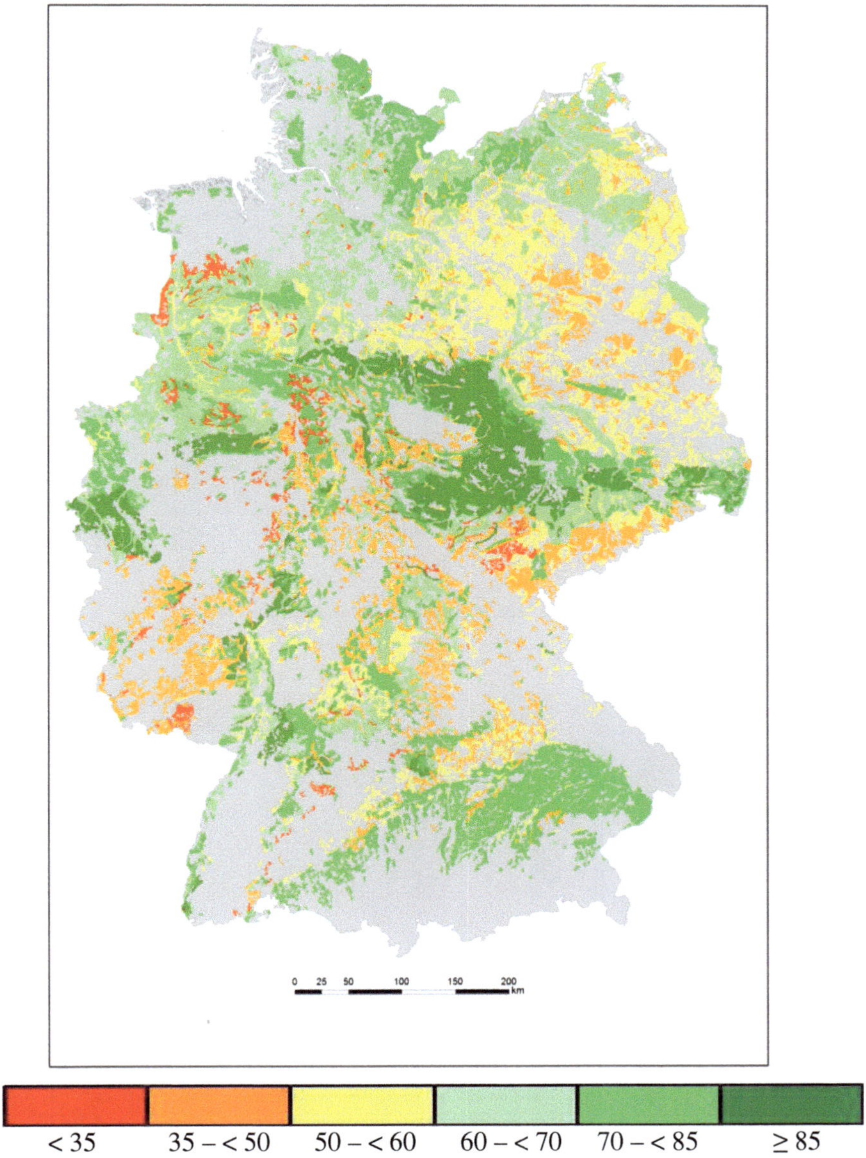

Fig. 6 Final M-SQR rating score for soils on arable land in Germany

$$\text{If AZ} > 30: \ \text{BS} = 17.47 + 0.168 \ \text{AZ}$$
$$\text{If AZ} < 30: \ \text{BS} = -6.72 + 83.08 \ \ln(\text{AZ})$$

Where there is no areal M-SQR field soil survey data but reliable Bodenschätzung AZ data and climate data for assessing the site-specific drought risk, the overall

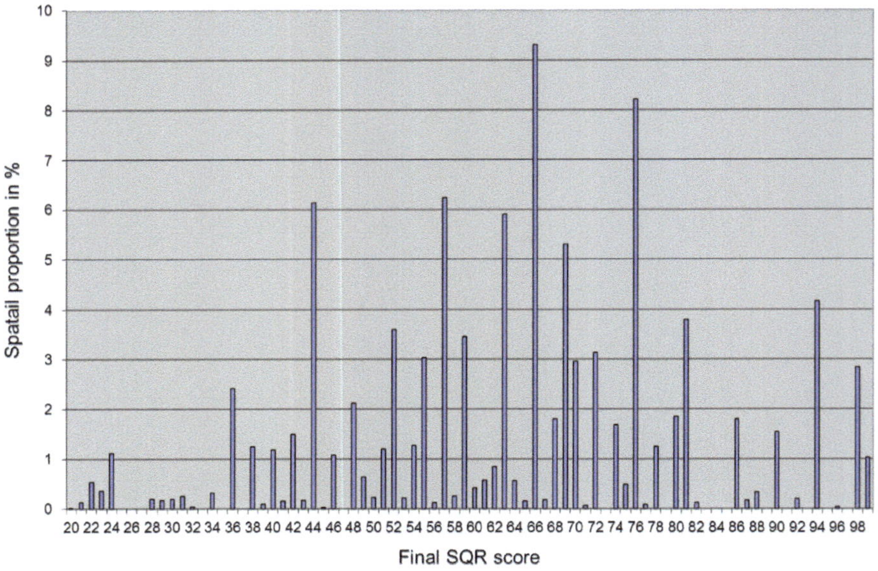

Fig. 7 Histogram of final M-SQR scores of soils on arable land in Germany

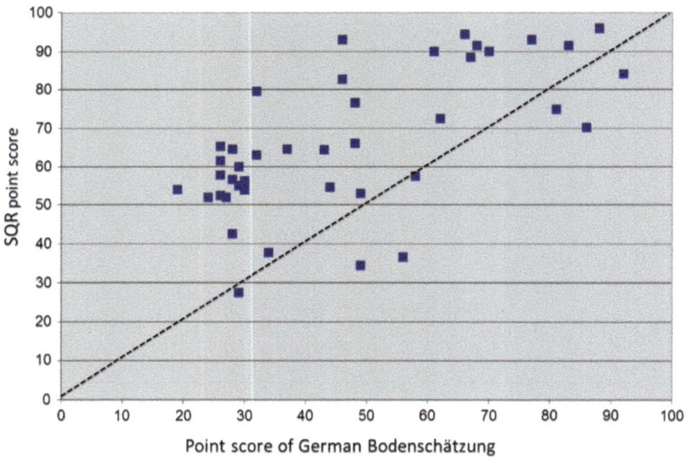

Fig. 8 Correlation between Bodenschätzung and Soil Quality Rating point scores on 47 sites from the LBEG database

M-SQR score can also be estimated for all regions in Germany on medium and large (field) scales.

Another interesting aspect that affects both evaluation schemes is the temporal variability of grain yields. Figure 10 shows the yields of winter rye in five consecutive years on five sites in Eastern Germany. In the case of Bodenschätzung,

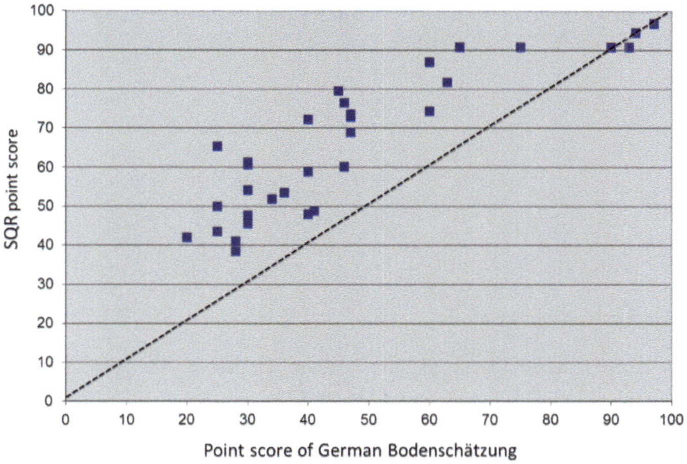

Fig. 9 Correlation between Bodenschätzung and Soil Quality Rating point scores on 32 sites from the ZALF database

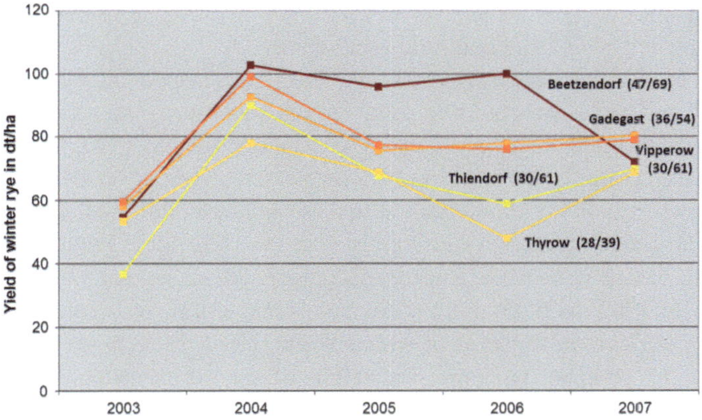

Fig. 10 Yields of winter rye in dt/ha from 2003 to 2007 at five selected sites (in *brackets*: Bodenschätzung/Soil Quality Rating point scores)

point scores range from 28 to 47 points, while in case of the M-SQR method they range from 39 to 69 points. The Beetzendorf site comes out best using both approaches, and in some years under consideration this site is characterized by the highest yields. On the other hand, the Thyrow site performs the worst and shows the lowest yields. In a year of optimal water supply during the period of growth, as in 2007, almost no differences in yield were recorded; by contrast, the broadest range of rye yields occurred in 2006, when only very little rainfall was registered between May and August. Natural site characteristics are evaluated using both approaches, but Bodenschätzung neglects climatic effects and Soil Quality Rating emphasizes

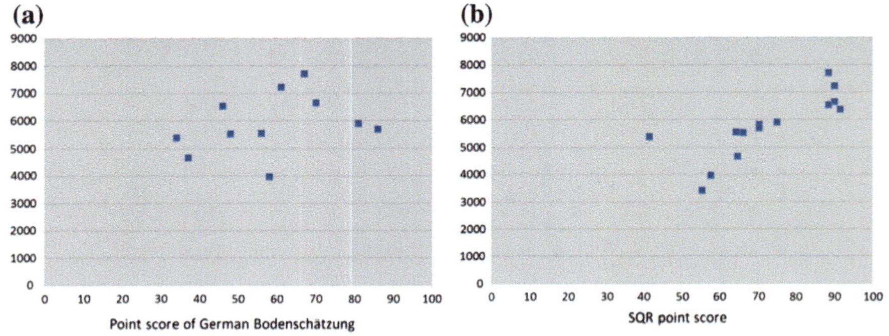

Fig. 11 Correlation between yields of winter wheat [dt/ha] in 2005 and point scores using Bodenschätzung (**a**, *left*) and Soil Quality Rating (**b**, *right*) methods on soil monitoring sites from Lower Saxony

them. Differences in natural site conditions are expressed by differences in yield, but only in a dry year such as 2006. In a wet year, as in 2007, differences in natural site conditions are leveled out. As a consequence, the degree of correlation between point scores and grain yield varies from year-to-year. These general conditions have to be taken into account, whenever the accuracy of any site assessment is to be evaluated on the basis of agricultural yield data.

For soil monitoring sites maintained by LBEG, no subset could be generated that fulfilled all demands and could be used for statistical analysis; against this background, M-SQR point scores could not be checked on the basis of multi-year means of reference crop yields. On sites from Lower Saxony, M-SQR results could only be compared with yields from individual years. For example, for 2005, yields of winter wheat are available for 14 soil-monitoring sites (see Fig. 11b). Figure 11a is based on only 11 data sets because Bodenschätzung point scores are not available for three sites. In 2005, M-SQR point scores are characterized by a smaller range and show better correlation to yields of winter wheat. Even if this is only a descriptive result, Fig. 11 provides an initial indication of the performance of the M-SQR method when an estimation of the crop yield potential is requested.

The inventory of the ZALF database allows M-SQR point scores to be compared with multi-year means of cereal yields, but the extent of the samples is limited ($n = 8$, see Figs. 12 and 13). Example sequences from 2003 to 2005 (Fig. 12) and sequences from 2006 to 2008 (Fig. 13) were chosen, with winter rye serving as a reference crop in both cases. The results differ considerably: while in 2003–2005 the degree of correlation to yields is higher for Bodenschätzung point scores, in 2006–2008 the M-SQR point scores show closer correspondence to measured yields of winter rye.

What is illustrated for two selected sequences of years in Figs. 12 and 13 is summarized for several sequences of years in Table 3: correlation statistics are presented if the ZALF database allows mean yields to be calculated on the basis of at least three years. In three cases, the results refer to yields of winter wheat, in six

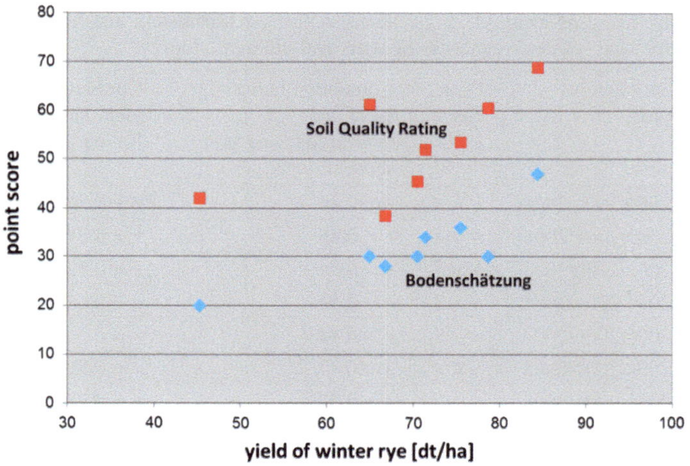

Fig. 12 Comparison of mean yields of winter rye [dt/ha] in 2003–2005 with Bodenschätzung and Soil Quality Rating point scores on eight sites in Eastern Germany

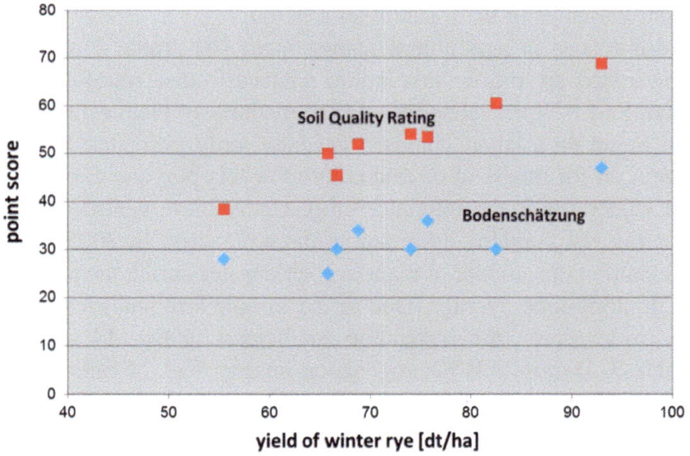

Fig. 13 Comparison of mean yields of winter rye [dt/ha] in 2006–2008 with Bodenschätzung and Soil Quality Rating point scores on eight sites in Eastern Germany

cases, the results refer to yields of winter rye. The contents of Table 3 do not paint a clear picture: in four cases Bodenschätzung point scores show closer correspondence to observed yields, while in the remaining five cases the M-SQR approach performs better. Even if results from Table 3 do not lead to universally valid conclusions, it can be stated that correlation coefficients r between M-SQR point scores and yields exceed 0.66 in all cases under consideration. In the majority of cases, the M-SQR approach provides a more precise and a more reliable evaluation of the productivity function of the soil than Bodenschätzung. The better

Table 3 Comparison of mean yields of cereals with Bodenschätzung and Soil Quality Rating point scores for nine selected sequences of years and different crops

Crop	Sequence of years	Data base	Correlation between yield and Bodenschätzung point score	Correlation between yield and Soil Quality Rating point score
Winter wheat	2004/2005/2006	$n = 6$	$r = 0.56$	$\boldsymbol{r = 0.90}$
	2003/2004/2006	$n = 6$	$r = 0.88$	$\boldsymbol{r = 0.89}$
	2005/2006/2007	$n = 7$	$\boldsymbol{r = 0.86}$	$r = 0.72$
Winter rye	2003/2004/2005	$n = 8$	$\boldsymbol{r = 0.86}$	$r = 0.66$
	2005/2006/2007	$n = 8$	$\boldsymbol{r = 0.84}$	$r = 0.76$
	2006/2007/2008	$n = 8$	$r = 0.77$	$\boldsymbol{r = 0.98}$
	2005/2006/2008	$n = 8$	$r = 0.87$	$\boldsymbol{r = 0.95}$
	2004/2005/2007	$n = 8$	$r = 0.61$	$\boldsymbol{r = 0.69}$
	2004/2005/2006	$n = 9$	$\boldsymbol{r = 0.83}$	$r = 0.80$

Bolditalics approach with better performance

performance of the M-SQR approach most likely originates in the incorporation of climatic input variables in the assessment scheme.

Due to differences in agricultural management (see Table 1) all yield-related results as presented up to now have either referred to data sets from LBEG or to data sets from ZALF. If the reliability of both methods of yield estimation are to be investigated based on a larger common database, only yield data from individual years can be used for this kind of evaluation. For this purpose the year 2005, was chosen and winter rye is the reference crop. Under these general conditions, the database for Fig. 14 contains information about 29 sites. In 2005, on most sites under consideration the amount of plant available water during the period of growth was above the long-time average. Due to the sample size and local differences in agricultural management, the correlation coefficients in Fig. 14 are smaller than those in Table 3. Despite this heterogeneous sample, Fig. 14 offers the chance to compare the performance of Bodenschätzung and Soil Quality Rating. As postulated by interpreting statistical measures from Table 3, the M-SQR approach provides the better performing soil productivity rating and the M-SQR point scores fit better to natural site conditions. While the degree of correlation between Bodenschätzung point scores and yields is minimal (Fig. 14a), M-SQR point scores at least reflect the general trend of rye yields (Fig. 14b).

4 Summary and Outlook

The M-SQR is an approach to evaluate how suitable soils are for agriculture; on many test sites in Germany the method has provided reliable estimates of the yield potential for cereals.

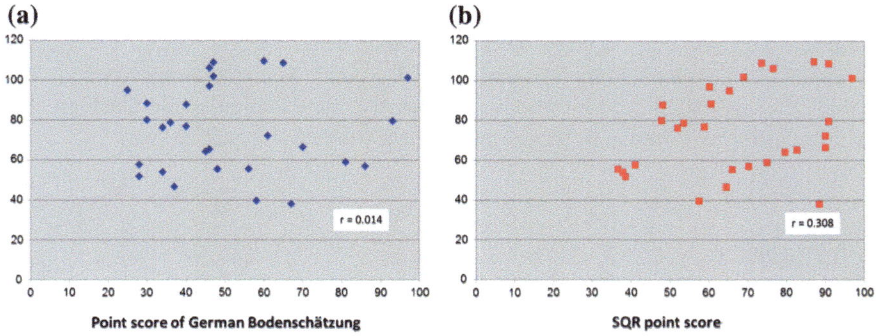

Fig. 14 Correlation between yields of winter rye [kg/ha] in 2005 and point scores using Bodenschätzung (**a**, *left*) and Soil Quality Rating (**b**, *right*) methods on 29 sites from different Federal States of Germany

The pedotransfer rules of the first M-SQR version, as published in the manual from 2007 (Mueller et al. 2007a), was improved in two details: the criterion of drought risk (hazard indicator 7) and the algorithms for evaluating the content of coarse texture fragments (hazard indicator 11) were modified.

Compared to soils in other countries, soils in Germany show a high yield potential for grain. The main determining factors in Germany are a moderate drought risk and limited rooting depth with shallow soils.

The M-SQR method performs best in assessing the agricultural yield potential when the water balance is used as a criterion of drought risk: this measure incorporates soil-related parameters as well as climatic parameters.

On 44.4% of all areas where factor scores of at least one hazard indicator fall below the maximum value of 3.0, the drought risk acts as the limiting indicator or relevant multiplier, respectively. On a nationwide map, the factor scores of this hazard indicator reach the maximum of 3.0 on 74 % of all areas of arable land.

Not only the final rating score, but also some of the M-SQR indicators such as the water balance during the period of growth can be used for political consulting, e.g., for the implementation of rules on agrarian policy such as the identification of less favored areas for agriculture.

Even though the Bodenschätzung system is a well-established assessment scheme, the M-SQR approach seems to be at least equivalent. In the majority of cases, M-SQR point scores correlate better to measured yields of cereals than Bodenschätzung scores. This can be explained easily: unlike Austria, Germany has failed to adapt the Bodenschätzung framework to the present-day state of scientific knowledge and lay more stress on climatic aspects (cf. Rust 2006).

A terminatory evaluation of the Soil Quality Rating in comparison to alternative methods should be based on a nationwide validation study, and should also include measured yields from sites in Southern Germany.

The good performance of the M-SQR method has been proven on the basis of grain yield data from various sites all over Northern and Eastern Germany. This may enable higher resolution crop yield potential maps to be created in these areas using the M-SQR methodology, digital soil maps on different scales and climate databases. As the M-SQR methodology was designed as a global approach, it also enables the consistent incorporation of the newly created small-scale German crop yield potential map into a potential small-scale global soil quality map.

References

Ad-hoc-Arbeitsgruppe Boden (2000) Methodendokumentation Bodenkunde. Auswertungsmethoden zur Beurteilung der Empfindlichkeit und Belastbarkeit von Böden, 2nd edn. Geol Jb, SG 1, Hannover, 232 p

Ad-hoc-Arbeitsgruppe Boden (2003) Methodenkatalog zur Bewertung natürlicher Bodenfunktionen, der Archivfunktion des Bodens, der Gefahr der Entstehung schädlicher Bodenveränderungen sowie der Nutzungsfunktion "Rohstofflagerstätte" nach BBodSchG. – Arbeitshefte Boden, Editors: Bundesanstalt für Geowissenschaften und Rohstoffe and Niedersächsisches Landesamt für Bodenforschung, vol 2003/2. Hannover, pp 4–73

Ad-hoc-Arbeitsgruppe Boden (2005) Bodenkundliche Kartieranleitung, 5th edn. Hannover, 438 p

Adler G, Eckelmann W, Hartwich R, Hennings V, Krone F, Stolz W, Utermann J (1998) The FISBo BGR soil information system: state of the art. In: Heineke H et al (eds) Land information systems. Developments for planning the sustainable use of land resources. Office for Official Publications of the European Communities, Luxembourg European Soil Bureau Research Report No. 4, pp 133–140

BBodSchG (1998) Gesetz zum Schutz vor schädlichen Bodenveränderungen und zur Sanierung von Altlasten (Bundes-Bodenschutzgesetz - BBodSchG). Date of issue: 17.03.1998. http://www.gesetze-im-internet.de/bundesrecht/bbodschg/gesamt.pdf. Accessed 09 Mar 2015

BÜK 1000N (2014) Nutzungsdifferenzierte Bodenübersichtskarte 1:1.000.000 (BÜK1000N). http://www.bgr.bund.de/DE/Themen/Boden/Informationsgrundlagen/Bodenkundliche_Karten_Datenbanken/BUEK1000/Nutz_BUEK/nutz_buek_node.html. Accessed 09 Mar 2015

FAO (1976) A framework for land evaluation. FAO Soils Bull 32, 72 p., Rome. http://www.fao.org/docrep/x5310e/x5310e00.htm. Accessed 09 Mar 2015

FAO (1980) Land evaluation guidelines for rainfed agriculture. World Soil Resour Rep 52: 118 (Rome)

Hennings V, Eckelmann W (2004) Das Fachinformationssystem Bodenkunde der Bundesanstalt für Geowissenschaften und Rohstoffe (FISBo BGR): Strategie und Status 2003. Bull BGS 27: 19–26 (Landwirtschaftliche Lehrmittelzentrale, Zollikofen)

IEEP (2006) An evaluation of the Less favoured Area measure in the 25 Member States of the European Union, http://ec.europa.eu/agriculture/eval/reports/lfa/index_en.htm. Accessed 09 Mar 2015

Institut für Länderkunde der Universität Leipzig (ed) (2003) Nationalatlas Bundesrepublik Deutschland, vol 2: Relief, Boden und Wasser. Spektrum Akademischer Verlag, Heidelberg-Berlin, 174 p

Mueller L, Schindler U, Behrendt A, Eulenstein F, Dannowski R (2007a) The Muencheberg Soil Quality Rating. Field manual for detecting and assessing properties and limitations of soils for cropping and grazing. Leibniz-Zentrum für Agrarlandschaftsforschung (ZALF), Müncheberg. 1st Draft, Nov. 2007, http://www.zalf.de/de/forschung/institute/lwh/mitarbeiter/lmueller/Documents/field_mueller.pdf. Accessed 09 Mar 2015

Mueller L, Schindler U, Behrendt A, Eulenstein F, Dannowski R (2007b) Das Müncheberger Soil Quality Rating (SQR): ein einfaches Verfahren zur Bewertung der Eignung von Böden als Farmland. Mitteilgn Dtsch Bodenkdl Ges 110/II: 515–516

Mueller L, Kay BD, Hu C, Li Y, Schindler U, Behrendt A, Shepherd TG, Bal BC (2009) Visual assessment of soil structure: evaluation of methodologies on sites in Canada, China and Germany. Part I: comparing visual methods and linking them with soil physical data and grain yield of cereals. Soil Tillage Res 103(1):178–187

Mueller L, Schindler U, Mirschel W, Shepherd TG, Ball BC, Helming K, Rogasik J, Eulenstein F, Wiggering H (2010) Assessing the productivity function of soils. A review. Agron Sustain Dev 30(3):601–614. doi:10.1051/agro/2009057

Mueller L, Smolentseva E, Rukhovich O, Lukin S, Hu C, Li Y, Schindler U, Behrendt A, Hennings V, Sauer S, Vorderbrügge T (2011) Zur Bewertung von Bodengüte und Ertragspotentialen in Agrarlandschaften. In: Böden verstehen – Böden nutzen – Böden fit machen: DBG-Jahrestagung, 3. 9 Sept 2011, Berlin, Deutsche Bodenkundliche Gesellschaft, Oldenburg, pp 1–4, http://eprints.dbges.de/680/1/LMüller_DBG11.pdf. Accessed 09 Mar 2015

Mueller L, Schindler U, Shepherd TG, Ball BB, Smolentseva E, Hu C, Hennings V, Schad P, Rogasik J, Zeitz J, Schlindwein S, Behrendt A, Helming K, Eulenstein F (2012) A framework for assessing agricultural soil quality on a global scale. Archiv Agro Soil Sci 58(sup1):76–82

Mueller L, Shepherd TG, Schindler U, Ball BC, Munkholm LJ, Hennings V, Smolentseva E, Rukhovic O, Lukin S, Hu C (2013) Evaluation of soil structure in the framework of an overall soil quality rating. Soil Tillage Res 127:74–84

Mueller L, Schindler U, Hennings V, Smolentseva EN, Rukhovich OV, Romanenkov VA, Sychev VG, Lukin S, Sheudshen A, Onishenko L, Saparov A, Pachikin K, Behrendt A, Mirschel W, Eulenstein F (2016) An emerging method of rating global soil quality and productivity potentials. In: L. Mueller et al. (eds) Novel methods for monitoring and managing land and water resources in Siberia. Springer, Cham, pp 573–595 (Chapter 26 of this book)

Müller U (2004) Auswertungsmethoden im Bodenschutz. Dokumentation zur Methodenbank des Niedersächsischen Bodeninformationssystems (NIBIS), 7th edn. Niedersächsisches Landesamt für Bodenforschung (ed) Arbeitshefte Boden, 2004/2, Hannover

Oelkers K-H, Sellge H (1993) Digitale Führung und Nutzung der Bodenschätzungsdaten. Nachrichten der Niedersächsischen Vermessungs- und Katasterverwaltung 4: 188–195

Pfeiffer E-M, Sauer S, Engel E (eds) (2003) Bodenschätzung und Bodenbewertung. Nutzung und Erhebung von Bodenschätzungsdaten. Chmielorz-Verlag Wiesbaden, 88 p

Richter U, Eckelmann W (1993) Das Ertragspotential ackerbaulich genutzter Standorte in Niedersachsen – Beispiel einer Auswertungsmethode im Niedersächsischen Bodeninformationssystem NIBIS. Geol Jb F 27:197–205

Richter A, Hennings V, Müller L (2009) Anwendung des Müncheberger Soil Quality Ratings (SQR) auf bodenkundliche Grundlagenkarten. In: Jahrestagung der DBG: Böden - eine endliche Ressource, Kommission V, September 2009, Bonn. Berichte der DBG, 4 p., Deutsche Bodenkundliche Gesellschaft, Oldenburg, http://eprints.dbges.de/243/. Accessed 09 Mar 2015

Roesch A, Kurandt F (1950) Bodenschätzung und Liegenschaftskataster. 3rd edn. Berlin, 148 p

Rust I (2006) Aktualisierung der Bodenschätzung unter Berücksichtigung klimatischer Bedingungen. Diss, Georg-August-Universität Göttingen, https://ediss.uni-goettingen.de/bitstream/handle/11858/00-1735-0000-0006-AB70-D/rust.pdf?sequence=1. Accessed 09 Mar 2015

Shepherd TG (2000) Visual soil assessment, Volume 1. Field guide for cropping and pastoral grazing on flat to rolling country. Horizons Regional Council, Landcare Research, 84 p, Palmerston North, 118 p

Storie RE (1933) An index for rating the agricultural value of soils. Agricultural Experiment, University of California Agricultural Experiment Station, Berkley, Station Bulletin 556, 44 p

Umweltministerium Baden-Württemberg (1995) Bewertung von Böden nach ihrer Leistungsfähigkeit. Luft, Boden, Abfall, vol 31, 34 p

Chapter 28
Balance of Nutrients and the Optimization of Their Use in Agroecosystems of the Russian Federation

Evgeny N. Yefremov, Victor G. Sychev and Vladimir A. Romanenkov

Abstract Nutrient balance, calculated at the national and regional level, is a useful tool to gain information on trends in nutrient depletion or enrichment which can be used to choose nutrient management strategies and to assess the undesirable effects of nutrient mining and environmental pollution. The balances of the main plant nutrients like nitrogen (N), phosphorus (P), and potassium (K) were calculated by taking into consideration the inputs of nutrients with mineral and organic fertilizers, seeds, biologically fixed N (symbiotic and non-symbiotic fixation) and rain, plus the outputs of nutrients through crop uptake, and losses through leaching, erosion, and denitrification. The increase in the scope of chemicalization in Russia led to the gradual elimination of the N and P deficit, and the surplus of these nutrients at the national level—from the approximate balance since the mid-1960s up to 37 and 25 kg/ha and between 1986 and 1990, respectively. The annual deficit of K was also gradually reduced in the same period from −15 to −2 kg/ha. Present-day agriculture has a serious annual excess of removal over input, with a long-term deficit of up to 30, 10, and 27 kg/ha for N, P, and K, respectively as the result of a drastic decrease in mineral and organic fertilization, since the 1990s. The regional balance also provides a link with monitoring data of nutrient availability in arable soils. For the regions with high-input agricultural production, the analysis of agrochemical survey data shows a consistent decrease in the weighted average content of available P and K forms in arable lands. This negative tendency became clear when the input of nutrients was inadequate to maintain soil fertility, being permanently lower than its removal from the agrolandscapes. The balance method helped to identify hot spots of unbalanced fertilization, where N consumption outstripped that of P and K. Siberia is one of these regions where arable lands are

E.N. Yefremov · V.G. Sychev · V.A. Romanenkov (✉)
Pryanishnikov All-Russian Research Institute of Agrochemistry (VNIIA),
Pryanishnikova St.31A, Moscow 127500, Russia
e-mail: geoset@yandex.ru

E.N. Yefremov
e-mail: info@vniia-pr.ru

V.G. Sychev
e-mail: info@vniia-pr.ru

L. Mueller et al. (eds.), *Novel Methods for Monitoring and Managing Land and Water Resources in Siberia*, Springer Water,
DOI 10.1007/978-3-319-24409-9_28

not only underfertilized, but mainly receive N fertilizers at the expense of P and K as a result of farmers concentrating on short-term decisions instead of long-term sustainability, while in the Russian Far East the situation is much more favorable. The development of policies and strategies relating to the fertilization requirements of Russian agriculture should be based on providing balanced nutrition conditions for sustaining agricultural systems and soil fertility conservation.

Keywords Plant nutrients · Fertilization · Balancing · Russia

1 Introduction

A permanent increase in the productivity of the crop farming sector of agrarian production is necessary to prevent an imbalance between the increasing population and the production of foods. Crops remove nutrients from the soil, and these losses need to be replaced by fertilization. This is a common rule of managing soils sustainably. Mineral fertilizers are an important factor for the intensification of arable and crop farming. Therefore, the consumption of mineral fertilizers is constantly growing. It is presently at 180 million t per year globally, including 58 % nitrogen fertilizers, 24 % phosphorus fertilizers, and 18 % potassium fertilizers (Heffer and Prud'homme 2013). In the 1970–1980s, the maximum amounts of fertilizers were consumed in the USA, European countries, and the USSR; however, 30 years later, the application focus was shifted to the Asian countries, primarily to the Eastern and South-Eastern Asian regions, where the growth of population, the change in food distribution, and, hence, the increase in demand for food products required a significant intensification of agricultural production. The unbalanced use of large amounts of fertilizers also has negative effects. During the last ten years, the efficiency of nitrogen fertilizers in China and India has decreased appreciably, and the environmental implications of intensive chemicalization in China are manifested not only in that country, but also in the border districts of the adjacent countries (Song et al. 2010). Moreover, in European and other countries, positive balances of nutrients and their negative impacts on water and air resources are of great concern. Because of intensive agriculture including high fertilization levels, the inputs and balances of nitrogen (Eulenstein et al. 2014), phosphorus (Chernenok and Barkusky 2014), and other plant nutrients and fertilizer by-products (Smidt et al. 2011) need to be strictly monitored and controlled.

Russia, which consumed 13.5 Mt of fertilizer nutrients in 1988, has still not overcome the decline in fertilizer application caused by the change in the politico-economic structure at the beginning of the 1990s. During the last 20 years, the use of mineral fertilizers was very low, not exceeding 1.3–2.0 Mt NPK. The application of organic fertilizers permanently decreased; only 50–60 Mt is used presently (Russia in figures 2012). The main cause for the low level of application of mineral fertilizers and other chemicals is the low profitability of most agricultural producers because of the unfavorable economic conditions for agricultural production and the nonequivalent exchange between agriculture and related sectors of the national economy.

2 Studies of Macronutrient Cycles with Balance Methods

Nutrients are mainly removed from the field with harvested products (agricultural crops and also weeds). Unproductive losses of nutrients (leaching from the root-inhabited layer, losses by erosion, and gaseous nitrogen losses) should be minimized but are sometimes inevitable and must be included in balances. As it was demonstrated in a series of our field experiments with fertilizers, losses of nitrogen may reach 25–30 kg/ha in zones of excessive moistening, and its loss into the air due to denitrification can reach 10–20 % of the applied nitrogen fertilizer rate (6 kg/ha soil on the average), while on eroded lands 18–20 kg nitrogen, 5–10 kg phosphorus, and 12–24 kg potassium are lost from 1 ha.

In Russia, large-scale calculations of the potassium balance were performed in stationary experiments on light soddy-podzolic soils at the Solikamsk Experimental Station, Perm Oblast (Belyaev 2005). During the 10–15-year period, a negative potassium balance was recorded in both the treatments with the application of NP fertilizers and the treatments with the additional application of potassium fertilizers. The content of exchangeable potassium did not change significantly in the NP treatments and increased in the NPK treatments.

Useful information is derived by calculating the nutrient balances in soils for very large areas of agricultural production, e.g., for the entire country, federal district, or region. It is believed that to obtain high and stable crop yields, the consumption of nitrogen and potassium should be compensated 100 %, and the phosphorus input items should exceed the output items by 1.5–2 times. In the forest-steppe zone, the balance intensity can be slightly lower: 85–90 % for nitrogen, 150–200 % for phosphorus, and 50–60 % for potassium. In the steppe zone, the corresponding values are 200–250, 60–75, and 25–30 % (Peterburgsky 1979).

Detailed balance calculations require a large number of input and output items to be considered. The balances of nitrogen, phosphorus, potassium, and humus are related to the original contents of mobile nutrient forms and organic matter in the soil. The output items include not only the removal of nutrients with crops and their migration to the groundwater, but also migration to the atmosphere with soil respiration and plant transpiration, fixation and nonexchangeable adsorption in the soil, the removal of soil with crops, the horizontal migration of substances in the catena, and losses by water and wind erosion. The input items should include not only the input of NPK with fertilizers, ameliorants, seeds, precipitation, and harvest residues, but also their input with rhizodeposition, migration from the overlying geomorphological elements, intake from the lower horizons with capillary fringe, uptake by plants and soil from the air, and the increase in the content of mobile biophilic elements because of climatic factors.

The coefficients of nutrient utilization by soils with different fertility levels vary in the same ranges as the yields of field crops depend on the weather conditions. On fields with low fertility, field plants can utilize 23–24 % of easily hydrolyzable nitrogen, 14–33 % of available phosphorus, and 10–15 % of exchangeable potassium from their reserves in the plow layer. The better cultivated and more enriched soil with available fertilizers, the lower its utilized portion. On well-cultivated fields, crops' utilization of phosphorus from the soil decreases to 2–10 % (Ionas et al. 1998; Kayumov 1977).

The migration of NPK beyond the soil profile with rain water is an important output item in the balance calculations of biophilic elements in the soil. The losses of nutrients with surface runoff and infiltrating water beyond the root-inhabited zone are determined by the wetting conditions, the presence of mobile biophilic elements in the soil, and the infiltration capacities of soils and sediments.

The soil releases gaseous products into the air, as well as water-soluble compounds, which volatilize with soil evaporation. This balance component is obviously characterized by different intensities depending on the soil, wetting conditions, and temperature, as well as the contents of components capable of migrating from the soil to the air. The greater the soil's supply of mobile and water-soluble nutrient forms, the more of them volatilize during evaporation from the soil (all other conditions being equal). The changes in the properties of soils, their fertility, and level of fertilization affect the compositions of gaseous emissions from the soils. The different forms of nitrogen fertilizers had different effects on the mineralization rate of soil organic matter (Budazghapov 1989).

The first suppositions about the structure of the NPK balance and the most detailed calculations of nutrient balances in Russian agriculture were made by D.N. Pryanishnikov in 1937. He posed questions about what shape of the balance structure for essential mineral nutrients in the national agriculture should take in order to balance out the removal of phosphorus with an excess of 100 % and more and the removal of nitrogen and potassium with an excess of 75–80 % (Pryanishnikov 1937). Later on, it was proposed that, because of the wide diversity of soil and climatic conditions in the country, the balances of nutrients should be calculated for separate agricultural areas characterized by common natural and economic factors for long time periods (10 years and more) in order to compare the balances of nitrogen, phosphorus, and potassium with the contents of their mobile forms in the soil recorded during periodical agrochemical surveys of agricultural lands.

3 Balancing of Nitrogen, Phosphorus, and Potassium

Roy et al. (2003) proposed an equation for calculating the balances of nutrients (N, P, and K) in soils of agroecosystems at a country or regional scale:

$$\mathrm{Rn}_{tn} = \sum^{tn} (\mathrm{AP}_t + \mathrm{AR}_{\Delta t} - \mathrm{RM}_{\Delta t} - \mathrm{L}_{\Delta t})$$

where Rn_{tn} is the quantity of inorganic and organic nutrients remaining in the soil at time *tn*; AP_t is the soil inorganic and organic nutrients present at time *t*; $AR_{\Delta t}$ is the inorganic and organic nutrients added or returned to the soil during the time interval Δt; the $RM_{\Delta t}$ estimate is the plant nutrients removed with the harvested product and residue management during the time interval Δt, and $L_{\Delta t}$ is the inorganic and organic nutrients loss during the time interval Δt; the value of *t* represents the beginning time period, *tn* represents the ending time period, and Δt is the time interval between *t* and *tn*.

Based on our own experience with agricultural experiments and nutrient analyses of plants and soils, and considering those methodical recommendations, we developed balancing methods and schemes for Russia (Methodological guidelines 2000; Shafran 2004).

Nutrient balancing for agricultural lands, farms, regions, and the whole country have been carried out on an annual basis. Balance sheets or computer programs have been developed. They contain the following components and factors valid for Russia:

Input items (kg nutrients/ha)

1. *Inputs of nutrients with mineral fertilizers*, kg N, P_2O_5, K_2O/ha of the corresponding agricultural land: $Q_{1(N)}$, $Q_{1(P_2O_5)}$, and $Q_{1(K_2O)}$, respectively.
2. *Inputs of nutrients with organic fertilizers*:

$$Q_{2(\mathrm{N})} = O \cdot c_N \cdot k \cdot k_1;$$
$$Q_{2(\mathrm{P_2O_5})} = O \cdot c_{\mathrm{P_2O_5}} \cdot k \cdot k_1;$$
$$Q_{2(\mathrm{K_2O})} = O \cdot c_{\mathrm{K_2O}} \cdot k \cdot k_1,$$

   where O is the rate of organic fertilizer, t/ha; c_{N}, $c_{\mathrm{P_2O_5}}$, and $c_{\mathrm{K_2O}}$ are the concentrations of N, P_2O_5, and K_2O, respectively, in the fertilizer, in %; *k* is the correction factor for the content of dry matter in manure; k_1 is the correction factor for the conversion of different types of organic fertilizer to litter manure as a standard (Reference book … 1989).

   In Russia, it is acceptable to take the following mean contents of nutrients in manure: N 0.5 %; P_2O_5 0.25 %; and K_2O 0.6 %.
3. *Inputs of nutrients with seeds*:

$$Q_{3(\mathrm{N})} = W \cdot c_{\mathrm{N}};$$
$$Q_{3(\mathrm{P_2O_5})} = W \cdot c_{\mathrm{P_2O_5}};$$
$$Q_{3(\mathrm{K_2O})} = W \cdot c_{\mathrm{K_2O}},$$

   where W is the sowing rate, kg/ha; c_{N}, $c_{\mathrm{P_2O_5}}$, $c_{\mathrm{K_2O}}$ are the concentrations of N, P_2O_5, and K_2O in seeds, respectively, kg/kg.

4. *Symbiotic nitrogen fixation*:

$$Q_{4(\mathrm{N})} = U \cdot a_{\mathrm{N}},$$

where U is the yield of crop, dt/ha; a_{N} is the symbiotic nitrogen fixation, kg N/dt crop.
If the data on the proportion of the legume component in the annual legume–cereal grass mixtures are unavailable, a conventional value of 6 kg N/t hay can be used. For the annual grain legumes, the corresponding value is 12.5–13 kg N/t hay.
5. *Nonsymbiotic nitrogen fixation*:
Depending on the soil and climatic conditions, 3–10 kg N/ha is fixed due to nonsymbiotic fixation:

$$Q_{5(\mathrm{N})} = na_{\mathrm{N}},$$

where na_{N} is the nonsymbiotic nitrogen fixation, kg N/ha.
6. *Inputs of N, P_2O_5, and K_2O with atmospheric precipitation*:

$$Q_{6(\mathrm{N})},\ Q_{6(\mathrm{P_2O_5})},\ \text{and}\ Q_{6(\mathrm{K_2O})}.$$

7. *Total inputs of N, P_2O_5, and K_2O*:

$$Q_{\mathrm{N}} = Q_{1(\mathrm{N})} + Q_{2(\mathrm{N})} + Q_{3(\mathrm{N})} + Q_{4(\mathrm{N})} + Q_{5(\mathrm{N})} + Q_{6(\mathrm{N})};$$
$$Q_{\mathrm{P_2O_5}} = Q_{1(\mathrm{P_2O_5})} + Q_{2(\mathrm{P_2O_5})} + Q_{3(\mathrm{P_2O_5})} + Q_{6(\mathrm{P_2O_5})};$$
$$Q_{\mathrm{K_2O}} = Q_{1(\mathrm{K_2O})} + Q_{2(\mathrm{K_2O})} + Q_{3(\mathrm{K_2O})} + Q_{6(\mathrm{K_2O})}.$$

Output items (kg nutrients./ha)

1. *Removal of nutrients with agricultural crop uptake*:

$$V_{1(\mathrm{N})} = U \cdot w_{\mathrm{N}};$$
$$V_{1(\mathrm{P_2O_5})} = U \cdot w_{\mathrm{P_2O_5}};$$
$$V_{1(\mathrm{K_2O})} = U \cdot w_{\mathrm{K_2O}},$$

where U is the crop yield, dt/ha; $w_{\mathrm{N,\,P_2O_5,\,K_2O}}$ denotes the removal of nitrogen, phosphorus, and potassium with crop, kg N, P_2O_5, K_2O/dt crop. If plant analyzes are not available, reference records per unit of harvested products can be assumed after Kayumov (1989).
2. *Removal of nutrients with weeds*.
If the specific data on the biomass of weeds and their contents of nutrients are available, the removal of NPK with weeds can be calculated from the equations

$$V_{2(N)} = U \cdot w_N;$$
$$V_{2(P_2O_5)} = U \cdot w_{P_2O_5};$$
$$V_{2(K_2O)} = U \cdot w_{K_2O},$$

where U is the weed biomass, dt/ha; $w_{N,\,P_2O_5,\,K_2O}$ denotes the removal of nitrogen, phosphorus, and potassium with weed biomass, kg N, P_2O_5, K_2O/dt biomass.

The weight of weeds in the plantations of row crops such as sugar beet, sunflower, and soybean is relatively small because of the intensive interrow tillage, and can be ignored. The weeds should be taken into consideration in the cereal crops calculations. The following statement can be used: the removal of nutrients with by-products should be increased by 50 % for winter crops and by 100 % for spring crops.

3. *Losses of nutrients through leaching (for soils with percolative water regime).* The available data on the loss of nutrients by leaching (e.g., from lyzimetric experiments) make the calculation of this balance item more accurate, otherwise reference records can be used (Lyzimetric research 2004; Postnikov 1999).

$$V_{3(N)},\ V_{3(P_2O_5)},\ V_{3(K_2O)}.$$

4. *Nitrogen losses through denitrification.*
Losses of nitrogen by denitrification are usually 10–20 % of the added fertilizer nitrogen. They make up about 10 % at application rates of 45–60 kg N/ha and increase to 15 % at the higher rates (Kayumov 1989). In addition, it should be taken into consideration that gaseous nitrogen losses can also be due to the soil nitrogen. The mean value of these losses is 6 kg N/ha.
5. *Losses of nutrients through erosion.*
The practical calculations are based on the supposition that the mean contents of nutrients lost from 1 ha of eroded soils $V_{5(N)}$, $V_{5(P_2O_5)}$, and $V_{5(K_2O)}$ vary depending on the degree of soil erosion (Table 1).
6. *Total losses of N, P_2O_5, and K_2O*:

$$V_{(N)} = V_{1(N)} + V_{2(N)} + V_{3(N)} + V_{4(N)} + V_{5(N)};$$
$$V_{(P_2O_5)} = V_{1(P_2O_5)} + V_{2(P_2O_5)} + V_{3(P_2O_5)} + V_{5(P_2O_5)};$$
$$V_{(K_2O)} = V_{1(K_2O)} + V_{2(K_2O)} + V_{3(K_2O)} + V_{5(K_2O)}.$$

Table 1 Mean contents of nutrients lost from eroded soils (kg/ha year)

	Degree of erosion		
	Slight[a]	Medium[a]	Strong[a]
N	18	25	30
P_2O_5	5	7.5	10
K_2O	12	20	24

[a]Degree of erosion according to Classification (1977)

7. *Overall Balances*

$$B_{(\mathrm{N})} = Q_{(\mathrm{N})} - V_{(\mathrm{N})},$$
$$B_{(\mathrm{P_2O_5})} = Q_{(\mathrm{P_2O_5})} - V_{(\mathrm{P_2O_5})},$$
$$B_{(\mathrm{K_2O})} = Q_{(\mathrm{K_2O})} - V_{(\mathrm{K_2O})}.$$

4 Assessment of Nutrient Balance in Russian Agriculture

The decrease in the fertilizer application rates significantly affected the balances of nutrients. The increase in the scope of chemicalization since the mid-1960s led to positive balances of nutrients, except for potassium. The latter is related to the good potassium supply of arable soils in the main natural zones; only 10 % of Russian arable soils have low or very low contents of exchangeable potassium. The inputs of nitrogen and phosphorus exceeded their removal with agricultural crops (Fig. 1). Presently, due to the significant excess of removal over input, the long-term negative balance is 30, 10, and 27 kg/ha for nitrogen, phosphorus, and potassium, respectively, and 60–70 kg NPK/ha annually on average (Table 2). In 2008, the record harvests for the past ten years ensured the highest removal of nutrients; therefore, these data were used in the calculations.

For the regions with high-input agricultural production, the excess of nutrient removal over input is significantly higher, as is evidenced by the data on the Stavropol krai (Table 3) (Stavropolskii TsAS 2012). Analysis of the long-term agrochemical survey data shows some tendencies in the dynamics of arable land areas with the optimum content of phosphates (30 mg/kg), due to the variation in the application level of phosphorus fertilizers to the soil during the 40-year period from 1968 to 2008. Until the early 1990s, the proportion of soils with an optimum content of available phosphorus progressively increased. This period corresponded to the maximum application of phosphorus fertilizers. Then, after the abrupt decrease in fertilization, the area of soils with an optimum phosphorus content decreased gradually. No significant changes were observed in the dynamics of arable lands with an optimum content of exchangeable potassium (300 mg/kg) during the 50-year period; however, a clear tendency toward a progressive, although insignificant, decrease in their area can be traced during the last 20 years. The changes in the content of exchangeable potassium in the arable lands are also due to the significant decrease in the application of potassium fertilizers. In the Belgorod oblast during the period from 2000 to 2004, the input of phosphorus was 2 times lower than its removal from the agrolandscapes; as a result, the weighted average content of its mobile forms in the soil decreased from 131 mg/kg in 1995–1999 to 121 mg/kg (by 7.6 %).

In 2005–2009, the weighted average content of available phosphorus was 116 mg/kg. Thus, a tendency for the available phosphorus supply of arable soils to decrease was revealed in the Belgorod oblast. The results of the agrochemical survey

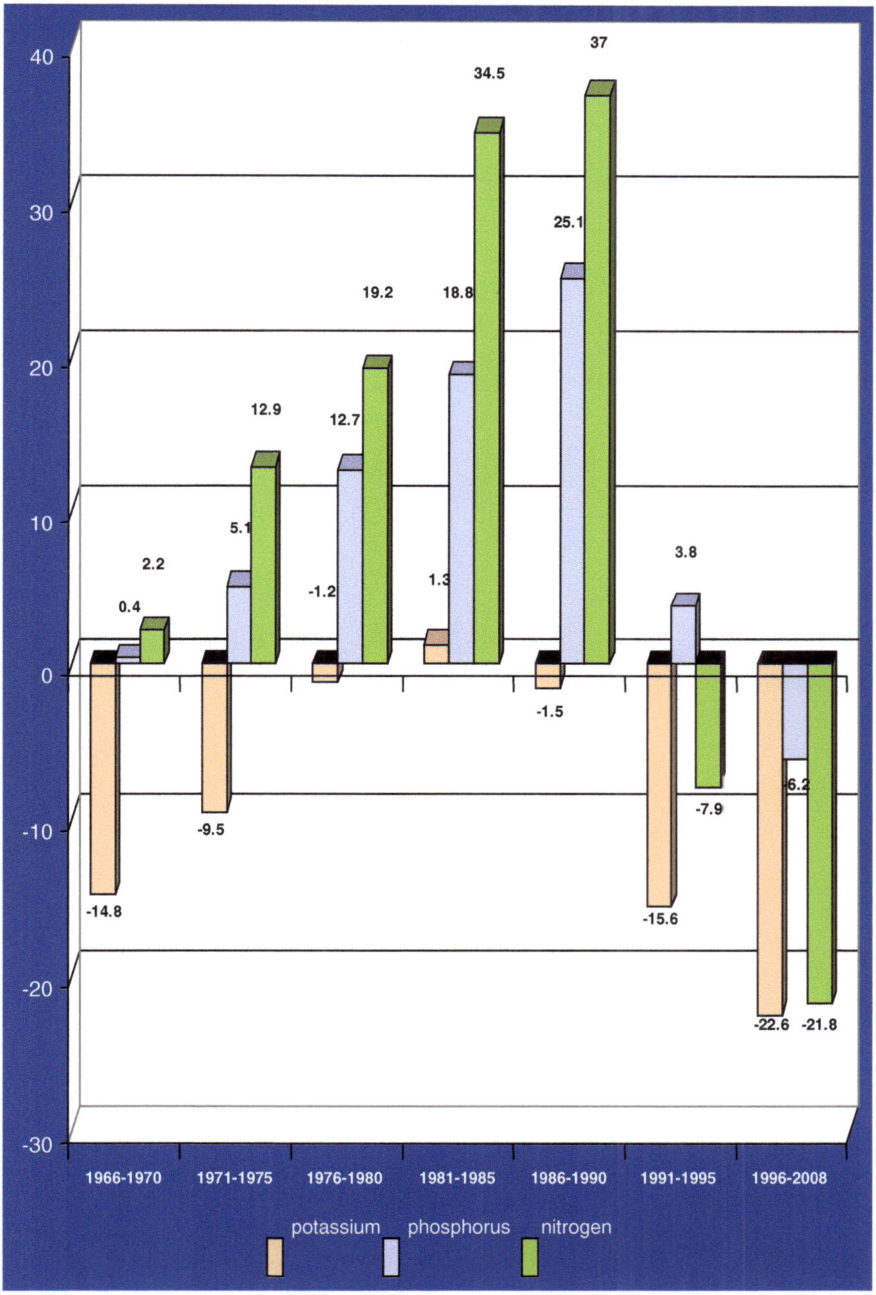

Fig. 1 Balance of nutrients in Russian agriculture (kg/ha cropland)

Table 2 Balance of nutrients in the agriculture of the Russian Federation in 2008

	Total	N	P	K
Added mineral fertilizers, Mt	2.2	1.4	0.5	0.3
Added organic fertilizers, Mt	0.7	0.25	0.13	0.32
Total added fertilizers, Mt	2.9	1.65	0.63	0.62
NPK return with straw, Mt	1.2	0.4	0.15	0.65
NPK return with plant residues, Mt	1.45	0.21	0.007	1.24
Total NPK input with fertilizers, straw, and plant residues, Mt	5.55	2.26	0.78	2.51
Nutrient removal, Mt	10.7	4.57	1.55	4.58
Nutrient balance, Mt	−5.15	−2.31	−0.77	−2.07
Nutrient balance, kg/ha crop area	−67	−30	−10	−27

Table 3 Balance of nutrients in the agriculture of the Stavropol Krai in 2011

Item	Total	Including		
		N	P	K
Added nutrients, 1000 t	204.6	118.0	60.2	26.4
With mineral fertilizers, 1000 t	166.1	103.2	56.9	6.0
With organic fertilizers, 1000 t	7.6	2.8	1.4	3.4
Nutrient return with straw and plant residues, 1000 t	30.9	12.0	1.9	17.0
Removal with crop, 1000 t	698.3	337.5	111.1	249.7
Nutrient balance, 1000 t	−493.7	−219.5	−50.9	−223.3
Nutrient balance, kg/ha crop area	−173	−77	−18	−78

of agricultural lands, local agroecological monitoring, and long-term field experiments of the Geographical Network revealed a slow but permanent decrease in the contents of mobile phosphorus and potassium forms (the rates of these processes vary among the regions, but a negative tendency is universal) (Sychev et al. 2011).

The crop yield does not depend on the absolute amount of fertilizers applied. The nutrients necessary for plants should be available in optimum proportions in order to supply the physiological need of plants at every development stage and ensure their high productivity. The conventional estimate of the optimum ratio of nutrients in fertilizers necessary for Russian agriculture is 1:0.9:0.7. The actual ratio established during the past seven years is 1:0.25:0.2. In agrochemical terms, this is an unfavorable ratio; however, it corresponds to the clearly traced general tendency: a permanent increase in the portion of nitrogen used for fertilizing agricultural crops (Table 4). The above ratio is based on data about the deliveries of fertilizers. The actual ratios of NPK applied for cereal, technical, and forage crops are slightly different: 1:0.24:0.15 for cereals, 1:0.55:0.44 for technical crops, and 1:0.08:0.1 for forage crops.

The use of mineral fertilizers in the Russian Federation varies significantly among the regions; most are used in European Russia, although large areas are

Table 4 Changes in the fertilizer nutrient ratios for the past 30 years

	1970	1980	1990	2000	2010
World	1:0.66:0.51	1:0.53:0.4	1:0.47:0.32	1:0.40:0.27	1:0.38:0.24
Western Europe	1:0.83:0.75	1:0.58:0.55	1:0.43:0.45	1:0.36:0.28	n.d.
Eastern Europe	1:0.68:0.72	1:0.65:0.64	1:0.45:0.45	1:0.28:0.28	n.d.
Russia	1:0.71:0.58	1:0.81:0.64	1:0.84:0.55	1:0.26:0.23	1:0.37:0.23

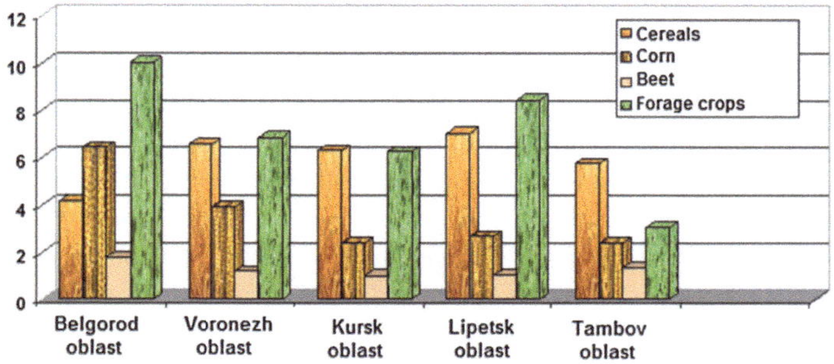

Fig. 2 The P:K_2O ratio in the mineral fertilizers used in the central chernozemic zone in 2012

occupied by cereals in the Asiatic part of the country. In 2012, the application of mineral fertilizers in the Central, Southern, and North-Caucasian federal districts covered 45–55 % of nitrogen requirements, 31–44 % of phosphorus requirements, and 21–40 % of potassium requirements.

The nitrogen/potassium ratio in the mineral fertilizers applied for the main crops in the Central Chernozemic zone in 2012 exceeded the recommended optimum ratios by 3.5–5 times for cereals, 1.4–4 times for grain corn, and by 2–8 times for forage crops; it was only for sugar beet that the increase did not exceed 2 times (Fig. 2).

The same approach applied for Siberia and the Russian Far East shows that the most favorable ratio of nutrients supplied with fertilizers is in the Far East Federal District, while in Siberia, where there is a demand for N and P application on at least 75 and 40 % of plowing lands, respectively, less than 20 % of arable lands receive fertilizers, with a predomination of N at the expense of P and K (Table 5).

Table 5 Nutrient ratios in fertilizers applied in the Asian part of Russia in 2013

Region, Federal District	N	P_2O_5	K_20	% of fertilized arable land
Siberian District	1	0.2	0.06	18
Krasnoyarskiy Krai	1	0.28	0.01	50
The Far East District	1	0.71	0.25	62
Amurskaya Oblast'	1	0.53	0.01	61

5 National Demand for Mineral Fertilizers Based on Balance Method

Fertilization strategies are based on nutrient balances and both the current and projected status of soil fertility (Kundler 1989). The demand of Russian agriculture for mineral fertilizers in the very near future was estimated based on compensation for the removal of essential nutrients with agricultural crops (Table 6) due to soil fertility and the application of 170–200 million tons of organic fertilizers (2 t/ha cropland annually on average) (Prognosis 2011; Sychev et al. 2013). Taking into consideration, the level of soil supply with available phosphorus and potassium established to 2003, the weighted average coefficients of compensation of nutrient removal by fertilizers are 0.87 for phosphorus and 0.63 for potassium (Table 6). According to agronomic studies, the return of nitrogen should make up 100 %. We suppose that given the annual demand for nitrogen is 5.8 Mt, about 0.9 Mt will be returned in the form of biological nitrogen by legumes, if they amount to 15 % of crop rotations and about 0.7 Mt nitrogen is added with organic fertilizers (Prognosis 2011).

The necessary volume of mineral fertilizers for the planned production of agricultural crops will be 8.5 Mt nutrients, including 4.2 Mt nitrogen fertilizers, 1.2 Mt phosphorus fertilizers, and 3.1 Mt potassium fertilizers (Table 7), (Sychev et al. 2013).

As the calculations show, the balance of nutrients in the agriculture (Table 8) ensures the sustainability of soil fertility, because the insignificant phosphorus and potassium deficits will be compensated due to the natural soil recourses, the decomposition and hydrolysis of soil minerals, and the utilization of nutrients from the subsurface layers by plants.

According to our estimates, the use of mineral fertilizers in Russia in the nearest future will remain at a level of 2.5–2.6 million t NPK, which is determined by the low effective demand of agricultural producers for any logistical support. An increase in the use of mineral fertilizers to 3 million t NPK can be expected by

Table 6 Calculation of the weighted average compensation coefficients of nutrient removal

Soil classification by the content of available phosphorus and potassium	P_2O_5		K_2O	
	Area of plowland with different supply (%)	Compensation coefficient	Area of plowland with different supply (%)	Compensation coefficient
Very low	4	1.3	1	1.2
Low	17	1.2	9	1.1
Medium	35	1.0	22	1.0
Increased	21	0.8	27	0.6
High	15	0.3	26	0.5
Very high	8	0	15	0
Weighted average		0.87		0.62

Table 7 Need for mineral fertilizers[a]

Nutrient	Removal, million tons	Weighted average compensation coefficient	Crop need for fertilizers, Mt	Input with organic fertilizers, Mt[b]	Biological nitrogen input, Mt	Needed mineral fertilizers, Mt
N	5.78	1.0	5.78	0.68	0.9	4.2
P_2O_5	1.76	0.87	1.53	0.34	–	1.19
K_2O	6.05	0.62	3.75	0.68	–	3.07
Total	13.59		11.06	1.70	0.9	8.46

[a]According to the methodological guidelines "Determination of the need of agricultural producers for mineral fertilizers," MSKh RF, Moscow, 2003
[b]The predicted value is 170 million t organic fertilizers calculated as litterless manure (2 t/ha); the factual value is 55–60 million t

Table 8 The optimistic scenario of the balance of nutrients in the agriculture

Items	Nutrients			
	N	P_2O_5	K_2O	Total
Removal with crops, Mt	5.78	1.76	6.05	13.59
Input with organic fertilizers, Mt	0.68	0.34	0.68	1.70
Biological nitrogen input, Mt	0.9	–	–	0.9
Input with mineral fertilizers, Mt	4.2	1.19	3.07	8.46
Nutrient input, Mt	5.78	1.53	3.75	11.06
Balance (±), Mt	0	−0.23	−2.30	−2.53
Nutrient removal, kg/ha	68	21	71	160
Mean rates of mineral fertilizers, kg/ha	49	14	36	99
Nutrients added with account for organic fertilizers, kg/ha	57	18	44	119
Overall added nutrients, kg/ha	68	18	44	130
Nutrient balance, kg/ha	0	−3	−27	−30
The N:P_2O_5:K ratio in the removal is 1:0.3:1.05				
The N:P_2O_5:K ratio in fertilizers (with account for biological nitrogen) is 1:0.26:0.65				

2020. The long-term prognoses for the development of the national economy traditionally consider several scenarios for the development of the agro-industrial complex. We developed a multi-scenario prognosis of the need and effective demand for mineral fertilizers. The optimistic innovation scenario allows us to predict the volume of applied mineral fertilizers sufficient for reaching the above parameters (Table 8).

References

Belyaev GN (2005) Potassium fertilizers from potassium salts of Verkghnekamskoe deposit and their efficiency (Калийные удобрения из калийных солей Верхнекамского месторождения и их эффективность). Perm book publishing house, Perm, 304 pp (in Russian)

Budazghapov L-ZV (1989) Effect of different forms of nitrogen fertilizers on cereals yield and speicifics of nitrogen transformation in soil (Влияние форм азотных удобрений на урожай зерновых культур и особенности трансформации азота в почве). Synopsis of PhD (Biol.) thesis, VIUA, Moscow, 26 pp (in Russian)

Chernenok V, Barkusky D (2014) Diagnosis and optimization of phosphorus nutrition conditions of grain crops in Northern Kazakhstan. In: Mueller L, Saparov A, Lischeid G (eds) Novel measurement and assessment tools for monitoring and management of land and water resources in agricultural landscapes of Central Asia. Environmental Science and Engineering 2014, Springer, Berlin, pp 667–679. http://link.springer.com/chapter/10.1007/978-3-319-01017-5_43. Accessed 09 Mar 2015

Classification (1977) Classification and diagnostics of USSR soils (1977) (Классификация и диагностика почв СССР). Kolos, Moscow, 220 pp (in Russian)

Eulenstein F, Tauschke M, Lana M, Sheudshen AK, Dannowski R, Schindler R, Drechsler H (2014) Nutrient balances in agriculture: a basis for the efficiency survey of agricultural groundwater conservation measures. In: Mueller L, Saparov A, Lischeid G (eds) Novel measurement and assessment tools for monitoring and management of land and water resources in agricultural landscapes of Central Asia. Environmental Science and Engineering 2014, Springer, Berlin, pp 263–273. http://link.springer.com/chapter/10.1007/978-3-319-01017-5_15. Accessed 09 Mar 2015

Heffer P, Prud'homme M (2013) Fertilizer outlook 2013–2017 www.fertilizer.org/imis20/images/Library_Downloads/2013_chicago_ifa_summary.pdf?WebsiteKey=411e9724-4bda-422f-abfc-8152ed74f306&=404%3bhttp%3a%2f%2fwww.fertilizer.org%3a80%2fen%2fimages%2fLibrary_Downloads%2f2013_chicago_ifa_summary.pdf. Accessed 09 Mar 2015

Ionas VA, Vil'dflush IR, Kukresh SP (1998) Fertilisation systems of agricultural crops (Система удобрения сельскохозяйственных культур). Urozhai, Minsk, 288 pp (in Russian)

Kayumov MK (1977) Reference book for crop yield programming (Справочник по программированию урожаев). Rosselkhozizdat, Moscow, 188 pp (in Russian)

Kayumov MK (1989) Crop yield programming (Программирование урожаев сельскохозяйственных культур). Agropromizdat, Moscow, 320 pp (in Russian)

Kundler P (1989) Erhöhung der Bodenfruchtbarkeit, 1st edn. VEB Deutscher Landwirtschaftsverlag, Berlin, 452 p

Lyzimetric research (2004) Lyzimetric research in Russia (2004) (Лизиметрические исследования в России). Institute of Central Regions of Non-Chernozem Zone, Moscow, 286 pp (in Russian)

Methodological guidelines (2000) Methodological guidelines on the determination of the budgets of nutrients, humus, and calcium (2000). TsINAO, Moscow, 26 pp (in Russian)

Peterburgsky AV(1979) Turnover and balance of nutrients in agriculture (Круговорот и баланс питательных веществ в земледелии). Nauka, Moscow, 168 pp (in Russian)

Postnikov AV (1999) Lyzimetric research of nutrient migration and balance in soddy-podzolic soil (Лизиметрические исследования миграции и баланса питательных веществ в дерново-подзолистой почве). VNIPTIKhIM Transacions, vol 1. Moscow. pp 77–86 (in Russian)

Prognosis (2011) Prognosis of the need and effective demand of the Russian agriculture for mineral fertilizers to 2020 (Прогноз потребности и платежеспособного спроса сельского хозяйства Российской Федерации на минеральные удобрения до 2020 года). VNIIA, Moscow, 52 pp (in Russian)

Pryanishnikov DN (1937) Chemical industry output and issue of crop rotation on the third fine-year plan (Размеры химической промышленности и вопрос севооборота в третьей пятилетке). USSR Acad Sci Bull (7–8): 33–40 (in Russian)

Reference book of organic fertilizers (1989) (Справочник по органическим удобрениям). Rosagropromizdat, Moscow, 110 pp (in Russian)

Roy FN, Misra RV, Lesschen JP, Smaling EM (2003) Assessment of soil nutrient balance, Approaches and methodologies. FAO fertiliser and plant nutrition bulletin 14, Rome, 87 p, ftp.fao.org/agl/agll/docs/fpnb14.pdf. Accessed 09 Mar 2015

Russia in figures (2012) (Россия в цифрах. 2012). Rosstat, Moscow, 573 pp (in Russian)

Shafran SA (2004) Dynamics of fertiliser application and the soil fertility. Moscow Agrokhimiya (1): 9–17

Smidt GA, Parks A, Pokharel R, Landes F, Schnug E, Koschinsky A (2011) Indications for phosphorus fertiliser derived uranium mobilization from arable soils to groundwater. Berichte der DBG, http://eprints.dbges.de/700/1/Smidt_Berichte_DBG_BERLIN_2011.pdf. Accessed 09 Mar 2015

Song CY, Zhang XY, Liu XB, Sui YY, Li ZL (2010) Impact of long term fertilization on soil water content in Haploborolls. Plant Soil Environ 56(9):408–411

Stavropolskii TsAS (2012) Nutrient balance in agriculture (Баланс питательных веществ в земледелии), www.stavagroland.ru/2012/01/24/balance-of-nutrients-in-agriculture/. Accessed 09 Mar 2015 (in Russian)

Sychev VG, Yefremov EN, Romanenkov VA (2011) Monitoring of agricultural lands and rational use of soil fertility potential: current status and future prospects. Probl Agrochem Ecol (4): 42–46 (In Russian)

Sychev VG, Yefremov EN, Romanenkov VA (2013) Agrochemical science: results and outlook. Probl Agrochem Ecol (4): 11–16 (in Russian)

Chapter 29
Assessing and Controlling Land Use Impacts on Groundwater Quality

Frank Eulenstein, Abdulla Saparov, Sergey Lukin, Askhad K. Sheudshen, Walter H. Mayer, Ralf Dannowski, Marion Tauschke, Olga V. Rukhovich, Marcos Lana, Roland Schindler, Konstantin Pachikin, Hartwig Drechsler and Nils Cremer

This chapter is based largely on the guidelines from the DWA—German association for water, wastewater and waste: "Efficiency of measures to control land use for groundwater conservation—the example of nitrogen" (DWA-M 911 2013). Roland Schindler was the author responsible; Frank Eulenstein, Nils Cremer and Hartwig Drechsler were also authors of this guideline.

F. Eulenstein (✉) · R. Dannowski · M. Tauschke · M. Lana
Leibniz Centre for Agricultural Landscape Research (ZALF), Eberswalder Str. 84, 15374 Müncheberg, Germany
e-mail: feulenstein@zalf.de

R. Dannowski
e-mail: rdannowski@zalf.de; danno@zalf.de

M. Tauschke
e-mail: mtauschke@zalf.de

M. Lana
e-mail: lana@zalf.de

A. Saparov · K. Pachikin
Kazakh Research Institute of Soil Science and Agrochemistry named after U.U. Uspanov, Al Faraby Ave 75b, 050060 Almaty, Kazakhstan
e-mail: ab.saparov@yahoo.com

K. Pachikin
e-mail: kpachikin@yahoo.com

S. Lukin
Institute for Organic Fertilization and Peat (VNIIOU), 601390 Vyatkino 492.2, Sudogotskiy Rayon, Vladimir Oblast, Russia
e-mail: vnion@vtsnet.ru

F. Eulenstein · A.K. Sheudshen
Kuban State Agrarian University, 13 Kalinin Str, 350044 Krasnodar, Russia
e-mail: kubagrohim@mail.ru

L. Mueller et al. (eds.), *Novel Methods for Monitoring and Managing Land and Water Resources in Siberia*, Springer Water,
DOI 10.1007/978-3-319-24409-9_29

Abstract Stewardship of fresh water resources is of paramount importance throughout Europe and for all environmental policies now and in the future. The problem has especially affected areas under predominantly agricultural use. The chapter provides an overview of methods used to assess soil, pore water and groundwater nutrient levels on farms and agricultural fields. We developed a zone monitoring model (ZMM) which is a basis for appropriate monitoring schemes in view of risks for the groundwater coming from agricultural lands. Based on this scheme, we describe various methods to monitor nitrate concentrations at different unit levels, from the farm to the soil zone and on to the groundwater. At farm level, nutrient balances are mandatory to identify the potentially remaining concentrations of nutrients in the soil. Nutrient balances provide an overview of nutrient levels, in particular to prevent surpluses which, as well as contaminating groundwater, could lead to environmental problems such as open water eutrophication, local air pollution and an increase in greenhouse gas emissions. Balances can be performed using operational records of nutrient application and other agronomic information (crops, yields, weather, etc.) at the farm or even field level. A catalogue of agricultural measures for groundwater conservation is available. It is to be supplemented by a methodology for the in situ monitoring of the groundwater quality as a basis for surveying the efficiency of those measures. General characteristics, the benefits and the disadvantages of recent monitoring methods are presented, summarised and rated under the heading "appropriateness for efficiency survey". The methods described here are groundwater sampling by means of a suction lance,

W.H. Mayer
PROGIS Software GmbH, Postgasse 6, 9500 Villach, Austria
e-mail: mayer@progis.com

O.V. Rukhovich
Pryanishnikov All-Russian Research Institute of Agrochemistry,
Pryanishnikova St., 31a, 127550 Moscow, Russia
e-mail: o_ruhovich@mail.ru

R. Schindler
NEW Niederrhein Wasser GmbH, Rektoratstrasse 18, 41747 Viersen, Germany
e-mail: info@niederrheinwasser.de

H. Drechsler
Dr. Drechsler Ingenieurdienst, Hannah-Vogt-Str.1, 37085 Göttingen, Germany
e-mail: post@drechsler-ingenieurdienst.de

N. Cremer
Erftverband, Bergheim, Germany
e-mail: nils.cremer@erftverband.de

F. Eulenstein · R. Schindler · H. Drechsler · N. Cremer
DWA, German Association for Water, Wastewater and Waste, Theodor-Heuss-Allee 17,
53773 Hennef, Germany

A.K. Sheudshen
Russian Scientific Research Institute of Rice, Byelosernoye settlem, 350921 Krasnodar,
Russia

soil sampling beneath the groundwater table, groundwater sampling using the direct-push method, sampling from observation wells, from multi-level observation wells and from production wells. Especially in the shallow, near-surface groundwater, the concentration of dissolved substances almost instantly mirrors the effect of agriculture on the aquatic environment. It can be considered a kind of early-warning system before the surface or drinking water quality are impaired. This survey is preferably focused on nitrate, an important substance for plant nutrition known to behave as a conservative tracer carried below the root zone with the subsurface water. Computer programs have been developed for achieving the locally specific optimal strategy for groundwater-protecting land management strategies and practices. Nutrient balances are included in the latest information and communication technology (ICT) and farm management information systems (FMIS). In Germany, the methods described in this chapter are recommended as a work basis and decision tool for all bodies required to assess the efficiency of agricultural operations in the framework of legal regulations or voluntary cooperation. They can be used by farmers, landscape planners, environmentalists, water associations, water companies, decision makers and others. The whole package of decision trees and monitoring methods in the ZMM, the FMIS and ICT computer programs also has the potential to be tailored and applied to other regions.

1 Introduction

Against the background of the paramount importance of fresh water resources for human well-being in Europe's industrialised societies, the protection of groundwater bodies is one of the main topics of the environmental ambitions of the European Union. This finds prominent expression in the European Water Framework Directive (EU WFD), which entered into force in December, 2000 (EU WFD 2000). It is flanked by the Groundwater Directive (EU GWD 2006), which was subsequently developed in response to the requirements of Article 17 of the EU WFD, formulating the EU policy on groundwater protection. By the end of 2015, the EU WFD aimed to achieve a good quantitative and qualitative state of all European surface water and groundwater bodies.

Abundant fresh, non-salty groundwater resources in particular are among the most valuable environmental goods ensuring life, ecosystem health and prosperity. Therefore, groundwater conservation and protection is a distinct environmental concern. In countries with a higher industrial level, the most prevalent impact on groundwater quantity and quality is exerted by agricultural production. Hence, conversely, agriculture is the most suitable field of action for implementing area-wide, large-scale groundwater conservation measures.

The aim of all the agricultural groundwater-protecting or -conserving measures is essentially—besides the quantitative aspects of preventing overuse—to reduce the amount of harmful substances introduced from agricultural land use into the groundwater, such as nutrients, pesticides, solutes, and other potential pollutants in general. In this context, systematically analysing the groundwater quality is an indispensable means and precondition for observing, inspecting, and evaluating these agricultural measures—in other words, checking their efficiency.

For some decades, the quantitative and qualitative conditions of surface and groundwater bodies have been of exceptional relevance for most of the water-related concepts and measures throughout many countries (Behrendt and Dannowski 2005; Nieder et al. 2007; FAO 2008; SRU 2013). However, the status and trends of water pollution are very different.

In Asian Russia, groundwater pollution by agriculture is currently still much less relevant as compared with Central and Western Europe. Apart from the fact that the level of public sensitivity and perception is different, this is because of different bioclimatic conditions, resulting different flow paths and patterns in the landscape, and the relatively low intensity of agricultural production in terms of fertilisation levels and livestock intensity (see Chap. 1, Mueller et al. 2016). However, the problem of water pollution by agriculture has already increased in European Russia (Lukin and Kozlova 2013) and parts of Central Asia (Saparov 2014; Mueller et al. 2014). In Siberia, climate change may alter the background conditions within the next decades. The potential of lands for agriculture will improve (see Chap. 11, Tchebakova et al. 2016), and subsequently the risk and level of water pollution will increase. It is time to think about monitoring systems and water conservation measures. This chapter is intended to support this cognitive process by presenting the knowledge and technologies available in Europe.

In many parts of its content, this chapter is based on methods that have already been presented in two chapters of another book (Eulenstein et al. 2014; Dannowski et al. 2014). The current chapter includes some updates about tools for optimising nutrient cycles in farming management.

2 Nutrients in Agro-Ecosystems

Agriculture is a multifunctional activity that goes way beyond the mere production of food and fibre. It acts as a source of fuels and raw materials, giving it a cultural value. In addition, agriculture increasingly plays an important role in agro-ecosystem conservation and recuperation. One of the tools that are relevant for the above-mentioned aspects is nutrient balances (Sutton et al. 2013).

Each plant requires nutrients for growing. One of the major nutrients is nitrogen (N), already present in the soil in certain organic or inorganic compounds. To achieve today's typical agricultural income, it is usually necessary to add nitrogen to the soil (fertilisation). The two predominant forms of fertiliser are manure (mostly from animal breeding) and mineral fertiliser. The latter arises from production processes in the chemical industry and its N content varies depending on the composition and type of product.

Nitrogen is of enormous importance as a major plant nutrient and as an ingredient of protein feedstock for animal breeding. The plant/soil system is fed by nitrogen fertilisation, whereas N is removed from the system by the removal of agricultural crops. Periodic replacement, by means of fertilisation, of the nitrogen removed with harvest products is a key prerequisite for high yields and maintaining soil fertility.

However, nutrient inputs by means of fertilisation exceeding the withdrawal by plants will result in significant damage to the environment, in particular by introducing nitrogen compounds and phosphate (Fig. 1). The resulting nitrogen balance (impact) of the operated area of a farm, based on its field of activity, or a region allows an estimate of the gross load potential for nitrogen discharges into waters and the atmosphere, and as well the efficiency of its application.

When nutrients are introduced into the environment in quantities exceeding the agro-ecosystems' carrying (extraction or absorption) capacity, they become polluting agents. Thus, even nitrogen and phosphate, though very important as agricultural nutrients, are commonly associated with pollution. They may contribute to a number of environmental concerns, including water eutrophication, species displacement, air pollution, groundwater contamination and increased greenhouse gas emissions.

Fig. 1 Example of a stable system for milk production with biogas plant for the production of electricity and heat (location near Frankfurt an der Oder, Germany). Surrounding lands and water bodies are subject to increased risk of nutrient pollution. Land management requires mandatory high standards for the protection of water resources

Regulations governing nutrient levels in soils or water are increasingly being used by government regulatory agencies as a tool for compliance with environmental regulations. The requirements for documenting agricultural production processes are currently very high. This applies, for example, to the documentation terms within fertiliser regulations, the EU's Water Framework Directive, or in connection with local environment acts governing drinking water protection zones or nature protection areas. At the same time, the demands placed on traders are rising: in addition to the traded goods, negotiators demand detailed documentation on the methods used for plant production.

3 Monitoring Nutrient Cycles to Protect Water Resources

3.1 The Zone Monitoring Model (ZMM)

The zone model of monitoring of nutrient cycles shown in Fig. 2 consists of different vertical compartments: the soil surface (farm zone) where nutrients are applied, the root zone, the drain zone, and the groundwater zone. It refers to a typical geo-hydrological situation in Europe, but could be modified and refined to suit other geo-hydrological situations as well.

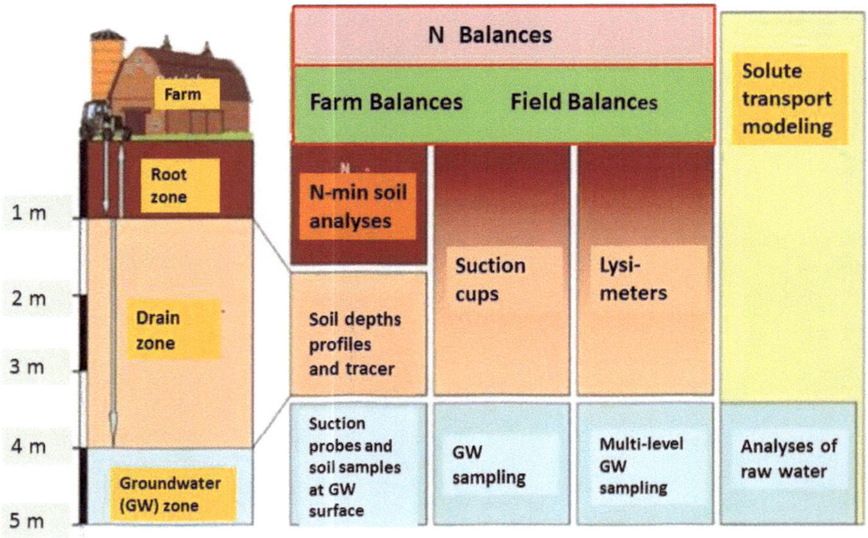

Fig. 2 Zone model of acknowledged methods for estimating agricultural nutrient surpluses for the purpose of groundwater protection, example of nitrogen. It demonstrates different monitoring levels and appropriate tools: the farm zone, the root and drain zone, and the groundwater zone (DWA M-9112013, modified)

Each zone is characterized by proven monitoring methods using analytical measurements and balancing calculations. Solute transport models enable an operator to analyse and predict nutrient flows over all these zones or even considering fluxes in all directions (Natkhin et al. 2013; Merz et al. 2009).

The following sub-sections of this chapter will focus on the farm zone above the soil, evaluating the performance of balances at field and farm level, and on sampling methods in the groundwater zone for monitoring purposes.

Methods applied to monitor the root and drain zone and the topic of solute transport modelling have already been characterized in concomitant papers (Meissner et al. 2014; Schindler 2014) or in previous chapters of this book (Schindler et al. 2016; Nendel 2016; Balykin et al. 2016).

3.2 Nutrient Balances at Farm and Field Levels

3.2.1 Data Requirements

Nutrient balances are acknowledged tools for estimating the nutrient loadings of nitrogen and phosphate in agricultural soils. European Community fertilisation law, for example, requires the preparation of an operational nutrient comparison for nitrogen and phosphate for the preceding agricultural year. This comparison can alternatively be performed by completing an "area balance sheet" or an aggregated field balance, which is then summarised in a long-term nutrient comparison. For this purpose, according to Sect. 6 of the German Directive of Fertilisation (Düngemittelverordnung, DüV 2007), the maximum nutrient surplus of phosphate is 20 kg/ha per year. Since 2011, a maximum surplus level of 60 kg/ha per year has been permitted for nitrogen. These surplus levels are extremely high, exceeding the overall consumption levels of most agricultural regions worldwide. In other regions of the world, however, permissible nutrient surplus levels may differ to those in the EU and its Member States, or are non-existent. Whilst thresholds differ, the same methods can be applied to detect and control levels.

Some direct and indirect methods of measuring, balancing and modelling nutrient cycles and recognising harmful quantities of nutrients for water bodies are acknowledged (Eulenstein et al. 2006, 2008; DWA M-911 2013). This chapter focuses on nutrient balances.

The nutrient balance, structured like an area balance sheet, is performed by calculating all of the nutrient inputs and those quantities that have been removed from the land. However, the accurate estimation of nutrient balances aiming at efficiency control and refined operational nitrogen management can be very complex, particularly in agro-ecosystems involving livestock production.

When the agronomic management of plant production (which is intrinsically associated with a flow of nutrients) is properly recorded and documented, it is easy to perform nutrient balances.

In nutrient balances, nutrient sources and sinks are compared in a system (Fig. 3). This can be carried out by completing operational balance sheets at farm or field level, depending on how detailed the available information is.

In order to prepare nutrient balance sheets, at least three years of farm/field monitoring or assessment are required. This is a mandatory requirement for properly identifying the contributions made by fertilisation, removal of agricultural crops and agricultural management, enabling the impact of weather on nutrient dynamics and balance to be identified. After three years of monitoring, moving averages are used as an indicator of the desired parameters.

Two main approaches can be used regarding nutrient balances: the farm balance and the field balance, both of which have advantages and disadvantages, which will be discussed below. Farm balances are usually easier to conduct since it is easy to gather the information required for the nutrient balance. Field balances, on the other hand, require greater precision in order to identify the quantity of inputs applied to the field, and hence outputs.

Balances at farm and field levels should be based on regional specific data. They are available for many regions. Major elements of the nitrogen balance, such as leaching, were carefully studied in parts of Europe several years ago (see Eulenstein and Drechsler 1992; Mueller et al. 2001, 2005; Eulenstein et al. 2003; Schindler et al. 2008). Thus, water flow pathways and data, and orientation values of nutrients can be used for balancing procedures in specific regions. Those values and coefficients are also available for nutrient balances over most countries including Russia and its Federal objects, and for Central Asian States such as Kazakhstan (Lukin

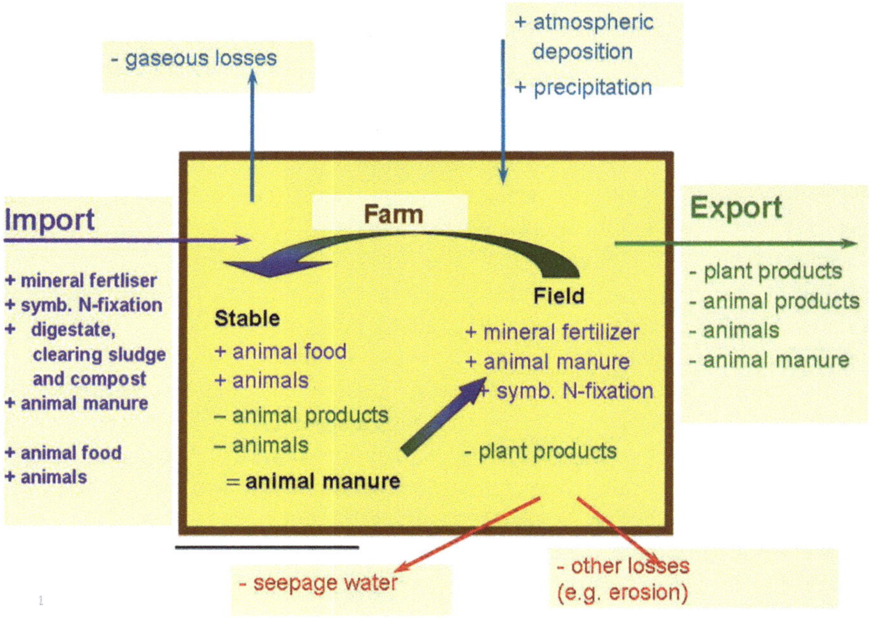

Fig. 3 Elements of nutrient balances

2013; Yefremov et al. 2016, Chap. 28; Saparov 2014), but not in detail for most specific sites.

Nutrient balancing methods can be transferred to other regions although more data is required to quantify balance input parameters there. For example, very different parameter values were found in sodic and saline soils under irrigation conditions (Devkota et al. 2013). In semi-arid climates, even very small nutrient surpluses become harmful agents for water bodies. Thus, carefully established balances in combination with advanced methods of field monitoring, such as lysimeters or soil hydrological set-ups, will provide improved information for ascertaining and controlling nutrient cycles in agro-ecosystems.

Soil hydrological studies including lysimeter experiments (Meissner et al. 2014; Hertel and von Unold 2014; Schindler 2014) and modelling approaches (Nendel 2014; Djanibekov and Sommer 2014) could help to gather more reliable data for typical landscape units of regions where the database is still unreliable.

3.2.2 Farm Balances

With farm balances, the input of nutrients for the whole establishment is compared with the amount of nutrients that leave the establishment from (animal and vegetable) production or residues (of fertilisers, fermenting remains, and so on). Gaseous nitrogen losses from livestock production or fertiliser storage (especially ammoniac) can be described and considered in the nitrogen balance as having been removed from the farm. When adequately described, individual balance sheets are a good representation of the nutrient balance, providing an insight into the nitrogen-disperse potential over the soil, water and air of the entire business. Table 1 and Fig. 2 show typical examples of inputs and outputs of nutrients; Eq. 1 is used to calculate the average annual nutrient surplus in kg/ha for a farm.

$$\text{Nutrient surplus}\left(\sum_{i=1}^{n}\left(\text{imports}_{\text{kg year}^{-1}}\right)-\sum_{i=1}^{n}\left(\text{exports}_{\text{kg year}^{-1}}\right)\right)\Big/\text{Farm area (ha)} \quad (1)$$

Table 1 Calculation of the nutrient balance at farm level

Nutrient imports due to:	Nutrient exports due to:
Mineral fertiliser	Plant products
Animal manure	Animal products
Digestate, clearing sludge and compost	Animal manure, crop residues
N-fixation by legumes	Losses (gas)
Animals	
Feedstock	

Using supporting reference information related to the location of the farm, the type of soil and other agronomic characteristics, the contamination potential can be evaluated and reported to decision-makers, who can take action to avoid contamination, if required.

Advantages of farm balances: The advantage of farm balances is that calculations of input and output factors are simple to perform. The results are also very accurate at farm level, providing an assessment of nutrient surpluses on a farm or field. Comparing different years, management strategies or agro-ecosystem settings enable potential improvements for nutrient use to be highlighted.

In addition, farm balances enable farms' inputs and outputs and their dynamic results to be compared. This feature also leads to the identification of the most successful practices which can possibly be adopted by other establishments.

Disadvantages of farm balances: Nitrogen-surplus results cannot easily be upscaled and attributed to similar regions.

The nitrogen surplus does not always provide absolute values, due to the complex temporal and spatial interactions of the nutrients and environmental factors. The nitrogen surplus cannot be used to determine the application of animal manure for crops, especially because it does not provide any information at the level of accuracy required for crop fertilisation, or recognise within-field spatial variability. For the above-mentioned reasons, nutrient balances at farm level cannot provide the adequate information required for water preservation purposes. In cases where governmental regulations determine average levels per farm, the potentiality of the method is also reduced due a relatively low degree of accuracy. Another shortcoming of nutrient balances is the data quality of inputs and outputs: nutrient concentrations of the manure used for fertilisation, yield and residues are usually estimated, whereas data concerning livestock production are defined arbitrarily. Both field and laboratory records are necessary in order to obtain a well-controlled system, which can increase monitoring costs significantly.

Suitability of farm balances to check the efficiency of groundwater protection: This method is especially suitable for developing a first-level risk assessment of the whole farm, providing information to analyse the nitrogen-surplus risk, enabling the prediction of potential action to be taken in order to mitigate the risk. In this case, the farm balance is to be considered as a component of the farm environment assessment. It is, in the first instance, merely a method for internal farming process control; it is not intended as a substitute for the more detailed inspections usually conducted by agents from regulatory agencies. The results of nutrient balances can be validated by using field samples to compare expected and observed nutrient levels. The same procedure can also help to calibrate the nutrient balance by reporting the initial soil conditions. In combination with the standard sampling procedure, nutrient balances assist with the long-term evaluation of farms or fields. The information yielded supports farm self-management, providing information about nutrient inputs and outputs.

3.2.3 Field Balances of Nitrogen

Nitrogen balances provide information about the management of individual fields. In this case, the amount of nitrogen fertilisers applied and the use of leguminous crops count as the nitrogen input. Additional inputs are also considered, such as organic and inorganic sources of nitrogen. In the case of organic fertilisation, no distinction is made between the various N fractions, i.e. 100 % of the nitrogen content of the source accounts for the balance. Gaseous losses are not considered in this case. For output calculations, all harvested yield or other materials removed must be taken into account. The nitrogen content of each fraction must be known in order to quantify the amount of nitrogen exported. To avoid errors, strict checks on the location and the size of the area are mandatory in this approach and require a Geographical Information System (GIS) to be in place.

Advantages of field balances: Nitrogen balances at field level help appraise the eutrophication risk considering different soils, crops and agronomic management practices. The information generated can be used in agronomic decision-making, leading to the correction of management errors that could otherwise cause inefficient nutrient use or contamination.

Disadvantages of field balances: As with farm balances, the temporal connection between nitrogen application and nitrogen demand is inadequately considered in field balances. As an example, nitrogen sources such as manure, biogas sludge, cover crops or other residues can release their nutrients after the cropping season. In this case, off-season nitrogen mineralisation can occur, even during the winter months, which could particularly affect the groundwater. Again, the input data quality plays a crucial role in the nutrient balance: the nitrogen levels of different nutrient sources—such as animal manure, digestate, clearing sludge and compost—are often assumed based on averages. In the same way, the concentration of nitrogen in harvested products is estimated, leading to probable inaccuracies in the balancing of accounts. In addition, the spatial balance of nitrogen requires records for each field, including inputs, agronomic operations and yield, as well as the production and removal of residues. This implies a great deal of effort for data collection and recording, which may call for extensive human and financial resources.

Suitability of farm balances to check the efficiency of groundwater protection: For the small-scale assessment and consideration of soil and climatic conditions, this balance can provide information which enables models to be run that calculate the nitrate-exportation potential and the potential nitrate concentration in the groundwater. In this case, depending on the quality and scale of application of the models, results can be very accurate, provided that the input data is sufficiently accurate. The validity and accuracy of field balances increases with the length of the balancing period. In order to carry out comparable analyses, it is recommendable to execute field balances over a period of at least one crop rotation (average of three years) in combination with additional farm balances for the same period. Figure 3 shows the schematic presentation of nutrient dynamics in an agro-ecosystem, also describing nutrient imports and exports. In order to appraise

Table 2 Examples of sources and extractors of nutrients for nitrogen balances at field level

Nutrient imports due to:	Nutrient exports due to:
Mineral fertiliser	Plant products
Animal manure	
Digestate, clearing sludge and compost	
N-fixation by legumes	
Quantity of imports	Quantity of exports
Nutrient surplus from field balance = quantity of imports − quantity of exports (kg/year)	

the nitrogen balance of a field, the parameters presented in Table 2 can be used, applying the same equation as that for farm-level assessment.

On the whole, balances can be used as a guide for agricultural farm management. The informative value of these balances basically depends on the accuracy of the input data, reflecting the quality of farm management. Thus, farms that are in a critical situation regarding water management can often only be appraised inadequately by field balances alone. Field balances are not recommended for large-scale analyses (e.g. entire water catchments) due to the large amount of work required for their parameterisation and the potential lack of accuracy involved.

3.2.4 Phosphate Field Balances as an Indicator of Organic Nitrogen Supply

The application of organic fertilisers, such as animal manure, digestate, clearing sludge and compost, harbours the risk of over-fertilising the soil over time. Limiting the quantity of nutrients must not focus on nitrogen alone, but must also involve phosphate. This problem is particularly important for farms where livestock production is performed alongside agriculture. For this reason, in addition to nitrogen, the adequate supply of soils with phosphate, in particular, has to be controlled according to the extraction capacity of this nutrient. When soil levels of phosphate are already sufficient or high, fertilisation should be calculated in order to maintain or reduce these levels. For this reason, soil samples must be taken and analysed prior to applying more fertilisers or other nutrient sources.

Pig manure, pig liquid manure and chicken dung contain high concentrations of phosphate, limiting the amount of these materials that can be applied to fields. For this reason, fertilisation based on one or several organic sources is usually not balanced in terms of phosphate and nitrogen: the level of the first nutrient is usually achieved due to a lack of the second. As a general recommendation, phosphate (in the form of P_2O_5) could be applied with a surplus of 20 kg/ha over a six-year period. For this reason, calculations to balance nitrogen and phosphate are more complex, and will usually include other nutrient sources in order to reach the desired nutrient levels for the field. Phosphate balances can also be performed using the same procedure as for nitrogen balances.

Advantages of phosphate field balances: Phosphate field balances are easy to establish, provided that the nutrient concentrations in the source materials are determined beforehand. For calculations, it can be assumed that 100 % of the phosphate in mineral and organic fertilisers is available. Losses by leaching or gaseous losses are usually not considered due to the relatively slow mobility of this nutrient (compared to nitrogen). Phosphate reference levels in the different exports (yield, residues) are relatively stable.

Disadvantages of phosphate field balances: As with nitrogen balances, phosphate balances also rely on the accuracy of input information to generate assessments with a relatively high degree of accuracy. The phosphate levels of organic fertilisers must also be known prior to application to the field. Unfortunately, the phosphate levels of inputs are usually estimated, causing inaccuracy.

Suitability of phosphate field balances for checking the efficiency of groundwater protection: On the basis of phosphate balance sheets, supplemented with data from soil analyses, farm managers can easily evaluate whether the field has been overcharged with phosphate in previous years or cropping seasons. Exaggeratedly high levels of phosphate in the soil usually indicate that organic fertilisation was used, albeit in amounts that surpassed adequate levels, at least for phosphate. This suggests that the next applications of fertiliser must be calculated in order to meet the demand for nutrients that are below the optimal level—nitrogen, for example—and to avoid over-applying nutrients that are already present in excess. This procedure can help to avoid the pollution and contamination of the soil and groundwater, preventing potential phyto-toxicity problems. Another advantage of phosphate field balances is that fertilisation costs can be rationalised.

3.3 Methods of In Situ Groundwater Quality Monitoring

3.3.1 Monitoring Rules

Monitoring the groundwater quality of an aquifer in the pre-development phase

In principle, any extraction well constructed and operated for drinking water supply or crop irrigation could equally be used for groundwater quality monitoring. However, drawing groundwater from those production wells will provide mixed water of average quality for the related groundwater catchment, without the possibility of further temporal or spatial differentiation. Therefore, regular groundwater sampling at various places within a subsurface catchment will provide more detailed information. In the unconsolidated rock region, which the following refers to, this should, in principle, be practicable at reasonable expense.

Prior to planning a monitoring program, the spatial interrelationships (flow paths) and temporal interrelationships (residence times) within the groundwater catchment ought to be unveiled; the hydrodynamic situation. Thus, in preliminary investigations, the aquifer structure is to be determined (stratigraphy, thicknesses, permeabilities), along with the hydrogeological characteristics (flow directions, flow velocities, depths to the groundwater, spatial extents of the aquifers and where appropriate the aquitards, properties of confining layers, rate of groundwater recharge, seepage dynamics, leakage conditions). Only then can a relationship be established between the implemented groundwater preserving measures and their effect on the groundwater body.

In the case of an allocated water protection area, as a rule, necessary data will exist, since the delineation of the protection zones is based on this information. For groundwater development projects whose water use permits have been granted recently, this data should also have been assembled in the course of the planning procedure.

3.3.2 Groundwater Sampling by Means of a Suction Lance

In regions with depths to the groundwater of less than 6 m, information can be obtained at relatively short notice about the impact of agricultural measures on groundwater quality by taking samples from the near-surface groundwater (Fig. 4a).

A drill hammer with a push rod is used to penetrate the seepage zone, and a lance complete with a suction cup is inserted down to at most 20 cm below the water table (which can be recognised from being saturated with water). This lance is joined to a suction unit via a hose. The application of low pressure causes water to be extracted from the groundwater surface. This procedure is repeated until a filter cake has built up around the suction cup to keep suspended materials out. Now a sample of

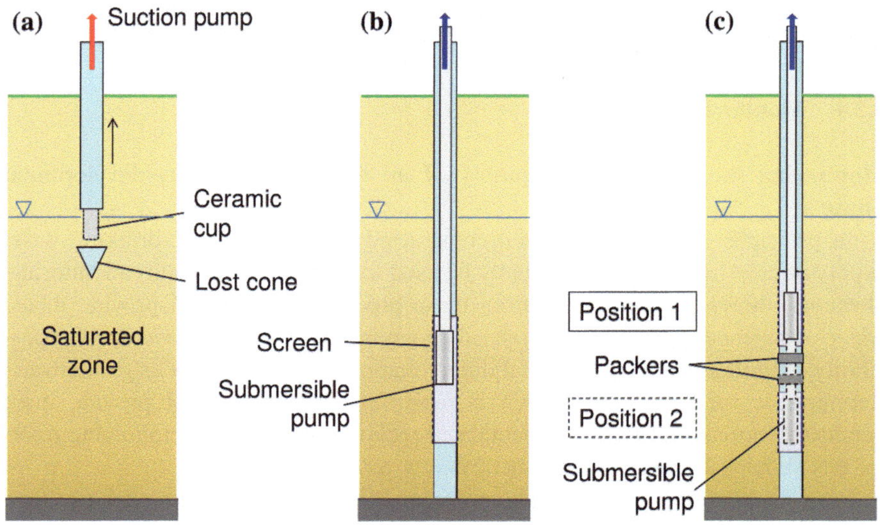

Fig. 4 Scheme of groundwater sampling methods. **a** Suction lance, **b** monitoring well, **c** multi-level sampling

relatively clear water can be abstracted to be analysed in a water laboratory in terms of its main ingredients and some physical-chemical parameters. In the case of peaty soils and sites with a higher content of fine silt, no effectual filter structure may develop outside the cup. Those samples are to be pressure-filtered prior to laboratory analysis.

In the best case, with regard to the probable temporal variability of solute concentrations over the seepage period and the effect of the lateral groundwater movement, sampling beneath the water table should be repeated several times during the period of seepage flow. If this does not happen to be feasible, water should at least be abstracted once a year until the end of the seepage period. Results obtained at that time at selected sampling points are best comparable with current solute inputs from the land surface because of the short transit time. In areas characterised by limited recharge, in the case of water sampling after prolonged periods of drought, and at higher lateral flow velocities in groundwater, the local displacement of solutes by groundwater flow must be taken into account when interpreting the results of the water analyses. Therefore, it is also advisable to choose groundwater sampling points a certain distance downstream of the particular area of survey.

Benefits: This relatively simple procedure does not require any especially high material or personnel expenses. Depending on the depth to the groundwater and the texture of the substrate, up to six samples can be taken within an 8-h day. Analysing the obtained water samples, compared with the analysis of soil samples (Sect. 3.3.3), is much simpler and less time- and cost-consuming. The impact of land use and agricultural protective measures on groundwater quality can be assessed at a relatively early stage.

Disadvantages: In the case of depths to the groundwater of more than 4 m, the procedure takes more and more time; at depths to the groundwater of more than 6 m the technical low pressure will not suffice to abstract enough water for analysis. Occasionally, water must be filtered prior to analysis, which can lead to errors in determining particular substances.

Appropriateness for efficiency survey: By taking water samples from beneath the water table, the quality of the recently constituted groundwater is examined. Thus, the results of water analyses can be related to matter inputs in the vicinity of the sampling point. Provided a reasonably fine sampling grid is established, hot spots of solute inputs can be identified within a groundwater catchment. The load balance of such matter inputs can be calculated from concentration, as long as the rate of groundwater recharge is known. Regular area-wide sampling repeated at an interval of one or two years can be used to assess the efficiency of agricultural measures in groundwater conservation. Compared with supervising based on the groundwater quality from deeper monitoring wells, where the interpretation is complicated by processes of solute displacement, mixing, and transformation in the related groundwater catchment, the described procedure is appropriate in providing results that can be definitely attributed to a particular cause-effect relationship within a clearly delineated sub-catchment.

3.3.3 Soil Sampling Beneath the Groundwater Table

This method based on taking soil samples from the near-saturation zone, as an alternative to the procedure described in Sect. 3.3.2, also enables one to draw conclusions about the effect of agricultural measures on groundwater conservation.

Using a drill hammer with a push rod, the unsaturated zone is penetrated down to the water table. The substrate material is sampled in the notch of a trenched rod. Knocking carefully against the rod makes the saturated zone visible (free water will appear), and the uppermost section of about 15 cm is taken from the saturated substrate. This way, per inspected unit (e.g., per field block), four or five soundings are executed and the substrate material is collected in one homogeneously mixed sample. The latter is analysed for nitrate and chloride in the laboratory after elution considering the water content (in Germany, ATV-DVWK 2004).

Benefits: Relatively simple soil sampling supports area-wide detection of nitrogen inputs at the groundwater table of a whole domain. By means of assembling mixed samples per inspected unit (field block), expenditure on analysis is minimised.

Disadvantages: At present, the application of the procedure is restricted to unconsolidated sediments and fine-grained substrates (silt, sand); coarse gravel material is out of its scope. At depths to the groundwater of more than 6 m, taking soil samples proves to become more and more time- and cost-consuming. In the case of a longer seepage period combined with higher recharge rates, the mean solute concentration of the recently constituted groundwater will tend to be under-estimated as a consequence of dilution.

Appropriateness for efficiency survey: Because of the fact that water eluted from the soil samples originates from a certain range of depths (approx. 15 cm), results from chemical analyses represent the nitrate input into the groundwater averaged over a period of up to several months. Compared with N_{min} soil analyses from beneath the root zone, they are less subject to short-term fluctuations caused by specific agricultural measures or weather events. Collecting mixed samples allows for the measurement-based detection of area-related nitrogen inputs into the groundwater. Just like taking water samples from the near-surface groundwater (Sect. 3.3.2), usage of this procedure indicates the effects of protective agricultural measures on the groundwater quality at an early stage.

3.3.4 Groundwater Sampling Using the Direct-Push Method

In areas sparsely equipped with monitoring wells, in recent years, groundwater soundings have been progressively executed by means of the direct-push method (EPA 2005; Hannappel and Braun 2010). This procedure allows groundwater samples to be gathered directly in the field and at specific depths. Likewise, an area-specific monitoring of the groundwater quality is feasible, even in the case of a deeper water table.

Direct-push groundwater sampling is performed by hammer-driven ramming a probe head connected to a liner rod down to the groundwater zone. In complex terrain, a small-sized drilling unit is adopted mounted on an off-road vehicle. After the intended depth is reached, the liner rod is lifted a few decimetres, pulling free a filter cone (screen) contained in the liner. This screen, as a rule, is 10–100 cm long and may consist of various materials (e.g., high-grade steel, PVC/HDPE, ceramics). Alternatively, an open filtering system can also be inserted. Afterwards, the height of the water table can be measured inside the liner by means of a light plummet. Finally, the groundwater sample is withdrawn via the liner rod by a pump system appropriate for the depth.

In the case of depth-specific groundwater sampling, the whole system is subsequently raised to the next position, and the sampling operation is executed as depicted before.

Benefits: As compared with conventional procedures, direct-push groundwater sampling allows the geologic and hydrochemical conditions underground to be surveyed faster and spatially more flexibly, combined with diversified local logging. Using lightweight automotive equipment makes sampling points accessible even on heavy terrain and supports fast relocation. Because of the groundwater observation points do not need permanent installation, no additional construction costs are incurred.

Disadvantages: As no regular groundwater observation wells are installed, the direct-push method only generates survey points for one-time usage. Repeated groundwater monitoring using the same installation is not possible.

Hammer-driven ramming for direct-push groundwater sampling is not feasible everywhere. Difficulties will occur depending on the conditions underground, especially for larger sampling depths and very deep water tables (>20 m), as well as in the case of stony or strongly cohesive sediments. To estimate the depths approachable for sounding and sampling, previous knowledge is needed from well-informed and skilled staff who are familiar with the hydrogeological situation. Under certain circumstances, preliminary investigations may be required. If a filter cake does not build up around the filter cone, groundwater samples may need to be filtered in place.

Appropriateness for efficiency survey: Since the direct-push sampling method offers considerable mobility, it is especially appropriate for application at hydro-geologically well-perceived areas which nonetheless have an inadequate density of observation wells, and require extra area- or usage-specific investigations into the groundwater quality. This is of concern, e.g., for groundwater bodies under the scope of the EU WFD, in the status of 'at risk', i.e. where it is actually unclear whether or unlikely that the water quality objectives will be achieved (EU WFD 2000). The procedure enables the investigator to specify the groundwater quality on a plot-by-plot basis. Knowledge obtained this way may provide the important completion of an existing groundwater-monitoring programme. In particular, potential locations of new points of inspection can be derived and measures to improve groundwater quality can be planned more selectively. By means of the depth-specific sampling provided by the direct-push method, the source and fate of pollutants can be traced in a more reasonable way. Further on, processes of matter (nitrate) transformation in the groundwater zone can be identified.

3.3.5 Groundwater Sampling from Observation Wells

In the catchments of groundwater abstraction wells for the public water supply, as a rule, observation wells are deliberately placed in the vicinity of the production wells (Fig. 4b). Based on those water level observations, or on pumping tests in the course of the pre-development phase of the well group, the shared groundwater catchment should have been delineated, the hydrogeological parameters of the aquifer should have been determined, and supervision of the hydrograph should be further ensured on a permanent basis. Such observation wells should also have been occasionally installed for the large-scale assignment of the subterranean catchment of a river or a canal. In addition, suppliers of potable water or water management authorities erect special observation networks to monitor the groundwater quality development, e.g. upstream of a well field or downstream of any polluters (Fig. 5).

These categories of observation wells, though installed for different purposes, all are suitable for investigations into the efficiency of agricultural groundwater-protecting measures provided they meet some requirements with regard to the site conditions and the level of their performance. In particular, this is important for assessing nitrogen inputs.

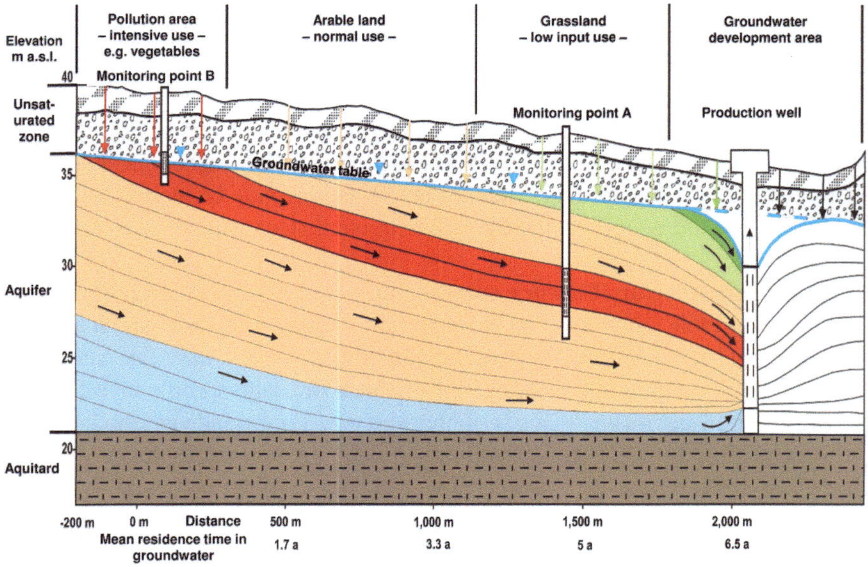

Fig. 5 Groundwater quality monitoring by means of observation wells in the upstream zone of a drinking water production well

These requirements are:

- Filter lining (screen) must be placed in the upper groundwater zone
- Short filter lines
- The related catchment area must be given a unique attribution

To interpret the measurement results, requirements regarding the construction and design of observation wells and monitoring networks must be taken into consideration, following the technical rules (in Germany, DVGW W 121 and DVGW W 108).

By systematic inspection of the results from observation wells, the effects of groundwater-protecting measures can be detected in the medium to long term. For this, the contributing subsurface catchment is to be attributed to the observation well according to the depth of the filter screen. This may be complicated in certain cases, and sometimes the catchment boundary may vary with time. To increase the trustworthiness of the efficiency assessment, if possible, several observation points should be used (Fig. 5).

Fingerprint parameters of the groundwater age can make sense when interpreting the analyses. In many cases, the identification of so-called environmental tracers (^{2}H, ^{3}H, ^{3}He, ^{18}O, ^{85}Kr, CFC, SF_6) has proven to be useful for groundwater age determination. To examine the origin of nitrates and to explore the quantitative schemes of their decay, radioisotope-hydrologic measurements (^{15}N–NO_3, ^{18}O–NO_3, ^{34}S–SO_4, ^{18}O–SO_4) are also appropriate (Schulenberg et al. 1990; Stadtwerke Viersen GmbH 1999). Recently, these methods have increasingly been applied when examining the efficiency of agricultural measures to mitigate groundwater pollution (Osenbrück et al. 2000; Schöpel 2000).

Groundwater quality should be monitored at selected observation wells over a longer period at adequate, regular intervals, paying attention to the groundwater flow velocity or residence time. Groundwater samples should be taken with moderate pumping power to minimise any water table drawdown and, thus, mixing with water from the adjacent (higher and lower) aquifer zones, which is indispensable in the case of regularly screened observation wells (in Germany, following DWA-A 909). Additionally, in interpreting the measuring results to narrow down the activity period of a source, the transit time across the seepage and groundwater zones upstream of an observation well is to be taken into consideration (Fig. 5).

One particular implementation of an observation well is the so-called multi-level well; its function and purpose are explained in Sect. 3.3.6.

Benefits: At waterwork catchment areas, as a rule, a groundwater observation network is found for which long-term records of groundwater quality already exist. Optimising this kind of network for the purposes of an efficiency survey allows the quality status of the upper groundwater layer to be captured at acceptable expense.

Disadvantages: The groundwater quality in the range of a well screen is dependent on a multitude of interacting factors, such as the spatial and temporal patterns of groundwater recharge, the flow velocity and solute transport across the aquifer, including dispersion, matter decay and transformation. Thus, the spatial and temporal attribution of groundwater quality data to the mostly non-point, distributed ('diffuse') sources will be relatively erroneous and complicated.

Appropriateness for efficiency survey: Groundwater observation wells are principally suitable to evaluate the efficiency and impact of agricultural groundwater protection measures. To be used for relatively short-term information on groundwater quality, however, they should be compatible with certain conditions. Their filter construction (screen) should not be placed too deep below the mean water level, and they must not be biased by larger-scale effects from the catchment. To assess the long-term groundwater quality development it is favourable to include regional main (background) parameters in the monitoring concept. Additionally, the residence time of the sampled groundwater should be estimated. Due to the uncertainties described above in the spatial and temporal allocation of matter inputs, if possible, the results should be re-evaluated and supplemented by independent detection methods.

Observation wells screened in the deeper range of the upper aquifer or in deeper groundwater layers do not provide reliable information about the impact of agricultural measures and are not suitable for the purpose of an efficiency survey.

3.3.6 Groundwater Sampling from Multi-level Observation Wells

Multi-level observation wells (Fig. 4c) are especially appropriate for the hydrochemical survey of thick aquifers or a system of aquifers hydraulically separated from each other by non- or semi-permeable confining layers. In general, they allow for the vertical differentiation of the hydrochemical conditions within one or between several groundwater-bearing units. In the context of the efficiency survey they are suitable for analysing the dynamics of the vertical matter exchange even in the near-surface groundwater zone. In recent times, multi-level observation wells have been often supplemented by probes located in the unsaturated zone at various depths.

In the case of the multi-level observation well "Ruhr University System", several minifilters are taken (the number depends on the permeability and the thickness of the aquifer and the specific intention of the groundwater hydrologist) and are fixed to a guide tube consisting of plastic (50 mm in diameter) by plastic clamps at arbitrary intervals. Each of the minifilters has its own riser tube (14 mm in diameter) leading through to the measuring equipment placed at the soil surface or in a well house. The minifilters are simultaneously pumped off by means of a rotary pump. This minimises vertical mixing between different aquifer zones and guarantees the intended depth-specific water sampling, provided the permeabilities of the individual layers induce an approximately horizontal flow towards the minifilters. For water sampling from probes >7 m below the surface and thus not accessible by suction pumps, so-called injectors can be mounted next to the minifilters. For water level measuring in a selected aquifer zone, a 1 m long filter tube is placed at the required depth within the 50 mm plastic guide tube.

While taking groundwater samples, the pumped discharge has to be adjusted to the water yield of the investigated aquifer zone. To prevent the hydraulic equilibrium over the whole depth of the aquifer from being disturbed too much during pumping, a certain water volume (ruled in DWA-A 909) drawn up before the actual sampling should not be exceeded.

As stated above with direct push, especially in the case of multi-level observation wells, it is indispensable, previous to drilling, to recognise the hydrogeological background within the groundwater catchment to be analysed. An important task is to determine the groundwater flow direction. Therefore, besides the extensive and costly multi-level equipment, a sufficient number of regular observation wells must also be installed within the catchment area.

The results from measuring the groundwater quality on various dates are assembled in a concentration depth profile. Provided the hydrogeological conditions within a given groundwater catchment are reasonably homogeneous, the depth-specific groundwater samples obtained from the multi-level wells can be allocated to their respective areas of origin, under consideration of the rate of groundwater recharge, the effective flow velocity, and the specific yield (actual porosity). In addition, deeper aquifer zones can be evaluated with respect to their denitrification potential, taking into account the obligatory regional background parameters such as NO_3^-, Cl^-, SO_4^{2-}, HCO_3^- or O_2. Additionally, isotope-hydrological methods can be used.

Benefits: The operation of multi-level observation wells provides vertical higher-resolution information about the hydro-chemical conditions of a pore aquifer (or even a system of aquifers and aquitards). Combined with the collection and analysis of additional hydrological, hydrogeological, hydro-chemical, and land use data, various impacts on groundwater quality can also be detected, as well as processes of matter transformation within the groundwater zone.

Disadvantages: Both the construction and operation of multi-level observation wells are very extensive and cost-consuming in comparison with regular groundwater observation wells. To avoid imperfections in groundwater sampling (evoking hydraulic short-circuits), only skilled personnel should be commissioned.

Appropriateness for efficiency survey: Multi-level observation wells provide an important tool to evaluate the efficiency of agricultural groundwater protection measures, as well as being an instrument for the long-term prognosis of the hydro-chemical conditions in an existing drinking water protection zone. They are suitable for comprehensively analysing and monitoring the hydro-chemical processes characterising the pore aquifer. Combined with compatible observation equipment installed in the overlying unsaturated zone, the effects of agricultural groundwater-conserving measures can be studied in their temporal and spatial development. This is of exceptional importance for groundwater catchments where the concentration of nitrate in the upper groundwater layer is approaching or exceeding the limit, or where concerns have been raised that the denitrification potential could decline and be exhausted within a relatively short period of time. Prior to establishing a multi-level observation well, however, sufficient information must be available on the hydrogeological conditions of the groundwater catchment to be able to localise the appropriate place.

3.3.7 Quality Monitoring of Raw Water from Production Wells

The effects of agricultural measures on diminishing nitrate loads are often quantified from time series of the untreated groundwater quality at drinking water works, especially as this method offers a specified set of target variables which can easily be determined. Interpreting the time series of raw water quality data, the following conditions ought to be reflected (Fig. 5):

- Water abstraction exhibits a markedly accelerating intervention into the natural groundwater dynamics of a catchment.
- In most of the water works, water abstraction takes place via several production wells. Their water quality will differ in general, based on the related sub-catchments characterised by differing groundwater quality parameters.
- Often the production wells have filter constructions located at different depths of an aquifer or even in different groundwater units.
- In larger-sized water works, the untreated groundwater from different well groups is mixed prior to treatment.

Therefore, when drawing conclusions about the effectiveness of agricultural measures for groundwater conservation from the evolution of single species concentrations in untreated groundwater, it is mandatory that the water originate at least from one particular groundwater catchment with stable ratios of water from different aquifers/depths throughout the monitoring period.

As a rule, however, it is preferable to monitor the raw water quality of selected production wells with definitely separated groundwater catchments. But again it must kept in mind that, in deeper wells, water is mixed from different recharge periods with different, non-natural residence times. Thus, the short-term development (less than 5 years) of matter concentrations in raw water is hardly to be spatially or temporally apportioned to specific agricultural measures. This is true in particular for spacious groundwater catchments with high aquifer thicknesses.

Benefits: Long-term records of untreated groundwater quality are already kept by suppliers of potable water, as regulated by law. These time series of critical substances, such as nitrates, pesticides or sulphates, are suitable for inexpensively assessing the effect of agricultural groundwater protection measures.

Disadvantages: Analysing untreated groundwater samples from production wells does not allow for the intended hydraulically undisturbed, highly differentiated spatial or temporal attribution of observed water quality to agricultural measures of groundwater conservation.

Appropriateness for efficiency survey: The evolution of substance concentrations in raw water from selected production wells can be used to evaluate efficiency under certain circumstances provided the substances are distinctly attributable to a particular groundwater catchment, and agricultural water protection measures have been practiced over a relatively long period of time. The duration of the monitoring period should be about one third of the average time of groundwater renewal (the quotient of the extractable groundwater volume and the annual recharge rate) in the developed aquifer, plus the mean transit time between the soil surface and the groundwater table. This method, however, will provide only rough information about larger-scale measures that cannot be locally and temporally differentiated.

Drinking water quality monitoring alone, as a rule, is not appropriate for efficiency assessment. Even long-term records are almost impossible to interpret, as they are usually dominated by the operational impacts of water extraction and groundwater treatment.

4 Advanced Tools for Optimising Farm Nutrient Cycles

As shown above, the requirements regarding reliable data are becoming more and more important for monitoring and controlling land use impacts on water quality. Data are multi-source and should be attributed to spatio-temporal coordinates. Along with basic requirements on specific procedures, such as nutrient balancing (see also Fernall 2010), agricultural processes and nutrient fluxes have to be optimised. On a global scale, cereal production could be raised by ~30 % at the same nitrogen application level but spatially redistributed and optimised (Mueller et al. 2014).

Specifically for the local and regional scale, this kind of optimisation is a promising realm for decision support systems (DSS, Djanibekov and Sommer 2014; Mirschel et al. 2016) and information and communication technology (ICT, PROGIS 2015) in agriculture.

Nutrient balances as a part of ICT are suitable to support the circular flow of natural elements and economical values within agriculture (PROGIS 2015).

Besides nutrient balance (farm and field balance) purposes, the latest **farm management information systems (FMIS)** with integrated GIS, including standard growth models for all crops, are also able to manage farms and fields with respect to other needs, such as logistics, precision farming, subsidies and more. They integrate cost calculation and possibly also CO_2 balancing, or even incorporate insurance or bank needs. Measures of precision farming help to save fertilisers, thus reducing nutrient surpluses.

Data requirements for these control measures are so high and complex (Figs. 6 and 7) that it is not reasonable to expect all farms to collect the full set of data on their own. This means basic data such as (1) the costs of different machines or information on (2) all fertilisers with nutrient contents and all organic fertilisers with their average or, better, precisely measured nutrient contents or (3) legally available pesticides including chemically active substances and the rules for their use or (4) the seeds and their varieties. Instead, it makes sense for a small number of experts within a region to start setting up a database that contains all these basic data, maintain it sustainably and make it accessible to interested farmers. Applying

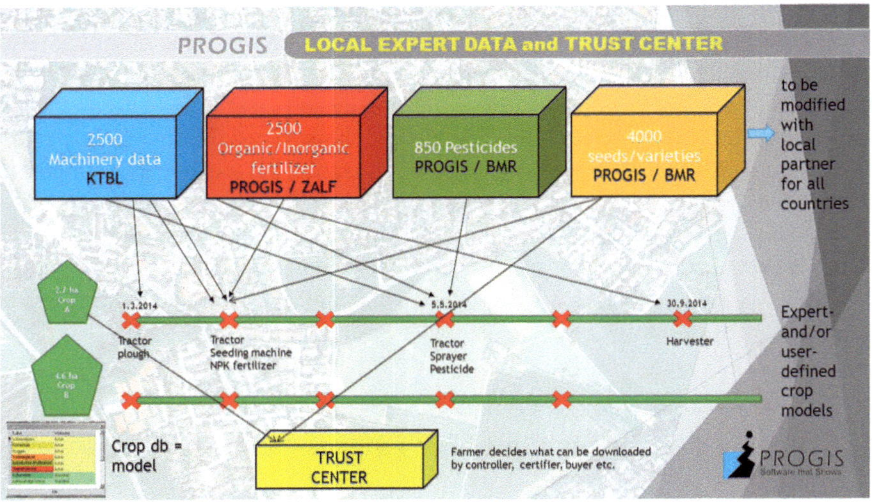

Fig. 6 Example of a database for optimising farmland management at field scale. Multi-source data need to be combined. Establishing and updating these complex databases requires cooperation between scientific institutions, public bodies and software companies (figure provided by PROGIS)

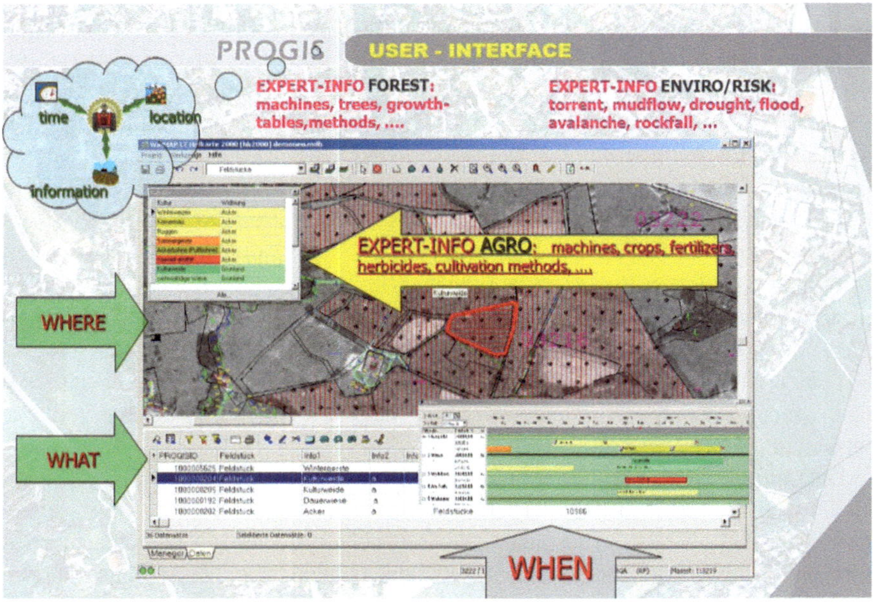

Fig. 7 Screen shot of the GIS map related to the example of Fig. 6. Expert data for setting up crop-related activity plans and a GIS tool allow for easy agricultural management planning (figure provided by PROGIS)

an appropriate growth model, this allows users to plan a crop on a specific field—the GIS provides the hectares—with a single click, with all needed data laid underneath. Everything is calculated with the standard data as a basis for the whole operated area of the farm, in terms of budget and for N, P, K, Ca and Mg also in kg/ha. All data are time related.

Once the plan data is available, farmers have to enter the real data, which might differ from the plan. This is done case by case as work progresses, manually or automatically. The system recalculates the targets including differences between the plan and reality for the fields or the farm as a whole. In the case of aggregation for a region or a country, this can also be organised within a trust centre that has to guarantee each farmer the exclusive ownership of information and that his data can be used only for specifically defined purposes.

For Germany, the expert database was created as a cooperation project of the Bundesmaschinenring (BMR), the Leibniz Centre for Agricultural Landscape Research (ZALF), both in Germany, and the ICT company PROGIS, Austria.

After this step to a field- and farm-oriented calculation of nutrient balances has been taken, we can go forward to optimise the system towards a higher level. **Agricultural Circular Flow Modelling (ACFM)** means evaluating nutrient contents in the soil with the latest laboratory technologies and starting a balance model from those precise monitoring data. This method is also used for nutrient contents in liquid manure and for livestock feed to work out precise feed models. To avoid financial loss and the occurrence of sludge gas, farmers equipped with a pH sensor can reduce the feed if the pH exceeds 7.0. This is helpful in meeting both economic and ecologic targets. Public or private organisations can support farmers and guide them to integrate feed, manure and soil nutrient data into a circular flow model. A 5-year long-term study in Luxemburg involving 250 farms has shown that based on feed, the best quarter could reach up to 30 % less feed costs with comparable performance.

Precision Farming (PF) is the next step towards minimising adverse environmental impacts. Of course, it will need more than only PF equipment, such as a fertiliser sprayer able to spray to m^2 precision or get a map from a satellite or a yield-map from the last harvest result to know in detail where to do what. Soils are heterogeneous and have different N supplies. They need a nutrient analysis from soil samples at least once every five years. An expert team can work for years with these basic data and other data gathered, preparing a fertiliser recommendation map that is precise and provides a further step to optimise the use of fertiliser.

Many more innovations are in progress in this field of operational control on the use of land and water resources. As an example, a PF farming map is shown (Fig. 8) that can be sent via an office-based logistics system to mobile equipment (e.g. tractors with tablet PCs or smart-phones) to trigger sprayers with m^2 precision according to the map. Additional meteorology sensors, as far as possible equipped with soil-moisture sensors at different depths, will not only allow meteorological monitoring but, together with expert information, can also guide the use of pesticides or even automatically trigger 'intelligent' irrigation equipment based on real-time data.

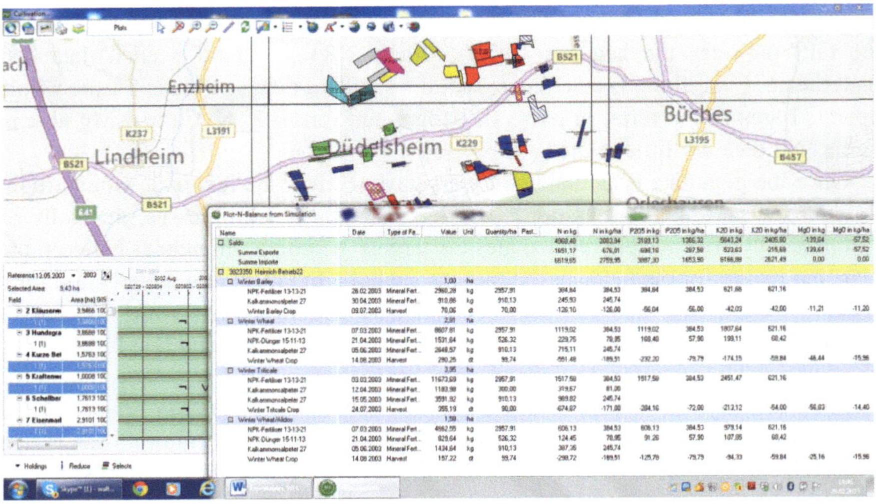

Fig. 8 Screen shot of a nutrient balance sheet and a GIS map of related fields in the background (figure of the PROGIS company)

For smallholder farms there is an additional need for ICT-equipped public or private advisors—one might serve up to 200 farmers, each with 25 ha = 5000 ha or more, depending on the land use structure—with access to high-end expert knowledge supporting them in complex cases. Machine cooperatives providing the shared use of agricultural equipment and also supporting additional social services could enable the use of the latest technologies even for smallholder farmers and optimise the use of fertilisers and pesticides, thereby guaranteeing economic and ecologic benefits for them and also integrating the idea of groundwater protection.

5 Conclusions

- On the basis of nutrient balance sheets, supplemented with data from soil analyses, farm managers can easily evaluate whether the field has been overcharged with nitrogen or phosphate in previous years or cropping seasons. The information generated can be used in agronomic decision-making, leading to the correction of management errors that could otherwise cause inefficient nutrient use or soil and water pollution and contamination. However, field balances are not recommended for large-scale analyses (e.g. entire water catchments) due to the large amount of work required for their parameterisation and the potential lack of accuracy involved.
- There are several methods of in situ groundwater quality monitoring suitable for the checking the efficiency of agricultural groundwater protection measures. Besides their general characteristics, the benefits and disadvantages of each

method were discussed and rated here under the summarised term "appropriateness for efficiency survey".

- The basics are outlined in the recently released German Technical Guideline DWA-M 911 (2013), "Efficiency of measures to control land use for groundwater conservation—the example of nitrogen", which was adopted for this contribution.
- It can be stated that there is no individual method that would manage all the practical problems at low cost and at the same place and time. Instead, a thorough selection, or even better a combination of the various methods, seems to be required and worthwhile depending on (for example) the agricultural, pedological, geological, climatic and hydrological site conditions.
- Some of the methods provide a relatively simple short-term indication of the local efficiency of specific agricultural groundwater protection measures. Others are more suitable for revealing and understanding the system behaviour of an entire complex of groundwater units in spacious catchments under anthropogenic impact.
- An efficiency survey should be carried out as close as possible to the actual agricultural area under reference to preclude incorrect or vague deductions. This will also raise the farmers' acceptance for conservation measures, since they are immediately informed and can understand the effect of their efforts on the groundwater resources under concern.
- The latest FMIS are also able to manage farms and fields with respect to other needs, such as logistics, precision farming, subsidies and more. They integrate cost calculation, possibly also CO_2 balancing, or even incorporate insurance or bank needs. Measures of precision farming help to save fertilisers, thus reducing nutrient surpluses.
- Their aim is to optimise both the farm economy and the ecology through land and water conservation measures. This is already under way on some projects and is the topic of ongoing work.

References

ATV-DVWK (2004) Weiterentwicklung einer Methode zur Messung des Stickstoffaustrages landwirtschaftlich genutzter Flächen – Teile 1 und 2. ATV-DVWK-Themen. ATV-DVWK-Arbeitsgruppe GB-6.2 „Effizienzkontrolle". Deutsche Vereinigung für Wasserwirtschaft, Abwasser und Abfall e. V. (ATV-DVWK), Hennef 177 p

Balykin D, Puzanov A, Meissner R, Stephan E (2016) Using the innovative lysimetric technology in the Russian-German Project KULUNDA. In: L. Mueller et al. (eds) Novel methods for monitoring and managing land and water resources in Siberia. Springer, Cham, pp 387–399 (Chapter 16 of this book)

Behrendt H, Dannowski R (eds) (2005) Nutrients and heavy metals in the odra river system—emissions from point and diffuse sources, their loads, and scenario calculations on possible changes. Weißensee Verlag, Berlin, 353 pp. ISBN 3-89998-046-8

Dannowski R, Schindler R, Cremer N, Eulenstein F (2014) Methods of in situ groundwater quality monitoring: basis for the efficiency survey of agricultural groundwater conservation measures.

In: Novel measurement and assessment tools for monitoring and management of land and water resources in agricultural landscapes of Central Asia. Environmental Science and Engineering 2014, pp 275–288. doi:10.1007/978-3-319-01017-5_16, http://link.springer.com/chapter/10.1007%2F978-3-319-01017-5_16. Accessed on 5 Feb 2015

Devkota KP, Manschadi A, Lamers JPA, Devkota M, Vlek PLG (2013) Mineral nitrogen dynamics in irrigated rice–wheat system under different irrigation and establishment methods and residue levels in arid drylands of Central Asia. Eur J Agron 47:65–76

Djanibekov N, Sommer R (2014) Integrated Decision Support for Sustainable and Profitable Land Management in the Lowlands of Central Asia. In: Novel Measurement and Assessment Tools for Monitoring and Management of Land and Water Resources in Agricultural Landscapes of Central Asia. Environmental Science and Engineering 2014, pp 407–422. doi:10.1007/978-3-319-01017-5_24, http://link.springer.com/chapter/10.1007/978-3-319-01017-5_24. Accessed on 5 Feb 2015

DüV (2007) Düngeverordnung: Verordnung über die Anwendung von Düngemitteln, Bodenhilfsstoffen, Kultursubstraten und Pflanzenhilfsmitteln nach den Grundsätzen der guten fachlichen Praxis beim Düngen in der Fassung der Bekanntmachung vom 27. Februar 2007, BGBl. I S. 221. Stand: zuletzt geändert durch Artikel 18 des Gesetzes vom 31. Juli 2009, BGBl. I S. 2585

DWA M-911 (2013) Möglichkeiten der Effizienzkontrolle von Maßnahmen zur Grundwasser schonenden Bodennutzung am Beispiel des Stickstoffs Deutsche Vereinigung für Wasserwirtschaft, Abwasser und Abfall e. V., Hennef, 33 pp

EPA (2005) Groundwater sampling and monitoring with direct push technologies. U.S. Environmental Protection Agency, Office of Solid Waste and Emergency Response, Washington DC, OSWER No. 9200.1-51, EPA 540/R-04/005, August 2005, 78 pp. http://www.clu-in.org/download/char/540r04005.pdf. Accessed on 5 Feb 2015

EU WFD (2000) Directive 2000/60/EC of the European Parliament and of the Council establishing a framework for the Community action in the field of water policy, OJ L 327, 22 Dec 2000, pp 1–73. http://eur-lex.europa.eu/LexUriServ/LexUriServ.do?uri=CONSLEG:2000L0060:20090625:EN:PDF. Accessed on 5 Feb 2015

EU GWD (2006) Directive 2006/118/EC of the European Parliament and of the Council of 12 December 2006 on protection of groundwater against pollution and deterioration, OJ L 372, 27 Dec 2006, pp 19–31. http://eur-ex.europa.eu/LexUriServ/LexUriServ.do?uri=OJ:L:2006:372:0019:0031:EN:PDF. Accessed on 5 Feb 2015

Eulenstein F, Drechsler H (1992) Ursachen, Differenzierung und Steuerung der Nitratkonzentration im Grundwasser überwiegend agrarisch genutzter Wassereinzugsgebiete. Institut für Bodenwissenschaft der Georg-August-Universität Göttingen. Eigenverlag, Göttingen, 269 pp

Eulenstein F, Schindler U, Müller L (2003) Feldskalige Variabilität von Frühjahrsfeuchte und mineralischem Stickstoff in der ungesättigten Zone sandiger Standorte Nordostdeutschlands. Arch Agron Soil Sci 49(2):187–201

Eulenstein F, Olejnik J, Willms M, Kersebaum KC (2006) Bestimmung tolerierbarer Stickstoff-Überschüsse in Agrarlandschaften mit Hilfe von Simulationsmodellen. In: Land- und Ernährungswirtschaft im Wandel – Aufgaben und Herausforderungen für die Agrar- und Umweltinformatik. Referate der 26. GIL Jahrestagung 2006, pp 57–60. Gesellschaft für Informatik, Bonn

Eulenstein F, Werner A, Willms M, Juszczak R, Schlindwein SL, Chojnicki BH, Olejnik J (2008) Model based scenario studies to optimize the regional nitrogen balance and reduce leaching of nitrate and sulfate of an agriculturally used water catchment. Nutr Cycl Agroecosyst 82(1):33–49

Eulenstein F, Tauschke M, Lana M, Sheudshen AK, Dannowski R, Schindler R, Drechsler H (2014) Nutrient balances in agriculture: a basis for the efficiency survey of agricultural groundwater conservation measures. In: Novel measurement and assessment tools for monitoring and management of land and water resources in agricultural landscapes of Central Asia. Environmental Science and Engineering 2014, pp 263–273. doi:10.1007/978-3-319-01017-5_15, http://link.springer.com/chapter/10.1007%2F978-3-319-01017-5_15. Accessed on 5 Feb 2015

FAO (2008) Current world fertilizer trends and outlook to 2011/12. Food and Agriculture Organization of the United Nations, Rome, 2008, 57 pp. ftp.fao.org/agl/agll/docs/cwfto11.pdf. Accessed on 5 Feb 2015

Fernall D (2010) Soil nutrient balances: reviewing and developing to meet customer needs. http://www.oecd.org/greengrowth/sustainable-agriculture/44793491.pdf. Accessed on 5 Feb 2015

Hannappel S, Braun P (2010) Planung und Durchführung eines Grundwasser-Monitorings mit Direct-Push in Nordbrandenburg. In: Grundwasser Bd. 15, H. 3, 191–201

Hertel C, von Unold G (2014) Third-Generation lysimeters: scientific engineered monitoring systems. In: Novel measurement and assessment tools for monitoring and management of land and water resources in agricultural landscapes of Central Asia. Environmental Science and Engineering 2014, pp 175–184. doi:10.1007/978-3-319-01017-5, http://link.springer.com/chapter/10.1007/978-3-319-01017-5_9. Accessed on 5 Feb 2015

Lukin SM (2013) The importance of organic fertilisation in the agronomic cycle and balance of biogenic elements (Роль органических удобрений в круговороте и балансе биогенных элементов в земледелии). In: Development and introduction of modern agricultural technologies and management systems (Развитие и внедрение современных технологий и систем ведения сельского хозяйства). Materials of the scientific-practical conference at Perm, NIISK 2013, vol 1, part 2, pp 17–24

Lukin SM, Kozlova MP (2013) Mitigation of environmental pollution by chemically active nitrogen—findings of a nitrogen expert group of countries from East Europe, the Caucasus region and Central Asia (Снижение загрязнения окружающей среды химически активным азотом: результаты работы экспертной группы по азоту стран Восточной Европы, Кавказа, Центральной Азии). Papers of the international scientific-practical conference dedicated to the 10th anniversary of the Sudogotsky long-term agricultural field experiment. Vladimir GNU VNIIOU 2023, vol 2, pp 191–198

Meissner R, Rupp H, Seyfarth M (2014) Advanced technologies in lysimetry. In: Novel measurement and assessment tools for monitoring and management of land and water resources in agricultural landscapes of Central Asia. Environmental Science and Engineering 2014, pp 159–173. doi:10.1007/978-3-319-01017-5_8, http://link.springer.com/chapter/10.1007/978-3-319-01017-5_8. Accessed on 5 Feb 2015

Merz C, Steidl J, Dannowski R (2009) Parameterization and regionalization of redox based denitrification for GIS-embedded nitrate transport modeling in Pleistocene aquifer systems. Environ Geol 58(7):1587–1599

Mirschel W, Wenkel K-O, Berg M, Wieland R, Nendel C, Köstner B, T"opazh AG, Terleev VV, Badenko VL (2016) A spatial model-based decision support system for evaluating agricultural landscapes under the aspect of climate change. In: L. Mueller et al. (eds) Novel methods for monitoring and managing land and water resources in Siberia. Springer, Cham, pp 519–540 (Chapter 23 of this book)

Mueller L, Augustin J, Eulenstein F, Seeger J, Meissner R, Behrendt A, Schindler U (2001) Einfluss unterschiedlicher Böden auf die Ausnutzung von Wasser und Stickstoff durch Weidelgras, Sommergerste und Mais auf einigen Grundwasserstandorten. Archiv für Acker- und Pflanzenbau und Bodenkunde 47:277–292

Mueller L, Wirth S, Augustin J, Tauschke M, Schindler U, Behrendt A, Eulenstein F, Meißner R (2005) Verlagerung von Agrochemikalien und Enzymen durch Dränsysteme (Lysimeterexperimente). - Mitteilungen der Deutschen Bodenkundlichen Gesellschaft (106): 45–46

Mueller L, Suleimenov M, Karimov A, Qadir M, Saparov A, Balgabayev N, Helming K, Lischeid G (2014) Land and water resources of Central Asia, their utilisation and ecological status. In: Novel measurement and assessment tools for monitoring and management of land and water resources in agricultural landscapes of Central Asia. Environmental Science and Engineering 2014, pp 3–59. doi:10.1007/978-3-319-01017-5_1, http://link.springer.com/chapter/10.1007%2F978-3-319-01017-5_1. Accessed on 5 Feb 2015

Mueller L, Sheudshen AK, Syso A, Barsukov P, Smolentseva EN, Khodzer V, Sychev VG, Romanenkov VA, Rukhovich OV, Puzanov A, Dannowski R, Schindler U, Eulenstein F

(2016) Land and water resources of Siberia, their functioning and ecological state. In: L. Mueller et al. (eds) Novel methods for monitoring and managing land and water resources in Siberia. Springer, Cham, pp 3–73 (Chapter 1 of this book)

Mueller ND, West PC, Gerber JS, MacDonald GK, Polasky S, Foley JA (2014) A tradeoff frontier for global nitrogen use and cereal production. Environ Res Lett 9:054002, (8 pp). doi:10.1088/1748-9326/9/5/054002

Natkhin M, Dannowski R, Dietrich O, Steidl J, Lischeid G (2013) Model-based impact analysis of climate and land use changes on the landscape water balance. In: Novel measurement and assessment tools for monitoring and management of land and water resources in agricultural landscapes of Central Asia. Environmental Science and Engineering 2014, pp 577–590. doi:10.1007/978-3-319-01017-5_36, http://link.springer.com/chapter/10.1007/978-3-319-01017-5_36. Accessed on Feb 5, 2015

Nendel C (2014) MONICA: a simulation model for nitrogen and carbon dynamics in agro-ecosystems. In: Novel measurement and assessment tools for monitoring and management of land and water resources in agricultural landscapes of Central Asia. Environmental Science and Engineering 2014, pp 389–405. doi:10.1007/978-3-319-01017-5_23 http://link.springer.com/chapter/10.1007/978-3-319-01017-5_23. Accessed on 5 Feb 2015

Nendel C (2016) Simulating temperature impacts on crop production using MONICA. In: L. Mueller et al. (eds) Novel methods for monitoring and managing land and water resources in Siberia. Springer, Cham, pp 503–518 (Chapter 22 of this book)

Nieder R, Köster WR, Kersebaum K-C (2007) Beitrag der Landwirtschaft zu diffusen N-Einträgen. Wasserwirtschaft 1–2:53–57

Osenbrück K, Heinrichs G, Göppel M, Eichinger L (2000) Nitratsanierung von Grundwassererschließungen: Konzept und Zielprognose durch isotopenhydrologische Untersuchungen. In: bbr 3/2000, pp 20–25

PROGIS (2015) PROGIS Software GmbH, www.progis.com. Accessed on 5 Feb 2015

Saparov A (2014) Soil resources of the Republic of Kazakhstan: current status, problems and solutions. In: Novel measurement and assessment tools for monitoring and management of land and water resources in agricultural landscapes of Central Asia. Environmental Science and Engineering 2014, pp 61–73. doi:10.1007/978-3-319-01017-5, http://link.springer.com/chapter/10.1007/978-3-319-01017-5_2. Accessed on 5 Feb 2015

Schindler U (2014) A field method for quantifying deep seepage and solute leaching. In: Novel measurement and assessment tools for monitoring and management of land and water resources in agricultural landscapes of Central Asia. Environmental Science and Engineering 2014, pp 185–198. doi:10.1007/978-3-319-01017-5_10, http://link.springer.com/chapter/10.1007/978-3-319-01017-5_10. Accessed on 5 Feb 2015

Schindler U, Müller L, Eulenstein F, Dannowski R (2008) A long-term hydrological soil study on the effects of soil and land use on deep seepage dynamics in Northeast Germany. Arch Agron Soil Sci 54(5):451–463

Schindler U, Mueller L, von Unold G, Durner W., Fank J (2016) Emerging measurement methods for soil hydrological studies. In: L. Mueller et al. (eds) Novel methods for monitoring and managing land and water resources in Siberia. Springer, Cham, pp 345–363 (Chapter 14 of this book)

Schöpel M (2000) Ergebnisse des Kooperativen Gewässerschutzes am Beispiel der Kooperation Holstershausen/Üfter Mark. In: Ber. IWW, Bd. 29, Mühlheim a. d. Ruhr

Schulenberg S, Schindler R, Rauert W, Wolf M, Weise S. Voerkelius S (1990) Hydrogeologische, hydrochemische und isotopenhydrologische Untersuchungen in den Grundwassereinzugsgebieten der Stadtwerke Viersen GmbH. Stadtwerke Viersen, 45 pp

SRU (2013) Wissenschaftliche Beiräte für Agrarpolitik (WBA) und für Düngungsfragen (WBD) beim Bundesministerium für Ernährung, Landwirtschaft und Verbraucherschutz (BMELV). Kurzstellungnahme: Novellierung der Düngeverordnung: Nährstoffüberschüsse wirksam begrenzen. Bundesministerium für Ernährung, Landwirtschaft und Verbraucherschutz (2013), 22 pp. http://www.umweltrat.de/SharedDocs/Downloads/DE/04_Stellungnahmen/2012_2016/2013_08_AS_Novellierung_Duengeverordnung.html. Accessed on 5 Feb 2015

Stadtwerke Viersen (1999) Hydrochemische und isotopenhydrologische Untersuchungen in den Grundwassereinzugsgebieten der Stadtwerke Viersen GmbH. Abschlussbericht, 189 pp

Sutton MA, Bleeker A, Howard CM, Bekunda M, Grizzetti B, de Vries W, van Grinsven HJM, Abrol YP, Adhya TK, Billen G, Davidson EA, Datta A, Diaz R, Erisman JW, Liu XJ, Oenema O, Palm C, Raghuram N, Reis S, Scholz RW, Sims T, Westhoek H, Zhang FS, with contributions from Ayyappan S, Bouwman AF, Bustamante M, Fowler D, Galloway JN, Gavito ME, Garnier J, Greenwood S, Hellums DT, Holland M, Hoysall C, Jaramillo VJ, Klimont Z, Ometto JP, Pathak H, Plocq Fichelet V, Powlson D, Ramakrishna K, Roy A, Sanders K, Sharma C, Singh B, Singh U, Yan XY, Zhang Y (2013) Our nutrient world: the challenge to produce more food and energy with less pollution. Global overview of nutrient management. centre for ecology and hydrology, Edinburgh on behalf of the global partnership on nutrient management and the international nitrogen initiative. http://www.initrogen.org/sites/default/files/documents/files/ONW.pdf. Accessed on 5 Feb 2015

Tchebakova NM, Chuprova VV, Parfenova EI, Soja AJ, Lysanova GI (2016) Evaluating the agroclimatic potential of Central Siberia. In: L. Mueller et al. (eds) Novel methods for monitoring and managing land and water resources in Siberia. Springer, Cham, pp 287–305 (Chapter 11 of this book)

Yefremov EN, Sychev VG, Romanenkov VA (2016) Balance of nutrients and the optimization of their use in agroecosystems of the Russian Federation. In: L. Mueller et al. (eds) Novel methods for monitoring and managing land and water resources in Siberia. Springer, Cham, pp 619–633 (Chapter 28 of this book)

Chapter 30
Principles of Conservation Agriculture in Continental Steppe Regions

Mekhlis Suleimenov, Zheksenbai Kaskarbayev, Kanat Akshalov and Aibek Tulegenov

Abstract In western Siberia, summer fallow-based crop rotation is practiced. These tillage-based cropping systems are not sustainable. They lead to a decline in soil fertility and damage to humans and the environment. The objective of this chapter is to analyze the principles of conservation agriculture (CA) and to replace wheat-fallow monocultures by stubble mulch farming. We conducted multi-factorial field experiments on crop rotations, tillage, and soil fertility management on three sites over more than five years. The research data include the results of studies in northern Kazakhstan, central Kazakhstan, and western Siberia. Furthermore, we analyzed results from other cool steppe regions such as Canada. In studies conducted in northern Kazakhstan, we found that cropland is most efficiently used in diversified crop rotations with no fallow. Summer fallow can be replaced by food legumes or legume forages. No-till has an advantage in terms of crop yields over traditional tillage on light textured soils of the Kostanai province, thanks to better moisture conservation. On heavy textured soils of the Akmola province, traditional tillage has an advantage in some cases, thanks to better snowmelt water intake and more active nitrogen mineralization. On Leached Chernozems of Trans-Ural Siberia, no-till is feasible only with the application of higher rates of nitrogen fertilizer. In the forest-steppe zone of western Siberia, no-till in the autumn provided the same grain yields as ploughing only when it was combined with the application of fertilizers, herbicides, fungicides, and growth regulators. Research in northern Kazakhstan

M. Suleimenov (✉) · Z. Kaskarbayev · K. Akshalov · A. Tulegenov
Scientific Production Center of Grain Farming named after A.I. Barayev, Shortandy 021601, Kazakhstan
e-mail: mekhlis@yahoo.com

Z. Kaskarbayev
e-mail: tsenter-zerna@mail.ru

K. Akshalov
e-mail: kanatakshalov@mail.ru

L. Mueller et al. (eds.), *Novel Methods for Monitoring and Managing Land and Water Resources in Siberia*, Springer Water,
DOI 10.1007/978-3-319-24409-9_30

shows that for soil fertility conservation, one should combine a reduced to minimum area under summer fallow with the replacement of summer fallow by pulses or legume forages and the application of nitrogen fertilizer, and thus avoid soil tillage.

Keywords Conservation agriculture • Soil tillage • No-till • Kazakhstan • Siberia

1 Objectives

After 42 million hectares of grasslands in northern Kazakhstan and western Siberia were ploughed back in 1954–1956, these lands were initially used for continuous spring wheat production. This was similar to practices in Canadian prairies of the early 1950s, where monocultures of spring wheat rotated with summer fallow were common (Hill 1954). For Canadian farmers, the advantage of this cropping system was avoiding crop failure in dry years. In 1956, when Barayev (1960) visited Canada, he returned with the idea that frequent summer fallows (with no crop for one year) was the only way to farm in dryland conditions, while stubble mulch farming was the best method to control wind erosion.

This is why the crop rotation study established at Shortandy in 1961 included wheat-fallow rotations, with fallow occupying 50, 33, and 25 % of the rotation area. There was also continuous wheat treatment with no fertilizer applied. A radical change in the methodology of the crop rotation study was made in 1984, when continuous wheat plots were started to be given necessary fertilizers. Since then, grain production from the total area including fallow has been at its highest in continuous wheat (Suleimenov 1988).

Advocates of wheat-fallow rotations emphasize the following advantages of summer fallow: better moisture accumulation, better weed control, and better accumulation of nitrates. According to our data, summer fallow accumulates 15–20 mm more moisture compared to stubble fields (Suleimenov and Akshalov 2007). As compared to controlling weeds efficiently in summer fallow, cereal cropping regions in Europe have no more fallow at all, and thus farmers control weeds efficiently. As to nitrate accumulation, this is true but this practice results in a more rapid loss of organic nitrogen. This was the main reason for stopping wheat-fallow rotation in western Canada (Gan et al. 2003; Larney et al. 2004).

In perennial grass-based cropping systems, which were widely practiced in the Soviet Union before 1954, the main tillage equipment was the moldboard plough. However, after testing Canadian tillage equipment it became obvious that conservation tillage using blades and sweeps was more suitable for steppe regions. The major advantage of conservation tillage was that it left standing stubble in the field to trap snow and accumulate more moisture, and protect the soil against wind erosion. Various tillage depths were identified for different soils, and in some cases no tillage was recommended in the autumn (Barayev 1984).

Meanwhile, in the USA, Canada, Australia, and in some countries of Latin America the idea of no-till was widely adopted (Derpsch 2008; Kassam et al. 2011). This is the idea of stopping any tillage except minimum soil disturbance when sowing seeds, leaving all residues on the soil surface, crop rotations, cover crops, and the application of fertilizers based on soil tests. Using no-till systems, the yield can be stabilized and soils can be protected from wind erosion in an effective way (Meinel et al. 2014). No-till systems and other soil fertility protecting measures, such as the elimination of fallow and providing crop rotations instead of monocultures, are part of Conservation Agriculture (CA).

CA is defined as "a production system in which crop, soil, nutrient, pest, water and energy management components and operations are based on a sustainable ecological foundation provided by three interlinked principles of: (1) minimum soil disturbance (no-till direct seeding); (2) maintenance of soil cover (mulch cover from crop residues and cover crops); and (3) diversification (rotation and/or association) of crops, including cover crops" (Kassam et al. 2011). Conservation agricultural systems create an ecologically protective interface between the soil profile and the atmosphere. This protects the soil surface from natural physical forces that can cause degradation by wind, water, and traffic, and allows soils to function to their highest potential (Franzlubbers 2009).

Current global studies show that expectations about possible crop yield are increasing (Pittelkow et al. 2014) or the climate mitigation (Powlson et al. 2014) thanks to no-till is limited, in humid regions in particular. A global meta-analysis study showed that no-till reduces yields by about 5–10 % in humid regions (Pittelkow et al. 2014). However, this yield deficit was lowest in combination with crop residue retention (mulch) and crop rotation. In dry climates, a significant increase in crop productivity of about 5–10 % was observed, though only if the cropping system comprised all three principles of CA (Pittelkow et al. 2014).

Those systems and principles have been tested in field trials in comparison with traditional systems. Since, the beginning of this century scientists in northern and central Kazakhstan and western Siberia have conducted no-till studies. The objective of this chapter is to provide an overview of the methodology of research on CA and to present some results of this research.

2 Materials and Methods

Our studies have been conducted in the continental steppe regions of Northern Kazakhstan and West Siberia. Northern Kazakhstan is a territory between 50° and 54°N latitude and between 60° and 78°E longitude. It covers an area of 57 million hectares, and comprises 4 provinces: Akmola, Kostanai, Pavlodar, and North Kazakhstan. The locations of the sites referred to in this chapter are as follows: Shortandy—51.7°N and 71°E, Kostanai Research Institute of Agriculture (RIA)—53.1°N and 63.8°E, Kokshetau RIA—53.2°N and 69.2°E, Karaganda RIA—49.8° N and 72.8°E. The Western Siberia area is located adjacent to and north of Northern

Kazakhstan and includes the provinces of Omsk, Altai, Novosibirsk, Kurgan, and Kemerovo. The dryland area is located between 50–55°N latitudes and 62–86°E longitudes. The locations of the sites referred to in this chapter are as follows: Omsk —58.6°N and 73.2°E, Kurgan—55.3°N and 65.2°E.

At Shortandy, CA has been studied in a number of multifactorial trials. The soil is a heavy clay loam Southern Calcareous Chernozem with an organic matter content of about 3.5 % in its arable layer. The long-term average annual precipitation is 322 mm. The distribution of precipitation is characterized by monthly falls of 20–25 mm throughout the autumn, winter, and spring and more rainfall in June (40 mm), July (54 mm), and August (35 mm). Snowfall makes up one-third of total annual precipitation and plays an important role in soil moisture accumulation. The annual average air temperature is 2.1 °C, with the highest temperatures in July (20.1 °C). This type of weather is typical for all the northern part of Kazakhstan and adjacent areas of western Siberia. In all parts of northern Kazakhstan, the application of phosphorus fertilizers at a rate of 15–20 kg ha^{-1} along with the seeds is the generally adopted practice, while nitrogen is applied based on soil analyses (Chernenok and Barkusky 2014).

At the Kostanai site, the soil is a Chernozem, sandy loam, with an organic matter content of 4.7 %, the average annual precipitation is 320 mm, and the average annual temperature is 2.5 °C. At the Karaganda site the soil is a Kastanozem, heavy clay loam, with an organic matter content of 2.5–2.8 %, the annual precipitation rate is 280 mm, and the temperature is 2.8 °C. At the Kurgan RIA site, the soil is a Leached Chernozem, clay loam, with an organic matter content of 4–5.2 % and a topsoil pH of 5.0–5.4, the annual precipitation is 350 mm, and the temperature is 1.8 °C. At the Omsk site of Siberian RIA, the soil is a Leached Chernozem, heavy clay loam, with an organic matter content of 6–7 %, the annual precipitation rate is 340 mm, and the temperature is 1.8 °C.

Our goal is not only to justify continuous wheat but to look for the best crop rotations with no fallow. For this purpose, back in the late 1990s, new crop rotation studies started at Shortandy. The traditional crop rotation of "fallow-wheat-wheat-barley-wheat" was taken as a control variant (Suleimenov et al. 2014). The new crop rotations were established by replacing summer fallow by crops: oats, dry pea, or rapeseed. These crop rotations were compared with three cropping technologies: simple, traditional, and no-till. Simple technology means that no intensification was used. Under traditional technology, tillage in the autumn and seedbed preparation was used, as well as snow ridging, fertilizer application, and chemicals to control weeds. Under no-till, direct seeding with tine was used as well as fertilizer being applied and chemicals to control weeds being sprayed (Fig. 1). In this chapter, only the yields of two crops of the rotations are presented.

In the second trial at Shortandy, three tillage methods in the autumn (deep, shallow and no-till) were studied in a 4-course rotation "3 times wheat-fallow" with a background of 4 tillage treatments during the summer fallow period: traditional (tillage to control weeds), minimum-d (one deep tillage + chemicals), minimum-s (one shallow tillage + chemicals), and chemical fallow (only herbicides to control weeds).

Fig. 1 No-till wheat

In the third trial, three factors were tested: (1) tillage, (2) sowing time, (3) wheat varieties (Suleimenov et al. 2014). The tillage treatments were traditional, minimum, and zero. The sowing dates were 19 May, 27 May, and 2 June. The wheat varieties were 3 varieties of bread wheat: Astana 2, Tselina 50, and Tselinnaya 2007, as well as 2 varieties of durum wheat: Korona and Damsinskaya yantarnaya. The total number of treatments was 45, and there were 135 plots. Plot size: width—4 m, length—50 m.

At Karaganda, no-till and traditional tillage has been applied in a five-year rotation "summer fallow—4 times wheat," since 2001 (Yushchenko 2012).

At Kurgan RIA a trial was conducted in a "fallow-wheat-wheat" rotation (Gilev et al. 2011). Two tillage treatments were tested: moldboard plough at a depth of 20–22 cm and no-till. In the no-till treatment, weeds were controlled by glyphosate at a rate of 2 litres ha^{-1} and sowing was direct, using tine sowing machines. Two fertilization treatments were tested: no fertilizer and 60 kg of N ha^{-1}.

At Omsk a trial was conducted in a 5-course crop rotation "fallow-wheat-maize-wheat-barley," which was established in 1972 (Kholmov and Shulyakov 2006). Four treatments of tillage in the autumn were tested: plough, mixed, blade, and zero. Moldboard plough was used at 20–22 cm depth. Mixed tillage treatment included plough in the fallow year and used for maize, while blades at 10–12 cm were used for grains. Three treatments of chemical application were tested: no, herbicides + fertilizer and complex. Fertilization included 60 kg ha^{-1} of P_2O_5 in fallow, 60 kg ha^{-1} of N and P_2O_5 for wheat, and 45 kg ha^{-1} of N and P_2O_5 for barley. Complex application of chemicals included fertilizers, herbicides, fungicides, and plant-growth regulators.

Soil moisture was determined by drying off soil samples at temperature 105 °C. Assessment of organic matter content in soil was made by method based on dissolving organic matter to carbonic dioxide and water (Tyurin method modified by TSINAO, standard GOST 26213-91 (1991)). Nitrates content in soil samples was determined by standard disulfurphenol method after Grandval Lyazhu (Piskunov 2004).

3 Results

3.1 Crop Rotations

In a trial at Shortandy, the yield of different crops was compared in two fields of four crop rotations. This comparison shows the advantage of replacing summer fallow by sowing various crops grown under different tillage technologies (Table 1).

Simple technology involves growing crops with no intensification. Traditional technology involves the application of all recommended intensification which means under the traditional tillage system. No-till involves the application of the same intensification means under the no-till system.

The best crop to replace summer fallow proved to be dry pea. The grain yield of wheat sown after fallow and after pea under both tillage methods was at the same level. However, dry pea itself provided 2.34 and 2.42 t ha^{-1} under no-till and traditional tillage methods, respectively. Replacing the summer fallow by rapeseed was less profitable because of the reduced yield of wheat sown after rapeseed, while the rapeseed yield was also rather low. Oats look much more advantageous than fallow, providing 3.76 t ha^{-1} of grain instead of a reduction in the wheat yield by 0.29–0.31 t ha^{-1}. However, it is less profitable than dry pea, which has much better market prices. No-till provided a reduced grain yield in wheat and dry pea because of lower nitrate availability, and a lower moisture storage in some years because of the lower snowmelt water permeability in early spring.

Table 1 Grain yield (t ha^{-1}) of preceding crops and of subsequent spring wheat under different tillage technologies at Shortandy (average for 2009–11)

Field	Preceding crop			
	Fallow	Oats	Pea	Rapeseed
Simple				
First	–	2.00	0.87	0.46
Wheat	1.32	1.25	1.11	0.98
Traditional				
First	–	3.76	2.42	1.16
Wheat	2.92	2.61	2.85	2.27
No-till				
First	–	3.76	2.34	1.15
Wheat	2.78	2.49	2.74	2.28

Crop rotation studies in northern Kazakhstan have shown that there is a considerable advantage in replacing summer fallow by other crops, first of all food legumes (Gilevich 2013; Sagalbekov et al. 2013). A number of studies have supported the idea that summer fallow can successfully be replaced by a variety of crops in dryland agriculture in a great deal of research carried out in Canada (Gan et al. 2003; Larney et al. 2004).

Diversified crop rotations are being adopted by the farmers of northern Kazakhstan, but not all farms are ready to reduce the summer fallow area and replace it by oilseeds or pulses. There are farms where alternate fallow-wheat rotation is preferred (Akayev 2011). However, there is a general trend toward considerably reducing the area under summer fallow.

3.2 Soil Tillage

In the second trial at Shortandy, three tillage methods (deep, shallow and no-till) were studied in a 4-course wheat-fallow rotation on the background of 4 tillage treatments during summer fallow: traditional, minimum-d, minimum-s, and chemical fallow. The results of the study showed that the tillage method had a significant influence on the wheat grain yield in the third year after fallow (Table 2).

Traditional tillage in the fallow includes 4–5 tillage operations during the summer to control weeds. Minimum-d treatment includes one tillage at 20–22 cm plus herbicide application. Minimum-s includes one shallow tillage at 12–14 cm plus herbicide application. Chemical fallow includes two applications of herbicides in summer to control weeds.

In the second year after fallow, in most cases the tillage methods in the fallow and in the stubble land did not affect spring wheat yields significantly. There was no need for tillage in the autumn with a background of tillage in the fallow. In the third year after fallow, no-till negatively affected the wheat yield with a background of

Table 2 Spring wheat yield (t ha^{-1}) as affected by tillage method in fallow and autumn tillage on stubble at Shortandy (average for 2007–09)

Tillage in fallow	Tillage in the autumn		
	Deep	Shallow	No-till
Second year after fallow			
Traditional	2.02	2.14	1.93
Minimum-d	1.93	1.93	1.98
Minimum-s	2.05	2.07	2.02
Chemical	2.09	2.08	2.11
Third year after fallow			
Traditional	1.75	1.97	1.76
Minimum-d	1.93	2.01	1.64
Minimum-s	1.85	1.89	1.69
Chemical	1.74	1.94	1.72

LSD_{05} t ha^{-1}: second year—0.08, third year—0.06

tillage or no tillage in the fallow. The best treatment was shallow tillage in the autumn, especially with a background of deep tillage in the fallow. This can be explained by two factors: better infiltration of thawing water in early spring and more active mineralization of nitrogen due to soil loosening.

In the third trial during the same period, a comparison of the two tillage methods has shown the advantage of zero tillage in bread wheat and durum wheat grain yields (Table 3).

The grain yield of two wheat types was higher with the delayed sowing date of 26 May. This is explained by the higher rainfall in July, which was better utilized by wheat sown later. At all sowing dates, no-till had an advantage over traditional tillage, thanks to better moisture conservation. The yield gains were higher at optimal sowing dates. Both bread wheat and durum wheat produced more nitrogen in grain under the traditional tillage method, which once more demonstrated the advantage of traditional tillage for organic matter decomposition.

In northern Kazakhstan, the best results in favor of no-till were obtained at Kostanai RIA (Aksagov 2010). In the trial, three tillage methods (traditional, minimum and no-till) were compared in a 4-course fallow-wheat rotation. In this study, the wheat yield advantage was in favor of no-till in the second and third years after fallow. The main reason was the advantage in water storage prior to sowing. Moisture storage in the 0–100 cm soil layer under traditional tillage in the first, second, and third years after fallow was 174, 121, and 118 mm, respectively, while under no-till it was 213, 161, and 140 mm, respectively. This is very different to the data obtained at Shortandy, which can be explained by the fact that this research was conducted on sandy loam black soil with an organic matter content of 4.7 %, while the soil at Shortandy is a heavy clay loam black soil with an organic matter content of 3.5 %. Additionally, in the study at the Kostanai site, traditional tillage did not include snow ridging to accumulate more snow and harvested straw was removed from the plots.

At Kurgan (Gilev et al. 2011), the moisture content prior to sowing after summer fallow was 102 mm on traditional tilled fallow and 109 mm on chemical fallow. On stubble land, no-till had a noteworthy advantage: 70 mm on plough and 110 mm on no-till. Traditional ploughing had a notable advantage on nitrate availability both after fallow and on stubble land. N-NO_3 availability on plough and no-till was 128 and 82 kg ha^{-1}, respectively after fallow and 71 and 56 kg ha^{-1}, respectively on stubble land. Spring wheat yields were affected by both the tillage method and the application of a nitrogen fertilizer (Table 4).

Table 3 Grain yield (t ha^{-1}) of bread wheat (average of 3 varieties) and durum wheat (2 varieties) as affected by tillage methods and sowing dates at Shortandy (average for 2009–2011)

Sowing date	Bread wheat		Durum wheat	
	Traditional	No-till	Traditional	No-till
19 May	1.32	1.38	1.20	1.38
26 May	1.39	1.58	1.46	1.73
2 June	1.32	1.42	1.25	1.42

Table 4 Spring wheat grain yield (t ha^{-1}) as affected by tillage method and fertilizer application in fallow—3 times wheat rotation in Kurgan, Trans-Ural central forest-steppe (average for 2007–10, Gilev et al. 2011)

Crop sequence	No fertilizer		60 kg of N ha^{-1}	
	Plough	No-till	Plough	No-till
Wheat after fallow	1.70	1.42	1.53	1.51
Wheat	1.05	1.10	1.12	1.59
Wheat	0.96	0.95	0.93	1.35

Wheat sown on fallow gave the highest grain yield when the soil was ploughed, which was obviously associated with more nitrate availability. In the second year after fallow, the wheat yield was a little higher under no-till thanks to the advantage in moisture availability. When nitrogen fertilizer was applied, ploughed fallow no longer had a yield advantage over no-till. The most noteworthy data on the advantage of no-till were obtained on stubble fields with nitrogen fertilizer applied. This was achieved, thanks to better moisture availability.

At Omsk, no-till treatment was not studied. There was a main tillage in the autumn tested together with chemical application treatments in a 5-course crop rotation. In spring, all plots were planted with a cultivator drill. The table below presents the average grain yield of wheat and barley (Table 5).

Plough—moldboard ploughing at 20–22 cm in the autumn. Blade—shallow tillage in the autumn at 12–14 cm. Mixed—plough for 2 crops and blade for 3 crops. Complex—application of fertilizers, herbicides, fungicides, insecticides, and plant-growth regulators. With no chemicals applied, tillage with a moldboard plough had a definite advantage in grain yield. When fertilizer and herbicides were applied, the best yields were obtained on plots tilled by plough annually or twice in five years. Only when all chemicals were applied, including fertilizer, herbicides, fungicides, and growth regulators, did all tillage treatments have comparable grain yields.

Comparative studies of no-till and traditional conservation tillage with sweeps and blades at the Shortandy site with a variety of combinations of crops, varieties, snow management, and sowing dates have shown that the results obtained provided a great number of data allowing a careful assessment of tillage treatments. In general, no-till on heavy clay loam soils may not be advantageous in some years when the intake of snowmelt water in early spring calls for tillage in the autumn.

Table 5 Grain yield in 5-course crop rotation as affected by tillage method and chemical application at Omsk (average for 2001–05) (Kholmov and Shulyakov 2006)

Chemicals	Tillage in the autumn			
	Plough	Mixed	Blade	Zero
None	1.94	1.76	1.69	1.64
Fertilizer + herbicides	2.97	2.90	2.73	2.73
Complex	3.95	3.97	3.96	3.91

LSD_{05} t ha^{-1}: tillage—0.15, chemicals—0.13

The most successful data in favor of no-till has been obtained at the Kostanai site on sandy loamy Chernozems. This can be explained by the fact that there is no problem of snowmelt water infiltration in early spring on light textured soils.

3.3 Soil Fertility Management

As found by Canadian scientists, major soil fertility losses happened during the summer fallow period (Renni et al. 1976). Moreover, frequent summer fallow reduced potential crop yields because of reduced soil fertility. Most soils in Saskatchewan lost some 35–44 % of their original nitrogen and organic matter. It was calculated that only 30–33 % of nitrogen in the lost organic matter was taken up by plants. By 1974, the nitrogen lost was equivalent to 36 million tons. These scientists also calculated that about 600 kg N ha^{-1} was leached below the plant root system layer, mainly during the summer fallow period.

According to our observations at Shortandy, snowmelt water runoff and soil losses occur in summer fallow fields on southern slopes almost every year. Soil losses go up to 0.5 t ha^{-1} even after winters with little snow. In one year, the soil losses on summer fallow fields may reach up to 292 m^3 ha^{-1}. Our research has shown that the intake of snowmelt water on summer fallow does not exceed 17.4 %. After a second winter of summer fallowing, up to 92 % of the snowmelt water is lost as runoff and through evaporation.

According to data obtained at Shortandy, in 50 years the organic matter content in 0–20 cm soil layer was reduced from 3.90–3.26 % under continuous wheat due to the common decay of organic matter. At the same time, in alternate fallow-wheat rotation, the organic matter content dropped dramatically to 2.48 %. That is, our new data is in line with Canadian findings. This is why CA should include diversified crop rotations with no fallow or a minimal area under fallow. Summer fallow might be used occasionally for certain purposes, but it should not be a systematic, frequent fallow. According to Kokshetau RIA (Sagalbekov et al. 2013) replacing the summer fallow by sweet clover harvested for green mass, followed by ploughing, increased the soil organic matter in four-year rotation by 10 % compared with the original content.

The second direction for soil fertility improvement is direct sowing. It is obvious that if there is lack of nitrates under no-till because of slow nitrogen mineralization, this leads to soil fertility conservation. At the Shortandy site there was a trial with soil mulching with straw at rates of 2 and 4 t ha^{-1} during three times of 4-course wheat—fallow rotation. At the end of the third rotation, after 12 years the organic matter content in the 0–10 cm soil layer at 2 and 4 t ha^{-1} rates increased from 3.52 to 3.69 and 4.10 %, respectively. However, nobody wants to carry straw from one field to another for mulching purposes. This is why no-till itself will affect soil fertility slowly, and it should be combined with the introduction of diversified crop rotations with no fallow.

The effect of no-till on organic matter content was obtained in the longest no-till study conducted at a Karaganda site in the dry steppe zone of Central Kazakhstan (Yushchenko 2012). After ten years of continuous no-till, an increase in the organic matter was observed. The organic matter content increased notably in the top 0–10 cm layer from 3.11 % under traditional tillage to 3.26 % under no-till as a result of continuous no-till application over ten years in a five-year rotation "fallow-4 year grains" at the expense of less organic matter losses through nitrification processes in no-tilled soil. In larger farm fields, these losses might be increased by soil losses due to runoff water, especially on sloping lands.

Not only on research fields but also on many working fields in Siberia, there has been evidence of significant soil degradation and desertification due to traditional farming and tillage methods on silty and sandy soils (Meinel 2002; Schreiner and Meyer 2014). If cereal-cropping systems are to be maintained in steppe regions, and soil fertility is to be preserved for coming generations, there is no alternative to installing CA.

4 Conclusions

1. The principles of CA include crop rotations with no fallow, no-till or minimum tillage, leaving all residues on the soil surface and applying recommended rates of nitrogen and phosphorus.
2. Compared with fallow-based wheat monoculture, crop rotations provide crop diversification and a more economical use of land resources, and guarantee better soil fertility management, thanks to the removal of summer fallow from rotations.
3. In western Siberia on Leached Chernozems no-till is feasible only with the application of increased rates of nitrogen fertilizers.
4. No-till enables better soil fertility conservation. On light textured soils it guarantees a crop yield increase, thanks to better moisture accumulation and conservation. On heavy textured soils, in some cases, soil loosening in the autumn is necessary to facilitate a better intake of snowmelt water in early spring.
5. For soil fertility conservation, what is most important is the removal or considerable reduction of summer fallow and its replacement by food legumes such as peas and sweet clover. In addition to this, no-till or minimum tillage should be applied with nitrogen fertilizer.
6. Soil degradation by nonsustainable tillage systems must be stopped. Conservation agriculture including no-till is practicable and must become the dominating practice in cereal cropping systems of steppe regions.

References

Akayev T (2011) No-Till technology and crop rotation with short rotation cycle at Tukym farm. J Agrarniy Sector 4:52–55 (in Russian)

Aksagov TM (2010) No-till technology as instrument of restructuring soil fertility conservation of southern chernozems in northern Kazakhstan. In: Proceedings of republican conference of young scientists "contribution of young scientists into agricultural science"—Shortandy, pp 17–20 (in Russian)

Barayev AI (1960) Cropping structure and crop rotations in grain farming of Northern Kazakhstan. J Zemledeliye, Moscow 7:15–21 (in Russian)

Barayev AI (1984) Further development of conservation agriculture—most important challenge. J Vestnik selskokhozyastvennoi nauki, Moscow 12:18–23 (in Russian)

Chernenok V, Barkusky D (2014) Diagnosis and optimization of phosphorus nutrition conditions of grain crops in Northern Kazakhstan. In: Mueller L, Saparov A, Lischeid G (eds) Novel measurement and assessment tools for monitoring and management of land and water resources in agricultural landscapes of Central Asia. Springer International Publishing, Cham, pp 667–683. http://link.springer.com/chapter/10.1007/978-3-319-01017-5_43. Accessed 10 Mar 2015

Derpsch R (2008) No-tillage and conservation agriculture: a progress report "no-till farming system". In: Goddard T, Zoebisch MA, Gan YT, Ellis W, Watson A, Sombatpanit S (eds) No-till farming systems. Special publication, #3, World Association of Soil and Water Conservation, Bangkok. http://betuco.be/CA/No-tillage%20%20Farming%20System.pdf. Accessed 10 Mar 2015

Franzlubbers AJ (2009) Soil organic carbon sequestration with conservation agriculture in the Southeastern USA: potential and limitations. In: Annual meeting of the soil science society of America. http://www.fao.org/ag/ca/Carbon%20Offset%20Consultation/CARBONMEETING/3FULLPAPERSBYCONSULTATIONSPEAKERS/PAPERFRANZLUEBBERS.pdf. Accessed 10 Mar 2015

Gan YT, Miller BG, McConkey BG, Zentner RP, Stevenson FC, McDonald CI (2003) Influence of diverse cropping sequences on durum wheat yield, and protein in semiarid Northern Great Plains. Agron J 95:245–252

Gilev SD, Kurlov AN, Zamyatin AA, Stepnykh NV (2011) Efficiency of water saving agro-technologies in trans-ural central forest-steppe zone. In: Proceedings of conference "crop diversification and no-till as basis of conservation agriculture and food security", Astana-Shortandy, pp 215–221 (in Russian)

Gilevich SI (2013) Crop rotations of sustainable farming. J Agrarniy Sector 2:62–68 (in Russian)

GOST 26213-91 (1991) State standard. Soil. Methods of estimation of organic matter (ГОСТ 26213-91. Почвы. Методы определения органического вещества), 8 p (in Russian)

Hill KW (1954) Wheat yields and soil fertility on the Canadian prairies after a half century of farming. Soil Sci Soc Am Proc 18:182–184

Kassam AH, Friedrich T, Shaxson F, Reeves T, Pretty J, de Moraes Sa JC (2011) Production systems for sustainable intensification—integrating productivity with ecosystem services. Technikfolgenabschätzung- Theorie und Praxis, Special Issue on Feeding the World, July 2011, pp 38–44. http://www.fao.org/ag/ca/CA-Publications/TATuP_2011.pdf. Accessed 10 Mar 2015

Kholmov VG, Shulyakov MI (2006) Improvement of grain agro-technologies in forest-steppe of western Siberia. In: Conference dedicated to 50 years anniversary of scientific production center of grain farming named after A.I. Barayev, Shortandy, Kazakhstan, pp 102–106 (in Russian)

Larney FH, Yanzen E, Smith, Anderson D (2004) Dryland agriculture on the Canadian prairies: current issues and future challenges. In: Rao S, Rayan J (eds) Challenges and strategies for dryland agriculture. CSSA Special Publication, Madison, pp 113–139

Meinel T (2002) Die geoökologischen Folgewirkungen der Steppenumbrüche in den 50-er Jahren in Westsibirien. Diss. Halle 2002. http://sundoc.bibliothek.uni-halle.de/diss-online/02/03H082/t1.pdf. Accessed 10 Mar 2015

Meinel T, Grunwald L-C, Akshalov K (2014) Modern technologies for soil management and conservation in Northern Kazakhstan. In: Mueller L, Saparov A, Lischeid G (eds) Novel measurement and assessment tools for monitoring and management of land and water resources in agricultural landscapes of Central Asia. Springer International Publishing, Cham, pp 455–464. http://link.springer.com/chapter/10.1007/978-3-319-01017-5_27. Accessed 10 Mar 2015

Piskunov AS (2004) Methods of agrochemical research. Moscow, Kolos, 312 pp (in Russian)

Pittelkow CM, Liang X, Linquist BA, van Groenigen KJ, Lee J, Lundy ME, van Gestel N, Six J, Venterea RT, van Kessel C (2014) Productivity limits and potentials of the principles of conservation agriculture. Nature. doi:10.1038/nature13809, http://www.nature.com/nature/journal/vaop/ncurrent/full/nature13809.html. Accessed 10 Mar 2015

Powlson DS, Stirling CM, Jat ML, Gerard BG, Palm CA, Sanchez PA, Cassman KG (2014) Limited potential of no-till agriculture for climate change mitigation. Nat Clim Change 4:678–683. doi:10.1038/nclimate2292, http://www.nature.com/nclimate/journal/v4/n8/full/nclimate2292.html. Accessed 10 Mar 2015

Renni DA, GJ Racz, DC McBeath (1976) Nitrogen losses. In: Proceedings of the western Canada nitrogen symposium—Alberta agriculture, Edmonton, Alberta, pp 325–353

Sagalbekov UM, Sagalbekov EU, Seitmaganbetova GT, Bogdanov IM (2013) Sweet clover semi-fallow as alternative to summer fallow. In: Proceedings of international conference "crop diversification and no-till technologies in dry areas", Astana, Shortandy, pp 114–118

Schreiner V, Meyer BC (2014) Indicators of land degradation in Steppe regions: soil and morphodynamics in the Northern Kulunda. In: Mueller L, Saparov A, Lischeid G (eds) Novel measurement and assessment tools for monitoring and management of land and water resources in agricultural landscapes of Central Asia. Springer International Publishing, Cham, pp 535–548. http://link.springer.com/chapter/10.1007/978-3-319-01017-5_33. Accessed 10 Mar 2015

Suleimenov M (1988) About theory and practice of crop rotations in Northern Kazakhstan. J Zemledeliye Moscow 9:14–20 (in Russian)

Suleimenov M, Akshalov K (2007) Eliminating summer fallow on black soils of northern Kazakhstan. In: Lal R, Suleimenov M, Stewart B, Hansen D, Doraiswamy P (eds) Climate change and terrestrial carbon sequestration in Central Asia. Taylor and Francis, London, pp 267–279

Suleimenov M, Kaskarbayev Z, Akshalov K, Yushchenko N (2014) Conservation agriculture for long-term soil productivity. In: Mueller L, Saparov A, Lischeid G (eds) Novel measurement and assessment tools for monitoring and management of land and water resources in agricultural landscapes of Central Asia. Springer International Publishing, Cham, pp 441–454. http://link.springer.com/chapter/10.1007/978-3-319-01017-5_26. Accessed 10 Mar 2015

Yushchenko NS (2012) No-till technologies in central Kazakhstan. J Agrarniy Sector 2:16–18 (in Russian)

Chapter 31
Modern Cropping Systems and Technologies for Soil Conservation in Siberian Agriculture

Lars-Christian Grunwald, Vladimir I. Belyaev, Malte Hamann, Patrick Illiger, Eckart Stephan, Norbert Bischoff, Nikolay V. Rudev, Nikita A. Kozhanov, Gerd Schmidt, Manfred Frühauf and Tobias Meinel

Abstract This chapter presents a concept and practicable solutions for effective and environmentally sound cereal-based cropping systems in Siberia. The work behind this was carried out in the framework of the international research project "KULUNDA, how to prevent the next global dust bowl?" We start with an analysis of the socio-economic and bio-climatic conditions and current practice for cereal cropping in Siberia. The climatic conditions of West Siberia were compared with those in the steppe regions of Canada. The adaption of current cropping practices based on South Canadian experiences, such as minimum tillage and direct seeding, seems to be promising for Siberia too. More reliable machinery for soil and crop management, exact seeding and spraying machinery in particular, will help to

L.-C. Grunwald (✉)
Amazone Company, 49205 Hasbergen, Germany
e-mail: LarsChristian.Grunwald@bbg-leipzig.de

V.I. Belyaev
Faculty Agricultural Engineering, Barnaul Altai State Agrarian University, Krasnoarmeyskiy Av. 98, 656049 Barnaul, Altai Krai, Russia
e-mail: belyaev@asau.ru

M. Hamann
Amazone Company, 27798 Hude, Germany
e-mail: Malte.Hamann@amazone.de

P. Illiger · E. Stephan · G. Schmidt · M. Frühauf
Martin-Luther-University Halle-Wittenberg, Institute of Geosciences and Geography, Von-Seckendorff-Platz 4, Haus Geographie, 06120 Halle, Germany
e-mail: patrick.illiger@geo.uni-halle.de

E. Stephan
e-mail: eckart.stephan@geo.uni-halle.de

G. Schmidt
e-mail: Gerd.Schmidt@geo.uni-halle.de

L. Mueller et al. (eds.), *Novel Methods for Monitoring and Managing Land and Water Resources in Siberia*, Springer Water,
DOI 10.1007/978-3-319-24409-9_31

establish more robust crops which provide higher yield stability and better water use efficiency, and maintain soil fertility by protecting the soil against wind erosion. We designed and conducted field trials on three experimental sites in the Altai Krai of Siberia. These trials represent the outcomes of our analysis by combining the local expertise of Siberian scientists and farmers with experience from Canadian and Kazakh cropping practices and with proven and newly developed German technology for their realisation. The focus is on innovative technologies for soil conservation tillage operations and direct drill seeders. On 70 randomised plots, most modern no-till methods were compared with other variants such as modernized and conventional agricultural machinery, and with extremely out-dated methods of conventional technology. The latter variants include practices still dominating in Siberia: no crop protection and fertilisation as well as cereal monoculture and fallow. The initial results under very dry weather conditions confirm our concepts that conservation practices perform better when considering all aspects of soil protection, the performance of machinery, plant establishment and yield. Wider crop rotation can give higher economic stability and will help to get the cropping systems closer to natural conditions. Soils can be protected from wind erosion in an effective way using no-till systems, and the yield can be stabilized. Most modern seeding machines with wide row spacings and narrow, single depth-adjusted hoe opener systems are able to carry out shallow tillage in the seeding furrow only and to place the seed and fertiliser exactly at the adjusted depth. The running experiments will deliver more detailed data about crop management and yields, soil fertility and micro-economy in the coming two years. Though those more detailed data are still required, we may already conclude that our new approaches seem to confirm our hypotheses. They are soil conserving, practicable and effective, and have great potentials for broader application in Siberian agriculture.

Keywords Siberia · Semi-arid climate · Soil erosion · Minimum tillage · Direct seed · Crop yield

---

M. Frühauf
e-mail: Manfred.Fruehauf@geo.uni-halle.de

N. Bischoff
Institute for Soil Sciences, Leibniz University Hannover (LUH), Herrenhäuser Str. 2, 30419 Hannover, Germany
e-mail: bischoff@ifbk.uni-hannover.de

N.V. Rudev · N.A. Kozhanov
Farm OOO KX "Partner", Poluyamki, Sovetskaya 31, Michailovskiy Rayon, Altai Krai, Russia
e-mail: Rudev.nikolaj@mail.ru

N.A. Kozhanov
e-mail: Kozhanov-partner@yandex.ru

T. Meinel
Amazone Company in Kazakhstan, Saryarka District, 010000 Astana, Kazakhstan
e-mail: Dr.Tobias.Meinel@amazone.de

1 Introduction

Since 2011, the scientists involved in the international KULUNDA research project (BMBF 2014) have been working on the challenge of improving the situation in the agricultural sector of the Altai region in West Siberia. This region is representative for all West Siberian and North Kazakh agricultural regions, which were subject to an extreme transformation of the natural landscape during the course of the Virgin Lands Campaign, aiming to achieve food security in post-war Soviet Union.

The associated subsequent damages have still not been remediated today: on the contrary, economic constraints, political uncertainties and handed-down conventions still currently impede the urgently required modernization of the agricultural sector. Within the scope of the KULUNDA joint research project, ten scientific sub-projects are collaborating to record the qualitative and quantitative causes and consequences of inadequate land cultivation in the marginal climatic regions of the Eurasian steppes, using the example of the Altai region, and to remediate these problems.

This project has involved investigations into soil degradation as well as into the dynamics of the soil and landscape water regimes depending on the intensity of land use. Field trials lasting several years aim to develop climate-adapted, soil-conserving agricultural methods which, more importantly, are also manageable for the farms' production. Socioeconomic studies are assessing the problems that have so far impeded extensive conversion to minimally invasive farming methods. The main focus is on overcoming monetary, administrative and mental hurdles and working on new approaches. An important aspect of the KULUNDA research project is to process and implement the knowledge gained effectively for stakeholders directly in the agricultural sector, from the worlds of business and politics, and in educational establishments. Thus, KULUNDA can take part in the row of scientific and practise orientated efforts to prevent the destruction of soil resources and the associated obliteration of the food source and livelihood of entire regions.

2 Initial Cropping Situation in the Steppes of Siberia

The cool, temperate grass steppes are one of the most widespread and populated types of ecosystem on earth. The Eurasian steppes from the south of Ukraine to the north of Kazakhstan and West Siberia represent the largest contiguous type of an ecosystem on earth (von Poletika 1936). On the other hand, they are among the most heavily utilised and destroyed landscapes on earth (Remmert 1997).

From the second half of the twentieth century, the West Siberian and Kazakh steppes played a vital role in terms of food production for the states of the former USSR, which were weakened by World War II. After the introduction of intensified arable farming on existing fields, new arable regions were to be taken into production (Georgiev 1955).

To expand industrial agriculture, the ploughing up of a total of 41.8 million ha of largely unused steppe land began in 1954, following the decision made by the

Central Committee of the CPSU (Eule 1962 p. 115; Wein 1980 p. 9). This effort was mainly concentrated on the driest regions of the Eurasian steppes (Fig. 1). Before the so-called Virgin Lands Campaign (*Russian*: *Tselina*), 70 % of the cultivated areas were in forest steppe regions; after the Tselina, however, 70 % of the cultivated areas were in arid steppe regions, and only 20 % were still in forest steppe regions (Eule 1962, p. 78).

Here, Kazakhstan and West Siberia are the regions with the largest proportion of newly ploughed steppe land, namely 25.5 and 6.2 million ha (Eule 1962, p. 115; Wein 1980, p. 7). The extensive expansion and the speed at which the ploughing of natural steppe soils took place were unique. The Tselina represents one of the most severe interferences in an ecosystem worldwide (Meinel and Frühauf 2003).

In the first year alone after the decision to implement Tselina, a total of 17 million ha of steppe soil were ploughed (Georgiev 1955). Within a short time, huge farm structures in the form of state farms (sovkhozy) were created in the virgin land regions. As soon as 1959, these farmed 27.9 million ha of arable land and therefore 27 % of the entire arable land of the former Soviet Union (Eule 1962, p. 74).

Several years of high precipitation with high wheat yields were followed by years of extreme drought. The tillage intensity was very high; the soil was deeply ploughed in preparation for the crops.

Because of the weak economy in the post-war period, mineral fertilisers and pesticides were not available (Georgiev 1955). In many Tselina regions, the cereal yields dropped dramatically from 1957 on. On the one hand, this was due to the monoculture of spring wheat and the associated depletion of the nutrient pool in the soil.

On the other hand, the highly intensive mechanical soil cultivation led to an increased mineralisation rate and therefore to a biological reduction in the humus content.

Fig. 1 Expansion of land use between 1900 and 1960—most of this area was first tilled during the Tselina Virgin Lands Campaign in 1954–1962

The mechanical destruction of the aggregate stability during the multiple passes of mechanical tillage peaked in a massive deflation of the bare topsoil during the sometimes severe dry storms in spring (Kaschtanov 1974). The events of human-induced soil deterioration occurring in the steppe regions of West Siberia and northern Kazakhstan as of 1957 perfectly mirror the so-called "dirty thirties" of the North American Great Plains in the 1930s (see Kostrovsky 1959; McLeman et al. 2013). Every year, 187,500 ha of arable land were destroyed in the Tselina regions by wind erosion during devastating dust storms (Wein 1980, p. 11). The irreversible damage of the soil as a means of livelihood by erosion and depletion-related degradation caused by inappropriate agricultural use is called the Dust Bowl Syndrome (WBGU 1994). After the first years with severe erosion events, measures for preventing erosion and intensifying agriculture were implemented, beginning in the 1960s in the former virgin land regions. These included erosion-reducing measures such as the introduction of non-inversion soil tillage, strip cultivation, reseeding grass on eroded fields and planting windbreaks against the prevailing wind direction (Baraev 2008). Furthermore, the use of pesticides and mineral fertiliser to increase yields and to reduce soil depletion was stepped up. The average cereal yields increased by 0.3 t/ha between the pre-Tselina period and 1980 (Jaehne 1981). In principle, the situation improved, although serious deficiencies due to insufficient supplies of fertiliser, pesticides, good seeds and reliable technology, as well as high losses caused by poor transport and storage logistics, strongly reduced the efficiency of the farms (Jaehne 1981; Peskaitis 1987).

During the course of the Perestroika and the deteriorating political and economic situation in the states of the former USSR as of 1991, the agricultural sector in particular fell into a period of deep recession (Viehrig 2005). Until the end of the 1990s, constantly dropping investments had drastic effects on agricultural productivity in the states of the former USSR, and especially in the dry farming regions of the former virgin land regions.

The annual wheat production of Ukraine, Kazakhstan and the Russian Federation (RF) in the year 1998 was 55 % of the volume from 1992 (FAOSTAT 2014).

The farm technology became ever more outdated, while important resources, chemical pesticides and mineral fertilisers could not be acquired because of the almost complete lack of government subsidies and bad credit conditions in the 1990s (Fellmann and Nekhay 2012). The highly intensive technique of mechanical soil tillage experienced a real renaissance in crop production in the states of the former USSR. This is shown in Fig. 2 using the example of bare fallow, the most intensive form of mechanical soil tillage.

Many farmers chose short-term profit-generating strategies at the expense of the soil resources over more sustainable farm management (Meinel 2002; Kostyayev 2003). At the trough of the economic depression, the proportion of areas with mechanical weed control through bare fallow reached its peak in the RF and the application of pesticides and fertiliser strongly decreased (Meyer et al. 2008; FSS ROSSIIJ 2014, Fig. 2). Figure 3 shows recent examples of massive wind erosion events between 2007 and 2014 from the Altai region in West Siberia. The

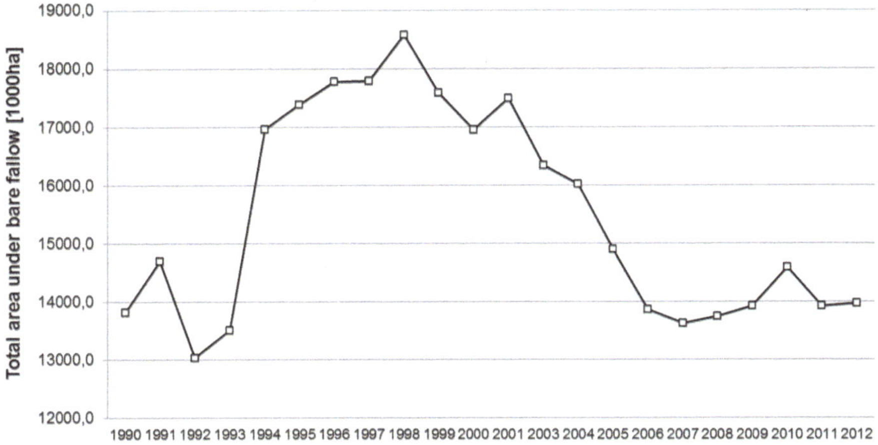

Fig. 2 Total area under bare fallow management in Russia 1990–2012 (author's illustration based on data from FSS ROSSIIJ 2014)

Fig. 3 Recent examples of disastrous events of wind erosion and soil deflation in the Altai region and northern Kazakhstan from 2007 to 2014 (photo bottom right courtesy of V. Sidorenko, Amazone Company)

hopeless situation in the 1990s led to a huge rural exodus. In particular, failed farmers, who had tried to establish their own private farms (*Russian*: *fermer*), but had to give up because of financial hurdles and non-profitable cultivation, left the villages of the West Siberian and north Kazakh steppe regions in their thousands (Meinel 2002).

Since the early 2000s, Russia's agricultural sector has been gaining strength. However, the development of Siberia's arable regions is considerably behind that of European regions. On the one hand, this becomes evident when looking at the machine park. On the other, significantly fewer pesticides and mineral fertilisers are used on average per unit area in the West Siberian regions. Interestingly, during the course of the last serious crisis starting in 2008/2009, significant investments in technology and active substances from the "Precision Farming" sector were made in Europe. In Russia, the proportion of bare fallow considerably increased during the economic crisis (Fig. 2). Another step back to soil-damaging conventional systems was taken.

3 Canadian Cultivation Systems—Adaptation for West Siberia?

All of these problems were found, and are still found today, in some areas of the North American Prairies. Particularly the Canadian provinces of Alberta, Saskatchewan and Manitoba have been facing this challenge for the last 50 years: developing sustainable agricultural concepts while maintaining or even increasing yields. Various scientific investigations have shown that solutions are to be found primarily in the minimization of soil tillage intensity and a diversification of the crop rotations (Lafond et al. 1992; Larney et al. 1994; Campbell et al. 2002; Hilliard et al. 2002). Starting in the 1990s, methods for weed control, seeding technology, fertilizer application and crop rotation were developed on a scientific basis. The establishment of the no-till concept for the northern Great Plains at the turn of the millennium provided a conclusive agricultural concept that both guaranteed soil conservation and gave farmers an economically viable option. With government support, this knowledge was also made accessible to farmers in the form of brochures and web pages. However, it has to be emphasized that each farmer had to adapt the system to the natural conditions and specific crop rotations in each case. Figures 4 and 5 show the rapid development of no-till at the turn of the millennium and the diversification of the crops. The relatively late conversion to no-till methods can be explained by the late availability of suitable herbicides and the delayed introduction of high-quality seeding technology.

Figure 5 clearly shows the trend towards conserving soil tillage in the Canadian Prairies. As the areas cultivated with cereals (mainly spring wheat) are reduced, the proportion of areas cultivated with oilseed and legumes, which are necessary for a sustainable crop rotation, increases. The proportion of traditional, but erosion-promoting, bare fallow decreases.

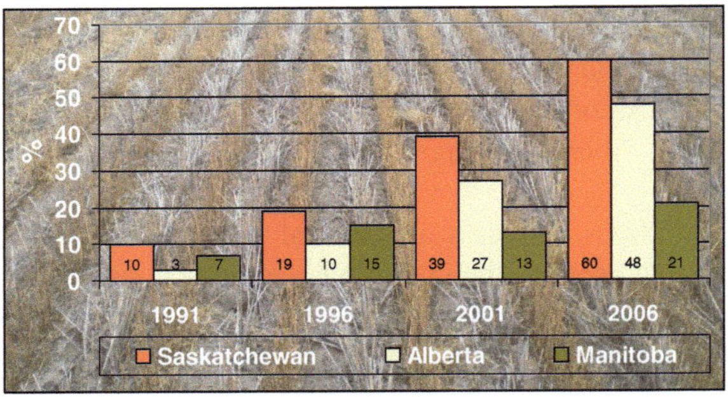

Fig. 4 Development of no-till areas in three provinces of the Prairie region of Canada (author's illustration based on data from Statistics Canada 2014)

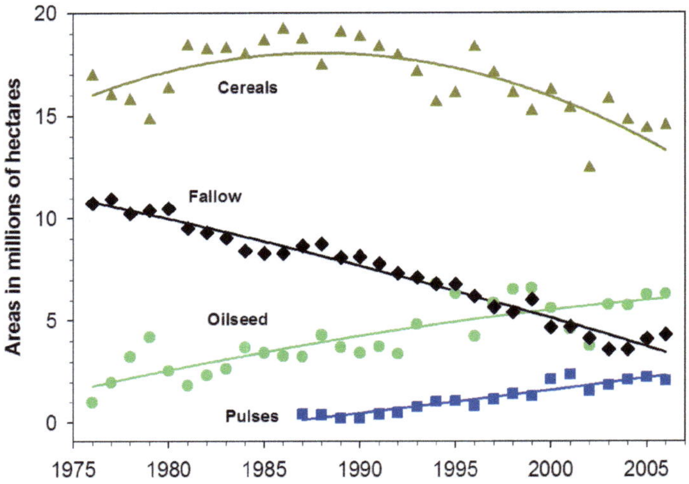

Fig. 5 Development of the diversification of crop rotations and steps towards conservation cropping in Canada (author's illustration based on data from Statistics Canada 2014)

The very successful and sustainable no-till agricultural concept, which has also already been introduced in real life in the Canadian Prairies, also seems to be very promising for the Altai region, which has a similar agricultural geography. However, these methods should not be introduced in the project area without precise prior analyses because of several factors. On the one hand, there are technical and agronomic reasons:

- Chemical plant protection plays a vital role in the no-till system. For historical reasons, however, real-life conditions concerning the methodology and the required application rates are not yet up to date.
- A wide-ranging crop rotation is indispensable for a functioning no-till system. However, the market is today still strongly oriented towards the high-cereal crop rotations in the region. The market is only slowly opening up to other crops.

- Technology that enables an intensification of the soil tillage intensity, field sprayers and seed drills capable of direct seeding are rarely found on the farms.
- The farm structures are different than in Canada. While farms in Canada are mostly family-run and the average farm size in the Prairies is about 3000 ha, the farms in southern Siberia are often much bigger. This makes a conversion of the production system more difficult and expensive.

In addition to the agronomic reasons, there are also agro-geographic reasons calling for an investigation of the Canadian no-till method in southern Siberia:

- In Canada, rainfed arable farming is performed down to a total annual precipitation of about 300 mm (Sauchyn 2009), whereas crops are produced on the Siberian steppes down to 230 mm (Atlas Altaiskovo Kraya 1978). Therefore, with assumed equally high evapotranspiration and also a very high probability of summer drought, the project region deals with drier conditions.
- Furthermore, the precipitation distribution in the Canadian Prairies looks more favourable for the cultivation of spring crops than in the Kulunda steppe. Figure 6 clearly shows the lower average precipitation in May and June in Siberia. However, the water in the early summer is particularly important for yield development. The high precipitation in July is often no longer relevant for yields, because the water deficit during the often dry month of June causes too much damage to the crops.
- In the Canadian no-till system, seeding is generally performed as early as possible after the snow has melted. This is usually not a problem, because the crops can develop well into the wet month of June. With optimal crop protection, this method results in the best growth development. In Siberia, field weeds are generally only moderately developed by the end of May, so chemical crop protection in spring is less effective than in southern Canada. Later seeding could offer better control concerning pre-emergence treatment.

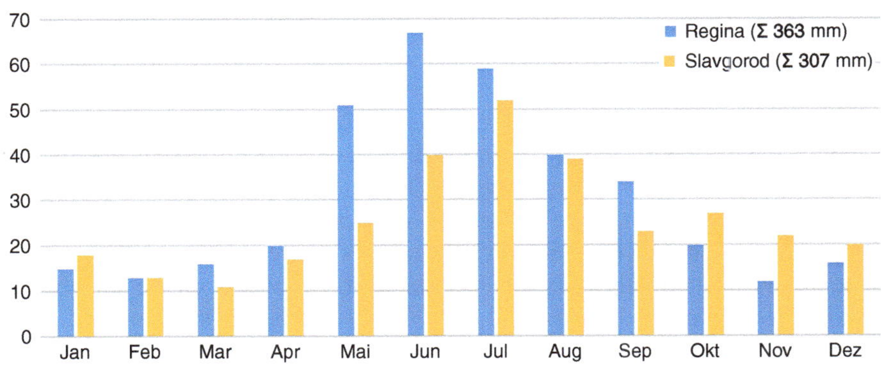

Fig. 6 Rainfall distribution in Regina (Canada) and Slavgorod (West Siberia). Data differ in the very important months of May and June (author's illustration based on data from the FAO (2006))

In addition to the agronomic and agro-geographic factors, traditions and stereotypes in crop production are also obstacles to conversion to no-till methods. But from an international viewpoint, this is nothing special. Everywhere in the agricultural sector, there are problems with the transition from traditional to modern methods.

"The main barriers to its adoption continue to be: knowledge on how to do it (knowhow), mindset (tradition, prejudice), inadequate policies such as commodity-based subsidies, the availability of adequate machines and the availability of suitable herbicides to facilitate weed management" (Derpsch et al. 2010, p. 3).

However, the positive effects of the no-till method on the soil and yield stability only become apparent after several years of consistent implementation (Lafond et al. 2006). Converting the production method is associated with great effort, and requires a complete reorientation, mainly to do with the renouncing of traditional cropping methods. However, the new acquisition of suitable technology, investment in active substances and the internal conversion of farm processes also create obstacles that make it hard for farmers to take this step. In this context, the risk of a moral and financial setback caused by possible errors in the method or by unforeseeable weather and market events, especially in the period directly following the conversion, is a major inhibiting threshold.

To scientifically test the theoretical assumptions about the positive potential of consistently converting the cropping method, long-term trials were established at three sites within the framework of the Kulunda project, which naturally also served the purpose of demonstration. In terms of their scope, these field trials are the first scientifically monitored tests on the introduction of no-till in the Kulunda steppe.

4 The Investigation Area—the Trial Sites

The Kulunda region and the Kulunda steppe are actually the largest fraction of the former virgin land regions in West Siberia (Meinel 2002) and were thus used as sites for experiments.

The investigation area stretches from the forested steppes of the pre-Ob plateau in the northeast, close to the Ob River, over the typical long grass steppes in the central Kulunda region, all the way to the dry steppe areas of the Kulunda steppe, which is geographically delimited by the Irtysh River (cf. Milkov 1977; Atlas Altaiskovo Kraya 1978). The climate is characterised by high continentality. Temperature amplitudes of 80 K over the course of the year are normal. The lion's share of precipitation falls in the summer half-year. Depending on the latitude, the growth period is short: between 110 and 90 days (Batalov 1980). In 2012, field trials were started at three sites in the Altai Krai. The selected farms are located along a climatic and natural landscape gradient from the northeast to the southwest (Fig. 7). This allows the field trials to be performed on a regional scale in three sub-types of the West Siberian steppe landscape and enables the effects of the regional climate and the prevailing soil types to be taken into account.

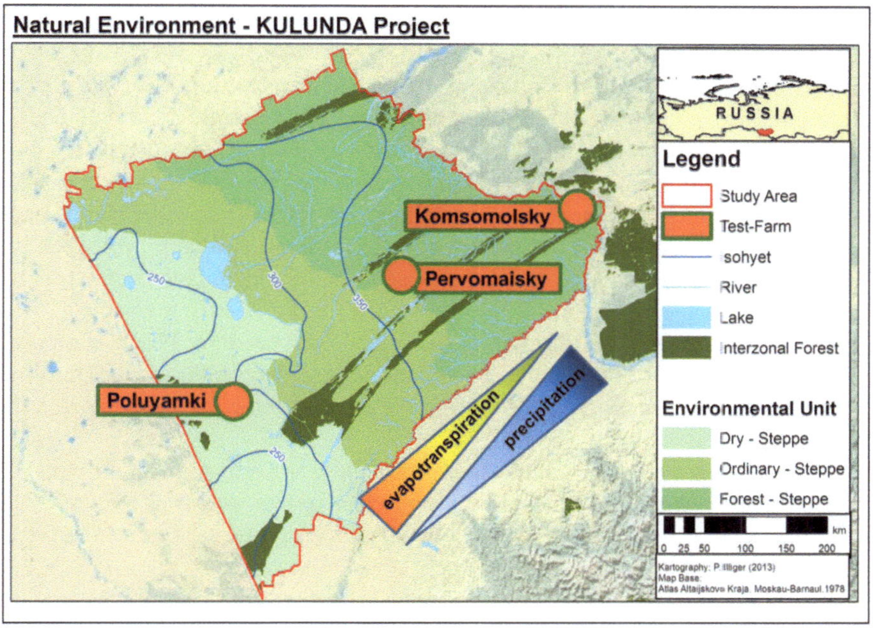

Fig. 7 Test farms of the KULUNDA project are located along a strong climatic gradient in the Altai region (Map based on Atlas Altaiskovo Kraya 1978, modified)

The selected trial sites have the following characteristics:

Komsomolsky:
Mean total annual rainfall: 419 mm
Mean annual temperature: 2.1 °C
Forest steppe zone with Chernozems (Haplic Chernozems)

Pervomaisky:
Mean total annual rainfall: 360 mm
Mean annual temperature: 2.1 °C
Forest steppe zone with Chernozems and Solonetz (Calcic Chernozems, Humic Solonetz)

Poluyamki:
Mean total annual rainfall: 280 mm
Mean annual temperature: 3.6 °C
Forest steppe zone with Chernozems, Kastanozems and Solonchaks (Calcic Chernozems, Haplic Kastanozems, Haplic Solonchaks)
Atlas Altaijskovo Kraja (1978), Burlakova et al. (1988) Stolbovoi (2000)

5 Experimental Design of the KULUNDA Trials

In 2012, the trials in Pervomaisky and in Poluyamki started. They focus on the soil tillage intensity in the context of the regional water regime as the main influencing factor on the yield levels. When planning these trials, we considered not only experience gained from Canada and local expertise, but also experience from conservation tillage experiments in the neighbouring Kazakhstan (Suleimenov et al. 2014; Meinel et al. 2014). The strategies for fertilisation and crop protection are adapted according to the crops, and remain unchanged throughout the entire trial for all of the soil tillage variants.

In all of the variants, fertilisation is performed with seeding. Fertilisation with seeding according to the "Single Shoot" principle: deposition of the fertiliser together with the seed in the seed furrow. The crop protection measures are the same on all plots.

Before seeding, treatment with a total herbicide was performed on all plots.

In the plots, wheat, rapeseed and peas are treated against weeds as required. The trials consist of 4 main blocks. Each block contains 6 plots, each of which corresponds to a tillage variant.

The crop rotation in the trial is a four-stage rotation with cereals, oilseed and legumes. Spring wheat (Ws)—field peas (FP)—spring wheat (Ws)—spring rapeseed (Rs). They rotate on blocks A to D (Fig. 8).

More details of the experimental design and agrotechnology are given in Tables 1 and 2.

These trials are not repeated multiple times; however, because of their extensive dimension, the trials are quite robust against outliers. Because of the large plot sizes of 1.4 and 0.7 ha, the trials can be integrated into the running production processes on the farms very well.

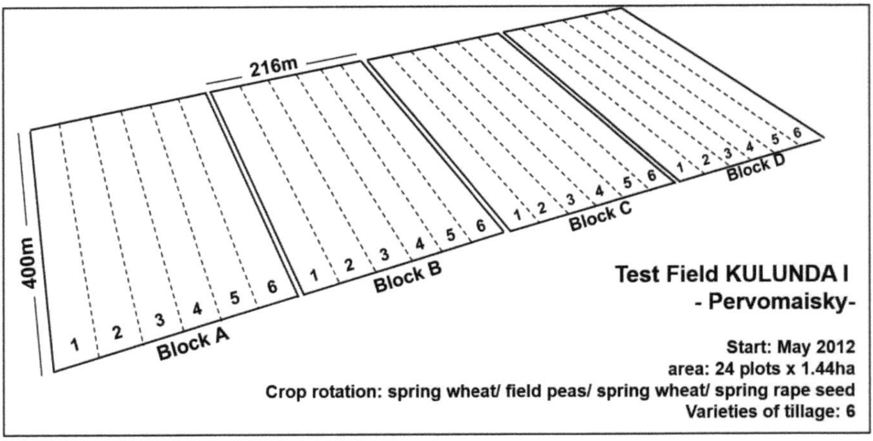

Fig. 8 KULUNDA - I - Test plots in Pervomaisky, the KULUNDA - I - Test in Poluyamki looks the same

Table 1 Field trials—scheme of treatments and techniques—Pervomaisky

No	Seedbed preparation	Coulter technology for seeding	Fertilising	Crop protection	Autumn cultivation
1	None	Individually depth-controlled narrow chisel tine coulter with trailing depth control/press roller, row spacing 25 cm [6]	Mineral with seeding (N fertiliser/MAP fertiliser)	Pre-emergence (glyphosate), in the crop as required	None
2	Disc cultivator 8–10 cm [1]				Disc cultivator 8–10 cm [1]
3	Sweep opener cultivator unit in drill combination 12–14 cm [2]	Non-individually depth-controlled double disc coulter with trailing steel roller packer and chain harrow, row spacing 15 cm [7]			None
4					Disc cultivator 8–10 cm [1]
5					Sweep opener cultivator 12–14 cm [2]
6					Chisel plough 20–22 cm [3]

Table 2 Field trials—scheme of treatments and techniques—Poluyamki

No	Seedbed preparation	Coulter technology for seeding	Fertilising	Crop Protection	Autumn cultivation
1	Disc cultivator 4 cm [5]	Individually depth controlled narrow chisel tine coulter with trailing depth control/press roller, row spacing 25 cm [6]	Mineral with seeding (N fertiliser/MAP fertiliser)	Pre-emergence (glyphosate), in the crop as required	Disc cultivator 4–6 cm [5]
2	None				None
3	Disc cultivator 4 cm [5]	Not individually depth-controlled sweep opener with trailing steel roller packer, row spacing 21 cm [8]			Disc cultivator 4–6 cm [5]
4	None				None
5					Chisel plough 14 cm [3]
6					Chisel plough 20–22 cm [4]

6 Agricultural Technology in the KULUNDA Trials

Items of machinery and their specifications are shown in the following figures. Numbers in angular parentheses refer to same numbers in Tables 1 and 2.

1. Heavy 4-row disc harrow (Fig. 9)
   - 560 mm disc diameter for tillage down to a depth of 18 cm
   - 5.7 m working width
   - Spiral roller packer for breaking clods
   - Good incorporation of large amounts of organic harvest residues (esp. maize)
   - Ultra-shallow tillage is not possible, because the implement is heavy and the disc diameter is very large.
2. 2-beam sweep opener cultivator (Fig. 10)
   - 42 cm line distance
   - 7.2 m working width

Fig. 9 Heavy disc harrow DISKATOR

Fig. 10 Sweep opener cultivator АПК 7.2

- Tandem flat bar roller
- Medium depth and deep tillage possible
- Shallow and ultra-shallow tillage not possible.

3. 1-beam chisel plough (no photo)

   - Cutting over the full area
   - Non-fracturing/non-inversion
   - Autumn tillage to 14 cm
   - Capillary break, water storage after harvest, weed control, reduction of erosion
   - 9 chisel shares
   - 1.11 m line distance
   - High fuel requirement because of the deep tillage and the wide coulter.

4. 2-beam chisel plough (Fig. 11)

   - Cutting over the full area
   - Non-fracturing/non-inversion
   - Autumn tillage to 20–22 cm
   - Capillary break, water storage after harvest, weed control, reduction of erosion
   - 5 chisel shares
   - 1.11 m line distance
   - High diesel and working time requirement because of deep tillage and narrow working width.

5. 2-row compact disc harrow (Fig. 12)

   - 460 mm disc diameter
   - Non-serrated discs for shallow and medium-depth tillage
   - Stubble breaking and seedbed preparation
   - Wedge ring roller for capillary connection of the tillage horizon with the deeper layers

Fig. 11 Chisel plough ПГ 3-5

Fig. 12 Compact disc harrow CATROS 6001-2

- Easy to pull soil tillage implement, high work rates
- Can be used from 2–12 cm working depth (extremely shallow to medium-depth mulch seeding).

6. Pneumatic air seeder (Fig. 13)

   - No previous tillage of the soil required
   - Individually depth-controlled narrow chisel (11 mm) with trailing depth control/press wheel
   - 25 cm row spacing
   - Fertiliser application with seed in the seed furrow (single shoot)
   - Minimally invasive soil movement when opening the seed furrow
   - Low erosion susceptibility and low water losses in the soils thanks to standing old stubble (low wind velocity)
   - Extremely low tractive force requirement for this working width (12 m width, 2–4 l/ha diesel).

7. Mechanical drill double disc opener (Fig. 14)

   - Double disc coulter without individual depth control
   - Row spacing 15 cm
   - Trailing steel roller packer and chain harrow for reconsolidation and refilling of the seed furrow
   - Previous mechanical tillage of the soil absolutely necessary
   - Fertiliser application together with seed in the seed furrow possible
   - Imprecise seed placement, poor coulter guidance on soils with poorly prepared seedbed—uneven field germination.

8. Mechanical sweep opener seed drill (Fig. 15)

   - No previous tillage of the soil required
   - 21 cm row spacing
   - Suitable for seeding directly into the stubble, intensive work pattern
   - Sweep opener without individual depth control

Fig. 13 No-Till—Airseeder CONDOR 15001-C Wind-stopper effect due to standing stubble after seeding with single guided narrow hoe opener systems (author's photos 2009/2012, shank scheme courtesy of Amazone company, Germany

Fig. 14 Mechanical drill СЗП 3.6—double disc opener only just cuts through straw, for use after seedbed preparation only

- Trailing steel roller packer
- Fertiliser with seed in the seed furrow (single shoot)
- Intensive mechanical tillage of the soil > mechanical weed control, higher soil moisture losses through soil tillage.

Fig. 15 Mechanical chisel opener drill C3C 2.1, no single guided shanks, heavy soil disturbance while opening the furrow

7 Influence of the Seeding Technology on Emergence and Crop Establishment

The KULUNDA I field trials investigate the effect of different cropping and agricultural technology variables on the key parameters "Field germination" and "Yield". In this context, the focus of the investigation is mainly on the parameters framed in green in the following scheme (Fig. 16).

Field germination stands for the germinated plants, as a relative or absolute fraction of all of the planted seeds per unit area. It is therefore the basis for the later crop development and the crop density, which is determined by the number of

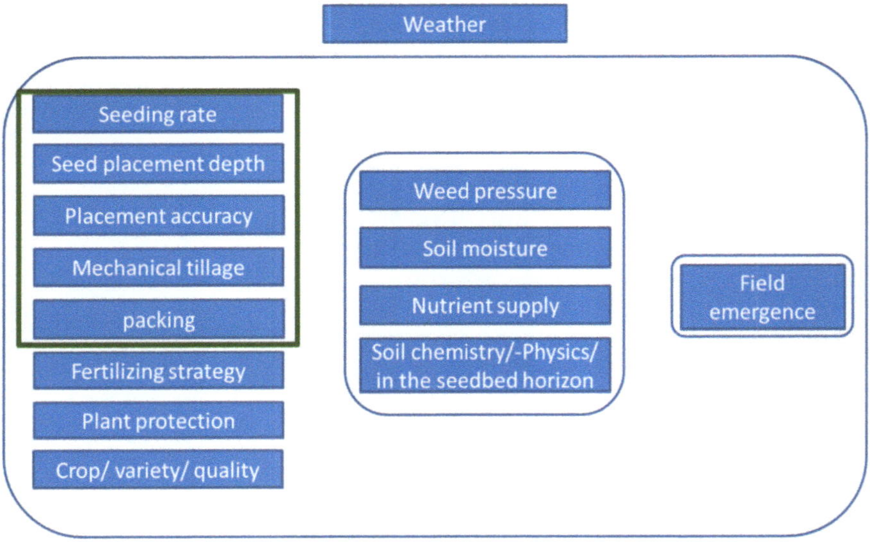

Fig. 16 System of factors influencing field germination

plants and the number of the ear-carrying stalks per unit area. Together, these parameters are decisive for yield development and are influenced by the variety selection, the nutrient availability and the weather conditions during the growing phase.

The course of the weather conditions and the nutrient availability determine whether a cereal crop is capable of fully ripening all of the ear-bearing stalks, or if deficiency stress (usually water deficits) causes premature ripening. In the traditional crop production of West Siberian regions, a very high seeding rate is generally used. On the one hand, this is intended to compensate for any inaccuracy of the seeding technology. The high deviation of the desired seed placement depth leads to a reduction in field germination. On the other hand, it is also designed to compensate for the low germination capacity of the seed, as more than 5 reproductions are often used in these regions.

It is not rare that the crops are very dense, especially in the early growing phase. This is because of the favourable germinating conditions of a tilled seedbed free of organic residues, which warms up quickly in the spring and also provides a larger amount of plant-available nitrogen without fertilisation. Furthermore, the weed pressure is much lower after the intensive tillage. During the dry period after seeding, however, the crops often suffer from massive water deficits and are significantly damaged. New precipitation, which can usually be expected as of July, is generally not capable of saving the crops, so they ripen prematurely, which leads to significant yield losses.

Therefore, a very dense crop stand is not necessarily a guarantee of high yields. However, the size of the relative field germination is a clear indication of the technical precision of the seeding technology implemented. Seed placement accuracy, the intensity of the mechanical tillage when opening the furrow, and the type of reconsolidation are directly dependant variables resulting from the seed drill construction.

For this reason, this trial aims to investigate their influence on field germination so as to be able to draw conclusions on the accuracy of the seeding technology used. For this purpose, the seed drills introduced in the trial description are used on the field. While variants one and two use minimally invasive seeding technology, in a pure direct-seeding system or in a system with minimal soil tillage, the tillage intensity increases progressively from variant three to variant six. For those comparisons, we also used seeding technology that is commonly found in the regions' conventional farming systems. Experience from cultivation regions in the south of Canada have shown that a high seed placement accuracy can be achieved with precisely guided seeding coulters, which enables a significant reduction in the seeding rate without suffering yield losses (PAMI and SSCA 1998).

Taking account of this, variants 1 and 2 are seeded at a lower rate than the conventional variants 3–6. The reduced soil tillage with the opening of the furrow using narrow chisel openers leaves a large amount of organic materials at the surface, and enables seeding in standing stubble rows.

The seeding rate and placement depth are classic agronomic parameters that primarily depend on the crop, the weather conditions and the germination capacity of the seed. Secondarily, the placement depth and the seeding rate are determined by the technical quality of the seeding procedure. The more precise the coulter guidance of the seed drill, the more accurate the placement of the seed at the preset placement depth. The seeding rate directly depends on the precision of the seed placement. The greater the fraction of seed that does not germinate or germinates late due to poor placement, the greater the required seeding density to compensate for the losses.

Figure 17 demonstrates the difference in crop establishment depending on the seeding coulter. On the left is an unevenly developed spring wheat crop due to differences in the time of field germination. This was caused by imprecise placement with a double-disc coulter without individual depth control. In the figure on the right, a very even crop can be seen, which was achieved through even field germination resulting from seed placement with chisel openers with individual depth control. This crop can be managed much better in terms of well-timed crop protection and ripens more evenly, which brings significant advantages for harvesting.

Furthermore, the chisel opener helps to prevent the so-called "*hairpinning effect*". In Fig. 18 the problem can be seen clearly. Due to their low coulter pressure, conventional double-disc coulters generally only work on pre-tilled soil, otherwise they could not cut through the organic harvest residues. But even with mulch seeding, the double disc coulter can fail and press the straw into the seed furrow. This is often the case with poor straw distribution by the harvester (swath formation) and with shallow pre-tilling. In contrast, a chisel tine clears the soil and therefore keeps the harvest residues out of the seed furrow. This ensures optimal closure between the seed and the soil. The moist fine earth fraction is also higher when seeding with a chisel. This also promotes germination. Furthermore, the coulter shape directly determines the intensity of soil movement when opening the furrow. Although the intensive soil movement may achieve better mechanical weed control, the water losses during soil tillage often lead to desiccation of the seedbed and therefore to poor field germination and a moderate establishment of the crop. This problem is mainly observed in very dry spring seasons.

Fig. 17 Uneven germination of seeds leads to uneven plant establishment—problems with the timing of plant protection treatments may occur (*left*) Regular establishment on *right*

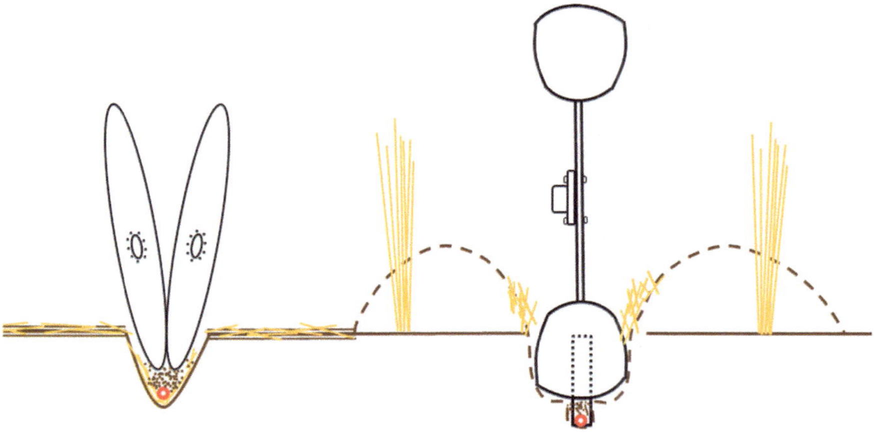

Fig. 18 Hair pinning is one of the biggest problems for disk openers under no-till systems, but also poor seedbed conditions due to poor straw distribution may lead to serious problems in crop establishment (author's illustration after PAMI 1987)

A special feature of the chisel opener system is the "*furrow cave-in*" principle. In Fig. 19 the scheme and the picture below graphically illustrate the working mode.

Here, it is not the distance between the level of the untilled soil and the seed that is important for seed placement, but rather the actual soil covering height directly above the seed grain. It is therefore possible to seed very shallowly, because the seeding chisel penetrates directly into the moist soil and the soil that is above and

Fig. 19 Hoe openers with packer wheels, guiding the depth placement of the seed according to the furrow cave-in principle (scheme by Amazone company, author's photo)

around the seed is also moist. This saves the energy of the seedlings, which must reach the soil surface as rapidly as possible.

This technique was proven to reduce evapotranspiration and to increase the soil water available for the crops in southern Canada's agricultural regions (AWWPC 2013 among others). The trial aims to demonstrate whether this fact is readily transferable to West Siberia.

Regarding the minimal placement depth, several coulter shapes also have clear limits. A sweep opener, as shown on the seed drill in Sect. 5 [8, Fig. 15], has a minimum placement depth of approx. 5 cm. This is due to the design, since a more shallow setting would not provide sufficient earth flow to cover the seed. This makes it difficult to use a conventional sweep opener seed drill for seeding crops such as flaxseed or rapeseed. These need to be able to reach the soil surface as rapidly as possible after germination.

8 Fertilization and Crop Protection in the KULUNDA I Trial

The fertilization, crop protection and the crop/variety selection should be the same for all variants in the trial. In all of the variants, 100 kg/ha Selitra (NH_4NO_3, 34.6 % pure *N*) were directly metered in the furrow with the seed. Because of the low precipitation rates and the low air humidity, fertilization performed with the spreader would have been only just or not at all effective (Lafond et al. 1999, p. 5). For the arid regions of West Siberia, the same can be assumed as for the agricultural areas in southern Canada. The mineral fertilizer would stay lying on the soil surface and would not become plant-available in the solution. In Canada, liquid fertilizer is often used, but it could not be considered in this trial because of technical reasons. Crop protection was performed with total herbicides at pre-emergence, and also in the crop as required using herbicides. For pre-emergence application, a common total herbicide based on glyphosate was used.

The in-crop application was different in terms of time, chemical composition and amount according to the demand. Usually, a selective herbicide was used to control monocots or dicots.

9 Results of the Tillage Trials in the Dry Year 2012

The weather conditions in the spring of 2012 were drier than average in the entire Altai region. At the Poluyamki site the precipitation in May and June was only 65 % of the precipitation of the long-term average. At the Pervomaisky site the May precipitation was only 75 % of the long-term average. As a result, soil moisture and precipitation were the most limiting factors, even more than in other years. As can be seen in Fig. 20, the conventional variants A3/A4 and C3/C4 exhibited lower

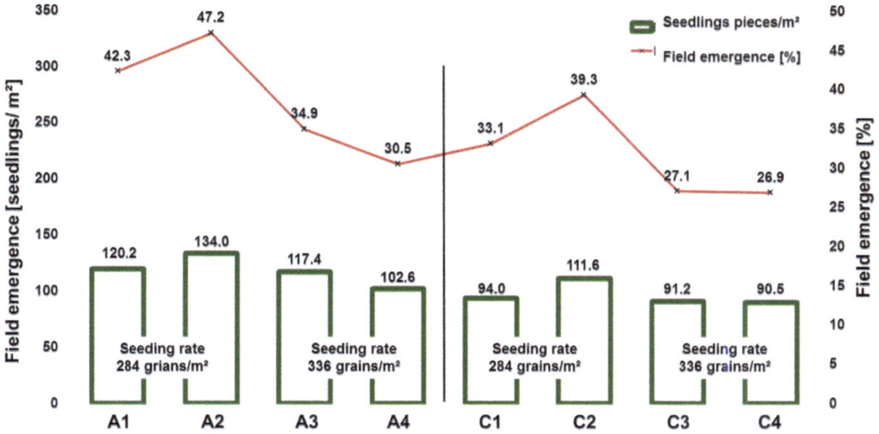

Fig. 20 Seeding rate and field emergence of spring wheat in Poluyamki. High seeding rates cannot compensate for the loss of seedlings due to moisture losses after intensive mechanical seedbed preparation and seeding with chisel openers—especially in extremely dry years

germination rates, despite higher seeding rates with the sweep opener, than plots A1/A2 and C1/C2, where direct seeding with narrow chisel tines was used.

The soil on the trial field had been kept under chemical fallow in the previous year, 2011, and then tilled several times in preparation for the trial. Although this created the same initial conditions for all of the plots, there was almost no soil cover to protect against desiccation.

The repeated strong soil movement during furrow opening with the wing coulter caused additional strong moisture losses in the topsoil on plots A3/A4 and C3/C4. Furthermore, the placement precision with the sweep opener is lower, because it has no individual depth control. For these reasons, there is considerably lower field germination despite higher seeding rates.

Here too, two extensively managed variants were compared to two intensively managed variants in 2012. Because of technical problems with the tractor technology, the required pressure to achieve optimal soil closure for the seeds could not be reached when seeding with the direct seeding technology and a narrow chisel opener. This is achieved with the trailing wheel, which both controls the depth of the coulter and is important for refilling and reconsolidating the seed furrow.

For the required reconsolidation, conventional seeding technology relies on a heavy steel roller packer, which is pressed onto the soil only through its weight, without hydraulic pressure. However, without good reconsolidation, there is no replenishment of soil water through capillary rise into the seed horizon.

Especially in dry years such as 2012, however, soil water is extremely important, since precipitation cannot be expected for a long time. For this reason, the more intensively managed plot A3 has better field germination than the extensively managed plots A1 and A2 (Fig. 21). The low field germination in plot A4 is notable. This is most likely due to the high soil moisture losses during the intensive seedbed preparation with the sweep opener cultivator. In 2012, another direct seed

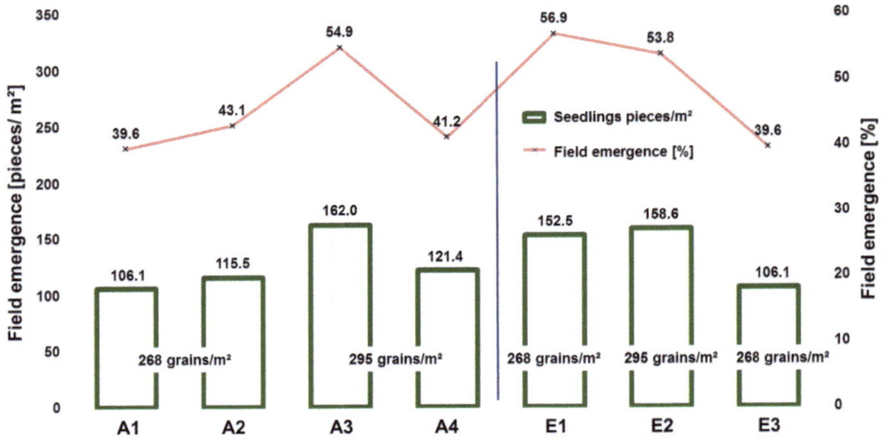

Fig. 21 Seeding rate and field emergence of spring wheat in Pervomaisky. Due to technical problems, the shank pressure to properly place and pack the seeds was not reached using no-till techniques in A1/A2—this led to poor germination under dry conditions

Fig. 22 DMC individually-controlled hoe opener system (Amazone company, Germany)

drill (DMC system) was also tested in Pervomaisky, which is also equipped with individual depth control on the chisel openers (Fig. 22).

Variant E1 (direct seeding) produced the best field germination. The precision of the individually controlled chisel opener via a parallelogram and the reduced soil movement when opening the furrow paid off under the extremely dry conditions in 2012.

The medium-depth tillage (12 cm) with the sweep opener cultivator and the subsequent seeding with the DMC deliver comparable results for E2 to those for E1.

However, it delivers considerably better results than on plot E3 (very shallow tillage—4 cm with compact disc harrow and subsequent seeding with the DMC).

It is conceivable that the reason lies in the rapid germination of weeds after the shallow tillage and the late treatment with crop protection products, which was performed more than three weeks after seeding, in the middle of June 2012.

As a matter of principle, Minkey et al. (2001) also come to the conclusion that for the arid cropping regions in Western Australia, an intensive mechanical soil tillage paired with a high seeding rate is clearly counterproductive for field germination. The following could apply for the high field germination rates on E2.

The medium-depth tillage with the cultivator on E2 brought up more moist soil from deeper layers for the seed horizon. Together with the warm weather conditions and the good seeding with the DMC, this led to very rapid field germination of the wheat, which then effectively suppressed the weeds. However, there can be no claim to statistical reliability because there is only one year of observations.

Because of the drought in June, the crops suffered major water deficits, which was exasperated by the low snow levels in the winter of 2011/12, which only provided negligible soil moisture reserves. Among all of the plots, only roughly 16 %-vol. soil moisture could be measured using FDR hand probes. At the Poluyamki site, there was higher desiccation stress under the intensively managed variants due to the stronger soil tillage, which resulted in lower yields (Fig. 23). The elevated desiccation stress in the acute growth period was particularly problematic. At the end of June 2012, the soil moisture values on the intensively managed plots were up to 21 % below those on the extensively managed plots.

Effectively, the yields on the trial plots behave proportionally to the field germination in the early summer. In terms of the economic balance of the trial, however, it is significant not only that the field germination under more intensive management was lower, but also that the seeding rates and the financial expenses due to the greater power requirement of the tractors were greater. Overall, without

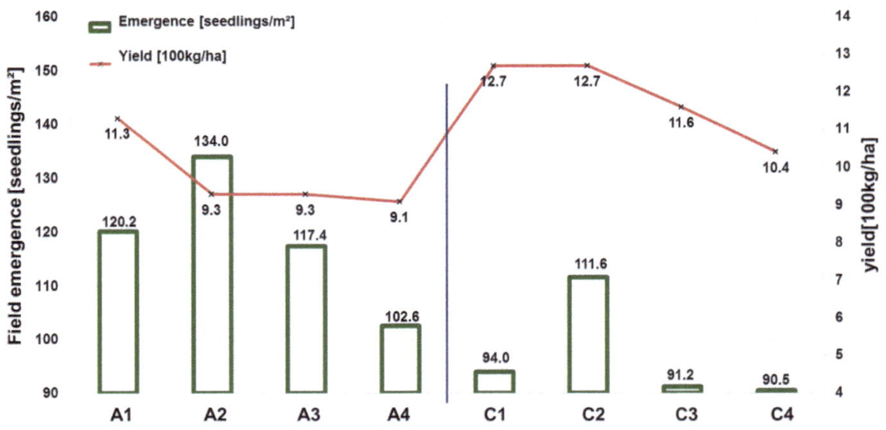

Fig. 23 Field emergence and yield of spring wheat (after spring wheat) in Poluyamki 2012. The more intensive the soil tillage, the higher the moisture loss and the lower the yields of spring wheat under very dry conditions in the dry steppe

precise balancing, it can already be seen that the less intensive cropping system offers savings potential.

In Pervomaisky (Fig. 24), it has been demonstrated that the thin crops under the most extensive soil tillage were capable of achieving the best yields.

Even the direct seeding and minimum till plots A1 and A2, where the field germination was significantly lower, were capable of establishing the best crops up to harvest. The auxiliary trial with the coulter system of the DMC also demonstrates that the most extensive variant achieved the best yields.

As in Poluyamki, the very low precipitation in Pervomaisky in the winter of 2011/2012 generated only very little or even no additional soil moisture. On average, the soil moisture on the trial field was approx. 15 %-vol. At the end of the growing period, the differences in soil moisture were within the measuring error of the manual measuring device (±5 %) across all of the variants. Precipitation in August 2012 then compensated for the differences. In the phenologically relevant phases in June/July, however, the soil moisture contents of the intensively tilled variants were approx. 10 % lower than those of the extensive variants. As can be readily seen in Fig. 25 the maturity of the crops on 16.07.2012 varies greatly.

The crops that were established under very extensive methods (A2/E1 Compact disc harrow for seedbed preparation/seeding with narrow chisel opener with individual depth control) still exhibit dark green colouring and are therefore still assimilating. After shallow soil tillage with the disc harrow, the crop A3 was seeded through the poorly guided double disc coulter.

The soil moisture losses through the weed pressure due to the shallow soil tillage forced the plants to mature early. Seeding with the precise individual depth control on the chisel openers and the thinner crops have proven to be beneficial under the very dry weather conditions in 2012.

The crop that was established with mulch seeding using a double disc coulter without individual depth control (A3 Sweep opener cultivator for seedbed

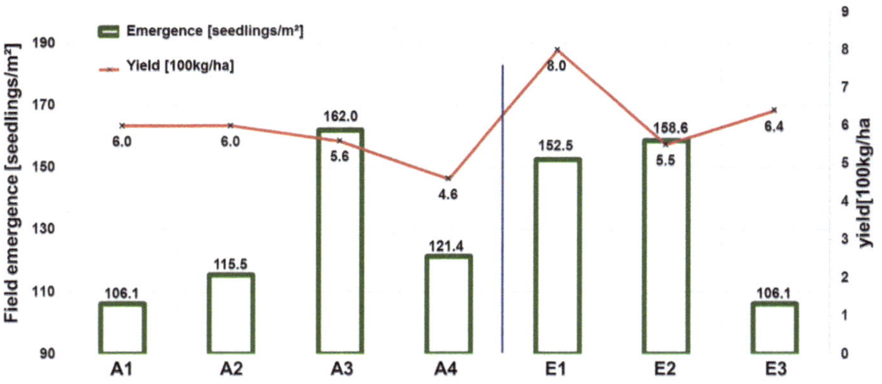

Fig. 24 Field emergence and yield of spring wheat in Pervomaisky 2012. Fewer plants per m^2 can help the established plants to use soil water and water from sparse precipitation more effectively, and this leads to higher yields in extensive plots in the very dry year 2012 at the tall grass steppe trial field

Fig. 25 Stands of spring wheat in Pervomaisky on July, 16, 2012. A2 and E1 spring wheat were seeded using individually controlled narrow hoe openers with less intensive soil disturbance than A3. The higher water availability and the better water use efficiency is shown by the more *greenish color*

preparation, seeding without individual depth control on the double disc coulter) was much denser at the beginning of the growth phase. The water stress caused the crop to mature more rapidly, as can be seen from the yellow colour.

The case example of the extremely dry year of 2012 demonstrated that extensive modern cropping systems offer many advantages. The minimized soil tillage can reduce the risk of erosion through deflation, because the organic cover effectively reduces the wind velocity at the soil surface. The evaporation and therefore the soil water losses are also reduced. The increased shading provided by the organic cover decreases the surface temperature. The reduced wind velocity decimates the suction pump effect, and therefore the constant replenishment of soil water from deeper layers through capillary rise (Cutfort and Mcconkey 1997).

The minimized tillage of the soil in combination with precise seeding technology enable a significant reduction in the seeding rate, because the deviation from the desired seed placement depth is much lower due to the individual depth control. At the same time, however, there should be an investment in better seed quality. A higher germination capacity of the seed also enables a further reduction in the seeding rate.

The conversion to less tillering cereal varieties helps to prevent drought stress and premature ripening of the side shoots, thus enabling the main shoot to develop optimally under the given weather problems. As was demonstrated by the trial in 2012, great importance must be placed on chemical crop protection measures. The weed pressure increases because of the minimized soil tillage. Chemical crop protection can effectively act against the weed pressure. A broad crop rotation helps to optimize the application of means and products and can contribute to increased resistance to market fluctuations due to the wider range of marketable products.

The results represent the case example of a very dry year, of the kind which occurs about every five years in the Altai region and also on the scale of West Siberia (Charlamova and Silanteva 2011).

It is conceivable that in a few individual cases during the cycle of the overall crop rotation, especially in years with very high precipitation, increased soil tillage could lead to higher yields. If less water stress occurs, more mechanical soil tillage would mean reduced competition with weeds and more nutrients available through mineralisation stimulus. However, the economic aspect must also be considered during the entire time period. It can be assumed that the additional expenses in conventional systems caused by higher intensity of autumn tillage, seedbed preparation as well as seed and weed control strongly reduce the profit from higher yields in isolated favourable years, or even make it not worthwhile. The four-year test period of the field trials will help to clarify this situation.

In addition to the general test of the method in Siberia, the focus is on the specific conditions, in particular the climatic conditions in the project region. For this reason, trials were established for the seeding date and the seeding depth of different crops. It is definitely conceivable that due to the pronounced drought in June, a later seeding date could bring higher yields.

10 Trials for Crop-Dependent Optimal Seeding Dates and Placement Depths

The seeding period for spring crops in West Siberia is short and often characterized by strong aridity (Belyaev and Volnov 2011). Important precipitation can be expected later in the season, when it rarely has a yield-relevant effect on the important cereal crops. In general, they fall on relatively limited areas in the summer months. The frequency varies extremely from one year to the next. The optimal seeding date is decisive for the course of crop development.

Seeding too late can result in high losses, because seed placement has to be very deep during the dry period. It is possible that the plants mature too late or not at all because of their retarded development. Seeding too early can result in crop stress during the dry period after seeding.

Another aspect is the timing of the weed management. If the seeding date is delayed, there can be the advantage that the field weeds are further developed and the treatment with crop protection products is therefore more effective. For example, late seeding provides good chances of effectively controlling wild millet. The mandatory glyphosate treatment in the middle of May mainly acts against shepherd's purse (*Capsella bursa-pastoris*), field pennycress (*Thlaspi arvense*), wild millet (*Echinocloa* species) and thistle (*Salsola* and *Cirsium* species) (Suleimenov 1991, quoted in Ullrich 2011).

This significantly minimises the competition pressure for growing space and water for the further growing period. The optimal seeding date is decisive for the logistics and the schedule organisation in the seeding period. In this context, it is important to make use of scheduling advantages resulting from the plants' physiological requirements of old and new crops in terms of the seeding date. This enables the seeding period to be better organized even though there are several crops in the crop rotation. This potential can, for example, be used for the integration of technical crops and oilseed.

In this trial, spring wheat and spring rapeseed are seeded at two different depths and at two dates (Fig. 26).

The goal is to determine the optimal seeding date together with the best placement depth under the site-specific conditions. Regarding the placement, the best possible compromise must be found, resulting from the depth at which germination is possible even under very dry weather conditions, and at which the sprout is able to reach the soil surface as rapidly as possible and to begin assimilation.

The trial is established in a four-year crop rotation consisting of spring wheat—field peas—spring wheat—spring rapeseed and replicated three times. In each row, the crops rotate on four blocks (Fig. 26). The plots with the combinations covering height/seeding date are randomized. The plots with field peas are not investigated and are only used for the design of the crop rotation. The method corresponds to the modern minimally invasive no-till method. The seed drill with a CONDOR chisel opener system is also used.

In the following, the seeding dates and covering height that are used in this trial will be listed. In this case, the covering height was determined using empirical values from the field. Seeding date and placement depth vary by a few days or a few millimetres according to the weather conditions.

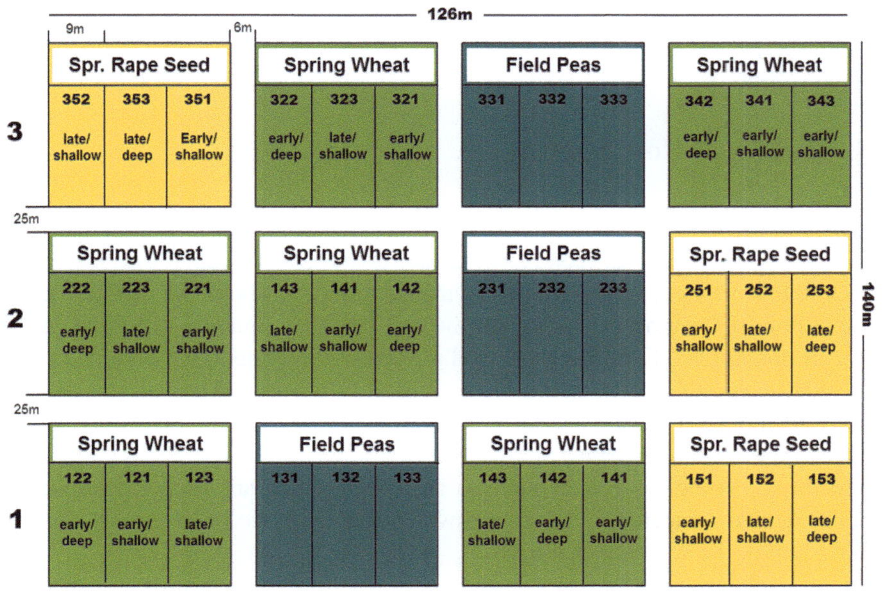

Fig. 26 Test plots for investigating the optimal seeding date and seed placement depth of spring wheat and spring rapeseed under dry steppe conditions in Poluyamki, beginning in 2013

Regarding the seeding dates in the practical course of the trial, it is ensured that the time differences between the scheduled dates for a crop are as great as possible.

Spring wheat		Spring rapeseed	
Shallow	3–4 cm	Shallow	2–3 cm
Deep	5–7 cm	Deep	3–5 cm
Early	10/05–20/05	Early	Beginning of May—10/05
Late	20/05—Beginning of June	Late	10/05–25/05

Field peas are seeded early at about the same time as the rapeseed variants. The covering height is usually about 6 cm. Fertilisation is performed using the single-shoot method with seeding. Crop protection is solely chemical. The treatments are performed at pre-emergence and, as required, at post-emergence.

The trial was established in the spring of 2013 and is scheduled to run for a total of three years. In terms of the crop establishment, the weather conditions are directly linked with the effect of the seeding date and the placement depth. At this point, no results will be presented, because at least two years of observations should be considered.

In addition to the question about the coulter shape, the optimal row spacing has not yet been determined. If other row spacings are possible (Lafond 1994), the work rate of the seed drills could be increased significantly, diesel could be saved, and due to the reduced soil movement, more moisture could also be conserved in the soil during seeding. Trials to address this were also established in 2013 within the scope of the project.

11 Trials on Row Spacing

Especially in the very dry areas of the northern Kazakh and West Siberian agricultural regions, there has been increased discussion about optimal row spacing in the last few years. There is the fundamental question about the optimal space distribution of the crop depending on the soil moisture situation, the trend in the weather conditions, the seeding rate and the fertilisation rate. Crop protection also has to be readjusted for changing row spacings. In addition, the questions of crop establishment, crop closure and the associated evaporation during the course of the season are also central points.

In the following, several hypotheses are listed, which are known assumptions among farmers concerning changing row spacings, and are to be proven right or wrong within the scope of this trial.

- With increasing row spacing, the water stress increases, because the space between the seeded rows receives more sunlight, resulting in increased evapotranspiration.

- Greater row spacing also significantly increases the weed pressure, because crop closure occurs later and is less complete. The chemical crop protection measures need to be more precise.
- Greater row spacing decreases fuel consumption per metre of working width (at the same speed). The newly available power is converted to seeding speed and area efficiency, in order to use the short seeding window more effectively. This results in economically and agronomically interesting aspects, which justify the investigation of the ideal row spacing.

The trial aims to determine which row spacing is optimal in the long run for spring wheat, spring rapeseed or field peas under the climatic conditions in West Siberia. The crops are seeded with row spacings of 25, 33.3, 37.5 and 50 cm. All of the row spacings are drilled accordingly at two different seeding rates. The cropping method essentially corresponds to the modern no-till method used for the KULUNDA I trials and for the trials for the optimal seeding date/covering height. No separate mechanical soil tillage is performed. Crop protection is performed chemically at pre-emergence and as required post-emergence. The mineral fertilisation is performed using the single shoot method with seeding. A four-year crop rotation is established: spring wheat—field peas—spring wheat—spring rapeseed.

The trial was randomised for statistical reliability. In doing so, the crops are rotated within four blocks for each row. The trial is replicated three times. This trial started at the same time as the trial for the optimal seeding dates and placement depths in 2013 at the Poluyamki site. Conclusive results are expected in 2015.

12 Summary and Outlook

The central question that was posed at the beginning was that of whether Canadian cropping systems could be adapted for use in West Siberia and in northern Kazakhstan. Based on the trial results from 2012, this can basically be endorsed. Given the very dry conditions in the first year of the project in 2012, the presented field trial showed that the intensity of soil tillage and especially the seeding technology used can have a great influence on crop establishment and ultimately on the yield expectations. Because the presented case example was only based on the results from one year, there can be no full claim for statistical reliability.

However, the trend towards positive effects on crop establishment, uniform crop development and yields is clearly visible.

High seeding rates were not capable of producing higher field germination in any of the (regular) cases. On the contrary, it was demonstrated that thinner crops achieve higher water use efficiency, especially under no-till, and therefore stronger assimilation.

Accordingly, the yields were highest on the plots with the most extensive variants. Given the dry weather conditions, the high soil moisture losses after mechanical seedbed preparation and repeated soil movement when opening the furrow led directly to yield depressions on the intensively tilled plots. For the sites and their natural landscape-climatic differences, it was not possible to make a qualitative distinction for the year 2012 in terms of the intensity of the effect of mechanical tillage and the precision of the technology on yields. At its full intensity the drought extended over the entire Altai region, meaning that the same disastrous weather conditions prevailed over great distances. The KULUNDA I field trial will be continued at least until the year 2015. In the further course of the project, there need to be investigations into the effect which the tillage intensity and the seeding technology used have on the yields of all crops in the crop rotation. The long-term average and the effects of different weather conditions are particularly interesting here. This will enable the investigation of the effects of crop rotation and the influence of the technology used on the economic balance of the trial. It is conceivable and to be expected that the different frequencies and intensities of use of the entirely different technologies will have an effect on the time expenditure, the tractive force and therefore fuel requirements. This was already proven in Canadian field trials with different cropping systems (Zentner et al. 1998). This observation should also include the expenses for the seed and fertiliser as well as a balance of the crop protection management.

Still, the question as to whether the Canadian no-till method can be agronomically adapted to West Siberia can already be agreed upon. In terms of the effects of extensive tillage and cropping methods, the same trends are observed concerning the soil water balance and crop development. The trends in terms of economic savings potentials go in the same direction as the one that scientists have proven right for modern minimally invasive cropping methods in the south of Canada.

However, it is important to look up from the individual factors and consider the overall system. A central expansion of the trial took place in 2013. In addition to the KULUNDA-I trials, where the focus of the investigation is on the soil tillage and the agricultural machinery, the KULUNDA-II trials in Poluyamki and in Komsomolsky systematically compare three complete cropping methods according to a multi-factorial approach.

A modern no-till method, a modernised cropping method according to modern agronomic standards but with conventional agricultural machinery, and an extremely out-dated method with conventional technology, without crop protection and fertilisation as well as cereal monoculture and fallow (both black and chemical fallow) are compared on 70 randomised plots.

This is to enable the agro-ecological comparison and the balancing of the economic parameters in the overall cropping system under the influence of all systematic parameters. It is conceivable that it will not be possible to quantitatively prove an individual factor to be the cause. However, it will be possible to assess the overall method in terms of its influence on the soil water and nutrient balance, on the agronomic parameters, especially the yield, as well as on the economic balance on a realistic basis. The establishment on two sites in the climatic boundary regions of the Altai steppes (both semihumid forest steppe at the Komsomolsky site and semiarid dry steppe at the Poluyamki site) enables a comparison that covers a broad region.

Their scope, the duration of their implementation and the depth of the scientific monitoring make the field trials of the KULUNDA project in West Siberia virtually unique. Within the scope of the sister project SASCHA in the Tyumen region, in-depth field trials are also being performed (Uni Münster 2014). In addition to the trials presented in this paper, the scope of the KULUNDA field trials also includes trials on strip-wise soil tillage in row crops and effective technologies for crop protection. Together, the results should serve to inform farmers in the steppe regions of West Siberia and northern Kazakhstan about the conversion and the extensification of the cropping systems. In addition to agri-ecological questions, technical solutions should also be generated to make agriculture more effective and resource-conserving.

These projects mainly aim to protect the economic foundation that is the soil and to stabilise the income situation through the yields and marketing of the fruit. This can make a valuable contribution to the stabilisation of the rural communities of the West Siberian steppes and to international food security.

References

Atlas Altaiskovo Kraya (1978) (Атлас Алтайского края) Moscow-Barnaul (1978), p 222 (in Russian)

AWWPC (2013) Winter wheat production manual. URL: http://www.growwinterwheat.ca/documents/winter-wheat-production-manual.pdf. Accessed 7 Jan 2015

Baraev AI (2008) Isbranniye Trudi (Избранные научные труды), Tom 2, 1963–1971. Almaty (in Russian)

Batalov FZ (1980) Selskochosaistvennaya produktivnost klimata dlya yarovich sernovich kultur (Сельскохозяйственная продуктивность климата для яровых зерновых культур). Leningrad 1980, p 32 (in Russian)

Belyaev VI, Volnov VV (2011) Conception of resource-saving technologies of cereal cropping in the Altai Krai (Концепция формирования ресурсосберегающих технологий возделывания зерновых культур в Алтайском крае). J Altai State Agri Univ Barnaul 11(85):92–97 (in Russian)

BMBF (2014). URL: www.kulunda.eu. Accessed 7 Jan 2015

Burlakova LM, Tatarinzev LM, Rassipnov VA (1988) Potshvuij Altaiskovo Kraja (почвы Алтайского края). Barnaul p 72

Campbell CA, Zentner RP, Gameda S, Blomert B, Wall DD (2002) Production of annual crops on the Canadian prairies: trends during 1976–1998. Can J Soil Sci 82(1):45–57. 10.4141/S01-046. URL: http://pubs.aic.ca/doi/pdf/10.4141/S01-046. Accessed 7 Jan 2015

Charlamova NF, Silanteva MM (2011) Savisimost uroshainosti sernovich kultur v raiyonach Kulundui ot klimatitsheskich faktorov. In: Isvestiya AltGU, Barnaul, vol 3/(71), pp 88–94

Cutforth HW, Mcconkey BG (1997) Stubble height effects on microclimate, yield and water use efficiency of spring wheat grown in a semiarid climate on the Canadian prairies. Can J Plant Sci, Ottawa 77:359–366

Derpsch R, Friedrich T, Kassam A, Hongwen L (2010) Current status of adoption of no-till farming in the world and some of its main benefits. Int J Agri Biol Eng 3(1):1–26

Eule W (1962) Das Problem der Neulandgewinnung in der Sowjetunion. Diss, Uni Bonn

FAO (2006) New_LocClim 1.10. Local Climate Estimator. Current Download–Address: http://www.juergen-grieser.de/downloads/ClimateInterpolation/ClimateInterpolation.htm. Accessed 7 Jan 2015

FAOSTAT (2014): URL http://faostat.fao.org/site/291/ default.aspx. Accessed 7 Jan 2015
Fellmann T, Nekhay O (2012) Agricultural sector and market developments: a special focus on Ukraine, Russia and Kazakhstan. Workshop proceedings. In: European commission JRS scientific and policy reports [publ.] report EUR 25554 EN. Seville. http://ipts.jrc.ec.europa.eu/publications/pub.cfm?id=5599. Accessed 7 Jan 2015
Frühauf M, Meinel T (2003) Die Neulandgewinnung in der südsibirischen Kulundasteppe und ihre geoökologischen sowie agronomischen Folgen. Geo-Öko. Darmstadt 1(2003):86–107
FSS ROSSIIJ (2014) EMISS—Jedinnaja Meshvedomstvennaja informazionno-statis-titsheskaja sistema (Единая межведомственная информационно-статистическая система) Moscow. URL: http://fedstat.ru/ about.do. Accessed 7 Jan 2015
Georgiev AV (1955) God rabota po osvoyeniyu zelinikh semel v Altaiskom Kraye (Год работа по освоению зелиних земел в Алтайском крае). Celchozgiz, Moskva
Hilliard C, Scott N, Lessa A, Reedyk S (2002) Agricultural best management practices for the Canadian prairies. Canada-Saskatchewan Agri-Food Innovation Fund. Agriculture and Agri-Food Canada—PFRA, p 72
Jaehne G (1981) Sowjet-Landwirtschaft 1981. In: Giessener Abhandlungen zur Agrar- und Wirtschaftsforschung des Europäischen Ostens, vol 110. Gießen
Kaschtanov AN (1974) Sashchita potshv ot vetrovoy i vodnoy erosii (Защита почв от ветровой и водной эрозии). Rosselchozizdat, Moskva 140 pp
Kostrovsky S (1959) Pylnye buri v Altaiskich stepakh (Пыльные бури в Алтайских степах). In: Selskoye chozyaistvo Sibirii 2, Omsk, p 6
Kostyayev AI (2003) Basic Reasons for the Failure of Agrarian Sector Transformation and the Problems on its Future Development. In: Schulze E, Knappe E, Serova E, Wehrheim P (2003) Success and failures of transition—The Russian agriculture between fall and Resurrection, Halle an der Saale, pp 145–157
Lafond GP (1994) Effects of row spacing, seeding rate and nitrogen on yield of barley and wheat under zero till management. Can J Plant Sci 74:703–711
Lafond GP, Loeppky H (1992) Derksen DA (1992) The effects of tillage systems and crop rotations on soil water conservation, seedling establishment and crop yield. Can J Plant Sci 72(1):103–115. doi:10.4141/cjps92-011
Lafond GP, Irvine B, Clayton G, Derksen DA, Johnson E, Johnston A, McConkey BG, Miller P, Rourke D (1999) Wide row spacing: Is it really possible with no-till—The Canadian Prairie experience. http://pnwsteep.wsu.edu/directseed/conf99/dsprogl2.htm. Accessed 7 Jan 2015
Lafond GP, May WE, Stevenson FC, Derksen DA (2006) Effects of tillage systems and rotations on crop production for a thin Black Chernozem in the Canadian Prairies. Soil Tillage Res 89(2006):232–245
Larney FJ, Lindwall CW, Izaurralde C, Moulin AP (1994) Tillage systems for soil and water conservation on the Canadian Prairie. In carter MR. Conserv Tillage Temperate Agro Ecosyst 1994:305–328
McLeman RA, Dupre J, Ford LB, Ford J, Gajewski K, Marchildon G (2013) What we learned from dust bowl: lessons in science, policy, and adaption. Popul Environ 35(4):417–440. URL: http://link.springer.com/article/10.1007%2Fs11111-013-0190-z. Accessed 7 Jan 2015
Meinel T (2002) Die geoökologischen Folgewirkungen der Steppenumbrüche in den 50er Jahren in Westsibirien. Ein Beitrag für zukünftige Nutzungskonzepte unter besonderer Berücksichtigung der Winderosion. Diss. Halle 2002. Online: http://sundoc.bibliothek.uni-halle.de/diss-online/02/03H082/t1.pdf. Accessed 7 Jan 2015
Meinel T, Grunwald L-C, Akshalov K (2014) Modern technologies for soil management and conservation in Northern Kazakhstan. In: Mueller L, Saparov A, Lischeid G (eds) Novel measurement and assessment tools for monitoring and management of land and water resources in agricultural landscapes of Central Asia. Springer International Publishing, p 455–464. http://link.springer.com/chapter/10.1007/978-3-319-01017-5_27. Accessed 7 Jan 2015
Meyer BC, Schreiner V, Smolentseva EN, Smolentsev BA (2008) Indicators of desertification in the Kulunda Steppe in the south of Western Siberia. Arch Agron Soil Sci 54:585–603

Milkov FN (1977) Natural zones of the USSR (Природные зоны СССР). Mysl, Moscow, p 296. Quoted in http://www.geo-site.ru/index.php/2011-01-11-14-47-10/133-2011-09-26-18-56-26/682-geographechesky-pojas.html. Accessed 7 Jan 2015

Minkey DM, Riethmuller G, Hashem A (2001) Effect of row spacing and seeding rate of wheat on the emergence and competitive ability of annual ryegrass in a no-tillage seeding system. Agriculture Western Australia, Merredin Dryland Research Institute, Crop Updates. URL: http://archive.agric.wa.gov.au/PC_91472.html. Accessed 7 Jan 2015

PAMI (1987) Evaluation report 519, Tye series V 114-5360 no till drill. URL: http://www1.agric.gov.ab.ca/$department/deptdocs.nsf/all/eng3079/$FILE/519.pdf. Accessed 7 Jan 2015

PAMI and SSCA (1998) Direct seeding manual: a farming system for the new millennium. Prairie Agricultural Machinery Institute (PAMI). ISBN 0969689640, 162 p.

Penkaitis N (1987) Neuausrichtung der sowjetischen Agrarpolitik unter Gorbatschow: Eine Übersicht über die Massnahmen im Agrarbereich. Unter Mitarb. von Luis Uribe Osses. In: Schinke E (eds) [Hrsg. von d. Komm. für Erforschung d. Agrar- u. Wirtschaftsverhältnisse d. Europ. Ostens e.V.] Editorial: Berlin: Duncker u. Humblot, (1987) pp 92. ISBN 10: 3428062965 /ISBN 13: 9783428062966

Remmert H (1997) Spezielle Ökologie/Terrestrische Systeme. Springer, Berlin GmbH. doi:10.1007/978-3-642-58975-1

Sauchyn D (2009) Saskatchewan's natural capital in a changing climate: an assessment of impacts and adaptation. Prairie Adaptation Research Collaborative. http://www.parc.ca/pdf/research_publications/pp2009-02_web.pdf. Accessed 7 Jan 2015

Statistics Canada (2014) URL:http://www5.statcan.gc.ca. Accessed 7 Jan 2015

Stolbovoi V (2000) Soils of Russia: correlated with the revised legend of the FAO soil map of the world and world reference base of soil resources. International Institute of Applied Systems Analyses (IIASA), Laxenburg. Research report, RR-00-13, p 123

Suleimenov M, Kaskarbayev Z, Akshalov K, Yushchenko N (2014) Conservation agriculture for long-term soil productivity. In: Mueller L, Saparov A, Lischeid G (eds) Novel measurement and assessment tools for monitoring and management of land and water resources in agricultural landscapes of Central Asia. Springer International Publishing, pp 441–454. http://link.springer.com/chapter/10.1007/978-3-319-01017-5_26. Accessed 7 Jan 2015

Ullrich SE (ed) (2011) Barley: production, improvement, and uses. Wiley, Blackwell, p 637

Uni Münster (2014). http://www.uni-muenster.de/SASCHA/. Accessed 7 Jan 2015

von Poletika WP (1936) Russische Steppen im klimatischen Rahmen der Großebene Eurasiens. In: Die Naturwissenschaften. Berlin/Heidelberg, vol 24, no 19, p 290–296

Viehrig H (2005) Russlands Landwirtschaft und ländliche Siedlungen in der Transformation. In: Praxis Kultur—und Sozialgeographie. Potsdam. PKS 34. 2005, P 128. ISBN 3-937786-53-8

WBGU (1994) Welt im Wandel. Die Gefährdung der Böden. Bonn, p 263. http://www.wbgu.de/fileadmin/templates/dateien/veroeffentlichungen/hauptgutachten/jg1994/wbgu_jg1994.pdf. Accessed 7 Jan 2015

Wein N (1980) Fünfundzwanzig Jahre Neuland. In: Geographische Rundschau. Hamburg. 32, vol 1, p 32–38

Zentner RP, McConkey BG, Stumborg MA, Campbell CA, Selles F (1998) Energy performance of conservation tillage management for spring wheat production in the Brown soil zone. In: Canadian Journal of Plant Science, Ottawa, vol 78, pp 553–563

Part IV
Synopsis and Overall Conclusions

Chapter 32
Potential of Applying Novel Monitoring and Management Methods to Siberian Landscapes

Lothar Mueller, Askhad K. Sheudshen, Victor G. Sychev, Vladimir A. Romanenkov, Ralf Dannowski and Frank Eulenstein

Abstract This chapter reviews and summarises the overall content of the book "Novel Methods for Monitoring and Managing Land and Water Resources in Siberia". The book starts with an extended analysis of water and land resources, characterising the natural conditions of Siberian landscapes, their ecosystems, crucial processes and human impacts on soil and water quality. The status of research and monitoring is characterised in another chapter, pointing both on substantial progress achieved during the past decades, but also on gaps in our knowledge. Both chapters reveal the Siberian landscapes' great potential for economically and ecologically viable business activities, but also inefficient and unsustainable land and water management practices and the decay of the rural infrastructure. Sustainable practices should be introduced soon, and this must be based on modern monitoring and management technologies. Some more studies show that thorough and innovative research and monitoring of water and land quality is provided by Siberian institutes and their leading researchers. Addressing climate change requires innovations in landscape research. Further book chapters deal with modern monitoring

L. Mueller (✉) · R. Dannowski · F. Eulenstein
Leibniz Centre for Agricultural Landscape Research (ZALF) e.V.,
Eberswalder Strasse 84, 15374 Müncheberg, Germany
e-mail: mueller@zalf.de

R. Dannowski
e-mail: rdannowski@zalf.de

F. Eulenstein
e-mail: feulenstein@zalf.de

A.K. Sheudshen
Kuban State Agrarian University, 13 Kalinin Str, 350044 Krasnodar, Russia
e-mail: kubagrohim@mail.ru

V.G. Sychev · V.A. Romanenkov
Pryanishnikov All-Russian Research Institute of Agrochemistry,
Pryanishnikova St. 31a, 127550 Moscow, Russia
e-mail: info@vniia-pr.ru

V.A. Romanenkov
e-mail: geoset@yandex.ru

L. Mueller et al. (eds.), *Novel Methods for Monitoring and Managing Land and Water Resources in Siberia*, Springer Water,
DOI 10.1007/978-3-319-24409-9_32

and management methods developed outside Siberia but having clear potential for application. We depict some highlights which could (a) lead to a significant knowledge shift, (b) initiate sustainable soil resource use and (c) trigger substantial improvement of the ecosystem status, if introduced into Siberia or applied there very soon on a wide scale. These are (1) soil and hydrological laboratory measurement methods, (2) process-based field measurement and evaluation methods of land and water quality, (3) remote sensing and GIS technology-based landscape monitoring methods, (4) process and ecosystem modelling approaches, (5) methods of resource and process evaluation and functional soil mapping and (6) tools for controlling agricultural land use systems such as nutrient balancing methods, conservation agriculture and their technologies. More than 15 concrete monitoring and management tools could immediately be introduced into research and practice, some of them without monetary investment. We conclude that strengthening international and national research cooperation in these fields will be a key for making novel methods operational. Agri-environmental research projects should have high priority as gaps in our knowledge are particular high, and a particularly large amount of novel measurement, evaluation, modelling and management tools are available. Various tools are ready for immediate introduction into Siberian landscapes in the framework of mutual pilot projects: state-of-the-science field monitoring technologies for soil and forest hydrology (EEM-HYPROP, virtual and real lysimeters), agro-ecological models and DSS (MONICA, LandCare-DSS), soil and land quality classification and evaluation tools (WRB 2014, Muencheberg-SQR), nutrient balancing tools, and technologies of conservation agriculture. The role of internationally linked monitoring capacities is particularly emphasised, with some existing stations established in the vast agri-environmental monitoring network and others to be newly built in remote regions of Siberia and the Far East, and supported by the latest remote sensing technologies. The book contributors represent an immense innovation network which should be employed to achieve both significant disciplinary and synergetic outreach effects. This should be imbedded into more sustainable strategies aiming at research cooperation between partners from EU countries, the Russian Federation and countries of Central Asia. Maintaining the functions of great landscapes for future human generations will be the reward of those efforts.

Keywords Land · Soil · Water · Ecosystems · Sustainability · Research cooperation · Monitoring · Methods · Siberia

1 Objectives

Siberia, the Asian part of Russia, has apparently inexhaustible natural resources of fossil fuels, minerals, land, water, forests and more. Where they have been already capitalised by the mining industries, forestry, fishery and agriculture, this has been done in a wasteful way. Due to the huge dimension of the region and the low population density, most ecosystems are still intact, and the overall pressure is low to moderate. However, several factors such as common land use practices, accelerated human activities, climate change and a lack of knowledge about the significance of trends in the drift of ecosystems harbour the risk that the dynamics of processes could diverge beyond the limits of our control. Public and official awareness about the value of the natural treasures must grow and measures for sustainably handling these resources must be initiated very soon.

This should be done based on profound knowledge about the most crucial resources for the prosperity and welfare of Russia: land and water. Russian scientists and the international scientific community have recognised that the land and water resources of Siberia underlie a significant trend of alterations and are increasingly affecting global cycles and energy fluxes.

Our book is intended to provide scientific information about modern methods of measuring and evaluating the status of soil and water as a basis for updated monitoring and sustainable management concepts.

Potential readers will come from different disciplines and should have the opportunity to pick out their topics of interest. Thus, the editors have kept the book's structure quite simple and non-hierarchical, with discrete chapters by different authors in three parts.

Both the longer chapters in Part 1, "Environmental and Societal Framework for Monitoring and Managing Land and Water Resources", analyse the status of land and water resources in Asian Russia and point out gaps in our knowledge. Part 2, "Methods and Case Studies for Understanding and Monitoring the Landscapes of Siberia", presented further advanced research studies from Siberia. Their authors came from leading Siberian research institutes. Some have worked abroad or cooperate with leading international scientific groups. The focus in this second part of the book was on explaining advanced tools and procedures for analysing, evaluating and estimating the status of soil and water. Their concrete results underlined the urgent need to maintain and deepen advanced research and monitoring technologies in their specific discipline.

Part 3, "Novel Approaches and Technologies of Application Potentials for Siberia", comprised methods developed outside Siberia, but with great potential to be applied there in the near future. The methods presented have a particular degree of novelty. Most of them represented "State of the Science" approaches. The authors of this section are their inventors, creators or protagonists. In some cases pilot projects are already underway, but more activities are needed to achieve significant effects for the landscapes of Siberia.

2 Thematic Clusters and Their Novelty

The separate chapters can be allocated by type to different thematic clusters which have similarities in terms of their scale, methodology or scientific discipline (Table 1). These clusters and individual chapters shall be reviewed in this subsection.

Table 1 Thematic clusters

Cluster	Chapters and authors
(1) Land and water resources, status and trends (2 chapters)	Both chapters of Part 1, Chaps. 1 and 2 (Mueller et al.)
(2) Measurement methods for soil and water properties (7 chapters)	Chapter 3 (Khodzer et al.); Chap. 4 (Parfenova et al.); Chap. 5 (Yermolaeva and Dvurechenskaya); Chap. 13 (Pankratova et al.); Chap. 14 (Schindler et al.); Chap. 15 (Müller); Chap. 16 (Balykin et al.)
(3) Process-based field measurement and evaluation methods (4 chapters)	Chapter 6 (Pokrovsky); Chap. 7 (Chumbaev and Tanasienko); Chap. 17 (Urbaniak et al.); Chap. 18 (Funk)
(4) Remote sensing methods and GIS technologies for monitoring and modelling landscape processes (4 chapters)	Chapter 8 (Peregon et al.); Chap. 9 (Zolnikov et al.); Chap. 19 (Urban et al.); Chap. 20 (Eberle et al.)
(5) Methods of ecosystem modelling (5 chapters)	Chapters 10 and 11 (Tchebakova et al.); Chap. 21 (Shary et al.); Chap. 22 (Nendel); Chap. 23 (Mirschel et al.)
(6) Methods of resource and process evaluation and functional mapping (5 chapters)	Chapter 24 (Sychev et al.); Chap. 12 (Mikheeva); Chap. 25 (Schad); Chap. 26 (Mueller et al.); Chap. 27 (Hennings et al.)
(7) Methods and technologies for controlling agricultural land use systems (4 chapters)	Chapter 28 (Yefremov et al.); Chap. 29 (Eulenstein et al.); Chap. 30 (Suleimenov et al.); Chap. 31 (Grunwald et al.)

Cluster 1, "Land and water resources, status and trends", characterised the current status of water and land resources and the framework conditions for monitoring and management options. It includes two introducing chapters:

1. "**Land and Water Resources of Siberia, their Functioning and Ecological State**" (Chap. 1, Mueller et al.)
2. "**Status Report about Monitoring, Understanding and Controlling Landscape Processes in Siberia**" (Chap. 2, Mueller et al.)

(1) The introductory analysis (Chap. 1) revealed that not only the resource-extracting and processing industries, but also forestry, agriculture and fishery have capitalised on the land and water resources of Siberia with implications for local and global processes of nature and society. Thanks to the exploitation of natural resources, the living standards of Siberia's urban populations have been enhanced. On the other hand, the status of ecosystems has worsened. Peatland and Tundra ecosystems are endangered by the resource-extracting industries and industrial air pollution. Mining and industrial activity have damaged soil and vegetation and accelerated thermokarst processes. Forest ecosystems suffer increasingly from fires, insect infestations and improper management. Past and recent mining and industrial activity has polluted soils and water seriously and persistently in many regions. Permafrost melting has revealed cases of old and inherited pollution. The impact of agriculture on water quality is still low but will increase.

Fig. 1 Seeding of spring wheat in West Siberia at the end of May. The seedbed has been prepared using reduced tillage, applying cultivators instead of ploughs. High amounts of stubble on the surface demonstrate this. However this technology is still sub-optimal and cannot prevent wind erosion. As the surface is dry, the risk of wind erosion is very high on loessy soils even at this time of high overall moisture. Wind erosion is a main threat to soils in Siberia not only in its southern steppe zone but also in the Arctic. Modern monitoring systems are needed as a basis for process quantification, risk assessment and soil conservation measures. Chapter 18 presents those tools

Agriculture is in a recession and operates inefficiently, destroying the soil (Fig. 1). The rural infrastructure is on the verge of collapse, in the High North and the Far East in particular. Siberian regions without resource-extracting industries have lost ground in the development indexes for human capital and rank far below the average of the Russian Federation. State natural reserves (*Zapovedniks*) are endangered by illegal activities and are not sufficiently integrated into scientific monitoring. Climate change will put much additional pressure on Siberian landscapes, but hard data are required, and monitoring systems need to be modernised. Siberian landscapes have great potential for mitigating climate change through carbon sequestration and for improving people's livelihoods. Environmentally friendly business activities such as controlled or organic food production, environmental tourism and recreational fishing are still underdeveloped. This chapter concludes that the status of food production and the disintegration of rural areas are risks for the food and national security of Russia. Modern technologies for understanding, monitoring and controlling ecosystems are needed to generate sustainable development in managing the land and water resources of Siberia.

(2) The second introductory chapter (Chap. 2) reviewed the role of science and technology for understanding, monitoring, and developing Siberian landscapes. Russia has experienced significant transformations over the past 70 years. This has caused implications not only for landscape processes and for the status of terrestrial and aquatic ecosystems, but also for research strategies and their outcomes.

Russia has great traditions in landscape research disciplines such as geography, soil science, hydrology and agronomy. They are based mainly on the economic revival of the country in the post-war period. Further substantial progress has been achieved in all these fields of science over the past 25 years. Particular progress in landscape research has been made based on international projects in the fields of Arctic research and climate changing processes such as soil carbon loss. Other fields such as agricultural research have remained traditional and lack modern technologies in Siberia.

In the 1990s, there was a great shift of knowledge and technology in the better cross-linked English-speaking European scientific community. In Russia, at the same time, the introduction of the market economy accelerated environmental problems, caused a greater discrepancy between the livelihoods of the urban and rural populations, created new knowledge gaps and enlarged the gap between theory and practice in landscape research. The decay of the infrastructure of rural landscapes produced an inhospitable environment for applications of science and technology. From this background, landscape research in Siberia and in the Far East remained very traditional. Other deficits are based on a lack of communication with the international community due to language barriers. Cooperation between leading Russian and European scientists is still poorly developed and underfunded.

The Russian academic scientific system was highly organised until 2013. However, efficiency was relatively low and scientific outputs did not meet decision makers' requirements. The ongoing reform of the academic system is aimed at increased innovation and higher efficiency. However, it harbours risks of precisely the opposite of the desired effects of higher efficiency coming true, such as accelerated brain-drain and loss of objectivity.

This chapter concludes that Trans-Eurasian research cooperation is becoming very important in stabilising the current critical transition phase. Some more modern analytical methods, sophisticated technologies, models and evaluation schemes for landscape research and environmental friendly soil management technologies are available in the English-speaking community and in leading Russian institutes outside Siberia. Substantial progress in monitoring, understanding and controlling landscape processes in the framework of international research projects could be achieved by applying these new research methods in Siberia. Some of them are presented in further chapters of this book.

Cluster 2 describes "Measurement methods for soil and water properties". This thematic cluster deals with novel laboratory methods or measurement methods at the interface between the laboratory and the field, and with their results. It includes the following chapters:

1. "Methods for Monitoring the Chemical Composition of Lake Baikal Water" (Chap. 3, Khodzer et al.)
2. "Microbiological Monitoring of Lake Baikal" (Chap. 4, Parfenova et al.)
3. "Developing the Regional Indicator Indexes of Zooplankton for Water Quality Class Determination of Water Bodies in Siberia" (Chap. 5, Yermolaeva and Dvurechenskaya)

4. "Study of the Suitability of NIR Spectroscopy for Monitoring the Contamination of Soils with Oil Products" (Chap. 13, Pankratova et al.)
5. "Emerging Measurement Methods for Soil Hydrological Studies" (Chap. 14, Schindler et al.)
6. "Methods for Measuring Water and Solute Balances in Forest Ecosystems" (Chap. 15, Müller)
7. "Using the Innovative Lysimetric Technology in the Russian-German Project KULUNDA" (Chap. 16, Balykin et al.)

The first three chapters dealt with measuring water quality, the fourth (Chap. 13) is an example of measuring soil quality (SQ), and the remaining chapters deal with soil hydrology. Analyses were done in the laboratory or as a combination of field and laboratory measurements. Two chapters (Khodzer et al. and Parfenova et al.) come from the Limnological Institute (LIN) in Irkutsk, a globally acknowledged scientific institute of the Siberian Branch of the Russian Academy of Sciences. Scientists of this Institute have developed sophisticated methods for monitoring the water quality of the deepest and purest freshwater lake on the globe. Another chapter (**Yermolaeva and Dvurechenskaya**) comes from the Institute for Water and Environmental Problems (IWEP) Barnaul, Novosibirsk Department. This Institute is responsible *inter alia* for methodologies of water quality monitoring in other waters of Siberia, and it operates mainly in the Ob-Irtysh watershed. The chapter by **Pankratova et al.** dealt with soil analytics in the laboratory. This chapter comes from the Pryanishnikov All-Russian Research Institute of Agrochemistry (VNIIA) in Moscow, the leading Russian institute in agrochemical research and monitoring. Another three chapters of this thematic cluster dealt with soil hydrology, a discipline at the interface between landscape hydrology and soil science. Analytical methods in this discipline are still underdeveloped in Russia. Thus, the methods presented here are based on German technologies. The chapter by **Schindler et al.** represented state-of-the-science technology developed in the Leibniz Centre for Agricultural Landscape Research (ZALF) in Müncheberg, in cooperation with the UMS Company in Munich and the Technical University of Braunschweig. The chapter by **Müller** about forest monitoring systems comes from the Thünen-Institute, Department of Forest Ecosystems, Eberswalde, Germany. **Balykin** et al. give an encouraging example of the introduction of the latest soil hydrological measurement technology into Siberian ecosystems, based on the Russian-German project KULUNDA. The first author comes from IWEP Barnaul, and the co-authors come from the Helmholtz Centre for Environmental Research (UFZ) in Halle/Leipzig.

(1) **Khodzer et al.** showed in Chap. 3 that in the early 1990s, a system of comprehensive monitoring, including hydrophysical, chemical and biological investigations, was developed at the LIN to estimate the current environmental state of Lake Baikal (Fig. 2). Because of the unparalleled significance of its dedicated object of investigation, international cooperation has been sustained at an outstanding level in this research institute. Chemical monitoring of the Baikal water includes checking the dynamics of chemical components in the

Fig. 2 Lake Baikal, the oldest, deepest and cleanest freshwater lake on the globe, underlies a careful and modern monitoring programme, developed, operated, refined and updated by the LIN in Irkutsk. The photo shows the research vessel "German Titov", the smaller of two vessels operated for sampling chemical and biological water quality parameters in different zones of the lake. Basic sample proceeding and analyses can be provided in a laboratory on the vessel. *Photo* LIN Irkutsk

pelagic and littoral areas of the lake, as well as their input from the atmosphere and with waters of tributaries. The monitoring system allows historical trends of chemical components in the lake to be assessed, and possible changes in the biota habitat to be forecast. Low concentrations of most components in the Baikal water initiated the development of more sensitive methods. New methods for analysing anions and persistent organic pollutants (POPs) (PAHs and PCBs) were elaborated at the Institute. The reliability of methods and the quality of analytical analyses undergo annual controls according to the international and Russian programmes on inter-laboratory calibration. Based on long-term data with the application of high-precision methods, scientists were able to assess the current chemical composition of the Baikal water. Present concentrations of pollutants in Lake Baikal, such as POPs and heavy metals, are low and do not directly affect the composition of the water and biota. Moreover, the ecosystem of Lake Baikal has a self-purifying ability. In its deep area, the water of Lake Baikal is one of the purest natural waters in the world and usable for drinking without any processing. However, external loads from rivers and the air are increasing. The authors recommended an emphasis on checking the littoral zone with intensive development of the coast and near-mouth areas of the lake's large tributaries, analysing parameters such as sanitary-microbiological characteristics, nutrients and biota. It is also necessary to monitor the concentration of persistent organic compounds in the air, water, bottom sediments and biota of the lake. Many of them display mutagenic and carcinogenic activity, and they are also able to accumulate in the food chains and transfer from one organism to another.

(2) Microbial communities are strong indicators of the ecological state of water bodies. In Chap. 4 **Parfenova et al.** provided an overview of the main regulatory documents used in the Russian Federation and explained the main analytical methods to assess the microbiological quality of water resources. This team has developed methods and monitored the microorganisms of Lake Baikal. Results of systematic analyses since 2005 of the microbiological monitoring of coliform bacteria, thermotolerant coliforms, coliphages, as well as *Pseudomonas*, *Clostridium* and *Enterococcus* in Lake Baikal have been presented. It was shown that the spatial distribution of allochthonous organotrophic as well as opportunistic bacteria associated with the local anthropogenic impact: settlements, deltas of the main tributaries or domestic wastewater discharge. In the deep layers of the pelagic zone of the lake groups of opportunistic bacteria were not found. Moreover, with increasing water depth changes in the structure of microbial community were observed. Oligotrophic and psychrotrophic microorganisms were predominant there, while organotrophic microorganisms do not exceed the background level. Microbiological indicators of the pelagic zone of the lake are quite constant, preserving the excellent drinking water quality. Microbiological monitoring methods to Lake Baikal and other water bodies need to be continued and refined to preserve the unique aquatic ecosystem and to avoid potential hazards to human health.

(3) **Yermolaeva and Dvurechenskaya** (Chap. 5) showed that the indicator values of species of zooplankton used to determine water quality classes in West Siberia differ from those in Europe. They recalculated the values of the indicator significance (s) and indicator weight (J) for 111 species of zooplankton taking into account the regional peculiarities of West Siberia. Comparing example calculations of saprobity indexes from the literature on one hand and those calculated for the specific region on the other hand showed that one can define the water quality class more exactly using indexes obtained from regional features of the hydrochemical background of reservoirs and rivers. Thus, using regional indexes for zooplankton species was appropriate because of their more objective assessment of the state of the ecosystem.

(4) Soil pollution by oil is a serious problem in many regions of Russia. In Chap. 13 **Pankratova et al.** studied the suitability of Near Infrared (NIR) spectroscopy for monitoring the contamination of soils with oil products and developed a calibration method for analytical devices. They conducted calibration experiments analysing various types of soils, differing in texture,

humus content and nutrients. Samples were artificially contaminated with commercial oil products (gasoline, kerosene, diesel fuel and motor oil). Laboratory-scale scanning diffusion-reflectance NIR analyzers were used. The authors quantified the influence of all these factors on NIR spectra and developed calibration functions. The use of separate calibrations for two soil groups (organomineral and mineral soils) gave better results than the common calibration for all soil types. The effect of particle-size distribution can be reduced by unifying the procedure of sample preparation used for the calibration of the instrument and for the analysis of unknown samples, and by using spectral derivation. The level of main nutrients (P, K, Ca and Mg) in the soil had no effect on the results of analysis. The content of a selected oil product in the soil can be determined in the presence of other oil products, if the calibration set of the NIR analyzer included all expectable oil products. The NIR analyzer calibrated on a single oil product will determine the content of total oil products in the soil. The obtained results showed that NIR spectroscopy is a promising technique for monitoring the contamination of soils with oil products. When introducing the NIR method into other Russian laboratories, including those in Siberia, standardised procedures for sampling, calibration and analysis can be developed and utilised.

(5) **Schindler** reported on innovations in soil hydrological measurement technology in Chap. 14. They are an integral part of the scale concept in hydrological research, starting with laboratory and lysimeter measurements, followed by investigations in the field and modelling of the whole watershed. Knowledge of a soil's hydraulic properties, such as its water retention curve and unsaturated hydraulic conductivity function, is required for soil water modelling and various soil hydrological studies. For measuring soil hydrological properties the Extended Evaporation Method (EEM) and the HYPROP device have been developed. Using new cavitation tensiometers and applying the air-entry pressure of the tensiometer's porous ceramic cup as the final tension value allows both hydraulic functions to be quantified even up close to the wilting point. Additionally, soil shrinkage dynamics and soil water hysteresis can be quantified. The experimental setup employed the HYPROP system, which is a commercial device with vertically aligned tensiometers that is optimised to perform evaporation measurements.

Another method for quantifying deep seepage and solute leaching under field conditions was developed, tested and applied at more than 40 soil hydrological field plots in Germany. This hydrological field setup can provide lysimeter-like results (Virtual Lysimeters, Kastanek 1995; Schindler and Mueller 2005) for water and solute transport at several sites. The method presented is designed to estimate deep seepage and solute leaching in the field based on soil hydrological measurements below the zero flux plane, based in turn on a calibrated hydraulic conductivity function. The method is simple to apply, flexible and cost saving compared with lysimeters. The required soil hydraulic properties are derived from tension and water content field recordings at the measurement depth. After calibration no further information on soil properties, weather, management and land use or other data is required. Lysimeter experiments were conducted to test the soil hydrological field method. A comparison between lysimeter discharge measurements and discharge calculations confirmed the validity of this method on sandy to loamy soils with a deep water table and a zero flux plane which did not fluctuate excessively.

(6) In Chap. 15 **Müller** presented soil hydrological innovations methods in the framework of the German and European Forest Monitoring Network (Fig. 3). Those results are required to assess future threats to today's forests and to develop strategies for adapting to anticipated climate change. Monitoring plots consist of open field plots and plots in forest stands. An open-field automatic weather station records meteorological and air quality parameters. In the forest stands, soil hydrological conditions and the quality of pore water are monitored by tensiometers, soil moisture sensors and suction probes. Tree growth increments are measured with dendrometers. Large lysimeters are used to measure the influence of trees of various ages and species on the groundwater recharge and evapotranspiration. High-precision weighable lysimeters allow

Fig. 3 Well-engineered forest monitoring plots operating in the German Forest Monitoring Network, Eberswalde site. The open field plot **a** consists mainly of meteorological instruments, while the plots in needle-leaf stands **b** and broadleaf stands **c** consist of tensiometers, TDR and FDR moisture probes, suction probes, lysimeters, dendrometers and more

the evapotranspiration of ground vegetation and young trees to be measured at high resolution. The results showed that water balance components of forest ecosystems on sandy soils in the sub-humid zone are very sensitive to climate variables, tree species and stand parameters. Groundwater recharge by seepage is a crucial target variable in these landscapes. It is required for drinking water and providing a base flow to rivers and wetlands. The monitoring indicates that the seepage water below the forest is clean but affected by periods of summer drought, which also reduce tree increments. It is influenced by the age and species of forest trees, the vertical structuring and heterogeneity of forests and the way they are managed. Broad-leaved forests are found to have more groundwater recharge than coniferous forests due to the differences in the interception between the evergreen canopies of coniferous forests and winter-bald broad-leaved forests. Field setups for measuring water and solute transport in forest ecosystems have potential for application in other landscapes.

(7) **Balykin et al**. (Chap. 16) reported on some initial lysimeter experiments in the interdisciplinary KULUNDA project (BMBF 2014). Russian and German scientists are trying to understand and tackle soil degradation and water scarcity in the Kulunda steppe of Siberia. The region under study is highly vulnerable to wind erosion, resulting in decreased topsoil thicknesses and humus contents and therefore in decreased concentrations of soil carbon. The assessment and management of the soil water, solute and matter balance are of great importance for crop yield potentials and the sustainable development of the territory. In 2013, the first weighable gravitation lysimeter station in Siberia was successfully installed in the Altai Krai under Kulunda dry steppe conditions. Weighable lysimeters allow the continuous monitoring of changes in soil monolith mass. This is the precondition for calculating actual evapotranspiration (ETa—a major component in the terrestrial water cycle) with high precision. Knowledge regarding the development of ETa is essential to evaluate the impact of climate change on the future water balance. Soil water balance and ecosystem models, which are required to estimate changes in water, matter and energy fluxes over large regions, need to be calibrated using these lysimeters (Meissner et al. 2014).

Cluster 3 "Process-based field measurement and evaluation methods" consisted of the following chapters:

1. "Measuring and Estimating Fluxes of Carbon, Major and Trace Elements to the Arctic Ocean" (Chap. 6, Pokrovsky)
2. "Measuring Snowmelt in Siberia: Causes, Process and Consequences" (Chap. 7, Chumbaev and Tanasienko)
3. "Measuring Major Components of the Terrestrial Carbon Balance" (Chap. 17, Urbaniak et al.)
4. "Assessment and Measurement of Wind Erosion" (Chap. 18, Funk).

The first chapter in this section estimated matter fluxes to the Arctic Ocean. It combined hydrological monitoring data with expert-based knowledge and semi-empirical modelling and balancing approaches at a large river catchment scale. The author **Pokrovsky** has worked in Siberia at Tomsk State University, Russia, and at Geoscience and Environment Toulouse, France. The second contribution informed readers about measurement technology and long-term monitoring results of snowmelt erosion processes at the field and small watershed scale. The authors **Chumbaev and Tanasienko** come from the Institute of Soil Science and Agrochemistry (ISSA) Novosibirsk. In the third and fourth paper they presented advanced measurement technologies at the interface between soil/vegetation and atmosphere. Their methodologies and data are important for estimating matter fluxes, soil functioning and degradation potentials. The authors **Urbaniak et al.** come from the Meteorology Department of the Poznan University of Life Sciences, Poland, a leading group in this research field in Europe. It cooperates closely with scientists at the Leibniz Centre for Agricultural Landscape Research (ZALF) Müncheberg, Germany. The author **Funk**, Germany's leading specialist in wind erosion, has also worked at ZALF.

This third cluster was overlapping with the second one described above but focussed on in situ measurement technologies and utilising field-monitored data.

(1) In Chap. 6 **Pokrovsky** described the methods for estimating element fluxes in large rivers in the Russian boreal and subarctic zone. The majority of existing flux measurements is based on a combination of daily discharges from Russian Hydrological Survey gauging stations with punctual or year-round sampling of the dissolved and particulate load following the chemical analysis. The author presented a new, geochemical-based means of examining the functioning of aquatic boreal systems. This took into account the following factors related to riverine element fluxes: (1) the specificity of the lithological substrate; (2) the importance of organic and organo-mineral colloidal forms, notably during the spring flood; (3) the role of permafrost presence within the small and large watersheds and (4) the governing role of terrestrial vegetation in element mobilisation from the rock substrate to the rivers. Two novel dimensions which added to the existing knowledge on element transport from the land to the Arctic Ocean by the Russian boreal and subarctic rivers were (i) evaluation of colloidal flux of dissolved substances and low molecular weight fraction and (ii) assessing, for the first time, the isotopic signatures of Ca, Mg, Si and Fe in several watershed cases with varying lithology and permafrost coverage. Thanks to this multiple approach, it is possible to make a first-order prediction of element fluxes under a scenario of progressive warming in high latitudes.

The results showed that, while climate warming will certainly affect the winter-time element fluxes and speciation, it is unlikely to change the nature and magnitude of the main fraction of trace element flux to the ocean. This fraction of the flux occurs in colloidal form during several weeks of the spring flood. At the present time it is not strongly affected by climate change, or this influence is within the uncertainty of the flux measurements. Overall, the major changes in the chemical and isotopic nature of riverine fluxes to the Arctic Ocean from Northern Eurasia under a climate warming scenario are likely to be linked to the change in vegetation (species, biomass) and geographical extension. The increase in the depth of the active layer has an influence of secondary-order importance on the riverine fluxes.

(2) **Chumbaev and Tanasienko** (Chap. 7) conducted long-term research on water erosion from agricultural lands. This is a main process of soil degradation in West Siberia which can lead to a catastrophic decrease in fertility. It poses a threat to food production, and it pollutes rivers and lakes. The purpose of this work was to show the main methods and devices used to analyse the quantity and quality of surface snowmelt water runoff and soil erosion. To quantify the overall snowmelt erosion process, the following parameters need to be measured: the total pre-winter water reserve of soil, snow depth, the snow water equivalent, the depth of soil frost penetration, the volume of snowmelt water runoff, the runoff coefficient, the water stream temperature, and soil loss with surface snowmelt water runoff. Experimental work requires 3 stages: (1) preparatory stage, during which the pre-winter soil moisture status is defined and the runoff and thermometric plots are constructed; (2) studying the process of accumulation of solid atmospheric precipitation, the nature of its distribution over the territory, and also the influences of snow depth on the frost penetration in soils; and (3) monitoring the snowmelt process in spring, during which the intensity of snowmelt, the volumes of a superficial drain of snowmelt waters and the damage caused by them to a soil cover are defined. One special feature of the Siberian soils during the cold period of the year is the intrasoil ice sheet, which is largely impenetrable to melting water and heat flux. This ice sheet in Siberian soils is one of the reasons for snowmelt water runoff forming. Over a period of 45 years the authors measured a mean annual soil loss of 6 t/ha caused by snowmelt erosion on arable land in West Siberia.

(3) **Urbaniak et al.** have developed novel technologies for quantifying the carbon balance of ecosystems and explained this in Chap. 17. The presence of carbon dioxide and methane in the atmosphere is the key factor of the so-called greenhouse effect. In this chapter, the authors focussed on the most common and practicable techniques for measuring this, such as chambers, eddy covariance and relaxed eddy accumulation (Fig. 4). The authors have gained experience in designing, building and using these techniques over the past decades. These systems have been successfully applied in wetlands, forests and crop ecosystems all over Europe, and have been reliable sources of ecological data until now. The presented overview of the measurement methods focused on providing an insight into the theoretical basis, as well as the advantages and limitations of all these techniques. This chapter is intended to help potential users decide what approach could be applied in their own investigations.

Fig. 4 Automatically operating eddy-flux and chamber measurement stations for analysing and modelling the carbon balance on different kinds of peatlands. *Left* eddy-flux station developed by the Meteorology Department of the Poznan University of Life Sciences, Poland. *Right* chamber measurement station operated by ZALF Müncheberg, Germany. Peatlands created in temporal ponding phases are characterised by particularly high methane emissions

(4) **Funk** reviewed the latest technologies in wind erosion research. Wind erosion has become and is recognised as an important soil degradation process around the world. The focus of the chapter was on assessing and measuring soil deflation from arable lands. The author created and implemented measurement technologies, risk assessment methods and models of wind erosion prediction in Germany and in the international community. Highly sophisticated technologies for measuring flux components are available and have been tested. Risk assessment methods such as the German standard DIN 19706 "SQ—Determination of the soil exposure risk from wind erosion" have also been developed. The wind erosion risk of every single field in Germany has been computed based on these assessment rules and is available for erosion control on a GIS platform. Wind tunnels have been created for refining erosion models. The paper revealed some focal points for better understanding, monitoring and controlling wind erosion. There is a need to refine, harmonise and standardise methods of wind erosion assessment, measurement methods and technologies, including on a trans-national scale. For more objective economical trade-offs about the efficiency of protection and mitigation measures, wind erosion risk assessment should be implemented in DSS (Chap. 23) and their underlying assessment frameworks such as the Muencheberg Soil Quality Rating (M-SQR) method (Chap. 26). Representative field monitoring plots should be installed at different locations across Europe. Measuring the horizontal fluxes or depositions at different heights is the most highly recommended method, because it is easy to carry out and further important parameters can be derived. Measurement technologies have to meet advanced international rules and standards, as do data storage and processing systems. These methods have the potential of being transferred to Siberian landscapes.

Cluster 4, "Remote sensing methods and GIS technologies for monitoring and modeling landscape processes", contains the following 4 chapters:

1. "Estimation of Biomass and Net Primary Production (NPP) in West Siberian Boreal Ecosystems: In Situ and Remote Sensing Methods (Chap. 8, Peregon et al.)
2. "GIS and Remote Sensing Data (RSD) Based Methods for Monitoring Water and Soil Objects in the Steppe Biome of Western Siberia" (Chap. 9, Zolnikov et al.)
3. "Multi-Scale Vegetation and Water Body Mapping of the Northern Latitudes in Siberia with Optical Remote Sensing" (Chap. 19, Urban et al.)
4. "Multi-Source Data Integration and Analysis for Land Monitoring in Siberia" (Chap. 20, Eberle et al.).

Monitoring and trend analyses over vast and inaccessible landscapes like many parts of Siberia would be impossible without modern information systems. GIS and remote sensing technologies are crucial to research and application studies.

The first chapter described the methodology of biomass estimation in boreal ecosystems as a combination of terrestrial and remote sensing methods. This chapter, by **Peregon et al.**, is a joint contribution of authors from Siberia (ISSA Novosibirsk), from the CEA Centre de Saclay, Laboratoire des Sciences Du Climat et de l'Environnement (LSCE), Gif-sur-Yvette, France, and from the Center for Global Environmental Research (CGER), National Institute for Environmental Studies (NIES), Tsukuba, Japan. The second chapter of this cluster explained methods for monitoring steppe biomes in Siberia using remote sensing. The authors of this study, **Zolnikov et al.**, came mainly from the Sobolev Institute of Geology and Mineralogy, Siberian Branch of Russian Academy of Sciences, with the participation of the ISSA Novosibirsk. Remote sensing vegetation, water body and land monitoring over Siberia were the topics of the next two chapters. The authors of the chapters by **Urban et al.** and **Eberle et al.** came mainly from the Department for Earth Observation, Institute of Geography, Friedrich-Schiller-University Jena, Germany. Their chapters tackled particularly challenging problems of remote sensing methods such as scale issues and complex data integration.

(1) **Peregon et al.** reported in Chap. 8 on the current state of in situ observations and remote sensing methods for assessing biomass and NPP in West Siberia. The natural ecosystems of the boreal region consist of two classes of vegetation: wetlands and forests. This requires different methods for biomass estimation. Basically, two methods are available for estimating NPP and biomass: (1) extrapolating field measurements up to a larger region, using the vegetation or land cover maps and (2) modelling productivity and plant biomass with or without the use of RSD. The first method was predominantly used for estimating wetland biomass, using extensive datasets of direct in situ measurements in both the above- and below-ground fractions of biomass. In forest ecosystems, the biomass can be estimated by processing satellite data from high-resolution radiometers measuring NDVI. Radar or LIDAR remote

sensing approaches are promising for direct observations of the three-dimensional structure of above-ground vegetation that can be used to make relatively straightforward calculations of carbon storage. However, this method works only in low to medium biomass ecosystems. The SAR-based biomass retrievals were found to be fairly uncertain in mature forests with high biomass values as the SAR signal often saturates at ~70 tonnes/ha. Common estimation errors take place at 25–30 % of the mean biomass. Remote sensing methods have to be refined to reduce uncertainties and to make them operational over the vast region of Siberia.

(2) **Zolnikov et al.** (Chap. 9) developed procedures based on GIS and RSD that allow short-term and retrospective monitoring of lakes (water bodies) and Solonchaks (soils). These components of landscapes in the steppe biome of West Siberia (WS) are indicators of ecosystem dynamics. They tested their methods at 4 key sites with a series of multi-temporal satellite images. Plots were chosen to reflect the change of climatic conditions from north to south, especially the influence of moisture on the parameters of the objects. Data from various satellite systems were utilised depending on the research scale. On the medium-scale level they used Landsat and Spot imagery at chronological intervals of 20 years. The dynamics of indicators were studied on a small-scale level from 250 m Moderate Resolution Spectroradiometer images (MODIS). To delineate the Solonchaks and water bodies on MODIS images, special indices were developed. For the purposes of retrospective monitoring, the ancient lake basins (50–60 thousand years old) were mapped. This was based on a combination of various data: geologic and topographic maps and RSD. The morphometric analysis was carried out on the basis of a Shuttle radar topographic mission (SRTM) digital elevation model. Moreover, the lake sizes during the mid-1970s to 1980s were digitised from topographic maps. The results of lake delineation on MODIS images were also used.
The results show a trend towards the degradation of lakes and the increasing area of Solonchaks in the steppe biome of WS, as a consequence of increased aridity. Paleogeographic data based on the analysis of SRTM, topographic maps from the last century and more recent satellite imagery led to the conclusion that large bodies of water had fragmented in arid areas, and then gradually dried. The results obtained illustrate that the methods based on GIS analysis of digital elevation models, in conjunction with the processing of RSD on all scales, are effective for retrospective and current monitoring of ecosystem dynamics under the influence of global environmental changes. The proven technologies are promising tools for environmental monitoring.

(1) The paper by **Urban et al.** (Chap. 19) presented the potential of using Earth observation data from various sources and time steps to monitor land cover characteristics and changes in the Arctic regions. Information on vegetation structure types and physiognomy is commonly incorporated into spatial models predicting the permafrost distribution. The MODIS land cover, the GlobCover land cover map, SYNMAP and MODIS VCF (vegetation

continuous field) have been combined in a product describing the fractional vegetation cover. The dataset, with a spatial resolution of 1 km, consists of four layers providing percentage cover information for trees, shrubs, herbaceous areas and barren areas. Additional information, such as the CAVM (Circumpolar Arctic Vegetation Map), has been integrated into the harmonisation approach. Local land cover and water body changes have been analysed using high-spatial-resolution earth observation information from Landsat, RapidEye and Corona Keyhole. This analysis was carried out for a test site in central Yakutia and the Lena river delta system in Siberia, Russia. High-resolution land cover information was mapped using an object-oriented classification approach. Object characteristics, such as the shape, spectral properties and information within different hierarchical object levels, were utilised to identify individual vegetation class properties for assignment to a thematic class. Water body changes were identified using historical earth observation data from the 1960s and recent RapidEye data.

(4) **Eberle et al.** (Chap. 20) developed data-processing middleware as a technical solution to improve interdisciplinary research using multi-source time-series data and standardised data acquisition, pre-processing, updating and analyses. This solution is being implemented within the Siberian Earth System Science Cluster (SIB-ESS-C), which combines various sources of Earth Observation and climate data with a focus on vegetation and temperature data. Products from the Moderate Resolution Imaging Spectroradiometer (MODIS), in situ data from meteorological stations and high-spatial-resolution Landsat data are available in the processing middleware, which is connected to different data providers. Analytical tools have been integrated and can be used for time-series plotting, phenological data, trend calculations, break-point detection and data comparison using existing open-source software packages. The development of this spatial data infrastructure (SDI) is based on the definition of automated and on-demand tools for finding, ordering and processing data, implemented along with standard-compliant web services. Therefore, open-source software is used to build this system. The tools developed, consisting of a user-friendly means of data access, download, analysis and interpretation, are available within SIB-ESS-C for operational use.

Cluster 5, "Methods of ecosystem modelling" included 5 chapters:

1. "Significant Siberian Vegetation Change is Inevitably Brought on by the Changing Climate" (Chap. 10, Tchebakova et al.)
2. "Evaluating the Agroclimatic Potential of Central Siberia" (Chap. 11, Tchebakova et al.)
3. "Analytical and Cartographic Predictive Modeling of Arable Land Productivity" (Chap. 21, Shary et al.)
4. "Simulating Temperature Impacts on Crop Production using MONICA" (Chap. 22, Nendel)

5. "A Spatial Model-Based Decision Support System (DSS) for Evaluating Agricultural Landscapes under the Aspect of Climate Change" (Chap. 23, Mirschel et al.).

The authors of the first three chapters of this cluster developed algorithms and utilised GIS technologies, remote sensing and other areal data to model vegetation change and productivity potentials over large landscapes. Two chapters of **Tchebakova et al.** came from the Sukachev Institute of Forest, Krasnoyarsk in cooperation with the Krasnoyarsk State Agricultural University, Russia, the NASA Langley Research Center, USA, and the Sochava Institute of Geography in Irkutsk. The chapter by **Shary et al.** focused on predictive modelling of crop yields. It originates from the Institute of Physicochemical and Biological Problems in Soil Science in Pushchino, in cooperation with the VNIIA in Moscow and the Institute of Ecology of the Volga Basin in Togliatti.

The topic of another two chapters in this cluster is the highly temporally resolved modelling of crop growth processes in agroecosystems. This kind of modelling, leading to a better understanding of plant biomass formation and matter cycling, is not yet adequately developed in Siberia. The main authors of these chapters by **Nendel**, and **Mirschel et al.** were from ZALF Müncheberg, Germany. Crop modelling can be a basis of DSS applicable up to the optimisation of agricultural systems at farm or regional level. The chapter about LandCaRe-DSS has co-authors from St. Petersburg (Agrophysical Research Institute and State Polytechnical University), Russia.

(1) **Tchebakova et al.** (Chap. 10) developed scenarios of the progression of potential vegetation and forest types in Siberia by the end of the twenty-first century, by coupling large-scale bioclimatic models of vegetation zones and major conifer species with climatic variables. They used the B1 and A2 Hadley Centre HadCM3 climate change scenarios. Their results show that in the predicted warmer and drier climate, Siberian taiga forest would decrease and shift to the north-east, and forest-steppe, steppe and novel temperate broadleaf forests are predicted to dominate most of Siberia by 2090. The permafrost would not retreat sufficiently to provide favourable habitats for dark taiga species (*Pinus sibirica*, *Abies sibirica*, and *Picea obovata*). The permafrost-tolerant larch taiga (*Larix dahurica)* would remain the dominant forest type in many current permafrost-underlain areas. Tree species resistant to water stress and fire (*Pinus sylvestris* and *Larix* spp.) would have an increased advantage over moisture-loving tree species (*Pinus sibirica*, *Abies sibirica*, and *Picea obovata*) in a new climate. Increased tree mortality from drought, insects and other factors, especially at the southern forest border and in Yakutia, would cause surface fuel loads to accumulate. Together with an increase in severe fire weather, this would also lead to increases in large, high-severity fires, which are expected to facilitate vegetation progression towards a new equilibrium with the climate. The adaptation of the forest types and tree species to climate change in the south may be based on the genetic means of individual species, and on human impacts on the vegetation

structure. Additionally, useful and viable crops could be established in agricultural lands, instead of failing forests.

(2) In the following Chap. 11 **Tchebakova et al.** focused on the future potential for agriculture in Siberia taking into account the soil carbon balance. They analysed current carbon (C) fluxes in agro-ecosystems, constructed envelope models that determine crop range and regression models that estimate crop yields. These data were applied to climate change scenarios for several time frames: 1960–1990, using historic data; and data taken from HadCM3 B1 and A2 scenarios for 2020 and 2090. Migration of the forest towards the north will create forest steppes which have greater agricultural potential. Between 50 and 85 % of central Siberia is predicted to be climatically suitable for agriculture by the end of this century, and only the soil potential would limit crop advance and expansion to the north. Crop production could increase twofold. Future Siberian climatic resources could provide the potential for a great variety of crops to grow which previously did not exist on these lands. Traditional Siberian crops could gradually shift as far as 500 km northward (about 50–70 km per decade) if soil conditions are suitable, and new crops may be introduced in the dry south that would necessitate irrigation. Agriculture in central Siberia would likely benefit from climate warming. Adaptation measures would sustain and promote food security in a warmer Siberia. Analyses of carbon fluxes in agro-ecosystems showed that plant phytomass and soil humus serve as a potential C sink but the current C balance is slightly deficient at a loss of 0.25 ton C ha^{-1} $year^{-1}$.

(3) **Shary et al.** created a model of crop yield prediction on the basis of environmental variables which is presented in Chap. 21. This predictive modelling is based on statistical procedures and is data driven. The authors present an inventory-based method to form spatial models of agricultural crop yields for extended regions (about 4° × 5° in size). The method is based on the analysis of links between the long-term characteristics of crop yields and environmental variables, such as the climate, topography and soils. The environmental variables used were climatic data on the long-term annual means of temperature and precipitation for each month and certain periods, the digital elevation model and 18 basic topographic attributes taken from it, plus soil type data. The topographic attributes were non-linearly transformed to obtain normally distributed residuals. Multiple regression models included validation using an empirically founded criterion, tests on multicollinearity, and determining the statistical significance of each environmental variable. The response variable in multiple regressions was the maximal crop yield surplus by fertilisation. The method was tested on long-term data of winter wheat yields for the western part of the Oka River basin. Analysis showed that the topography-generated microclimate is a major factor determining the maximum addition to yield. Based on the model, a gridded map was constructed for the whole region (4° × 5°). The results of the analysis indicate that among the environmental factors, topography may have the largest influence, and it should be taken into account, including in the case of crop yield prognoses in

the context of a changing climate. The method of predictive modelling could also be applied to Siberian landscapes but requires more detailed data from this region.

(4) **Nendel** (Chap. 22) developed the process-based simulation model MONICA (**MO**del of **NI**trogen and **C**arbon dynamics in **A**gro-ecosystems), which predicts crop growth, evapotranspiration, nitrate leaching and other environmental variables. Temperature relations are the most crucial for the success of individual models to capture crop growth and yield formation at a specific site. For Siberia, where the annual temperature fluctuations can easily exceed 80 K at some locations, temperature extremes are the most important challenge for the application of agro-ecosystem models. This chapter presented temperature-related algorithms of the dynamic simulation model MONICA, including the temperature dependencies of soil organic matter turn-over, plant photosynthesis and respiration, ontogenesis and the impacts of extremely high or low temperatures on crop growth. MONICA was developed to demonstrate the impact of the climate and management on crop yields and environmental variables on the plot scale and in smaller regions in Central Europe. The chapter showed that, based on known bio-physical processes, MONICA has the potential to assess the impacts of climate change and land management on crop yields, carbon balance and nitrogen efficiency in Siberia.

(5) In Chap. 23 **Mirschel et al**. presented a brand-new DSS for developing practicable, resilient climate change adaptation strategies for the sustainable use of agro-landscapes. The authors created the model-based interactive spatial information and DSS LandCaRe-DSS (Fig. 5). This chapter described the conceptual framework, the structure, the methodology and the basic principles. LandCaRe-DSS uses different ecological impact models, including models for crop yield, erosion risk, regional evapotranspiration, total water discharge and irrigation water demand. At the local level, a farm economy model is directly coupled with both the biophysically based agro-ecosystem model MONICA (Chap. 22) and the statistically based crop yield model YIELDSTAT to simulate the economic consequences of regional climate change and of proposed agricultural adaptation strategies. Due to the modular architecture and innovative design of LandCaRe-DSS, alternative or new impact models can easily be incorporated into the system. Scenario simulation runs can be realised in a reasonable amount of time. The interactive LandCaRe-DSS prototype offers a variety of data analysis and visualisation tools and an information system for agricultural adaptation to the climate. This system does not only support interactive scenario simulations and multi-ensemble and multi-model simulations at the regional level by providing information about the complex long-term impacts of climate change. It also helps different stakeholders to find suitable, sustainable agricultural adaptation strategies to climate change (crop rotation, soil tillage, fertilisation, irrigation, price and cost changes) at the local or farm level. LandCaRe-DSS is constantly being developed, updated and adapted. The planned integration of the M-SQR procedure (Chap. 26) and data will provide an improved prediction of soil functions and crop yield

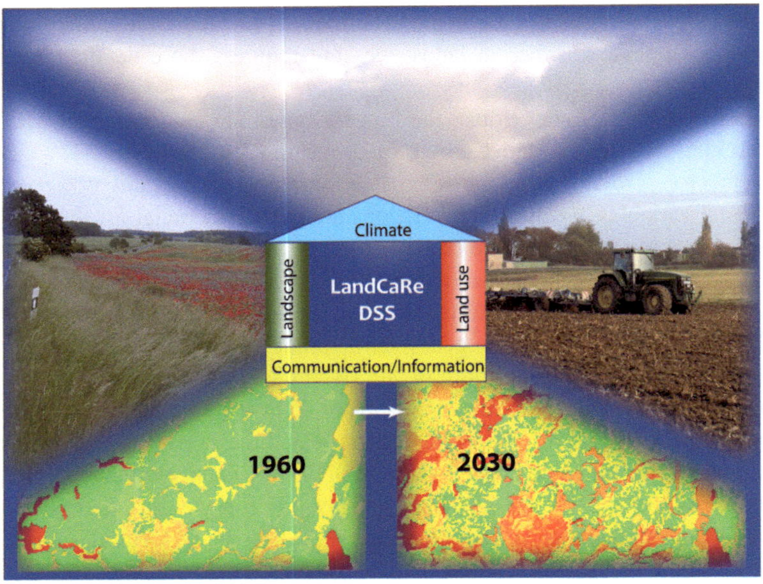

Fig. 5 LandCaRe-DSS is a model-based interactive spatial information and DSS for developing climate change adaptation strategies for the sustainable use of agro-landscapes. Picture from the authors' homepage (ZALF-LSA 2015)

potentials consistently over spatial scales. LandCaRe-DSS has already been successfully tested in different regions of Germany and in Brazil; ongoing work for its application in Poland and Russia is running.

Cluster 6, "Methods of resource and process evaluation and functional mapping", concentrated on evaluating and monitoring soil and land resources. The cluster contains the following chapters:

1. "Monitoring of Soil Fertility (Agroecological Monitoring)" (Chap. 24, Sychev et al.)
2. "Probabilistic Assessment of Contemporary Soil Evolution (CSE) in the South of Western Siberia Based on Analysis of Soil Monitoring Data" (Chap. 12, Mikheeva)
3. "The International Soil Classification System WRB, third edition, 2014" (Chap. 25, Schad)
4. "An Emerging Method of Rating Global SQ and Productivity Potentials" (Chap. 26, Mueller et al.)
5. "Small-Scale Soil Functional Mapping of Crop Yield Potentials in Germany" (Chap. 27, Hennings et al.)

The first chapter of this cluster (**Sychev et al.**) was a contribution of the VNIIA in Moscow, describing agro-ecological monitoring programmes which continue to run successfully. The author **Mikheeva** from ISSA Novosibirsk informed readers about a theoretical concept of soil state monitoring. The next two chapters dealt with soil classification. International efforts to protect soils from mismanagement and degradation have failed *inter alia* because of "Babylonian confusion" in terminology and the definition of soil types, processes and functions. Both chapters depicted solutions to overcome this situation. The author **Schad** (Technical University of Munich, Germany) presented the new International Soil Classification System WRB 2014 in brief. He headed an international working group who had elaborated this new system. It designates every soil unit on earth with a meaningful name. The approach used by **Mueller et al.** about the M-SQR allocates fertility numbers which describe the crop yield potentials of these soil units. The authors of this chapter came from ZALF Müncheberg, Germany, and a number of internationally acknowledged research institutions in Russia and Kazakhstan such as ISSA Novosibirsk, VNIIA Moscow, Kuban State Agrarian University in Krasnodar, the Institute of Peat and Organic Fertiliser in Vladimir, and the Kazakh Research Institute of Soil Science and Agrochemistry in Almaty. **Hennings et al.** created a small-scale SQ map of Germany by utilising the M-SQR approach and parameterising it with data from the soil information system of the Geo-Center in Hannover, Germany.

(1) In Chap. 24 **Sychev et al.** analyzed fundamentals, developments, trends and present results of agroecological monitoring in Russia. Monitoring the ecological status of agricultural land is a fundamental precondition for controlling its sustainable functions for the human society and for maintaining ecosystem capacity. This system has been developed and operated by the Pryanishnikov Institute of Agrochemistry (VNIIA) in Moscow. Agroecological monitoring in Russia was installed in the 1970s and is based on a regular 5-year agrochemical survey of agricultural lands all over the country, more than 300 field experiments in all bioclimatic zones of the country and more than 1000 reference monitoring plots. In trials with different inputs of fertilisers the focus is on analysing soil fertility indicators and their impact on productivity. Some of these experiments are long-term experiments and part of international networks. The authors found significant tendencies towards decreasing soil fertility in some regions, such as decreased contents of humus and plant-available minerals, and topsoil acidification. Nutrient withdrawals must be made up for by regular fertilisation regimes, and nutrient mining must be avoided. The authors detected some knowledge gaps in the topic of balancing elements and modelling the agro-ecosystem's response to climate and land use changes. They concluded that there is a need to implement modern measurement and modelling systems in some key long-term trials. The VNIIA has taken responsibility for coordinating running programmes in different regions and administrative units of the Russian Federation and for elaborating methodical guidelines and most advanced monitoring technologies. National and international cooperation, research programmes and networks are key for the agroecological monitoring systems of the twenty first century in addressing

challenges for highly productive, stable, sustainable, and environmentally safe food production.

(2) **Mikheeva** (Chap. 12) developed a theoretical approach to analysing soil monitoring data over a period extending from ten to one hundred years. Contemporary Soil Evolution (CSE) is considered a continuous block process of change in soil conditions leading to changes in the probability distribution functions (pdfs) of soil properties at different time steps. The author developed a method that consists of: (1) identifying the pdfs of soil properties, which means a quantitative evaluation of different kinds and parameters of pdfs according to data samples resulting from soil investigations; (2) calculating probabilistic indicators such as the statistical entropy of pdfs as probabilistic characteristics of soil status; and informational divergence that is a measure of pdf difference. A case study carried out on the large territory in the south of Western Siberia revealed new findings: changes in the probability structure of Kastanozem soil properties during CSE under natural processes and anthropogenous influences. Distinctions in pdfs were evaluated from the values of informational divergence and increment of statistical entropy, which were quantitatively different for soils of different granulometric composition; that is useful to identify the most vulnerable soils in the territory under investigation. It may be concluded that it is necessary to use probabilistic indicators to assess CSE from pdf alterations. They characterise a degree of influence of soil-forming factors and anthropogenous influences on the probability structure of the properties of the ground and its stability. Thus, they could be reliable indicators of environmental transformation; that is important for land resources research, land use policy planning, and basic research.

(3) **Schad** (Chap. 25) reported on the third edition (2014) of the International Soil Classification System WRB. The first level of the WRB allocates soils to 32 Reference Soil Groups (RSGs), which are identified using a key. Many RSGs represent specific soil-forming processes, are representative of major soil regions or reflect special parent materials. In the second level, a set of qualifiers can be added to the name of the RSG. The principal qualifiers characterise the most significant properties of the particular RSG, and supplementary qualifiers give some further details about the soil. The WRB uses diagnostic horizons, diagnostic properties, and diagnostic materials. The definitions of many RSGs and qualifiers refer to the presence or absence of certain diagnostics at a certain depth. In addition, many definitions refer to individual features such as the base saturation or clay content. For naming a soil, the RSG has to be provided with all applicable principal and supplementary qualifiers. For map legends, the number of qualifiers depends on the scale. Detailed rules are established to create comprehensive names for map units at different scale levels. The major changes in WRB 2014 compared to WRB 2006 are the replacement of Albeluvisols by Retisols at RSG level. Retisols have a broader definition and include the former Albeluvisols. Fluvisols have moved down in the key to be the second last RSG, which changes some former Fluvisols into Solonchaks, Gleysols and others. This gives soil properties (salinity, redox

conditions, etc.) priority over the mere position in the landscape. Three new diagnostic horizons have been defined. The chernic horizon is now required for Chernozems. The WRB supplements national classification systems. This is an important step towards uniform soil naming based on the criteria that characterise soil processes.

(4) Chapter 26, by **Mueller et al.**, provided information about an emerging approach for rating agricultural Soil Quality (SQ) and crop yield potentials consistently over a range of spatial scales. The authors developed and tested the Muencheberg Soil Quality Rating (M-SQR), an indicator-based straightforward method of assessing the overall agricultural SQ. It is a framework covering aspects of soil texture, structure, topography and climate which is based on eight basic indicators and more than 12 hazard indicators. Ratings are performed by visual methods of soil evaluation and supported by monthly climate data. A field manual is then used to provide ratings from tables based on indicator thresholds (Fig. 6). Finally, overall rating scores are given, ranging from 0 (worst) to 100 (best) to characterise crop yield potentials. The current approach is valid for grassland and cropland. Field tests in the main global agricultural regions confirmed the practicability and reliability of the method. At the field scale, soil texture and structure (Fig. 7) are most

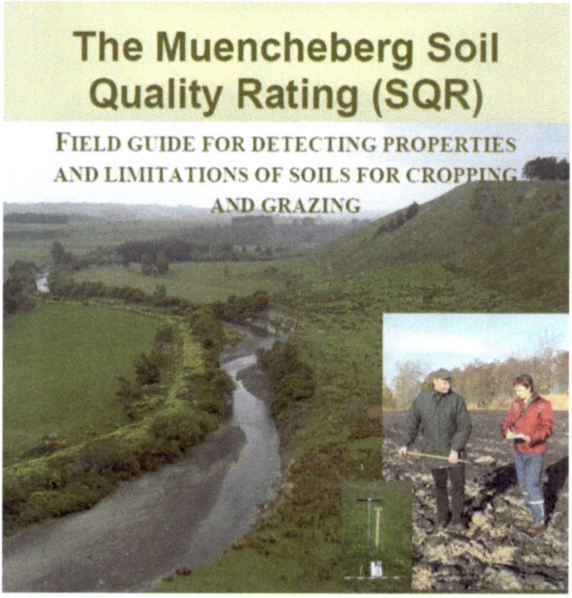

Fig. 6 Front page of the field manual of the Muencheberg Soil Quality Rating (*M-SQR*). This indicator-based assessment method of soil quality of global lands has proven its applicability to Siberia at single sites. It has potential for mapping crop yield potentials over all Eurasian lands. The methodology is explained in Chap. 26, and the soil quality map of Germany, based on this approach, in Chap. 27. The field guide is available as internet source (ZALF-LWH 2013)

Fig. 7 Soil structure expresses the influence of soil management by farmers on agricultural soil quality. Types of structure and biological pores are significantly correlated with soil functional parameters of water and gas transport. Visual soil structure serves thus as a quickly recognisable indicator of soil quality within modern soil evaluation frameworks such as M-SQR. Favourable soil structure under conservation tillage [rotations including ley (*left*) show a clearly better structure than monocultures (*right*)]

important criteria of agricultural SQ. At the global scale, climate-controlled hazard indicators of drought risk and the soil thermal regime are crucial for soil functioning and crop yield potentials. The final rating scores are well correlated with crop yields. Some tests were already carried out in Siberia (Smolentseva et al. 2014). The authors concluded that the combination of the M-SQR with the World Reference Base of soil resources (WRB 2014, Chap. 25) provides key information about main soil functions and processes. This system could be evolved for ranking and controlling agricultural SQ on a global scale as a basis for sustainable land use. Current concepts and data have led to a new crop yield potential map of Germany (Chap. 27) and could help create a similar map of Russia using the same methodology.

(5) In Chap. 27 **Hennings et al**. explained the method of creating a SQ map based on the M-SQR approach (Chap. 26). It covers the arable lands of Germany. M-SQR rules and algorithms were adapted to the terminology and classification of soil parameters as defined by German soil mapping guidelines and were applied on soil-profile-related data sets of the land use stratified soil map of Germany at a scale of 1:1,000,000. According to the resulting thematic map, soils in Germany show a high yield potential for grain; the nationwide mean score is 64 out of 100 possible points. A moderate drought risk is the main crop yield limiting factor in Germany, while shallow soil depth and other crop yield-limiting factors may also be locally important. The authors concluded that the M-SQR methodology also enables the consistent incorporation of the newly created small-scale German crop yield potential map into a potential global SQ map.

Cluster 7 "Methods and technologies for controlling agricultural land use systems" contains the following four chapters:

1. "Balance of Nutrients and the Optimization of their Use in Agroecosystems of the Russian Federation" (Chap. 28, Yefremov et al.)
2. "Assessing and Controlling Land Use Impacts on Groundwater Quality" (Chap. 29, Eulenstein et al.)
3. "Principles of Conservation Agriculture in Continental Steppe Regions" (Chap. 30, Suleimenov et al.)
4. "Modern Cropping Systems and Technologies for Soil Conservation in Siberian Agriculture" (Chap. 31, Grunwald et al.).

The first two chapters explained methods of nutrient balance. These are important tools for evaluating the sustainability of agricultural practices towards maintaining soil fertility and preserving water and air from pollution. The chapter by **Yefremov et al**. came from VNIIA in Moscow, the chapter by **Eulenstein et al**. originated in ZALF Müncheberg with inputs from colleagues from some other research institutions and scientific companies in Russia, Kazakhstan, Germany and Austria. The next two chapters dealt with a crucial problem of agricultural land management in Siberia and throughout Russia: the lack of knowledge and experience with conservation agriculture. The first authors from the chapters explaining their principles and results thus came from abroad. The chapter by **Suleimenov et al**. came from the Scientific Production Center of Grain Farming named after A.I. Barayev, Shortandy, Kazakhstan. The chapter by **Grunwald et al**. originates from the Amazone Company and Halle University, both in Germany, with participation from the Barnaul Altai State Agrarian University and further institutions working on the KULUNDA project.

(1) In Chap. 28 **Yefremov et al**. calculated the balances of the main plant nutrients nitrogen (N), phosphorus (P), and potassium (K) in certain regions of Russia and for the whole country over different time periods. They took into consideration all the main elements of the balances such as inputs of nutrients with mineral and organic fertilisers, seeds, biologically fixed N (symbiotic and non-symbiotic fixation) and rain, and outputs through crop uptake and losses through leaching, erosion and denitrification. The increase in the scope of chemicalisation in Russia led to the gradual elimination of the N and P deficit and a surplus of these nutrients at the national level—from an approximate balance since the mid-1960s to 37 and 25 kg/ha/year between 1986 and 1990, respectively. Annual deficits of K were also gradually reduced in the same period from −15 to −2 kg/ha. Today, agriculture has a serious annual excess of removal over input with a long-term deficit of up to 30, 10, and 27 kg/ha/year for N, P, and K, respectively as the result of a drastic decrease in mineral and organic fertilisation since the 1990s. The regional balance also provided a link with monitoring data on nutrient availability in arable soils. For the regions with high-input agricultural production the analysis of agrochemical survey data shows a consistent decrease in the weighted average content of available

P and K forms in arable lands. The negative tendency was clear when the input of nutrients was inadequate to maintain soil fertility, as it was permanently lower than the amount of nutrient removal from soils. Negative hotspots were identified. Siberia is one such region where arable lands are not only under-fertilised but mainly receive N at the expense of P and K as a result of the farmers' short-term decisions instead of long-term sustainability; in the Far East the situation is much more favourable. The development of policies and strategies relating to the fertilisation requirements of Russian agriculture should be based on providing balanced nutrition conditions for sustaining agricultural systems and soil fertility conservation.

(2) Chapter 29 shows that protecting aquifers and open water bodies against pollution caused by ineffective and excessive fertilisation is a very important issue in modern agriculture (Fig. 8). **Eulenstein et al**. explained different balancing and measurement methods for the identification of nitrate problems and risks to water bodies. This chapter describes various methods for monitoring the efficiency of groundwater conservation by optimising soil and fertiliser management (Fig. 9) at different levels, from the farm through the soil zone to the groundwater. Nutrient balances are mandatory for all farms and fields in Germany. Depending on the spatial and temporal scale and the severity of the problem, different balancing and measurement methods are also mandatory or preferentially recommended. Balancing programmes differ between farm balances and field balances. Special methods are required in areas under particular groundwater protection for drinking water exploitation. Recommended sampling methods in aquifers included the direct push method, sampling from observation wells and sampling from multi-level observation wells.

Fig. 8 Monitoring the quality of soil water by balancing and direct measurements (pore water, seepage water, runoff, groundwater) is particularly important on lands belonging to bioenergy farms. In biogas plants, liquid manure is a valuable fertiliser but may lead to excessive contents of nitrate and other agents in the soil and water at the local scale

Fig. 9 Methods of precision agriculture can help to optimize fertiliser demands according to the spatio-temporal specific conditions of crops. The photo shows an "N-sensor" mounted on a tractor. It is combined with a GPS system and can be a basis for the operational control of fertiliser application

Nutrient balancing is managed using information and communication technology (ICT) and the latest farm management information systems in agriculture. In Germany, the methods described in this chapter are recommended as a work basis and decision tool for all bodies that have to assess the efficiency of agricultural operations in the framework of legal regulations or voluntary cooperation. The methods described here can be used by farmers, landscape planners, environmentalists, water authorities and other stakeholders as a basis for taking agricultural groundwater conservation measures. Decision trees, monitoring methods and information technologies have the potential of being transferable to other regions. Modern GIS-based software is available for this purpose.

(3) In chap. 30, **Suleimenov et al**. reported on the principles of conservation agriculture and experimental results in Siberia and Kazakhstan. It is time to put an end to the non-sustainable summer-fallow-based cropping practiced in West Siberia. Conservation agriculture consists of three principles: (i) no-tillage or reduced tillage, (ii) permanent plant- or plant-residue-covered soil (mulch), and (iii) crop rotations. The authors conducted multi-factorial field experiments on crop rotations, tillage and soil fertility management on sites in Kazakhstan over more than 5 years and reviewed analogous studies from West Siberia. In studies conducted in northern Kazakhstan they found that cropland is most efficiently used in diversified crop rotations without black fallow. Summer fallow can be replaced by food legumes or legume forages. No-Till had an advantage in terms of crop yields over traditional tillage on the light-textured soils of the Kostanai province thanks to the better moisture conservation. On the heavy-textured soils of the Akmola province, traditional tillage had an advantage in some cases thanks to the better snowmelt water

intake and more active nitrogen mineralisation. On the leached Chernozems of Trans-Ural Siberia, No-Till is feasible only with the application of higher rates of nitrogen fertiliser. In the forest-steppe zone of West Siberia, No-Till in the autumn provided the same grain yields as ploughing only when it was combined with the application of fertilisers, herbicides, fungicides and growth regulators. The major advantages of conservation agriculture are that it maintains soil fertility and long-term soil productivity, and prevents on-site and off-site damage by wind and water erosion. The authors concluded that conservation agriculture should become the dominating cropping system in Siberia on soils prone to erosion.

(4) The chapter by **Grunwald et al.** (Chap. 31) analyzed topics of conservation agriculture from another perspective and presented the most advanced technologies for productive and sustainable cropping. No-till and direct seeding is a basic backbone of conservation agriculture. Applying this concept in agricultural practice does not work without modern technical equipment (Meinel et al. 2014). The authors reported on their first experiments in Siberia within the framework of the international research cooperation project KULUNDA. More reliable machinery for soil and crop management, exact seeding and spraying machinery in particular, will help to establish more robust crops which provide higher yield stability and better water use efficiency, and maintain soil fertility by protecting the soil against wind erosion. Modern drills have been designed for specialised direct seeding on stubble under steppe conditions. Wide row spacings and narrow, single depth-adjusted hoe opener systems are able to carry out shallow tillage only in the seeding furrow and to place seed and fertiliser exactly at the adjusted depth. The running experiments will deliver more detailed data about crop management and yields, soil fertility and micro-economy in the coming two years. Though those more detailed data are still required, authors may already conclude that their new approaches seem to confirm their hypotheses. Initial results in very dry weather conditions confirmed that conservation practices perform better when all aspects of soil protection, performance of machinery, plant establishment and yield are considered. Soils can be protected from wind erosion in an effective way using no-till systems, and the yield can be stabilised. They are soil conserving, practicable and effective, and have great potentials for broader application in Siberian agriculture.

3 Availability of Novel Methods and Conditions for Their Application

Part 3 of the book contains a number of methods with potential for application in Siberia. Their introduction and broader application would provide a shift in knowledge and technology towards better monitoring and the sustainable use of land and water resources. Table 2 provides an overview of novel tools. There are

Table 2 Characteristics of novel methods presented in Part 3 of the book

Tool designation	Chapter no, author	Addressees	Accessibility
Calibration of NIR spectrometers for monitoring the contamination of soils with oil	13, Pankratova et al.	Research laboratories	Calibration recommendations potentially for free
Soil hydrological measurement method (EEM, HYPROP device)	14, Schindler et al.	Research laboratories	Commercial product
Soil hydrological field setup for estimation of leaching below the root zone	14, Schindler et al.	Research and monitoring stations	Consists of commercial products, secondary investments in laboratory water analytics
Monitoring field setup for measuring water and solute transport in forest ecosystems	15, Müller	Research and monitoring stations	Consists of commercial products, secondary investments in laboratory water analytics
Lysimetric technology, (designed for steppe conditions)	16, Balykin et al.	Research and monitoring stations	Consists of commercial products, secondary investments in laboratory water analytics
Carbon flux measurement technology	17, Urbaniak et al.	Research and monitoring stations	Consists of commercial products
Measurement technologies for the quantification of wind erosion, algorithms of site-specific risk assessment	18, Funk	Research and monitoring stations	Measurement technologies consist of commercial products; algorithms and models are for free; training advisable
Multi-scale vegetation and water body mapping procedures	19, Urban et al.	Research and monitoring groups	Data access, investments in GIS technologies, method for free, training advisable
Middleware for land monitoring	20, Eberle et al.	Research and monitoring groups	Data access, investments in GIS technologies, method partly for free, training advisable
Predictive crop yield modelling	21, Shary et al.	Research and monitoring groups	Data access, investments in GIS technologies, method for free, training advisable
MONICA model for simulating impacts on crop production	22, Nendel	Research and planning groups, consultants	Method for free, training advisable

(continued)

Table 2 (continued)

Tool designation	Chapter no, author	Addressees	Accessibility
LandCaRe-DSS for simulating and evaluating agricultural landscapes	23, Mirschel et al.	Research and planning groups, consultants	Method for free, training advisable
Agroecological monitoring of Russia	24, Sychev et al.	Research groups, monitoring authorities	Method for free, training advisable
International Soil Classification System WRB	25, Schad	Soil scientists, landscape planners, consultants	Method for free, training advisable
Global soil quality rating procedure (Muencheberg Soil Quality Rating, M-SQR)	26, Mueller et al.	Research and planning groups, consultants	Method for free, training advisable
Procedure of mapping crop yield potentials based on M-SQR	27, Hennings et al.	Research and planning groups, consultants	Data access, investments into GIS technologies, method for free, training advisable
Nutrient balancing of agricultural land in Russia	28, Yefremov et al.	Research and planning groups, consultants	Method for free, training advisable
Guidelines for assessing and controlling land use impacts on groundwater quality	29, Eulenstein et al.	Research and planning groups, consultants	Investments in measurement devices and GIS, ICT programmes are commercial products, training advisable
Principles of conservation agriculture, cropping systems and technologies for Siberian agriculture	30, Suleimenov et al. 31, Grunwald et al.	Farmers, research and planning groups, consultants	Investments in soil tillage and modern crop management technologies

various potential addressees for these methods. Most tools could be applied by research and monitoring groups, but landscape planners, consultants, farmers, decision makers and other people could also benefit from them.

The novel methods presented have a different degree of accessibility. Some require investment, others are available for free. We may distinguish between:

(1) Well-engineered commercial products. The measurement devices presented in this group are useful for application in research and monitoring and may provide a knowledge shift in understanding soil hydrological processes (EEM-HYPROP, lysimeters and hydrological field setups for arable lands and

forest ecosystems). Water and gas flux measurement devices are also in high demand as research tools. Machinery for conservation tillage, and special agrochemicals, are very important for the functioning of conservation agriculture but costly for farmers.

(2) Knowledge-based, virtual tools for the recognition, assessment and evaluation of states and processes in agricultural landscapes. This includes land evaluation and mapping frames and guidelines, agro-ecosystem models, GIS and remote sensing tools. Some addressees of these evaluation methods and models are also researchers, but others are monitoring and planning groups, authorities or advanced farmers. These methods are freely available, but they also require qualifications and a lot of creativity. Access to geospatial data, especially at high resolution, is sometimes commercially restricted by costs and licensing.

What all the new methods presented here have in common is that their introduction requires the people dealing with them to have particular qualifications. Also, the application of commercially available measurement devices and systems needs to be embedded in an advanced and creative environment, dominated by motivated and skilled scientific staff.

It should be mentioned that further transferrable tools are available as well. They have already been described in a book about monitoring and management of soil and water resources in Central Asia (Mueller et al. 2014a, b, c) but are applicable in Siberia, too. Most of these methods are listed in Table 3.

The authors are aware that presenting novel methods in a book can only set slight impulses for their potential application. The key to their real application is mental access to available research tools, including "non-physical" ones. This clearly requires open minds among scientists and decision makers, plus high-level training and qualification in their handling, based on a standard level of scientific education.

4 Addressing Climate Change Requires Innovations in Landscape Research

Climate change in Siberia is significant and progressing. Chaps. 1, 2, 6, 10 and 11 presented data of significant warming, and indicated some possible consequences for nature and civilisation. Climate change will lead to a shift in the natural vegetation northwards and increase the potential for agriculture as a whole. Its impact on water resource availability, however, is largely uncertain at the time being.

Can this potential be utilised to improve food security? Dronin and Kirilenko (2011) doubt this, mainly because of drought risk and the lower soil fertility of Taiga soils. On the other hand, drought risk can be diminished by irrigation, and Taiga soils of limited inherent fertility can be improved by proper soil reclamation and management. In North Germany, soils of the Podsol type which are

Table 3 Further novel methods with potential for application in Siberian landscapes

Methods	Authors	Links
Field methods for measuring and evaluating the quality of grasslands	Mueller et al. (2014b)	http://link.springer.com/chapter/10.1007/978-3-319-01017-5_11
Method of Impact Assessment for multi-functional land use	Helming (2014)	http://link.springer.com/chapter/10.1007/978-3-319-01017-5_12
Method of mapping groundwater recharge	Hennings (2014)	http://link.springer.com/chapter/10.1007/978-3-319-01017-5_14
Method of assessing groundwater risk from trace elements	Godbersen et al. (2014)	http://link.springer.com/chapter/10.1007/978-3-319-01017-5_17
Statistical tool for detecting structures and hidden processes in series data	Lischeid (2014)	http://link.springer.com/chapter/10.1007/978-3-319-01017-5_21
ZEPHYR model for sprinkler irrigation control	Michel and Dannowski (2014)	http://link.springer.com/chapter/10.1007/978-3-319-01017-5_22
FLEOM model for site-specific planning and operating of surface irrigation systems	Djanibekov and Sommer (2014)	http://link.springer.com/chapter/10.1007/978-3-319-01017-5_24
Water desalinisation by *Lemnaceae*	Balla et al. (2014)	http://link.springer.com/chapter/10.1007/978-3-319-01017-5_25
Phosphogypsum application for rehabilitation of high-magnesium soils	Qadir et al. (2014)	http://link.springer.com/chapter/10.1007/978-3-319-01017-5_28
Automatic, intelligent sprinkler irrigation system	Evans (2014)	http://link.springer.com/chapter/10.1007/978-3-319-01017-5_29
Multi-species grazing systems in deer farms	Behrendt et al. (2014)	http://link.springer.com/chapter/10.1007/978-3-319-01017-5_30

traditionally considered infertile may provide cereal yields of 10 t/ha thanks to suitable climate and careful soil cultivation over decades or centuries. Soils do not only degrade as a result of the climate and human impacts. They may also be improved (aggraded) in their productivity function by human activity (Geßl and Nestroy 2008; Mueller et al. 2014c). Framework conditions determined by society are decisive factors for utilising the potential of land for farming or not.

Climate change will create new bio-physical facts. How will the newly created landscapes develop? All scenarios considering factors of societal behaviour are very uncertain because of the apparently unforeseeable reaction of human civilisation. Even a simple, basic question such as "Can soil reclamation and sustainable use be achieved on land under varied ownership?" lies beyond the topic of this book and beyond our expertise.

Instead, we concentrated on other subtopics associated with climate change and more related to questions which can largely be answered from the viewpoint of applied natural science:

"Can soil reclamation and management be performed in an effective and sustainable way by applying obsolete technologies without public monitoring and control?" Of course, they cannot. Chaps. 7, 30 and 31 show that on soils prone to erosion by wind and water this will not be possible. Under those conditions there is no alternative to conservation agriculture if those soils need to be cropped. On the other hand, in more detail, a dualism in terms of "yes" or "no" will always fail when assessing landscapes and preparing planning decisions. The recognition of the ecosystem's behaviour has to follow dialectical principles and consider back coupling processes, disturbances by boundary and neighbour systems, memory effects and other aspects. For example, in the case of successful soil reclamation by drainage, irrigation or other management technologies, undesired side effects or subsequent damage may occur besides the desired effects of improved fertility (Zaydelman 2009). The same principle holds for conservation agriculture, which is not practicable without increased herbicide application. One cannot rule out possible consequential damage to agro- or aquatic ecosystems.

It is very likely that the utilisation of the enhanced agroclimatic potential of a warming Siberia can only be achieved by drainage and irrigation in many parts of the region. In the south, measures against soil degradation by salinisation and sodification have to be taken. Improved knowledge and modern technologies for soil reclamation are available (Berezin et al. 2013; Evans 2014; Michel and Dannowski 2014; Qadir et al. 2014; Semendyaeva et al. 2015), many of them again outside of Siberia (Fig. 10). However, drainage, water table control and irrigation will have implications for soil development, vegetation and biodiversity (Mueller et al. 2000, 2002). They need to be planned and handled with more care than formerly. If planned for larger regions or watersheds they will require impact assessment procedures (Helming 2014) including economic and ecological trade-offs. The M-SQR (Chap. 26) could serve as an important decision tool within those procedures by making them more objective and comparable.

Permanent research and monitoring about climate change and landscape development will increase our knowledge and allow new conclusions to be drawn for optimising the environmentally sound use of resources. Consequently, all decisions about resource allocation have to be based on the latest state of the science. This is our claim and mission, and we can already answer a key question in this context: "Can we develop strategies for the optimum use of landscapes which have been modified by climate change?" Yes, we can. Chapters 13–29 give examples of new monitoring, assessment and modelling tools that are required for those thought experiments. A further example of a question might be: "Is a recently observed altered water quality caused by climate change or caused by poor practices in agricultural and forest management?" This question is not yet very relevant to Siberia because mining and industrial activities have been identified as the main source of soil and water pollution there. However, it is a permanent issue in Western Europe, has growing relevance to European Russia and could reach Siberia

Fig. 10 Examples of modern drainage and irrigation technology: **a** Well-constructed ditch for land drainage and water table control, **b** Laser-controlled machine for trenchless subsurface drainage, **c** Automatically operating low-pressure sprinkler irrigation machine. It is likely that these and further technology and knowledge about their operation will be needed to exploit the higher land use potential of a warming Siberia. Their application requires complex planning, careful construction, operation and maintenance to achieve their main effects and to minimise negative side effects

soon. It can be answered using the monitoring programmes (Fig. 11), modelling studies and technologies shown in several chapters of this book, such as Chaps. 3–6 , 14, 28 and 29.

Summarising these considerations, it may be concluded that climate change to Siberia is evident and may be a threat or an opportunity. Society needs to find solutions and make mandatory decisions. Introducing and enabling climate-strategic agriculture requires a permanent dialogue between scientists on the one hand and policy makers and managers on the other hand to convert research data into flexible policy and action plans (Lal 2013). There is no alternative to permanent research and monitoring and to applying the latest knowledge and technologies in adaptation programmes. Not doing this, and neglecting the state of the science, would have negative implications for nature and future generations. The degree of application of our monitoring and management tools for land and water resources in Siberian research and practice could be a first good indicator of the willingness and ability of responsible decision makers to address important aspects of climate change.

Fig. 11 Agricultural field experiments, and in particular long-term experiments (*LTEs*), are a very valuable source of information about land use strategies, management systems, nutrient cycling, carbon sequestration, climate change implications and other topics and data. The monitoring network of the Pryanishnikov All-Russian Research Institute of Agrochemistry (*VNIIA*) in Moscow provides the derivation of scientific recommendations for research and practice based on this information. The network needs to be extended by including more Siberian sites and modernized by installing automatically operating monitoring stations at key sites. The figure depicts the LTE on the "Effect of mineral fertilization and liming on crop productivity and fertility of soddy-podzolic soil" carried out by the Krasnoyarsk State Agricultural Research Institute (*NIISKH*). *Source* VNIIA

5 Suggestions for Initiating Sustainable Use of Land and Water Resources

Analyses of the first chapters showed that agriculture has operated in a relatively endemic and unsustainable manner in Siberia. Research and monitoring in agriculture and related disciplines are also traditional and require modernisation and better linkage with the European scientific community. The chapters reviewed above presented some novel methods which could help to improve the situation at different levels of science and technology significantly. However, how to make them operational?

Reading the book chapters and understanding their key messages can be only a first step towards achieving our goals: initiating joint actions for sustainably handling land and water resources in rural landscapes.

On a global scale, international cooperation in science and technology has been a proven key to practical progress based on new ideas developed at any location in the world. It is high time to strengthen cooperation in scientific research and teaching in agricultural and agri-environmental disciplines on a Eurasian scale, with a strong focus on Russia and Siberia. An emerging generation of Russian scientists

Fig. 12 Ecological summer schools and excursions have been excellent initiatives for teaching students about ecosystem functioning and sustainable landscape handling. The *left photo* shows a group working in a mountain tundra landscape, and the *right photo* a group in a mountain desert of the Altai. Photos courtesy of Christian Siewert

well educated in their scientific discipline and in English, the language of science, will be involved.

The contributors to this book are willing to integrate their inventions and other novel methods developed into these initiatives. They represent an immense international research network with great potential for supporting activities of agri-environmental research, education and outreach in Siberia in the coming years. Those actions need to be better interlinked with currently running activities. Soil-ecological field courses across West Siberia (Siewert et al. 2014, Fig. 12), international summer schools at leading Siberian research institutes and the international research project KULUNDA (BMBF 2014) are some initial promising activities on the path to more durable cooperation projects.

Strategies of cooperation in science, technology and education between the EU (and its constituent countries) and Russia must become more sustainable. For example, pilot project groups for field-testing novel methods and technologies could be installed at Siberian research institutes. Their work could be supported by external scientific consultants. In the area of education, it is worth thinking about "Schools of Environment and Natural Resources". A single Siberian scientific research and education centre for "Rural Landscape Research" would express the required focus on the sustainable use of land and water resources associated with problems of agriculture, forestry, inland fishery, environmental protection and further related topics. Employing external lecturers would mean that new ideas and methods could have an innovative effect. Also, professors and academic teachers at universities in Russia have a particular responsibility to consider the latest findings of science and technology in their lessons for students.

Maintaining rural landscapes and their basics (fertile, clean soils and pure water) requires some effort in the urban century (Watson 2014). New initiatives for Siberia could be effectively interlinked with efforts to improve the situation in Central Asia. Former Soviet Republics there share similar problems related to the need for innovation in soil and water monitoring and management (Mueller et al. 2014a; Meyer and Lundy 2014). Large regions of Kazakhstan border Siberia. The Kulunda steppes and their desertification problems, transboundary river water management, research and monitoring methodologies and many other issues are common with Siberia. All contributing partners will benefit from those innovative initiatives and from a durable scientific and technological partnership. Maintaining the functions of great landscapes for future human generations will be the reward of those efforts.

6 Overall Conclusions

1. Huge natural resources of Siberia, the Asian part of Russia, are the basis for the welfare of modern urban Russia. The land and water resources of Siberia are of crucial importance for global cycles and show great potential for effective, sustainable and ecologically viable activities around their handling and utilisation. This would provide a solid basis for maintaining natural landscapes, developing rural cultural landscapes, and achieving food security and a high living standard of the population, rural regions included.
2. Not only the resource-extracting and processing industries, but also forestry, agriculture and fishery have capitalised land and water resources in a wasteful way. Peatlands, Tundra, Taiga, and most aquatic ecosystems are endangered by these industries, industrial air pollution and uncontrolled natural and human-induced hazards (fires, thermokarst processes, floods, droughts, insect outbreaks). Agriculture is in a recession and operates inefficiently and soil-destructively. The impact of agriculture on water quality is still low but will increase. The rural infrastructure is on the verge of collapse.
3. Measures for sustainably handling land and water resources should be initiated soon. This has to be managed based on a profound knowledge and reliable data about the significance of trends in ecosystem drift. Russia has great traditions in landscape research disciplines such as geography, soil science, hydrology and agronomy. They experienced a great development during the economic revival of the country in the post-war period. Further substantial progress has been achieved in all these fields of science over the past 25 years. Respectable basic monitoring programmes for soil and water quality have been installed but need to be modernised and interlinked with the scientific community.
4. Addressing climate change requires innovations in landscape research. Novel methods and technologies developed outside Siberia for monitoring and controlling land and water resources, meeting international standards, should be set in operation. The above book chapters present an array of methods which can provide a knowledge shift, particularly in agri-environmental research, monitoring and agricultural practice.
5. More than 15 concrete monitoring and management tools could immediately be introduced into research and practice, some of them without monetary investment. The agro-ecosystem and plant growth model MONICA, the DSS LandCaRe-DSS and the M-SQR for estimating SQ and crop yield potentials are some of these freely available tools with a focus on preserving and improving rural landscapes.

6. If any cropping practices are required for natural steppe regions, the depicted principles and modern technologies of conservation agriculture are needed to protect soils from erosion and degradation by wind and water.
7. Strengthening international and national cooperation will be key to substantial progress in monitoring, understanding and controlling landscape processes with a focus on the soil and water resources of Siberia. A new era of innovative, durable cooperation should be initiated. The book contributors are part of an immense innovation network which should be employed to achieve both significant disciplinary and synergetic effects.
8. New initiatives for monitoring and managing the land and water resources of Siberia could be effectively interlinked with efforts to improve the situation in parts of European Russia and in Central Asia. These regions share similar problems of natural conditions, landscape development and the need for innovation. All contributing partners, future human generations and great landscapes will benefit from those innovative initiatives.

References

Balla D, Omar M, Maassen S, Hamidov A, Khamidov M (2014) Efficiency of duckweed (*Lemnaceae*) for the desalination and treatment of agricultural drainage water in detention reservoirs. In: Mueller L, Saparov A, Lischeid G. (eds) Novel measurement and assessment tools for monitoring and management of land and water resources in agricultural landscapes of Central Asia. Springer International Publishing, pp 423–440. http://link.springer.com/chapter/10.1007/978-3-319-01017-5_25. Accessed 27 Feb 2015

Behrendt A, Fischer A, Kaiser T, Eulenstein F, Ortmann S, Berger A, Mueller L (2014) Multi-species grazing on deer farms. In: Mueller L, Saparov A, Lischeid G (eds) Novel measurement and assessment tools for monitoring and management of land and water resources in agricultural landscapes of Central Asia. Springer International Publishing, pp 491–501. http://link.springer.com/chapter/10.1007/978-3-319-01017-5_30. Accessed 27 Feb 2015

Berezin LV, Saparov AS, Kan VM, Shayakmetov MR (2013) Technology of integrated ecosystem reclamation of Russia and Kazakhstan (Технология комлексной мелиорации экосистем России и Казахстана). Almaty-Omsk 1st edition 2013, p 215 (in Russian)

BMBF (2014) www.kulunda.eu. Accessed 27 Feb 2015

Djanibekov N, Sommer R (2014) Integrated Decision Support for Sustainable and Profitable Land Management in the Lowlands of Central Asia. In: Mueller L, Saparov A, Lischeid G (eds) Novel Measurement and Assessment Tools for Monitoring and Management of Land and Water Resources in Agricultural Landscapes of Central Asia. Springer International Publishing, pp. 407–422. http://link.springer.com/chapter/10.1007/978-3-319-01017-5_24. Accessed on 30 Jan 2015

Dronin N, Kirilenko A (2011) Climate change, food stress, and security in Russia. Reg Environ Change 11(Suppl 1):S167–S178. doi:10.1007/s10113-010-0165-x

Evans RG (2014) Advanced technologies for irrigated cropping systems. In: Mueller L, Saparov A, Lischeid G (eds) Novel measurement and assessment tools for monitoring and management of land and water resources in agricultural landscapes of Central Asia. Springer International Publishing, pp 475–489. http://link.springer.com/chapter/10.1007/978-3-319-01017-5_29. Accessed 27 Feb 2015

Geßl A, Nestroy O (2008) Veränderungen landwirtschaftlich genutzter Böden in Österreich, dargestellt an den Erhebungen der Bodenschätzung. Mitt. naturwiss. Ver. Steiermark Band 137 pp 5–24 Graz 2008. http://www.landesmuseum.at/pdf_frei_remote/MittNatVerSt_137_0005-0024.pdf. Accessed 27 Feb 2015

Godbersen L, Utermann J, Duijnisveld WHM (2014) Methods in the exploratory risk assessment of trace elements in the soil-groundwater pathway. In: Mueller L, Saparov A, Lischeid G (eds) Novel measurement and assessment tools for monitoring and management of land and water resources in agricultural landscapes of Central Asia. Springer International Publishing, pp 289–314. http://link.springer.com/chapter/10.1007/978-3-319-01017-5_17. Accessed 27 Feb 2015

Helming K (2014) Impact assessment for multifunctional land use. In: Mueller L, Saparov A, Lischeid G (eds) Novel measurement and assessment tools for monitoring and management of land and water resources in agricultural landscapes of Central Asia. Springer International Publishing, pp 223–234. http://link.springer.com/chapter/10.1007/978-3-319-01017-5_12. Accessed 27 Feb 2015

Hennings V (2014) Use of pedotransfer functions for land evaluation: mapping groundwater recharge rates under semi-arid conditions. In: Mueller L, Saparov A, Lischeid G (eds) Novel measurement and assessment tools for monitoring and management of land and water resources in agricultural landscapes of Central Asia. Springer International Publishing, pp 249–262. http://link.springer.com/chapter/10.1007/978-3-319-01017-5_14. Accessed 27 Feb 2015

Kastanek F (1995) Kritische Bemerkungen zur Verwendung von Lysimetern. 5. Gumpensteiner Lysimetertagung „Stofftransport und Stoffbilanz in der ungesättigten Zone" BAL Gumpenstein. 25-26 April 1995. BAL Bericht, pp 93–102

Lal R (2013) Climate-strategic agriculture and the water-soil-waste nexus. J Plant Nutr Soil Sci 176:479–493. doi:10.1002/jpln.201300189

Lischeid 2014 Non-linear approaches to assess water and soil quality. In: Mueller L, Saparov A, Lischeid G (eds) Novel measurement and assessment tools for monitoring and management of land and water resources in agricultural landscapes of Central Asia. Springer International Publishing, pp 365–378. http://link.springer.com/chapter/10.1007/978-3-319-01017-5_21. Accessed 27 Feb 2015

Michel R, Dannowski R (2014) Using soil–water–plant models to improve the efficiency of irrigation. In: Mueller L, Saparov A, Lischeid G (eds) Novel measurement and assessment tools for monitoring and management of land and water resources in agricultural landscapes of Central Asia. Springer International Publishing, pp 379–387. http://link.springer.com/chapter/10.1007/978-3-319-01017-5_22. Accessed 27 Feb 2015

Meinel T, Grunwald L-C, Akshalov K (2014) Modern technologies for soil management and conservation in northern kazakhstan. In: Mueller L, Saparov A, Lischeid G (eds) Novel measurement and assessment tools for monitoring and management of land and water resources in agricultural landscapes of Central Asia. Springer International Publishing, pp 455–464. http://link.springer.com/chapter/10.1007/978-3-319-01017-5_27. Accessed 27 Feb 2015

Meissner R, Rupp H, Seyfarth M (2014) Advanced technologies in Lysimetry. In: Mueller L, Saparov A, Lischeid G (eds) Novel measurement and assessment tools for monitoring and management of land and water resources in agricultural landscapes of Central Asia. Springer International Publishing, pp 159–173. http://link.springer.com/chapter/10.1007/978-3-319-01017-5_8. Accessed 27 Feb 2015

Meyer BC, Lundy L (eds) (2014) Integrated water cycle management in Kazakhstan. Al-Farabi Kazakh National University, Publishing House, Almaty, p 320. ISBN: 978-601-04-0900-2

Mueller L, Frielinghaus M, Schindler U, Behrendt A, Schmidt W (2000) Bodenschutz auf Standorten der Flussniederungen und Niedermoore. Archiv für Acker- und Pflanzenbau und Bodenkunde 45:239–262. http://www.zalf.de/de/forschung/institute/lwh/mitarbeiter/lmueller/Documents/L7_60gruen.pdf. Accessed 27 Feb 2015

Mueller L, Schindler U, Behrendt A, Frielinghaus M (2002) Protecting lowland soils by agricultural practices. In: Proceedings 12th international soil conservation organization conference, 26–31 May 2002. Beijing, China, pp 328–333. http://www.zalf.de/de/forschung/institute/lwh/mitarbeiter/lmueller/Documents/L7_65gruen.pdf. Accessed 27 Feb 2015

Mueller L, Suleimenov M, Karimov A, Qadir M, Saparov A, Balgabayev N, Helming K, Lischeid G (2014a) Land and water resources of Central Asia, their utilisation and ecological status. In: Mueller L, Saparov A, Lischeid G (eds) Novel measurement and assessment tools for monitoring and management of land and water resources in agricultural landscapes of Central Asia. Springer International Publishing, Environmental Science and Engineering 2014, pp 3–59. Accessed 27 Feb 2015

Mueller L, Behrendt A, Shepherd TG, Schindler U, Ball BC, Khudyaev S, Kaiser T, Ralf Dannowski R, Eulenstein F (2014b) Simple field methods for measurement and evaluation of grassland quality. In: Mueller L, Saparov A, Lischeid G (eds) Novel measurement and assessment tools for monitoring and management of land and water resources in agricultural landscapes of Central Asia. Springer International Publishing, Environmental Science and Engineering 2014, pp 199–222. http://link.springer.com/chapter/10.1007/978-3-319-01017-5_11. Accessed 27 Feb 2015

Mueller L, Schindler U, Ball BC, Smolentseva E, Sychev VG, Shepherd TG, Qadir M, Helming K, Behrendt H, Eulenstein F (2014c) Productivity potentials of the global land resource for cropping and grazing. In: Mueller L, Saparov A, Lischeid G (eds) Novel measurement and assessment tools for monitoring and management of land and water resources in agricultural landscapes of Central Asia. Springer International Publishing, Environmental Science and Engineering 2014, pp 115–142. http://link.springer.com/chapter/10.1007/978-3-319-01017-5_6. Accessed 27 Feb 2015

Qadir M, Vyshpolsky F, Mukhamedjanov K, Bekbaev U, Ibatullin S, Yuldashev T, Noble AD, Karimov A, Mirzabaev A, Aw-Hassan A (2014) Enhancing the productivity of high-magnesium soil and water resources in Central Asia. In: Mueller L, Saparov A, Lischeid G (eds) Novel measurement and assessment tools for monitoring and management of land and water resources in agricultural landscapes of Central Asia. Springer International Publishing, Environmental Science and Engineering 2014, pp 465–474. http://link.springer.com/chapter/10.1007/978-3-319-01017-5_28. Accessed 27 Feb 2015

Schindler U, Mueller L (2005) Comparison of deep seepage estimations of a virtual with a real lysimeter by means of TDR measurements. Int Agrophysics Lublin 19:69–73. http://www.old.international-agrophysics.org/artykuly/international_agrophysics/IntAgr_2005_19_1_69.pdf. Accessed 27 Feb 2015

Semendyaeva NV, Lomova TG, Utenkov GL (2015) Scientific grounds of agricultural alcaline soil development in the south of the Western Siberia from 1969 to 2014 (Научное обеспечение сельскохозяйственного освоения солонцовых почв юга западной сибири за период с 1969 по 2014 г.) Vestnik Novosibirsk State Agrarian University NGAU 1(34)/2015, pp 7–21

Siewert C, Barsukov P, Demyan S, Babenko A, Lashchinsky N, Smolentseva EN (2014) Teaching soil science and ecology in West Siberia: 17 years of field courses. Environ Edu Res 20 (6):858–876. http://dx.doi.org/10.1080/13504622.2013.839778. Accessed 27 Feb 2015

Smolentseva E, Smolentsev B, Pachkin K, Mueller L (2014) Assessing the soil quality and crop yield potentials of some soils of Eurasia. In: Mueller L, Saparov A, Lischeid G (eds) Novel measurement and assessment tools for monitoring and management of land and water resources in agricultural landscapes of Central Asia. Environmental Science and Engineering 2014, pp 505–517. http://link.springer.com/chapter/10.1007%2F978-3-319-01017-5_31. Accessed 27 Feb 2015

Watson EE (2014) The continuing significance of rural landscapes in the urban century. Rural Landscapes 1(1): 3:1–3. doi:http://dx.doi.org/10.16993/rl.ab

ZALF-LWH (2013) The Muencheberg Soil Quality Rating. URL:http://www.zalf.de/de/forschung/institute/lwh/mitarbeiter/lmueller/Documents/field_mueller.pdf. Accessed 27 Feb 2015

ZALF-LSA (2015) URL:http://www.zalf.de/en/forschung/institute/lsa/forschung/Folgenab/landcare/Pages/default.aspx. Accessed 27 Feb 2015

Zaydelman FR (2009) Genesis and ecological basics of soil and landscape reclamation (Генезис и экологические основы мелиорации почв и ландшафтов) KDU Moscow 3rd edition 2009, p 720. ISBN 978-5-98227-554-7 (in Russian)

MIX
Papier aus verantwortungsvollen Quellen
Paper from responsible sources
FSC® C105338

If you have any concerns about our products,
you can contact us on
ProductSafety@springernature.com

In case Publisher is established outside the EU,
the EU authorized representative is:
Springer Nature Customer Service Center GmbH
Europaplatz 3, 69115 Heidelberg, Germany

Printed by Libri Plureos GmbH
in Hamburg, Germany